# Stereochemistry of Organic Compounds

V. K. Ahluwalia
Department of Chemistry (Retd.)
University of Delhi
New Delhi, Delhi, India

ISBN 978-3-030-84963-4          ISBN 978-3-030-84961-0   (eBook)
https://doi.org/10.1007/978-3-030-84961-0

Jointly published with ANE Books Pvt. Ltd.
In addition to this printed edition, there is a local printed edition of this work available via Ane Books in
South Asia (India, Pakistan, Sri Lanka, Bangladesh, Nepal and Bhutan) and Africa (all countries in the
African subcontinent).
ISBN of the Co-Publisher's edition: 978-1-910-39018-4

This Springer imprint is published by the registered company Springer Nature Switzerland AG
The registered company address is: Gewerbestrasse 11, 6330 Cham, Switzerland

V. K. Ahluwalia

# Stereochemistry of Organic Compounds

**Ane Books**
**Pvt. Ltd.**

# Contents

# About the Author

**Dr. V. K. Ahluwalia** was a Professor of Chemistry at Delhi University for more than three decades teaching Graduate, Postgraduate and M.Phil. Students. He was also a Postdoctoral Fellow between 1960 and 1962 and worked with renowned global names from prestigious international universities. He was a visiting professor of Biomedical Research at University of Delhi. Dr Ahluwalia is widely regarded as a leading subject expert in chemistry and allied subjects along with being a "Choice Award for an Outstanding Academic Title" winner. He has published more than 100 titles. Apart from books, he has published more than 250 research papers in national and international journals.

# Part I
# Introduction

# Chapter 1
# Introduction

## 1.1 Introduction

Stereochemistry is the chemistry of organic compounds in three dimensions. The term stereochemistry has been derived from the Greek word '*steros*' meaning solid. In fact, stereochemistry is an important branch of chemistry. It is not only concerned with the geometry of the molecules but is also of immense use in understanding the pathway of chemical transformations.

The origin of stereochemistry stems from the discovery of plane-polarised light by French physicist Malus (1809). Subsequently, another French physicist Biot (1815) discovered the existence of two types of quartz crystals, which rotated the plane-polarised light in opposite directions. It was found that this property was not only associated with the crystalline structure but some compounds in solution also exhibited this property. It was Pasteur who studied various salts of tartaric acid and observed that optically inactive sodium ammonium tartarate is actually a mixture of two different kinds of crystals which were mirror images of each other and rotated the plane-polarised light in opposite directions. He was able to separate them by using a hand lens and a pair of tweezers. It was concluded that the optical activity in solution is due to some molecular property which is retained in solution. It is now well known that this property is due to the presence of asymmetric carbon in the compound. A detailed discussion on this forms the subject matter of a sub-section chapter (see Chap. 2).

It is now well known that a very large number of compounds can be represented in more than one form. The phenomenon of the existence of two or more compounds having the same molecular formula is known as isomerism. Such compounds are referred to as isomers. Thus, it can be said that the isomers have the same molecular formula. However, they differ from each other in their physical and chemical properties.

© The Author(s), under exclusive license to Springer Nature Switzerland AG 2022  3
V. K. Ahluwalia, *Stereochemistry of Organic Compounds*,
https://doi.org/10.1007/978-3-030-84961-0_1

Isomerism is basically of two types. These are structural isomerism and stereoisomerism. The structural isomerism, as we know, is due to the difference in the structures of the molecules. These structural differences can be further classified into, *i.e.*, chain isomerism, position isomerism and functional group isomerism. Besides these, we also come across metamerism and tautomerism. On the other hand, stereoisomerism is not structural isomerism; it is due to constituent atoms or groups differing in their arrangement in space. The different types of isomerisms are represented below:

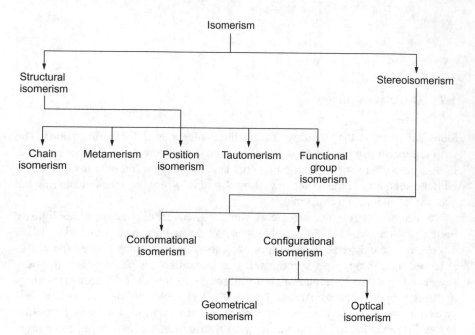

The stereoisomers can be either conformational isomers (leading to conformational isomerism); they arises due to rotation about a carbon–carbon sigma bond (single bond) or configurational isomers, which are of two types, viz., geometrical isomers (leading to geometrical isomerism) and optical isomers (leading to optical isomerism).

**Chain isomerism** arises due to the different carbon skeletons of the isomers (called chain isomers). Some examples are

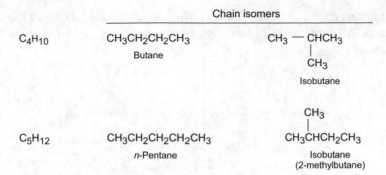

Chain isomers

$C_4H_{10}$   $CH_3CH_2CH_2CH_3$        $CH_3 — CHCH_3$
                    Butane                                    |
                                                            $CH_3$
                                                          Isobutane

$C_5H_{12}$   $CH_3CH_2CH_2CH_2CH_3$     $CH_3$
                    n-Pentane                        |
                                           $CH_3CHCH_2CH_3$
                                               Isobutane
                                            (2-methylbutane)

**Position isomerism** as the name implies is due to the difference in the position of the substitutions of the isomers (called position isomers). As an example, $C_3H_8O$ can be 1-propanol or 2-propanol.

$CH_3CH_2CH_2OH$                    OH
                                     |
   1-Propanol              $CH_3CHCH_3$
(n-Propyl alcohol)            2-Propanol
                        (isopropyl alcohol)

**Functional group isomerism** arises due to different functional groups. The isomers are called functional isomers. As an example, $C_3H_6O$ can be either acetone or propanal and $C_3H_8O$ can be either an ether or an alcohol.

                                O
                                ‖
$C_3H_6O$          $CH_3CCH_3$                  $CH_3CH_2CHO$
                       Acetone                        Propanal

$C_3H_8O$          $CH_3CH_2OCH_3$              $CH_3CH_2CH_2OH$
                   Methoxyethanel                     Propanal
                 (Ethyl methyl ether)            (n-Proyl alcohol)

**In case of metamerism,** the isomers (known as metamers) have the same functional group. Some examples include

$C_4H_{10}O$      $CH_3OCH_2CH_2CH_3$           $CH_3CH_2OCH_2CH_3$
                   Methoxy propane                  Ethoxyethane
                 (Methylpropyl ether)             (Diethyl ether)

                        O                              O
                        ‖                              ‖
$C_5H_{10}O$    $CH_3CH_2CCH_2CH_3$          $CH_3CCH_2CH_2CH_3$
                     3-Pentanone                    2-Pentanone
                   (Diethyl ketone)           (Methylpropylketone)

In **tautomerism** a compound can exist in two interconvertible forms known as tautomers. It is also called dynamic isomerism or Keto-enol tautomerism. Some examples are

$$\underset{\substack{\text{Acetone}\\\text{(Keto form)}\\93\%}}{CH_3-\overset{\overset{O}{\|}}{C}-CH_3} \rightleftharpoons \underset{\substack{\text{Acetone}\\\text{(Enolic form)}\\7\%}}{CH_3-\overset{\overset{OH}{|}}{C}=CH_2}$$

$$\underset{\substack{\text{Ethylacetoacetate}\\\text{(Keto form)}}}{CH_3\overset{\overset{O}{\|}}{C}CH_2COOC_2H_5} \rightleftharpoons \underset{\substack{\text{Ethylacetoacetate}\\\text{(Enolic form)}}}{CH_3\overset{\overset{OH}{|}}{C}=CHCOOC_2H_5}$$

**Stereoisomerism**, as has already been stated, is due to differences in the arrangement of atoms or groups in space. Stereoisomerism is of two types, viz., conformational isomerism and configurational isomerism. A discussion on these forms is the subject matter of subsequent chapters (*see* Chap. 2, Sect. 2; Chaps. 3 and 4).

Stereochemistry plays a special role in drugs. It is now known that only one enantiomer of a drug is useful for the effective treatment of a disease. As an example, the well-known drug ibuprofen (which contains a stereogenic centre) exists as a pair of enantiomers. However, only (S)-ibuprofen is effective as an anti-inflammatory agent. The (R)-ibuprofen shows no anti-inflammatory activity but is slowly converted into the (S)-enantiomer in vivo. Another drug, fluoxetine, is used as an antidepressant; in this case, only the (R)-enantiomer is the active component. The most interesting example is the case of the drug thalidomide, which was taken during pregnancy to avoid morning sickness. This drug (which was a mixture of (R) and (S) enantiomers) caused catastrophic birth defects to children born to women who took thalidomide. It was subsequently formed that only the (R) enantiomer has the desired effect and the (S)-enantiomer was responsible for birth abnormalities. This has been discussed in detail in Sect. 18.2 of Chap. 18.

Stereochemistry plays a vital role in the outcome of products obtained in various reactions like addition reactions, elimination reactions, substitution reactions, rearrangement reactions, free radical reactions and pericyclic reactions. All these form the subject matter of Part-III of this book.

# Part II
# Stereochemistry of Organic Compounds

# Chapter 2
# Stereochemistry of Organic Compounds Containing Carbon–Carbon Single Bonds (Hydrocarbons)

## 2.1 Introduction

Organic compounds containing carbon–carbon single bonds are called alkanes. Since there is free rotation about the single bonds (sigma bond), various spatial arrangements are obtained by rotation about the single bonds. These spatial arrangements are called **conformations** of a molecule. Out of all the conformations, the stable ones are known as **conformers** or **conformational isomers**.

## 2.2 Projection Formula of Conformers

The projection formula of conformers are useful for studying the conformations of simple molecules. Two projections, *viz.,* Newman projection and Sawhorse representation are possible.

In order to write the **Newman Projection** of a molecule (say ethane), it is viewed along a carbon–carbon bond [Fig. 2.1a]; in this case, the ethane molecule can be represented in wedge and dash drawings. For drawing Newman's projection, the carbon near the observer is represented by a point and the three groups attached to it are represented by three lines emerging from this point [Fig. 2.1b]. The other carbon (rear carbon) is shown by a circle and the three substituents attached to this carbon are shown by three lines emerging from the edge of the circle. The angle, θ, between the H—C—C plane and the C—C—H plane of the H—C—C—H unit is called the **dihedral angle.**

In **Sawhorse representation,** the carbon–carbon single bond of ethane is represented by a line, which is oriented diagonally backwards, i.e., the left-hand carbon projects towards the viewer and the other carbon (right-hand carbon) projects away from the viewer (Fig. 2.2). As in case of Newman projections, the substituents on each carbon are also shown by lines.

© The Author(s), under exclusive license to Springer Nature Switzerland AG 2022
V. K. Ahluwalia, *Stereochemistry of Organic Compounds,*
https://doi.org/10.1007/978-3-030-84961-0_2

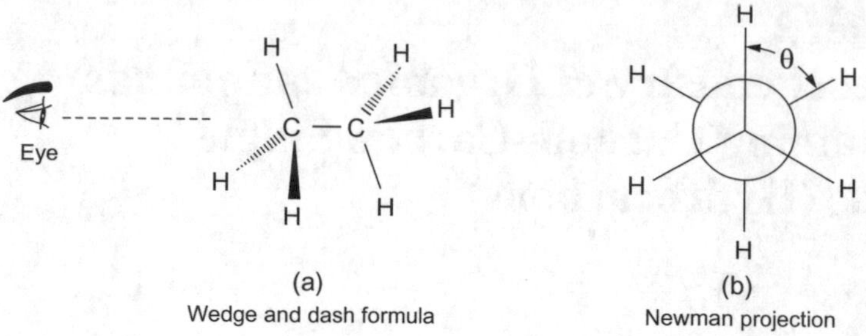

**Fig. 2.1** **a** Wedge and dash formula and **b** Newman projection of ethane

**Fig. 2.2** Sawhorse
representation of ethane

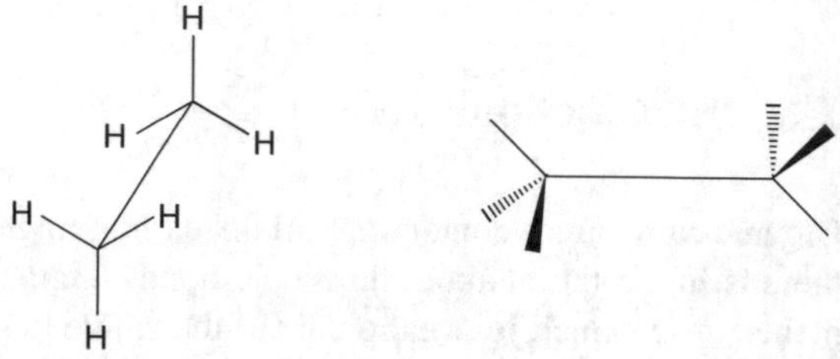

The projection formulae (as mentioned above) can be best understood by using molecular models. The four substituents attached to the carbon atoms in ethane are arranged in a tetrahedral fashion.

It was J. H. van't Hoff who first postulated that certain formations of molecules are favoured. He was the winner of the first Nobel Prize in Chemistry (1901) for his work on Chemical Kinetics.

## 2.3   Conformations of Ethane

As already stated, in case of ethane a large number of conformations are possible resulting from the rotation of $CH_3$ group about carbon–carbon single bond. The different conformations in ethane can be either staggered or eclipsed. These are represented in wedge and dash forms as shown in Fig. 2.3.

The scattered and eclipsed conformations of ethane are best represented by Newman projections. These projections represent the view along the $C_1$—$C_2$ bond. Carbon 1 is represented by a small black circle and carbon 2 is represented by a large circle. The staggered and eclipsed conformations of ethane are represented by the Newman projection as shown in Fig. 2.4.

When the dihedral angle is 0°, then the hydrogens on the two carbon atoms are parallel and the conformation is called **eclipsed conformation**. However, when the dihedral angle is 60°, the hydrogens on two carbons are at maximum distance

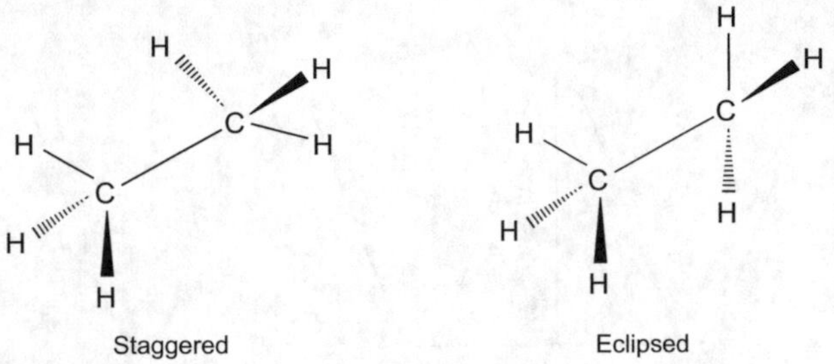

Fig. 2.3 Wedge and dash representations of staggered and eclipsed conformations of ethane

**Fig. 2.4** Newman projections of staggered and eclipsed conformations of ethane

Staggered
(More stable)

Eclipsed
(Less stable)

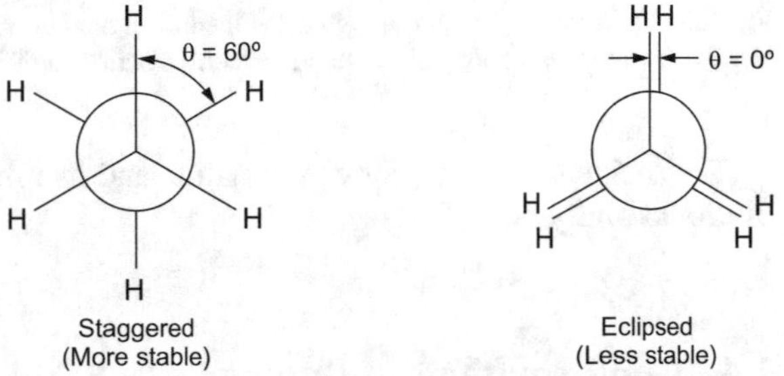

and the conformation is called staggered conformation (*see* Fig. 2.4). It has been found that there is an energy difference of 12.13 kJ mol$^{-1}$ between the eclipsed and staggered conformations and that the staggered conformation has the lower energy. This is due to the reason that there is maximum separation of bonded pairs in the staggered conformation leading to maximum repulsion between them. On the other hand, in the eclipsed conformation, the C—H bonds are very close to each other and this results in repulsion between the electrons forming the bonds. So the staggered conformation is more stable than the eclipsed conformation. Though the energy difference between the two conformations is very small (12.13 kJ mol$^{-1}$) in comparison to the kinetic energy of the molecule due to molecular motions. Even at low temperature, a molecule can pass through one staggered conformation to another staggered conformation, although in between it has to pass through an eclipsed conformation at a rate of about $10^{11}$ times per second.

This interconversion, as seen, is very rapid though it is not completely 'free' in the sense that the energy barrier of 12.13 kJ mol$^{-1}$ is there to be overcome. In fact, ethane molecule spends most of its time in staggered conformation, passing only transiently through its eclipsed form. At ordinary temperature, the rotation about the C—C bond is restricted. The energy required to rotate the molecule about the C—C bond is called **torsional energy**.

The variation of potential energy for various conformations of ethane is shown in Fig. 2.5.

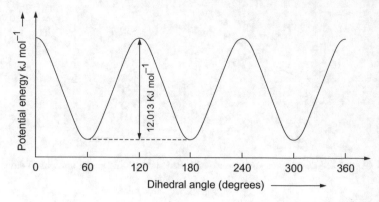

**Fig. 2.5** Variation of potential energy with dihedral angle. The molecule is in eclipsed conformation at 0°, 120°, 240° and 360° and in staggered conformation at 60°, 180° and 300°

The analysis of molecular conformation and its relative energy is called **conformational analysis.**

## 2.4  Conformations of Butane

The Sawhorse and Newman projections of butane are represented as

As in the case of ethane, in butane also various conformations are possible by the rotation of the C—C bond formed by carbon atoms 2 and 3. These conformations are represented as follows:

| I | II | III | IV |
|---|---|---|---|
| Conformations: Eclipsed | Gauche or Skew | Eclipsed | Anti |
| Dihedral angle: 0° | 60° | 120° | 180° |

| V | VI | VII |
|---|---|---|
| Conformations: Eclipsed | Gauche | Eclipsed |
| Dihedral angle: 240° | 300° | 360° =0 |

When the dihedral angle is 0°, the conformation (I) is called **eclipsed conformation.** As the dihedral angle increases to 60°, another conformation (II) called **gauche** or **skew conformation** is obtained. When the dihedral angle is 120°, another **eclipsed conformation** (III) results. In this conformation (III), $CH_3$ and H are eclipsed, whereas in the earlier eclipsed conformation (I) the two methyl groups were eclipsed. Hence, this eclipsed conformation (III) is at a lower energy level than the earlier conformation (I). When the two methyl groups are maximum apart (when the dihedral angle is 180°), the conformation is called **anti-conformation** (IV). This conformation (IV) is the most stable conformation of butane since it has the lowest energy value. Further rotation gives another eclipsed (V) and gauche (VI) conformations. The difference in energy between the anti- and gauche conformations is about 3.14 kJ mol$^{-1}$. It is found that at room temperature, butane is a mixture of 72% anti-conformation and 28% gauche conformation. As in the case of ethane, in this case also the interconversion is quite rapid. In case one wants to separate the conformations, it is necessary to slow down the interconversion by working at a low temperature of about 43 K. Figure 2.6 shows the potential energy variations for various conformations of butane.

**Conformational analysis** or the study of conformations is helpful in explaining the specificity of reactions, particularly the reactions observed in living systems.

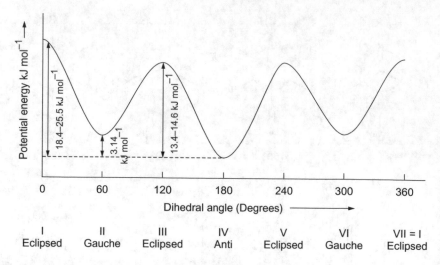

**Fig. 2.6** Potential energy diagram for the conformations of butane

## 2.5   Conformations of Cycloalkanes

Cycloalkanes are cyclic compounds and contain carbon–carbon single bonds as in the case of alkanes. Examples include cyclopropane, cyclobutane, cyclopentane and cyclohexane. However, cycloalkanes having more than six carbons are also known.

### 2.5.1   Stability of Cycloalkanes

Cycloalkanes do not have the same relative stability. Their stabilities are determined on the basis of their **heats of combustion.** The heat released on complete combustion of one mole of a substance is called its heat of combustion. Higher the heat of combustion per methylene group, lower will be its stability. The cycloalkanes, as we know, constitute a homologous series, each member differing from the one immediately preceding it by one —CH$_2$— group. On this basis, the general equation for the combustion of a cycloalkane is given by the equation

$$(CH_2) + \frac{3}{2}nO_2 \rightarrow nCO_2 + nH_2O + heat$$

In case of cycloalkanes, it is possible to calculate the amount of heat evolved per CH$_2$ group, on the basis of which the stabilities of cycloalkanes are directly compared. Table 2.1 gives the heat of combustion per CH$_2$ group in cycloalkane and the ring strain. It is also possible to determine the bond angle in cycloalkanes, on which their stabilities depend. The bond angle is determined by using the formula:

**Table 2.1**  Bond angle and ring strain of cycloalkanes

| Cycloalkane $(CH_2)_n$ | $n$ | Bond[1] angle | Heat of combustion cycloalkane $(kJ\ mol^{-1})$ | Heat of combustion[2] per $CH_2$ group $(kJ\ mol^{-1})$ | Ring strain[3] $(kJ\ mol^{-1})$ |
|---|---|---|---|---|---|
| Cyclopropane | 3 | 60° | 2091 | 697.0 | 115 |
| Cyclobutane | 4 | 90° | 2744 | 686.0 | 109 |
| Cyclopentane | 5 | 108° | 3320 | 664.0 | 27 |
| Cyclohexane | 6 | 120° | 3592 | 658.7 | 0 |
| Cycloheptane | 7 | 128.6° | 4637 | 662.4 | 27 |
| Cyclooctane | 8 | 135° | 5310 | 663.8 | 42 |
| Cyclononane | 9 | 140° | 5981 | 664.6 | 54 |
| Cyclodecane | 10 | 144° | 6636 | 663.6 | 50 |
| Cyclopantadecane | 11 | 147.2° | 9885 | 659.0 | 6 |
| Unbranched alkane | – | – | – | 658.6 | – |

[1] Bond angle determined by the equation given above
[2] Heat of combustion per $CH_2$ group is determined by dividing the heat of combustion of the cyclohexane (as determined experimentally) by the number of $CH_2$ groups
[3] Data taken from Organic Chemistry, Graham Solomons and Craig Fryhle, 7th Edn. p. 155

$$Bond\ angle = \left(\frac{2n-4}{n}\right) \times 90°$$

An example of the bond angle of cyclopropane ($n = 3$) is

$$= \left(\frac{2 \times 3 - 4}{3}\right) \times 90° = 60°$$

The stability of cycloalkanes in some cases is, in fact, related to the deviation of the bond angle from tetrahedral bond angle of 109.5°. This deviation causes a strain in the molecule leading to decreased stability. This type of instability is referred to as **angle strain**. As the deviation from the tetrahedral value decreases, the stability decreases. Thus, stability should increase from cyclopropane to cyclobutane to cyclopentane. The rule of angle strain is valid only for the first three members. On this basis, cyclopentane (in which case the deviation from the tetrahedral angle of 109.5° is minimum) is most stable. Thus, it is more appropriate to use the data of heat of combustion per $CH_2$ group (Table 2.1).

On the basis of the data obtained from the heat of combustion per $CH_2$ group, it is found that cyclohexane has the lowest heat of combustion per $CH_2$ group ($658.7\ kJ\ mol^{-1}$); this value is approximately of the same order in case of unbranced alkane. It is, therefore, assumed that cyclohexane has no ring strain and is thus used as a standard for comparison with other cycloalkanes. The ring strain can be calculated for other cycloalkanes by multiplying $658.7\ kJ\ mol^{-1}$ by $n$ and then subtracting the result from the heat of combustion of the cycloalkane. As an example, the heat of combustion of cyclopentane is found to be 3320 kJ mol$^{-1}$. So, the ring strain is = 3320 − 658.7 × 5 = 3320 − 3293.5 = 26.5 = 27 approximately.

As seen (Table 2.1), in case of cyclopropane the heat of combustion per $CH_2$ group evolves the greatest amount of heat and so cyclopropane has maximum ring strain (115 kJ mol$^{-1}$). On this basis, it can be said that in cyclopropane there is the greatest amount of potential energy per $CH_2$ group. Thus, what is called ring strain is a form of potential energy that the cycloalkane contains. It is, therefore, concluded that the more the ring strain a molecule has, the more the potential energy it will have and the less stable the compound will be.

In case of cyclobutane, the heat of combustion per $CH_2$ group is the second largest amount (Table 2.1) and so cyclobutane has the second largest ring strain (109 kJ mol$^{-1}$).

### 2.5.2 Conformations of Cyclopropane

In cyclopropane, all the carbons are in one plane and the hydrogen atoms are situated below and above the plane of the ring and is a flat molecule. So there are no conformational isomers in cyclopropane [Fig. 2.7a].

In cyclopropane (a molecule with the shape of a regular triangle), the internal angle is 60°, a value which is less by 49.5° from the tetrahedral angle of 109.5°. This results in **angle strain.** This is because the $sp^3$ orbitals of the carbon atoms are unable to overlap as effectively [Fig. 2.7b] as they do in alkanes. In fact, the carbon–carbon bonds of cyclopropane are described as 'bent'. The angle strain in cyclopropane is responsible for the ring strain. The hydrogen atoms of the ring are all eclipsed [Fig. 2.7c] and the molecule also has torsional strain. A Newman projection formula (6c) as viewed along a carbon–carbon bond shows the eclipsed hydrogens.

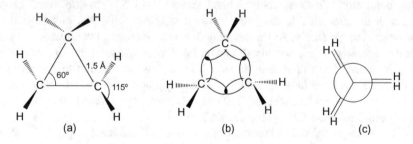

**Fig. 2.7** Angle strain in cyclopropane

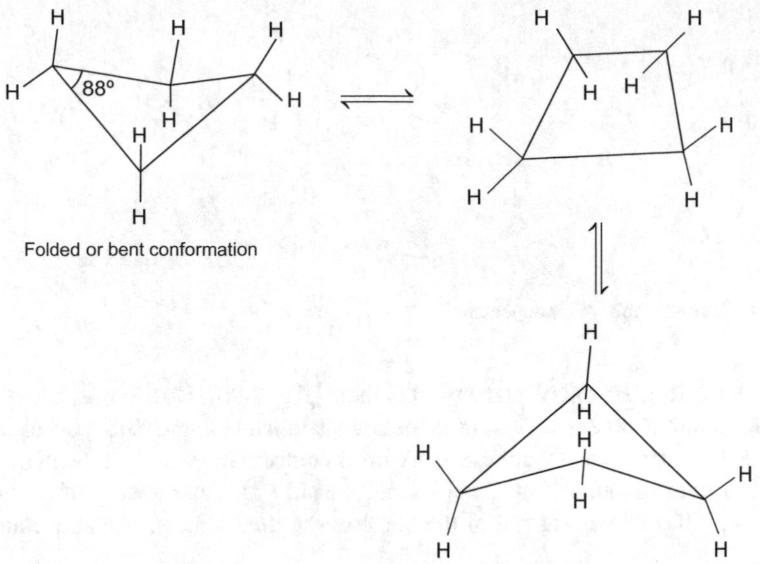

**Fig. 2.8**  Conformations of cyclobutane

### 2.5.3  Conformations of Cyclobutane

Like cyclopropane, cyclobutane also has angle strain. In cyclobutane, the internal angles are 88°, a departure of more than 21° from the tetrahedral bond angle. Unlike cyclopropane, cyclobutane is slightly folded. In case the cyclobutane ring is planar, the internal angles will be 90° instead of 88°, but the torsional strain will be considerably larger due to eight hydrogens being eclipsed. By bending slightly or folding, the cyclobutane ring relieves most of its torsional strain than it gains in the slight increase in its angle strain. The folded or bent conformations of cyclobutane are shown in Fig. 2.8.

### 2.5.4  Conformations of Cyclopentane

Like cyclobutane, cyclopentane has three conformations. A completely planar conformation of cyclopentane [Fig. 2.9a] has all C—H bonds eclipsed.

The internal angles of cyclopentane are 108°, a value close to the tetrahedral bond angle 109.5°. Thus, if cyclopentane molecule is planar, it would have little angle strain. However, planarity would introduce considerable torsional strain due to all 10 hydrogen atoms being eclipsed. Therefore, cyclopentane (like cyclobutane) assumes a slightly bent conformation (envelope conformation) in which two or more of the

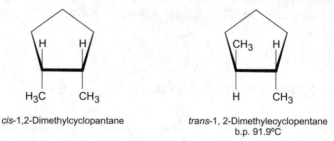

|       |          |            |
|-------|----------|------------|
| (a)   | (b)      | (c)        |
| Planar| Envelope | Half chair |

**Fig. 2.9**  Conformations of cyclopentane

atoms of the ring are out of plane of the others [Fig. 2.9b]. Due to this, some of the torsional strain is relieved. With little torsional strain and angle strain, cyclopentane is virtually as stable as cyclohexane. A third conformation of cyclopentane is the half-chair conformation [Fig. 2.9c] which is nearly of equal energy as that of the envelope; it has two atoms out of the plane of the three others, one above and one below.

In case two substituents are present in the cyclopropane ring, there is a possibility of *cis–trans* **isomerism**. Let us consider the example of 1,2-dimethylcyclopentane. It can be represented in two forms, *cis*-1,2-dimethylcyclopentane (the methyl groups are on the same side of the ring) and *trans*-1,2-dimethylcyclopentane (the methyl groups are on the opposite side of the ring) (Fig. 2.10).

The *cis* and *trans*-1,2-dimethylcyclopentane are stereoisomers. These differ from each other in the arrangement of atoms in space. These forms cannot be interconverted without cleavage of carbon–carbon bond.

In a similar way, 1,3-dimethylcyclopentane also exhibits *cis–trans* isomerism (Fig. 2.11).

A discussion of *cis–trans* isomerism forms the subject-matter of Chap. 3.

*cis*-1,2-Dimethylcyclopantane

*trans*-1, 2-Dimethylecyclopentane
b.p. 91.9°C

**Fig. 2.10**  *cis* and *trans*-1,2-dimethylcyclopentane

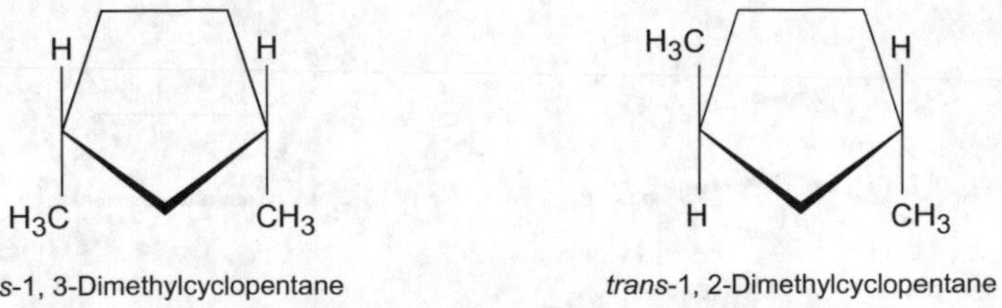

cis-1, 3-Dimethylcyclopentane

trans-1, 2-Dimethylcyclopentane

**Fig. 2.11** *Cis* and *trans* forms of 1,3-dimethylcyclopentane

## 2.5.5 Conformations of Cyclohexane

Cyclohexane is known to exist in two main conformations known as the **chair** and **boat conformations**. The chair conformation of cyclohexane is most stable. It is a non-planar structure; the carbon–carbon bond angles are all 109.5°. The chair conformation is free of angle strain and also of torsional strain. On viewing along any carbon–carbon bond, the atoms are perfectly staggered and the hydrogen atoms at opposite corners of the cyclohexane ring are maximally separated (*see* Fig. 2.12).

The boat conformation of cyclohexane is formed by flipping one end of the chair form (upward or downward). The flipping requires only rotation about a carbon–carbon single bond. The boat conformation like the chair conformation is also free of angle strain. However, the boat conformation is not free of torsional strain (*see* Fig. 2.13).

As seen, the H substituents in the boat conformation are eclipsed [*see* Fig. 2.13c], Newman Projection). Also, the two hydrogens on C1 and C4 are close enough to cause van der Waals repulsion [Fig. 2.13d]; this effect is called the 'flagpole' interaction of the boat conformation. Due to torsional strain and flagpole interactions, the boat conformation has higher energy than the chair conformation.

**Fig. 2.12** Chair conformation of cyclohexane. **a** Line drawing, **b** Illustration of large separation between H atoms at opposite corners and **c** Newman projection of chair conformation

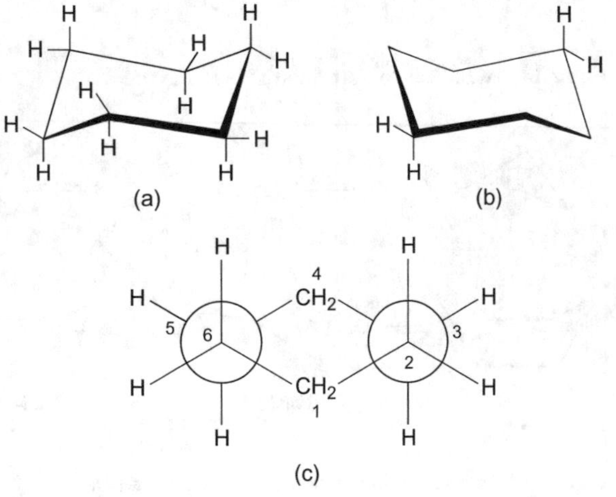

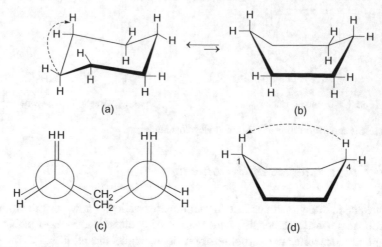

**Fig. 2.13** Boat conformations of cyclohexane. **a** Chair conformation and **b** its conversion into boat conformation, **c** Newman projection of boat conformation and **d** Flagpole interaction of the C1 an C4 hydrogen atoms of boat conformation

The conversion of boat form into the chair form involves the pulling down of the uppermost carbon down so that it becomes the bottommost carbon atom. However, in case the uppermost carbon is pulled just a little, another conformation, known as twist boat or skew boat conformation, results (Fig. 2.14). During this conversion, some of the torsional strain of boat conformation is relieved.

The various conformations of cyclohexane are chair, half-chair, twist boat and boat conformation. The interconversion of these formation are shown in Fig. 2.15.

**Fig. 2.14** Twist boat conformation of cyclohexane

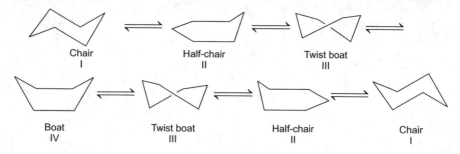

**Fig. 2.15** Interconversion of conformers of cyclohexane

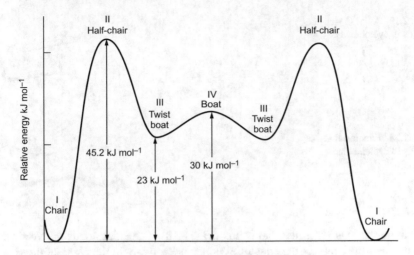

**Fig. 2.16** Relative energies of the various conformations of cyclohexane

The relative energies of various conformations of cyclohexane are depicted in Fig. 2.16.

As seen (Fig. 2.16), the energy barriers between the chair, boat and twist boat conformations of cyclohexane are low. This makes the separation of the conformations rather impossible at room temperature.

At room temperature, about 1 million interconversions occur each second and due to greater stability of the chair, more than 99% of the molecules are believed to be in the chair conformation at any time.

The NMR spectra of cyclohexane shows a sharp peak at $\delta 1.5$–$2.0$ (the region of methylene absorption), due to the fact that the two types of protons are indistinguishable due to rapid reversible interconversion of one form of cyclohexane into another. However, at the temperature, when its equilibrium is frozen, two signals are observed—one upfield (due to axial protons) and the other downfield (due to equatorial protons).

For the development and applications of the principles of conformation in chemistry, Derek H. R. Barton and Odd Hassel shared the Nobel Prize in 1969. Their work was useful for the understanding of not only the conformations of cyclohexane rings but also the structures of other compounds (like steroids) containing cyclohexane rings.

### 2.5.5.1 Axial and Equatorial Bonds in Cyclohexane

The chair form of cyclohexane has two types of carbon–hydrogen bonds, viz., axial (*a*) and equatorial (*e*). The C—H bonds which are parallel to the axis are axial bonds (Fig. 2.17) and the remaining six C—H bonds which extend outward at an angle of

**Fig. 2.17**  Axial C—H
bonds in cyclohexane

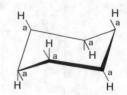

**Fig. 2.18**  Equatorial C—H
bonds in cyclohexane

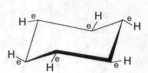

109.5° to the axis are equatorial bonds (Fig. 2.18). The axial bonds are vertical to
the plane of the ring and the equatorial bonds are roughly in the plane of the rings.

When ring flipping takes place from one chair form to another, all the axial bonds
become equatorial bonds and equatorial bonds become axial bonds.

In case of substituted cyclohexanes (mono or di), there will be two cyclohexanes,
axial or equatorial. It is interesting to know why one form is more stable than the
other. This forms the subject matter of the following sections.

### 2.5.5.2  Conformations of Monosubstituted Cyclohexane

Monosubstituted cyclohexane, for example, methylcyclohexane has two chair
conformations. In one form, methyl group is axial and in the other the methyl group
is equatorial (Fig. 2.19). Both these forms are interconvertible by ring flipping.

On the basis of studies, it has been shown that methyl cyclohexane with the methyl
group equatorial is more stable than the conformation with methyl group axial by
about 7.6 kJ mol$^{-1}$. In fact, the methylcyclohexane with equatorial methyl is present
to the extent of 95% in the equilibrium mixture. The greater stability of methylcyclohexane with equatorial methyl is because in axial methyl group (in position 1),

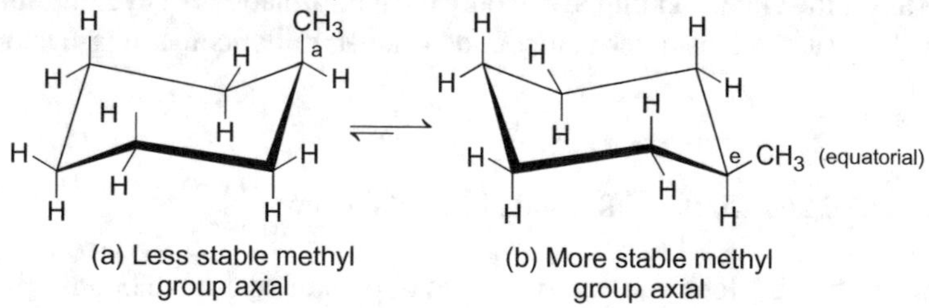

|  (a) Less stable methyl | (b) More stable methyl |
|---|---|
| group axial | group axial |

**Fig. 2.19**  Conformations of methylcyclohexane

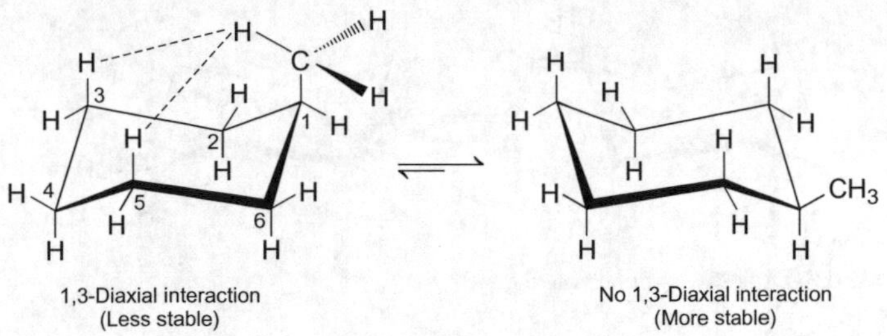

1,3-Diaxial interaction
(Less stable)

No 1,3-Diaxial interaction
(More stable)

**Fig. 2.20**  Equatorial methylcyclohexane is more stable than axial methylcyclohexane

Equatorial *tert* butylcyclohexane
(More stable)

Axial *tert* butylcyclohexane
(Less stable)

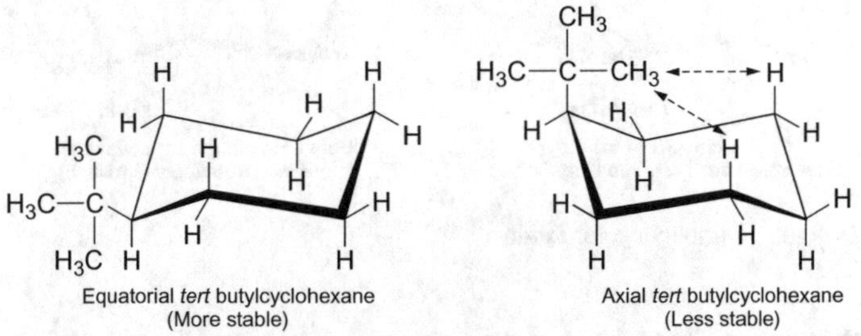

**Fig. 2.21**  Equatorial *tert* butylcyclohexane is more stable than axial *tert* butylcyclohexane

there is non-bonded interaction between $CH_3$ group and H atoms at positions 1 and 3, commonly known as **1,3-diaxial interaction.** In this case, the distance between H of $CH_3$ and H at C-3 and C-5 is less than the sum of van der Waals radii of two hydrogens. This repulsion destabilises axial conformation of methylcyclohexane, whereas no such interaction is possible with equatorial conformation of methylcyclohexane making it a more stable conformation (Fig. 2.20).

In case, monosubstituted cyclohexane has a large alkyl substituent (e.g., tertiary butyl group), the strain caused by 1,3-diaxial interaction is much more pronounced. The conformation of *tert*-butylcyclohexane with *tert*-butyl group in equatorial is found to be more than 21 kJ mol$^{-1}$ more stable than the axial form (Fig. 2.21).

At room temperature, 99.99% of equatorial *tert*-butylcyclohexane is present in the equilibrium mixture.

### 2.5.5.3   Conformations of 1,2-Dimethylcyclohexane

As in the case of 1,2-dimethylcyclopentane (*see* Sect. 2.5.4), in case of 1,2-dimethylcyclohexane also there is a possibility of *cis–trans* isomerism.

Let us first consider the case of 1,2-dimethylcyclohexane, in which one $CH_3$ group occupies axial position at C-1 and the second methyl group occupies an equatorial position at C-2. As both the methyl groups are on the same side, this is the *cis* arrangement (Fig. 2.22).

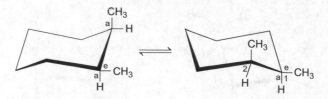

**Fig. 2.22**   *Cis*-1,2-Dimethylcyclohexane

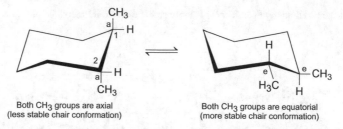

Both CH₃ groups are axial          Both CH₃ groups are equatorial
(less stable chair conformation)   (more stable chair conformation)

**Fig. 2.23**   *Trans*-1,2-Dimethylcyclohexane

In the above case, the two $CH_3$ groups are closer to each other; there is a crowding between the hydrogens of the two methyl groups. Ring flipping of the *cis* form gives another equivalent *cis* form. However, this does not lead to any change as far as interactions between the hydrogens of the two methyl groups are concerned.

In case both the methyl groups occupy axial position, the arrangement is a *trans* arrangement of the groups. This *trans* conformation on ring flipping gives another *trans* configuration in which both the methyl groups occupy equatorial positions (Fig. 2.23).

In the above configuration, the axial —$CH_3$ group at C-1 faces van der Waals repulsion by axial hydrogens at C-3 and C-5 carbon atoms. Similar repulsion for the axial C-2 methyl group with hydrogens at C-4 and C-6 carbon atoms makes this diaxial *trans* conformation less stable compared to the diequatorial *trans* conformation. The equatorial positions are free from such interactions as the substituents project outward. As a general rule, any substituent is more stable in the equatorial position than in the axial position. Thus, *trans* diequatorial conformation of 1,2-dimethylcyclohexane is more stable than the *trans* diaxial cyclohexane.

### 2.5.5.4   Conformations of 1,3-Dimethylcyclohexane

In 1,3-dimethylcyclohexane, when both the methyl groups are axial, the arrangement is *cis*. This arrangement on ring flipping gives another *cis* form in which both the methyl groups occupy equatorial positions (Fig. 2.24).

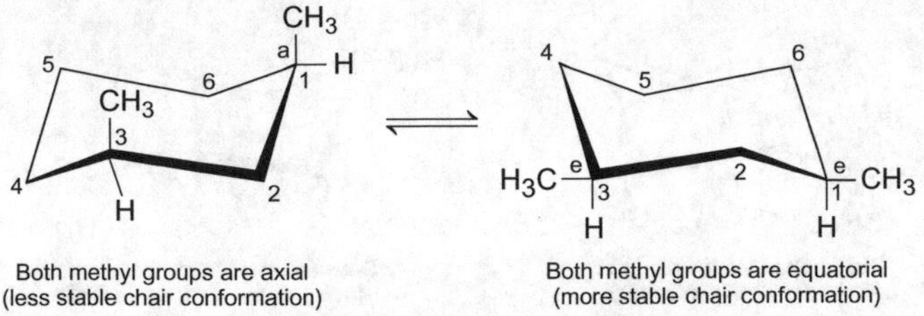

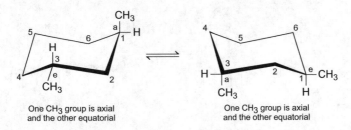

Fig. 2.24 *Cis*-1,3-Dimethylcyclohexane

Fig. 2.25 *Trans*-1,3-Dimethylcyclohexane

However, in case C-1 methyl is axial and C-3 methyl is equatorial (i.e., both the CH$_3$ groups are *trans* to each other), ring flipping gives another equivalent *trans* form (Fig. 2.25).

As seen, the *trans* form has one methyl group in equatorial position and the other methyl group in axial position compared to the *cis* form, which is its more stable conformation and both the methyl groups are in equatorial positions. So in this case, the *cis* form with diequatorial substituents is more stable than the *trans* form.

### 2.5.5.5 Conformations of 1,4-Dimethylcyclohexane

In 1,4-dimethylcyclohexane, when both the methyl groups are axial, it leads to *trans* arrangement of methyl groups. The resulting diaxial arrangement of methyl groups on ring flipping is changed into another chair form having diequatorial arrangement of methyl groups (Fig. 2.26).

If one methyl at C-1 is axial and the other methyl at C-4 is equatorial, it leads to *cis* arrangement of methyl groups. The ring flipping gives another equivalent chair conformation (Fig. 2.27).

The *cis*-1,4-dimethylcyclohexane (Fig. 2.27) has two equivalent chair conformations having one axial and equatorial methyl substituent.

Comparison of the *cis* and *trans* conformations shows that in the *trans* conformation, both the methyl groups are equatorial and so it is more stable than the *cis* form which has one substituent each in the axial and equatorial positions.

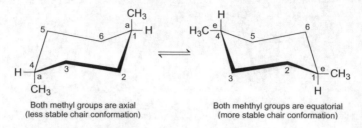

Both methyl groups are axial          Both mehthyl groups are equatorial
(less stable chair conformation)        (more stable chair conformation)

**Fig. 2.26**  *Trans*-1,4-Dimethylcyclohexane

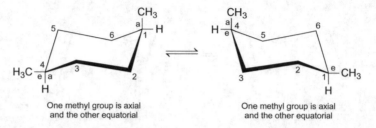

One methyl group is axial            One methyl group is axial
and the other equatorial              and the other equatorial

**Fig. 2.27**  *Cis*-1-,4-Dimethylcyclohexane

**Table 2.2**  Conformations of dimethylcyclohexane

| Compound | *cis* isomer | *Trans* isomer |
|---|---|---|
| 1,2-dimethylcyclohexane | a, e or e, a | **e, e** or a, a |
| 1,3-dimethylcyclohexane | **e, e** or a, a | a, e or e, a |
| 1,4-dimethylcyclohexane | a, e or e, a | **e, e** or a, a |

The different conformations of 1,2-, 1,3- and 1,4-dimethylcyclohexanes are summarised in Table 2.2. The more stable conformation, where one exists, is shown in bold type.

As seen, the conformation in which both the methyl groups are in equatorial positions is more stable.

In case, the disubstituted cyclohexanes have different substituents, the isomer with the larger substituent occupying the equatorial position is more stable.

### 2.5.5.6    Conformations of Cycloheptane

Cycloheptane exists in non-planar conformation and may be considered to be derived from chair cyclohexane where one apex is extended by one carbon atom. This is the chair conformation, since it is derived from chair cyclohexane. This conformation is not the most stable conformation due to excessive eclipsing of the H atoms at C-4 and C-5 [Fig. 2.28a]. Twisting leads to a more stable conformation 'twist cycloheptane', where some of the strain due to eclipsing along C-4 and C-5 is eliminated, Fig. 2.28b.

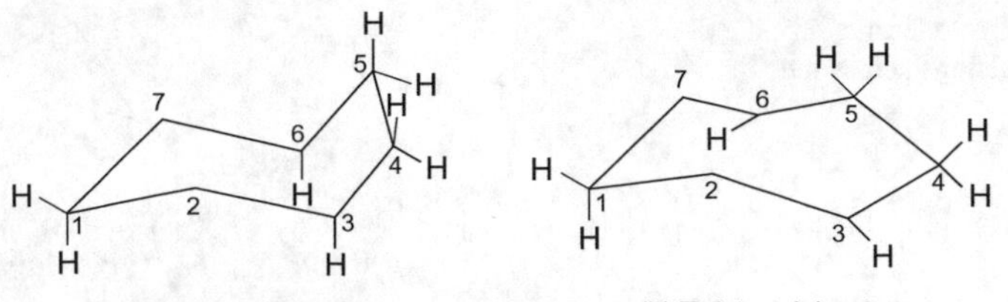

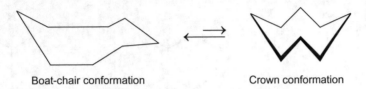

**Fig. 2.28** Conformations of cycloheptane

Boat-chair conformation          Crown conformation

**Fig. 2.29** Conformations of cyclooctane

### 2.5.5.7 Conformations of Cyclooctane

Like cycloheptane, cyclooctane also exists in a non-planar conformation. The X-ray analysis of cyclooctane–*trans*-1,2-dicarboxylic acid and other cyclohexane derivatives establishes a chair–boat conformation. The other conformation, called crown conformation, is a highly symmetrical form (Fig. 2.29).

### 2.5.5.8 Conformation of Cyclodecane

Cyclodecane like cycloheptane and cyclooctane exists in non-planar configuration. Its smaller unstability is caused by torsional stress and van der Walls repulsions between hydrogen atoms across rings, called **transannular strain**. The non-planar conformation is free of angle strain. X-ray crystallographic studies of cyclodecane reveal that the most stable conformation has a carbon–carbon–carbon bond angle of 117°, indicating some strain. The wide bond angle, of course, permits the molecule to expand thereby minimising unfavourable repulsions between hydrogen atoms across the ring (Fig. 2.30).

## 2.5.6 Conformations of Fused Six-Membered Rings

Common example of fused six-membered rings contains a bicyclic system. An example of fused bicyclic six-membered rings is decalin, which is known to exist in two forms, viz., *cis*- and *trans*-decalins (Fig. 2.31).

**Fig. 2.30** Cyclodecane
(H—H repulsions across the
ring)

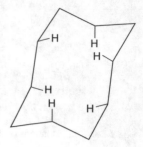

**Fig. 2.31** *Cis-* and
*trans*-decalins

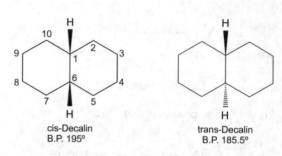

cis-Decalin
B.P. 195°

trans-Decalin
B.P. 185.5°

Decalin is the common name. It is described as bicyclo (4.4.0) decane (*see* also Fig. 8.10 in Chap. 8).

A knowledge of the conformations of 1,2-dimethylcyclohexane (Sect. 2.5.5.3) is useful for the conformational analysis of fused saturated ring. As an example, *trans*-decalin can be visualised to be derived from the most stable conformation of *trans*-1,2-dimethylcyclohexane by extending the methyl groups with two additional carbons into another ring. Alternatively, the two methyl groups can be visualised to be the two ends of a four carbon bridge, that is, —$CH_2$—$CH_2$—$CH_2$—$CH_2$— (*see* Fig. 2.32).

The *trans* ring fusion is understood by the relative positions of H atoms at the bridge head positions (the carbon atoms common to both the rings). However, unlike *trans*-1,2-dimethylcyclohexane, *trans*-decalin cannot flip into another stable chair–chair form.

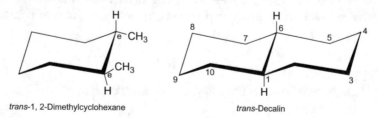

trans-1, 2-Dimethylcyclohexane

*trans*-Decalin

**Fig. 2.32** *Trans*-Decalin visualised to be derived from *trans*-1,2-dimethylcyclohexane

In a similar way, *cis*-decalin can be visualised to be formed from *cis*-1,2-dimethylcyclohexane (one $CH_3$ group axial and the other equatorial). In this isomer, chair–chair ring flipping is easy (Fig. 2.33).

The *trans*- and *cis* fusion of chair cyclohexane units in decalins do not involve any change in the dihedral angle. However, *cis*-decalin has relatively higher enthalpy due to more severe non-bonded interactions in its *cis* form than in the *trans* form.

In case there is a methyl group in position 6, it introduces increased non-bonded interaction; this is more serious in *trans* isomers. Also, a trigonal centre such as carbonyl group, on the other hand, reduces axial interactions, which are more serious in the *cis*-decalin series. Table 2.3 gives the enthalpy differences for *cis* and *trans* forms of various fused ring structures.

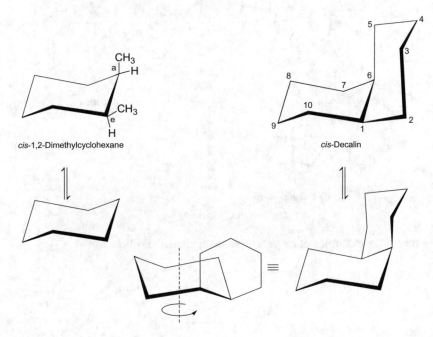

*cis*-1,2-Dimethylcyclohexane

*cis*-Decalin

**Fig. 2.33** *Cis*-Decalin visualised to be derived from *cis*-1,2-dimethyl cyclohexane and its ring flipping to give another chair–chair conformation

**Table 2.3** Enthalpy differences for *cis* and *trans* forms of various fused ring structures

| Compounds | (XXIII) | Enthalpy differences kJ/mole | |
|---|---|---|---|
| Decalin | R = R' = H | *trans → cis* | 11.5 |
| 10-Methyldecalin | R = CH₃; R' = H | *trans → cis* | 5.9 |
| 10-Methyl-1-decalone, | R = H; R' = O | *trans → cis* | 0.8 |
| Hydrindane | (structure given below) | *cis → trans* | 4.2 |
| Perhydroazulene | (structure given below) | *cis → trans* | small |
| Bicyclo[3,3,0]octane | (structure given below) | *cis → trans* | 29.6 |

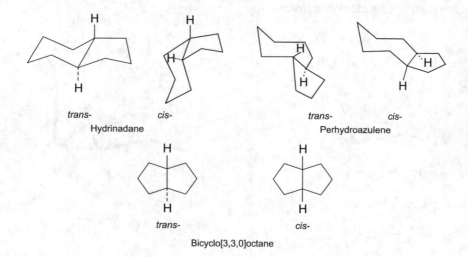

*trans-*          *cis-*                              *trans-*          *cis-*
  Hydrinadane                                  Perhydroazulene

*trans-*                      *cis-*
        Bicyclo[3,3,0]octane

Some other examples of bicyclic systems include the following:

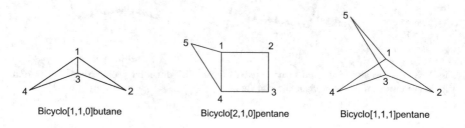

Bicyclo[1,1,0]butane          Bicyclo[2,1,0]pentane          Bicyclo[1,1,1]pentane

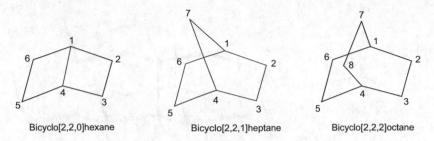

Bicyclo[2,2,0]hexane          Bicyclo[2,2,1]heptane          Bicyclo[2,2,2]octane

The conformation of the fused ring system is of particular interest in the determination of the stereochemistry of steroids. A typical example is that of cholesterol.

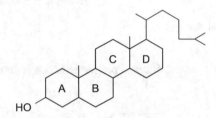

It has been found that in cholesterol
Rings A and B are *cis* fused.
Rings B and C are *trans* fused and
Rings C and D are *trans* fused.

So far we have seen conformations of bicyclic rings. Let us now consider a tricyclic system containing a three-dimensional cyclohexane ring [Adamantane, tricyclo $(3.3.1.1^{3,7})$ decane].

The three-dimensional structure of diamond is derived from the structure of adamantane (Fig. 2.34).

The conformation of a six-membered ring is useful for the structure elucidation of complex natural products like cholesterol.

A twist boat of adamantane is twistane. In twistane, the six-membered ring (chair conformation) is forced into twist boat conformation by the bridging carbon atoms.

Twistane

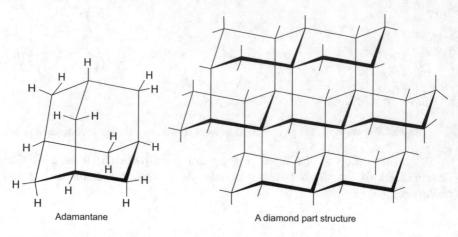

Adamantane                              A diamond part structure

**Fig. 2.34** Adamantane and diamond

An example each of a tetracyclic and pentacyclic compounds are given below.

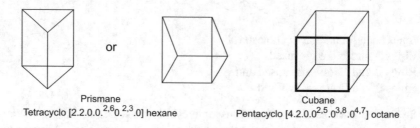

or

Prismane                                        Cubane
Tetracyclo [2.2.0.0.$^{2,6}$0.$^{2,3}$.0] hexane          Pentacyclo [4.2.0.0$^{2,5}$.0$^{3,8}$.0$^{4,7}$] octane

## Key Concepts

- **Angle Strain**: Important in small ring cycloalkanes, where it results from compression of bond angles to less than tetrahedral value. An example in cyclopropane internal angle is 60°, a value which is less by 49.5° from the tetrahedral angle of 109.5°. This results in angle strain.
- **Anti-Conformation**: The geometric arrangement around a C—C single bond, in which two largest substituents are 180° apart as viewed in the Newman projection.

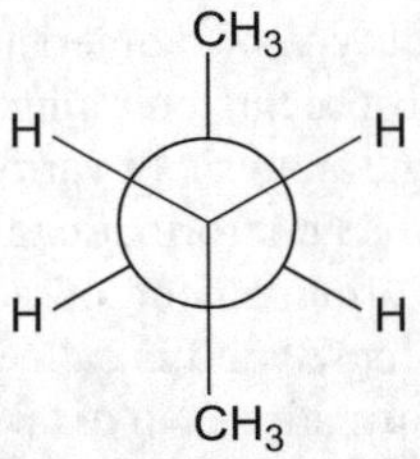

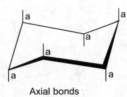

Anti conformation of
butane (Newmann Projection)

- **Axial Bond**: A bond in the chair cyclohexane which lies along the ring axis perpendicular to the rough plane of the ring.

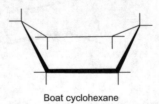

Axial bonds

- **Boat Cyclohexane**: A three-dimensional conformation of cyclohexane resembling the shape of a boat. It has no angle strain but has a large number of eclipsing interactions that make it less stable than chair cyclohexane.

Boat cyclohexane

- **Chair Cyclohexane**: A three-dimensional conformation of cyclohexane resembling the shape of a chair. It has no angle strain and represents the lowest energy conformation of the molecule.

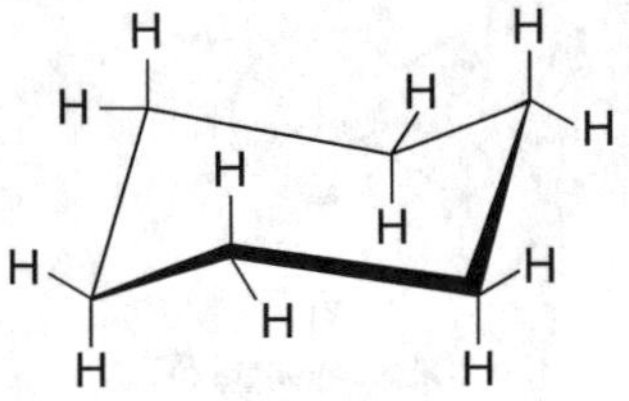

Chair cyclohexane

- **Cis–trans Isomerism**: Special type of isomerism due to the presence of double bond in a molecule or two substituents on a ring.
- **Conformational Analysis**: The analysis of various conformations in a molecule, and their relative energy is called conformational analysis.
- **Conformation**: The spatial arrangement of atoms or groups in a molecule containing a carbon–carbon single bond is called conformation.
- **Conformers**: The stable conformations out of all the conformations of a molecule are known as conformers or conformational isomers.
- **Decalin**: It is bicyclo (4.4.0) decane and exists in two forms, viz., *cis*-decalin (the two H atoms attached to the bridgehead atoms lie on the same side of the ring and in *trans*-decalin they are on opposite side).

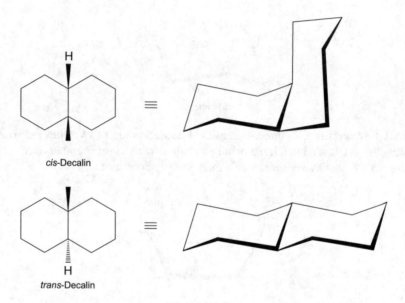

- **Dihedral Angle**: The angle, θ, between H—C—C plane and the C—C—H plane of the H—C—C—H unit in the Newman projection formula (e.g., ethane) is called dihedral angle.

- **Eclipsed Conformation**: The arrangement around carbon–carbon single bond in which the bonds to substituents on one carbon are parallel to the bonds to the substituents on the neighbouring carbon as viewed in a Newman projection. Thus, in the Newman projection of ethane, the C—H bonds on one carbon are lined up with the C—H bonds on the neighbouring carbon atom.

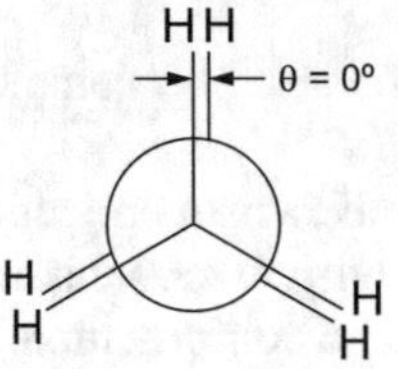

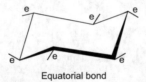

Eclipsed conformation θ = 0°

- **Equatorial Bond**: A bond in the chair cyclohexane that lies along the rough equator of the ring.

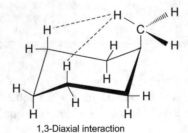

Equatorial bond

- **1,3-Diaxial Interaction**: In chair conformation of methyl cyclohexane, there is non-bonded interaction between $CH_3$ group at position 1 and hydrogens at positions 1 and 3. This is related to as 1,3-diaxial interaction.

1,3-Diaxial interaction

- **Gauche Conformation**: The conformation of butane in which the two methyl groups lie 60° apart as viewed in Newman projection.

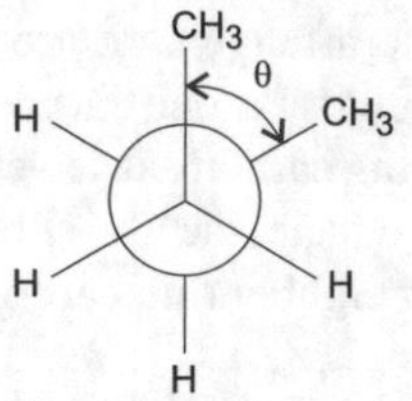

CH3

H

CH3

θ

H

H

H

Guache conformation
θ = 60°

- **Heat of Combustion**: Heat released on complete combustion of one mole of a substance.
- **Staggered Conformation**: In Newman projection (of ethane), when the dihedral angle is 60°, the hydrogens on two carbons are at maximum distance and the conformation is called staggered conformation.

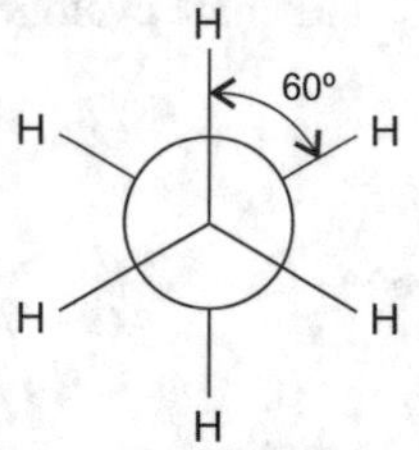

H

H

60°

H

H

H

H

Staggered conformation
θ = 60°

- **Stereochemistry**: Branch of chemistry concerned with three-dimensional arrangement of atoms or groups in molecules.
- **Torsional Energy**: The energy required to rotate the molecule about the C—C bond is called torsional energy.

## Problems

1. What is Newman projection? How does it differ from Sawhorse formula?
   **Hint**: In Sawhorse representation, the C—C bond is viewed sideways and in Newman projection, the C—C bond lies along the line of vision and so cannot be seen.
2. What is conformation?

3. Describe the conformation of butane arising due to rotation about the central bond. Which of the conformations is most stable? Explain.
4. Explain the term torsional energy.
5. Equatorial methylcyclohexane is more stable than axial methylcyclohexane. Explain why?
6. Equatorial *tert*-butyl cyclohexane is more stable than axial *tert*-butyl cyclohexane. Explain why?
7. *Trans* diequatorial conformation of 1,2-dimethyl cyclohexane is more stable than the *trans* diaxial cyclohexane.
8. Which of the conformationS in the following pairs is more stable?

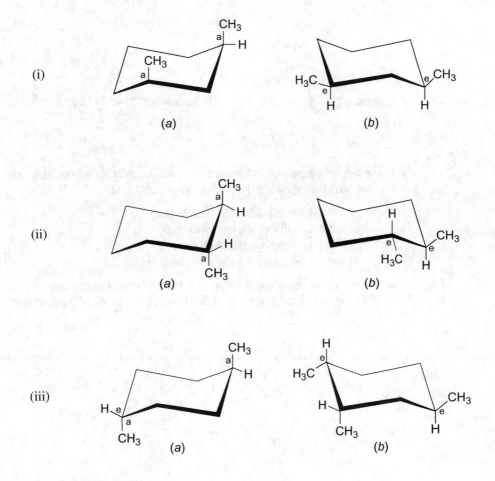

**Ans.** (*b*) in all the cases.

9.    In disubstituted cyclohexanes, when is the conformation *cis* or *trans*.

| Ans | Position | trans | cis |
|---|---|---|---|
| | 1,2 | (e, e) or (a, a) | (e, a) or (a, e) |
| | 1,3 | (e, a) or (a, e) | (e, e) or (a, a) |
| | 1,4 | (e, e) or (a, a) | (e, a) or (a, e) |

10.   In case of *trans*-1-t-butyl-3-methylcyclohexane, which conformation is predominant? Explain.

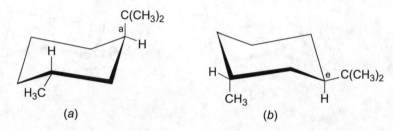

11.   Which of the following pairs of isomers is conformationally more stable, and draw a line structure showing the preferred conformation?

   (i)    *cis* or *trans*-1-t-butyl-2-methylcyclohexane
   (ii)   *cis* or *trans*-1,4-diisopropylcyclohexane
   (iii)  *cis* or *trans*-1,3-dibromocyclohexane
   (iv)   *cir* or *trans*-1-t-butyl-3-ethylcyclohexane

12.   Draw *trans*- and *cis*-decalin using the chair form of cyclohexane ring.
13.   The *trans* isomer of decalin is more stable than the *cis* isomer. Explain why?

# Chapter 3
# Stereochemistry of Organic Alicyclic Compounds Containing Carbon–Carbon Double Bonds (Alkenes and Cycloalkenes)

## 3.1 Introduction

The simplest compound containing a carbon–carbon double bond is ethylene. Such compounds are called alkenes. In ethylene, a sigma bond between two carbon atoms is formed by using one $sp^2$-orbital from each atom. Besides this sigma bond, one unhybridized $p$-orbital on each carbon atom overlaps each other sideways producing a new type of bond called π-bond, and the electrons involved in the formation of π-bond are called π-electrons. The π-electrons are distributed above and below the plane of the sigma bond. Along with these, two sigma bonds on each carbon atom are formed by the overlapping of the $sp^2$-orbitals in each carbon atom with 1 $s$-orbital of two hydrogen atoms (Fig. 3.1).

It can thus be said that a carbon–carbon double bond is made up of a σ-bond and a π-bond. The π-bond is weaker than a σ-bond. On the basis of thermochemical calculations, it is found that the strength of the π-bond is 264 kJ mol$^{-1}$ compound to 13–26 kJ mol$^{-1}$ for a carbon–carbon single bond. Also, in ethene, the carbon–carbon bond length is shorter (1.34 Å) than the carbon–carbon bond length in ethane (1.54 Å). The three $sp^2$-orbitals resulting from $sp^2$-hybridization are directed towards the corners of a regular triangle with an angle of 120° between them (Fig. 3.2).

## 3.2 Restricted Rotation Around a Carbon–Carbon Double Bond

The presence of a double bond consisting of a σ-bond and a π-bond accounts for an important property of the double bond. There is a large energy barrier to rotation associated with the atoms or groups joined by a double bond. There is maximum overlap between the $p$-orbital of a π-bond when the axes of the $p$-orbitals are exactly parallel. As already stated, the strength of a π-bond is 264 kJ mol$^{-1}$. This, in fact, is the barrier to the rotation of the double bond. This energy barrier is markedly

© The Author(s), under exclusive license to Springer Nature Switzerland AG 2022
V. K. Ahluwalia, *Stereochemistry of Organic Compounds*,
https://doi.org/10.1007/978-3-030-84961-0_3

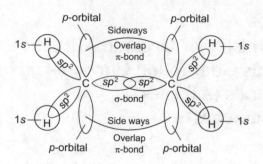

**Fig. 3.1**  Ethylene formation of a double bond

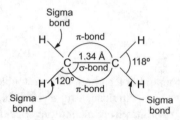

**Fig. 3.2**  Representation of Ethylene

higher than the rotational barrier of groups joined by a carbon–carbon single bond (13–26 kJ mol$^{-1}$). Thus, the groups joined by a single bond rotate comparatively freely at room temperature, and those joined by double bond do not.

It is, in fact, the restriction to rotation around a carbon–carbon double bond that gives rise to *cis–trans* isomerism.

Besides alkanes, there are a number of other cyclic alkanes which contain carbon–carbon single bonds, and there is a restriction of rotation also exhibiting *cis–trans* isomerism (see Chap. 2, Sects. 2.5.5.3, 2.5.5.4 and 2.5.5.6).

## 3.3  *Cis–Trans* Isomerism

As already stated, the restriction in the rotation of groups joined by a carbon–carbon double bond gives rise to another type of isomerism, called *cis–trans* isomerism. As an example, 1,2-dichloroethene can be represented by the following two structures:

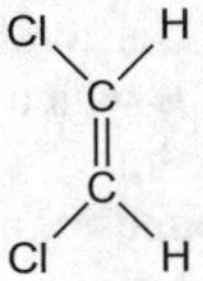

Two representations of 1,2-dichloroethene.

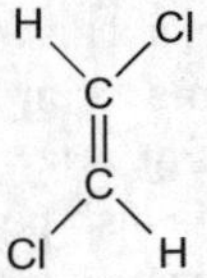

The two compounds I and II are isomers. These are different compounds having the same molecular formula. The two compounds differ in the arrangement of their atoms in space. In structure I, the two chlorine atoms are on the same side of the double bond and is called *cis* isomer. The structure I is thus *cis*-1,2-dichloroethene. On the other hand in structure II, the two chlorine atoms are on the opposite side and is called the *trans* isomer. Thus, structure II is *trans*-1,2-dichloroethene. These isomers are called geometrical isomers, and the isomerism is called geometrical isomerism. It is, however, more appropriate to call the isomer *cis*- and *trans* isomers and the isomerism *cis–trans* isomerism.

Some other examples of *cis–trans* isomerism are (Fig. 3.3).

It can, however, be said that *cis–trans* isomers are of the type $baC == Cab$.

In molecules of this type, the $\overset{a}{\underset{a}{>}}C=C\overset{b}{\underset{b}{<}}$ or $\overset{a}{\underset{a}{>}}C=C\overset{b}{\underset{b'}{<}}$ where the carbon atoms forming the double bond carry identical substituents, such as isomerism is impossible.

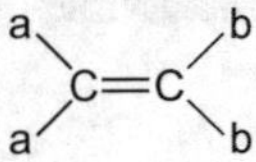

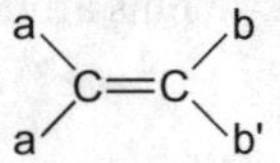

**Fig. 3.3**  Some examples of *cis–trans* isomerism

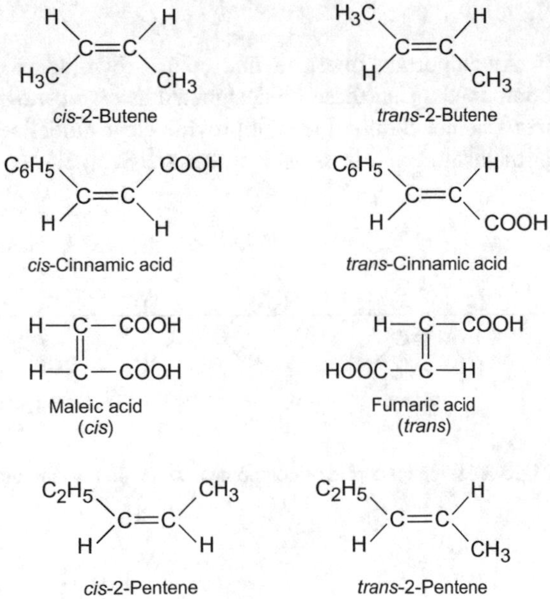

*Cis–trans* isomerism is not possible if one of the unsaturated carbons is attached to two identical groups. Some examples of compounds which do not exhibit *cis–trans* isomerism are shown in Fig. 3.4.

*Cis–trans* isomerism is also possible in compounds containing $C == N$ bonds. One such example is oximes of aldehydes or ketones.

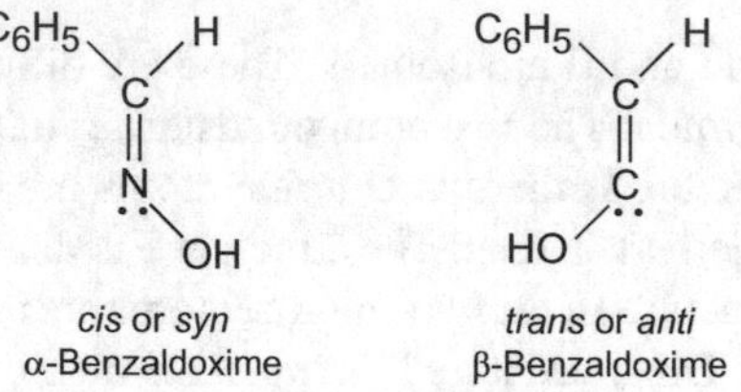

*cis or syn*
α-Benzaldoxime

*trans or anti*
β-Benzaldoxime

For details, see Sect. 15.1.4 in Chap. 15.

As already mentioned, the presence of two substituents on the ring of cycloalkane (alicyclic compounds) also exhibits *cis–trans* isomerism. Some examples include as shown in Fig. 3.5.

In case all the four substituents around the double bond are different, the following arrangements are possible.

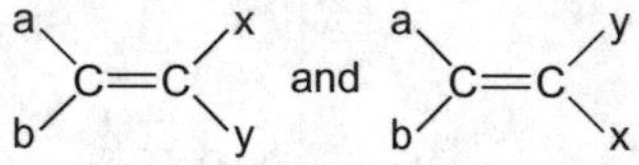

An important question that comes to mind is how to differentiate these two compounds. Can these be designated as *cis* or *trans*? The answer is no as the *cis–trans* nomenclature does not provide clear guidelines for their designation. For the nomenclature of *cis–trans* isomers, see Sect. 3.4.

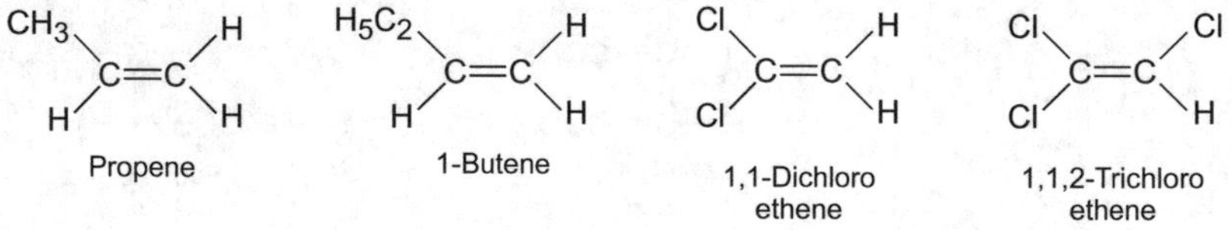

Propene          1-Butene          1,1-Dichloro
ethene
1,1,2-Trichloro
ethene

**Fig. 3.4**  Some examples of compounds which do not exhibit *cis–trans* isomerism

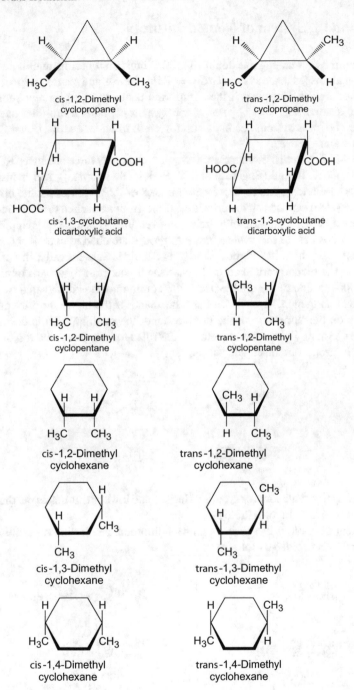

**Fig. 3.5**   *cis–trans* isomerism in cycloalkanes

## 3.4   *E* and *Z* System of Nomenclature

For compounds containing one double bond having different substituents attached to the carbon atoms (as in the system $baC == Cab$), the *cis*- and *trans* system of nomenclature works well. However, if the compound contains four different substituents (as in the system $abC == Cxy$) or if the system contains more than one double bond, the *cis–trans* system becomes complicated. In all cases, the *E* and *Z* system of nomenclature is used.

The *E* and *Z* system of nomenclature was developed by three chemists, R. S. Cahn (England), C. K. Ingold (England) and V. Prelog (Switzerland), and is also called Cahn–Ingold–Prelog system. This system is part of IUPAC rules for the nomenclature. This system depends on the priorities of the groups attached to a carbon–carbon double bond as per Cahn–Ingold–Prelog convention. This system is applicable to all types of alkenes. In this system, the two groups attached to one carbon atom are considered and which has higher priority is decided. Subsequently, the process is repeated to the second carbon atom. In case the higher priority groups or atoms are on the opposite side of the double bond, the configuration is designated as *E* (from the German, Estgenen, across). On the other hand, if the higher priority groups or atoms are on the same side of the double bond, the configuration is designated *Z* (from the German, Zusammen, together) (the rules for the determination of priority of atoms or groups are discussed subsequently).

(*H* and *L* represent atoms or groups of higher and lower priorities as per the system developed by Cahn–Ingold–Prelog).

A typical example is 2-bromo-1-chloro-1-fluoroethene. In this case, the *Z* and *E* conformation is decided as follows:

| Cl > F; Br > H | (Cl > F; Br > H) |
| (*Z*)-2-Bromo-1-chloro-1-fluoroethene | (*E*)-2-Bromo-1-chloro-1-fluoroethene |

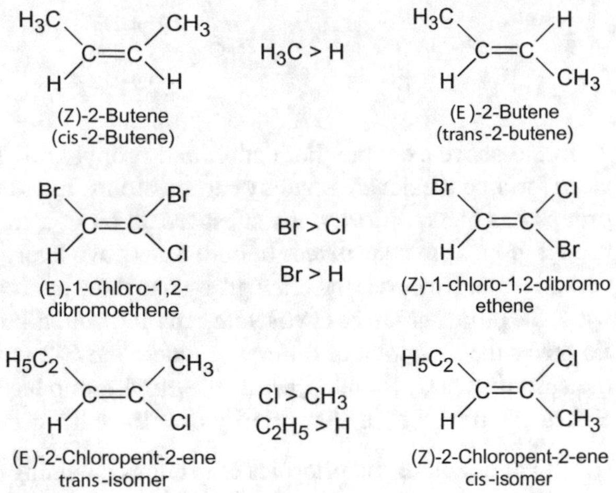

**Fig. 3.6** *E* and *Z* nomenclature of some compounds

(Z)-2-Butene
(cis-2-Butene)

$H_3C > H$

(E)-2-Butene
(trans-2-butene)

(E)-1-Chloro-1,2-dibromoethene

$Br > Cl$

$Br > H$

(Z)-1-chloro-1,2-dibromo ethene

(E)-2-Chloropent-2-ene
trans-isomer

$Cl > CH_3$
$C_2H_5 > H$

(Z)-2-Chloropent-2-ene
cis-isomer

Some other examples are given as shown in Fig. 3.6.

***Chan–Ingold–Prelog Priority Rules***

The following are the priority rules as determined by Chan–Ingold–Prelog:

(i) Each group attached to the carbon atom attached to the double bond is assigned priority or preference. Higher atomic number atom gets priority. For example, O (atomic number 18) gets a priority over N (atomic number 7), which in turn has a higher priority than H (atomic number 1). (In the case of a group, the priority of the first atom is considered.)

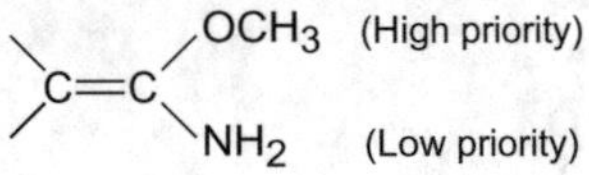

OCH₃ (High priority)

NH₂ (Low priority)

(ii) Among the isotopes of the same element, the isotope of higher atomic mass gets priority over the isotope with lower atomic mass.

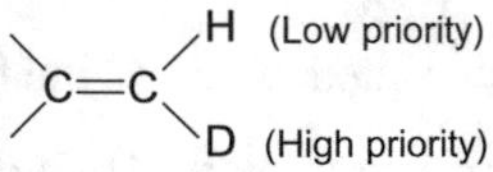

H (Low priority)

D (High priority)

(iii) If the two groups attached to the carbon atom involved in the formation of the double bond have the same atom as points of attachment (as in the case of ethyl, $CH_3CH_2$—, and propyl groups, $CH_3CH_2CH_2$—), the priorities are assigned on the basis of the first point of difference, using the same considerations of atomic number and atomic mass. Thus, as an example, the two groups attached to a carbon atom of a double bond are ethyl and propyl groups as shown below.

$$=C\overset{\overset{1\ \ 2}{CH_2CH_3}}{\underset{\underset{1\ \ 2\ \ 3}{CH_2CH_2CH_3}}{}}$$

In the above example, both ethyl and propyl groups are attached to the carbon atom forming the double bond by carbon atoms. In order to decide which of the two groups has higher priority, the substituents on C-1 carbon atoms of the ethyl and propyl groups are considered. In both cases, two hydrogens are attached to the C-1 carbon atom. However, the ethyl group has three hydrogens attached to C-2 carbon, while the propyl group has two hydrogens and one carbon attached to the C-2 carbon. So this is the first point of difference where the C-2 carbon of the propyl group has the substituents C, H and H, while the ethyl group has the substituent H, H and H. Hence, the propyl group has priority over the ethyl group.

(iv)   For assigning the priorities of groups containing double or triple bonds, these bonded groups are imagined in such a way that the bonded atoms are duplicated or triplicated as the case may be. Thus, for example, the group-$HC == CH_2$, a carbon atom attached to another carbon atom by a double bond, is considered as bonded to two carbon atoms. On this basis, the group —$HC == CH_2$ is

considered to be as $\overset{-HC-CH_2}{\underset{\ \ C\ \ \ \ C}{|\ \ \ \ \ |}}$.

Similarly, the group $\overset{O}{\underset{-C-H}{\|}}$ is treated as $-\overset{O}{\underset{O}{\overset{|}{\underset{|}{C}}}}-H$.

In the case of the group $-\overset{1}{C} \equiv \overset{2}{C}H$, the C at C-1 is written as $-\overset{C}{\underset{C}{\overset{|}{\underset{1|}{C}}}}-\overset{2}{C}H$ and the

C at $C_2$ is written as $-\overset{C}{\underset{C}{\overset{|}{\underset{1|}{C}}}}-\overset{C}{\underset{C}{\overset{|}{\underset{2|}{C}}}}H$ ; this is equivalent to —$C\equiv CH$ group. The group

—$C\equiv N$ is considered as $-C\overset{N}{\underset{N}{\overset{\diagup}{\diagdown}}}N$

On the basis of what has been stated, we have the following:

(a)   The priority of common groups on the basis of the first atom attached to the carbon of the double bond is

$-I > Br > -Cl > -Sb > -SO_3H > -NH_2 > -COOCH_3 >$
$-COOH > -CH_3$

(b)  In case the first atom is the same, the priority is determined on the basis of the second atom. This is demonstrated by

$$-N(CH_3)_2 > -NHCH_3 > -NH_2$$

(v)  If the second atom is the same, the priority is assigned on the basis of the third atom, as in a carboxylic acid and its ester

$$-COOCH_3 > -COOH$$

On the basis of guidelines provided by the sequence rules, some of the commonly occurring groups are arranged in order of their priority as follows:

$$I \rightarrow Br \rightarrow Cl \rightarrow S \rightarrow F \rightarrow CH_3 \overset{O}{\overset{\parallel}{C}} -O \rightarrow H \overset{O}{\overset{\parallel}{C}} -O \rightarrow CH_3CH_2O \rightarrow$$

$$HO \rightarrow H_2N \rightarrow CH_3O \overset{O}{\overset{\parallel}{C}} \rightarrow HO \overset{O}{\overset{\parallel}{C}} \rightarrow CH_3 \overset{O}{\overset{\parallel}{C}} \rightarrow H - \overset{O}{\overset{\parallel}{C}} \rightarrow HOCH_2 \rightarrow$$

$$(CH_3)_3C \rightarrow CH_2 == CH \rightarrow (CH_3)_2CH \rightarrow CH_3CH_2 \rightarrow CH_3 \rightarrow H$$

## 3.5  Relative Stabilities of *Cis* and *Trans* Alkenes

*Cis* and *trans* alkenes do not have the same stability. Their relative stabilities can be determined by the heat of hydrogenation or heat of combustion.

### 3.5.1  From Heat of Hydrogenation

The hydrogenation of alkene is an exothermic reaction; the enthalpy changes involved is called the heat of hydrogenation. The differences in the heat of hydrogenation permit the measurement of relative stabilities of alkene isomers when hydrogenation converts them to the same product. The heat of hydrogenation of the *cis* and *trans* butenes (along with that of 1-butene for comparison) is given below (Fig. 3.7).

As seen, 1-butene evolves the greatest amount of heat on hydrogenation and *trans*-2-butene evolves the least. So, 1-butene must have the greatest energy (enthalpy) and is the least stable isomer. *Trans*-2-butene has the lowest energy and is the most stable isomer. Thus, it can be said that *trans* isomers are more stable than the *cis* isomer. According to the energy diagram (Fig. 3.8) of the three butene isomer, the order of stability is *trans*-2-butene > *cis*-2-butene > 1-butene.

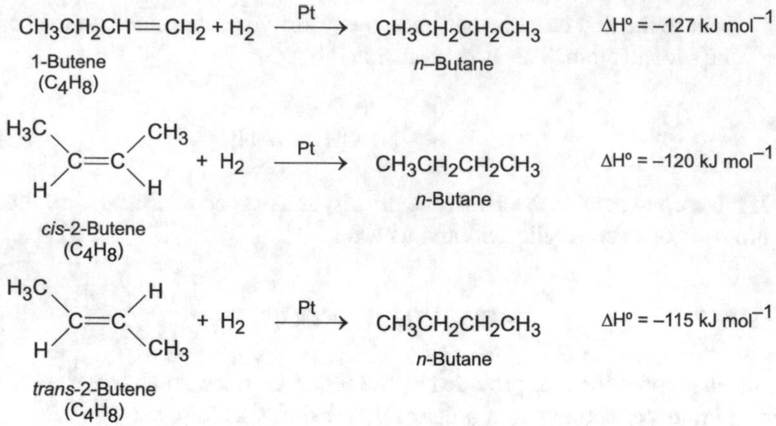

CH3CH2CH=CH2 + H2
Pt
CH3CH2CH2CH3
ΔH° = −127 kJ mol−1
1-Butene
(C4H8)
n–Butane
H3C
CH3
C=C
H
H
+ H2
Pt
CH3CH2CH2CH3
ΔH° = −120 kJ mol−1
n-Butane
cis-2-Butene
(C4H8)
H3C
H
C=C
H
CH3
+ H2
Pt
CH3CH2CH2CH3
ΔH° = −115 kJ mol−1
n-Butane
trans-2-Butene
(C4H8)

**Fig. 3.7** Relative stability of *cis* and *trans* alkenes, on the basis of heat of hydrogenation

CH3CH2CH=CH2
+
H2
7 kJ mol−1
ΔH° = −127 kJ mol−1
H3C
CH3
C=C
H
H
+
H2
ΔH° = −120 kJ mol−1
H3C
H
C=C
H
CH3
+
H2
5 kJ mol−1
ΔH° = −115 kJ mol−1
CH3CH2CH2CH3

**Fig. 3.8** Energy diagram for three butene

The greater stability of *trans*-2-butene when compared to *cis*-2-butene is a general pattern found in *cis–trans* alkene pairs. In the case of 2-pentenes, also the *trans* isomer is more stable than the *cis* isomer as collaborated by their heat of hydrogenation (Fig. 3.9).

The greater enthalpy of *cis* isomer is attributed to the strain caused by the crowding of two alkyl groups on the same side of the double bond Fig. 3.10.

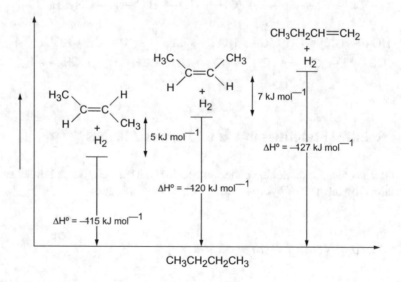

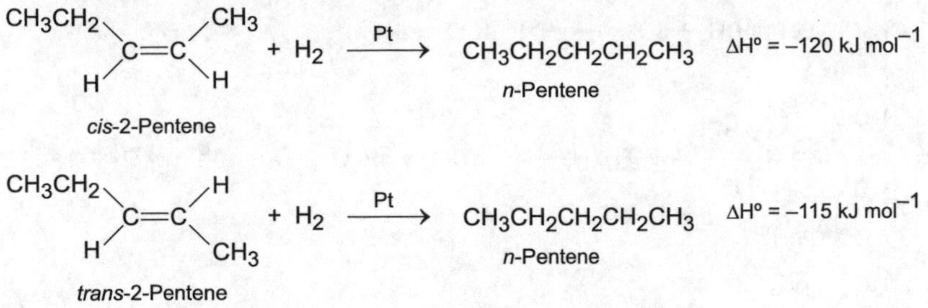

$$CH_3CH_2\text{---}C\text{==}C\text{---}CH_3\ (cis) + H_2 \xrightarrow{Pt} CH_3CH_2CH_2CH_2CH_3 \quad \Delta H^\circ = -120\ kJ\ mol^{-1}$$

cis-2-Pentene

n-Pentene

$$CH_3CH_2\text{---}C\text{==}C\text{---}H + H_2 \xrightarrow{Pt} CH_3CH_2CH_2CH_2CH_3 \quad \Delta H^\circ = -115\ kJ\ mol^{-1}$$

trans-2-Pentene

n-Pentene

**Fig. 3.9**   Relative stability of *cis*- and *trans*-2-pentenes

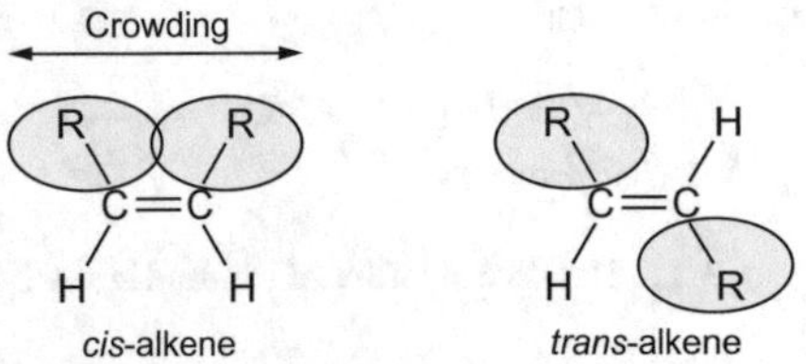

**Fig. 3.10**  *Cis* and *trans* alkenes. Less stable *cis* isomer has the greater strain

Crowding

cis-alkene                    trans-alkene

## 3.5.2   Heat of Combustion

Each isomer of a particular compound (e.g., of *cis* and *trans* butene) consumes six molar equivalents of oxygen and produces four molar equivalents of $CO_2$ and four molar equivalents of $H_2O$. On the basis of the comparison of heat of combustion, it is found that *trans*-2-butene is more stable than *cis*-2-butene. For the sake of comparison, the heat of combustion of 1-butene and 2-methylpropene are also included (Fig. 3.11).

On the basis of the above, the stability of butene isomers is

$$CH_3\underset{CH_3}{\overset{\phantom{CH_3}}{C}}\text{==}CH_2 > trans\ CH_3CH\text{==}CHCH_3 > cis\ CH_3CH\text{==}CHCH_3 > CH_3CH_2CH\text{==}CH_2$$

The same order was also obtained on the basis of heat of hydrogenation.

## 3.6   Synthesis of *Cis* and *Trans* Alkenes

Both *cis* and *trans* alkenes can be synthesised from appropriate alkynes by syn addition of hydrogen (to give *cis* alkene) or anti-addition of hydrogen (to give *trans* alkenes).

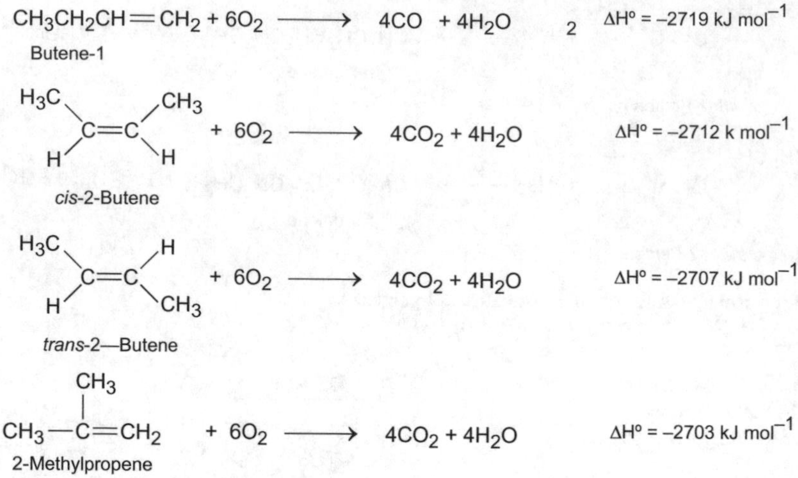

$$CH_3CH_2CH = CH_2 + 6O_2 \longrightarrow 4CO + 4H_2O \quad {}_2 \quad \Delta H° = -2719 \text{ kJ mol}^{-1}$$

Butene-1

cis-2-Butene

$$+ 6O_2 \longrightarrow 4CO_2 + 4H_2O \qquad \Delta H° = -2712 \text{ k mol}^{-1}$$

trans-2—Butene

$$+ 6O_2 \longrightarrow 4CO_2 + 4H_2O \qquad \Delta H° = -2707 \text{ kJ mol}^{-1}$$

$$\underset{\text{2-Methylpropene}}{CH_3 - \overset{\overset{\displaystyle CH_3}{|}}{C} = CH_2} \quad + 6O_2 \longrightarrow 4CO_2 + 4H_2O \qquad \Delta H° = -2703 \text{ kJ mol}^{-1}$$

**Fig. 3.11**  Relative stability of *cis-* and *trans*-2-butene

## 3.6.1  Synthesis of Cis Alkenes (Catalytic Reduction)

*Cis* alkenes are obtained by syn addition of hydrogen to alkynes using nickel boride compounds known as P-2 catalyst. The catalyst is prepared from nickel acetate by reduction with sodium borohydride (Fig. 3.12).

The reduction of alkynes in the presence of P-2 catalyst involves the syn addition of hydrogen. The alkene obtained from an alkyne with an internal triple bond is *cis* alkene (Z-configuration) (Fig. 3.13).

In syn addition, both the hydrogen atoms add from the same side of the molecule (Fig. 3.14).

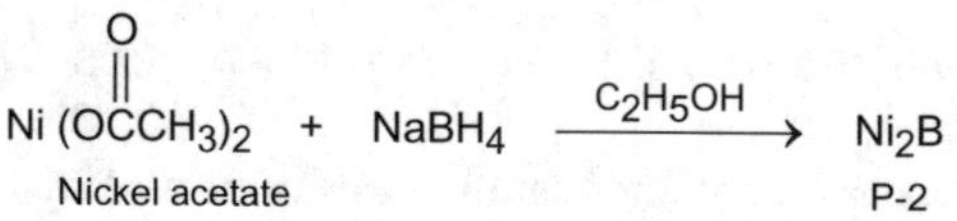

$$\underset{\text{Nickel acetate}}{Ni \, (O\overset{\overset{\displaystyle O}{||}}{C}CH_3)_2} \; + \; NaBH_4 \; \xrightarrow{\; C_2H_5OH \;} \; \underset{\text{P-2}}{Ni_2B}$$

**Fig. 3.12**  Preparation of P-2 catalyst

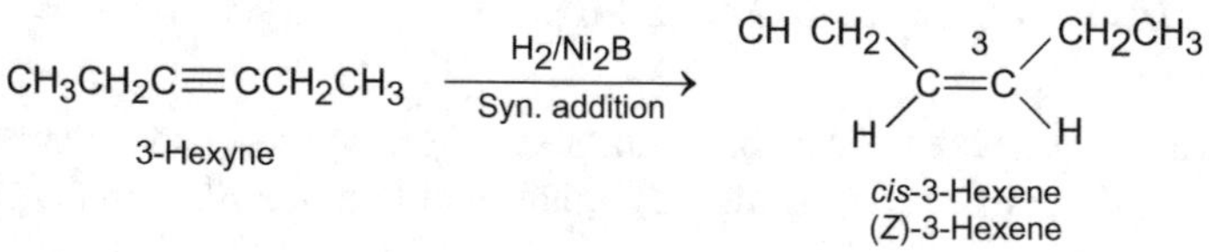

$$\underset{\text{3-Hexyne}}{CH_3CH_2C \equiv CCH_2CH_3} \; \xrightarrow[\text{Syn. addition}]{H_2/Ni_2B} \;$$

cis-3-Hexene
(Z)-3-Hexene

**Fig. 3.13**  Syn addition of H$_2$

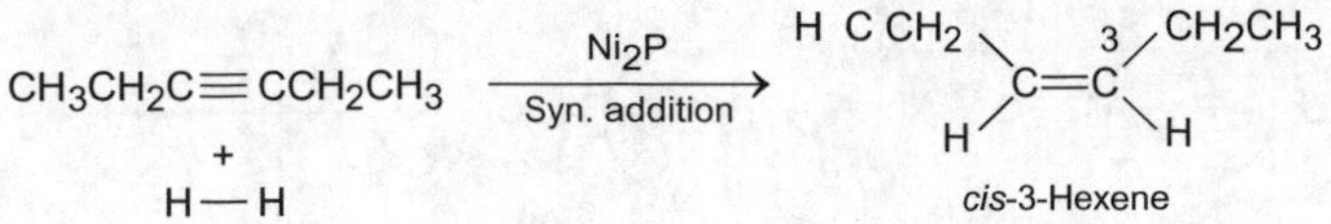
CH3CH2C≡CCH2CH3
+
H—H
Ni2P
Syn. addition
H C CH2
3
CH2CH3
C=C
H
H
cis-3-Hexene

**Fig. 3.14** Synthesis of *cis*-3-hexene

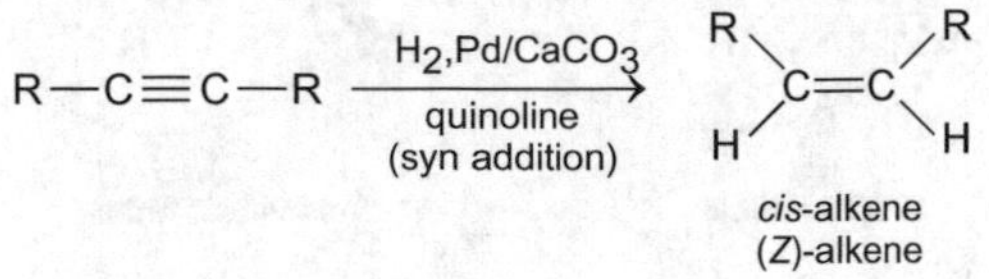
R—C≡C—R
H2,Pd/CaCO3
quinoline
(syn addition)
R
R
C=C
H
H
cis-alkene
(Z)-alkene

**Fig. 3.15** Reduction of alkynes by Lindlar's catalyst

In place of $Ni_2P$ (P-2 catalyst), **Lindlar's Catalyst** (metallic palladium deposited on calcium carbonate conditioned with lead acetates and quinoline) (Fig. 3.15).

### 3.6.2 *Synthesis of* Trans *Alkenes (Chemical Reduction)*

These are prepared by anti-addition of hydrogen atoms to triple bonds. The catalyst used in this case is Lithium or sodium metal or ethylamine at low temperature. This reaction known as **dissolving metal reduction** yields *trans* or (*E*) alkene (Fig. 3.16).

The mechanism of anti-addition of hydrogen to alkyne involves the following steps:

(i) Lithium atom donates an electron to the π-bond of the alkyne to give a radical anion (*A*).
(ii) The radical anion acts as a base and removes a proton from ethylamine to give a vinylic radical (*B*).
(iii) The vinylic radical receives an electron from a second lithium atom to give *trans*-vinylic anion (*C*).
(iv) The *trans*-vinylic anion acts as a base and removes a proton from a second molecule of ethylamine.

Various steps involved are shown below (Fig. 3.17).

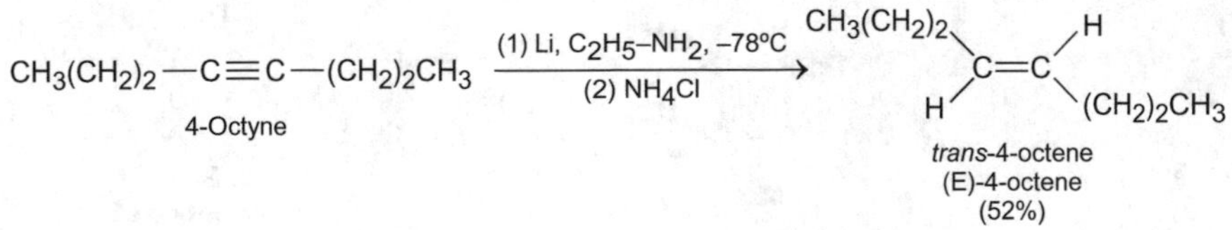
CH3(CH2)2—C≡C—(CH2)2CH3
4-Octyne
(1) Li, C2H5–NH2, –78°C
(2) NH4Cl
CH3(CH2)2
H
C=C
H
(CH2)2CH3
trans-4-octene
(E)-4-octene
(52%)

**Fig. 3.16** Synthesis of *trans*-4-octene

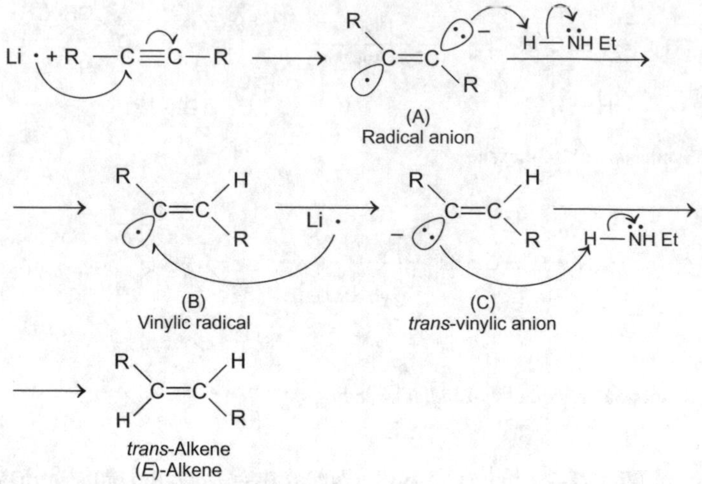

(A)
Radical anion

(B)
Vinylic radical

(C)
trans-vinylic anion

trans-Alkene
(E)-Alkene

**Fig. 3.17**  Mechanism of anti-addition of H$_2$ to alkyne

Both *cis* and *trans* alkenes can be obtained by the **Wittig Reaction** of an aldehyde with alkylidene phosphorane (R CH = PR$_3$). Normally, the *cis* or Z alkene is obtained. However, in the presence of lithium salt, the *E* or *trans* alkene is obtained as shown below (Fig. 3.18).

*E* and *Z* alkene can also be obtained by the Peterson reaction, silicon version of Wittig reaction. For details, see Sect. 20.10 in Chap. 19.

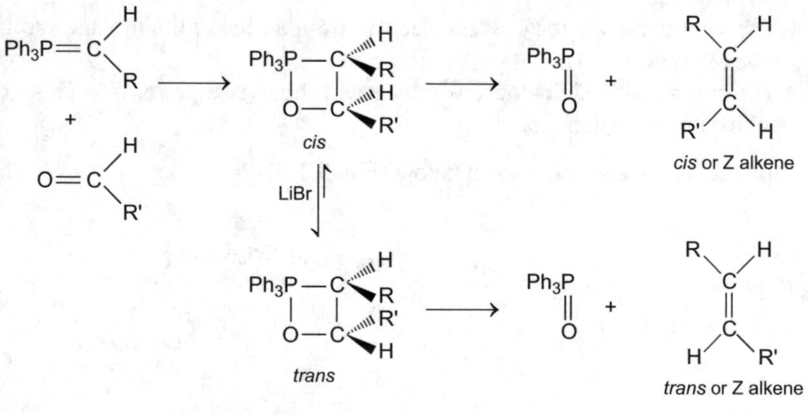

cis

cis or Z alkene

LiBr

trans

trans or Z alkene

**Fig. 3.18**  Wittig reaction

## 3.7 Characterisation of *Cis* and *Trans* Isomers

### 3.7.1 Physical Properties

The *cis* and *trans* isomers generally differ in their physical properties like melting point, boiling point, solubility, dipole moment and spectral properties.

(i) **Melting Point:** In general, *trans* isomers have a higher m.p. than the *cis* isomers. This is because the *trans* isomer is more symmetrical and fits into the crystal lattice more easily.

Some examples include as shown in Fig. 3.19 and for more examples, see Table 3.1.

(ii) **Boiling Point:** The correlation of boiling points with the configuration of the isomer is not as exact as in the case with melting points. This is because the b.p. depends on molecular volume and so b.p. is not of much use for such determinations. For examples, see Table 3.1.

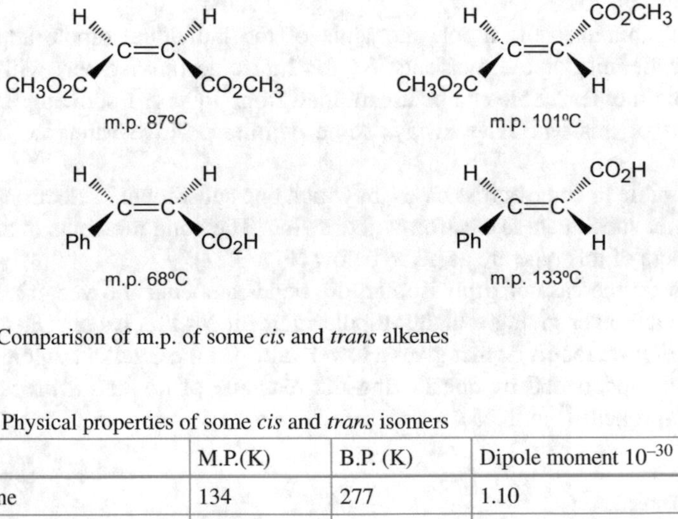

**Fig. 3.19** Comparison of m.p. of some *cis* and *trans* alkenes

**Table 3.1** Physical properties of some *cis* and *trans* isomers

| Isomer | M.P.(K) | B.P. (K) | Dipole moment $10^{-30}$ (cm) |
|---|---|---|---|
| *cis*-2-butene | 134 | 277 | 1.10 |
| *trans*-2-butene | 167 | 274 | 0 |
| *cis*-1,2-dichoroethene | 193 | 333 | 6.17 |
| *trans*-1,2-dichloroethene | 223 | 321 | 0 |
| *cis*-1,2-dibromoethene | 220 | 383 | 4.5 |
| *trans*-1,2-dibromoethene | 267 | 381 | 0 |
| *cis*-1,2-diiodoethene | 259 | 345 | 2.50 |
| *trans*-1,2-diiodoethene | 461 | 465 | 0 |

(iii)   **Solubilities:** The *cis* isomers have in general higher solubility than the *trans* isomer. As an example, the solubilities of maleic acid (*cis* isomer) and fumaric acid (*trans* isomer) are 79.0 g and 0.79 g, respectively, in 100 mL water at 20°.

(iv)    **Dipole Moment:** Dipole moment is most dependable for differentiation between *cis* and *trans* isomers. In general, the *cis* isomers have a greater dipole moment than the *trans* isomers. In compounds of the type $abC == Cab$, the *trans* isomer has zero dipole moments. Some examples are given in Table 3.1. This is because in the *trans* isomer, the same substituents are in opposite directions and so whatever is the magnitude of dipole moments due to one bond in one direction is cancelled by an equal moment operating in the opposite direction and so the resultant dipole moments are zero.

Depending on whether the substituents are electron donating or electron withdrawing, the directions of the dipole moments due to individual bonds for *trans* isomer are as given below (Fig. 3.20).

In both the cases, the resultant dipole moments is zero.

However, in the *cis* isomer, depending upon whether the groups are electronwithdrawing or electron-donating, the direction of individual bond moments is as given below (Fig. 3.21).

In both situations, the dipole moments of the individual dipole add vectorially giving definite dipole moments. So the molecule (*cis* isomer) will have the same dipole moment. This can be ascertained from Table 3.1 showing that the *cis* compounds of this type have always some definite positive value for the dipole moment.

It is possible to come across cases, in which one substituent is electron-donating and the other substituent is electron-withdrawing. The bond moments in the *cis* and *trans* isomers of this type are as given below (Fig. 3.22).

As seen, in the case of *trans* isomer, the bond moments add vectorially and so reinforce each other giving a higher dipole moment for this isomer. However, the vectorial addition for *cis* isomer gives a lower value for the resultant dipole moment. This is well understood by considering the example of *cis* and *trans* isomers of 1-chloro-1-propene (Fig. 3.23).

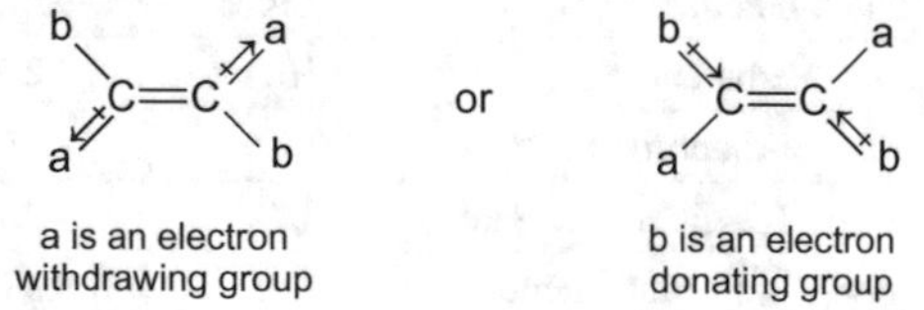

**Fig. 3.20** Direction of bond moment

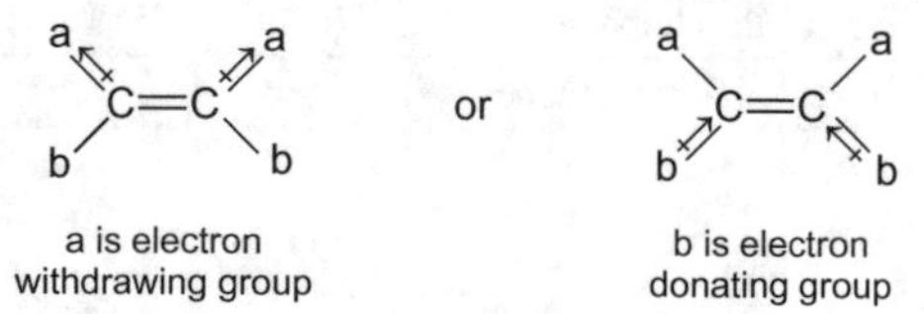

a is an electron withdrawing group

b is an electron donating group

**Fig. 3.21** Dipole moment depends on the nature of the group

a is electron withdrawing group

b is electron donating group

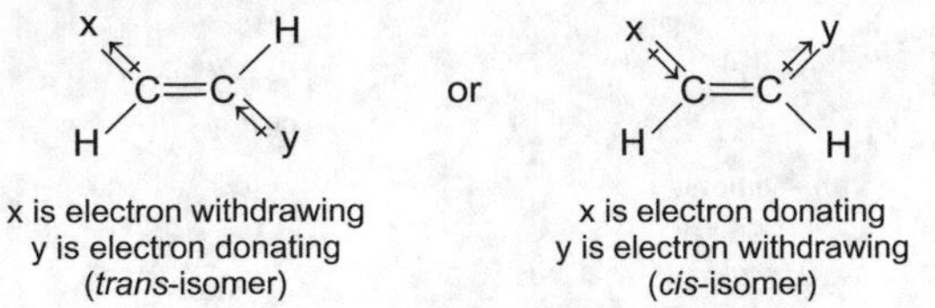

x is electron withdrawing
y is electron donating
(*trans*-isomer)

x is electron donating
y is electron withdrawing
(*cis*-isomer)

**Fig. 3.22** Dipole moment

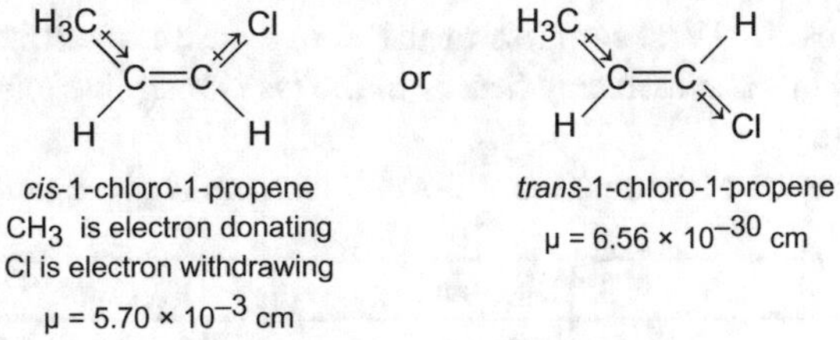

*cis*-1-chloro-1-propene
$CH_3$ is electron donating
Cl is electron withdrawing
$\mu = 5.70 \times 10^{-3}$ cm

*trans*-1-chloro-1-propene
$\mu = 6.56 \times 10^{-30}$ cm

**Fig. 3.23** Dipole moment of *cis*- and *trans*-1-chloro-1-propene

The m.p., b.p. and dipole moments of some *cis* and *trans* isomers are given in Table 3.1.

(v) **Spectral Properties:** The important spectral properties which are used to differentiate between *cis* and *trans* isomers are infrared spectrum, UV spectrum and NMR spectrum.

(a) *Infrared Spectrum*: Infrared absorptions occur in those molecules which produce a change in dipole moment. Thus, *trans*-1,2-dichloroethene shows no IR absorption due to $C = C$ stretching since its dipole moment is zero. However, *cis*-1,2-dichloroethene shows a strong $C = C$ stretching vibration at 1590 cm$^{-1}$, since the molecule has a dipole moment. In a similar way, there is a difference in the IR spectra of fumaric and malic acids and *cis*- and *trans*-3-hexenes. The differences in the IR absorptions are much less marked when the substitution at the olefinic double bond is not symmetrical; in such a case, even the *trans* isomer has a small dipole moment. As an example, *trans*-2-hexene, (E) $CH_3CH = CHC_3H_7$ shows a $C = C$ stretching frequency at 1670 cm$^{-1}$ and is much less intense than that of *cis* (Z) isomer at 1656 cm$^{-1}$. The *cis* and *trans* isomers of 1,2-dichloropropene, $ClCH = CClCH_3$, both show IR absorption in the $C = C$ stretching region at 1614 and 1615 cm$^{-1}$, respectively.

(b) *Ultraviolet Spectrum*: Simple alkenes show UV absorption at about 180 nm (*trans* isomer) or 183 nm (*cis* isomer). However, when the double bond is conjugated with another double bond in an aromatic ring, there are appreciable differences between the *cis* and *trans* isomers, the *trans* isomer displaying longer wavelength absorption and higher intensity than the corresponding *cis* isomer. A typical example is the UV absorption of *cis*- and *trans*-stilbenes (Fig. 3.24).

trans-stilbene
$\lambda_{max}$ 295 nm
$\varepsilon_{max}$ 27000

cis-stilbene
$\lambda_{max}$ 280 nm
$\varepsilon_{max}$ 13500

**Fig. 3.24** UV absorption of *cis*- and *trans*-stilbenes

The differences in UV absorption in the above case are attributed to effective π-orbital overlap in *trans* isomer, whereas in the *cis* isomer due to steric effects, the coplanarity is lost.

The UV absorption of some other *cis–trans* isomers is given below.

| Compound | $\lambda_{max}$ (nm) | $\varepsilon_{max}$ Lmol$^{-1}$ cm$^{-1}$ |
|---|---|---|
| *trans*-cinnamic acid | 273 | 21,000 |
| *cis*-cinnamic acid | 264 | 9,500 |
| *trans*-α-methylstilbene | 272 | 21,000 |
| *cis*-α-methylstilbene | 267 | 9,340 |
| *trans*-1-phenylpropene | 265 | 18,200 |
| *cis*-1-phenylpropene | 249 | 16,600 |

(c)  *Nuclear Magnetic Resonance Spectrum*: NMR spectra is most useful to differentiate *cis*- and *trans* isomers of the type RCH = CHR′. The two protons have different coupling constants. The coupling constant of *trans* form is higher than that of *cis* form. A typical example is that of cinnamic acid (Fig. 3.25).

In *cis–trans* isomers, if a proton can couple with a proton or fluorine, in such a case the coupling constants are different (Fig. 3.26).

Also in *cis–trans* isomers, the protons which are coplanar (or near coplanar) with an aromatic ring (as in the case of stilbene) are deshielded. However, protons which lie above them are shielded. Thus, in *trans*-stilbene, each of the *trans* olefinic protons ($H_a$) is deshielded by both aromatic rings while in the *cis* form each of the hydrogens ($H_b$) is deshielded by only one adjacent aromatic ring. Therefore, $H_a$'s appear at a lower field than $H_b$'s (Fig. 3.27).

cis-cinnamic acid
$J_{H,H}$ = 12 Hz

trans-cinnamic acid
$J_{H,H}$ = 16 Hz

**Fig. 3.25** Coupling constants of *cis* and *trans* cinnamic acids

Fig. 3.26 Coupling constants between H, H and H, F in *cis* and *trans* isomers

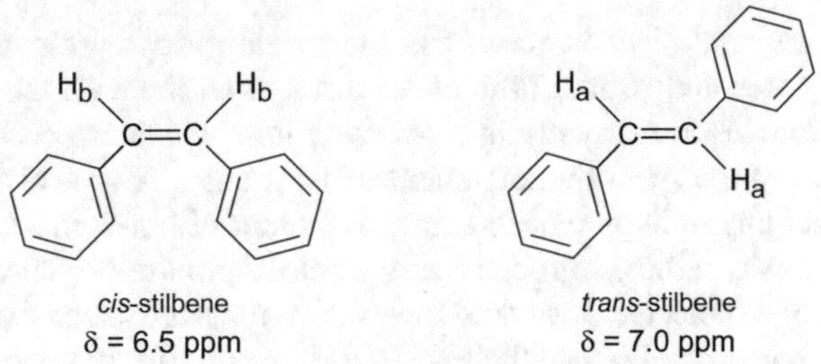

| X, Y | $J_{cis}$ (Hz) | $J_{trans}$ (Hz) |
|------|----------------|------------------|
| H, H | +4 to +12 | +12 to +19 |
| H, F | −4 to +20 | +12 to +50 |

*cis*-stilbene
δ = 6.5 ppm

*trans*-stilbene
δ = 7.0 ppm

Fig. 3.27 NMR absorptions in *cis*- and *trans*-stilbenes

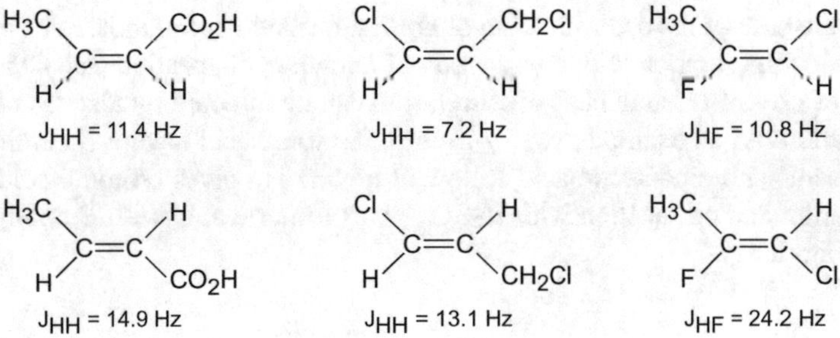

$J_{HH}$ = 11.4 Hz

$J_{HH}$ = 7.2 Hz

$J_{HF}$ = 10.8 Hz

$J_{HH}$ = 14.9 Hz

$J_{HH}$ = 13.1 Hz

$J_{HF}$ = 24.2 Hz

Fig. 3.28 Coupling constants in some alkenes

In the case of trisubstituted alkenes RR′C = CHR″, neither IR nor NMR coupling constants are useful for configurational assignments. This problem has been solved by $^{13}$CNMR spectroscopy. The only condition is that both R and R′ (or either of them) must have a carbon atom at the point of attachment to the C=C group. In such a case, the carbon nucleus *cis* to R″ group will be shifted upfield relative to the same carbon nucleus positioned *cis* to H. Some examples are given below (Fig. 3.28).

The *cis* and *trans* isomers can also be determined on the basis of $^{13}$C shifts (ppm). Some examples are given below (Fig. 3.29).

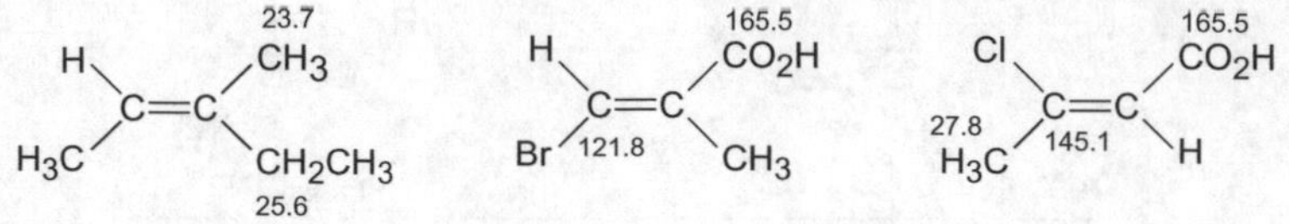

**Fig. 3.29**  $^{13}$C shifts in some *cis* and *trans* alkenes

## 3.7.2   Chemical Reactions

In the case of *cis*- and *trans* isomers, the functional groups present are the same in most of the cases and so it is difficult to distinguish them on the basis of their chemical reactions. However, only in a few reactions which are possible with one isomer only due to the spatial arrangement of its groups. A typical reaction is the formation of an anhydride by maleic acid (*cis* isomer of but-2-ene-1,4-dioic acid). In this case, the two carboxyl groups being in close proximity yield an anhydride by elimination of a molecule of water. However, in the *trans* isomer (fumaric acid), no such reaction is possible since the two COOH groups are in opposite directions (Fig. 3.30).

It is, however, found that on strong heating, fumaric acid (*trans* isomer) forms maleic anhydride.

As in the case of the determination of configuration in chiral molecules by correlating their configuration with compounds of known configuration (see Chap. 4), in the case of *cis–trans* isomers also, correlation can be affected by alkenes of known configuration. As an example, γ, γ, γ-trichlorocrotonic acid having *E* configuration on treatment with zinc-acetic acid followed by Na–Hg gives crotonic acid with *E* configuration and on reaction with $H_2SO_4$ gives fumaric acid with *E* configuration (*trans*) (Fig. 3.31).

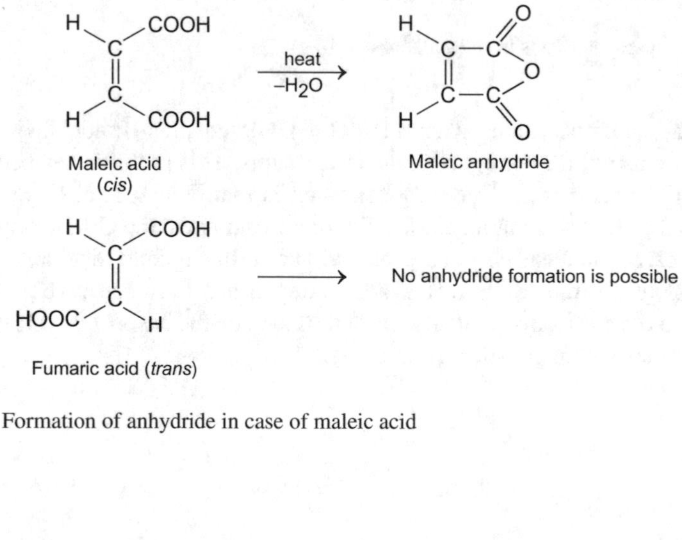

**Fig. 3.30**  Formation of anhydride in case of maleic acid

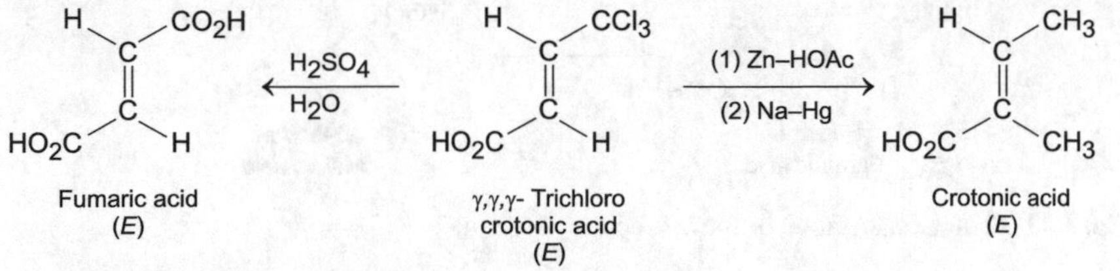

Fumaric acid (E) ← H₂SO₄ / H₂O — γ,γ,γ- Trichloro crotonic acid (E) → (1) Zn–HOAc / (2) Na–Hg — Crotonic acid (E)

**Fig. 3.31** Correlation of crotonic acid with fumaric acid

**Fig. 3.32** Synthesis of *cis* and *trans* β-methyl styrene

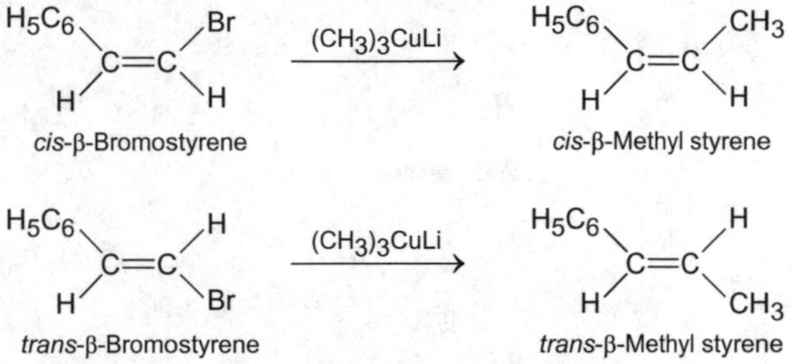

cis-β-Bromostyrene — (CH₃)₃CuLi → cis-β-Methyl styrene

trans-β-Bromostyrene — (CH₃)₃CuLi → trans-β-Methyl styrene

Nucleophilic displacement (with retention) of an olefinic halide of known configuration has also been used to get *cis* or *trans* olefins (Fig. 3.32).

## 3.8 Interconversion of *Cis*–Trans Isomers

The following methods are commonly used for the interconversion of *cis–trans* isomers.

### 3.8.1 Photoisomerisation of Cis and Trans Isomers

Olefins undergo isomerisation on irradiation with UV light in the presence of a sensitizer. In the case of simple olefins, *E*-isomers absorb energy more effectively and at a slightly different wavelength than the *Z* isomer. Usually, *E*-isomer is partially converted into its thermodynamically less stable *Z*-isomer. An example is the interconversion of fumaric and maleic acids (Fig. 3.33).

The ratio of the product formed depends on the substrate and the sensitizer.

As an example, photoisomerisation of *cis* pentene in a *cis–trans* mixture using benzene as a photosensitizer gives a *cis–trans* ratio of about unity. However, the use of benzophenone (or acetophenone) as a photosensitizer gives a *trans/cis* ratio of about 5.5. The mechanism (G. O. Schenck and R. Steinmtz, Bull. Soc. Chem. Belg, 1962, **71**, 781) of the isomerisation is shown below (Fig. 3.34).

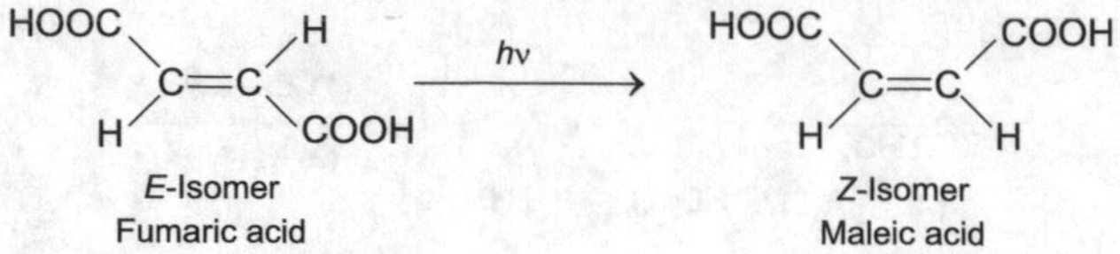

**Fig. 3.33** Photoisomerisation of fumaric acid

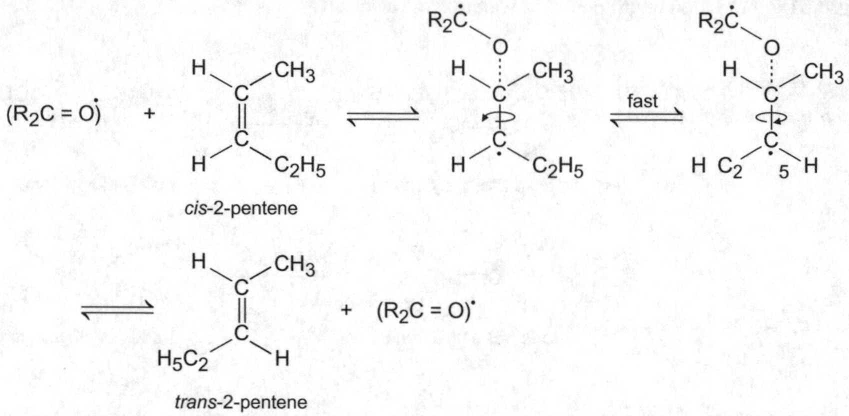

**Fig. 3.34** Mechanism of photochemical isomerisation of *cis*-2-pentene

Another interesting example of photoisomerisation is that of stilbene. *Trans*-stilbene on irradiation in hexane in UV light results in the formation of *cis* isomer. After some time, the *cis–trans* ratio in UV light does not change and becomes constant. This condition called photostationary state is also achieved in the case of *cis* isomer. The equilibrium favours the formation of a less stable *cis* form. At equilibrium, the *cis–trans* isomers are in relative amounts of 10:1 (Fig. 3.35).

The mechanism of the reaction is similar to that given above.

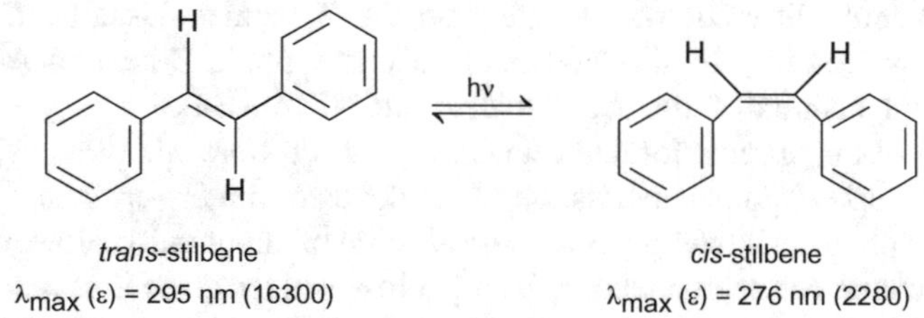

*trans*-stilbene
$\lambda_{max}$ ($\varepsilon$) = 295 nm (16300)

*cis*-stilbene
$\lambda_{max}$ ($\varepsilon$) = 276 nm (2280)

**Fig. 3.35** Photoisomerisation of *trans*-stilbene

### 3.8.2 Conversion of One Isomer (Cis or Trans) into Another Isomer (Trans or Cis)

We have seen that the photoisomerisation of *cis* or *trans* isomers gives a mixture of *cis* and *trans* forms. The separation of the mixture is difficult and so this method is of not much use. The conversion of one isomer into another isomer is called directed *cis–trans* interconversion.

As an example, *trans*-β-ional on irradiation with UV light in the presence of 2-acenaphthone is converted into a *cis–trans* mixture containing 65% of *cis* isomer (Fig. 3.36).

The conversion of *trans* alkene to *cis* alkene can be effected by a sequence of reactions involving *anti*-addition of chlorine followed by syn elimination (P. E. Sonnet and J. E. Oliver, J. Org. Chem., 1976, **41**, 3284) (Fig. 3.37).

Another route for conversion of *trans* into *cis* olefin is given below (Fig. 3.38).

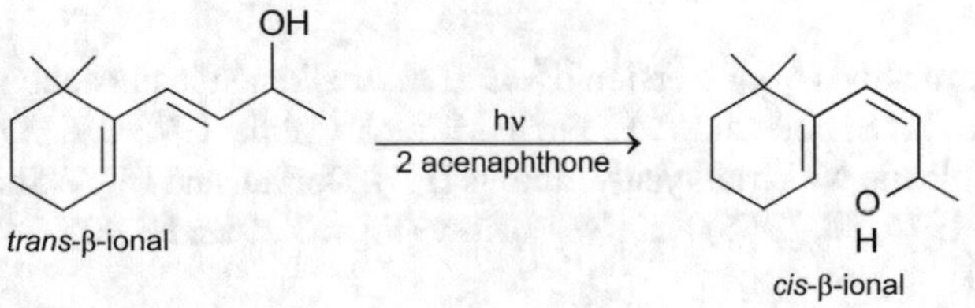

**Fig. 3.36** Photoisomerisation of *trans*-β-ional

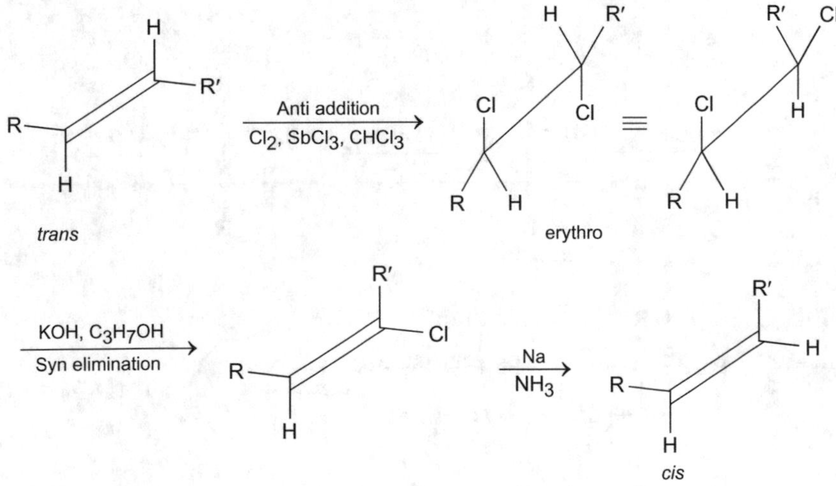

**Fig. 3.37** Conversion of *trans* alkene into *cis* alkene

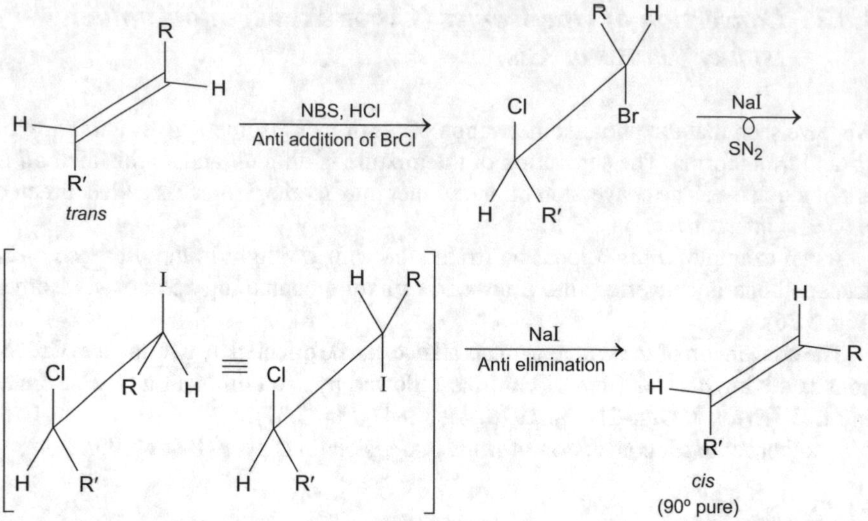

**Fig. 3.38**  Another route for the conversion of *trans* alkene to *cis* alkene

Two other methods of conversion of *cis* to *trans* alkene via phosphonium betaines (E. Vedejs, K. A. Snoble and P. L. Fuchs, J. Org. Chem, 1973 **38**, 1178) and *trans* alkene to *cis* alkene with trialkylsilyl anions (P. B. Dervan and M. A. Shippey, J. Am. Chem. Soc., 1976, **98**, 1265) are given below (Figs. 3.39 and 3.40).

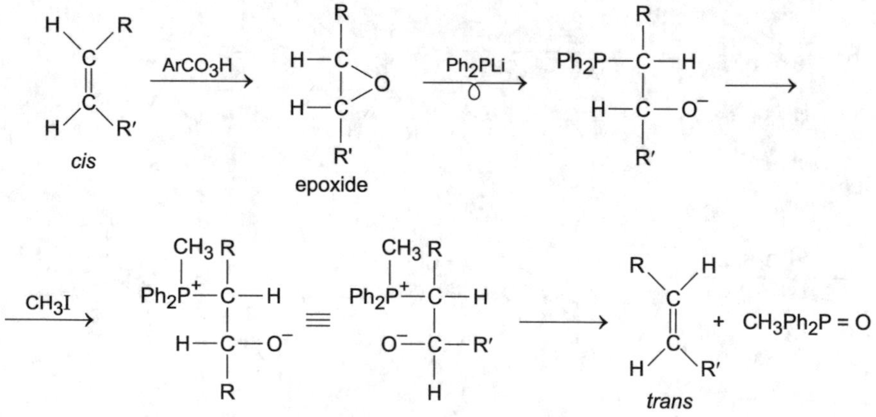

**Fig. 3.39**  Conversion of *cis* alkene into *trans* alkene via phosphonium betaines

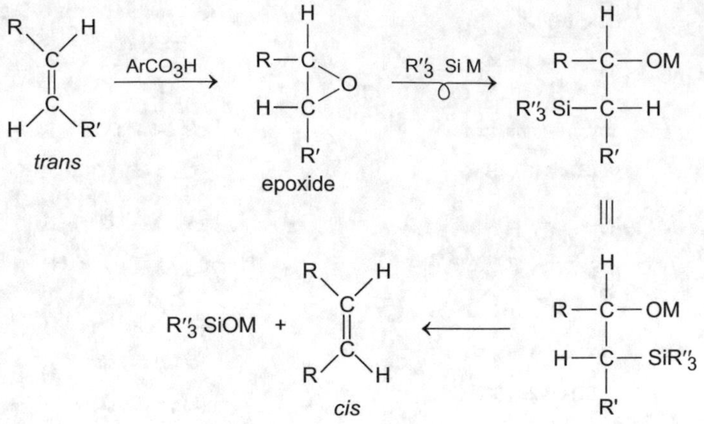

**Fig. 3.40**  Conversion of *trans* olefin into *cis* by trialkylsilyl anions

## 3.9  *Cis–Trans* Isomerism in Conjugated Dienes

In conjugated dienes (which have alternate single and double bonds in a chain which has at least four carbons), e.g., 2,4-hexadiene, besides configuration of the double bonds, the two double bonds can get oriented in *cis-* and *trans* manner. This is known as *s-cis-* and *s-trans* isomerism (*s* stands for a single bond). The *s-cis* and *s-trans* isomerism is in addition to *cis-* and *trans* isomerism exhibited by molecules around a double bond.

In the case of 2,4-hexadiene, there can be as many as six geometrical isomers as given below (Fig. 3.41).

## 3.10  *Cis–Trans* Isomerism in Cumulenes

Besides conjugated dienes *cis–trans* isomerism also exists in cumulenes (J. H. van 't Hoff, Die Lagerung der Atom in Raume, Viehweg and Sohn, Braunchweig, Germany, 1877, p. 14) of the type which have an odd number of double bonds [ab (C = )$_n$ Ccd, where *n* is odd]. This type of isomerism is due to the fact that successive planes of $\pi$-bonds are orthogonal to each other. In butatriene, the *cis–trans* isomerism was observed in 1959 (R. Kuhn and B. Schulz., Chem. Ber., 1959, **92**, 1483).

The stereo isomerism in cumulenes is shown below (Fig. 3.42).

In the case of substituted butatriene, the *cis–trans* isomerism is shown below (Fig. 3.43).

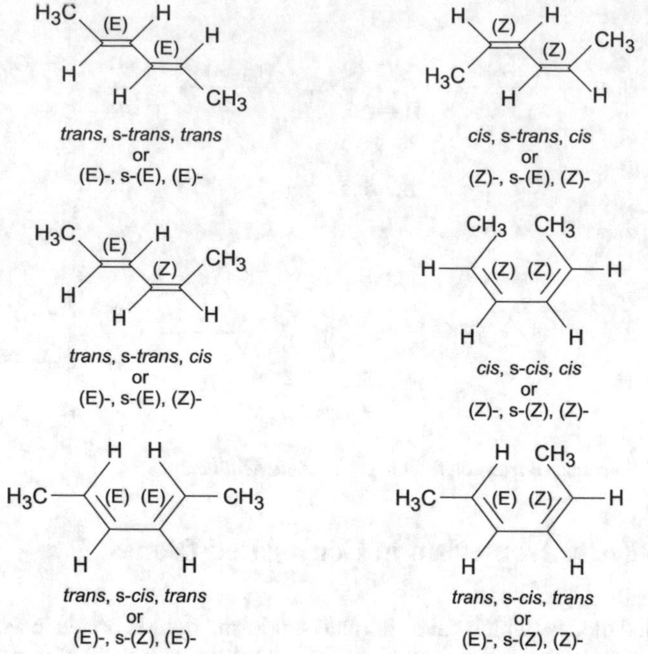

trans, s-trans, trans
or
(E)-, s-(E), (E)-

cis, s-trans, cis
or
(Z)-, s-(E), (Z)-

trans, s-trans, cis
or
(E)-, s-(E), (Z)-

cis, s-cis, cis
or
(Z)-, s-(Z), (Z)-

trans, s-cis, trans
or
(E)-, s-(Z), (E)-

trans, s-cis, trans
or
(E)-, s-(Z), (Z)-

**Fig. 3.41** Geometrical isomers of 2,4-hexadiene, 's' refers to single bond

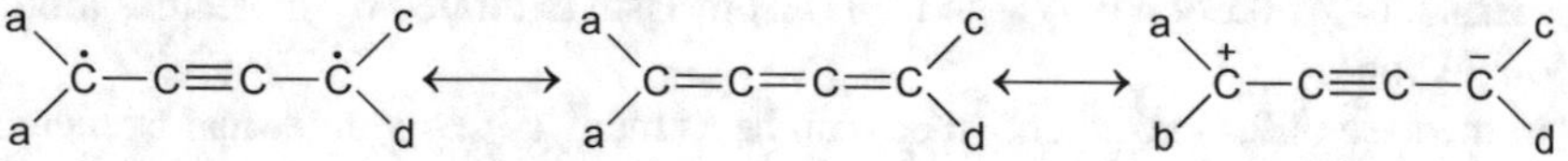

**Fig. 3.42** Stereoisomerism in cumulenes

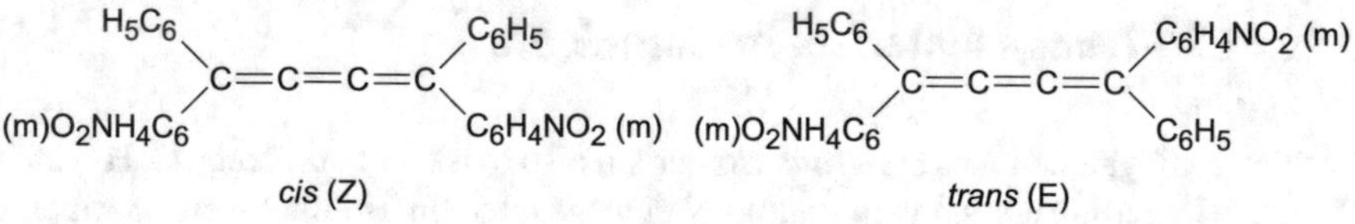

cis (Z)

trans (E)

**Fig. 3.43** *Cis–trans* isomerism in substituted butatriene

## 3.11 *Cis–Trans* Isomerism Due to Restricted Rotation About C–N Bonds

Geometrical isomers of compounds of the types I and II have been isolated by Adams et al. (1950). In these cases, geometrical isomerism is possible due to restricted rotation about the C–N bonds (Fig. 3.44).

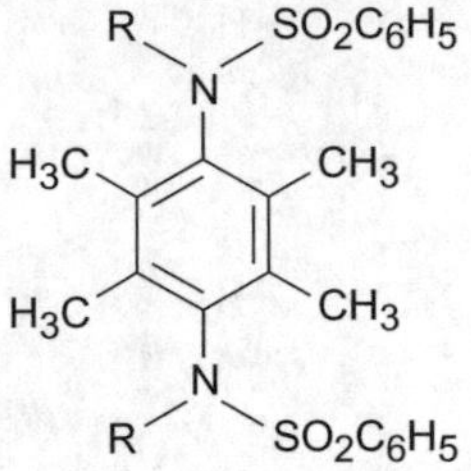

**Fig. 3.44** Geometrical isomerism due to restricted rotation about C–N bonds

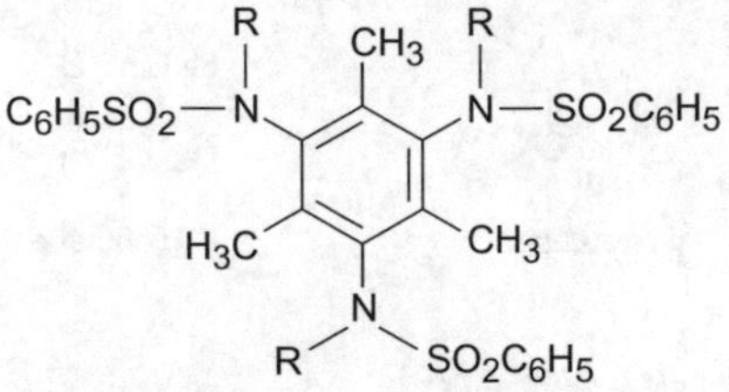

**Fig. 3.45** *Cis–trans* isomerism in terphenyl compounds

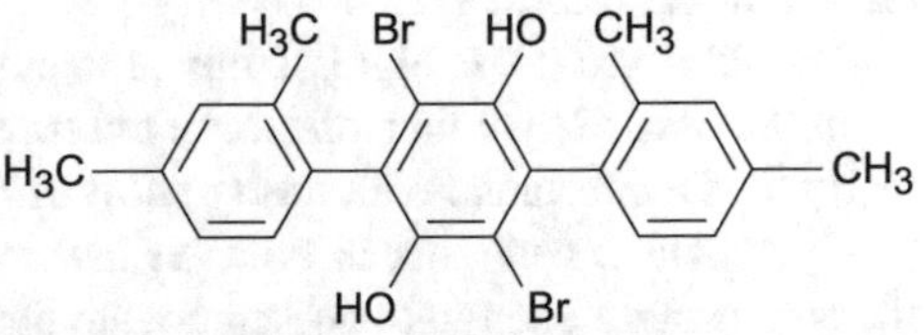

## 3.12 *Cis–Trans* Isomerism in Terphenyl Compounds

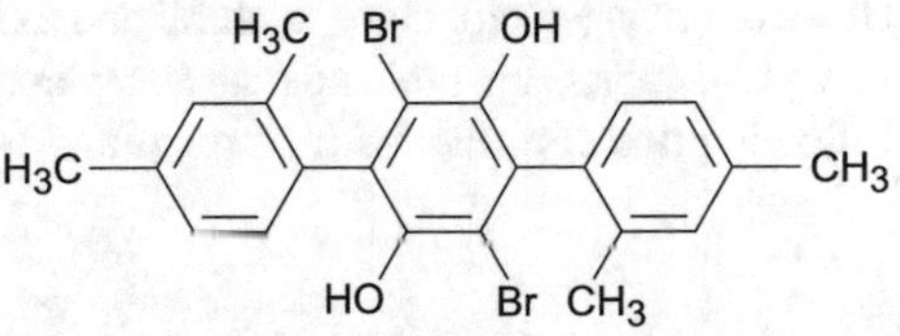

Suitably substituted terphenyl compounds exhibit geometrical isomerism when the substituents present to prevent the free rotation about single bonds.

As an example, *cis* and *trans* forms of (3) was prepared (Browning and Adams, 1930) (Fig. 3.45).

## 3.13 Stereochemistry of Cycloalkenes

Cycloalkenes, like alkenes, also contain a double bond. Cycloalkenes up to five carbon atoms exist only in the *cis* form. In cycloalkenes, the introduction of a *trans* double bond will introduce much greater strain than the bonds of the ring system can accommodate (Fig. 3.46).

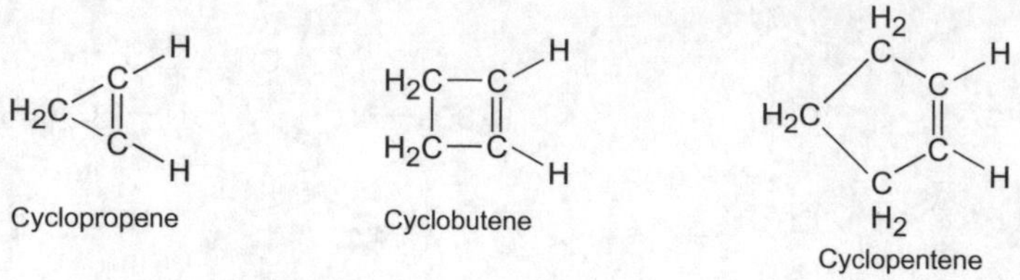

Fig. 3.46  *cis* cycloalkenes

The cyclohexene (A) might resemble the structure (B). It is believed that the structure (B) can be formed as a very reactive short-lived intermediate in some chemical reactions (Fig. 3.47).

The structure (B) is highly strained to exist at room temperature.

In the case of cycloheptene, only the *trans* form has been observed spectroscopically. It is a substance with a very short lifetime and it is not possible to isolate it.

Cyclooctene can exist in both *cis* and *trans* forms. In the case of cyclooctene, the ring is large enough to accommodate the geometry necessary for a *trans* double bond. *Trans*-cyclooctene is stable at room temperature (unlike other cycloalkenes). However, *trans*-cyclooctene is chiral and exists as a pair of enantiomers (Fig. 3.48).

Cycloheptatriene molecule is non-planar with a rapid inversion between the following two equivalent structures A and B.

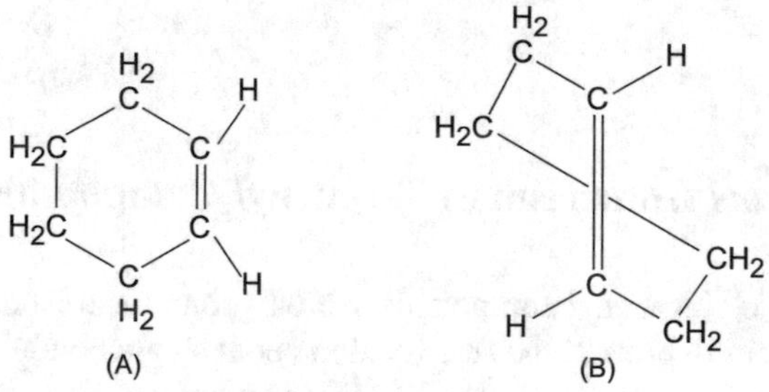

Fig. 3.47  Cyclohexene

Fig. 3.48  *cis*- and *trans*-cyclooctene

*cis*-cyclooctene

*trans*-cyclooctene

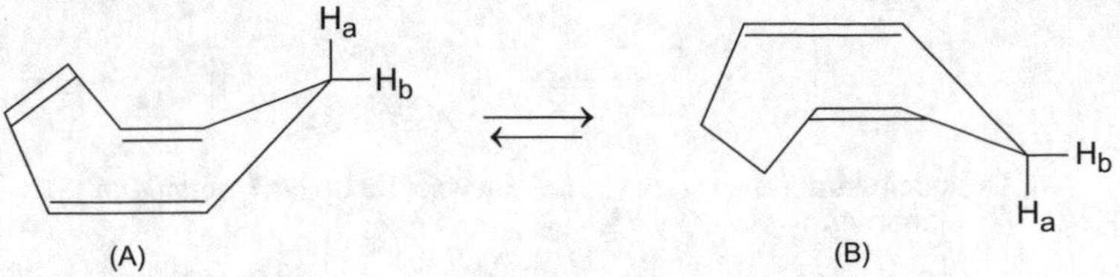

(Two equivalent structures of cycloheptatriene)

The inversion at low temperature is seen by NMR proton signals for $H_a$ and $H_b$ (which are resolved) at low temperature.

Cycloheptatriene also exhibits thermal isomerism due to 1,5-hydrogen transfer across the molecule. This is supported on the basis of change in NMR signal for allylic and vinylic protons.

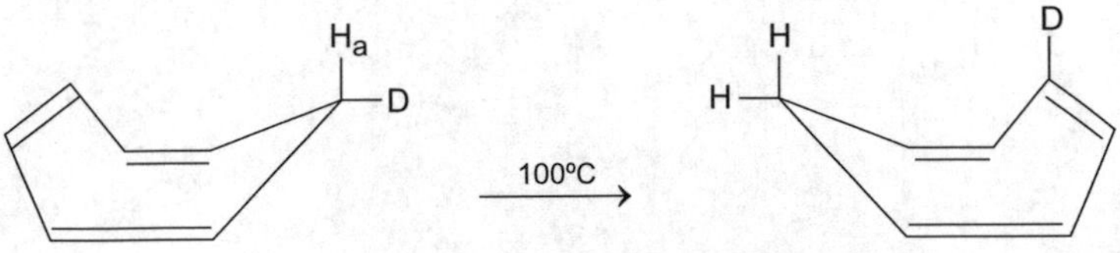

Thermal isomerism in cycloheptatriene

Cyclooctatetraene has been shown to exist in tub form with alternating single and double bonds. The bonds of cyclooctatetraene are shown to be alternatively long (1.48 A) and short (1.34 A), respectively (on the basis of X-ray studies) (Fig. 3.49).

Cyclooctatetraene is found to exist as bicyclo [4.2.0] −2,4,7-octatriene. This has been shown on the basis of adduct formation with maleic anhydride (Fig. 3.49) (Fig. 3.50).

**Fig. 3.49** Cyclooctatetraene

Cyclooctatetraene
planar form

Tub form
of cyclooctatetraene

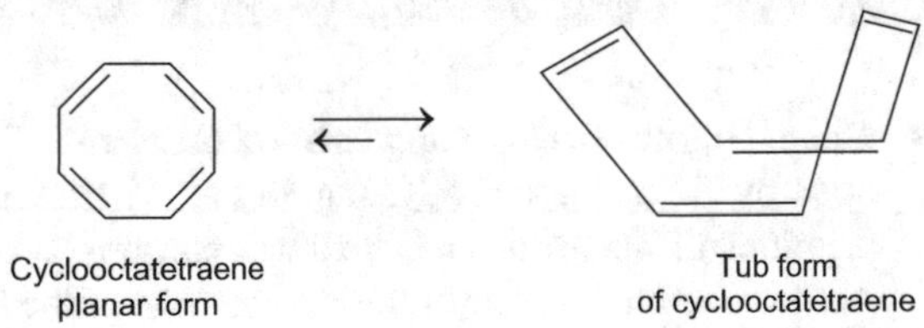

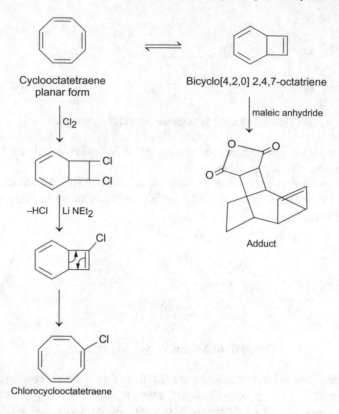

**Fig. 3.50** Adduct of cyclooctatetraene with maleic anhydride

**Key Concepts**

- **Chan–Ingold–Prelog Convention:** An *E* and *Z* system of nomenclature developed by R. S. Cahn, C. K. Ingold and V. Prelog. This system depends on the priorities of the groups attached to a carbon–carbon double bond. The priorities are decided on the basis of the atomic mass of the first atom attached to the double bond.
- *Cis–trans* **Isomerism:** Also known as geometrical isomerism, it is generally encountered in compounds containing one or more double bonds and arises due to restriction in the rotation due to the presence of double bonds.
- *Cis–trans* **Isomers:** Stereo isomers that differ in their stereochemistry about a double bond. These are also called geometric isomers.
- **Conjugated Dienes:** Olefinic compounds having alternate single and double bonds, e.g., 2,4-hexadiene. These also exhibit *cis–trans* isomerism known as s-*cis* and s-*trans* isomerism (s stands for a single bond).

- **Cumulenes:** Polyenes containing continuous double bonds. Cumulene containing an odd number of double bonds exhibits *cis–trans* isomerism. An example is butatriene

$$
\underset{b}{\overset{a}{>}}C=C=C=C\underset{b}{\overset{a}{<}} \qquad \underset{b}{\overset{a}{>}}C=C=C=C\underset{a}{\overset{b}{<}}
$$

cis (Z)          trans (E)

- **Dipole Moment:** It is a measure of the polarity of a molecule and is useful for differentiation between the *cis* and *trans* isomers.
- **E Configuration:** In case of alkenes if the higher priority groups or atoms are on the opposite side of the double bond, the configuration is designated as E (from German, Estgenen, across): An Example is E-2-bromo-1-chloro-1-fluoroethene

$$
\begin{array}{c}
F \qquad Cl \\
\diagdown \ \diagup \\
C \\
\| \\
C \\
\diagup \ \diagdown \\
Br \qquad H
\end{array}
\qquad
\begin{array}{l}
Cl > F \\
Br > H
\end{array}
$$

E-2-Bromo-1-chloro-1-fluoroethene

- **Heat of Combustion:** The amount of heat released when a compound is burned in a calorimeter according to the equation.

$$
C_nH_m + O_2 \rightarrow nCO_2 + m/2\ H_2O
$$

The comparison of the heat of combustion of different alkenes permits one to determine the relative stability of different alkenes.

- **Heat of Hydrogenation:** The amount of heat released when a carbon–carbon double bond is hydrogenated. Comparison of the heat of hydrogenation of different alkenes permits one to determine the relative stability of different double bonds. Alternatively, the enthalpy changes involved in the hydrogenation of an alkene are called heat of hydrogenation.
- **Lindlar's Catalyst:** It is metallic palladium deposited on calcium carbonate, conditioned with lead acetate and quinoline. Using internal alkynes and reduction with Lindlar's catalyst gives *cis* (Z) alkenes.
- **Nickel Boride:** Also known as P-2-catalyst, it is prepared from nickel acetate by reduction with sodium borohydride.

$$
\underset{}{\overset{O}{\overset{\|}{Ni(OCCH_3)_2}}} + NaBH_4 \xrightarrow{C_2H_5OH} \underset{P\text{-}2}{Ni_2B}
$$

The reduction of alkene using P-2 catalyst involves the syn addition of hydrogen to an internal triple bond to give *cis* (Z) alkene.

- **Photo Isomerisation:** Isomerisation of *cis* and *trans* alkene by UV light in the presence of a sensitizer.
- **Wittig Reaction:** The reaction of a carbonyl compound with alkylidene phosphorane gives *E* and Z alkene.
- **Z Configuration:** In case of alkenes, if the higher priority groups or atoms are on the same side of the double bond, the configuration is designated Z (from the German, Zusammen, together). An example is Z-2-bromo-1-chloro-1-fluoroethene

Cl > F
Br > H

Z-2-Bromo-1-chloro-1-fluoroethene

## Problems

1.  Explain the term *cis–trans* isomerism.
2.  Which of the following alkenes can exist as *cis–trans* isomers. Write their structures.

    (a)  1, 2-Difluoroethene
    (b)  1, 2-Dichloro-1,2-difluoroethene
    (c)  2-Methylpropene
    (d)  1-Chloro-1-butene.

3.  Explain the reason for alkenes exhibiting *cis–trans* isomerism.
4.  Discuss *E* and Z system of nomenclature.
5.  How is the priority of groups containing double and triple bonds determined?
6.  Arrange methyl, ethyl, propyl and isopropyl groups in order of their priorities.
7.  Using *E* and Z system of nomenclature, give IUPAC names of the following:

8.    Assign $E$ or $Z$ configuration to the following compounds:

(a) 
$$CH_3CH_2 \diagdown \qquad \diagup CH_3$$
$$C=C$$
$$H \diagup \qquad \diagdown CH_2CH_3$$

(b) 
$$Br \diagdown \qquad \diagup CH_2Br$$
$$C=C$$
$$Cl \diagup \qquad \diagdown CH_3$$

(c) 
$$\qquad\qquad\qquad O$$
$$\qquad\qquad\qquad \|$$
$$ClCH_2CH_2 \diagdown \qquad \diagup COH$$
$$C=C$$
$$H_3CH_2C \diagup \qquad \diagdown H$$

(d) 
$$CH_3CH_2 \diagdown \qquad \diagup C\equiv CH$$
$$C=C$$
$$H \diagup \qquad \diagdown CH_2CH_3$$

9.    How will you synthesis *cis-* and *trans*-3-hexenes and *cis-* and *trans*-4-octenes?
10.   Besides alkenes, which other types of compounds exhibit *cis–trans* isomerism?
11.   Using the Chan–Ingold–Prelog priority rules, arrange the groups Cl, Br, I, $SO_3$, SH, $NH_2$, COOH and $COOCH_3$ in order of their priority.
12.   Determine the priority of the groups $NMe_2$, —NHMe and —$NH_2$.
13.   Arrange —$CH_2CH_3$, $CH_2CH_2CH_3$, —$CH_2$—$CHMe_2$ in order of their priority.
14.   How will you determine the relative stabilities of *cis* and *trans* alkene?
15.   Give two methods for the synthesis of *cis* and *trans* alkenes.
16.   Write a note on the characterisation of *cis* and *trans* isomers.
17.   Discuss methods for the interconversion of *cis* and *trans* isomers.
18.   Write a note on a *cis–trans* isomerism in conjugated dienes and cumulenes.
19.   Discuss the stereochemistry of cycloalkenes.

# Chapter 4
# Stereochemistry of Organic Compounds Containing Asymmetric Carbon

## 4.1 Introduction

Carbon is tetravalent, that is it can form four bonds, as in the case of methane. In case, a carbon atom is attached to four different groups or atoms it is said to be chiral or asymmetric. An example of a simple compound containing four different groups attached to a carbon is lactic acid, the four groups/atoms are H, OH, $CH_3$, COOH. The presence of asymmetric carbon atom in an organic molecule makes the compound optically active. A substance that rotates the plane of polarisation of plane-polarised light is said to be optically active. An optically active sample must contain a chiral carbon but all compounds containing chiral carbon are not necessarily optically active. As an example, a racemic sample is not optically active. The rotation of the plane polarised light is referred to as **optical rotation**. Compounds that have similar physical and chemical properties but differ in their behaviour towards plane polarised light are known as **optical isomers** and the phenomenon is known as **optical isomerism**.

The isomer that rotates the plane polarised light in the clockwise direction (right) is called **dextrorotatory** (Latin, *dexter* = right) and is indicated by *d* or (+), sign. On the other hand, the isomer that rotates the plane polarised light in anticlockwise direction (left) is called **laevorotatory** (Latin, *lavus* = left) and is indicated by *l* or (–) sign.

**What is plane polarised light and how it is obtained?**: Ordinary light is an electromagnetic vibrations, which is of different wave lengths. The vibrations occur in all planes at right angles to the line of propagation when such a light is passed through a '**Nicol prism**', it is converted into plane polarised light, which is a monochromatic ray of light with uniplanar vibrations only in a plane which is perpendicular to the direction of propagation.

The Nicol prism is a doubly reflecting crystal of calcite ($CaCO_3$). For more details about Nicol prism see key concepts at the end of the chapter). The formation of plane polarised light from ordinary light is shown in Fig. 4.1.

© The Author(s), under exclusive license to Springer Nature Switzerland AG 2022
V. K. Ahluwalia, *Stereochemistry of Organic Compounds*,
https://doi.org/10.1007/978-3-030-84961-0_4

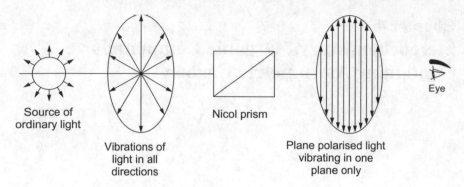

Source of
ordinary light

Vibrations of
light in all
directions

Nicol prism

Plane polarised light
vibrating in one
plane only

Eye

**Fig. 4.1**    Formation of plane polarised light

## 4.2    Origin of Optical Activity

It was Biot (1850) who discovered optical activity. He found that there are two types
of quartz crystals, which rotate the plane of polarised light in opposite direction.

It was substiquently found that even some compounds in solution also exhibit
the same property. It was found that the optical activity in solution is due to some
molecular property (*viz.,* arrangement of groups attached to carbon), which is also
[retained in solution]. Subsequently, Pasteur studied the salts of tartaric acid. He
found that sodium ammonium tartrates, which is optically inactive, is infact, a mixture
of two kinds of crystals which were mirror image of each other and rotated the plane
polarised light in opposite direction. In fact, Pasteur was successful in separation of
the two kinds of crystals using a pair of tweezers and a hand lens.

According to van Hoff and Le Bell (1874), the optical activity is caused due to
four valencies of carbon being directed in three dimensional space towards the four
corners of a regular tetrahedron, the carbon atom being at the centre. In case, the
four different groups (*e.g., a, b, c, d*) are attached to carbon, the molecule becomes
asymmetric. Thus, for a compound *Cabcd*, there are two ways of arrangement of
groups.

The two structural arrangements are non-superimposable and are mirror image at
each other (Fig. 4.2). Thus, the presence of four different atoms or groups attached to
a carbon atom is the cause of optical activity. This carbon atom is called asymmetric
carbon atom (Fig. 4.2).

The carbon atom, which has four different groups or atoms attached to it, is called
an **asymmetric carbon atom** and is generally indicated by an asterisk (*).

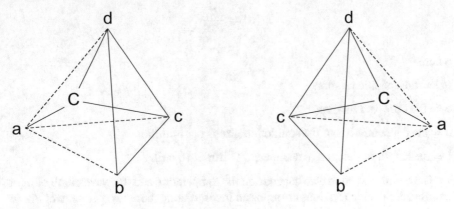

**Fig. 4.2**  Non-superimposable mirror images of Cabcd

## 4.3   Measurement of Optical Activity

The optical activity is measured by an instrument called, a **polarimeter**. It consists of a source of light (usually sodium light of fixed wavelength is used). This light is narrowed down by using a slit and this passes through a 'Nicol prism' called a polariser. The plane polarised light, thus obtained is passed through a solution of an organic compound. In case, optically active compound is used, the plane of polarisation gets rotated to right or left. The angle θ through which the plane polarised light gets rotated is measured by using another Nicol prism (called the analyser) and a scale for measuring the number of degrees that the plane of polarised light is rotated. As already mentioned, a substance their rotates the plane-polarised light in clockwise direction is said to be dextrorotatory, and the one that rotates the plane-polarised light in a counter clockwise direction is called laevorotatory. Figure 4.3 shows the schematic diagram of a polarimeter.

It has been shown that the degree of optical rotation, θ (theta), due to an optically active compound depends on the length of the tube and the concentration of the optically active compound and is given by the formula

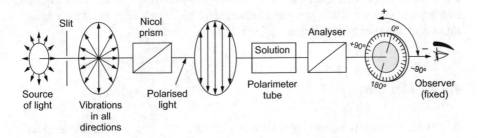

**Fig. 4.3**  Schematic representation of a polarimeter

$$[\alpha] = \frac{\alpha}{c \cdot l}$$

where

$[\alpha]$ = the specific rotation

$\alpha$ = the observed rotation

$c$ = the concentration of the solution in grams per milliliter

$l$ = the length of the tube in decimeters (1 dm = 10 cm).

The specific rotation also depends on the temperature and the wavelength of light employed. Specific rotations are reported incorporating these two parameters. As an example, a specific rotation is given by:

$$[\alpha]_D^{25} = +3.12°$$

This implies that the D line of sodium lamp ($\lambda = 589.6$ nm) is used for the light at a temperature of 25 °C and the sample containing 1.00 g mL$^{-1}$ of the optically active substance in a 1 dm tube produced a rotation of 3.12° in a clockwise direction. The specific rotation also depends on the solvent used and due to this reason, this solvent is also specified when rotation is reported. Thus,

$$[\alpha]_D^{25°C} = +13.5°(C = 0.25 \text{ g}/100 \text{ mL, solvent ethanol})$$

## 4.4   Optical Isomerism

Optical isomerism is exhibited by compounds which show optical activity, i.e., which rotate the plane of polarised light. Such compounds, as has already been stated are called optically active compounds. The property of optical activity is due to the presence of chiral carbon. In case a tetrahedral carbon is attached to four different substituents, it results in the existence of mirror-image molecules which are non-superimposable. Such non-superimposable molecules which are mirror image molecules are called **enantiomers** (Fig. 4.4).

### 4.4.1   Chirality

A molecule, which is non-superimposable with its mirror image is known as chiral. Such a molecule can exist as two enantiomers. The word chiral is derived from the

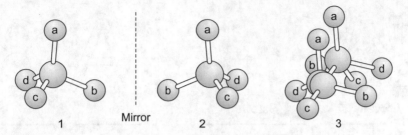

**Fig. 4.4**  Structure 2 is mirror image of structure 1. In Fig. 4.4(3) it is seen that the structures 2 and 1 are not superimposable

Greek word '*Cheir*', which means hand. Thus, it can be said that chiral molecules show handedness. This means that as our left hand and right hands are not super-imposable with their mirror images (Fig. 4.5), in a similar way, the enantiomers of chiral molecule are not superimposable with each other.

As in the case of hands, a number of other objects are also chiral; these include gloves, shoes, screws etc. On the other hand, achiral objects are those, which are superimposable on their mirror images. Some example include objects like pencil, letter A, alphabet 8 etc. (Fig. 4.6).

According to Cahn, Ingold and Prelog (1966), a molecule is chiral, when it has no elements of symmetry (plane of symmetry, centre of symmetry and alternating axis of symmetry). A discussion on these forms the subjects-matter of a subsequent section (Chap. 5).

The chiral centre (asymmetric carbon atom) is also called a **stereogenic centre**. A stereogenic centre is defined as an atom, in which interchange of any two atoms or groups result in a new stereoisomer (Fig. 4.7). A molecule that cannot be superimposed on its mirror image is called chiral or dissymmetric.

The carbon atom which is responsible for chirality is the asymmetric carbon atom, which is $sp^3$ hybridised. As already stated final conformation of chirality in

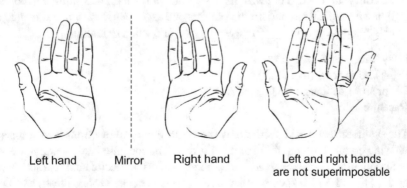

**Fig. 4.5**  Mirror image relationship of left hand and right hand and their non-superimposibility

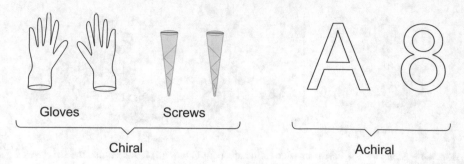

Gloves          Screws                         A  8

Chiral                                              Achiral

**Fig. 4.6**  Some chiral and achiral objects

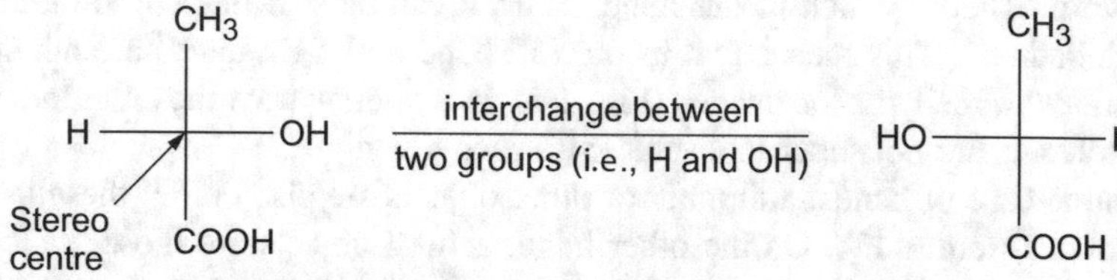

$$
\begin{array}{ccc}
& CH_3 & \\
H \rule[0.5ex]{3em}{0.4pt} OH & \underset{\text{two groups (i.e., H and OH)}}{\overline{\text{interchange between}}} & HO \rule[0.5ex]{3em}{0.4pt} H \\
\underset{\text{centre}}{\overset{\text{Stereo}}{}} \; COOH & & COOH
\end{array}
$$

**Fig. 4.7**  Non-superimposable pairs of enantiomers

a molecule is obtained by considering the super imposibility of a molecule with its mirror image (*see* Fig. 4.7). Besides these, the elements of symmetry have also to be considered.

### 4.4.2   Optical Isomerism in Compounds Having One Stereogenic Centre

(a)  **Lactic Acid**: Lactic acid was the first compound which was studied in connection with optical isomerism. It contains one asymmetric carbon atom and so it exhibits optical activity. Three forms of lactic acid are known.

L (+) Lactic acid, m.p. 53°C.
D (–) Lactic acid, m.p. 52.8°C.
DL or (±) Lactic acid, m.p. 16.8°C.
(racemic acid).

The synthetic lactic acid obtained by any of the chemical reactions is racemic acid (±) having equal amounts of (+) and (–) forms. The racemic lactic acid, as we will see subsequently can be resolved into (+) and (–) forms. Lactic acid extracted from meat is L (+) and is also known as sarco lactic acid (Greek, *Sarkos*, Flesh) and D (–) Lactic acid is obtained by the fermentation of sucrose *by Bacillus acid laevolactic*.

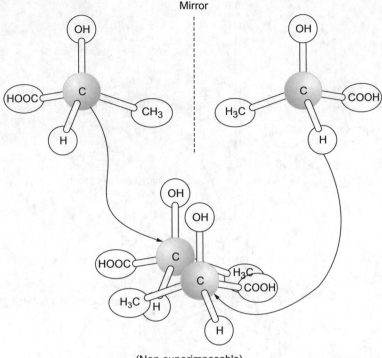

Mirror

(Non-superimposable)

**Fig. 4.8** Enantiomers of Lactic acid

In lactic acid, the carbon is attached to four different substituents, viz., H, $CH_3$, OH and COOH and are arranged tetrahedrally around the carbon. These isomers being non-superimposable mirror image isomers are enantiomers (Fig. 4.8).

It should, however, be noted that the molecule (lactic acid) is chiral rather than one of its atoms. It has been suggested that it is more appropriate to call carbon atom of this type as **stereocentre**. According to IUPAC Rules, the term chiral centre is used for stereochemical notation.

(b) **2-Butanol**: The enantiomers of 2-butanol are represented as shown in Fig. 4.9. As seen the two enantiomers are non-superimposable.

(c) **Other examples of compounds** containing one asymmetric centre are given below (Fig. 4.10).

It should be noted that—the chiral centre is shown by an asterisk mark.

– A carbon atom forming a double bond cannot be a chiral centre since it cannot have four different substituents.

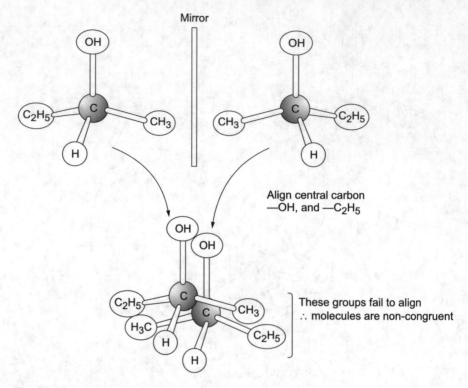

**Fig. 4.9** Enantiomers of 2-butanol

**Fig. 4.10** Some examples of compounds containing one stereogenic centre

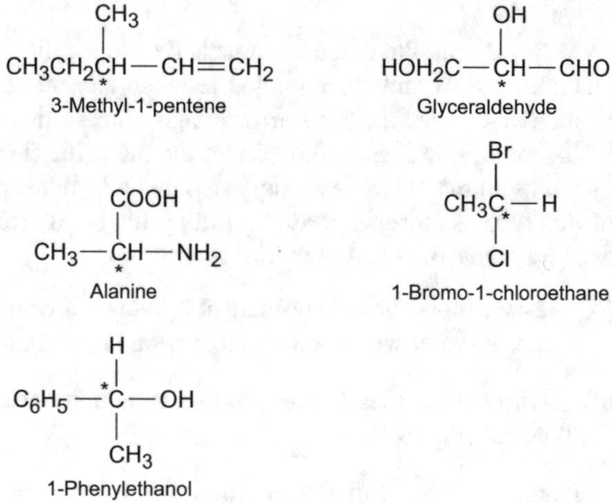

### 4.4.3 Optical Isomerism in Compounds Having Two Asymmetric Centres (Stereogenic Centres)

The number of stereoisomers in a molecule having $n$ chiral centres is given by $2^n$. Since tartaric acid has two chiral centres, it can have four stereoisomers.

In case of compounds having two stereogenic centres, there are two possibilities, i.e., either the two stereogenic centres are similar or different.

(a) **Optical Isomerism in compounds having two similar stereogenic centres**: A compound with two similar stereogenic centre does not have four possible stereoisomers; instead they have only three stereoisomers. This is due to the fact that some molecules are achiral even though they contain stereocentres.

A typical examples is that of tartaric acid. The possible isomers of tartaric acid are given below (Fig. 4.11).

**Fig. 4.11** Stereoisomers of tartaric acid

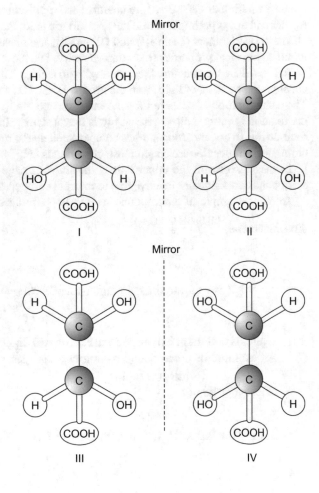

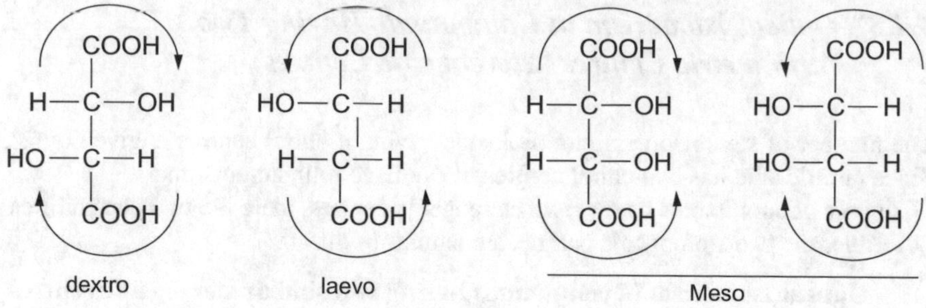

Fig. 4.12 (a) Dextro, laevo and meso forms of tartaric acid

In the Fig. 4.12, structures I and II are enantiomers. However, structures III and IV, although they are mirror image isomers, they are superimposable. This can be ascertained by making models of structures III and IV; these are convertible to each other by rotation of 180°, these are identical and represent two molecules of the same isomer and are superimposable. Thus, for tartaric acid, there are only three isomers. Of the above isomers (I and III) and (II and III) are diastereoisomers. These show identical physical properties (like, m.p., density) but show different sign of rotation.

In Fig. 4.12a, the structure I represents dextro isomer. This is well understood if we consider the rotation of light from the groups H, COOH and OH. This is clockwise. The same rotation is observed for the other carbon atom. However, in structure II, the rotation is counter clockwise and so is laevorotatory. In structures III and IV, the rotation due to upper carbon atom is compensated due to rotation of the lower carbon atom. Such molecules are called meso compounds (Fig. 4.12a).

In the meso form, the upper half is the mirror image of the lower half. Meso compounds are optically inactive due to internal compensation.

Another example of a compound having two similar stereogenic centre is 2,3-dibromobutane, $(CH_3 \overset{*}{C}H \overset{*}{C}H CH_3)$
$$\underset{Br\ \ Br}{\mid \ \ \mid}$$

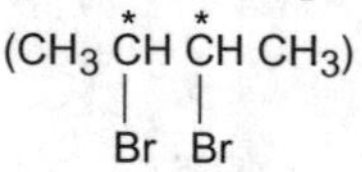

1,2,3,4-tetrahydroxybutane $(HOH_2C—\overset{*}{C}H—\overset{*}{C}H—CH_2OH)$
$$\underset{OH\ \ \ OH}{\mid \ \ \ \ \ \mid}$$

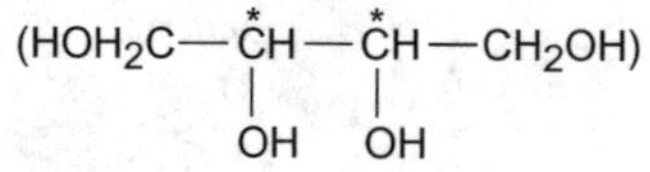

(b)   **Optical isomerism in compounds having two dissimilar stereogenic centre**: An example of a compound having two dissimilar stereogenic centre is 3-chloro-2-butanol    $CH_3 \overset{*}{C}H \overset{*}{C}H CH_3$
$$\underset{Cl\ \ OH}{\mid \ \ \mid}$$
3-chloro-2-butanol

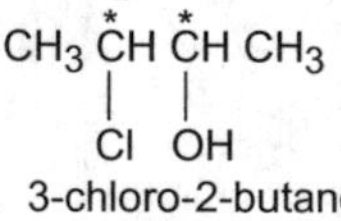

It exists in the following four optically active forms (Fig. 4.13).

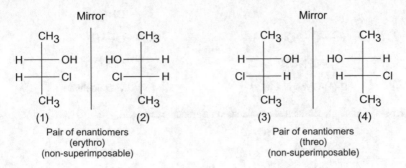

**Fig. 4.13**  Four stereoisomers of 3-chloro-2-butane

The structures (1) and (2) are not superimposable; these represent different compounds differing in the arrangement of the atoms/groups in space. These are called **stereoisomers**.

Structures (1 and 2), and (3 and 4) are mirror images of each other. These are called **enantiomers**. Structures (1) and (3) are stereoisomers and are not mirror images of each other. These are called **diastereomers**. In a similar way (1 and 4), (2 and 3) and (2 and 4) constitute pair of diastereomers. The diastereomers differ in physical properties like melting point, boiling point, solubilities and specific rotation. These can, however, be separated by fractional crystallisation, fractional distillation and chromatography etc. The chemical properties [of the diastereomers are not identical but] are similar. As an example the rates of their reactions with different reagents are different.

Another example of a compound containing two different asymmetric carbon is 2-bromo-3-chlorobutane $\begin{array}{c}[H_3C-CH-CH-CH_3]\\ \quad\quad |\quad\; |\\ \quad\quad Br\;\; Cl\end{array}$. Some other examples are given in Sect. 4.4.6.

### 4.4.4  Relative Configuration of Stereoisomers (D and L System of Nomenclature)

Relative configuration is assigned on the basis the configuration of a standard reference compound. As a matter of convention, glyceraldehyde is taken as a reference compound. In glyceraldehyde, if OH group is at the right side with CHO group at the top, the compound belongs to D-series. On the other hand, if the OH group is at the left side with CHO group at the top, the compound belongs to L series. The D and L-configuration in case of glyceraldehyde is as per accepted convention (Fig. 4.14).

Thus, any compound that can be prepared from or converted into D(–) glyceraldehyde without any disturbance in the bonds of the asymmetric carbon will belong to D-series (D-configuration). Alternatively, any compound that can be prepared

$$
\begin{array}{ccc}
& \text{CHO} & \\
& | & \\
\text{H} - & \text{C} & - \text{OH} \\
& | & \\
& \text{CH}_2\text{OH} &
\end{array}
\qquad\qquad
\begin{array}{ccc}
& \text{CHO} & \\
& | & \\
\text{HO} - & \text{C} & - \text{H} \\
& | & \\
& \text{CH}_2\text{OH} &
\end{array}
$$

D-(+)-Glyceraldehyde                     L-(–)-Glyceraldehyde

**Fig. 4.14**   D and L-glyceraldehyde as standard reference compound

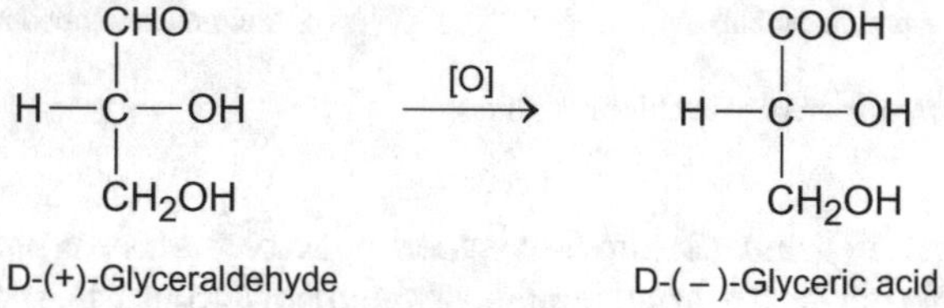

$$
\begin{array}{ccc}
& \text{CHO} & \\
& | & \\
\text{H} - & \text{C} & - \text{OH} \\
& | & \\
& \text{CH}_2\text{OH} &
\end{array}
\xrightarrow{[O]}
\begin{array}{ccc}
& \text{COOH} & \\
& | & \\
\text{H} - & \text{C} & - \text{OH} \\
& | & \\
& \text{CH}_2\text{OH} &
\end{array}
$$

D-(+)-Glyceraldehyde                     D-(–)-Glyceric acid

**Fig. 4.15**   Oxidation of D-(+)-glyceraldehyde to D-(–) glyceric acid

or converted to L-glyceraldehyde will belong to L-series (L-configuration). As an example [D-(+)-glyceraldehyde on oxidation gives D-(–)-glyceric acid (Fig. 4.15)].

In the example cited above (Fig. 4.15) both glyceric acid and glyceraldehyde have D configuration (i.e., OH group is at the right side). However, their specific rotations are different (glyceraldehyde is dextrorotatory) $[\alpha]_D + 3.82°$ while glyceric acid is laevorotatory, $[\alpha]_D - 8.25°$.

On the same analogy, the amino acids D-serine and L-serine are represented as shown below (Fig. 4.16).

It should be noted that there is no correlation between the D and L and (+) and (–) or $d$ and $l$. The capital D are L represent relative configuration while (+) and (–) represent the direction of rotation.

In case a compound contains more than one asymmetric carbon atom, the first asymmetric carbon atom from the bottom decides the configuration. As an example see the configuration of L-tartaric acid and D-glucose (Fig. 4.17).

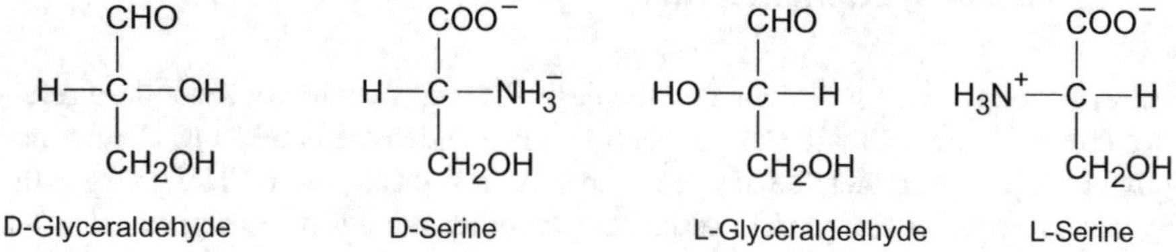

$$
\begin{array}{ccc}
& \text{CHO} & \\
& | & \\
\text{H} - & \text{C} & - \text{OH} \\
& | & \\
& \text{CH}_2\text{OH} &
\end{array}
\quad
\begin{array}{ccc}
& \text{COO}^- & \\
& | & \\
\text{H} - & \text{C} & - \text{NH}_3^- \\
& | & \\
& \text{CH}_2\text{OH} &
\end{array}
\quad
\begin{array}{ccc}
& \text{CHO} & \\
& | & \\
\text{HO} - & \text{C} & - \text{H} \\
& | & \\
& \text{CH}_2\text{OH} &
\end{array}
\quad
\begin{array}{ccc}
& \text{COO}^- & \\
& | & \\
\text{H}_3\text{N}^+ - & \text{C} & - \text{H} \\
& | & \\
& \text{CH}_2\text{OH} &
\end{array}
$$

D-Glyceraldehyde          D-Serine          L-Glyceraldedhyde          L-Serine

**Fig. 4.16**   Correlation of configuration of D- and L-serine with D- and L-glyceride

**Fig. 4.17** Configuration of
L-tartaric acid and D-glucose

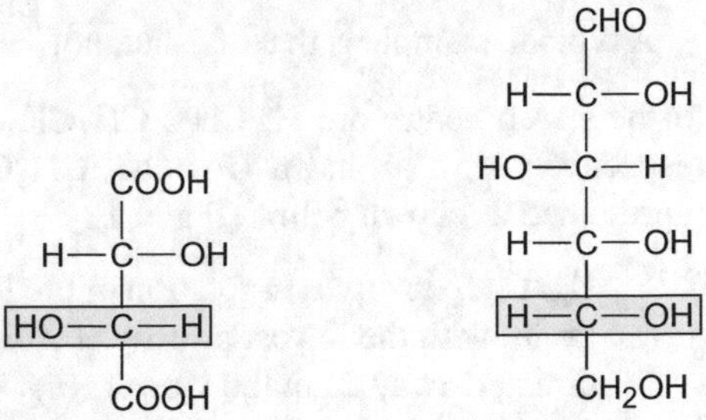

```
        CHO
         |
   H —— C —— OH
         |
  HO —— C —— H
         |
   H —— C —— OH
         |
   H —— C —— OH
         |
       CH₂OH
     D-Glucose
```

```
       COOH
        |
  H —— C —— OH
        |
 HO —— C —— H
        |
      COOH
   L-Tartaric acid
```

## 4.4.5 Absolute Configuration of Stereoisomers (R and S-system of Nomenclature)

The absolute configuration of a stereoisomer is assigned by the (R) and (S) system as developed by R.S. Cahn-C.K. Ingold-V. Prelog. This system is now commonly used and is part of the IUPAC system of nomenclature. According to this system, the R and S configuration is assigned on the basis of the following procedure:

(i) Each of the four groups attached to a stereo centre is assigned a priority or preference *a, b, c, d* on the basis of atomic number of the atom directly attached to the stereo centre. The group with lowest atomic number is given the lowest priority (*d*), the group with next higher [atomic number is given the next higher] priority (*c*) and so on. In case of isotopes, the isotope of greatest atomic mass is given the higher priority.

(ii) In case priority cannot be assigned on the basis of the atomic number of the atoms directly linked to the stereo centre, then the next set of atoms in the unassigned group are considered. The process is continued till a decision can be made. The priority is assigned at the first point of difference.

As an illustration out of methyl ($CH_3$) and ethyl group ($CH_2CH_3$), ethyl group gets the higher priority. This is because in both cases, the first point of attachment is the same (carbon). However, the next set of carbon consists of three hydrogen atoms (H, H, H) in case of methyl group and in case of ethyl group, the next sets of atoms consists of one carbon atom and two hydrogen atoms (C, H, H). Since carbon has a higher atomic number than hydrogen, so ethyl group gets the higher priority [(C, H, H) > (H, H, H)].

Another example is to find the priority of ethyl ($CH_3CH_2$) and isobutyl

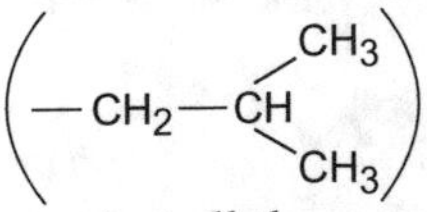

$$\left( - CH_2 - CH \begin{smallmatrix} CH_3 \\ \\ CH_3 \end{smallmatrix} \right)$$ groups. In this case, there is no difference at the point of attach-

ment as all the groups have C, as the second carbon atom in case of ethyl group is attached to three H atoms), whereas in case of isobutyl group.

It is attached to one H and two carbon atoms. So, isobutyl group gets a priority over ethyl group.

A typical example is that of 2-butanol $\overset{CH_3\ CH\ CH_2\ CH_3}{\underset{OH}{|}}$. The four groups attached to the stereo centre are H, $CH_3$, $CH_2CH_3$, OH the priorities of these groups are respectively $a, b, c$ and $d$ (for OH, $CH_2CH_3$, $CH_3$ and H respectively). This is represented as shown below (Fig. 4.18).

(iii)   Next step involves in rotation of the formula (or model) in such a way that the group with the lowest priority (H in the above example, having priority $d$) is directed away from the viewer (Fig. 4.19).

Finally, the path from $a$ to $b$ to $c$ is traced. In case the tracing is clockwise the enantiomer is designated (R). In case the direction of tracing is counter clockwise, the enantiomer is designated (S). Thus, on this basis the 2-butanol enantiomer is (R)-2-butanol (Fig. 4.20a).

On the basis of the above three rules of Cahn–Ingold-Prelog system, it is possible to assign (R) or (S) designation for most of the compounds containing single bonds.

(iv)   In case, the groups contain double bond or triple bond, these are assigned priorities as if both atoms are duplicated or triplicated. Thus,

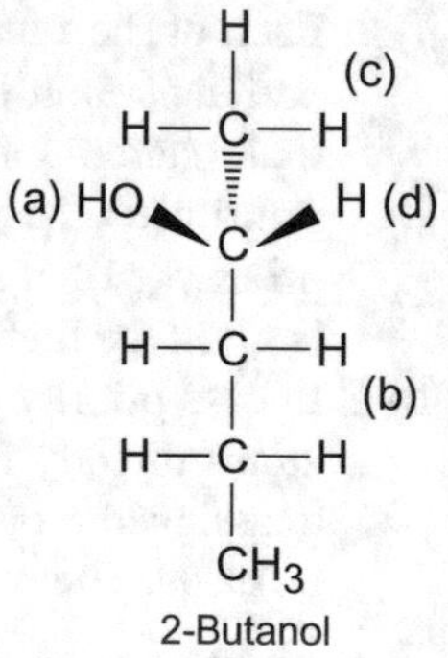

**Fig. 4.18**   Priorities assigned to four different groups in 2-butanol

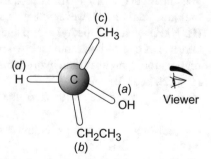

2-Butanol

**Fig. 4.19**   Viewing 2-butanol in the right perspective

**Fig. 4.20** (a). *(R)-2-butanol*
*(arrows are clockwise)*

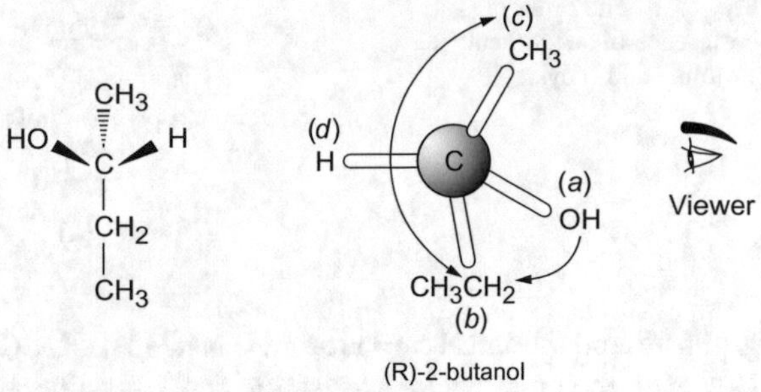

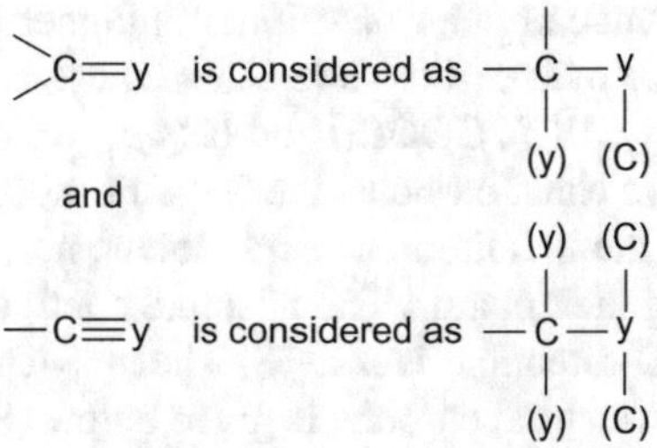

(R)-2-butanol

$$\underset{}{>}C{=}y \quad \text{is considered as} \quad -\underset{\underset{(y)}{|}}{\overset{|}{C}}-\underset{(C)}{y}$$

and

$$-C{\equiv}y \quad \text{is considered as} \quad -\underset{\underset{(y)}{|}}{\overset{\overset{(y)}{|}\ \overset{(C)}{}}{C}}-\underset{(C)}{y}$$

where the symbols in parenthesis are the duplicate or triplicate representations of the atoms at the other end of the double or triple bond.

Thus, vinyl group, —CH==CH$_2$ is of higher priority than isopropyl group.

$$-CH{=}CH_2 \quad \text{is considered as} \quad -\overset{\overset{H}{|}}{\underset{(C)}{C}}-\overset{\overset{H}{|}}{\underset{(C)}{C}}-H$$

$$\text{and} \quad -CH(CH_3)_2 \quad \text{is considered as} \quad -\overset{\overset{H}{|}}{\underset{\underset{H}{\overset{|}{H{-}\overset{|}{C}{-}H}}}{C}}-\overset{\overset{H}{|}}{\underset{H}{C}}-H$$

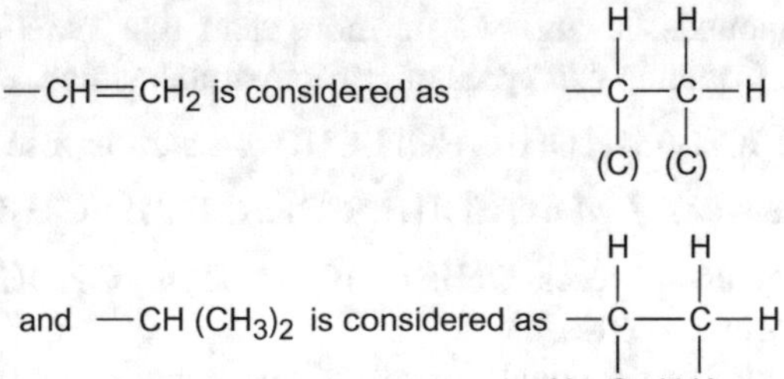

So Vinyl group has a higher priority than isopropyl group. This is because the second set of atoms in vinyl group are C, H, H and in isopropyl group these are H, H, H.

### 4.4.5.1 The R–S Notation in Amino Acids

In proteins, all amino acids are chiral and have L-configuration; all these correspond to S notation. This in because in most amino acids the order of preference of the

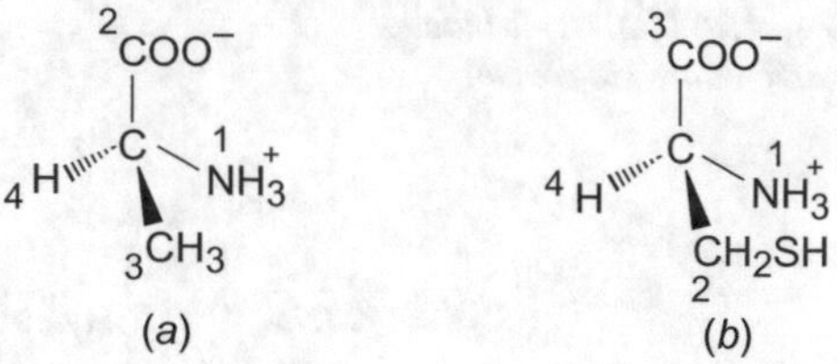

**Fig. 4.21** The order of
preference of substituents in
**a** alaline and **b** cysteine

(a)                                    (b)

group around the chiral carbon atom is $NH_3^+$(1), $CO\overline{O}$(2), R(3) and H(4) (Fig. 4.21a). However, the amino acids L-cysteine and L-cystine are exceptions as these have R notation. It is due to the presence of sulphur atom in the side chain (i.e., $CH_2SH$). In this case, since the atomic number of sulphur is higher than that of oxygen, the group R takes precedence over carboxylate ion in these amino acids (Fig. 4.15b). The order is $NH_3^+$(1), $CH_2SH$(2), $CO\overline{O}$(3) and H (4).

In amino acids with two chiral carbons, the D- and L-notations refer to the configuration of the α-carbon atom. Threonine and isoleucine both have a asymmetric centre at position 3 along their chains. The absolute configuration of L-threonine is 2S, 3R. Its enantiomer, D-threonine is 2R, 3S. The absolute configuration of some other amino acids with two chiral carbons is given below (Fig. 4.22).

### 4.4.6   Optical Isomerism in Compounds Having More Than Two Stereogenic Centres

Examples of compounds having two or more than two stereogenic centres are the carbohydrates. Common examples of carbohydrates which are optically active include tetroses, $HOCH_2 \cdot \overset{*}{C}H OH. \overset{*}{C}HOH.CHO$ (an aldrotetrose having two stereogenic centres), pentoses, $CHO. \overset{*}{C}HOH. \overset{*}{C}HOH. \overset{*}{C}HOH \cdot \overset{*}{C}H_2OH$ (having three stereogenic centres). and hexoses, $OHC. \overset{*}{C}HOH. \overset{*}{C}HOH. \overset{*}{C}HOH. \overset{*}{C}HOH.CH_2OH$ (having four stereogenic centres.)

The maximum number of optical isomers of a carbohydrate is given by $2^n$, where $n$ is the number of stereogenic centres.

#### 4.4.6.1   Optical Isomerism in Tetroses

The aldotetroses, $HOCH_2. \overset{*}{C}HOH. \overset{*}{C}HOH.CH_2OH$ has two stereogenic centres (marked *) and four optical isomers ($2^n$). These are represented in two pairs in simple fisher projections (Fig. 4.23).

The ketotetrose, $HOCH_2 - CO - \overset{*}{C}HOH - CH_2OH$ contains one stereogenic centre and has two optical isomers called D- and L-erythrulose (Fig. 4.24).

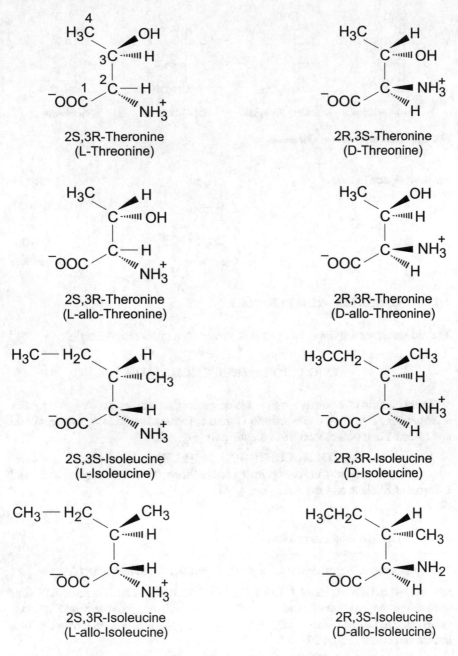

**Fig. 4.22** Absolute configuration of some amino acids having two chiral carbons

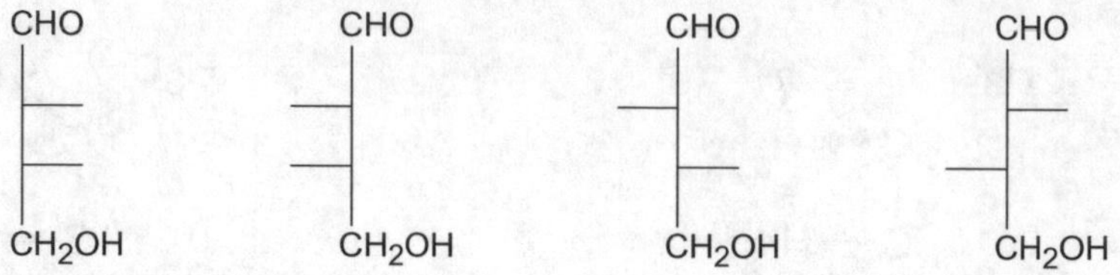

| CHO | CHO | CHO | CHO |
|---|---|---|---|
| CH₂OH | CH₂OH | CH₂OH | CH₂OH |
| D(−)-Erythrose | L(+)-Erythrose | D(−)-Threose | L(+)-Threose |

**Fig. 4.23** Optical isomers of aldotetrose

**Fig. 4.24** Optical isomers of ketotetrose

$$CH_2OH$$
$$C=O$$
$$CH_2OH$$

D-Erythrulose

$$CH_2OH$$
$$C=O$$
$$CH_2OH$$

L-Erythrulose

### 4.4.6.2  Optical Isomerism in Pentoses

The aldopentoses are important group of monosaccharides represented as

$$OHC.\overset{*}{C}HOH.\overset{*}{C}HOH.\overset{*}{C}HOH.CH_2OH$$

These contain three different stereogenic centres and can exist in eight optically active forms ($2^3$). They correspond to D- and L-forms of arabinose, xylose, ribose and lysose. Each of these exist as racemic pair (Fig. 4.25).

The ketopentoses, $HOCH_2.CO.\overset{*}{C}HOH.\overset{*}{C}HOH.CH_2OH$ have two different stereogenic centres and exist in four optically active forms. These correspond to D- and L-forms of ribulose and xylulose (Fig. 4.26).

### 4.4.6.3  Optical Isomers of Hexoses

The aldohexoses are the most important group of monosaccharides and are represented as $OHC.\overset{*}{C}HOH.\overset{*}{C}HOH.\overset{*}{C}HOH.\overset{*}{C}HOH.CH_2OH$. These contain four different stereogenic centres and so exist in sixteen optically active forms ($2^4$). These correspond to D- and L-forms of glucose, mannose, galactose, allose, altrose, gulose, iodose and talose (Fig. 4.27).

The ketohexoses $CH_2OH.CO.\overset{*}{C}HOH.\overset{*}{C}HOH.\overset{*}{C}HOH.CH_2OH$ have three stereogenic centres and so exist in eight optically active forms. Of these only six are known. These are D- and L-fructose, D- and L-sorbose, D-tagalose and D-psicose (Fig. 4.28).

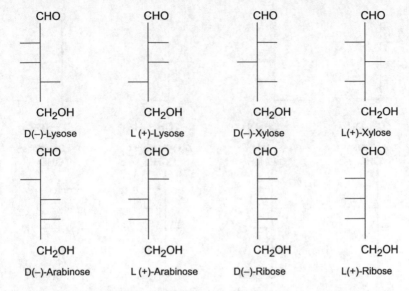

**Fig. 4.25** Optical isomers of aldopentoses

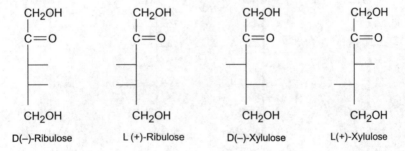

**Fig. 4.26** Optical isomers of ketopentoses

## 4.5 Racemic Mixture

An equimolar mixture of two enantiomers (+ and −) is known as a racemic mixture. A racemic mixture does not show rotation of plane polarised light and is designated as (±). A racemic form of (R)-(−)-2-butanol and (S)-(+)-2-butanol is represented as

$$(\pm)-2- \text{Butanol or } (\pm) \text{ - } CH_3CH_2\,CHOH\,CH_3$$

A racemic mixture is optically inactive due to external compensation, where as the meso compounds are optically inactive due to internal compensation.

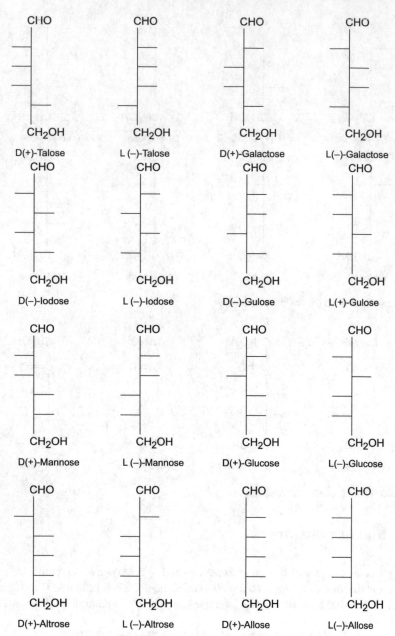

**Fig. 4.27**   Optical isomers of aldohexoses

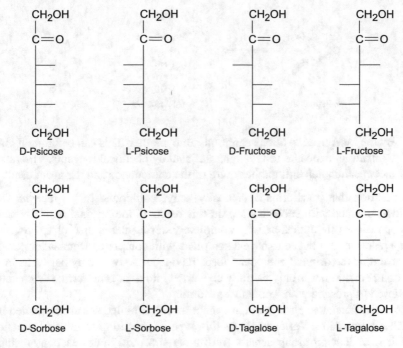

**Fig. 4.28** Optical Isomers of Ketohexoses

## 4.5.1  *Formation of Racemic Mixtures*

The racemic mixtures can be obtained by any of the following ways:

(a)  By mixing equimolar amounts of two enantiomers
(b)  Attempts to prepare a pure optically active compound results in the formation of a racemic mixture. Some examples include:

   (i)  Synthesis of α-amino acid by strecker synthesis. It involves the reaction of an aldehyde (e.g., acetaldehyde) with ammonia and hydrogen cyanide to give 2-aminopropanenitrile which on hydrolysis give (±) alanine.

$$CH_3CHO + NH_3 + HCN \xrightarrow{-H_2O} CH_3CH\begin{smallmatrix}\nearrow NH_2 \\ \searrow CN\end{smallmatrix} \xrightarrow[\text{(2) OH}^-]{\text{(1) heat, H}_2O} CH_3CH-\overset{+}{N}H_3$$

Acetaldehyde                                       2-Amino
                                                   propanenitrile                    COO$^-$
                                                                                     (±)-Alanine

   (ii)  Hydrogenation of 2-butanone gives a racemic mixture of (R)-2-butanol and (S)-2-butanol.

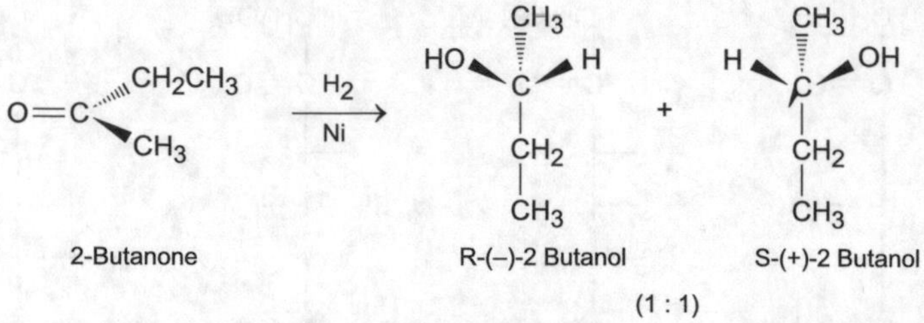

2-Butanone          R-(−)-2 Butanol          S-(+)-2 Butanol

(1 : 1)

(c)    By the racemisation of one eantiomer into another. This can be brought about
       by heating, exposure to UV light and also by chemical reagents. The rate of
       racemisatoin depends on the nature of the compounds and the agent used.

A reaction that transforms an optically active compound into a racemic form
is said to proceed with **racemisation**. In this reaction, the original optically active
compound loses its optical activity completely during the course of the reaction.
Such a reaction is believed to have taken place with complete racemisation. On the
other hand, if the original compound loses its optical activity only partly, as in the
case of an enantiomer is only partially converted into a racemic form; such reaction
is believed to proceed with partial racemisation.

Racemisation takes place when in a reaction chiral molecules are converted into
achiral intermediates. An example of this type of reaction are the $S_N1$ reactions
in which the leaving group departs from a stereocentre. Such reactions result in
extensive or sometimes complete racemisation. An example is heating an optically
active (S)-3-bromo-3-methylhexane with aqueous acetone resulting in the formation
of a racemic mixture of (S) and (R)-3-methyl-3-hexanol (Fig. 4.29).

The above reaction (Fig. 4.29) proceeds *via*, the formation of an intermediate
carbocation (which has trigonal planar configuration and so is achiral). It reacts with
water at equal rates from either side to form enantiomers of 3-methyl-3 hexanol in
equal amounts. Thus, the $S_N1$ reaction proceeds with racemisation (Fig. 4.30).

Another very interesting reaction involving racemisation is the treatment of a
solution of (R)-(+)-sec. butyl phenyl ketone in aqueous ethanol with acid or base.
Gradually, the compound loses its optical activity showing that the compound has
racemised (Fig. 4.31).

In the above case (Fig. 4.31), in presented acid or base, racemisation takes place to
form enol (achiral). The enol on reverting back to the keto form gives equal amounts
of the two enantiomer (Fig. 4.32).

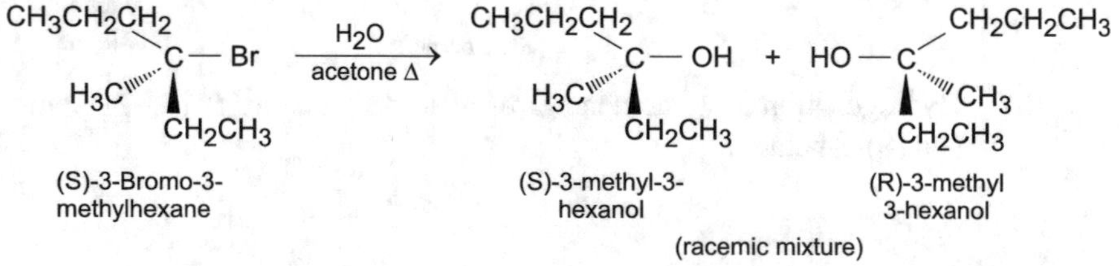

(S)-3-Bromo-3-          (S)-3-methyl-3-          (R)-3-methyl
methylhexane            hexanol                 3-hexanol

(racemic mixture)

**Fig. 4.29**  A typical $S_N1$ reaction

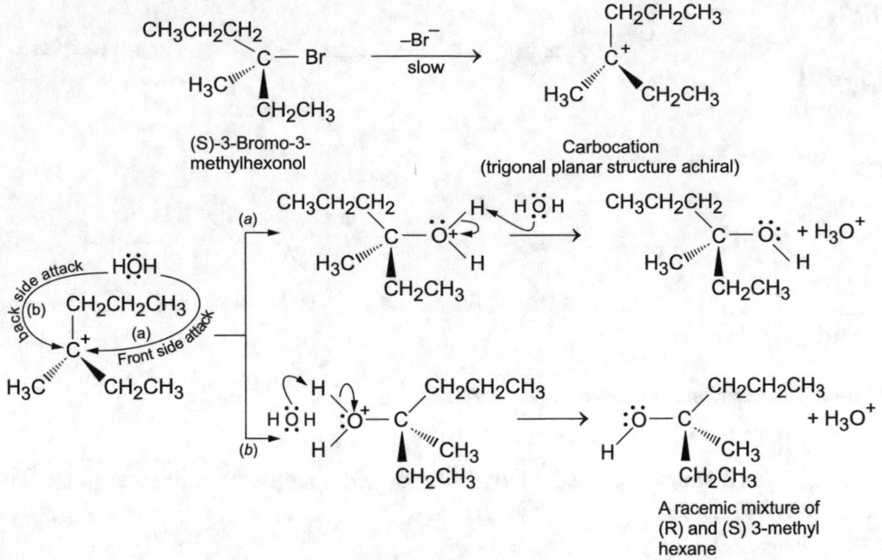

**Fig. 4.30**  Mechanism of $S_N1$ reaction

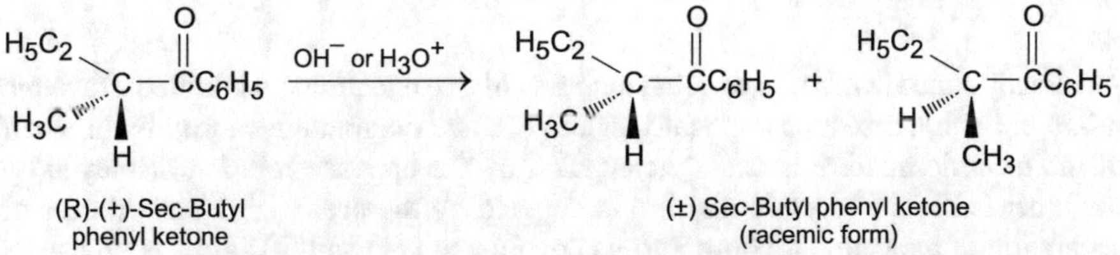

**Fig. 4.31**  Racemisation in presence of acid or alkali

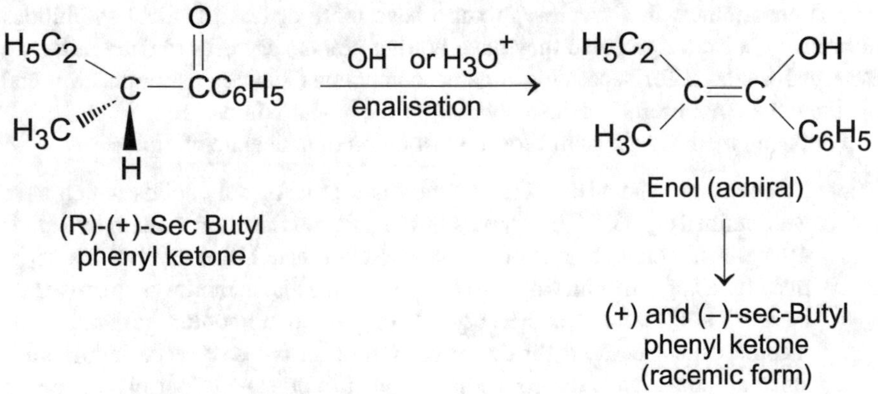

**Fig. 4.32**  Racemisation in presence of acid or alkali

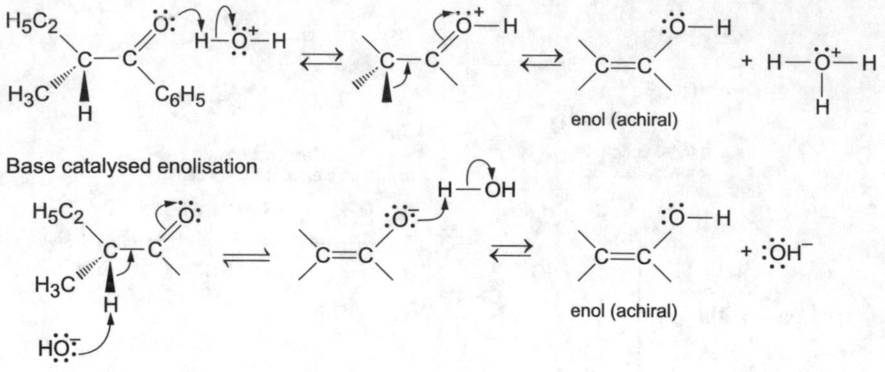

**Base catalysed enolisation**

**Fig. 4.33** Enolisation in presence of acid or alkali

The mechanism of the base catalysed and acid catalysed enolisation is shown below (Fig. 4.33).

## 4.5.2 *Resolution of Racemic Mixture*

As already stated, a mixture of equal amounts of two enantiomers is called a racemic mixture. Such a mixture is optically inactive, since the rotation caused by one half of the racemic mixture is exactly cancelled by the opposite rotation displayed by the other half. The racemic mixture is denoted by the prefix ($\pm$). The process of separation of a racemic mixture into its enantiomers (+) and (−) forms is known as resolution. It is extremely important to get the individual enantiomers in pure state since these have varied characteristic properties and biological reactions.

The enantiomers in a racemic mixture have more or less identical solubilities in ordinary solvents, and also they have boiling points very near to each other. So the usual methods for separating organic compounds such as crystallisation and distillation are not useful for the resolution of racemic mixtures.

Following methods are useful for the resolution of a racemic mixture:

(i)    **Mechanical Separation**: This method is useful for only solids which have well-defined crystals. The crystals of racemic modification contain two types of crystals, which are mirror image of each other. These crystals are separated by using a magnifying lens and tweezer. This method was first used by Pasteur (1848) who separated crystals of sodium ammonium tartarate. This method cannot be used for the resolution of all types of racemic mixtures, since all compounds do not form asymmetric crystals to permit separation mechanically. It is therefore important to use other methods.

(ii) **Biochemical Method**: This method consists in the treatment of a racemic mixture with certain microorganisms, which utilises one of the enantiomer leaving behind the other in solution. The separation by this method is almost quantitative. The enantiomer that is left in solution is isolated by fractional crystallisation. As an example, a solution of ($\pm$) tartaric acid is treated with an ordinary mould '*Penicillium glaucum*'. The mould selectively utilises dextro tartaric acid and leaves laevo tartaric acid.

   This method (described above) was developed by Pasteur (1858) is a better procedure. However, the drawback is that one of the enantiomers which may or may not be useful is sacrificed. Also, toxic racemic mixture kill the micro organisms and the expected reaction will not commence.

(iii) **Resolution of Racemic Mixture by Inclusion Complex Procedure**: An inclusion complex is formed by the treatment of one component called the host (which has a cavity or space in the crystal lattice) with a second compound called the guest. In inclusion complex, the guest fits into the cavity of the host. In the inclusion complex there are no covalent bonding but only the van der Waals forces. Chiral phenols like 1,1'-binaphthol, 2, 2'-binaphthols which are optically active act as good host for the separation of racemic mixture. In such separations, solvents like benzene or hexane is used.

   The procedure used for separation consists in keeping the solution of the racemic mixture (Guest) and the host (like 1,1'-binaphthol) in a solvent (benzene or hexane) for 5-10 hrs. During this process crystals of inclusion complex separate out, which are filtered. The mother liquor is used for the isolation of the second isomer. The inclusion complex is formed in the 1:1 or 1:2 host-guest ratio and is recrystallised from the same solvent (benzene or hexane). The enantiomer from the inclusion complex is obtained by distillation of the complex or recrystallisation from a different solvent. Alternatively, high pressure liquid chromatography can also be used. This method gives a convenient separation of enantiomers.

(iv) **Chemical Methods**: This method involves treatment of the racemic mixture with an appropriate optically active compound. For example, racemic lactic acid on treatment with optically active brucine (a base) gives the following two salts:

   + (Acid) (–) Base
   – (Acid) (–) Base

These salts can be separated by fractional crystallisation and the individual salts on treatment with sulphuric acid give optically pure acid (Fig. 4.34).

In a similar way, a racemic mixture of a base can be converted into diastereometric salts by treating with a chiral acid. The salts so obtained are separated by fractional crystallisation and treated with a strong base to yield pure enantiomers. The chiral bases and acid used in such separations are called resolving agents A list of resolving agents used for salt formation are given below.

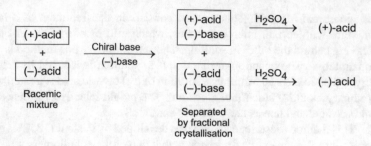

**Fig. 4.34** Separation of a racemic mixture (acid) using a chiral base

| Racemic mixture | Resolving agents (optically active) |
|---|---|
| Two enantiomers of carboxylic acid | Bases used:<br>(+)-strychnine, (−)-brucine (−)-quinine, β-picoline,<br>(−)ephedrine, (−) menthyl amine |
| Two enantiomers of bases | Acid used:<br>(+)-alanine or any amino acid; (+)-10-camphor<br>sulphonic acid |

The structures of some of the resolving agents are given below (Fig. 4.35).

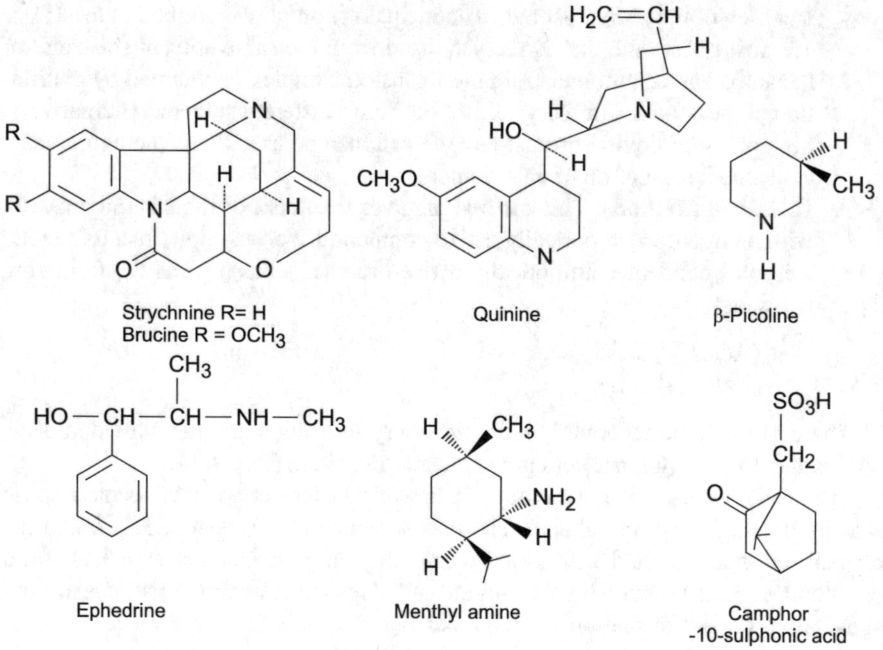

Strychnine R= H
Brucine R = OCH₃

Quinine

β-Picoline

Ephedrine

Menthyl amine

Camphor
-10-sulphonic acid

**Fig. 4.35** Structures of some resolving agents

### 4.5.2.1 Enantiometric Resolution of α-Amino Acids

All amino acids, except glycine synthesised by any chemical method are obtained as racemic mixtures. It is necessary to resolve these racemic mixtures to get pure D or L enantiomer. Following are given some of the methods used for their resolution:

(i) **Amine Salt Formation**: The method consists of reaction of the racemic mixture with an enantiomer (either R or S) of a naturally occurring amine like strychnine or brucine. The formed diastereomeric pair of salts are separated by crystallisation. On acidification of the salts, the respective amino acids are obtained in optically pure state. The separation is represented as follows (Fig. 4.36).

(ii) **Ester Formation**: The mixture of amino acids is converted into diastereomeric esters, which are separated by crystallisation and the amino acids recovered by hydrolysis (Fig. 4.37).

(iii) **Enzymatic Resolution of Amino Acids**: Enzymes (biological catalysts) can be used to resolve a mixture of amino acids. Certain enzymes (called *deacylases*) obtained from living organisms can selectively catalyse the hydrolysis of one of the enantiomeric N-acylamino acid. As an examples, the enzyme

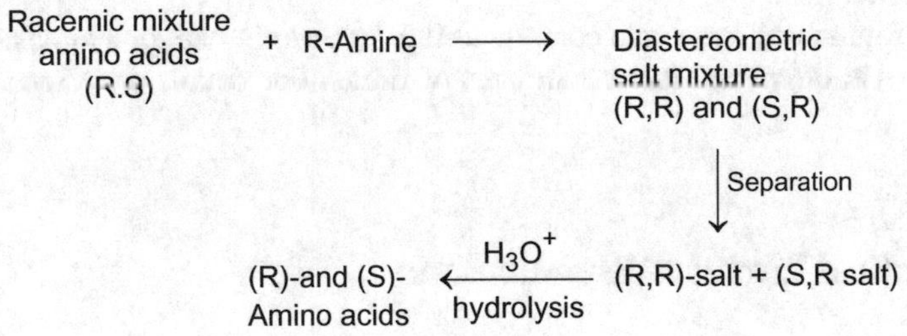

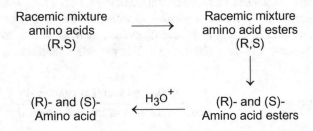

**Fig. 4.36** Enantiomeric resolution of amino acids by salt formation

Racemic mixture amino acids (R,S) ⟶ Racemic mixture amino acid esters (R,S)

(R)- and (S)- Amino acid ←$H_3O^+$— (R)- and (S)- Amino acid esters

**Fig. 4.37** Resolution amino acids by ester formation

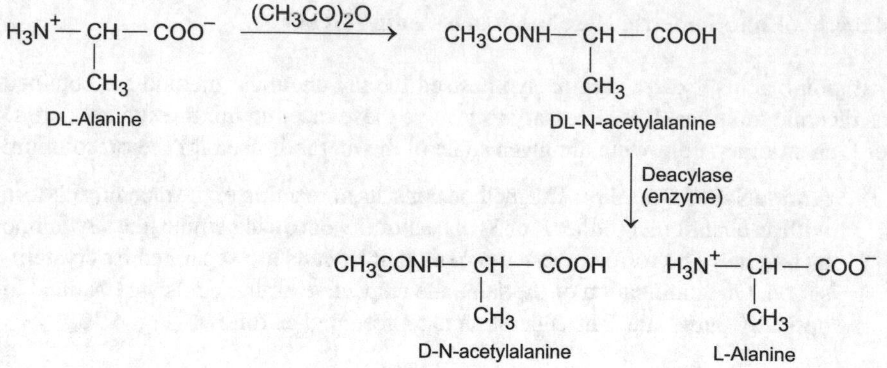

$$H_3N^+ - CH - COO^- \xrightarrow{(CH_3CO)_2O} CH_3CONH - CH - COOH$$

$$\underset{\substack{| \\ CH_3}}{} \qquad\qquad\qquad \underset{\substack{| \\ CH_3}}{}$$

DL-Alanine                                    DL-N-acetylalanine

Deacylase
(enzyme)

$$CH_3CONH - CH - COOH \qquad H_3N^+ - CH - COO^-$$

$$\underset{\substack{| \\ CH_3}}{} \qquad\qquad\qquad \underset{\substack{| \\ CH_3}}{}$$

D-N-acetylalanine                              L-Alanine

**Fig. 4.38** Enzymatic resolution of α-amino acids

obtained from hog kidney cleaves the acyl group from the L-enantiomer while the D-enantiomer remains uneffected. Thus to resolve a racemic mixture of amino acids, it is converted into an N-acyl derivatives and the mixture is hydrolysed with the help of deacylase enzyme (Fig. 4.38).

The liberated amino acid is then precipitated from ethanol while N-acyl derivative remains in solution.

The resolution of a racemic compound into its enantiomers or synthesis of pure enantiomer is of special interest in case of drugs. For details (*see* Sect. 18.2. in Chap. 2).

## 4.6   Optical Purity of Enantiomers

An enantiomerically pure optically active substance (containing a single enantiomer) is said to have **enantiometric excess of 100 present**. A enantiometrically pure sample of (S)-(+)-2-bromobutane has a specific rotation of + 23.1°. If a sample contains both the enantiomers (R and S) having a positive specific rotation but whose value is less than + 23.1°, say 11.55° then we say that it contains more of S-enantiomer and has enantiometric excess less than 100 per cent. The per cent enantiomeric excess (ee) is defined as

$$\% \text{ Enantiomeric excess} = \frac{\text{Observed specific rotation}}{\text{Specific rotation of the pure enatiomer}} \times 100$$

$$= \frac{+11.55°}{+23.1°} \times 100$$

$$= 50\%$$

On the basis of the above calculations, it can be said that of the total mixture 50% consists of racemic form (which contains equal number of the two enantiomers). So, half of 50% (or 25%) is the (–) enantiomers and 25% is the (+) enantiomer. The remaining 50% of the mixture is also the (+) enantiomer. So the mixture is 75% (+) enantiomer and 25% (–) enantiomer.

It should be noted that some compounds which do not have asymmetric carbon can also be optically active and exhibit optical isomerism. Examples include appropriately substituted biphenyls, allenes and spiranes. This forms the the subject-matter of Chap. 6.

## Key Concepts

- **Absolute Configuration**: Also known as R- and S-system of nomenclature as developed by Cahn-Ingold-Prelog. It is assigned on the basis of a number of rules.
- **Chiral Centre**: Also known as stereogenic centre, usually a carbon atom that is bonded to four different atoms or groups and is therefore chiral.
- **Chirality**: Property of an optically active substance. The chiral molecules show handedness. In such cases the object and its mirror image are not superimposable.
- **Dextrorotatory**: An optically active substance that rotates the plane of polarisation of plane polarised light in clockwise direction. It is indicated by d or (+) sign.
- **Diastereoisomers**: The stereoisomers that are not mirror images and are non-superimposable. As an example the following two stereoisomers of 3-chloro-2-butanol are diastereoisomers.

Pair of diastereomers

- **Enantiomers**: Stereoisomers of a chiral substance that have a mirror image relationship. These have opposite configuration at all chiral centres in the molecule.
- **Enontiomeric Excess**: It is the observed specific rotation (of an optically active mixture) divided by the specific rotation of the pure enantiomer and multiplied by 100.

$$\% \text{ enantiomeric execess} = \frac{\text{Observed specific rotation}}{\text{Specific rotation of the}} \times 100$$
$$\text{pure enantiomer}$$

It is used to determine the percentage of an enantiomer in an enantiomeric mixture.

- **Inclusion Complex**: A complex formed by the teatment of one compound (called the host which has a cavity or space in the crystal lattice) with a second component (guest). In this complex the guest fits into the cavity of the host and there are no covalent bonding but only van der Waals forces.
- **Laevorotatory**: An optically active substance that rotates the plane of polarisation of plane polarised light in counter clockwise direction. It is indicated by l or (−) sign.
- **Meso Compound**: Optically inactive compound (though they contain two asymmetric carbons). The upper half of the molecule is the mirror image of the lower half. Meso compounds are optically inactive due to internal compensation. An example is meso tartaric acid.

Meso tartaric acid

- **Nicol Prism**: A device that produces plane polarised or the light in which the waves vibrate in one plane only. It is made by joining together two appropriately cut pieces of the mineral calcite. Calcite has the ability to split a ray of light in two parts one of which obeys normal laws of refraction and the other does not. These two, which vibrate along two mutually perpendicular planes, are called ordinary and extraordinary rays respectively. The ordinary ray undergoes total internal reflection while the extraordinary ray passes through. Thus, the light ray emerging through the prism is polarised in one plane only. The device was first made by William Nicol (1768–1851).
- **Optical Isomerism**: Isomerism exhibited by optically active molecules. The optically active isomers are called enantiomers.
- **Optical Isomers**: Known as enantiomers, these have a mirror image relationship.
- **Optically Active**: A substance that rotates the plane of polarisation of plane-polarised light. For a compound to be optically active it must have a chiral centre. This is not strictly true since all chiral molecules are not optically active. *See* racemic mixture and meso compounds.

- **Polarimeter**: An apparatus used for the measurement of the plane of polarisation of light by substances showing optical activity. It consists of a light source and two Nicol prisms called polariser and analyser placed at the end of a tube which is filled with a solution of an optically active substance. Any rotation of the plane of polarisation can be measured by turning the analyser.
- **Racemic Mixture**: An equimolar mixture of two enantiomers (+ and −). It does not rotate the plane of plane polarised light and is designated as (±).
- **Relative Configuration**: The configuration assigned on the basis of a standard reference compound. Glyceraldehyde is taken as a reference compound. In glyceraldehyde, if OH group is at the right side with CHO group at the top, the compound belongs to D-series.

However, if the OH group is at the left side with CHO group at the top the compound belongs L-series.

$$
\begin{array}{ccc}
\text{CHO} & \qquad & \text{CHO} \\
| & & | \\
\text{H}-\text{C}-\text{OH} & & \text{HO}-\text{C}-\text{H} \\
| & & | \\
\text{CH}_2\text{OH} & & \text{CH}_2\text{OH}
\end{array}
$$

D-Glyceraldehyde        L(−)-Glyceraldehyde

Any compound that can be prepared from or converted into D(−) glyceraldehyde without any disturbance in the bands of the asymmetric carbon will have D-configuration. The same reasoning is also true about L-configuration. Relative configuration is also known as D-and L-configuration.

- **Resolution**: The process of separation of a racemic mixture into its enantiomers (+ and − forms).
- **Specific Rotation**: It is the optical rotation caused by an optically active compound having a concentration of 1 g/mL in a tube of 1 dm of the polarimeter. It is denoted by $[\alpha]_D$.
- **Stereochemistry**: Branch of chemistry dealing with three dimensional arrangement of atoms or groups in a molecule.
- **Stereogenic Centre**: A chiral centre. It is defined as an atom, in which interchange of any two atoms or group result in a new stereoisomer.

$$
\begin{array}{ccc}
\text{CH}_3 & & \text{CH}_3 \\
| & \text{interchange between} & | \\
\text{H}-\text{C}-\text{OH} & \overline{\text{two groups (H and OH)}} & \text{HO}-\!\!\!-\!\!\!-\text{H} \\
| & & | \\
\text{COOH} & & \text{COOH}
\end{array}
$$

## Problems

1. Describe the origin of optical activity. How it is measured?
2. The specific rotation of a compound is represented as $[\alpha]_{25}^{D} = + 3.15°$. What do you understand from this?
3. What do you understand by the word 'chirality'? What are the necessary conditions necessary for a compound to exhibit chirality?
4. Write notes on:

   (a) dextrorotatory
   (b) laevorotatory
   (c) stereogenic centre
   (d) enantiomer
   (e) diastereomers

5. What is optical isomerism? Discuss optical isomerism exhibited by 3-chloro-2-butanol.
6. What do you understand by relative configuration? Give examples of relative configuration of amino acid and glucose.
7. How will you assign absolute configuration to an optically active substance taking the example of 2-butanol.
8. Discuss R and S notation in amino acids.
9. Discuss optical isomerism in compounds having more than two (five or six) stereogenic centres.
10. Arrange the groups in order at their priority from higher to lower.

    (a) Cl, OH, SH
    (b) $CH_3$, $CH_2Cl$, $CH_2Br$
    (c) CHO, COOH
    (d) $CH_3$, $OCH_3$, $N(CH_3)_2$

Assign (R) or (S) designation to the following compounds:

(a)

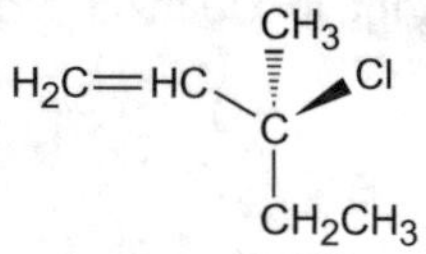

(b)

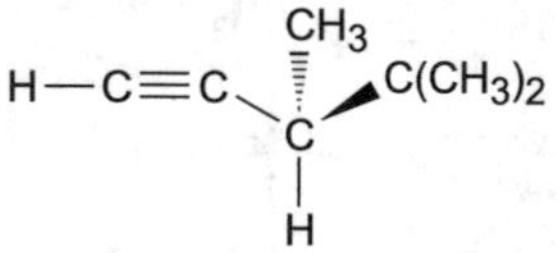

(c)

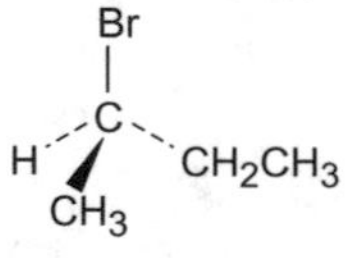

(d)

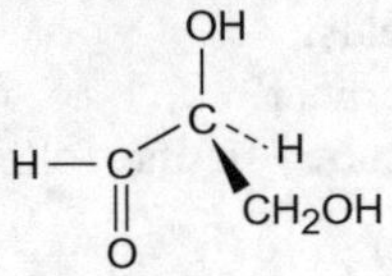

$$\begin{array}{c} OH \\ | \\ H-C \overset{\displaystyle C \cdots H}{\underset{\displaystyle CH_2OH}{\big|}} \\ \| \\ O \end{array}$$

(e)

$$\begin{array}{c} COOH \\ | \\ H \cdots C \diagdown OH \\ \diagup \\ CH_3 \end{array}$$

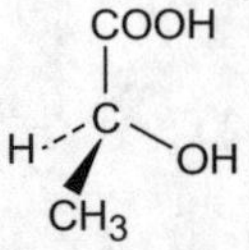

11.  Determine the R and S configuration of all the asymmetric centres, in the following compounds:

(a)

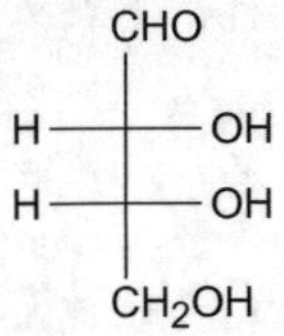

$$\begin{array}{c} CHO \\ H \!-\!\!\!\!-\!\!\!\!- OH \\ H \!-\!\!\!\!-\!\!\!\!- OH \\ | \\ CH_2OH \end{array}$$

(b)

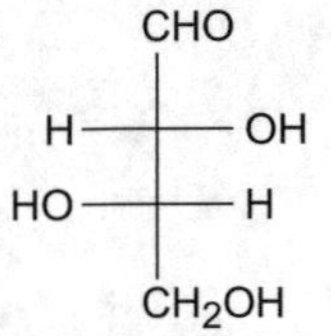

$$\begin{array}{c} CHO \\ H \!-\!\!\!\!-\!\!\!\!- OH \\ HO \!-\!\!\!\!-\!\!\!\!- H \\ | \\ CH_2OH \end{array}$$

(c)

$$\begin{array}{c} CHO \\ H \!-\!\!\!\!-\!\!\!\!- Br \\ Br \!-\!\!\!\!-\!\!\!\!- H \\ | \\ COOH \end{array}$$

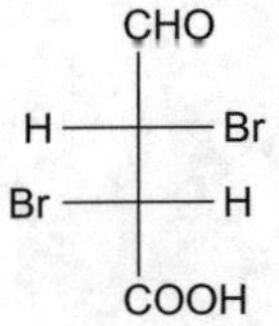

(d)

$$\begin{array}{c} CHO \\ H \!-\!\!\!\!-\!\!\!\!- NH_2 \\ H \!-\!\!\!\!-\!\!\!\!- OH \\ | \\ CH_3 \end{array}$$

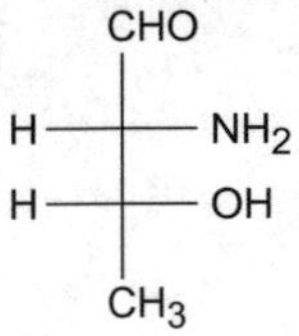

**Ans.**   (a) $C_2$ and $C_3$ both R      (b) $C_2$ is R and $C_3$ is S
(c) $C_2$ and $C_3$ both S      (d) $C_2$ and $C_3$ both R.

12.  Calculate the enantiomeric excess of a sample of (S)–(+)-2-butanol that has an specific rotation of +6.76° (given specific rotation of enantiometrically pure sample of (S)-(+)-2-butanol is + 13.52°). Also calculate the composition of the mixture of (S) and R-(2)-butanol.

   **Ans.** Enantiomeric excess = 50 per cent.
   The mixture contains 75% (+) enantiomer and 25%(–) enantiomer.

13.  What is a racemic mixture? How these are formed? How can a racemic mixture resolved into its compounds?

14.   Write a note an enantiomeric resolution of α-amino acids.
15.   How will you determine the optical purity of an enantiomer?
16.   Assign R and S configuration to all stereogenic centres in the following compounds.

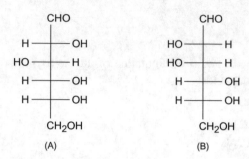

(A)                              (B)

**Ans.** Compound A configuration of $C_2$, $C_4$ and $C_5 = R$ configuration of $C_3 = S$
Compound B configuration of $C_2$, $C_3 = S$
configuration of $C_4$ and $C_5 = R$.

# Chapter 5
# Symmetry Elements

## 5.1 Introduction

It has already been stated (Chap. 4) that according to van 't Hoff and Le Bel, the optical activity in a compound is related to the presence of an asymmetrically substituted carbon atom (also called asymmetric atom or stereogenic centre). These authors also found that optical activity could also be in compounds having no asymmetric atoms. This aspect forms the subject matter of subsequent sections. Subsequently, Pasteur postulated that the optical activity is due to the presence of molecular dissymmetry. We have also known that a necessary condition for a compound to show optical activity is that an optically active molecule should not be superimposable with its mirror image (Sect. 4.4.2). A molecule which is superimposable with its mirror image does not show optical activity.

Cahn, Ingold and Prelog (1966) postulated that for a molecule to be optically active, it should not have any elements of symmetry (see Sect. 5.2). In fact, as per Group Theory, the superimposibility of a molecule with its mirror image is governed by the symmetry properties of the molecule.

## 5.2 Elements of Symmetry

There are three elements of symmetry. These include plane of symmetry, centre of symmetry and alternating axis of symmetry. A molecule is achiral if it has one or more elements of symmetry.

© The Author(s), under exclusive license to Springer Nature Switzerland AG 2022     107
V. K. Ahluwalia, *Stereochemistry of Organic Compounds*,
https://doi.org/10.1007/978-3-030-84961-0_5

### 5.2.1  Plane of Symmetry

A plane of symmetry is an imaginary plane which divides a molecule into two halves in such a way that the part of the molecule on one side of the plane is a mirror image on the other side of the plane. As an illustration, the isomers of tartaric acid are represented as given ahead (Fig. 5.1).

In Fig. 5.1, structures I and II do not have a plane of symmetry and so are optically active; this is also supported by the fact that I and II have a mirror image relationship and are enantiomers. On the other hand, structures III and IV have a plane of symmetry and so are optically inactive. This has already been discussed in connection with the optical isomerism of tartaric acid (Sect. 4.4.3). It should be noted that both structures III and IV have two asymmetric carbons but are optically inactive.

Two other examples of compounds having a plane of symmetry are 1,1-dichloro ethene and 1,1,2,2-tetrachloro ethene; both these are optically inactive (Fig. 5.2).

We also come across compounds which have an asymmetric substituent but are optically inactive. The asymmetric group is *sec*-butyl, $CH_3$— $C^1$H—$C_2H_5$. Thus,

**Fig. 5.1** Stereoisomers of tartaric acid

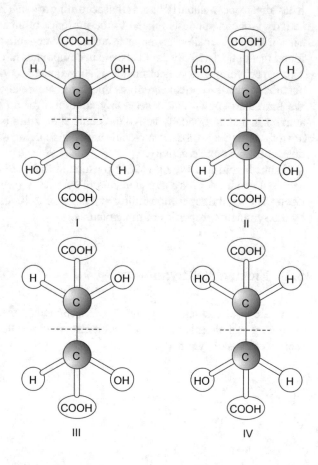

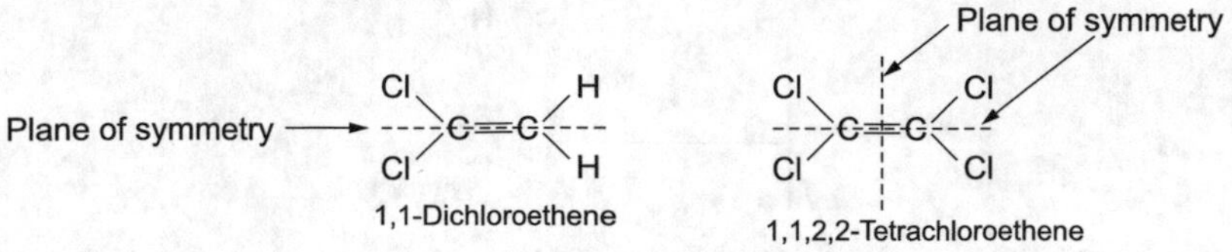

Fig. 5.2 1,1-dichloro and 1,1,2,2-tetrachloro ethene. Both are optically inactive

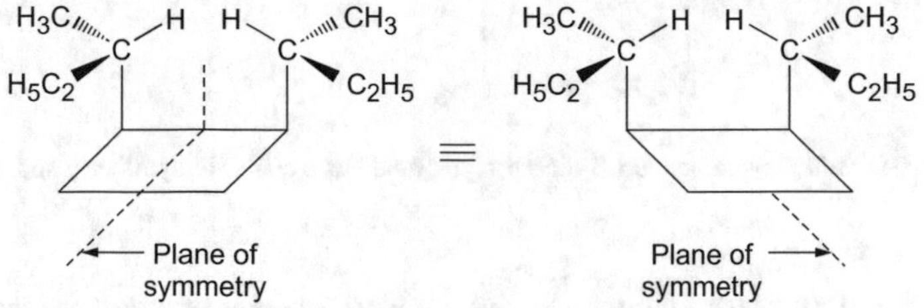

Fig. 5.3 1,2-di-sec-butylcyclobutane having plane of symmetry and is optically inactive

1,2-di sec-butylcyclobutane is optically inactive since it has a plane of symmetry, though it contains asymmetric carbons in the *sec*-butyl group (Fig. 5.3).

## 5.2.2 Centre of Symmetry

It is a centre in a molecule, from where identical atoms or groups in the molecule are located diametrically at equal distance. An example is the *meso* form of tartaric acid (Fig. 5.4).

Another example of a compound having a centre of symmetry is 2,4-dimethyl cyclobutane 1,3-dicarboxylic acid which is optically inactive (Fig. 5.5).

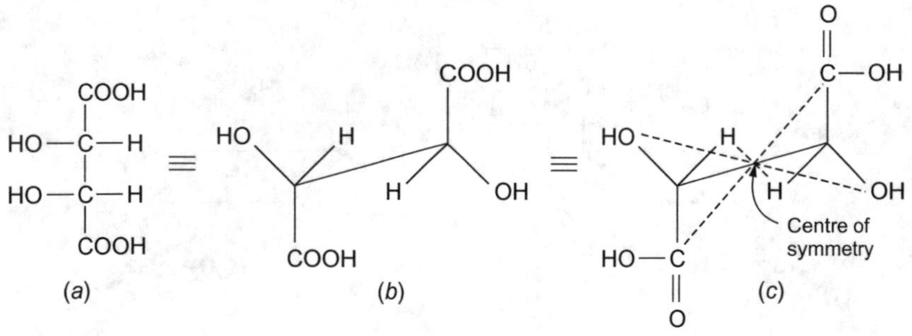

Fig. 5.4 Meso tartaric acid **a** normal representation **b** Sawhorse projection, **c** shows centre of symmetry

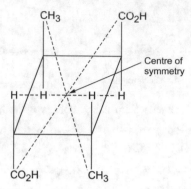

**Fig. 5.5** 2,4-Dimethylcyclobutane 1,3-dicarboxylic acid has a centre of symmetry and is optically inactive

Two other examples of compounds which have a centre of symmetry are ethane and *trans*-1,2-dichloroethene and are therefore optically inactive.

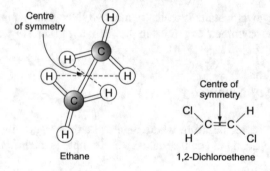

An interesting example is that of 1,3-di-sec-butylcyclobutane. This contains two substituents which have asymmetric carbons (*sec*-butyl group), but it is optically inactive as it has a centre of symmetry (Fig. 5.6).

**Fig. 5.6** 1,3-di-sec-butylcyclobutane has a centre of symmetry and is optically inactive

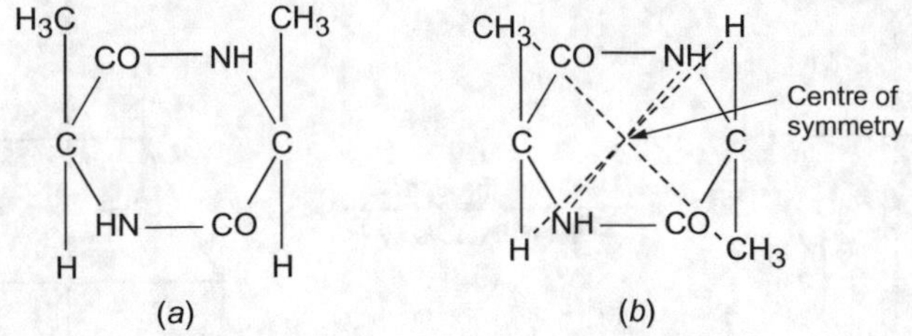

H₃C      CH₃

CO —— NH

C      C

HN —— CO

H      H

(a)

CH₃   CO —— NH   H

Centre of symmetry

C      C

NH —— CO

H      CH₃

(b)

**Fig. 5.7** Dimethyl diketopiperazine **a** *cis* form—optically active **b** *trans* form optically inactive

Another interesting example is that of dimethyl diketopiperazine, which exists in *cis* and *trans* forms. The *cis* isomer is optically active since it has no centre of symmetry. However, the *trans* isomer is optically inactive since it has a centre of symmetry (Fig. 5.7).

## 5.2.3 *Alternating Axis of Symmetry*

A structure possessing this axis on rotation around this axis results in another identical structure. In case the rotation of the molecule is by 360°, the identical structure is obtained twice; such an axis is called **two-fold axis of symmetry**. In a similar way, when identical structures repeat thrice, it is called a **three-fold axis of symmetry** and so on.

When such identical points alternate around an axis or a plane, then the axis of symmetry is called **alternating axis of symmetry**. An example is 1,3-dichlorobutane which has two-fold alternating axis of symmetry and not a simple axis of symmetry. This is because on rotation around this axis although the Cl group appears twice, one of them appears above the plane of the molecule and the other appears below the plane.

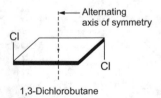

1,3-Dichlorobutane

In general, a molecule has an *n*-fold axis of symmetry, when on rotation through an angle of 360°/*n* about this axis and then reflected across the plane perpendicular to the axis, an identical structure results. Thus, as an example, we have 1,2,3,4-tetra-methyl cyclobutane (I) that contains a four-fold axis of symmetry if it is rotated through 90° about the axis *xy*, and the structure (2) results. Reflection of (2) in the plane of the ring

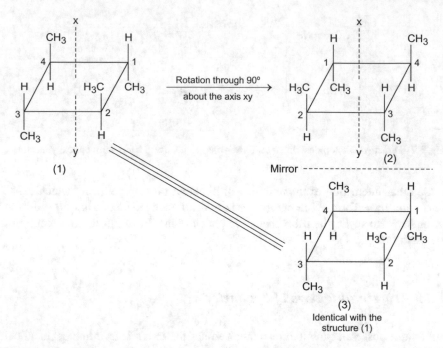

**Fig. 5.8** 1,2,3,4-Tetramethyl cyclobutane having four-fold axis of symmetry

gives (3), which is identical to the starting molecule (1). So, it contains a (360°/90° = 4) four-fold axis of symmetry (Fig. 5.8).

In a similar way, it can be shown that mesotartaric acid is optically inactive as it has an axis of symmetry (Fig. 5.9).

An interesting case is that of 1,3-di-sec-butylcyclobutane having the structure shown below (Fig. 5.10). This molecule has a two-fold (simple) axis of symmetry. It is not superimposable with its mirror image and so this molecule is optically active.

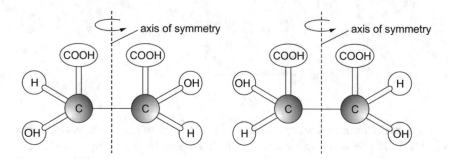

**Fig. 5.9** Mesotartaric acid having axis of symmetry

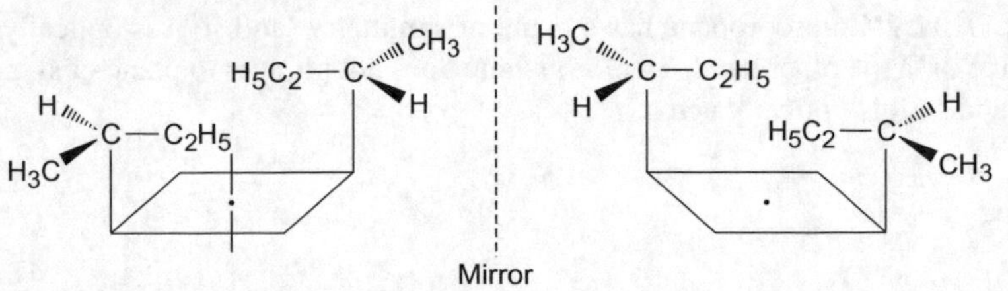

**Fig. 5.10** 1,3-di-sec-butyl cyclobutane with the simple axis of symmetry (two-fold) optically active

---

### Key Concept

- **Alternating Axis of Symmetry:** It is the axis on which identical points alternate around the axis. A molecule is said to have $n$-fold axis of symmetry when on rotation through an angle of $360°/n$ about this axis and then reflecting across the plane perpendicular to the axis, an identical structure results.
- **Centre of Symmetry:** A centre in a molecule from which identical atoms or groups in the molecule are located diametrically at equal distance.
- **Plane of Symmetry:** An imaginary plane which divides a molecule into two halves in such a way that the part of the molecule on one side of the plane is the mirror image on the other side of the plane.

---

### Problems

1. If an organic molecule has a plane of symmetry, will it be optically active? Explain with the help of an example.
2. Explain with the help of symmetry elements how mesotartaric acid is optically inactive.
3. Explain the term symmetry elements.
4. If a compound has any one of the elements of symmetry, will it be optically active?
5. In the case of 1,3-disec-butylcyclobutane, which structure will be optically active?
6. Draw the plane of symmetry in $CH_2BrCl$ and $CHBrClF$. Which one will be optically active?
   **Ans.** $CHBrClF$ is chiral
7. Draw a conformation of 2,3-dibromobutane which has a plane of symmetry.
8. Out of 2-chloropropane and 2-chlorobutane which is optically active on the basis of consideration of plane of symmetry?

**Ans.** 2-Chloropropane has a plane of symmetry (and so it is optically inactive). On the other hand, 2-chlorobutane does not possess to plane of symmetry (and so it is optically active).

# Chapter 6
# Stereochemistry of Optically Active Compounds Having no Asymmetric Carbon Atoms

We have known (Chap. 1) that for a compound to exhibit optical activity or optical isomerism, the presence of an asymmetric carbon is a must. However, there are a number of compounds that do not have asymmetric carbon atoms that exhibit optical activity. It is of interest to study the stereochemistry of such compounds, which include appropriately substituted biphenyls, allenes and spiranes.

## 6.1 Stereochemistry of Biphenyls

The optical activity was first observed in biphenyls in which the four ortho positions are substituted by a bulky group. Thus o,o′-dinitrodiphenic acid is optically active. This is because the rotation about the central bond does not occur (due to steric hindrance), and the two rings lie in different planes, i.e., the two rings are inclined to each other. In fact, the actual angle of inclination of the two rings depends on the substituent groups, but it is usually about 90°, i.e., the two rings are approximately perpendicular to each other. Such a molecule (as o,o′-dinitrodiphenic acid) is not superimposable on its mirror image, and therefore, it will be optically active (Fig. 6.1).

In the above case (Fig. 6.1), the two rings are not coplanar and so (*b*) is not superimposable on (*a*), i.e., (*a*) and (*b*) are enantiomers. As seen, in molecule (*a*), there is no chiral centre. In fact, it is the molecule as a whole that is chiral due to restricted rotation.

The evidence about the restriction (or lack) of rotation about the central bond is obtained by the failure to resolve biphenyl derivatives that are not substituted in ortho positions. Also, the resolution of ortho-substituted derivatives is possible only if the ortho substituents are sufficiently large. Thus, o,o′-difluorodiphinic acid, like the corresponding dinitro derivative (1*a*), is resolvable but is easily racemised than

© The Author(s), under exclusive license to Springer Nature Switzerland AG 2022    115
V. K. Ahluwalia, *Stereochemistry of Organic Compounds*,
https://doi.org/10.1007/978-3-030-84961-0_6

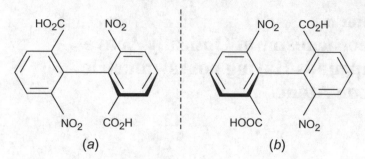

**Fig. 6.1**  o, o′-Dinitrodiphenic acid and its mirror image

1 *a*. This is attributed to the smaller size of the fluorine atom relative to the $NO_2$ group, with less interference from the ortho substituents. Once the compound passes through the planar conformation, the asymmetry is lost resulting in racemisation (Fig. 6.2).

It is not essential that all the four ortho positions be substituted by the bulky group in order to be optically active. The only prerequisite for a biphenyl to exhibit optical activity is that the substituents should be large enough to prevent free rotation and that each ring should be unsymmetrically substituted. Thus, compounds I and II (in Fig. 6.3a) are optically active and can be resolved. However, compound III is not resolvable as the ortho substituents are too small and one ring is symmetrically substituted (the molecule has a plane of symmetry when the rings are mutually perpendicular) (Fig. 6.3a).

Thus, it is seen (Fig. 6.3a) that even if two or three ortho positions are suitably substituted by bulky groups (so that the free rotation about the central bond is restricted), the compound can be optically active and is resolvable.

The enantiomers of an optically active biphenyl are called **atropisomers**. This type of isomerism is called **atropisomerism**. A discussion on atropisomerism forms the subject matter of a subsequent section.

A considerable amount of work has been carried out to study the effect of a 2,2′-bridge on the optical activity of compounds of the type (IV) (Fig. 6.3b). In case $n = 1$, the compound is a disubstituted fluorene, which being flat is not resolvable. When $n = 2$, the compound is a disubstituted 9, 10-dihydrophenanthrene, such compounds (e.g., V) has been resolved. When $n = 3$, the compounds can be resolved and are highly optically stable.

Some other such compounds containing 2,2′-bridge can be resolved and are given below (Fig. 6.4a).

The stereochemistry of biphenyls having 2,2′-bridge has been studied by UV and NMR spectroscopy and X-ray analysis. On the basis of X-ray diffraction study of a biphenyl derivative (VI), it was shown (Wahl, Jun. et al. 1972) that this compound exists in pseudo-chair form (VII) (Fig. 6.4b). The compound (VI) was prepared and resolved chromatographically (on cellulose acetate) (Lüttringhaus et al. 1967). On

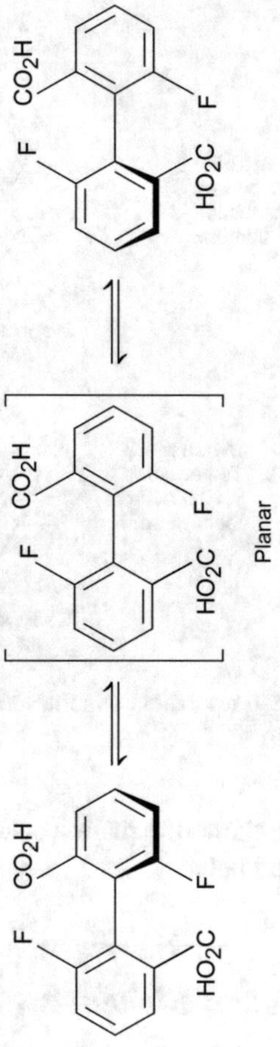

**Fig. 6.2**  o,o′-Difluorodiphinic acid (racemisation)

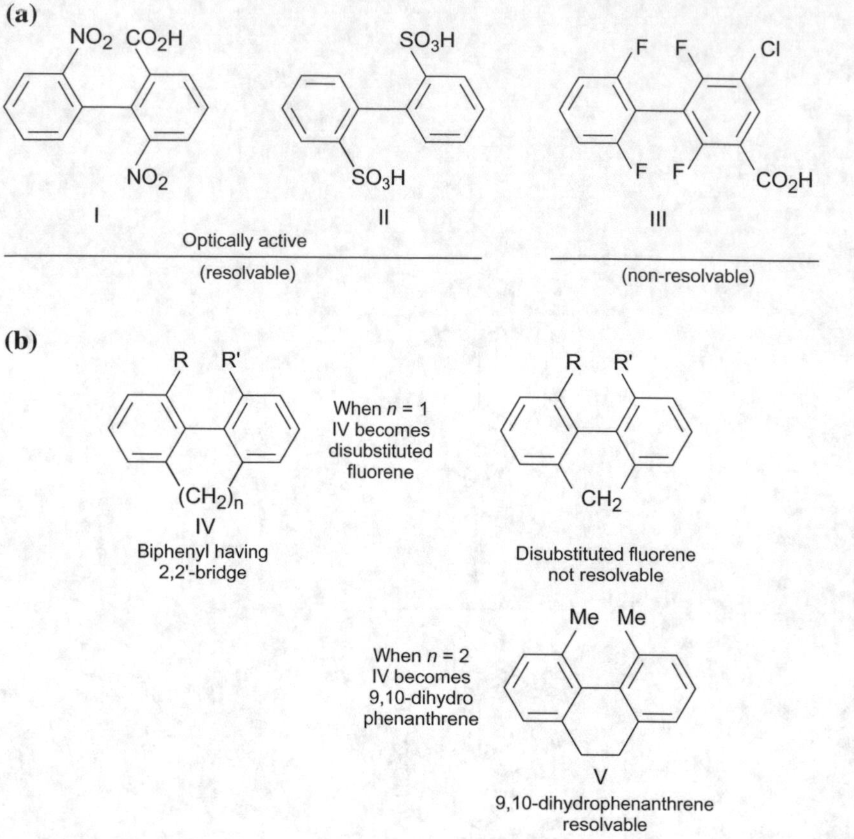
(a)
NO2 CO2H
NO2
I
SO3H
SO3H
II
F F Cl
F F CO2H
III
Optically active
(resolvable)
(non-resolvable)
(b)
R R'
(CH2)n
IV
Biphenyl having
2,2'-bridge
When n = 1
IV becomes
disubstituted
fluorene
R R'
CH2
Disubstituted fluorene
not resolvable
When n = 2
IV becomes
9,10-dihydro
phenanthrene
Me Me
V
9,10-dihydrophenanthrene
resolvable

**Fig. 6.3** **a** Examples of resolvable and non-resolvable of some substituted biphenyls. **b** Effect of 2,2'-bridge in biphenyls

the basis of considering bond angles, the conformation of VI was assigned pseudo-tub conformation, (VIII) which is more likely.

### *6.1.1 Absolute Configuration of Biphenyls*

The method used for the correlation of the absolute configuration of compounds containing chiral carbon cannot be used for determining the absolute configuration of biphenyls since these do not contain an asymmetric carbon. In the case of biphenyls, the absolute configuration is determined by asymmetric synthesis as developed by Mislow et al. (1957). The method consists in the reduction of (+) and (–) ketones (obtained from 6,6′-dinitro-2,2′-diphenic acid) by the Meerwein–Ponndorf–Verley method (*see* Sect. 19.3.11 in Chap. 19), using asymmetric alcohol of known absolute

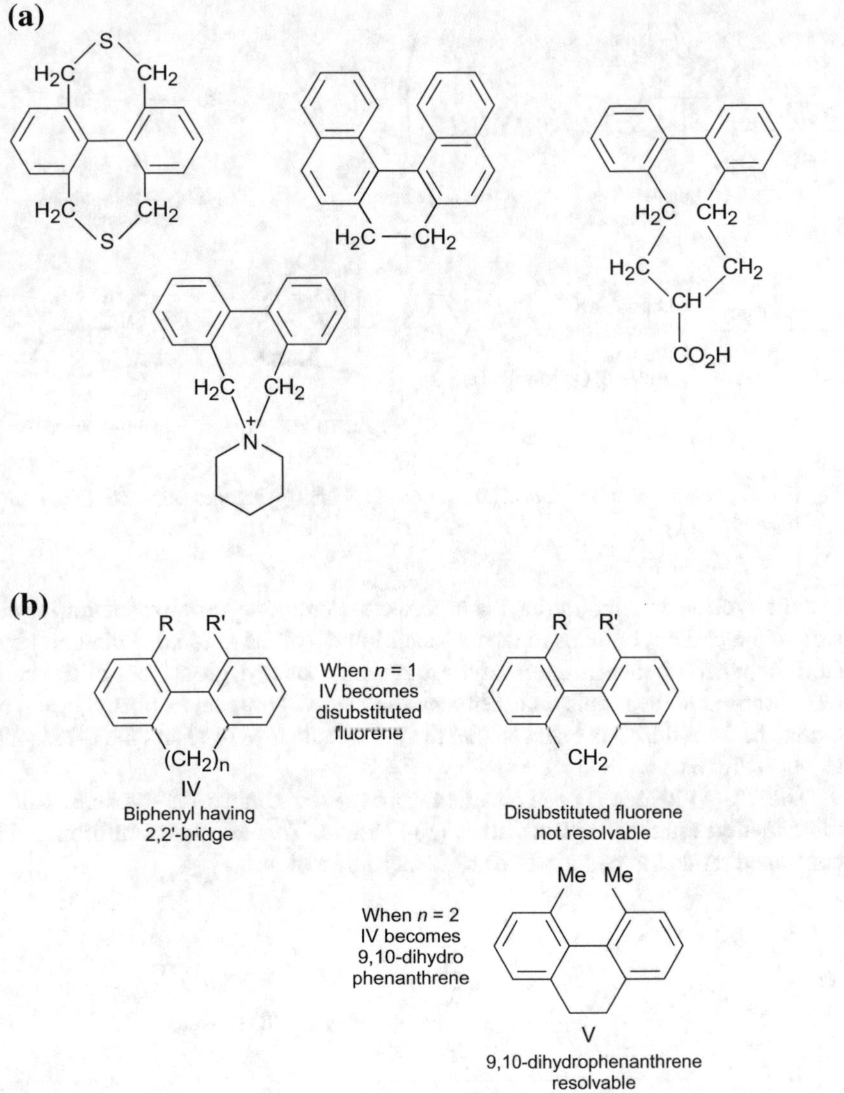
(a)
S
H2C
CH2
H2C
CH2
S
H2C — CH2
H2C CH2
H2C
CH2
CH
CO2H
H2C
CH2
N
+
(b)
R R'
(CH2)n
IV
Biphenyl having
2,2'-bridge
When n = 1
IV becomes
disubstituted
fluorene
R R'
CH2
Disubstituted fluorene
not resolvable
Me Me
When n = 2
IV becomes
9,10-dihydro
phenanthrene
V
9,10-dihydrophenanthrene
resolvable

**Fig. 6.4** **a** Some resolvable biphenyls having 2,2′-bridge. **b** X-ray diffraction study of a biphenyl derivative

configuration. As an example, (+) and (−) ketones (I) on reduction with (*S*)-(+)-methyl-t-butylmethanol in presence of aluminium-t-butoxide gave the unchanged ketone (in which the (+) form predominated) and the (+)- and (−)-alcohols (II) (in which the (−) form predominated). Since the unchanged ketone contained an excess of the (+) form, the method constituted a kinetic resolution (the alcohol (II) become enriched in the (–)-form). It is believed (on the basis of models of (+) and (–) alcohols)

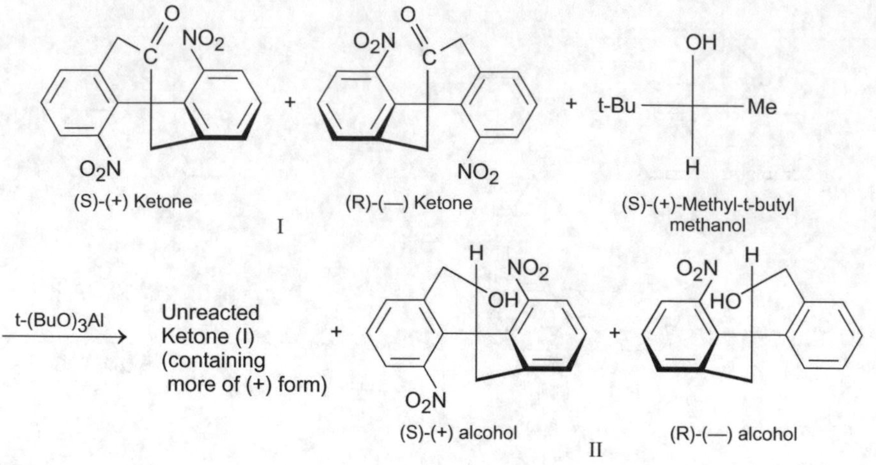


**Fig. 6.5** Determination of absolute configuration of (+)- and (–) ketones derived from 6,6'-dinitro-2,2'-diphenic acid

that the hydride transfer (during the Meerwein–Ponndorf–Verley reduction) to either side of the carbonyl groups in the (*S*)-enantiomer (of the ketone) is hindered by the steric repulsion between the t-butyl group and a phenyl group, but in the case of (*R*)-enantiomer, the repulsion is between the methyl group and a phenyl group only. On the basis of this, it is believed that the (+)-ketone is (*S*)-(+) and the (–) alcohol is (*R*)-(–) (Fig. 6.5).

The (*S*)-(+)-ketone (I) was obtained from (–)-6,6′-dinitro-2,2′-diphenic acid (via the dimethyl ester) (III). Thus, III is (*S*)-(–)-acid. This absolute conformation was confirmed (Ankemoto et al. 1968) by X-ray diffraction.

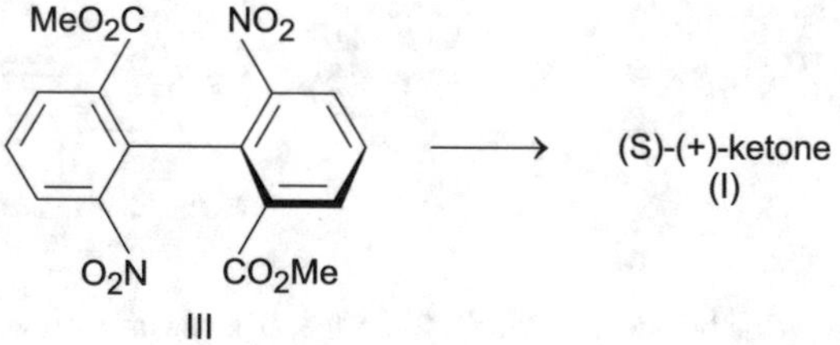


III
Dimethyl ester of (S)-(–)-6,6'-dinitro-2,2'-diphenic acid

The above method of determination of absolute configuration is dependable only if there is no change of configuration during the transformation; however, if the change in configuration occurs, it should be predictable. Using the (*S*)-(–)-acid as a standard, a number of transformations were carried out (Mislow et. al.). One such example is given below (Fig. 6.6).

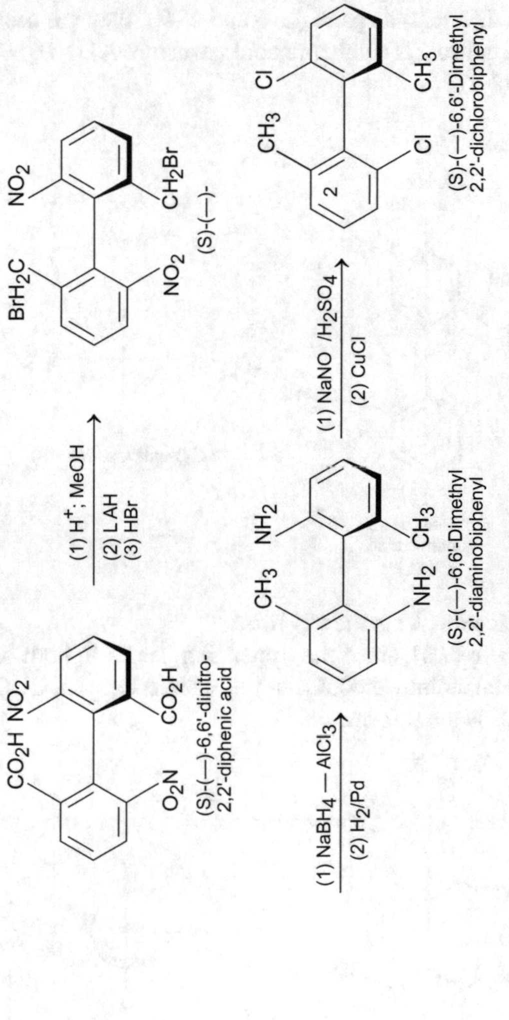

**Fig. 6.6** Absolute stereochemistry of (S)-(–)-6,6'-diphenyl 2,2'-dichlorobiphenyl

The absolute configuration of the biphenyl compounds has been confirmed (Mislow et al. 1960) by the rotary **dispersion method** (see Chap. 9). It was found that the ORD curves depended on the configuration and conformation of biphenyls.

The assignment of the absolute configuration ($R$ or $S$) in biphenyls is carried out in the following:

1.   In the case of biphenyl, the line joining the positions 1, 1′, 4, 4′ is the chiral axis (i.e., an axis of asymmetry).
2.   In the case of substituted biphenyls, the four ortho substituents are examined. In case, they are different, as pairs (2–6 and 2′-6′), they are used. Thus, in the compound 6, 6′-dinitro-2,2′-diphenic acid (structure A), $NO_2 = a$ and COOH $= b$ (priority $a > b$).

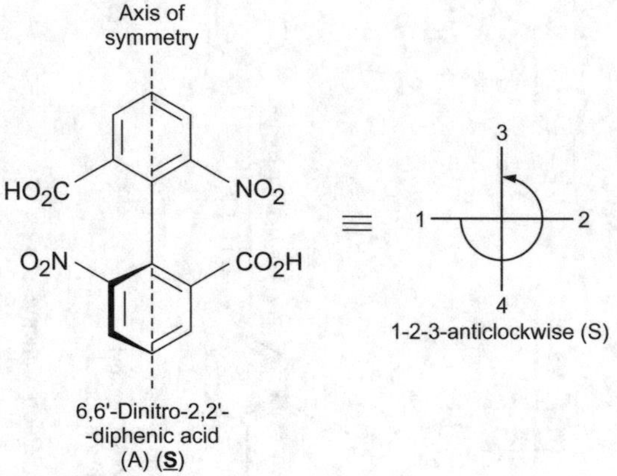

Axis of symmetry

$HO_2C$          $NO_2$

$O_2N$          $CO_2H$

6,6′-Dinitro-2,2′-
-diphenic acid
(A) (**S**)

1-2-3-anticlockwise (S)

Therefore, this molecule (A) is the (S)-form.

In the case of biphenyl (B), since the upper ring has Cl in both ortho positions, groups H and Me are taken into account. In this case, $NO_2 = a$, COOH $= b$, Me $= c$ and H $= d$. Hence, B is the S form.

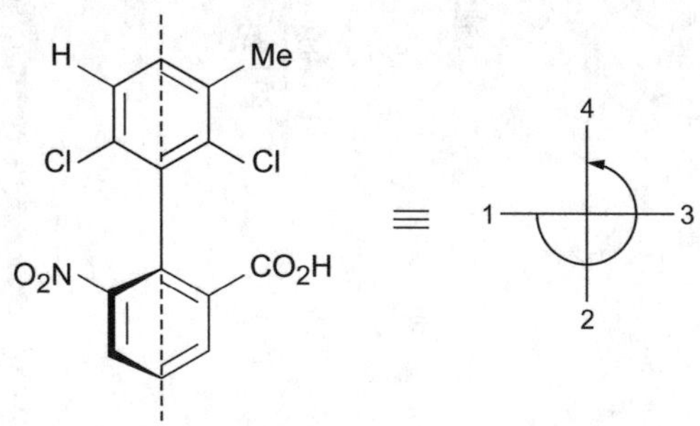

H          Me

Cl          Cl

$O_2N$          $CO_2H$

(B)

Hence, B is the (S) form. The ketone I (Fig. 6.5) is in the (S) forms; $NO_2 = a$ and $CH_2CO = b$.

## 6.1.2 Atropisomerism

As already mentioned (Sect. 6.1), the restriction in rotation around a single bond due to the presence of bulky groups in ortho positions of biphenyls gives rise to isomerism known as atropisomerism. The isomers involved are called atropisomers and are optically active. A number of examples of biphenyls exhibiting atropisomerism have already been discussed. In addition to the biphenyls, there are a number of other examples where the optical activity is due to restricted rotation about a single bond. Some examples of this type of atropisomerism are given below.

(a)  N-phenyl pyrroles (Adams et al. 1931)

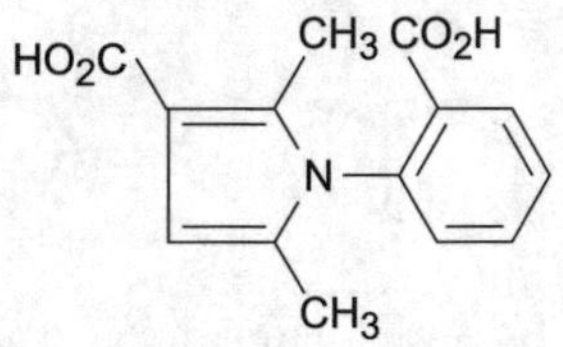

(b)  N, N′-Bipyrroles (Adams et al. 1932)

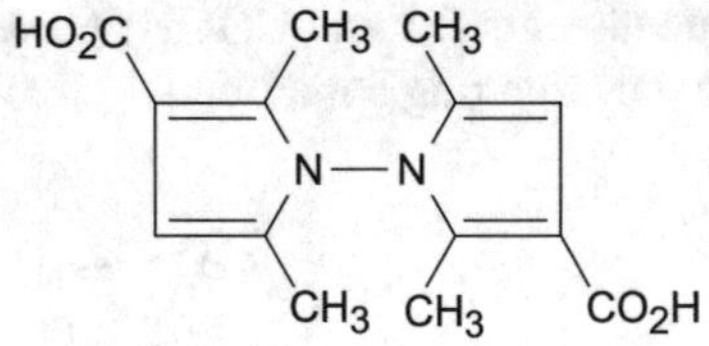

(c)  3,3′-Bipyridyls

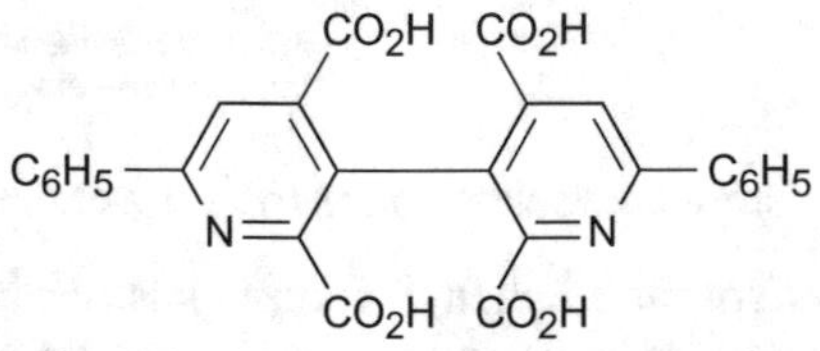

(d)  1,1′-Binaphthyl-8–8′-dicarboxylic acid (Stanley 1931)

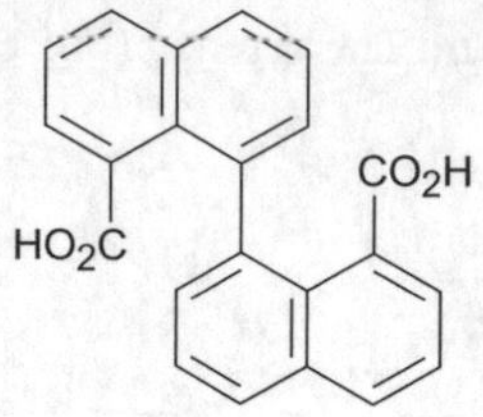

(e)  Atropisomerism due to restricted rotation about C–N bond, the carbon being the ring carbon to which N is attached, some examples are as follows:

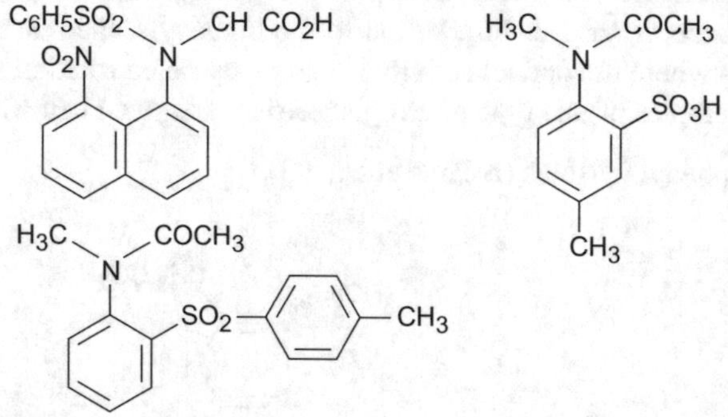

(f)  Atropisomerism in ansa compounds: Two optically active forms of 4-bromogentisic acid decamethylene ether (belonging to the group known as ansa compounds) were isolated by Lüttringhaus et al. (1940, 1947). In this compound, the methylene ring is perpendicular to the plane of the benzene ring and the two substituents (Br and COOH) prevent the rotation of the benzene nucleus inside the large ring compound.

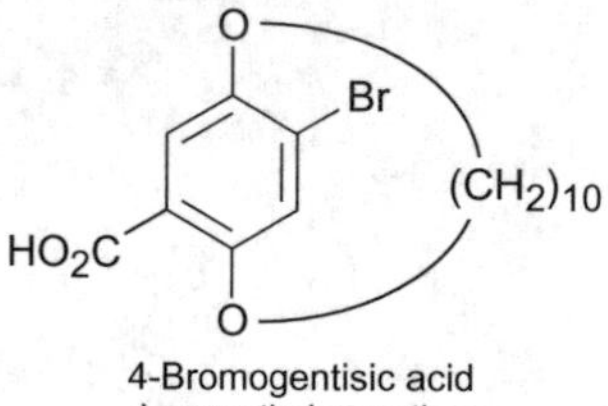

4-Bromogentisic acid
decamethylene ether

A discussion of ansa compound forms the subject matter of a subsequent section.

(g)  Atropisomerism in terphenyl compounds: Suitably substituted terphenyl compounds exhibit both geometrical and optical isomerisms. The substituents are such so that free rotation about single bonds is prevented. It was possible to obtain *cis* and *trans* forms of the terphenyl compound (I) (Shildneck and Adams 1931). In this case, the free rotation due to methyl and hydroxyl group in ortho positions is prevented and the two outside rings are perpendicular to the central ring. In case the centre ring does not possess a vertical plane of symmetry, then there is a possibility of optical activity. In fact, the *cis* and *trans* forms of (II) were prepared (Browning and Adams 1930). It was possible to resolve the *cis* form of (II); the *trans* form is non-resolvable since it has a centre of symmetry.

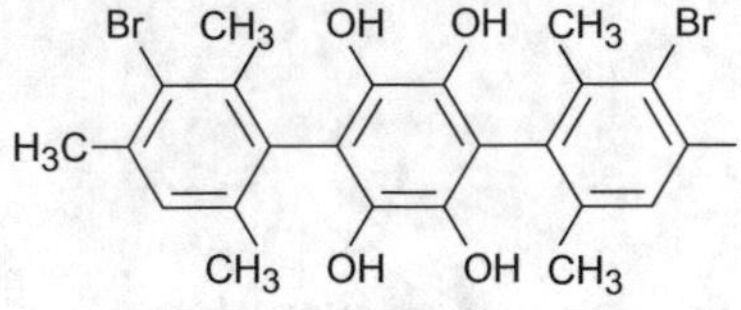

cis-(I)

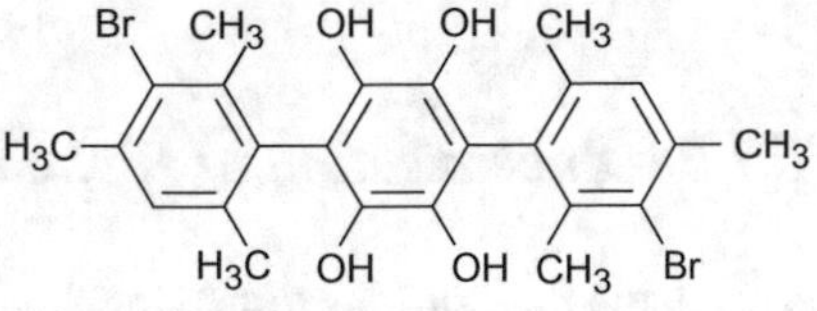

trans-(I)

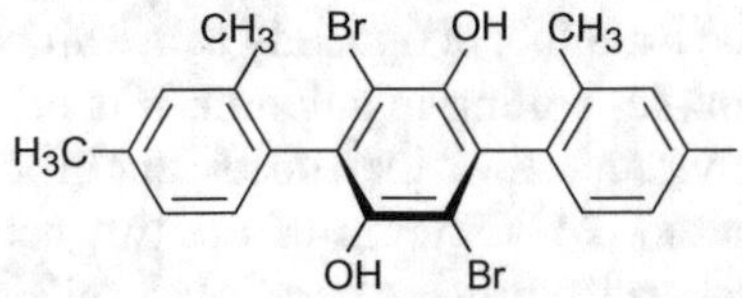

cis-(II)
(resolvable)

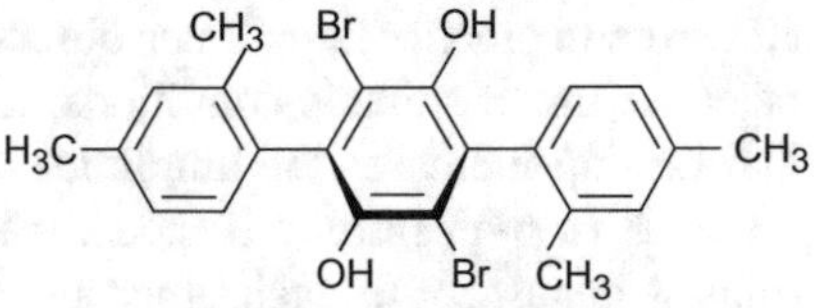

trans-(II)
(non-resolvable)

(h) Atropisomerism in 10-*m*-aminobenzylideneanthrone: 10-*m*-Aminobenzy-lideneanthrone was prepared by Ingram (1950). This compound is asymmetric, which is due to the restricted rotation of the phenyl group about the C-phenyl bond; the restriction in the rotation is brought about by hydrogen atoms (labelled H*) in ortho positions.

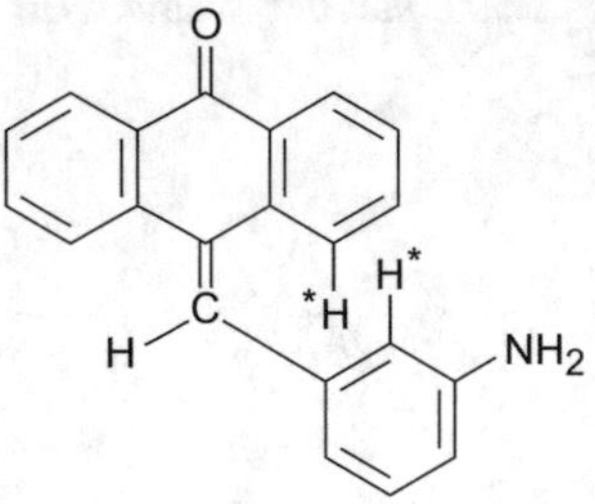

10-m-Aminobenzylideneanthrone

(i) Atropisomerism in appropriately substituted cinnamic acids: It has been found (Adams et al. 1940, 1941) that in substituted cinnamic acids, the benzene ring and the ethylenic double bond cannot become planar and so are optically active.

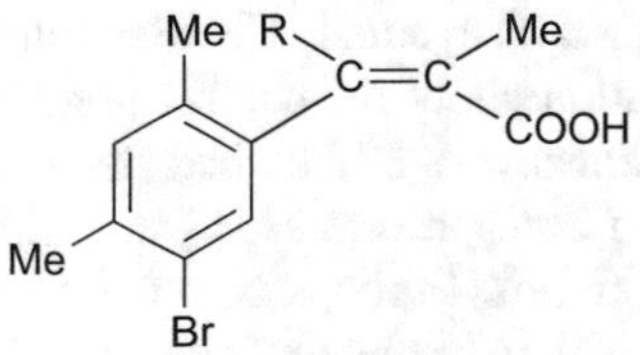

Substituted Cinnamic acid
R = Cl, Me, OMe

The order of stability to racemisation in the above case has been found to be
Cl > Me > OMe.

(j)     A number of other compounds also exhibit atropisomerism: These include
cyclophanes, benzocycloalkanes, helicenes and annulenes. A discussion on all
these forms the subject matter of subsequent sections.

### 6.1.3   Racemisation of Biphenyls

The optical activity of biphenyls is due to the restricted rotation about the central
bond. In view of this, it may be expected that these compounds will not racemise.
However, in practice, it has been observed that a number of optically active biphenyls
racemise under suitable conditions, for example, boiling in solution. It is believed
that heating increases the amplitude of the vibrations of the substituent groups in
the 2, 2' 6, 6'-positions. Besides, the amplitude of vibration of the two benzene
rings with respect to each other also increases. Both these permit the substituent
groups to slip by one another. The nucleus, thus, passes through a common plane
and so there is a possibility that the final product will be an equimolecular amount
of (+)-and (–)-form. This is referred to as the **obstacle theory**. Subsequently, it was
believed (Westheimer, 1946–1950) that in addition to the above bond stretchings, the
angles α, β, and γ are also deformed and the benzene rings are also deformed during
racemisation. There is good agreement between estimated and measured activation
energies of racemisation of some di-ortho-substituted biphenyls. The calculation is
based on known values of the van der Waals radii and stretching and bending-force
constants of the bonds.

On the basis of extensive work, it has been found that 2, 2', 6, 6'-tetra-substituted
biphenyls are of three types depending on the nature of the substituent groups.

(i)      **Non-resolvable**: These biphenyls contain the groups like hydrogen, methoxy
or fluorine, which have small effective volumes which in turn are too small
to prevent rotation about the central bond. As an example, 2,2'-difluoro-6,
6'-dimethoxybiphenyl-3, 3'-dicarboxylic acid (I) is non-resolvable.

(ii)     **Resolvable and easily racemised**: Such biphenyl must contain at least two
amino or two carboxyl groups or one amino and one carboxyl group. The
remaining groups can be any of those given in (*i*) above except hydrogen (i.e.,

OMe, F). As an example 6, 6′-diflurodiphenic acid (II) is resolvable and is easily racemised.

2,2′-Difluoro-6,6′-dimethoxy
biphenyl-3,3′-dicarboxylic acid

6,6′-Difluorodiphenic acid

(iii)   **Not racemisable:** Such biphenyls contain at least two nitro groups; the other group can be any of those given in (*i*) above except hydrogen (viz., OMe, F). As an example, 2, 2′-difluoro-6, 6′-dinitrobiphenyl (III) is resolvable but cannot be racemised.

The rate of racemisation also depends on the nature and position of the substituents besides the size of the group in ortho positions. As an example, the rate of racemisation of (IV) is much slower than that of (V) (Adams et al. 1932, 1934). Thus, it can be said that the nitro group in 3′-position has a much greater stabilising influence than in 5′-position. The reason for this is not well understood.

III

2,2′-Difluoro-6,6′-
dinitrobiphenyl

IV

2,3′-dinitro-
2′-methoxyl biphenyl
6-carboxylic acid

V

2,5′-Dinitro-2′-
methoxybiphenyl
6-carboxylic acid

The order of steric hindrance produced by different groups is of the order:
Br > > Me > Cl > $NO_2$ > $CO_2H$ > > OMe > F.

This order roughly corresponds to the order of the van der Waals radii of the groups.

It has also been found (Adams et al. 1954, 1957) that the rate of racemisation of (VI) is increased if R is an electron-withdrawing group (like $NO_2$ and CN) and decreased if R is an electron-releasing group (like Me and OMe).

PhSO$_2$ — N — CH$_2$CO$_2$H

Me

R

VI

Besides heating in a suitable solvent, racemisation can also be affected by irradiation (in ether solution) with UV light or by heating above 200°C in dark.

As already stated, the obstacle theory explains the resolution of optically active biphenyls (appropriately substituted in 2, 2', 6–6'-positions).

However, the two benzene rings in the biphenyls should be non-planar. Support for the non-planar configuration was provided by Meisenheimer et al. (1927). The method involved uniting the obstacle groups in optically active biphenyls resulting in the formation of five- or six-membered rings. Thus, an optically active enantiomer of 2, 2'-diamine 6, 6'-dimethylbiphenyl was subjected to the following sequence of reactions.

H$_3$C        NH$_2$    (1) Ac$_2$O    HO$_2$C        NHAc    H$_2$SO$_4$    OC        NH
H$_2$N        CH$_3$    (2) [O]        AcHN        CO$_2$H                HN        CO

| Opticaly active enanatiomer of 2,2'-diamino-6,6'-dimethylbiphenyl | Optically active | Optically inactive (Dilactam) |

In all optical biphenyls, the rings cannot be coplanar. In case the rings in biphenyls are planar, the molecule will possess a centre or plane of symmetry. If the dilactam is not planar, it will not possess elements of symmetry and so will be optically active. As this dilactam is optically inactive, it must be planar. This planarity is explained in terms of resonance. All double bonds in dilactam have double bond character and so the molecule is planar.

The fact that the two rings in o-substituted biphenyls are not coplanar is shown by the UV spectrum of biphenyl, $\lambda_{max}$ 248(19,000) nm, different from that of benzene, $\lambda_{max}$ 198(8000) nm. The shift to a longer wavelength is explained on the basis that biphenyl is a resonance hybrid, and one of the contributing structures is in extended conjugated form (II). Thus, the interannular bond is expected to have

Some double bond character and so the molecule will tend to be planar. However, when o-positions are occupied, the coplanarity is prevented and the spectrum will be different, the maximum absorption shifts to the shorter wavelength. Thus, 2-methylbiphenyl shows $\lambda_{max}$ 236(10,000) nm and 2, 2′-dimethylbiphenyl shows $\lambda_{max}$ 224(~700) nm. It has been shown (Picketl et al. 1936) that the UV absorption maximum of bimesityl $\lambda_{max}$ 267 (545) nm and Mesitylene $\lambda_{max}$ 266 (nm) are almost identical but different from the $\lambda_{max}$ of biphenyl. In the case of bimesityl, there is steric inhibition of resonance thereby preventing the coplanarity.

## 6.2 Stereochemistry of Optically Active Compounds Due to Intramolecular Crowding

We have known that for a compound to exhibit optical activity, it must have a chiral carbon (Chap. 1). Even compounds having no asymmetric carbon (like appropriately substituted biphenyls, allenes and spirans) can exhibit optical isomerism. There is, however, another way by which some steric factors may produce asymmetry. In general, it has been found that non-bonded carbon atoms cannot be closer to each other than about 3.0 Å. In case, the geometry of the molecule is such so as to produce 'intramolecular overcrowding', the molecule becomes distorted. An example of this type is 4, 5, 8-trimethyl-1-phenanthryl acetic acid (I). It is known that the phenanthrene nucleus is planar, and the substituents are lying in this plane. In case, reasonably large groups are present in positions 4 and 5, then there will be insufficient room to accommodate both the large groups in the plane of the nucleus. In such a case, strain is produced in the molecule by intramolecular overcrowding. This produced strain can be relieved by bending of the substituents out of the plane of the nucleus and or by bending (buckling) of the aromatic rings. Thus, the molecule will not be planar and will be asymmetric and so it can be resolved (in principle). In fact, the compound I was resolved by Newman et al. (1940, 1946). These authors also resolved the compound (II). Subsequently, Bell et al. (1949) resolved the compound (III). In fact, these authors introduced the term 'intramolecular overcrowding'.

I

II

III

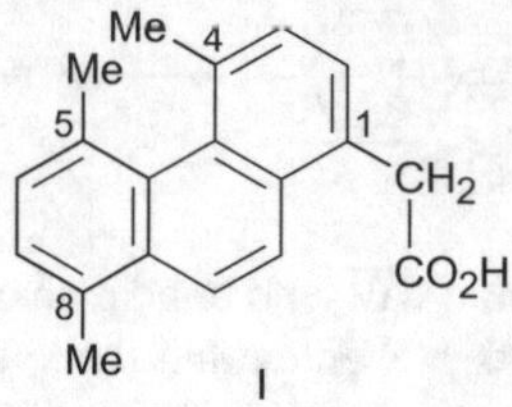

An heterocyclic analogue (IV) of phenanthrene was resolved by Theilacker et al. (1953). All the compounds (I to IV) had low optical stability. Two other compounds (V) and (VI), having good optical stability, were prepared by Newman et al. (1955, 1956); the compound VI is hexahelicene.

IV

V

VI
(Hexahelicene)

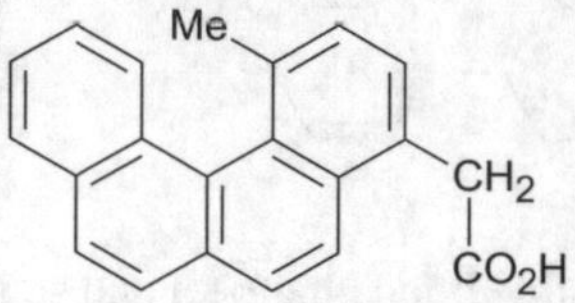

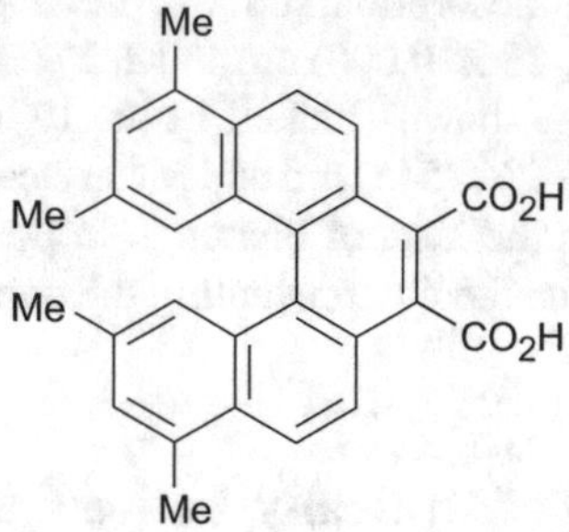

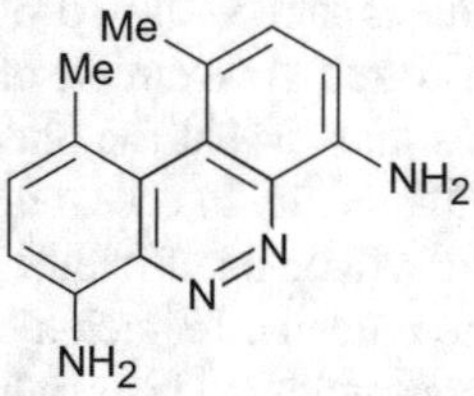

In the compound (VI), out-of-plane distortion occurs by buckling of the molecule. An example of a simple molecule that exhibits overcrowding and consequently out-of-plane buckling is 3, 4-benzophenanthrene (VII). On the basis of X-ray analysis, VII has been shown to be non-planar (Schmidt et al. 1954). In a similar way, Robertson et al. (1954) showed that (VIII) also exhibited out-of-plane buckling. It should be noted that in all compounds (like VI, VII and VIII), the buckling is distributed over all the rings in such a way so as to cause minimum distortion in any one ring.

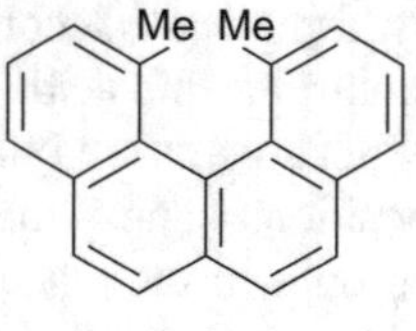

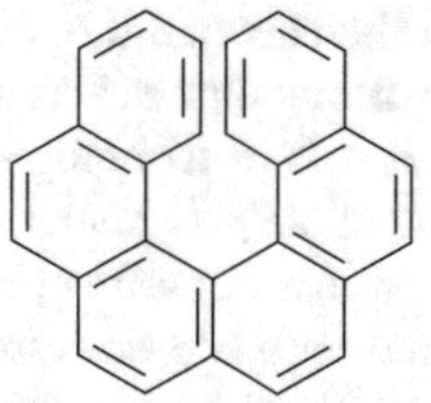

VII        VIII        IX

The distortion (as mentioned above) avoids the non-bonded carbon atoms (marked with dots in VII and VIII) to avoid being closer together than 3.0 Å and results in forcing some other carbon atoms to adopt an almost tetrahedral valency arrangement (the original hybridisation being trigonal). This in turn affects the chemical and physical properties of the molecule. The deformation in VIII has been found (Coulson et al. 1955) to produce a loss of resonance energy of about 75 kJ mol$^{-1}$.

On the basis of analysis of circular dichroism spectra of (–)-(VII$a$) and (+)-(VII$b$) (both are derivatives of VII), it has been concluded (Mason et al. 1965) that the former [(–)-(VII$a$)] has the M–(minus, left-handed) and the later [(+)-(VII)$b$] had P-(plus, right-handed) helical configuration as viewed in the direction perpendicular to the mean molecular plane.

(—)-VII$a$            (+)-VII b

[**Helical molecules** are those molecules in which the arrangement of the atoms or groups is an imaginary helix].

The helical molecules are optically active due to the presence of **helical dissymmetry** or **helicity**. In fact, helicity is a particular type of chirality.

## 6.3 Stereochemistry of Allenes

The 1, 2-dienes having cumulated double bonds one called allenes. It can also be said that allenes are compounds in which a carbon atom is bonded to two other carbon atoms by double bonds. The simplest member of the series is represented as $H_2C = C = CH_2$ (1,2-propadiene) and is called allene. In allene, the terminal carbons are $sp^2$-hybridised and use the three hybridised orbitals to form a σ bond with two hydrogens and a carbon. The central carbon atom in allene is $sp$-hybridised and is attached to two carbon atoms through these orbitals forming σ bonds. The central carbon has two $p$-orbitals which are involved in π bonding. Thus, any allene with the substitution pattern $abC = C = Cab$ will exist in two enantiomeric forms.

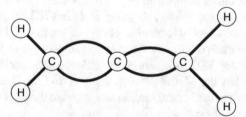

Examination of the ball-and-stick model of allene indicates that the two double bonds and so the terminal methylene groups are in different planes at right angles.

Ball-and-stick model of allene

The same stereochemistry has been predicted by the atomic-orbital representation of allene.

The allenes can be inferred on the basis of IR and NMR spectra. IR absorption due to double bond stretching vibration appears at about $1950\,\text{cm}^{-1}$ rather than $1650\,\text{cm}^{-1}$ (alkenes) or $2200\,\text{cm}^{-1}$ (alkynes). Such an absorption band ($1950\,\text{cm}^{-1}$) is normally observed with cumulated system and is also found in the spectra of ketenes ($R_2 = C = O$) and isocyanates ($R-N = C = O$). A special feature in the NMR spectra of allenes is the large J value for hydrogen on the double bonds (the hydrogens are separated by four bonds).

5-10 cps

H$\diagdown$ $\diagup$H
$\diagup$C$=$C$=$C$\diagdown$

It was as early as 1875 that the optical isomerism of allene was predicted by Van't Hoff, but experimental verification was obtained only in 1935 when catalytic asymmetric dehydration of 1,3-di-1-naphthyl-1,3-diphenyl prop-2-enol was carried out to given dinaphthyldiphenyl allene (Mills and Mailland).

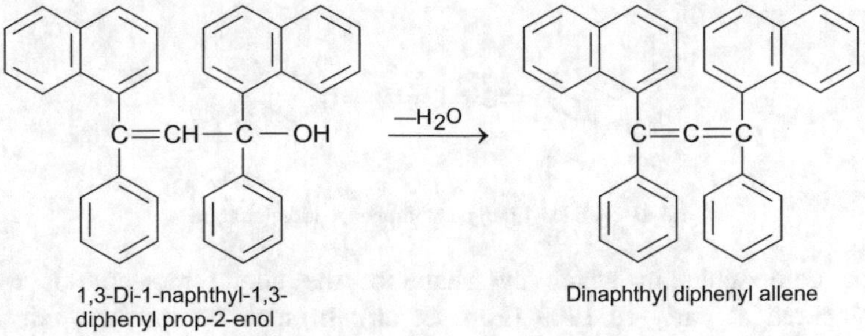

1,3-Di-1-naphthyl-1,3-
diphenyl prop-2-enol

Dinaphthyl diphenyl allene

In case dehydration in the above case was carried with an optically inactive catalyst (e.g., *p*-toluene sulphonic acid), the allene derivative was obtained in racemic modification. However, dehydration with boiling 1 per cent solution of (+)-camphorsulphonic acid gave the corresponding dextrorotatory allene, and dehydrating using (−)-camphoric acid gave the corresponding laevorotatory allene.

Another asymmetric synthesis of 1,3-diphenylallene was carried out by Jacobs et al. (1957) by rearrangement of 1,3-diphenylpropyne by absorption on alumina impregnated with brucine or quinoline. The use of brucine gave the (−)-allene and the use of quinoline gave the (+)-allene.

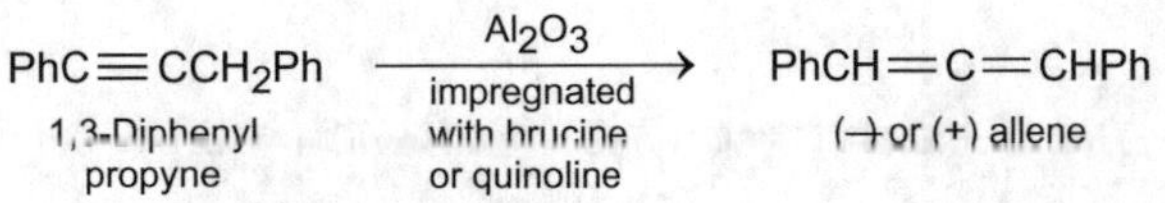

$$PhC \equiv CCH_2Ph \xrightarrow[\substack{\text{impregnated} \\ \text{with brucine} \\ \text{or quinoline}}]{Al_2O_3} PhCH = C = CHPh$$

1,3-Diphenyl                                            (−) or (+) allene
propyne

The first successful resolutions of asymmetric allenes (I) and (II) were carried out by Kohler et al. (1935).

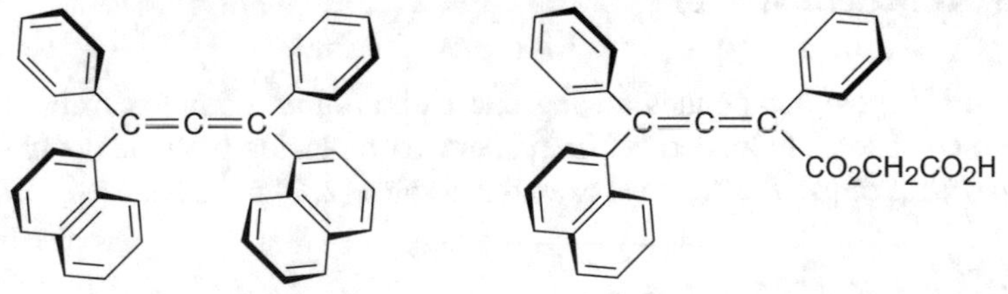

I

II

The allenes so far discussed contain two double bonds. Subsequently, Nakagawa et al. (1961) prepared 1,5-di-t-butyl-1,5-di-*p*-chlorophenyl pentatetraene, an cumelene with four double bonds.

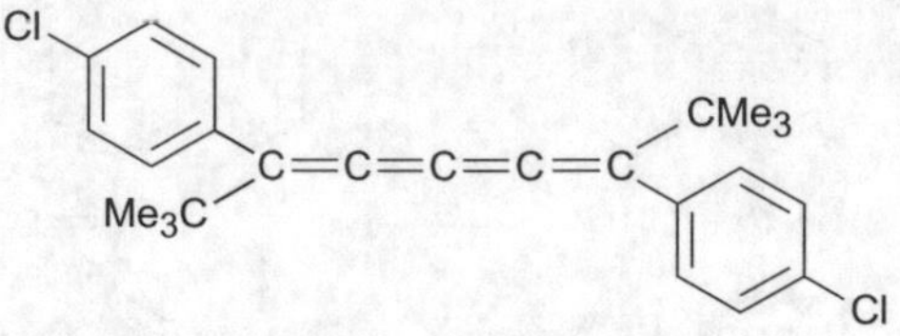

1,5-di-tert. butyl-1,5-di-p-chlorophenyl pentatetraene

Some compounds in which the shape of the allene molecule is retained were prepared as early as 1909 (Pope et al.). In such compounds, one of the double bonds of allene is replaced by a six-membered ring. A typical example is 4-methylcyclohexylidene-1-acetic acid which could be resolved.

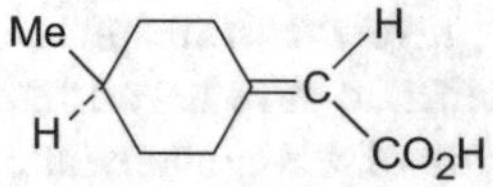

4-Methylcyclohexylidene-1-active acid

The importance of the allene group is evident from the observation that mycomycin an antibiotic and other polyacetylenes contain the allene group. The optical activity of mycomycin is due to the presence of the allene group.

$$HC \equiv C - C \equiv C - CH = C = CH - CH = CH - CH = CH - CH_2 - CO_2H$$

Mycomycin

## 6.4   Stereochemistry of spirans

Spirans are bicycle compounds having one carbon atom common to both rings. These are considered to be derived from allene by replacing both the double bonds (in allene) with rings. The simplest example is spiro [2, 2] pentane.

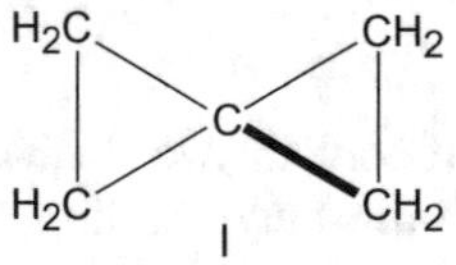

spiro [2,2] pentane

The spirans are named by selecting the root name from the number of carbons in the structure and then prefixing the term 'spiro' followed by numbers placed in square brackets indicating the number of carbon atoms joined to the 'Junction' carbon atom (higher number is placed first). The position of the substituents is indicated by the numbers, the numbering beginning with the smaller ring and ending on the Junction carbon atom; the numbering is continued on the bigger ring. The junction carbon atom gets the last number. As an example, the structure (I) is spiro [2.2]-pentane and the structure II is named 1-chlorospiro [5, 3]-nonane.

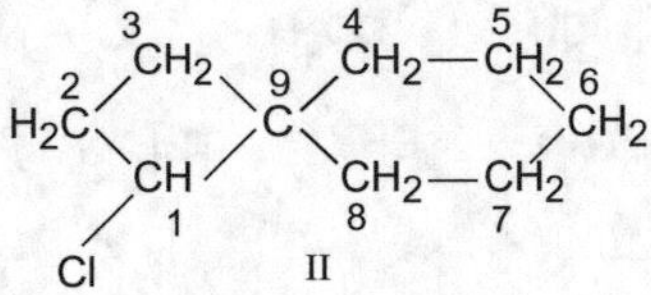

II
1-Chlorospiro[5,3]nonane

On examination of formulae I and II, it is seen that the two rings are perpendicular to each other. So, suitable substituents will give molecules that will have no elements of symmetry, and the forms, thus, obtained will be optically active. As an example, the dilactone (III) of benzophenone-2, 2′, 4 4′-tetracarboxylic acid could be resolved into two enantiomers (Mills and Nodder, 1920, 1921).

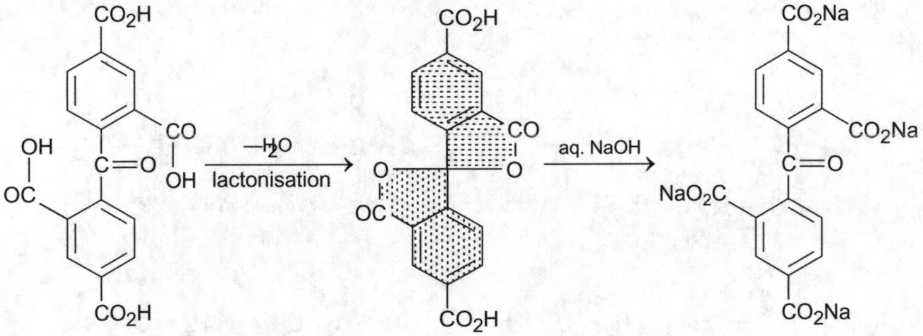

IV
Benzophenone
2,2',4,4'-tetracarboxylic

III
Dilactone-ol
benzophenone-2,2',4,4'
tetracarboxylic acid

V

In the dilactone (III), the two shaded portions being perpendicular to each other have no elements of symmetry. Treatment of the dilactone with alkali opens up the lactone rings to give (V), which is not optically active.

Some other spiro compounds have been prepared and resolved. These include the following:

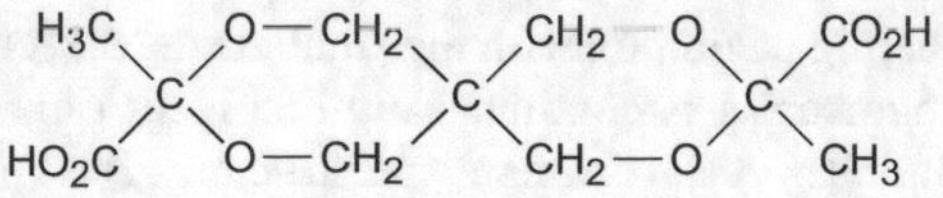

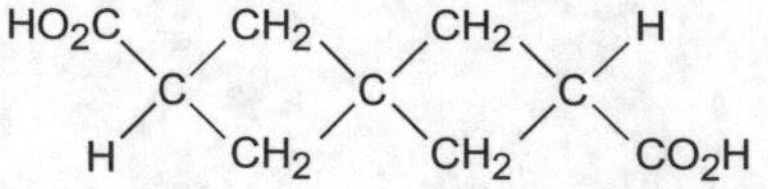

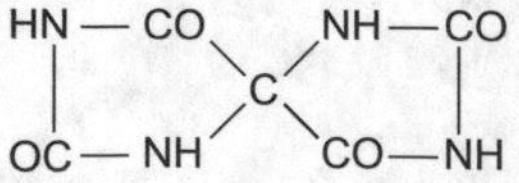

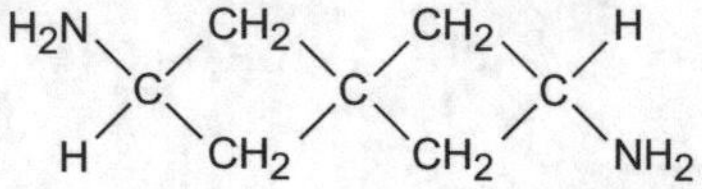

$$H_3C\!-\!\underset{HO_2C}{\overset{}{C}}\!\!<\!\!\underset{O-CH_2}{\overset{O-CH_2}{}}\!\!>\!\!C\!\!<\!\!\underset{CH_2-O}{\overset{CH_2\;\;O}{}}\!\!>\!\!\underset{CH_3}{\overset{CO_2H}{C}}$$

$$\underset{H}{\overset{HO_2C}{}}\!\!>\!\!C\!\!<\!\!\underset{CH_2}{\overset{CH_2}{}}\!\!>\!\!C\!\!<\!\!\underset{CH_2}{\overset{CH_2}{}}\!\!>\!\!\underset{CO_2H}{\overset{H}{C}}$$

$$\underset{OC-NH}{\overset{HN-CO}{}}\!\!>\!\!C\!\!<\!\!\underset{CO-NH}{\overset{NH-CO}{}}$$

$$\underset{H}{\overset{H_2N}{}}\!\!>\!\!C\!\!<\!\!\underset{CH_2}{\overset{CH_2}{}}\!\!>\!\!C\!\!<\!\!\underset{CH_2}{\overset{CH_2}{}}\!\!>\!\!\underset{NH_2}{\overset{H}{C}}$$

In all the above spirans (discussed so far), the optical activity is due to the asymmetry of the molecule as a whole and each gives one pair of enantiomers. In case a spiro compound contains also asymmetric carbons atoms, the number of optical forms will be more than two the actual number depending on the structure of the spiran. As an example, the spiro compound (VI) was prepared by Sutter and Wijkman (1935). As seen, this spiro compound (V) contains two similar asymmetric carbon atoms (marked *). In this case, three differential geometrical isomers (VII, VIII and IX) are possible.

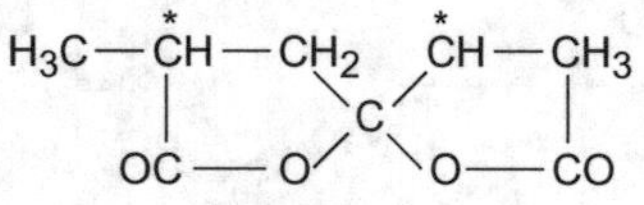

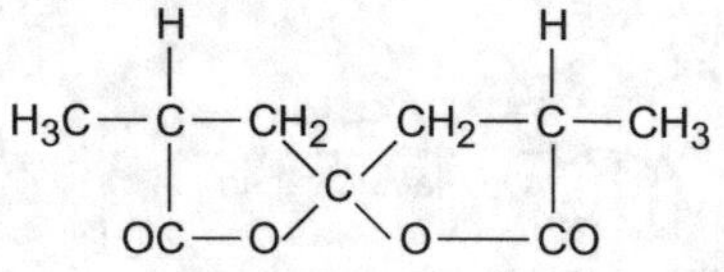

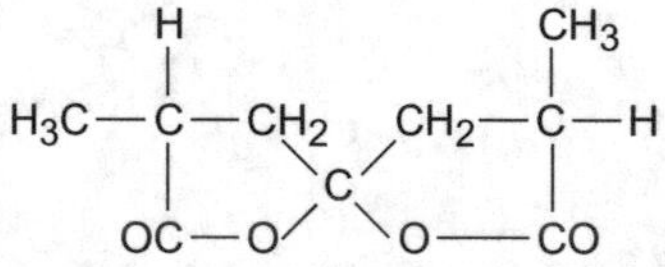

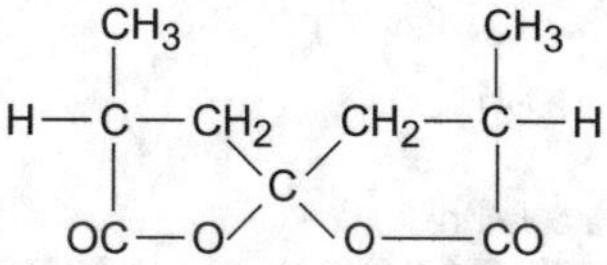

The three isomers can be easily visualised with the help of models. A number of spiro compounds containing nitrogen and phosphorus have been prepared. Some examples are given below.

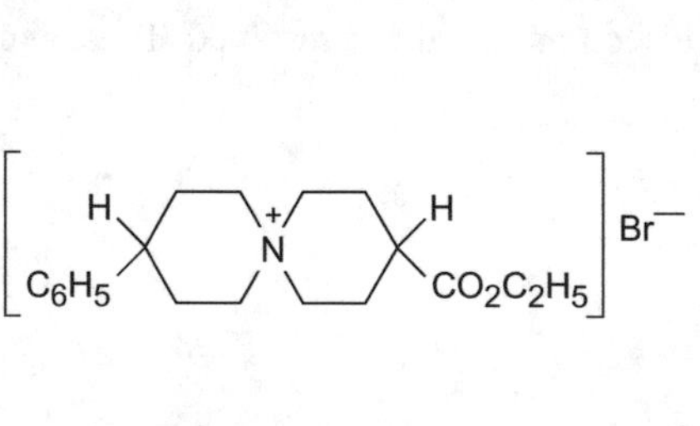

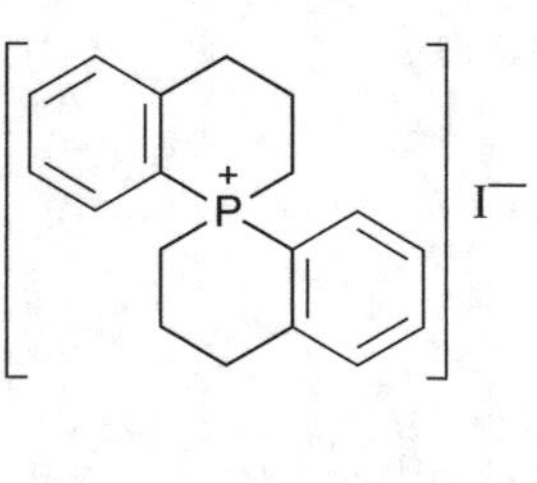

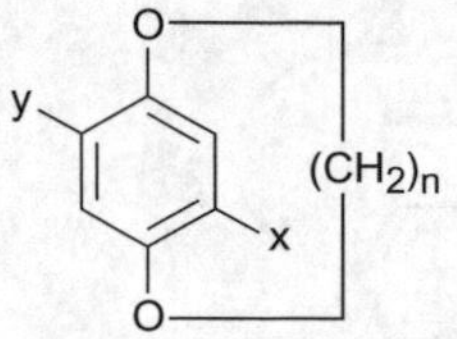

**Fig. 6.7** Some examples of ansa compounds having restricted rotation

I    x = $CO_2H$; y = Br; $n = 10$
II   x = y = Br; $n = 10$
III  x = $CO_2H$; y = H; $n = 10$
IV   x = $CO_2H$; y = H; $n = 9$
V    x = $CO_2H$; y = H; $n = 8$

It is interesting to know that a spiran terpenoid (**accerone**) has been found to occur in nature in sweet flag oil.

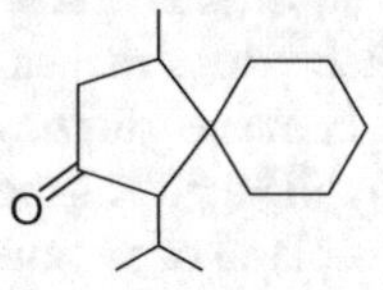

Acorone

## 6.5  Ansa Compounds

The carbon-bridged analogues of benzene rings are known as ansa compounds (*ansa* = Latin for handle). In these compounds (as in the case of appropriately substituted biphenyls), the restricted rotation of benzene ring is responsible for their chirality. However, to exhibit optical activity, such compounds (ansa compound) should be appropriately substituted. Thus, as an example, compounds I and II having two ortho substituents and a ten-membered methylene bridge are optically stable. In case there is only one substituent (as in the case of III), the optical activity disappears; the compound V with an eight-numbered methylene bridge can be resolved and is optically stable. However, in cases (when there is only one substituent) with a nine-membered bridge, the compound is of intermediate optical stability as in the case of IV (Fig. 6.7).

## 6.6  Cyclophanes

These are interesting compounds in which two benzene rings are joined by methylene bridges on either side between para positions and exhibit atropisomerisms. Such compounds are known as paracyclophanes and have the following structure.

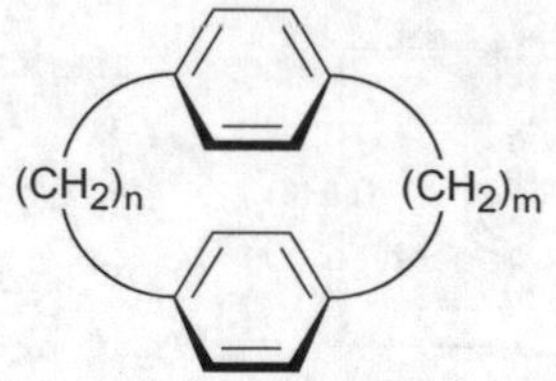

$$m = n = 2$$
$$m = 3, n = 4$$
$$m = n = 4$$

Paracyclophanes

Paracyclophane with $m = n = 2$ is optically stable. If $m = 3$ and $n = 4$, the compound is resolvable but racemises at 150–160 °C. However, if both $m$ and $n$ are 4, the compound cannot be resolved.

In the case of unsymmetrically substituted paracyclophanes, the presence of even one substituent in any of the benzene rings makes the structure chiral. An example is the substitution of one of the benzene rings with the COOH group. The (+)-antipode of paracyclophane carboxylic acid has the structure as shown below; these compounds have a structure with a plane of symmetry.

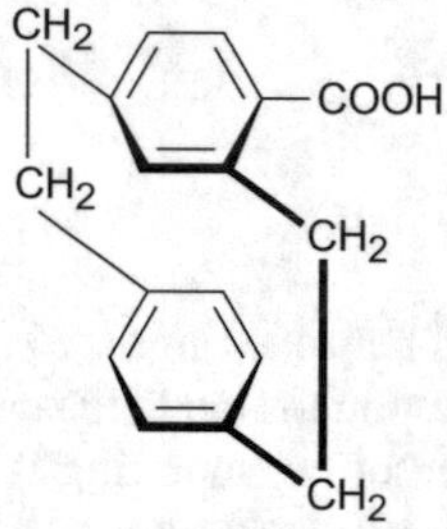

Paracyclophane carboxylic acid

In the above case, even if the COOH group is replaced by the amino group (by the Hofmann rearrangement) and by chlorine and hydroxyl (by diazo reactions), the optical activity is retained in all these compounds.

Some other optically active paracyclophanes have been prepared and have the following formula:

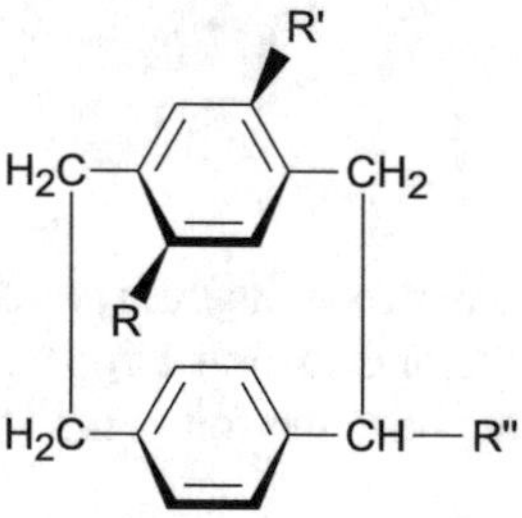

$$R' = CH_3 ; R = R'' = H$$
$$R' = R = CH_3 ; R'' = H$$
$$R' = R = H ; R'' = COOCH_3$$
$$R' = R = H ; R'' = CH_2OH$$

The absolute configuration of a paracyclophane, whose optical activity is due to the chirality of the paracyclophane structure and an asymmetric centre in the side chain ($R = CH_2-CHOH-CH_3$), is given below.

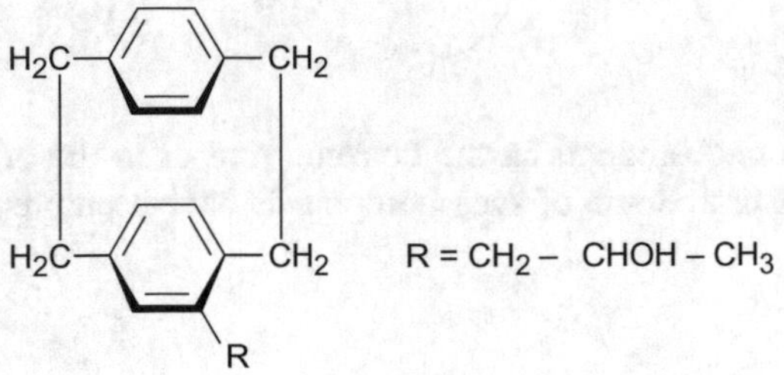

$$R = CH_2 - CHOH - CH_3$$

Paracyclophanes having sulphur-containing bridges are also known. These have the general formula as shown below.

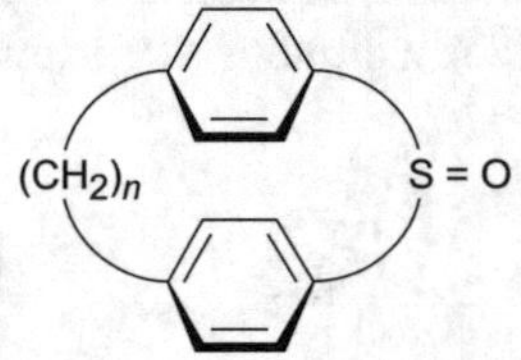

$(CH_2)_n$          $S = O$

Cyclophanes having bridges in meta position are called metacyclophanes, and such cyclophanes can exist in optically active forms. An interesting metacyclophane with three bridges in the meta positions is given below.

Metacyclophane

Metacyclophane
with three sulphur bridges
in meta-positions

Some paracyclophanes contain only one aromatic ring and an aliphatic bridge in the para position. Some examples of such compounds are given below.

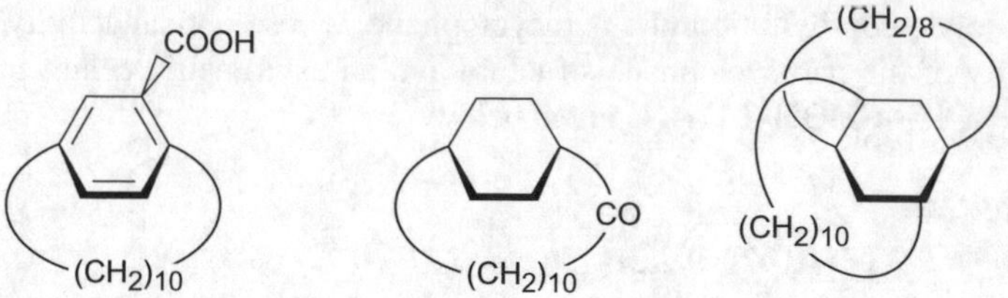

The presence of substituents in the benzene ring or in the bridge makes such paracyclophanes chiral. Some of such compounds have been prepared in optically active forms.

## 6.7   Benzocycloalkanes

These are non-planar cycloalkane ring compounds. Their conformation is 'frozen' by fused to benzene rings. A number of such compounds are optically active. One of the simple benzocycloalkane is a diphenyl derivative (I).

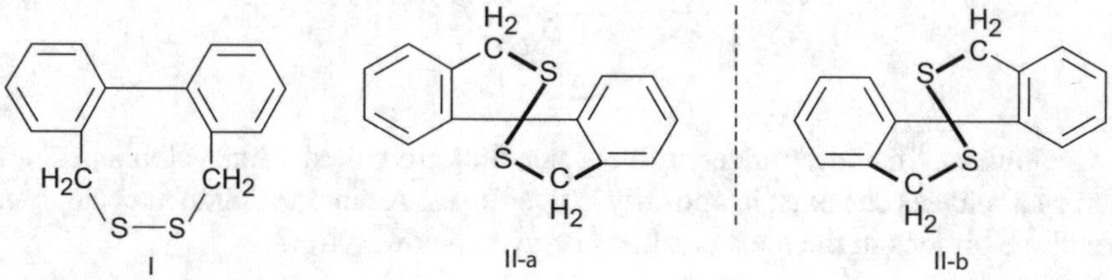

In compound (I), the central sulphur-containing eight-membered ring exists in asymmetric non-planar 'pseudo-chair' and 'pseudo-boat' forms as shown in structures IIa and IIb. It is believed (Lüttringhans and Rosenbaum) that the longer C–S and S–S bonds are responsible for the formation of stable asymmetric conformers.

The rigidity of the cycloheptane is increased by the presence of double bonds and fusion to benzene rings. Such molecules are asymmetric as the compound (III). In such compounds, the central seven-membered ring is rigidly fixed in the boat conformation as shown in structure IV.

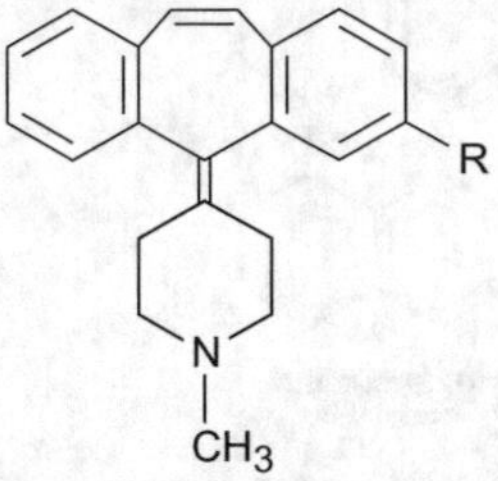

III

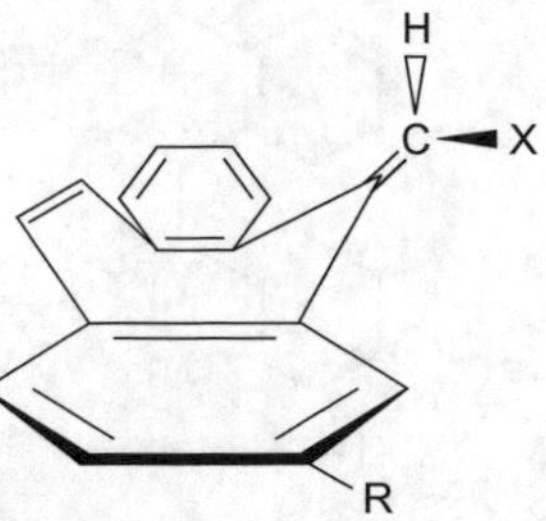

IV

The molecular asymmetry in compounds of type III can be due to the non-identity of benzene rings or due to non-identical substituents in the seven-membered ring as in structure IV. Also, the asymmetry in compound with a nine-membered central ring (V) in place of a seven-membered ring in III is also of the same nature.

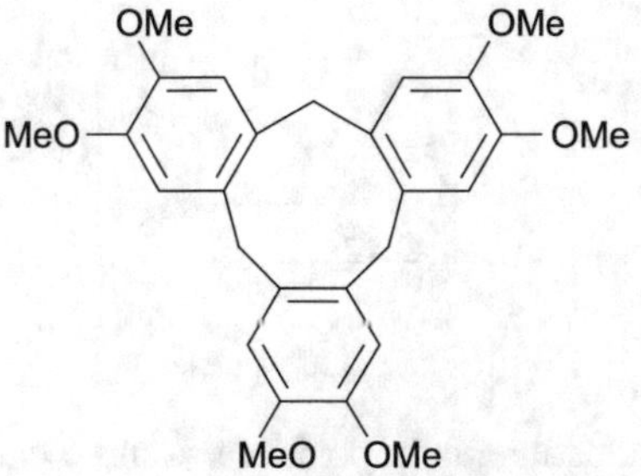

V

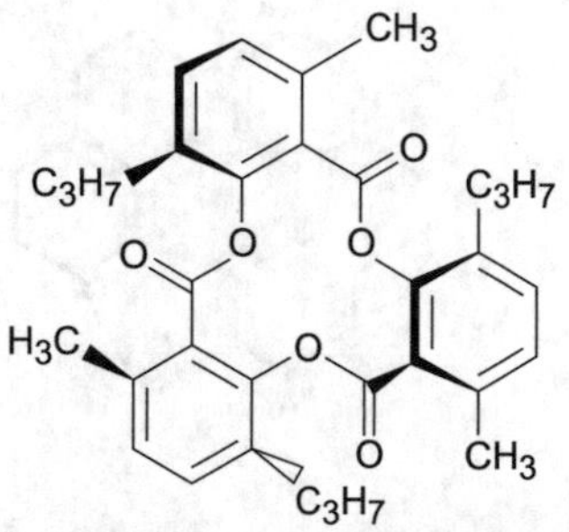

VI

A special case of asymmetry was described by Newman and Powell who prepared tri-*o*-thymotide (VI) in optically active form. In this case, the molecular asymmetry is due to the phenyl rings not lying in the same plane; they are in an inclined position relative to each other. However, the compound (VI) undergoes racemisation spontaneously.

## 6.8 Helicenes

These are compounds in which a number of aromatic rings are joined in ortho-junction to form a helical type of structure. Such compounds are optically active since they have no elements of symmetry; these may have left- or right-handed orientation. Stereoisomerism is possible in helicenes having a minimum of 5 benzene rings. some examples are as follows:

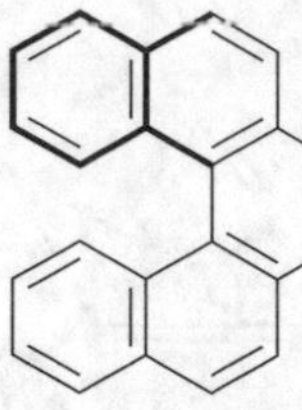

(+)-Pentahelicene 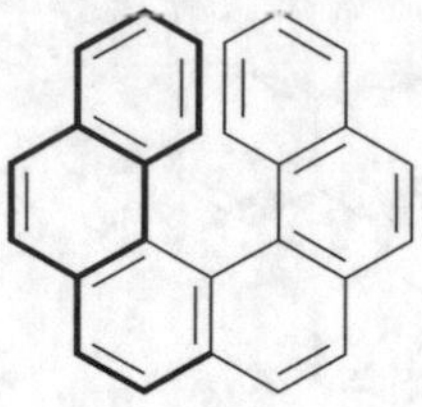                Hexahelicene

Helicenes having a larger number of benzene rings (7, 8 or 9) are also known. The helicenes can be racemised by heating above their melting points. Even the heterocyclic analogues of hexahelicene in active forms have also been prepared. Some examples I and II are given below.

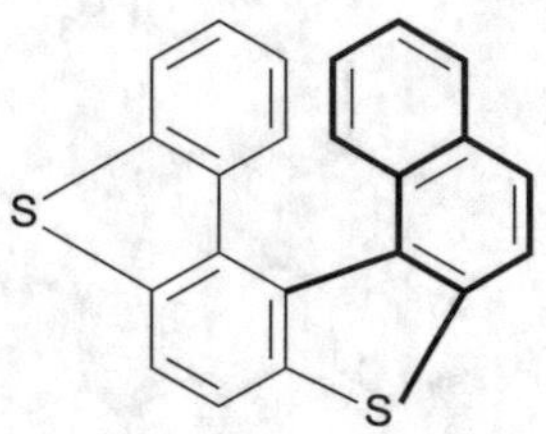

I 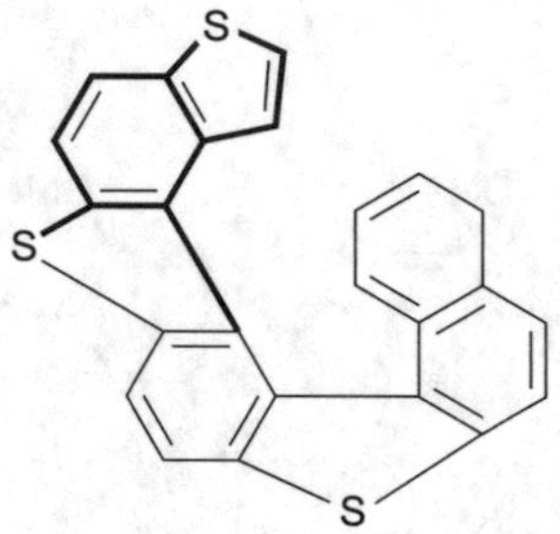                                II

Even phenanthrene derivatives having substituents in positions 4 and 5 can also have a helical shape as in helicenes. In such cases, the substituent can also be present in positions 4 and 5 by distortion of the phenanthrene ring. Such a system can be chiral and can be resolved into its antipodes. An example of such compounds is represented below [(IIIa) and (IIIb)] (Newman et al.).

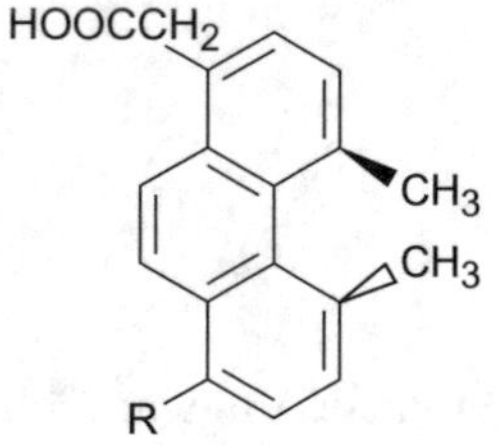

III-a 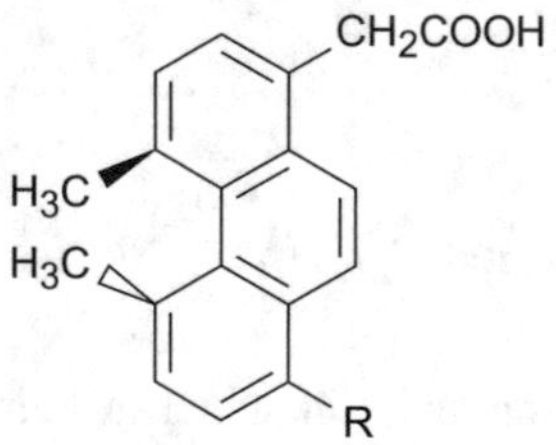                            III-b

Another optically active compound (IV) containing nitrogen has also been prepared. On heating at 60° C for one hour, complete racemisation occurs. Also, the phenanthraquone derivative (V) has been obtained in an optically active form.

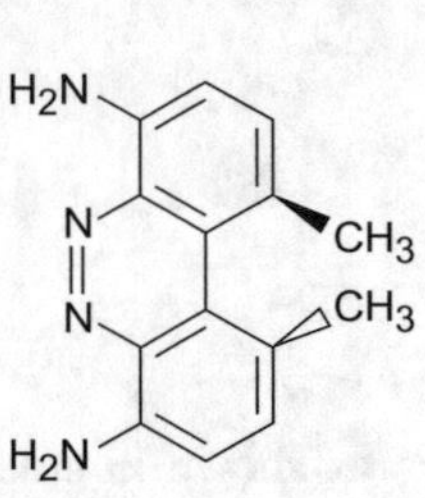

IV 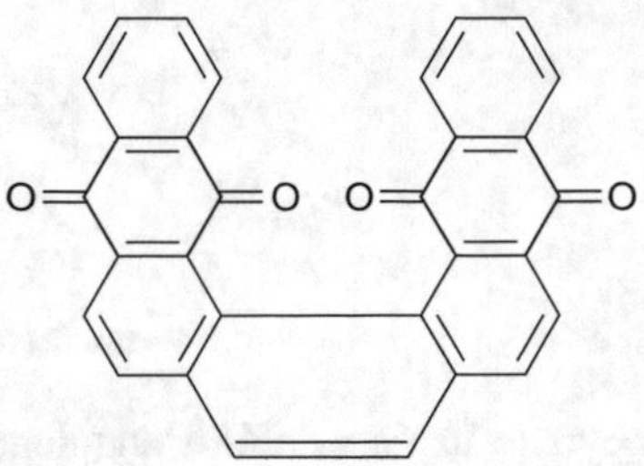                                              V

Optically active helianthrones also belong to a similar type of compounds. An example (VI) is given below.

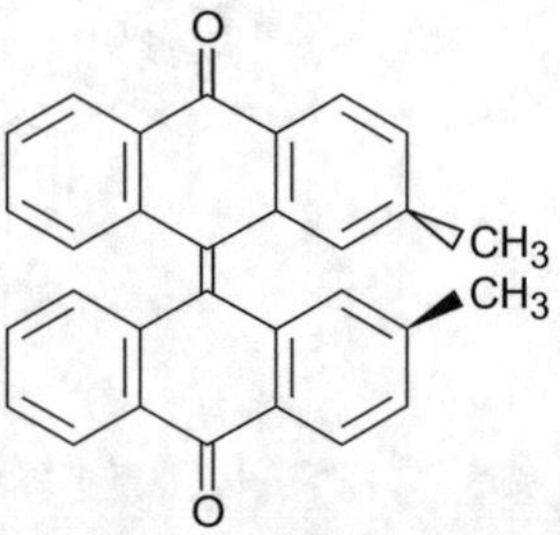

VI

## 6.9   Annulenes

These are macrocyclic compounds having conjugated double bonds and are known to display stereochemical character. As an example, [14] annulene can be separated into two rapidly interconvertible forms by chromatography on silica coated with silver nitrate. These two forms differ in the mode of overlapping of the intraannular hydrogen atoms as shown below.

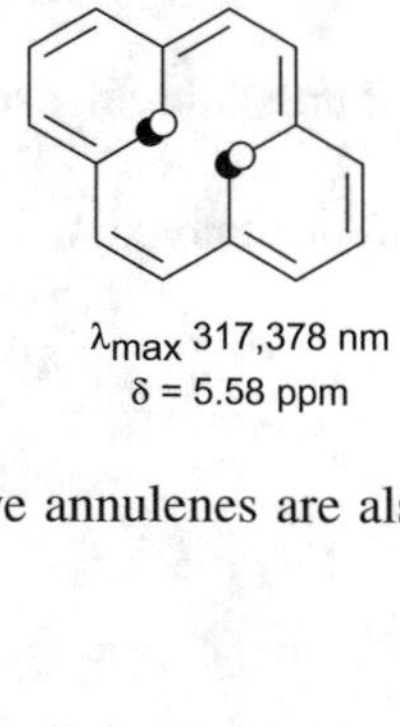

$\lambda_{max}$ 317,378 nm            $\lambda_{max}$ 317,380 nm
$\delta$ = 5.58 ppm                   $\delta$ = 6.07 ppm

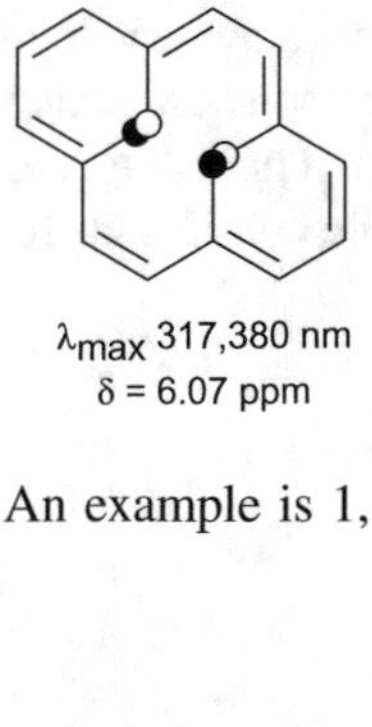

Optically active annulenes are also known. An example is 1, 6-methylene [10] annulene (I).

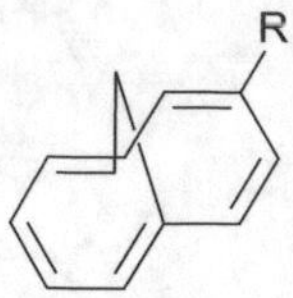

I

1,6-Methylene[10]annulene

Dihydropyrene has a peculiar 'annulene' structure as shown in structure (II). In case, there are two substituents to the central atoms of dihydropyrene and one substituent in the peripheral atom, the molecule (III) exhibits optical activity.

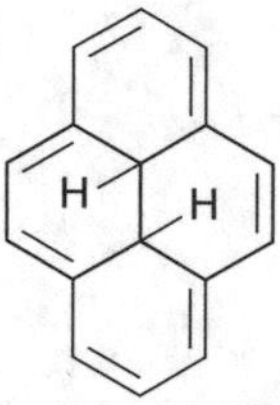

III 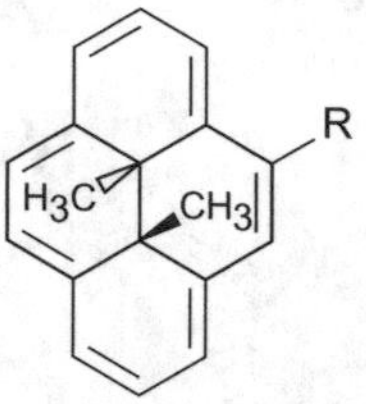                    IV   R = CH$_3$, COOH

## Key Concepts

- **Allenes:** 1, 2-dienes having cumulated double bonds, i.e., compounds in which a carbon atom is bonded to two other carbon atoms by double bonds.
- **Annulenes:** Macrocyclic compounds having conjugated double bonds.
- **Ansa compounds:** Carbon-bridged analogues of the benzene ring.
- **Atropisomerism:** The term atropisomerism (meaning no rotation) covers cases of spatial isomerism which is due to restriction in rotation about a single bond. It was first discovered in biphenyl derivatives.
- **Benzocycloalkanes:** Non-planar cycloalkene ring compounds.
- **Cyclophanes:** Compounds in which two benzene rings are joined by a methylene bridge on either side between para positions (known as paracyclophanes) or meta positions (known as metacyclophanes).
- **Helicenes:** Compounds in which a number of aromatic rings are joined in ortho junctions to form a helical type of structure.
- **Spirans:** Bicyclic compounds having one carbon atom common to both rings.

## Problems

1. Which of the following compounds are optically active:

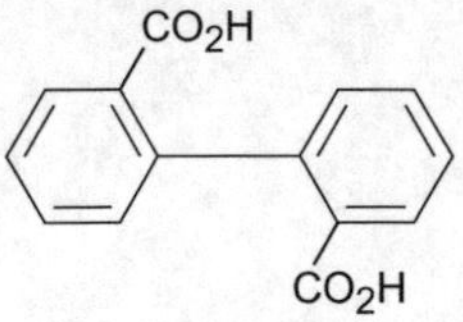

2. Discuss the stereochemistry of biphenyls.
3. Explain why the following biphenyl is not resolvable.

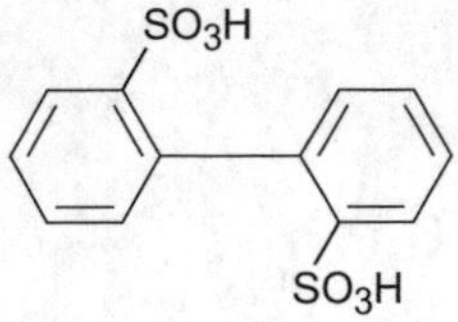

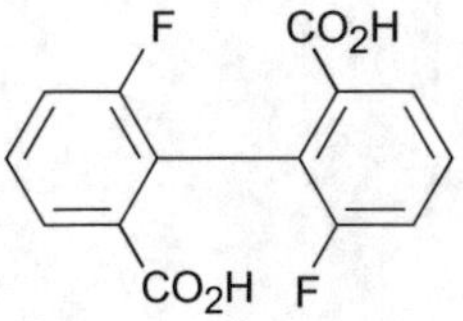

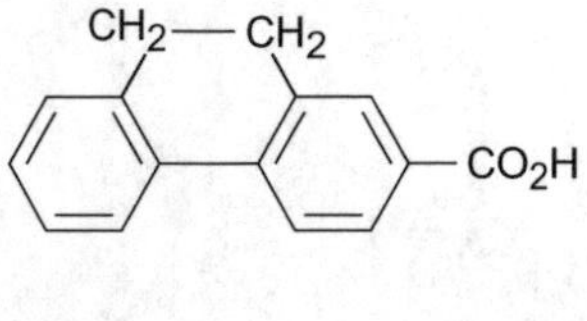

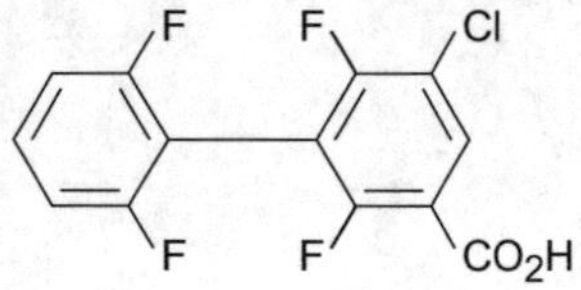

4. What do you understand by the term atropisomerism? Give examples of compounds that exhibit atropisomerism. Can heterocyclic compounds exhibit atropisomerism? If so give examples.
5. How do you show that the two rings in ortho-substituted biphenyls are not coplanar?
6. Explain the term intramolecular overcrowding.
7. What are allenes? Discuss their stereochemistry. How do you find that a compound is an allene?
8. What are spirans? How they are named? Can compounds containing N or P can be spirans? If so give examples.

9.   Write notes on the following:

   (a)   Ansa compounds.
   (b)   Cyclophanes.
   (c)   Benzocycloalkanes.
   (d)   Helicenes.
   (e)   Annulenes.

# Chapter 7
# Stereochemistry of Trivalent Carbon

## 7.1 Introduction

Carbon is tetravalent and normally forms four covalent bonds involving four bond orbitals each having two electrons. In case one of the orbitals does not form a bond, the carbon atom is trivalent. In such a case, there are three possibilities:

- The non-bonding orbital is empty. In this case, we are dealing with a carbocation $R\,R'R''\,C^+$.
- The non-bonding orbital contains one electron. In this case, there is the formation of a carbon radical (free radical), $R\,R'R''\,C^\bullet$.
- The non-bonding orbital contains two electrons. In this case, we are dealing with a carbanion, $R\,R'R''\,C^-$.

The trivalent carbocations, carbanions and free radicals are the most important reactive intermediates. A special common feature of all these reactive intermediates is the electron-deficient and electrophilic character. Due to their electron-deficient character, they undergo skeletalrearrangements. All these intermediates are of transitory existence and take part in a reaction as soon as they are formed. Being very reactive, they cannot be isolated under normal conditions, except in cases where the non-bonding orbital overlaps with other pi-orbitals in a molecule. One such example is the triphenylmethyl carbocation, radical and carbanion.

The stereochemistry of trivalent carbon intermediate is of importance since being short-lived these are quite common and their stereochemistry is reflected in the stereochemistry of the products obtained in reactions involving these intermediates.

© The Author(s), under exclusive license to Springer Nature Switzerland AG 2022    147
V. K. Ahluwalia, *Stereochemistry of Organic Compounds*,
https://doi.org/10.1007/978-3-030-84961-0_7

## 7.2   Carbocations

Carbocations are obtained by the heterolytic fission of a C—X bond in an organic molecule. In case, X is more electronegative than a carbon atom, X takes away the bonding pair of electrons and attains a negative charge (: $\overline{X}$) and an ion having a positive charge also results.

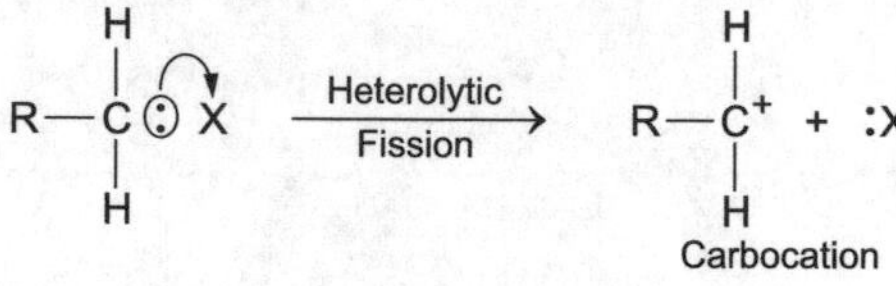

$$
R—\underset{\underset{H}{|}}{\overset{\overset{H}{|}}{C}}\!\!\overset{\frown}{\odot}\,X \quad \xrightarrow[\text{Fission}]{\text{Heterolytic}} \quad R—\underset{\underset{H}{|}}{\overset{\overset{H}{|}}{C}}{}^{+} \; + \; :X^{-}
$$

Carbocation

The positively charged species is called a carbocation. In fact, a carbocation is defined as a carbon atom that is trivalent, has an even number of electrons and has a positive charge. The carbon atom in a carbocation is $sp^2$-hybridised; it uses all its three hybridised orbitals for bond formation with other atoms, and the remaining $p_z$ orbital is empty and is perpendicular to the plane of the other three bonds. A carbocation is described as planar (trigonal coplanar) with a bond angle of 120°. Its formation is not possible in compounds (as in bridgehead compounds) which do not permit the attainment of a planar geometry (Fig. 7.1).

For the formation of carbocation, the reaction of RR′R″—X has to be carried out in a suitable solvent. In case, a good ionising solvent is used, the carbocation is associated with solvent molecules (solvation). However, if the solvent is poorly ionising, the carbocation forms an ion pair. For example, triphenylmethyl chloride. $Ph_3CCl$ will give an ion pair $Ph_3C^+Cl^-$. Thus, it is not possible to observe a free carbocation.

The reaction of optically active compounds RR′R″CX which proceed via a free carbonium ion RR′R″C$^+$ gives racemic products. As an example, phenyl-$p$-biphenylyl-α -naphthyl carbocation is obtained in solution by the treatment of phenyl-$p$-biphenylyl-α-naphthylmethyl thioglycolic acid with perchloric acid or sulphuric acid. The formed carbocation on treatment with water gives a racemic phenyl-$p$-biphenylyl-α-naphthylcarbinol (M. Gemberg and W. E. Gorden, J. Am. Chem. Soc., 1935, **57**, 119) (Fig. 7.2).

**Fig. 7.1**  Carbocation

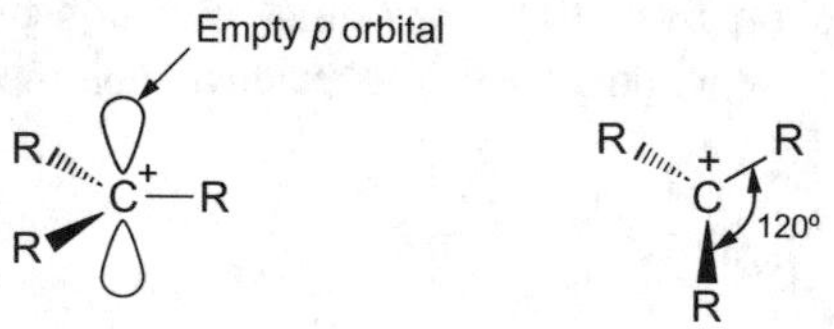

Empty $p$ orbital

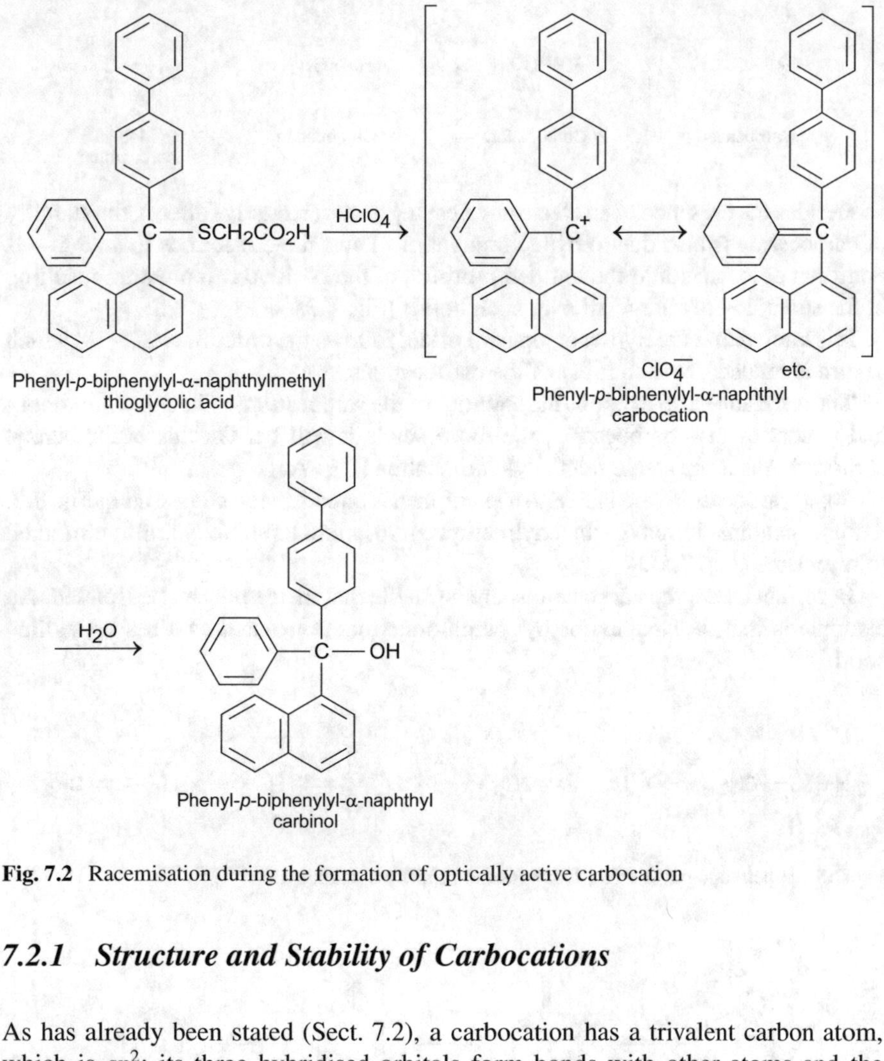
HClO4
SCH2CO2H
C
C+
C
ClO4−
etc.
Phenyl-p-biphenylyl-α-naphthylmethyl thioglycolic acid
Phenyl-p-biphenylyl-α-naphthyl carbocation
H2O
C — OH
Phenyl-p-biphenylyl-α-naphthyl carbinol

**Fig. 7.2** Racemisation during the formation of optically active carbocation

### *7.2.1 Structure and Stability of Carbocations*

As has already been stated (Sect. 7.2), a carbocation has a trivalent carbon atom, which is $sp^2$; its three hybridised orbitals form bonds with other atoms and the remaining $p_z$ orbital is empty and is perpendicular to the plane of the other three bonds. It is described as planar with a bond angle of 120° (see Fig. 7.1).

The presence of an electron-releasing group such as an alkyl group adjacent to a carbon atom bearing a positive charge increases the stability of the carbocation. This explains why tertiary carbocation is more stable than a secondary carbocation, which is more stable than primary carbocation.

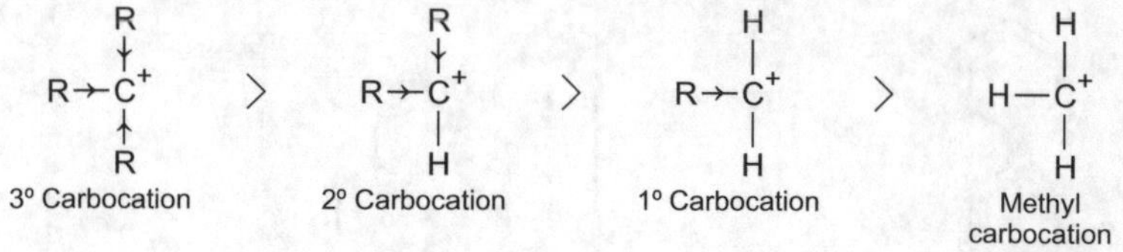

Besides the presence of an electron-releasing group (inductive effect), the stability of carbocation is also due to hyperconjugation. Thus, the σ electrons to an α C—H bond can delocalise into the unfilled *p*-orbital of the positive carbon atom resulting in the spreading of charge all over such bonds (Fig. 7.3).

It is known that the hyperconjugative effect follows the order 3° > 2° > 1°, which in turn is related to the stability of the carbocations.

The resonance also explains the stability of the carbocations. The more the canonical structures of a carbocation, the more stable it will be. On this basis, benzyl carbocation is more stable than allyl carbocation (Fig. 7.4).

Some carbocations are stable. An example has already been discussed in Fig. 7.2. Another example is that of triphenylmethyl carbocation. Its stability is also attributed to resonance (Fig. 7.5).

In certain cases, the carbocations are so stable that their salts can be isolated. An example is that of triphenylmethyl perchlorate that is isolated as a red crystalline solid.

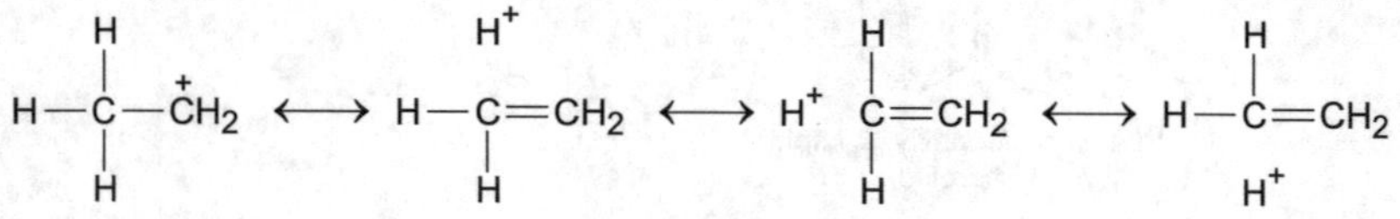

**Fig. 7.3** Hyperconjugative effect increases the stability of carbocation

Benzyl carbocation (4 canonical forms)

$$CH_2=CH-\overset{+}{C}H_2 \longleftrightarrow \overset{+}{C}H_2-CH=CH_2$$

Allyl carbocation (2 canonical forms)

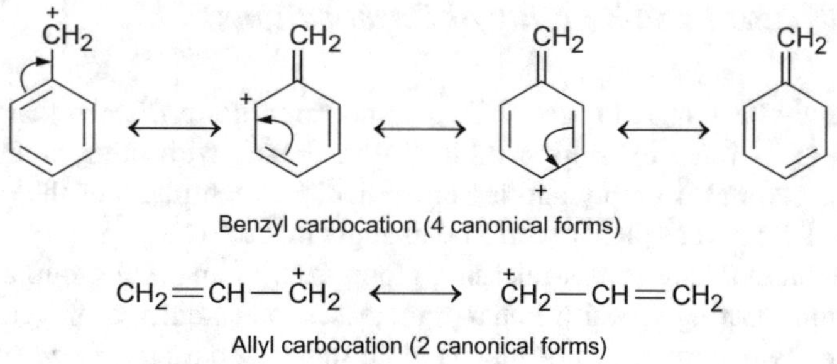

**Fig. 7.4** Canonical forms of benzyl carbocation and allyl carbocation

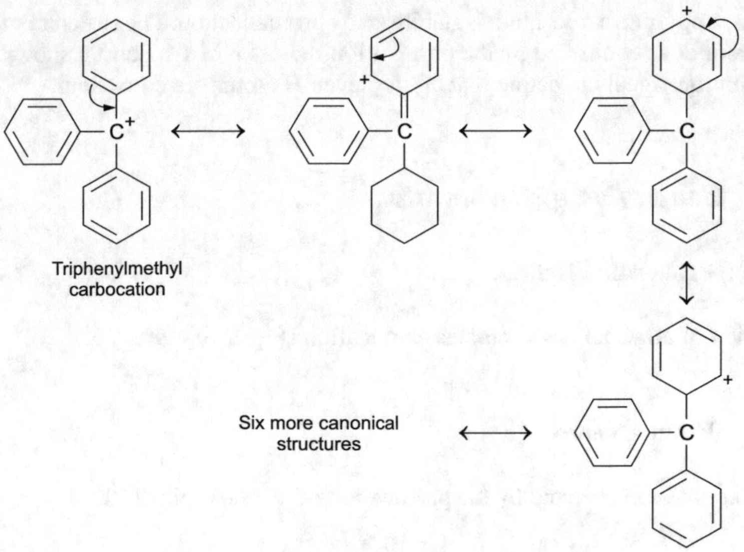

**Fig. 7.5** Resonance (canonical) structures of triphenylmethyl carbocation

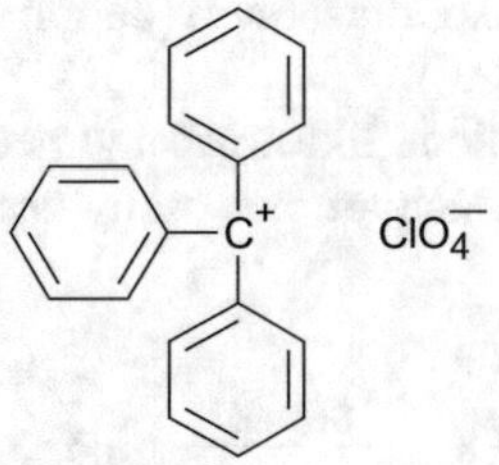

Triphenyl methyl perchlorate

The stabilisation of carbocations can also occur through aromatisation. As an example, the tropylium cation has 6 π electrons which are accommodated in three delocalised molecular orbitals (spread over seven atoms). Thus, it is a Hückel $4n + 2$ system ($n = 1$) and exhibits quasi-aromatic stability.

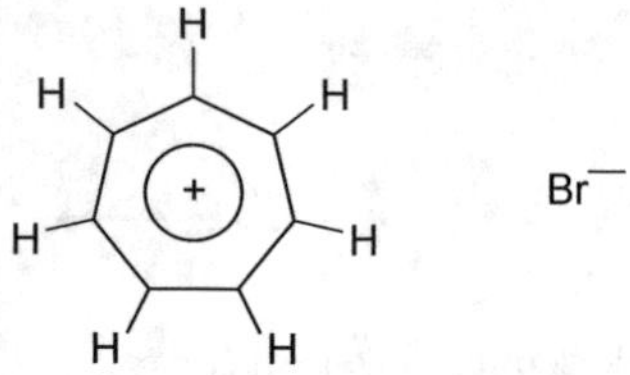

Tropylium bromide

Thus, tropylium carbocation is stabilised by aromatisation. The structure of tropylium bromide is confirmed by the observation that its NMR spectrum shows only a single proton signal indicating that all the seven H atoms are equivalent.

## 7.2.2 Generation of Carbocations

### 7.2.2.1 From Alkyl Halides

Ionisation of alkyl halides generates carbocation (Fig. 7.6).

### 7.2.2.2 From Alkenes

Carbocations are generated by the protonation of alkenes (Fig. 7.7).

### 7.2.2.3 From Other Cations

Decomposition of cations like diazonium cation (obtained by the action of $NaNO_2/HCl$) on amines (Fig. 7.8).

Some readily available cations like triphenylmethyl carbocations are used to generate other carbocations which are not easily accessible. An example is given in Fig. 7.9.

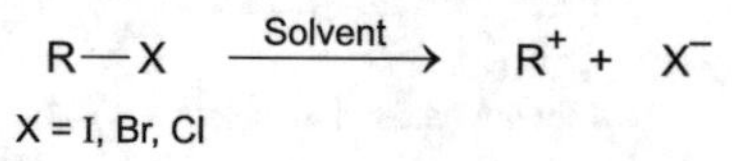

$$R-X \xrightarrow{\text{Solvent}} R^+ + X^-$$
$$X = I, Br, Cl$$

**Fig. 7.6** Generation of carbocations from alkyl halides

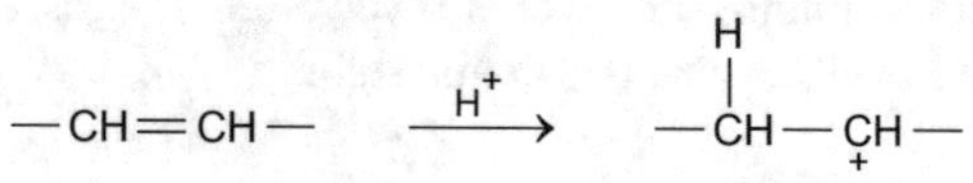

$$-CH=CH- \xrightarrow{H^+} -\overset{\overset{H}{|}}{CH}-\underset{+}{CH}-$$

**Fig. 7.7** Generation of carbocations from alkene

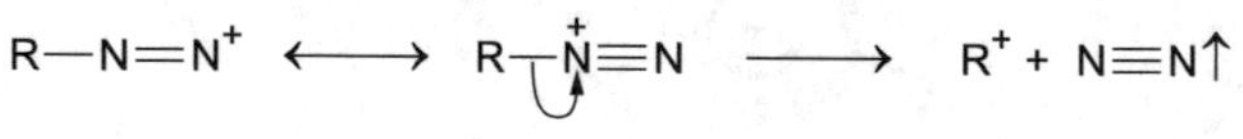

$$R-N=N^+ \longleftrightarrow R-\overset{+}{N}\equiv N \longrightarrow R^+ + N\equiv N\uparrow$$

**Fig. 7.8** Generation of carbocations from diazonium cation

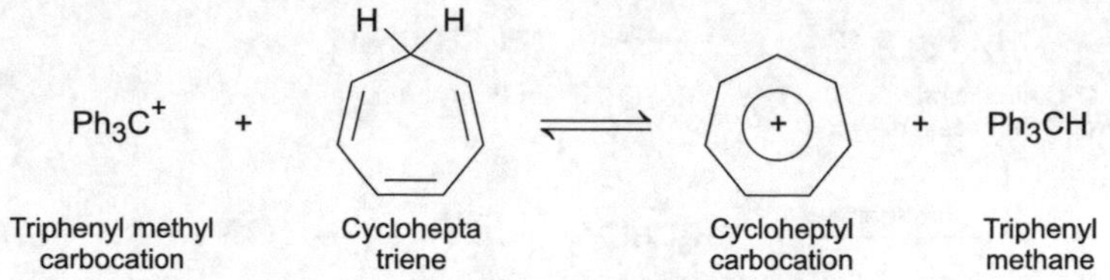

**Fig. 7.9** Generation of cycloheptyl carbocation

## 7.2.3   Reactions of Carbocations

Carbocations undergo the following reactions:

- Elimination of a proton.
- Reaction with nucleophiles.
- Addition to unsaturated compounds.
- Molecular rearrangements.

### 7.2.3.1   Elimination of a Proton

A carbocation can lose a proton to form an alkene. As an example, 1-propyl carbocation (which is generated by the diazotisation of 1-aminopropane followed by decomposition of the formed diazonium salt) can lose a proton to form propene. It is also possible that the 1-propyl carbocation may rearrange to a more stable secondary carbocation, which can lose a proton to give propene (Fig. 7.10).

### 7.2.3.2   Reaction with Nucleophiles

Carbocation reacts with nucleophiles to form a new bond. Thus, propyl carbocation or isopropyl carbocation on reaction with halides ($X^-$) yields the corresponding halides (Fig. 7.11).

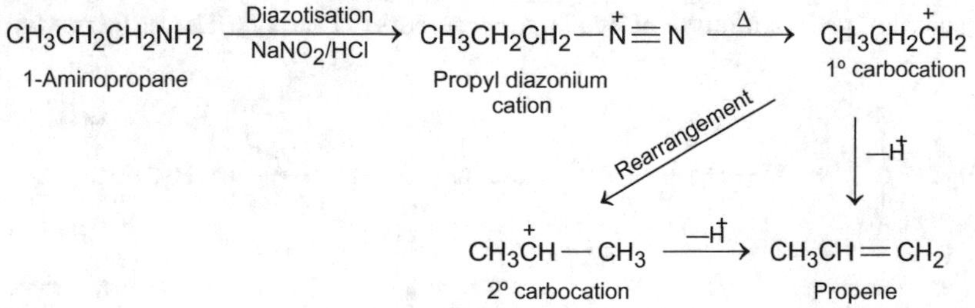

**Fig. 7.10** Elimination of a proton from carbocation

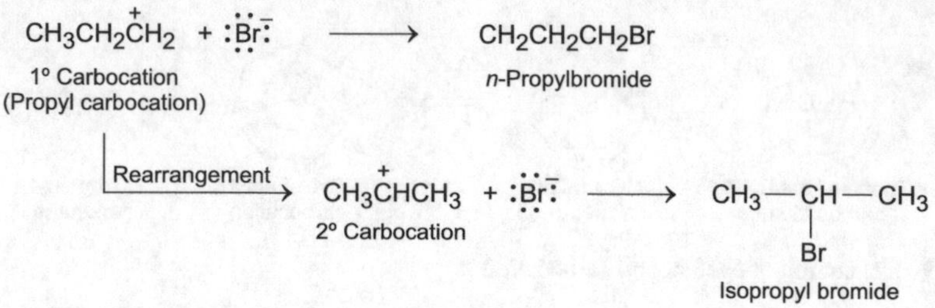

$$CH_3CH_2\overset{+}{C}H_2 \ + \ :\overset{..}{\underset{..}{Br}}:^{\,-} \ \longrightarrow \ CH_2CH_2CH_2Br$$

1° Carbocation
(Propyl carbocation)

n-Propylbromide

$$\xrightarrow{\text{Rearrangement}} \ CH_3\overset{+}{C}HCH_3 \ + \ :\overset{..}{\underset{..}{Br}}:^{\,-} \ \longrightarrow \ CH_3—CH—CH_3$$

2° Carbocation

|
Br

Isopropyl bromide

**Fig. 7.11** Reaction of carbocation with a nucleophile

A neutral nucleophile (like $H_2O$) on reaction with a carbocation gives a protonated alcohol, which gives an alcohol by elimination of a proton (Fig. 7.12).

### 7.2.3.3   Molecular Rearrangements

The molecular rearrangements can take place either with a change in carbon skeleton or without a change in carbon skeleton.

### 7.2.3.4   Molecular Rearrangements Without Change in Carbon Skeleton

This type of rearrangement has already been discussed (Figs. 7.10 and 7.11) in which 1-propyl carbocation rearranges to the 2-propyl carbocation (isopropyl carbocation) by migration of a hydrogen atom along with its electron pair (i.e., as $H^-$) as shown in Fig. 7.13.

The above rearrangement (Fig. 7.13) shows that the 2° carbocation is more stable than the 1° carbocation. The rearrangement can also take place by the migration of an alkyl group with its pair of bonding electrons as in the case of conversion of 2° carbocation into 3° carbocation (Fig. 7.14).

Another example of molecular rearrangement without change in carbon skeleton is **allylic rearrangement**. As an illustration, 3-chlorobut-1-ene on solvolysis in ethyl alcohol gives a mixture of two isomeric ethers (1) and (2). The same mixture of

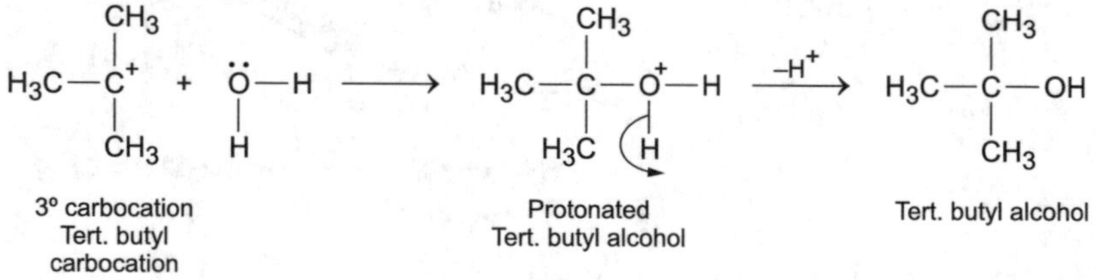

$$H_3C—\overset{\overset{\displaystyle CH_3}{|}}{\underset{\underset{\displaystyle CH_3}{|}}{\overset{+}{C}}} \ + \ \overset{..}{\underset{\underset{\displaystyle H}{|}}{O}}—H \ \longrightarrow \ H_3C—\overset{\overset{\displaystyle CH_3}{|}}{\underset{\underset{\displaystyle H_3C}{|}}{C}}—\overset{+}{O}—H \ \xrightarrow{\ -H^+\ } \ H_3C—\overset{\overset{\displaystyle CH_3}{|}}{\underset{\underset{\displaystyle CH_3}{|}}{C}}—OH$$

3° carbocation
Tert. butyl
carbocation

Protonated
Tert. butyl alcohol

Tert. butyl alcohol

**Fig. 7.12** Reaction of 3° carbocation with the neutral nucleophile

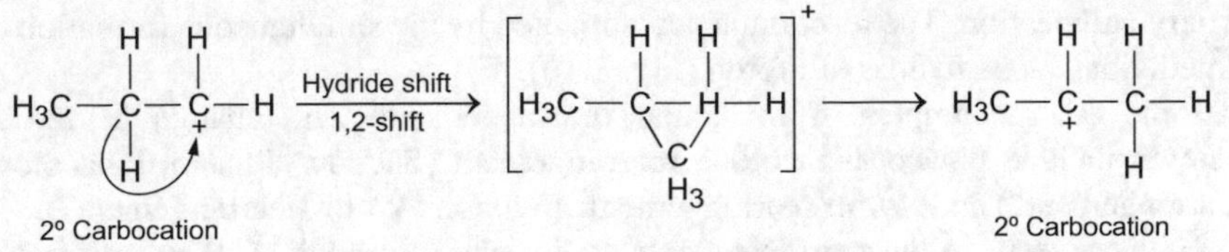

**Fig. 7.13** 1,2-Hydride shift in molecular rearrangement (without change in carbon skeleton)

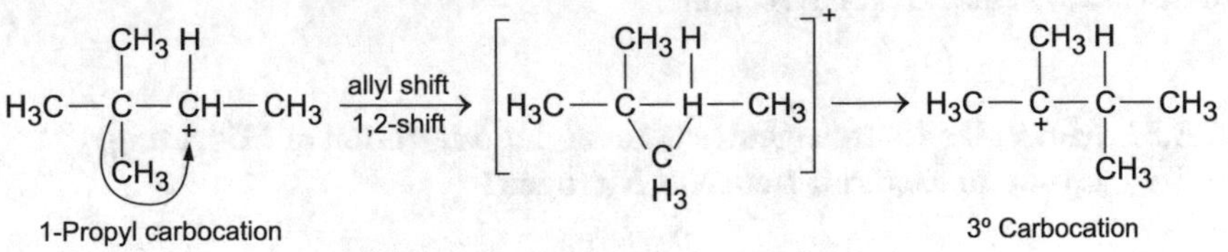

**Fig. 7.14** 1,2-Methyl shift in molecular rearrangement without change in the carbon skeleton

ethers (1) and (2) is also obtained in the solvolysis of 1-chlorobut-2-ene (Fig. 7.15). This clearly supports the formation of the same delocalised allylic cation (see also Fig. 7.4):

## Molecular Rearrangement with Change in Carbon Skeleton

A typical example of such rearrangement is **Neopentyl rearrangement**. Thus, neopentyl bromide on solvolysis gives the rearranged product, 2-methyl butan-2-ol, and not the expected 2,2-dimethyl propanol (neopentyl alcohol). Such a rearrange-ment is called **neopentyl rearrangement**. The 2-methyl butan-2-ol is obtained by the conversion of the initially formed primary carbocation ion into tertiary carboca-tion followed by a reaction with water. The driving force for the C—C bond-breaking involved in the migration of methyl group with its electron pair gives more stable

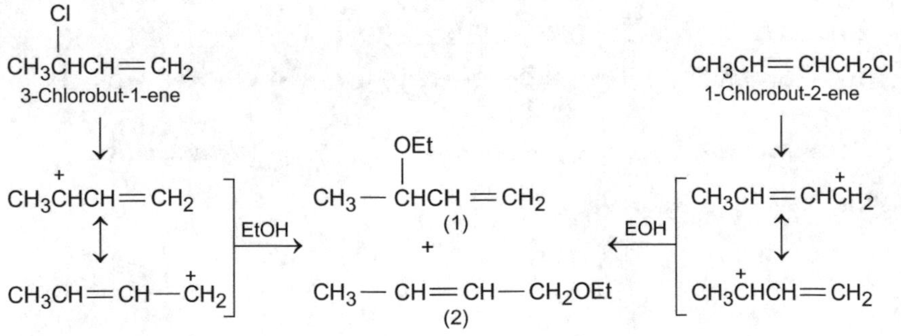

**Fig. 7.15** Solvolysis of 3-chlorobut-1-ene and 1-chlorobut-2-ene

tertiary carbocation. The reaction is accompanied by the simultaneous formation of 2-methylbut-2-ene by loss of proton (Fig. 7.16).

Some other examples of molecular rearrangements with change in carbon skeleton include pinacol-pinacolone rearrangement (Section "Pinacol-Pinacolone Rearrangement") and 'Wolff rearrangement' (Section "Wolff Rearrangement").

We have seen, in the rearrangements so far cited, that the alkyl or aryl group migrates to a carbon atom (carbocation). We also come across rearrangements in which the migrating group migrates to electron-deficient nitrogen or oxygen. These are discussed in the foregoing sections.

### 7.2.3.5  Molecular Rearrangements Involving Migration of Migrating Group to Electron-Deficient Nitrogen

Such rearrangement involves the well-known Beckmann rearrangements (Section "Beckmann Rearrangement") and the Hofmann rearrangement (Section "Hofmann Rearrangement").

### 7.2.3.6  Molecular Rearrangements Involving Migration of Migrating Group to Electron-Deficient Oxygen

Such rearrangements involve the well-known Baeyer–Villiger oxidation (Section "Baeyer-Villiger Oxidation") and hydroperoxide rearrangement (Section "Hydroperoxide Rearrangement").

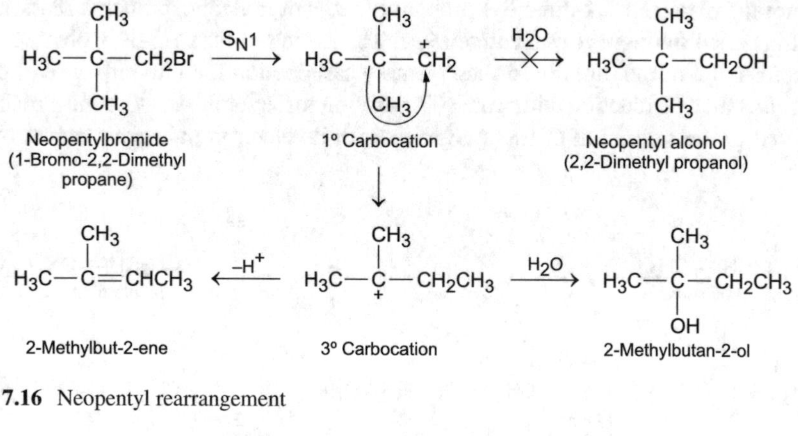

**Fig. 7.16**  Neopentyl rearrangement

### 7.2.4 Stereochemistry of Rearrangements Involving Carbocations

In rearrangements involving carbocations, there are three points of stereochemical interest:

- What happens to the configuration of the carbon atom from which migration has taken place (migration origin)?
- What happens to the configuration of the carbon atom to which migration has taken place (The migration terminus, the cationic carbon atom)?
- What happens to the configuration of the migrating group if it is chiral?

It has been shown that the migrating group is not detected and becomes free during the rearrangement. This has been conclusively proved by taking the example of two similar pinacols (1 and 2) that rearrange at nearly the same rate but these have different migrating groups (Me and Et in 1 and 2, respectively). If both 1 and 2 are rearranged simultaneously as a mixture in the same solution (a crossover experiment), no cross migration is observed (Fig. 7.17).

It should be noted that in unsymmetrical diols, the OH group that is lost is the one that gives a more stable carbocation. Thus, the stability of the carbocation is an important factor. The order of migratory aptitude is Ar > H > CH$_3$ (see also section "Pinacol–Pinacolone Rearrangement").

In a similar way, rearrangements reactions in which there is hydride shift (1,2-shift) (see section "Molecular Rearrangements Without Change in Carbon Skeleton", Fig. 7.13), the addition of a deuterated solvent (e.g., D$_2$O), no deuterium is incorporated in the new C—H bond in the rearrangement product (Fig. 7.18).

Both the experiments given above in Figs. 7.17 and 7.18 conclusively prove that the rearrangement is strictly intramolecular, i.e., the migrating group is not detached

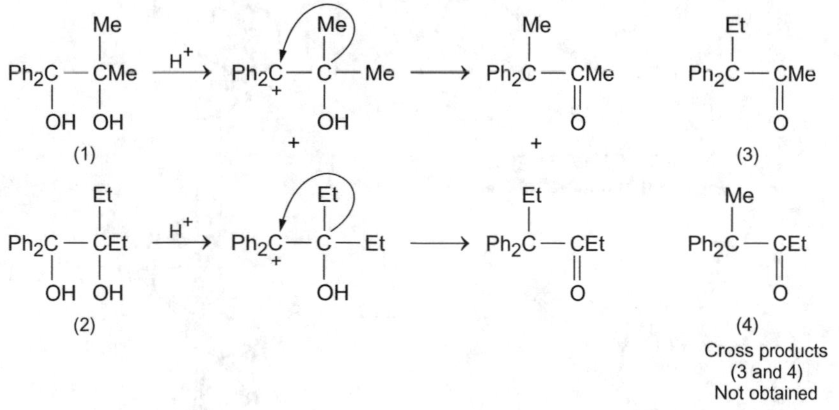

**Fig. 7.17** Crossover experiment shown that the migrating group is not detached and becomes free during rearrangement

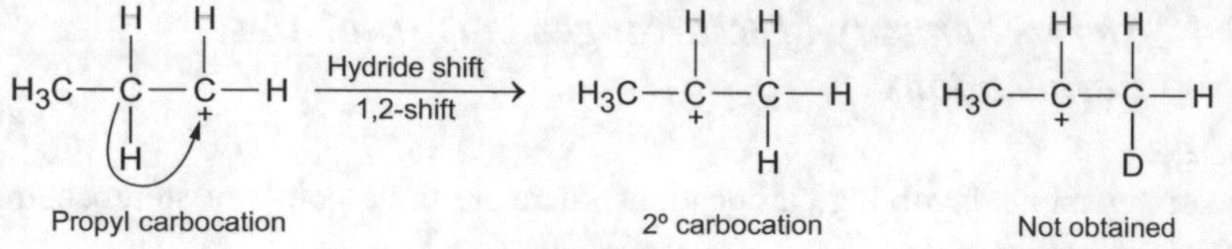

**Fig. 7.18** Crossover experiment during hydride shift. No crossover product obtained

from the synthon. In other words, one should not expect the configuration of the migration group to change, i.e., there is the retention of configuration if the migrating group is chiral. This has been confirmed by the reaction given in Fig. 7.19.

It has also been shown that during molecular rearrangement there is a predominant inversion of configuration at both the migration origin and migration terminus (Fig. 7.20).

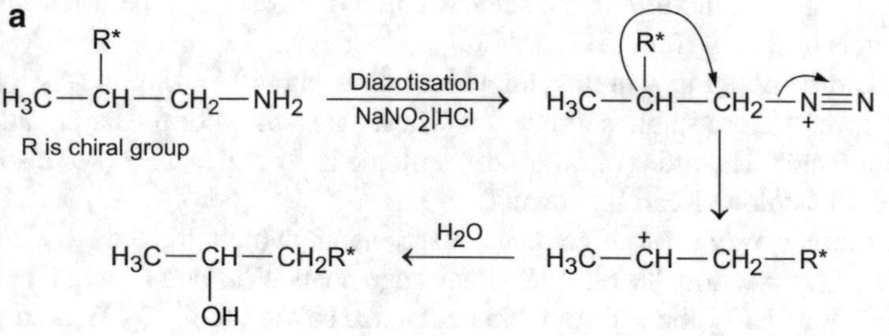

**Fig. 7.19** There is no change in the configuration of the migrating chiral group

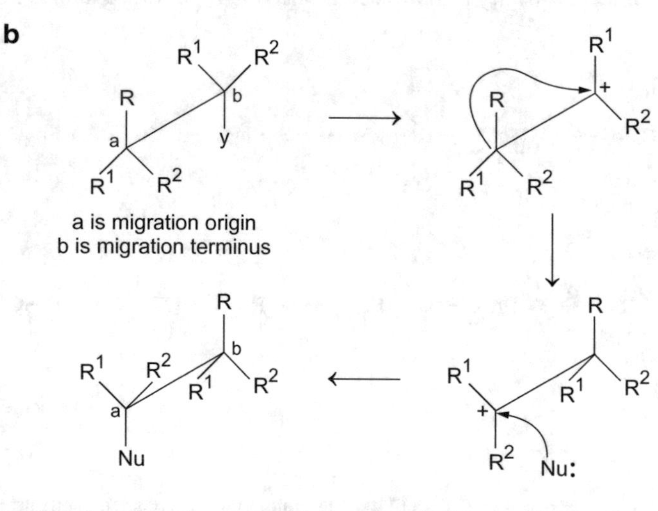

**Fig. 7.20** Inversion of configuration at migration origin and migration terminus

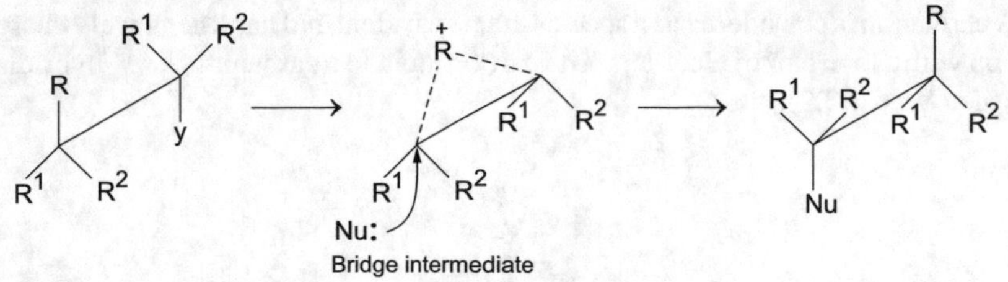

**Fig. 7.21** Bridged intermediates formation during molecular rearrangement

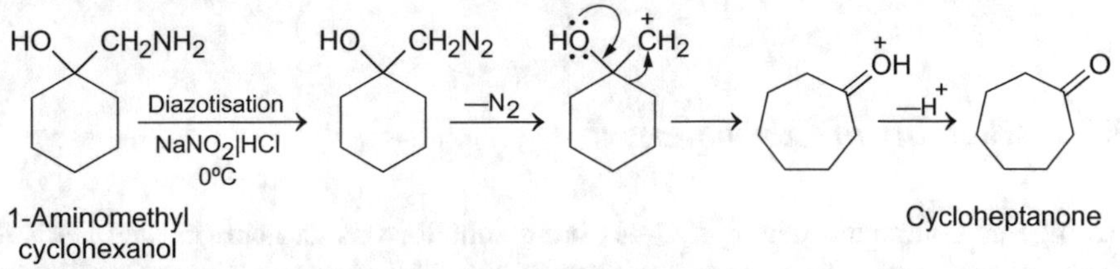

**Fig. 7.22** Ring expansion in molecular rearrangement involving carbocation

In cyclic compounds, (where $C_1$—$C_2$ rotation is prevented) the inversion is almost complete. However, in alicyclic compounds, the inversion is considerable. This is explained on the basis of a 'bridged' intermediate (Fig. 7.21).

In some cases, molecular rearrangement involving carbocation may lead to ring expansion. An example is given in Fig. 7.22.

## 7.3   Carbanions

Carbanions result from the heterolytic bond fission of a C—X bond, in which carbon is more electronegative than X. In this fission, the carbon atom takes the bonding electron pair and acquires a negative charge and X attains a positive charge. The ion carrying a negative charge and having a pair of the electron is called a carbanion (Fig. 7.23).

**Fig. 7.23** Heterolytic bond fission of a C—X bond

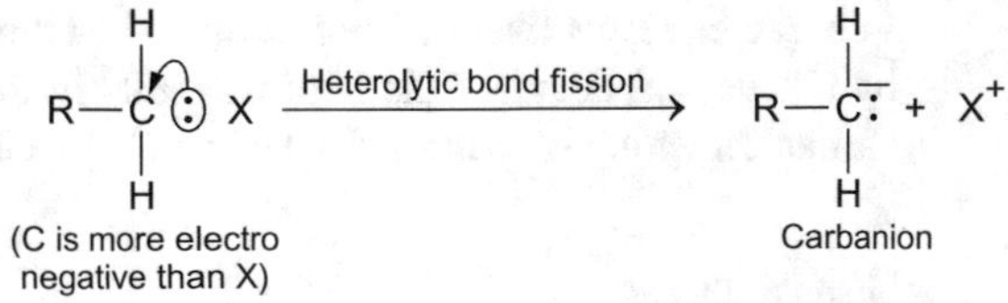

A carbanion is considered as a species that is trivalent and has a negatively charged carbon with a lone pair of electrons. The carbon atom in a carbanion is $sp^3$-hybridised (see also Sect. 7.3.2).

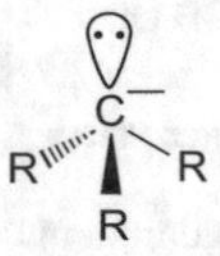

Carbanion

## 7.3.1 Stability of Carbonians

An organic compound having a C—H bond functions as an acid (in the classical sense) by donating a proton on treatment with a base forming a carbanion (Fig. 7.24).

Thus, a carbanion possesses an unshared pair of electrons and is considered as a base. The carbanion can accept a proton to give its conjugate acid. The stability of the carbocation, infect, depends on the strength of theconjugate acid. The weaker the acid, the more is its basic strength and less will be the stability of the carbanion. Table 7.1 gives the base (carbanion) obtained from the corresponding acid along with its $pKa$ value.

The following are the given main features on which the stability of carbanions depends:

(a)    Increase in the $s$-character of the carbanion carbon.
(b)    Electron-withdrawing inductive effect.
(c)    Conjugation of the lone pair of carbanion with a polarised multiple bond.
(d)    Aromatisation.

The understanding of the above features is clear from the following discussions:

(a)    *Increase in the s-character at the carbanion carbon*: There is an increase in the acidity of the hydrogen atom in the sequence $CH_3CH_3 < CH_2 = CH_2 < HC \equiv CH$. The increase in acidity is marked (*see* Table 7.1) as one goes from alkene to alkyne. This also reflects the increasing $s$-character of the hybrid orbital involved in the sigma bond to hydrogen, i.e., $sp^3 < sp^2 < sp^1$ (25, 33 and 50% $s$-character, respectively). The H atom is lost more easily from alkyne followed by alkene and alkane. This is turn results in the stabilisation of the resultant carbanion which is of the order $HC \equiv C^- > CH_2 = \bar{C}H > H_3C = \bar{C}H_2$.

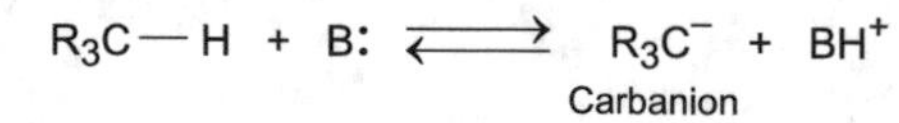

**Fig. 7.24** Formation of carbanion

$$R_3C-H + B: \rightleftharpoons R_3C^- + BH^+$$

Carbanion

**Table 7.1** Carbanions obtained from the corresponding acids along with their *pKa* values

| Acid (*pKa*) | Base (carbanion) |
|---|---|
| CH$_4$ (43) | $\overline{C}$ H$_3$ |
| CH$_2$=CH$_2$ (37) | CH$_2$=CH |
| C$_6$H$_6$ (37) | C$_6$H$_5$$^-$ |
| C$_6$H$_5$CH$_3$ (37) | C$_6$H$_5$$\overline{C}$H$_2$ |
| (C$_6$H$_5$)$_3$CH (33) | (C$_6$H$_5$)$_3$C$^-$ |
| CF$_3$ H (28) | C$^-$F$_3$ |
| HC $\equiv$ CH (25) | HC$\equiv$$\overline{C}$ |
| CH$_3$ CN (25) | $\overline{C}$H$_2$CN |
| CH$_3$COCH$_3$ (20) | CH$_3$CO$\overline{C}$H$_2$ |
| C$_6$H$_5$COCH$_3$ (19) | C$_6$H$_5$CO$\overline{C}$H$_2$ |
| CH$_2$ (CO$_2$Et)$_2$ (13.3) | $^-$CH(CO$_2$Et)$_2$ |
| CH$_2$ (CN)$_2$ (12) | $\overline{C}$H(CN)$_2$ |
| HC (CF$_3$)$_3$ (11) | $^-$C(CF$_3$)$_3$ |
| CH$_3$COCH$_2$CO$_2$Et (10.7) | CH$_3$CO$\overline{C}$HCO$_2$Et |
| CH$_3$NO$_2$ (10.2) | $^-$CH$_2$NO$_2$ |
| (CH$_3$CO)$_2$ CH (6) | (CH$_3$CO)$_2$C$^-$ |
| CH$_2$(NO$_2$)$_2$ (4) | $^-$CH(NO$_2$)$_2$ |

(b)  *Electron-withdrawing inductive effect:* The presence of an electron-releasing group (like an alkyl group) at the end of the chain decreases the stability of the carbanion. On the other hand, the presence of an electron-attracting group (like C $\equiv$ N, > C = O, etc.) at the end of the chain increases the stability of the carbanion. Thus,

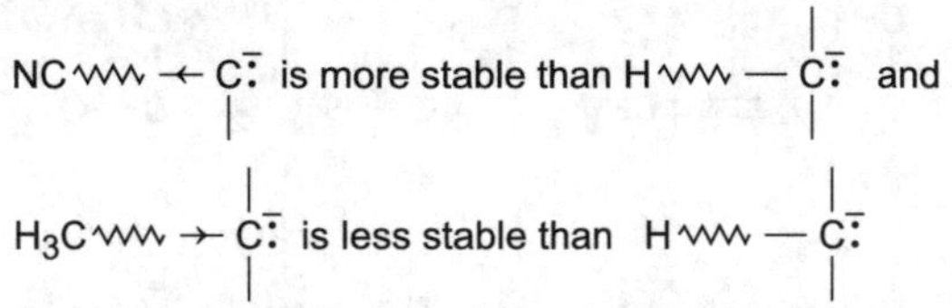

NC $\sim\!\!\sim\!\!\sim$ $\leftarrow$ $\overset{|}{\underset{|}{C}}$: is more stable than H $\sim\!\!\sim\!\!\sim$ — $\overset{|}{\underset{|}{C}}$: and

H$_3$C $\sim\!\!\sim\!\!\sim$ $\rightarrow$ $\overset{|}{\underset{|}{C}}$: is less stable than H $\sim\!\!\sim\!\!\sim$ — $\overset{|}{\underset{|}{C}}$:

the fact that the electron-withdrawing inductive effect is an important factor in determining the stability of the carbanions is well understood by considering the case of HCF$_3$ (*pKa* = 28) and HC(CF$_3$)$_3$ (*pKa* = 11). In both cases, the strong electron-withdrawing inductive effect of fluorine atoms makes the H atoms more acidic and also stabilises the resulting carbanions $^-$CF$_3$ and $^-$C(CF$_3$)$_3$ by electron withdrawal since $^-$C(CF$_3$)$_3$ has nine F atoms compared with only 3 in HCF$_3$, the former, i.e., $^-$C(CF$_3$)$_3$ is more stabilised than $^-$CF$_3$. It should be noted that in the formation of $^-$CCl$_3$ from HCCl$_3$, a similar electron-withdrawing inducting effect is operative. However, the electron-withdrawing inductive of Cl is less than that of F, and $^-$CF$_3$ is more stabilised than $^-$CCl$_3$.

Due to the deshielding effect of the electron-donating inductive effect of alkyl groups, the stability of the carbanions is of the following order.

It should, however, be noted that the stability sequence for carbocations is the exact reverse of the stability sequence for carbanions (see Sect. 7.2.1).

(c)  *Conjugation of the lone pair of carbanions with a polarised multiple bond:* The stability of the carbanion depends on the conjugation of the lone pair of carbanions with a polarised multiple bond as seen in the following examples (Fig. 7.25).

Though in each of the above examples, the electron-withdrawing inductive effect increases the acidity of the H atoms on the formed carbanions, but the stabilisation of the formed carbanion by the delocalisation is of greater significance. Thus, as expected, $NO_2$ is much more powerful.

(d)  *Aromatisation:* Carbanions are stabilised by aromatisation. As an example, cyclopentadiene (*pKa* 16 compared to about 37 for simple alkene) on treatment with a base gives cyclopentadiene anion (a 6 π electron system, a $4n + 2$ Hückel system, where $n = 1$). The 6 electrons in the anion are accommodated in three stabilised π molecular orbitals (like benzene). The cyclopentadiene anion is stabilised by aromatisation and shows quasi-aromatic stabilisation (Fig. 7.26).

In a similar way, cyclooctatetrane on treatment with potassium gives an isolable, crystalline salt of cyclooctatetraenyl dianion which is also a Hückel $4n + 2p$ electron

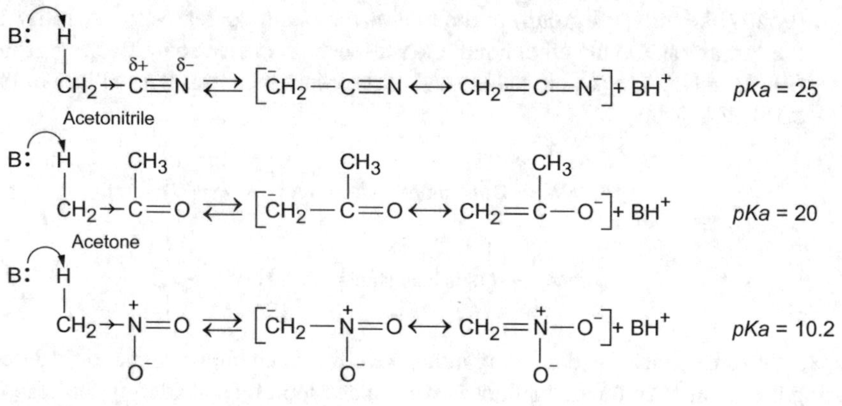

**Fig. 7.25** Stability of some carbocations

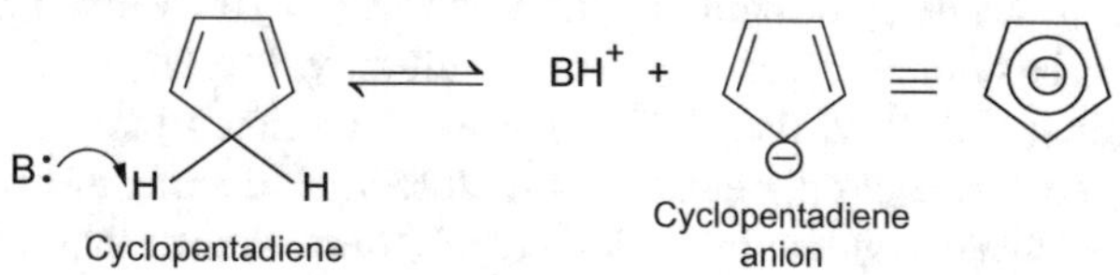

**Fig. 7.26** Cyclopentadiene anion

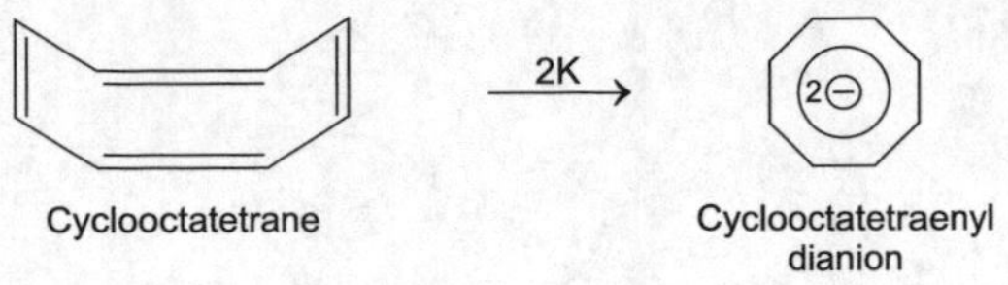

Fig. 7.27 Cyclotetradenyl dianion

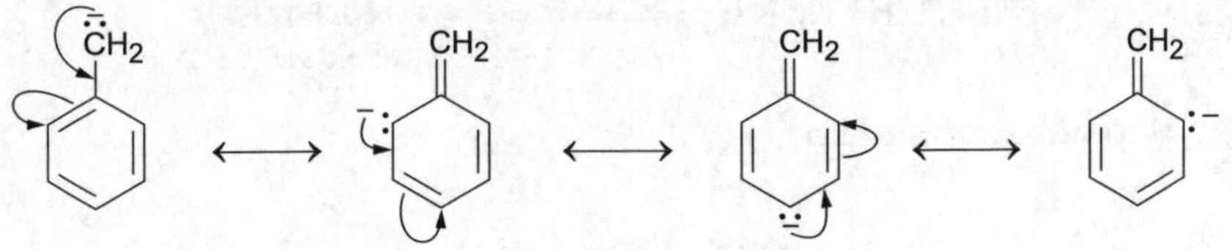

Fig. 7.28 Canonical forms of benzyl carbanion

system ($n$ being 2) and shows quasi-aromatic stability. In this case also, stabilisation is by aromatisation (Fig. 7.27).

As in the case of carbocations (Sect. 7.2.1), the carbanions are also stabilised by resonance, which is due to delocalisation of the negative charge and then distributed over other carbon atoms. Thus, benzyl carbanion $C_6H_5 - \ddot{C}H_2$ is more stable than ethyl carbocation $CH_3 - \ddot{C}H_2$. The canonical forms of benzyl carbanion are given in Fig. 7.28.

The stability of various carbanions is of the order:

Benzyl > vinyl > phenyl > cyclopropyl > ethyl > $n$-propyl > isobutyl > neopentyl > cyclobutyl.

The stability of carbanions carrying a functional group at α-position follows the order:

$NO_2$ > RCO > COOR > $SO_2$ > CN ≈ $CONH_2$ > halogen > H > R.

## 7.3.2  Structure of Carbanions

A simple carbanion of the type $R_3C^-$ can assume a pyramidal ($sp^3$) or a planar ($sp^2$) configuration depending on the nature of R. On the basis of energy grounds, it is believed that the pyramidal configuration is preferred since the unshared electron pair can be accommodated in an $sp^3$ orbital (Fig. 7.29).

The pyramidal configuration (Fig. 7.29) is similar to the one adopted by tertiary amines $R_3N$:, with which simple carbanions, $R_3C^\ominus$ are isoelectronic. In carbanions, ready inversion of configuration occurs (Fig. 7.29a ⇌ Fig. 7.29b) as in the case of amines.

**Fig. 7.29** Structure of
carbanions

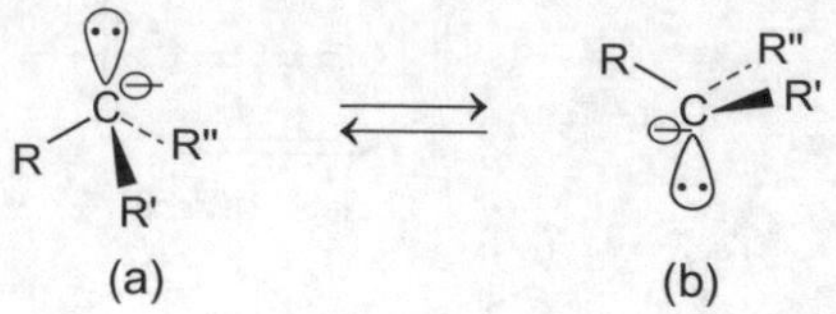

(a)                                    (b)

$$\overset{+}{Ph_3C} - H + Na\,\overset{-}{N}\,H_2 \xrightarrow{\text{liq.NH}_3} Ph_3\,\overset{-}{C}\,\overset{+}{Na} + NH_3$$

Blood red salt

**Fig. 7.30**  Generation of carbanion

The pyramidal $sp^3$ configuration finds support by the observation that the reactions involving the formation of carbanion intermediates at the bridgehead positions take place quite readily.

## 7.3.3  Generation of Carbanions

### 7.3.3.1  Abstraction of H by a Base

As already stated (Sect. 7.3.1), an organic compound containing a C—H bond on treatment with a base (like NaNH$_2$ in liq. NH$_3$) generates a carbanion (Fig. 7.30).

The formed salt, sodium triphenyl methyl, is a very strong organic base due to its proton-accepting ability.

An acidic hydrogen can be easily abstracted by using an appropriate base like NaOH/ether, aqueous NaOH, $t$-BuO$^-$/$t$-BuOH, RLi/DMF, etc. Some examples are given in Fig. 7.31.

### 7.3.3.2  From Alkyl Halides

Alkyl halides on treatment with a suitable base generate carbanions. Some examples are given in Fig. 7.32.

### 7.3.3.3  From Unsaturated Compounds

Unsaturated compounds on reaction with nucleophiles generate carbanions (Fig. 7.33).

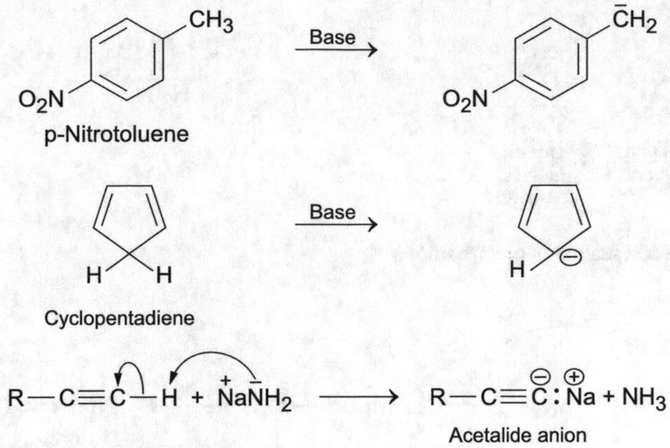

**Fig. 7.31**  Generation of carbanion from compounds containing acidic H

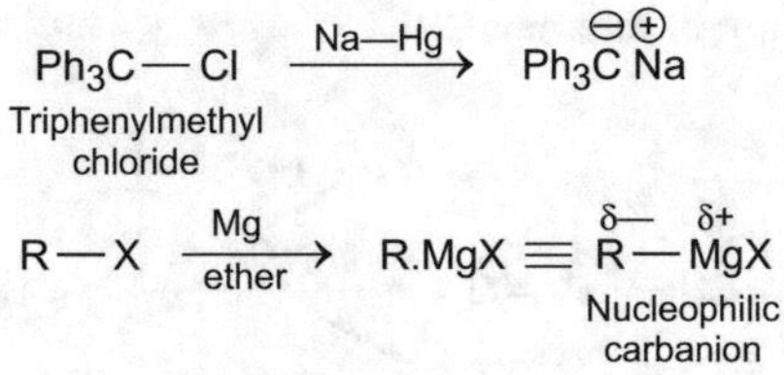

**Fig. 7.32**  Generation of carbanions from alkyl halides

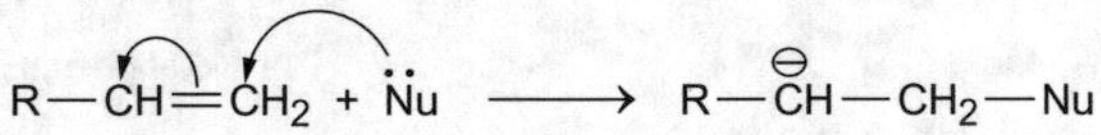

**Fig. 7.33**  Generation of carbanions from unsaturated compounds

## 7.3.4   Reactions of Carbanions

Carbanions undergo a number of reactions like addition reactions, elimination reactions, displacement reactions, rearrangement reactions, decarboxylations and oxidation.

### 7.3.4.1   Addition Reactions

Carbanion, being electron-rich, behaves like nucleophiles and adds to the carbonyl group of aldehydes or ketones (Fig. 7.34).

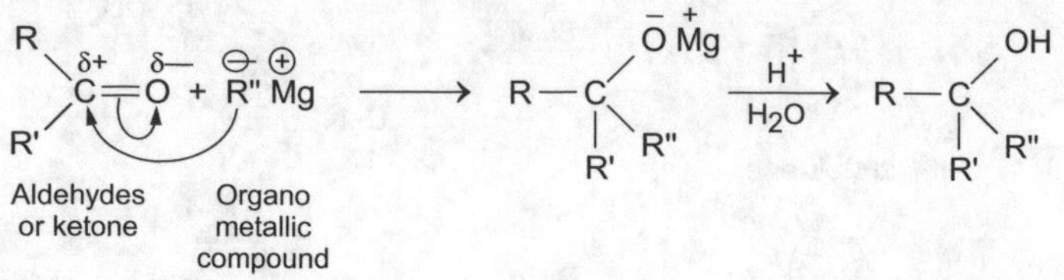

**Fig. 7.34** Addition reactions of carbanions

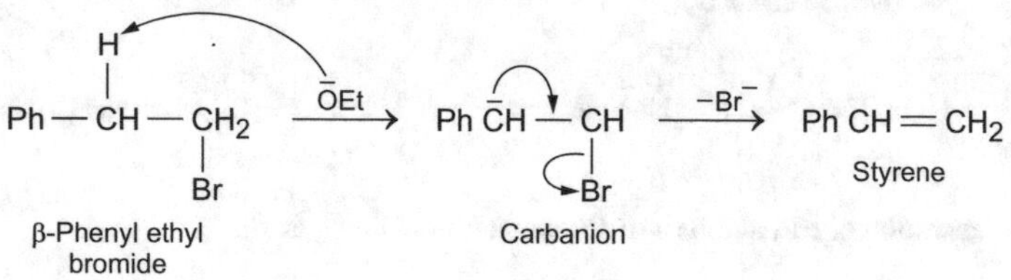

**Fig. 7.35** Elimination reaction of carbanion

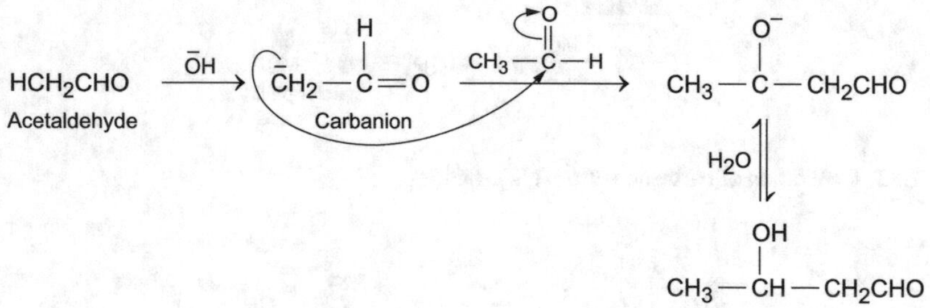

This is the well-known **Aldol condensation**

### 7.3.4.2    Elimination Reactions

Carbanions undergo elimination reactions. As an example, β-phenyl ethyl bromide on treatment with a base gives styrene via the intermediate formation of carbanion followed by elimination of Br⁻ (Fig. 7.35).

### 7.3.4.3    Displacement Reactions

A typical example of a displacement reaction is the formation of monoalkyl derivative of diethyl malonate or acetylacetone via the intermediate formation of carbanion (Fig. 7.36).

In these reactions, the formed carbanions act as nucleophiles.

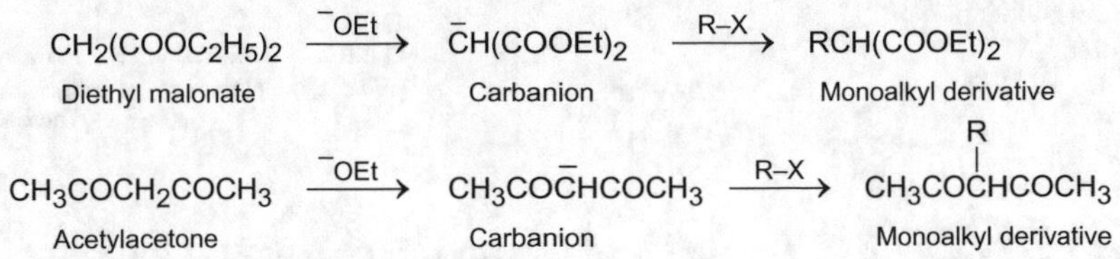

$$CH_2(COOC_2H_5)_2 \xrightarrow{\text{–OEt}} \bar{C}H(COOEt)_2 \xrightarrow{\text{R–X}} RCH(COOEt)_2$$

Diethyl malonate · Carbanion · Monoalkyl derivative

$$CH_3COCH_2COCH_3 \xrightarrow{\text{–OEt}} CH_3CO\bar{C}HCOCH_3 \xrightarrow{\text{R–X}} \overset{\overset{\displaystyle R}{|}}{CH_3COCHCOCH_3}$$

Acetylacetone · Carbanion · Monoalkyl derivative

**Fig. 7.36** Displacement reactions of carbanions

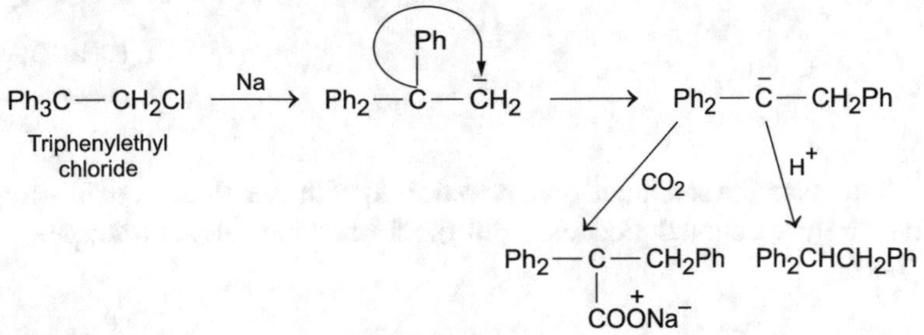

$$Ph_3C-CH_2Cl \xrightarrow{\text{Na}} Ph_2-\overset{\overset{\displaystyle Ph}{|}}{C}-\bar{C}H_2 \longrightarrow Ph_2-\bar{C}-CH_2Ph$$

Triphenylethyl chloride

$$Ph_2-\overset{|}{\underset{\overset{|}{COONa^-}}{C}}-CH_2Ph \qquad Ph_2CHCH_2Ph$$

with $CO_2$ and $H^+$ pathways

**Fig. 7.37** Rearrangements of carbanions

**Fig. 7.38** Decarboxylation of carboxylate anion

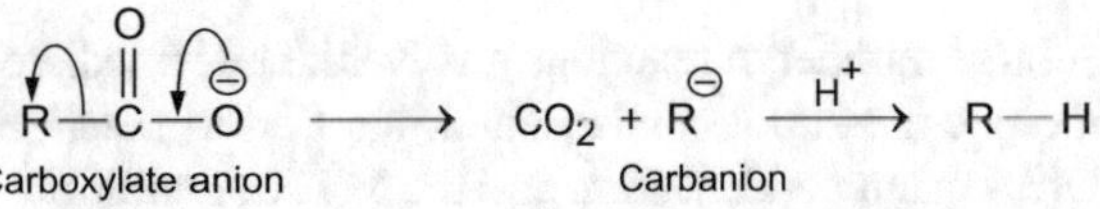

$$R-\overset{\overset{\displaystyle O}{||}}{C}-O^{\ominus} \longrightarrow CO_2 + R^{\ominus} \xrightarrow{H^+} R-H$$

Carboxylate anion · Carbanion

### 7.3.4.4 Rearrangement Reactions

The reactions involving rearrangement are comparatively less common. An example is given in Fig. 7.37.

### 7.3.4.5 Decarboxylation

The carboxylate anion loses $CO_2$ via the intermediate formation of carbanion followed by acquiring a proton from the solvent (Fig. 7.38).

### 7.3.4.6 Oxidation of Carbanions

The carbanion under suitable conditions undergoes oxidation. Thus, triphenylmethyl anion can be oxidised to triphenylmethyl radicals, which can be reduced back to carbanion (Fig. 7.39).

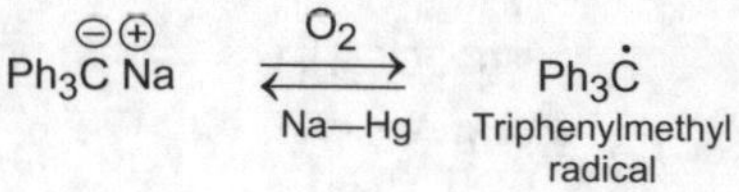

**Fig. 7.39** Oxidation of
carbanion

$$Ph_3\overset{\ominus}{C}\ \overset{\oplus}{Na} \quad \underset{Na\text{—}Hg}{\overset{O_2}{\rightleftarrows}} \quad \underset{\substack{\text{Triphenylmethyl} \\ \text{radical}}}{Ph_3\overset{\cdot}{C}}$$

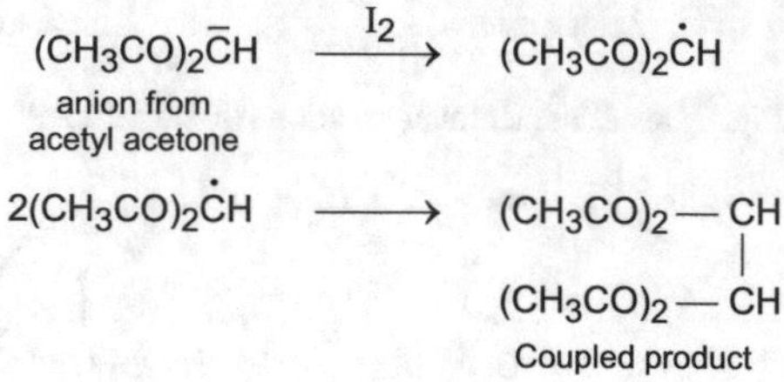

**Fig. 7.40** Oxidation of
carbanion with iodine

$$\underset{\substack{\text{anion from} \\ \text{acetyl acetone}}}{(CH_3CO)_2\overset{-}{C}H} \quad \overset{I_2}{\longrightarrow} \quad (CH_3CO)_2\overset{\cdot}{C}H$$

$$2(CH_3CO)_2\overset{\cdot}{C}H \quad \longrightarrow \quad \begin{array}{c} (CH_3CO)_2 \text{— CH} \\ | \\ (CH_3CO)_2 \text{— CH} \end{array}$$

$$\text{Coupled product}$$

In certain cases, the carbanions can be oxidised with one-electron oxidising agent like iodine to give coupled products from the formed radical. An example is given in Fig. 7.40.

## 7.3.5  Reactions Involving Carbanions

A large number of reactions involve the intermediate formation of carbanions. These include the Aldol condensation, the Perkin reaction, the Claisen condensation, the Dieckmann condensation and the Michael addition.

### 7.3.5.1  Aldol Condensation

For details, see Sect. 20.1.

### 7.3.5.2  Perkin Reaction

Aromatic aldehydes on heating with acetic anhydride and sodium or potassium acetate give cinnamic acid. The reaction proceeds via the intermediate formation of carbanion (Fig. 7.41).

### 7.3.5.3  Claisen Reaction

The reaction of a carbonyl compound (aldehyde or ketone) without α-hydrogen atom with an ester having active hydrogen in presence of a base gives α, β-unsaturated ester (Fig. 7.42).

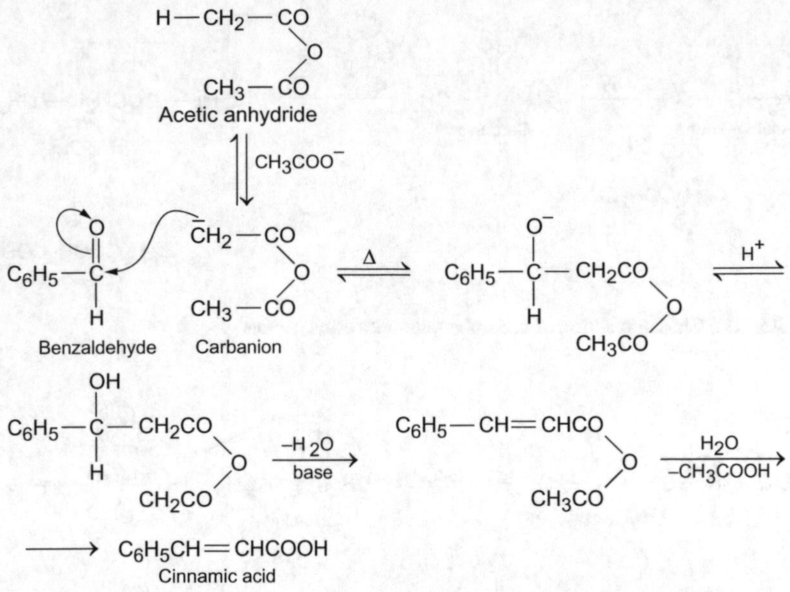

Fig. 7.41  The Perkin reaction

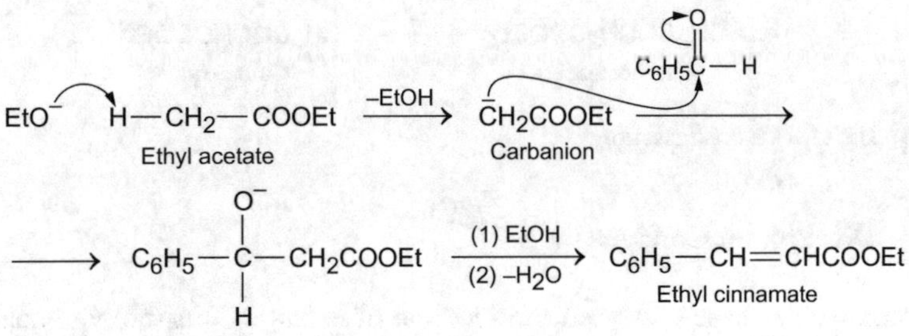

Fig. 7.42  The Claisen reaction

The Claisen reaction of aromatic ketones having α-hydrogen with ethyl acetate in presence of pulverised sodium gives β-diketones (Fig. 7.43).

### 7.3.5.4  Claisen Condensation

Base catalysed condensation of an ester containing a α-hydrogen with the same or different ester gives β-ketoester (Fig. 7.44).

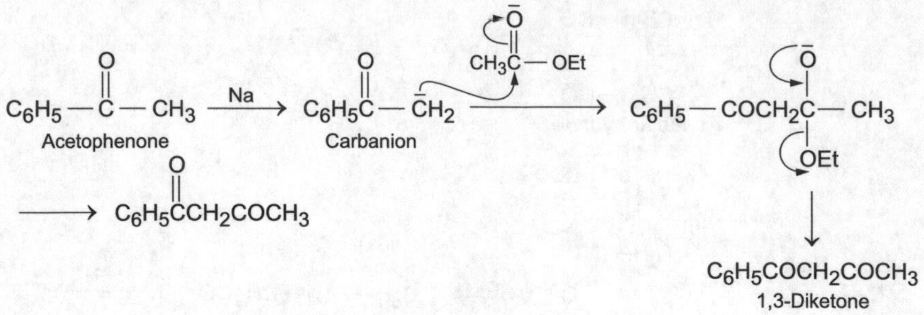

**Fig. 7.43**  The Claisen reaction of acetophenone with ethyl acetate

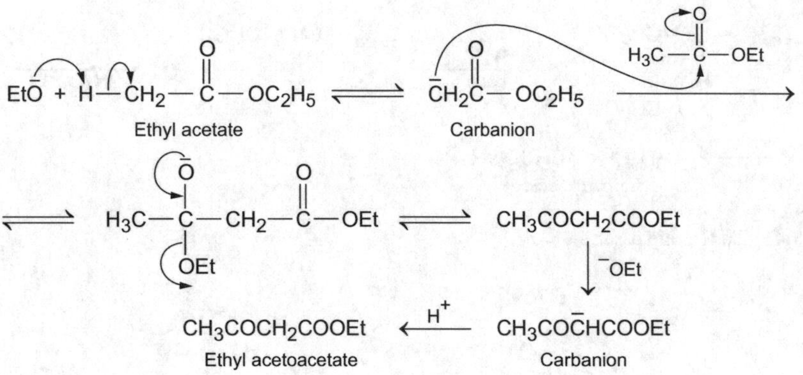

**Fig. 7.44**  The Claisen condensation

### 7.3.5.5  Dieckmann Condensation

An intramolecular base catalysed the reaction of esters of dicarboxylic acids and is useful for the synthesis of cyclic ketones. Thus, the reaction of diethyl adipate with sodium gives 2-carbethoxy cyclopentanone, which is used for the synthesis of cyclopentanone (Fig. 7.45).

In a similar way, diethyl pimelate $EtO_2C(CH_2)_5 COO_2 Et$ gives cyclohexanone.

### 7.3.5.6  Michael Addition

The reaction of an $\alpha$, $\beta$-unsaturated carbonyl compound with an active methylene group (e.g., malonic ester, acetoacetic ester and cyanoacetic ester) in presence of base gives the adduct as shown ahead (Fig. 7.46).

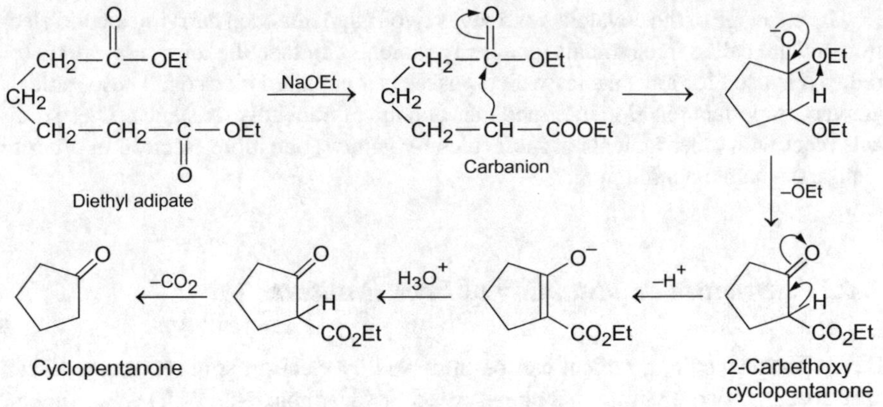

Fig. 7.45 The Dieckmann condensation

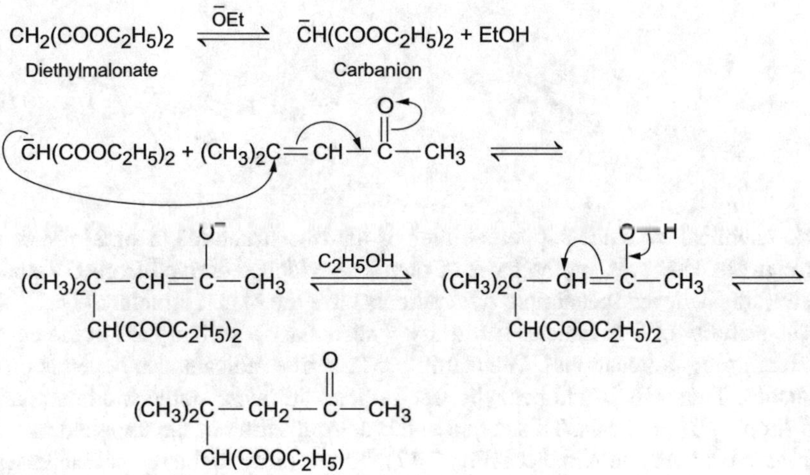

Fig. 7.46 The Michael addition

## 7.4 Free Radicals

In the homolytic bond cleavage of a covalent bond, the separating atoms take one electron each. Such a bond cleavage is symmetrical and is represented as a single electron shift and is represented by a half arrow (called fish arrow) (Fig. 7.47).

Fig. 7.47 Homolytic bond fission of a covalent bond

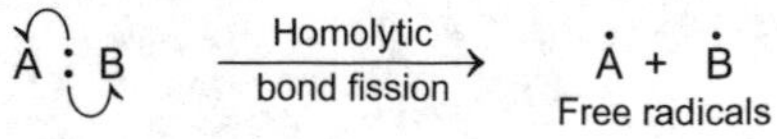

The cleavage of the covalent bond gives two fragments each carrying an odd electron and are called free radicals (or simply radicals). In fact, the term free radical (or radical) is used for any species which possess an unpaired electron. These radicals are very important reaction intermediates and are of transitory existence. These radicals react with other radicals or molecules by gaining one more electron in order to restore the stable bonding pair.

### 7.4.1   Structure and Stability of Free Radicals

Being paramagnetic, a radical can be observed by electron spin resonance (ESR) spectroscopy. Two structures have been proposed for simple radicals. These are planar $sp^2$-hybridised radical (1) (similar to carbocation) or a pyramidal $sp^3$-hybridised radical (2) (similar to carbanion).

Planar
(1)

Pyramidal
(2)

No chemical evidence supports either of the two structures (1 or 2) for simple free radicals. However, on the basis of physical evidence derived from UV and IR spectra, it is believed that simple free radicals have the planar structure (1).

The stability of free radicals is tertiary > secondary > primary (as in the case of the stability of carbocations). The stability of the free radicals also depends on the resonance. Thus, allylic and benzylic free radicals are more stable and less reactive than simple alkyl radicals. This is due to the delocalisation of the unpaired electron over the π orbital system in them (Fig. 7.47). The stability of the radical increases as the delocalisation increases. Thus, Ph2ĊH is more stable than PhCH$_2$•, and Ph$_3$C• is a reasonable stable radical (Fig. 7.48).

In the case of triphenylmethyl radical, Ph$_3$C•, the radical carbon is $sp^2$-hybridised; this implies that the bonds joining it to the three phenyl groups lie in the same plane. However, maximum stabilisation will be only if all the three phenyl groups can be coplanar simultaneously. On the basis of X-ray crystallographic studies, the Ph$_3$C• radical is found to be propeller-shaped, the phenyl groups being angled at about 30° out of the common plane.

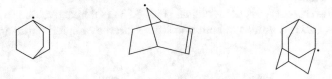

RCH $=$ CHCH$_2^\bullet$ $\longleftrightarrow$ [RCH $\cdots$ CH $\cdots$ CH$_2$]$^\bullet$
Allyl free radical

◯$-$ĊHR $\longleftrightarrow$ [◯$-$CHR]$^\bullet$

Benzylic
free radical

Ph$_3$C$\cdot$
Triphenyl
methyl radical

**Fig. 7.48**   Stability of some free radicals

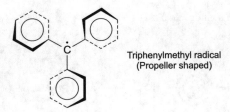

**Fig. 7.49**   Some bridged free radicals

Triphenylmethyl radical
(Propeller shaped)

Some radicals have rigid structures with fixed dihedral angles and bond angles. These are referred to as **bridged free radicals**. The structures of bridged radicals are supported on the basis of physical and chemical evidence. The following are given some bridged radicals which can be easily formed (Fig. 7.49).

## 7.4.2   Heteroradicals

Radicals involving atoms other than carbon are called heteroradicals. Such radicals have varying degrees of stability. Heteroradicals containing nitrogen, sulphur and oxygen are discussed below.

A nitrogen radical is formed by warming N, N, N', N'-tetraphenyl hydrazine in a non-polar solvent. The required tetraphenyl hydrazine is obtained by the oxidation of diphenylamine with $KMnO_4$ (Fig. 7.50).

Another stable nitrogen radical 1,1-diphenyl-2-picryl hydrazyl is obtained by the oxidation of triaryl hydrazine, which in turn is obtained from diphenyl hydrazine by reacting with picryl chloride (Fig. 7.51).

Radicals containing sulphur are known as **thiyl radicals**. An example of a thiyl radical is $PhS^•$. It is obtained by heating diphenyl sulphide. This radical becomes yellow on heating and the colour disappears on cooling (Fig. 7.52).

The thiyl radicals (Fig. 7.52) can be trapped with N free radicals such as 1,1-diphenyl-2-picryl hydrazyl (see Fig. 7.51).

Simple alkyl thiyl radicals such as $MeS^•$ have also been detected as reaction intermediates. Such radicals are highly reactive.

Stable oxygen-containing radicals are also known. As an example, 2,4,6-tert butyl phenoxy radical is obtained by oxidation of 2,4,6-tert butyl phenoxide with $K_3Fe$ $(CN)_6$ (Fig. 7.52).

The 2,4,6-tert. butyl phenoxy radical (Fig. 7.53) is unreactive due to hindrance by the bulky $(CMe_3)$ groups in both o-positions, thereby hindering the approach of

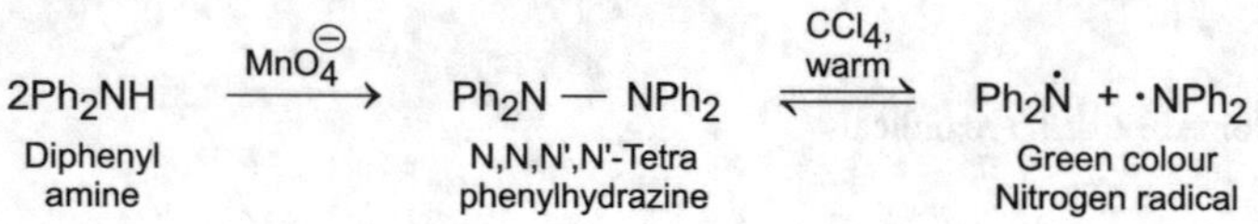

$$2Ph_2NH \xrightarrow{MnO_4^{\ominus}} Ph_2N - NPh_2 \underset{warm}{\overset{CCl_4,}{\rightleftharpoons}} Ph_2\overset{•}{N} + {}^•NPh_2$$

Diphenyl               N,N,N',N'-Tetra          Green colour
amine               phenylhydrazine        Nitrogen radical

**Fig. 7.50** Preparation of a nitrogen radical

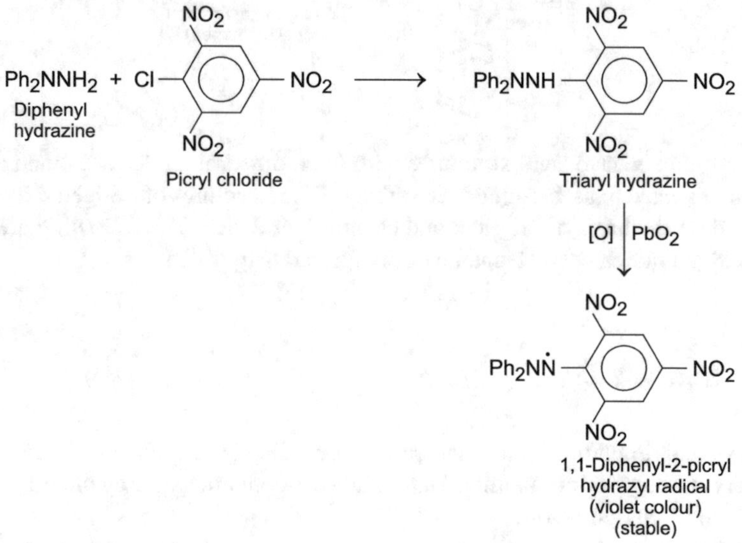

**Fig. 7.51** Preparation of 1,1-diphenyl-2-picryl hydrazyl radical

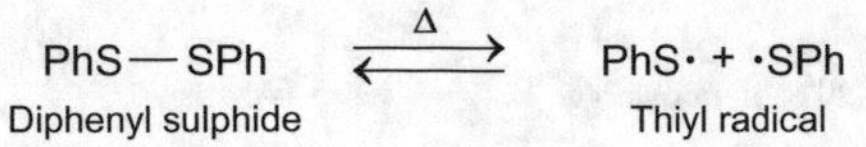

$$PhS-SPh \xrightarrow{\Delta} \text{⇌} \quad PhS\cdot + \cdot SPh$$

Diphenyl sulphide      Thiyl radical

**Fig. 7.52** Formation of thiyl radical

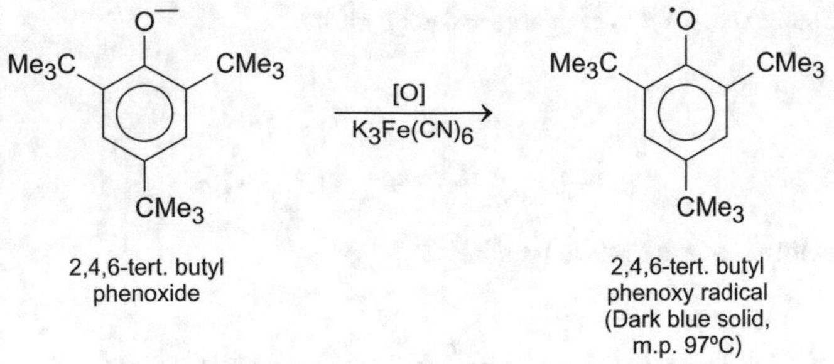

2,4,6-tert. butyl
phenoxide

2,4,6-tert. butyl
phenoxy radical
(Dark blue solid,
m.p. 97°C)

**Fig. 7.53** Synthesis of 2,4,6-tert butyl phenoxy radical

another molecule of the phenoxy radical or of other species to the radical oxygen atom.

## 7.4.3 Generation of Free Radicals

Some of the important methods for the generation of free radicals are photolysis thermolysis and redox reactions.

### 7.4.3.1 Photolysis

Only compounds that absorb radiations in the UV or visible range undergo photolysis. Thus, acetone in the vapour phase undergoes photolysis by light (wavelength 320 nm, 3200 Å) to yield two molecules of methyl free radicals (Fig. 7.54).

Alkyl hypochlorites and alkyl nitrites also undergo photolysis to give alkoxy radicals (Fig. 7.55).

Halogens also undergo photolysis to give free radicals (Fig. 7.56).

The chlorine or bromine free radicals (Fig. 7.56) can initiate halogenation of alkenes or addition to alkenes.

**Fig. 7.54** Photolysis of acetone

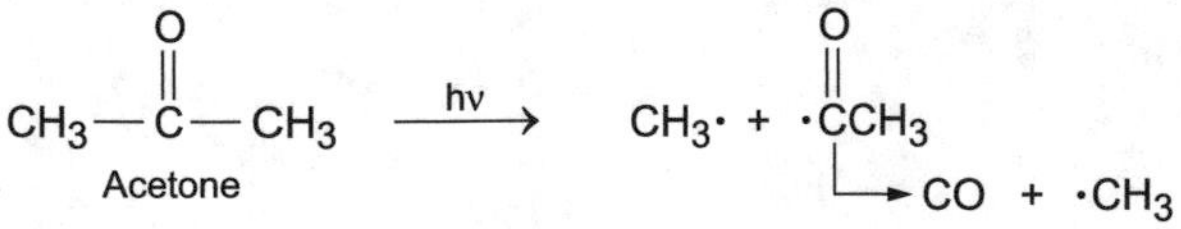

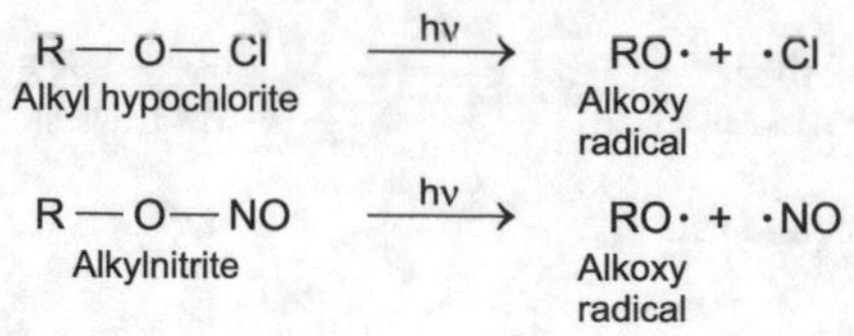

$$R-O-Cl \xrightarrow{h\nu} RO\cdot + \cdot Cl$$

Alkyl hypochlorite                     Alkoxy
                                       radical

$$R-O-NO \xrightarrow{h\nu} RO\cdot + \cdot NO$$

Alkylnitrite                           Alkoxy
                                       radical

**Fig. 7.55**  Photolysis of alkyl hypochlorite and alkyl nitrite

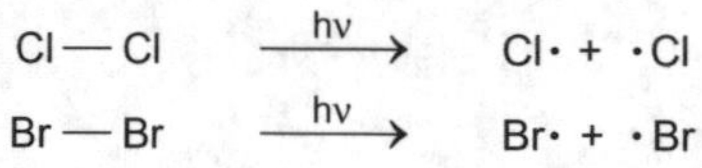

$$Cl-Cl \xrightarrow{h\nu} Cl\cdot + \cdot Cl$$

$$Br-Br \xrightarrow{h\nu} Br\cdot + \cdot Br$$

**Fig. 7.56**  Photolysis of chlorine and bromine

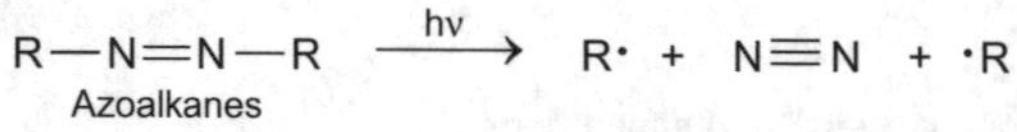

$$R-N=N-R \xrightarrow{h\nu} R\cdot + N\equiv N + \cdot R$$

Azoalkanes

**Fig. 7.57**  Photolysis of azoalkanes

$$R-\overset{\overset{\displaystyle O}{\|}}{C}-O-O-\overset{\overset{\displaystyle O}{\|}}{C}-R \xrightarrow{h\nu} 2R-\overset{\overset{\displaystyle O}{\|}}{C}-O\cdot \longrightarrow 2R\cdot + 2CO_2$$

Diacetyl peroxide                      Acyl free radical

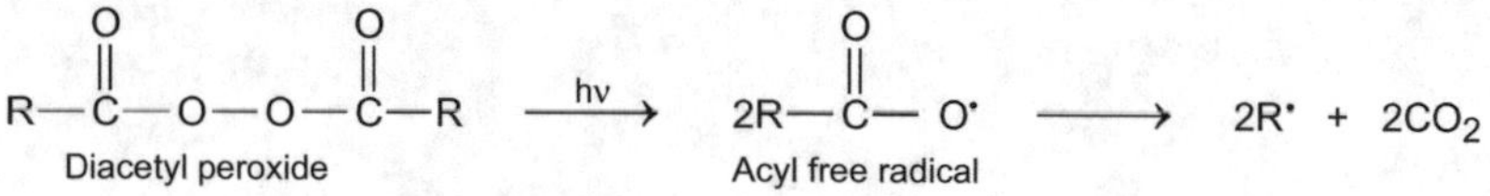

**Fig. 7.58**  Photolysis of deacetyl peroxides

Photolysis can cleave strong bonds that do not break readily. As an example, the photolysis of azoalkanes gives alkyl free radicals (Fig. 7.57).

In the above illustration (Fig. 7.57), energy at only one particular level is transferred to a molecule. In view of this, photolysis is a specific method for effecting homolysis compared to pyrolysis, where a number of byproducts are obtained. As an example, the photolysis of diacetyl peroxides occurs cleanly (Fig. 7.58).

## 7.4.3.2  Thermolysis

In thermolysis, the substrate is heated at a suitable temperature. Some examples are given in Fig. 7.59.

The free radicals obtained above (Fig. 7.59) find application in the manufacture of polymers.

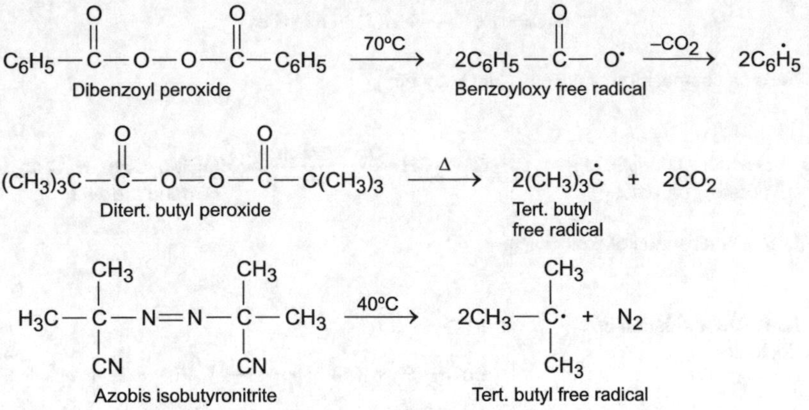

Fig. 7.59 Some examples of thermolysis

### 7.4.3.3 Redox Reactions

The reactions involve one-electron transfer for generating free radicals. In these reactions, metal ions such as $Cu^+/Cu^{2+}$ and $Fe^{2+}/Fe^{3+}$ are involved. As an example, $Cu^+$ ions accelerate the decomposition of acyl peroxides (Fig. 7.60).

Figure 7.60 is a convenient method for the generation of acyloxy free radical $(Ar-\overset{\overset{\displaystyle O}{\|}}{C}-O^{\bullet})$, since, in thermolysis, the acyloxy free radical is decomposed to $Ar^{\bullet}$ + $CO_2$.

$Cu^+$ also takes part in the Sandmeyer reaction (conversion of $ArN_2^+Cl^-$ to $ArCl$ + $N_2$), wherein $Ar^{\bullet}$ is formed transiently as an intermediate (Fig. 7.61).

The above two reactions (Figs. 7.60 and 7.61) are reductions. An oxidation reaction involves the use of $Fe^{2+}$ to catalyse the oxidation reaction of aqueous $H_2O_2$ (Fig. 7.62).

A mixture of $H_2O_2$ and $Fe^{2+}$ is the well-known **Fenton reagent**, wherein the effective oxidising agent is the hydroxy radical ($HO^{\bullet}$), which in turn can abstract a

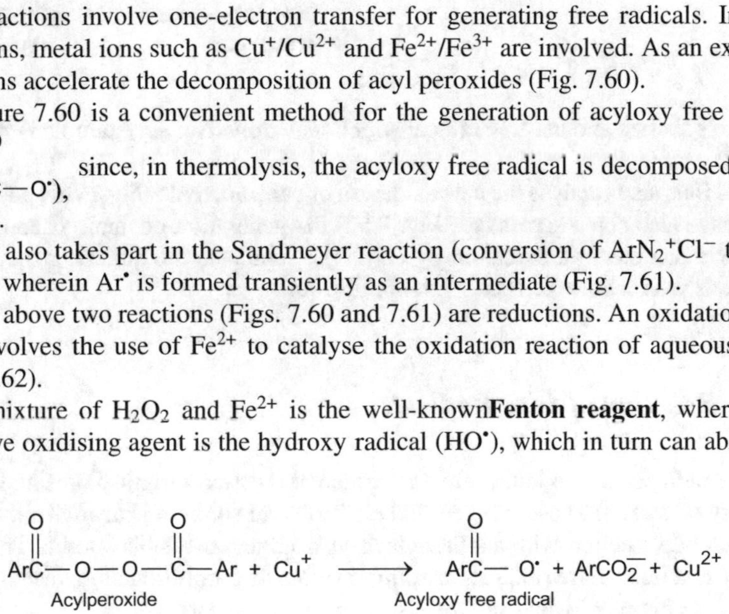

Fig. 7.60 Formation of acyloxy free radical from acyl peroxide by $Cu^+$

$$Ar\,N_2^+ + Cu^+ \longrightarrow Ar^{\bullet} + N_2 + Cu^{2+}$$

Fig. 7.61 Role of $Cu^+$ in the Sandmeyer reaction

$$H_2O_2 + Fe^{2+} \longrightarrow HO^{\bullet} + \bar{O}H + Fe^{3+}$$

**Fig. 7.62**  Decomposition of aqueous $H_2O_2$ by $Fe^{2+}$

$$HO^{\bullet} + H - CH_2CMe_2OH \longrightarrow {}^{\bullet}CH_2CMe_2OH \xrightarrow{\text{Dimerisation}} HOCMe_2CH_2CH_2CMe_2OH$$

Trimethyl carbinol                                                        Dimeric product

**Fig. 7.63**  Dimerisation of free radical

**Fig. 7.64**  Autooxidation of benzaldehyde

$$\underset{\text{Benzaldehyde}}{Ph\overset{\overset{\displaystyle O}{\|}}{C}-H} + Fe^{3+} \longrightarrow \underset{\substack{\text{Benzoyl} \\ \text{free radical}}}{Ph-\overset{\overset{\displaystyle O}{\|}}{C}\cdot} + H^{+} + Fe^{2+}$$

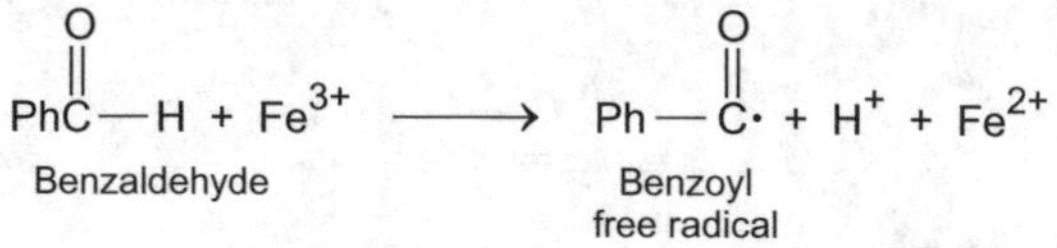

**Fig. 7.65**  Dimerisation of free radicals

$$\overset{\bullet}{C}H_3 + \overset{\bullet}{C}H_3 \longrightarrow CH_3 - CH_3$$

Methyl free radical                Ethane

$$2CH_3\overset{\bullet}{C}H_2 \longrightarrow CH_3CH_2CH_2CH_3$$

Ethyl free radical                Butane

proton to generate another free radical which may dimerise. An example is given in Fig. 7.63.

Metal ions also catalyse the autooxidation of benzaldehyde (Figs. 7.64 and 7.65).

We have also come across (see Fig. 7.53) the generation of stable phenoxy free radical via one-electron oxidation by $Fe(CN)_6^{3-}$ and also the dimeric oxidation of carbanions with iodine (see Sect. 7.3.4.6, Fig. 7.40).

## 7.4.4  Reaction of Free Radicals

The free radicals, as we know, are very reactive reaction intermediates and are of transitory existence. These have an odd electron and so there is always a tendency to pair its odd electron with another electron from any available source. The most common reactions of free radicals are dimerisation, disproportionation, reaction with olefins, reaction with iodine and metals and rearrangements.

### 7.4.4.1  Dimerisation

The free radicals may dimerise (or recombine) to give hydrocarbon (Fig. 7.65).

An interesting example of dimerisation of free radicals is discussed in Sect. 7.3.4.6, Fig. 7.40.

$$CH_3CH_2^{\bullet} + CH_3CH_2^{\bullet} \longrightarrow CH_2{=}CH_2 + CH_3CH_3$$

Ethyl
free radical

Ethylene        Ethane

**Fig. 7.66** Disproportionation of free radicals

$$CH_3^{\bullet} + CH_2{=}CH_2 \longrightarrow CH_3{-}CH_2{-}CH_2^{\bullet}$$
$$CH_2{-}CH_2{-}CH_2^{\bullet} + CH_2{=}CH_2 \longrightarrow CH_3CH_2CH_2CH_2CH_2^{\bullet}$$
$$2CH_3CH_2CH_2CH_2CH_2^{\bullet} \longrightarrow CH_3CH_2CH_2CH_2CH_2CH_2CH_2CH_2CH_2CH_3$$

**Fig. 7.67** Reaction of free radicals with olefins

**Fig. 7.68** Reaction of methyl radical with iodine and metals

$$2CH_3^{\bullet} + I_2 \longrightarrow 2CH_3I$$
$$2CH_3^{\bullet} + Zn \longrightarrow (CH_3)_2Zn$$
Dimethyl zinc
$$2CH_3^{\bullet} + Hg \longrightarrow (CH_3)_2Hg$$
Dimethyl mercury

### 7.4.4.2 Disproportionation

Alkyl radicals undergo disproportionation. As an example, one radical (e.g., $CH_3CH_2$) takes up a H from another free radical to give an unsaturated compound (Fig. 7.66).

### 7.4.4.3 Reaction with Olefins

The reaction of alkyl radicals with olefins generates a new radical. The reaction continues till the formed free radical couples with another free radical and the reaction gets terminated (Fig. 7.67).

### 7.4.4.4 Reaction with Iodine and Metals

The alkyl radicals combine with iodine (see also Fig. 7.40) and other metals to form alkyl derivatives. The same examples are given in Fig. 7.68.

### 7.4.4.5 Rearrangements

The free radicals (like carbocations and carbanions) undergo rearrangements involving 1,2-aryl shifts via a bridged transition state. An example is that of an aldehyde (1), which undergoes abstraction of H from the CHO group by the free

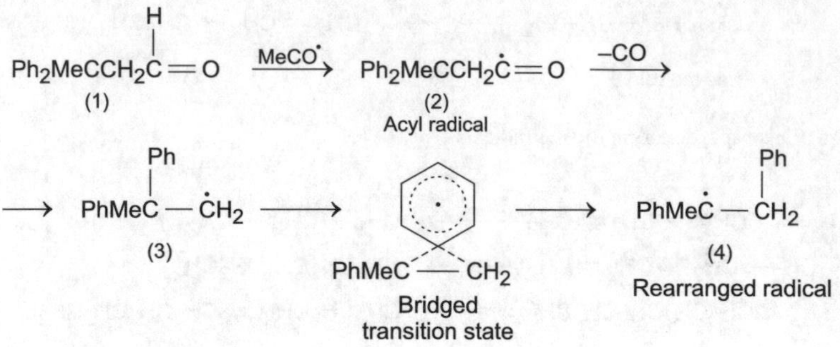

$$\underset{(1)}{Ph_2MeCCH_2\overset{\overset{\displaystyle H}{|}}{C}=O} \xrightarrow{MeC\overset{\bullet}{O}} \underset{\underset{Acyl\ radical}{(2)}}{Ph_2MeCCH_2\overset{\bullet}{C}=O} \xrightarrow{-CO}$$

$$\longrightarrow \underset{(3)}{PhMe\overset{\overset{\displaystyle Ph}{|}}{C}-\overset{\bullet}{C}H_2} \longrightarrow$$

PhMeC —— CH₂
Bridged
transition state

$$\longrightarrow \underset{\underset{Rearranged\ radical}{(4)}}{PhMe\overset{\overset{\displaystyle Ph}{|}}{\overset{\bullet}{C}}-CH_2}$$

**Fig. 7.69**  Rearrangements in free radicals

radical Me3CO. to yield an acyl radical (2), which in turn loses CO to form another radical (3). This radical undergoes rearrangement to give the radical (4).

In the formation of (4) from (3), there is the involvement of the migration of Ph via the bridged transition state (Fig. 7.69).

## 7.4.5  Reactions Involving Free Radicals

A large number of reactions involve the formation of free radicals as intermediates. Some of such reactions are the Sandmeyer reaction, the Gomberg free radical reaction, the Wurtz reaction, the Hunsdiecker reaction, Kolb's electrolytic reaction, halogenation and vinyl polymerisation.

### 7.4.5.1  Sandmeyer Reaction

It involves the decomposition of a diazonium salt in presence of a cuprous halide to give the corresponding aryl halide. The reaction proceeds via the formation of a free radical, $C_6H_5{}^{\bullet}$. Thus, benzenediazonium chloride on treatment with CuCl gives chlorobenzene (Fig. 7.70).

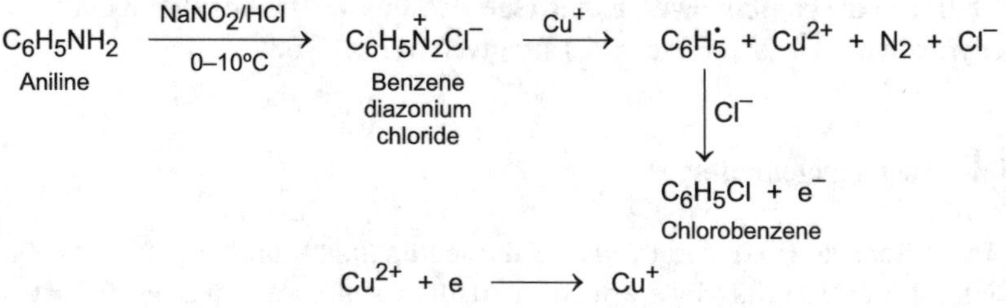

$$\underset{Aniline}{C_6H_5NH_2} \xrightarrow[0-10^\circ C]{NaNO_2/HCl} \underset{\underset{chloride}{\underset{diazonium}{Benzene}}}{C_6H_5\overset{+}{N_2}Cl^-} \xrightarrow{Cu^+} C_6H_5^{\bullet} + Cu^{2+} + N_2 + Cl^-$$

$$\Big\downarrow Cl^-$$

$$\underset{Chlorobenzene}{C_6H_5Cl + e^-}$$

$$Cu^{2+} + e \longrightarrow Cu^+$$

**Fig. 7.70**  The Sandmeyer reaction

**Fig. 7.71** The Gomberg free radical reaction

$$2(C_6H_5)_3CCl + Zn \longrightarrow 2(C_6H_5)_3\overset{\cdot}{C} + ZnCl_2$$

Triphenyl methyl chloride

Triphenyl methyl radical

Using this method, the $NH_2$ group can also be replaced by Br, CN, etc.

### 7.4.5.2 Gomberg Free Radical Reaction

This reaction is useful to prepare free radicals by the abstraction of halogen from triarylmethyl halides by reacting with metals (Fig. 7.71).

### 7.4.5.3 Wurtz Reaction

Alkyl halides on reaction with sodium in presence of ether yield hydrocarbons. The reaction takes place via the formation of free radicals (Fig. 7.72).

The reaction is useful for the preparation of hydrocarbons containing an even number of C atoms. An example is given in Fig. 7.73.

The use of aryl halides and alkyl halide in place of alkyl halides gives alkyl-substituted benzenes. The reaction is called the **Wurtz–Fittig reaction**. In this reaction, a small amount of diphenyl (from aryl halide) and a hydrocarbon (from alkyl halide) are obtained as byproducts (Fig. 7.74).

$$RX + Na \longrightarrow R\cdot + NaX$$

Free radical

$$R\cdot + R\cdot \longrightarrow R{-}R$$

Hydrocarbon

**Fig. 7.72** The Wurtz reaction

$$CH_3CH_2Br \xrightarrow[\text{ether}]{Na} CH_3CH_2CH_2CH_3$$

Ethyl bromide

$n$-Butane

**Fig. 7.73** The Wurtz reaction of ethyl bromide

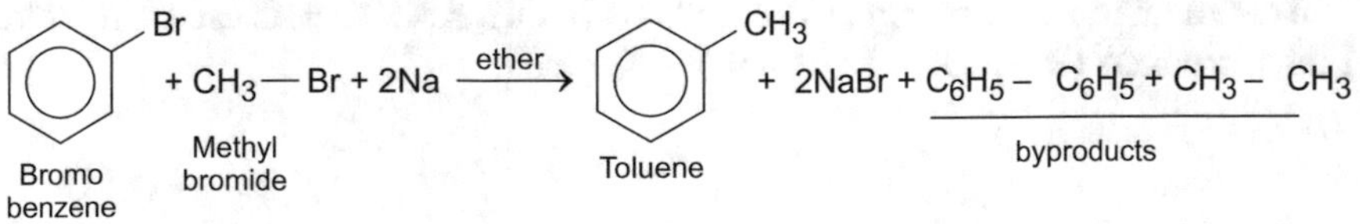

Bromo benzene + CH₃—Br + 2Na $\xrightarrow{\text{ether}}$ Toluene + 2NaBr + C₆H₅ − C₆H₅ + CH₃ − CH₃

Methyl bromide

byproducts

**Fig. 7.74** The Wurtz–Fittig reaction

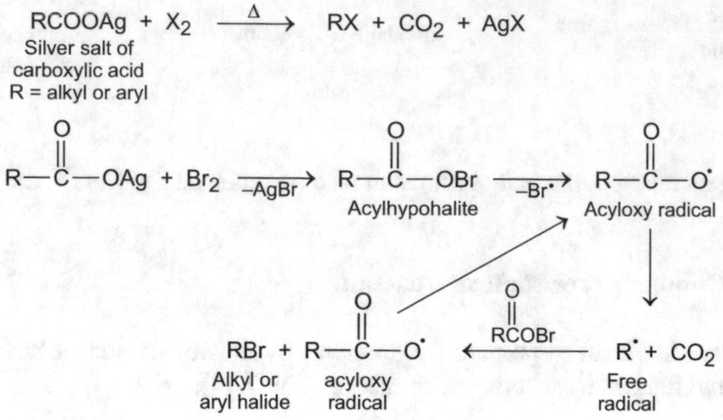

$$RCOOAg + X_2 \xrightarrow{\Delta} RX + CO_2 + AgX$$

Silver salt of
carboxylic acid
R = alkyl or aryl

Fig. 7.75 The Hunsdiecker reaction

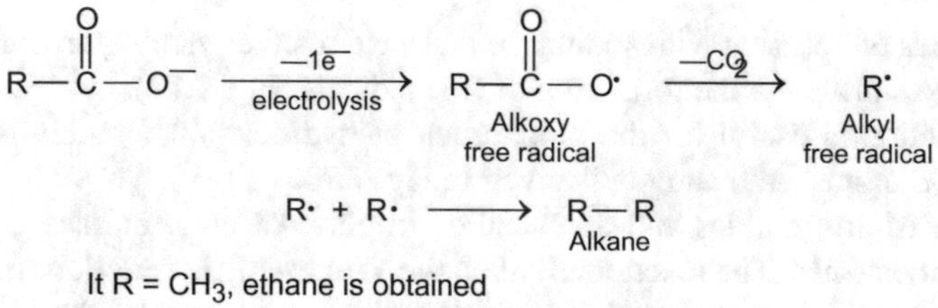

It R = CH$_3$, ethane is obtained

Fig. 7.76 The Kolbs electrolytic reaction

### 7.4.5.4   Hunsdiecker Reaction

This reaction is useful for the conversion of silver salts of carboxylic acids into alkyl or aryl halides by the thermal decomposition of the silver salt in presence of halogen. The reaction involves decarboxylative halogenation and proceeds via a free radical (Figs. 7.75 and 7.76).

### 7.4.5.5   Kolbs Electrolytic Reaction

Electrolysis of sodium or potassium salt of carboxylic acid gives alkanes. The reaction proceeds via the formation of free radicals (Fig. 7.32).

In case a mixture of two carboxylic acids (RCOOH, R$'$ COOH) is used, a mixture of three products (R – R, R$'$ – R$'$, R – R$'$) is obtained.

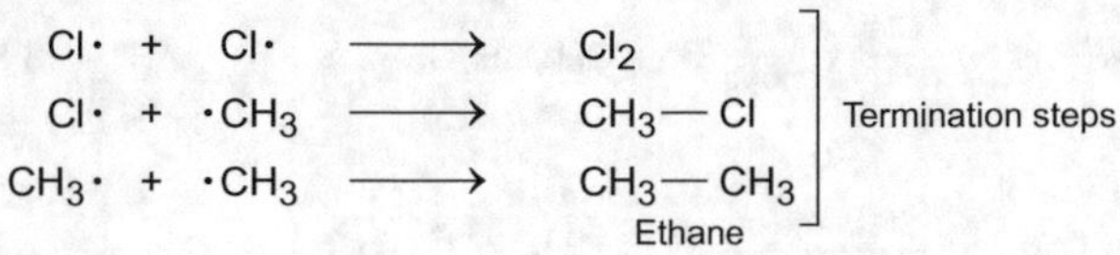

$$Cl\cdot \ + \ Cl\cdot \longrightarrow Cl_2$$
$$Cl\cdot \ + \ \cdot CH_3 \longrightarrow CH_3 - Cl$$
$$CH_3\cdot \ + \ \cdot CH_3 \longrightarrow CH_3 - CH_3$$

Termination steps

Ethane

**Fig. 7.77** Conversion of methane into carbon tetrachloride

### 7.4.5.6  Halogenation

A number of halogenations proceed via the formation of free radicals. Some of the important halogenations are given below.

Conversion of Methane into Carbon Tetrachloride

Methane on treatment with excess chlorine in presence of sunlight gives carbon tetrachloride. The reaction takes place via the formation of free radicals (Fig. 7.77).

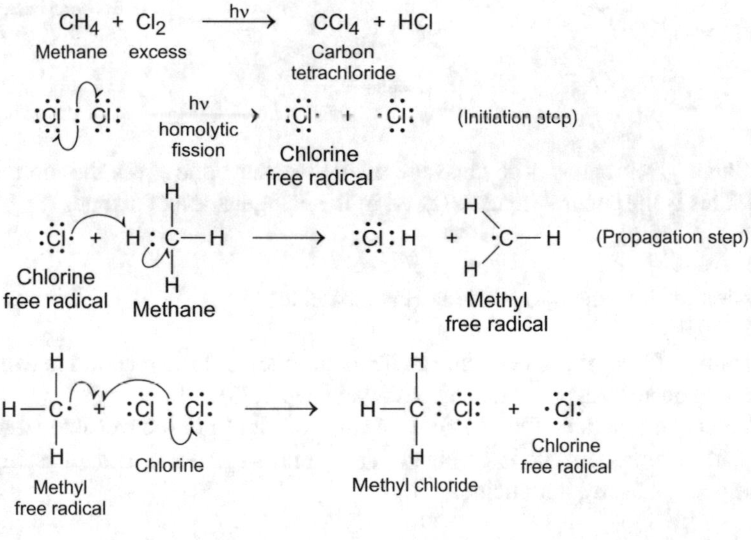

The last two steps are repeated to finally give carbon tetrachloride. Finally, the reaction is terminated by any of the following steps:

In general, the halogenation of alkane can take place either by cleavage of primary C–H bond (to form a 1° radical) or by cleavage of a secondary C–H bond to form a 2° radical. Thus, in the case of propane, the cleavage of a stronger 1° C–H bond to give 1° radical ($CH_3CH_2CH_2{}^\bullet$) requires more energy (98 kcal/mol) than the cleavage of the weaker 2° C–H bond to give a 2° radical [$(CH_3)_2$ $CH^\bullet$] (95 kcal/mol). Thus, the 2° radical is more stable since it requires less energy for its formation as shown

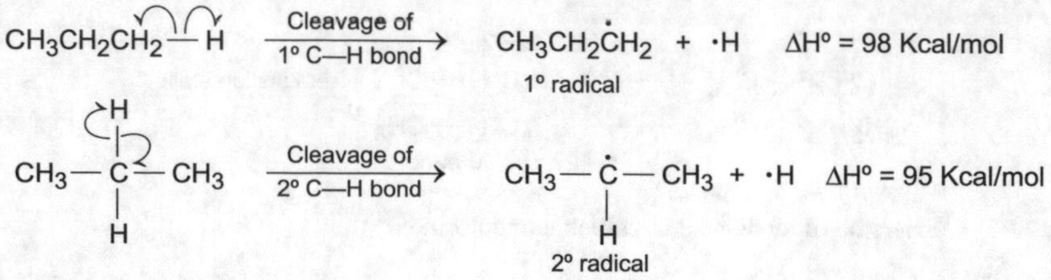

**Fig. 7.78** Halogenation of alkane

**Fig. 7.79** Conversion of benzene into benzene hexachloride

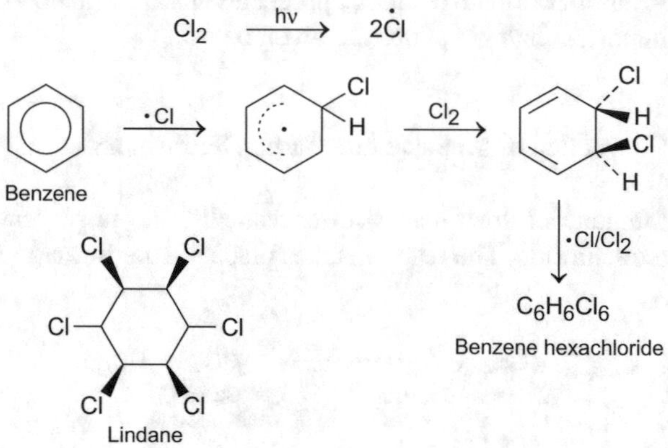

in the figure given below. The cleavage of the weaker bond gives the more stable radical. This is the general trend followed in the halogenation of alkanes (Fig. 7.78).

## Conversion of Benzene into Benzene Hexachloride

Chlorination of benzene gives benzene hexachloride, and the reaction is catalysed by light and occurs via a free radical pathway (Fig. 7.79).

In the above reaction (Fig. 7.79), a mixture of eight non-convertible stereoisomers of hexachlorobenzene is obtained. The gamma isomer is known as Lindane (gammexane), a useful insecticide.

## Conversion of Toluene into Benzyl Chloride

Toluene on reaction with chlorine in presence of sunlight gives benzyl chloride. The chlorine free radical, ˙Cl (obtained from chlorine by sunlight), abstracts a proton to give delocalised benzyl radical, $C_6H_5-CH_2$ (which is resonance stabilised) rather a hexadienyl radical in which the aromatic stabilisation of the starting toluene is lost. The formed benzyl radical on reaction with chlorine gives benzyl chloride (Fig. 7.80).

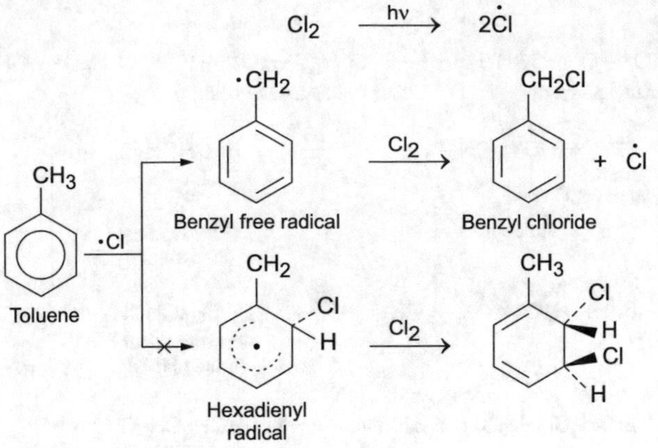

**Fig. 7.80** Conversion of toluene into benzyl chloride

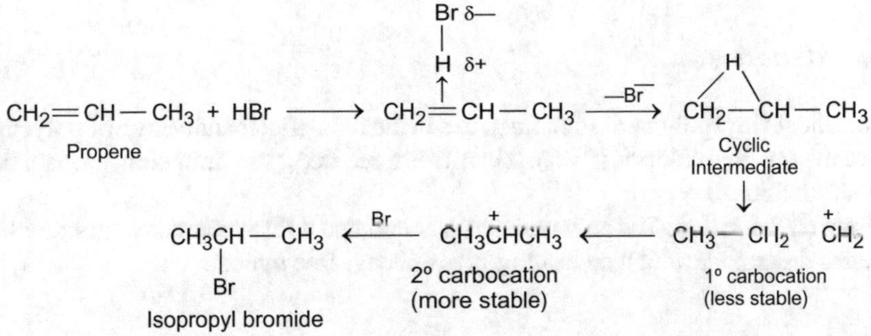

**Fig. 7.81** Addition to H Br to propene gives isopropyl bromide

## Conversion of Propylene into n-Propyl Bromide

Hydrogen bromide is known to add on to propene to give isopropyl bromide via the formation of a carbocation intermediate. The first formed 1° carbocation gets converted into more stable 2° carbocation, which reacts with the Br⁻ to give isopropyl bromide. The addition takes place as per Markownikoff's rule (Fig. 7.81).

In case the above reaction (Fig. 7.81) is carried out in presence of a peroxide, *n*-propyl bromide results. This is known as the **peroxide effect** and the addition takes place in anti-Markownikoff fashion (Fig. 7.82).

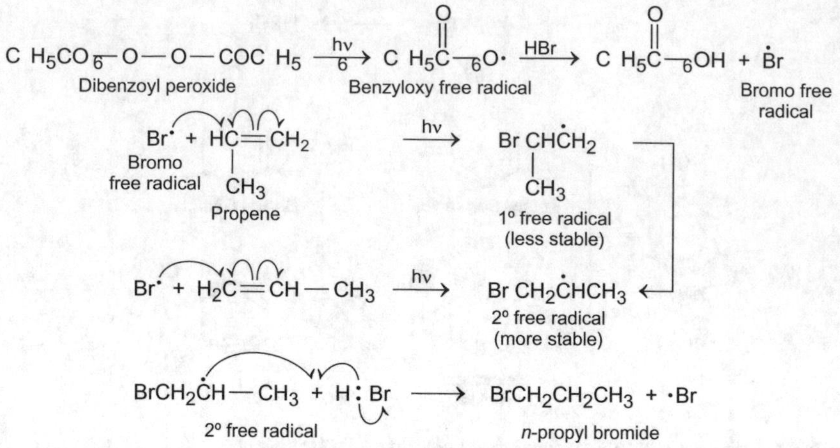

$$C\ H_5CO_6-O-O-COC\ H_5 \xrightarrow{hv} C\ H_5\overset{O}{\overset{\|}{C}}-_6O\cdot \xrightarrow{HBr} C\ H_5\overset{O}{\overset{\|}{C}}-_6OH + \dot{B}r$$

Dibenzoyl peroxide         Benzyloxy free radical               Bromo free radical

$$\underset{\substack{\text{Bromo} \\ \text{free radical}}}{Br^{\cdot}} + \underset{\substack{| \\ CH_3 \\ \text{Propene}}}{HC\!=\!CH_2} \xrightarrow{hv} \underset{\substack{| \\ CH_3 \\ \text{1° free radical} \\ \text{(less stable)}}}{Br\ CH\dot{C}H_2}$$

$$Br^{\cdot} + H_2C\!=\!CH-CH_3 \xrightarrow{hv} \underset{\substack{\text{2° free radical} \\ \text{(more stable)}}}{Br\ CH_2\dot{C}HCH_3} \longleftarrow$$

$$\underset{\text{2° free radical}}{BrCH_2\dot{C}H-CH_3} + H\!:\!Br \longrightarrow \underset{n\text{-propyl bromide}}{BrCH_2CH_2CH_3} + \cdot Br$$

**Fig. 7.82** Conversion of propylene into *n*-propylbromide

## Vinyl Polymerisation

Free radical vinyl polymerisation finds use in the industrial manufacture of polyethylene and polyvinyl chloride (PVC). It involves three steps, viz., initiation, propagation and termination.

**Step (*i*) Initiation**: The polymerisation is initiated with a free radical initiator like dibenzoyl peroxide which on heating gives phenyl free radical.

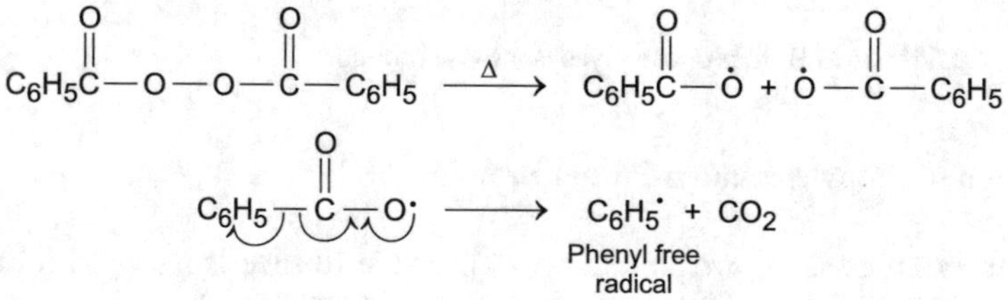

$$C_6H_5\overset{O}{\overset{\|}{C}}-O-O-\overset{O}{\overset{\|}{C}}-C_6H_5 \xrightarrow{\Delta} C_6H_5\overset{O}{\overset{\|}{C}}-\dot{O} + \dot{O}-\overset{O}{\overset{\|}{C}}-C_6H_5$$

$$C_6H_5\overset{O}{\overset{\|}{C}}\dot{O} \longrightarrow \underset{\substack{\text{Phenyl free} \\ \text{radical}}}{C_6H_5^{\cdot}} + CO_2$$

The radical so formed adds on to a molecule of a monomer (ethylene or vinyl chloride).

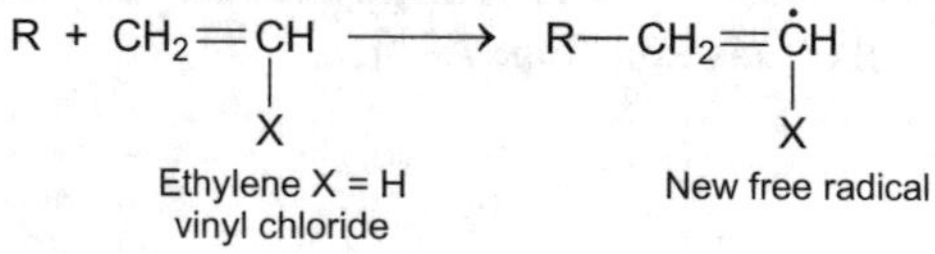

$$R + \underset{\substack{| \\ X \\ \text{Ethylene X = H} \\ \text{vinyl chloride}}}{CH_2\!=\!CH} \longrightarrow \underset{\substack{| \\ X \\ \text{New free radical}}}{R-CH_2\!=\!\dot{C}H}$$

**Step (*ii*) Propagation**: The new free radical generated in step (*i*) adds on to another free radical. Successive addition of monomer to these radicals gives a long chain radical.

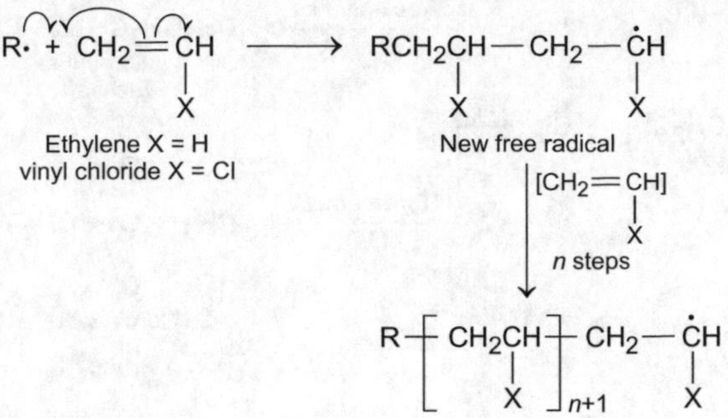

$$R\cdot + CH_2\!=\!\overset{\displaystyle |}{\underset{\displaystyle X}{C}}H \longrightarrow RCH_2\overset{\displaystyle |}{\underset{\displaystyle X}{C}}H\!-\!CH_2\!-\!\overset{\displaystyle |}{\underset{\displaystyle X}{\dot{C}}}H$$

Ethylene X = H
vinyl chloride X = Cl                 New free radical

$$\Big|[CH_2\!=\!\overset{|}{\underset{X}{C}}H]$$

$n$ steps

$$R\!-\!\!\left[CH_2\overset{\displaystyle |}{\underset{\displaystyle X}{C}}H\right]_{n+1}\!\!CH_2\!-\!\overset{\displaystyle |}{\underset{\displaystyle X}{\dot{C}}}H$$

**Step (*iii*) Termination**: The long chain radical formed above in step (*ii*) is terminated by radical coupling or disproportionation as shown below.

*Radical coupling*: Coupling of two radicals formed in step (*ii*) gives long chain polymer.

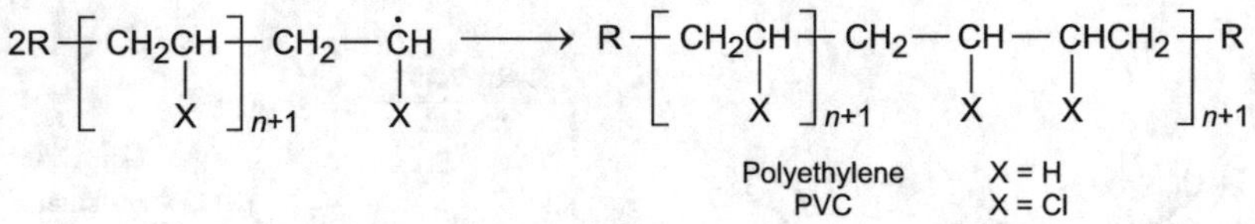

$$2R\!-\!\!\left[CH_2\overset{|}{\underset{X}{C}}H\right]_{n+1}\!\!CH_2\!-\!\overset{|}{\underset{X}{\dot{C}}}H \longrightarrow R\!-\!\!\left[CH_2\overset{|}{\underset{X}{C}}H\right]_{n+1}\!\!CH_2\!-\!\overset{|}{\underset{X}{C}}H\!-\!\overset{|}{\underset{X}{C}}HCH_2\!\!\left]\right._{n+1}\!\!R$$

Polyethylene    X = H
PVC            X = Cl

*Disproportionation*

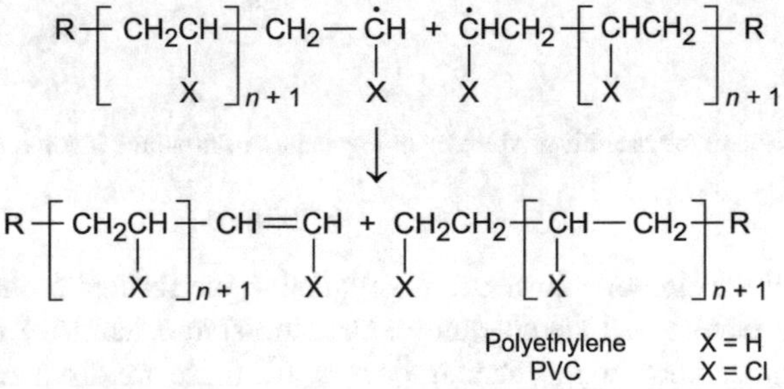

$$R\!-\!\!\left[CH_2\overset{|}{\underset{X}{C}}H\right]_{n+1}\!\!CH_2\!-\!\overset{|}{\underset{X}{\dot{C}}}H + \overset{|}{\underset{X}{\dot{C}}}HCH_2\!\!\left[\overset{|}{\underset{X}{C}}HCH_2\right]_{n+1}\!\!R$$

$$\downarrow$$

$$R\!-\!\!\left[CH_2\overset{|}{\underset{X}{C}}H\right]_{n+1}\!\!CH\!=\!\overset{|}{\underset{X}{C}}H + CH_2\overset{|}{\underset{X}{C}}H_2\!\!\left[\overset{|}{\underset{X}{C}}H\!-\!CH_2\right]_{n+1}\!\!R$$

Polyethylene    X = H
PVC            X = Cl

As seen, depending on the type of termination step, the initiator residue (R) is incorporated into the polymer chain at one or both ends.

Stereochemistry of Halogenation

During halogenation, the stereochemistry of the reaction product depends on whether halogenation occurs at a stereogenic centre of another atom. It is found that on halogenation of an achiral starting material invariably gives either an achiral or a racemic product. In case, the halogenation does not occur at a stereogenic centre, the

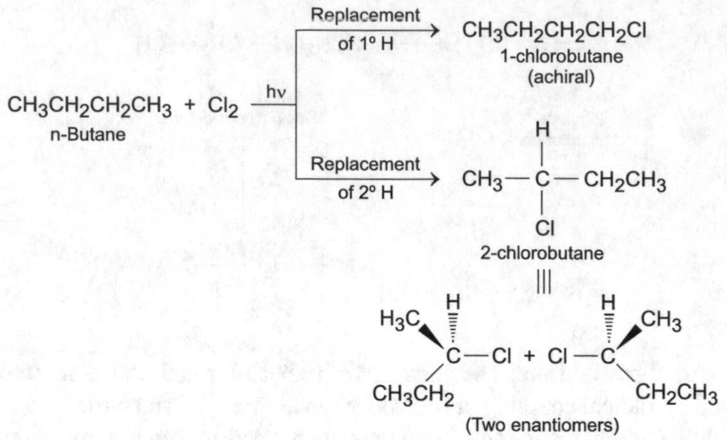

**Fig. 7.83** Chlorination of *n*-butane

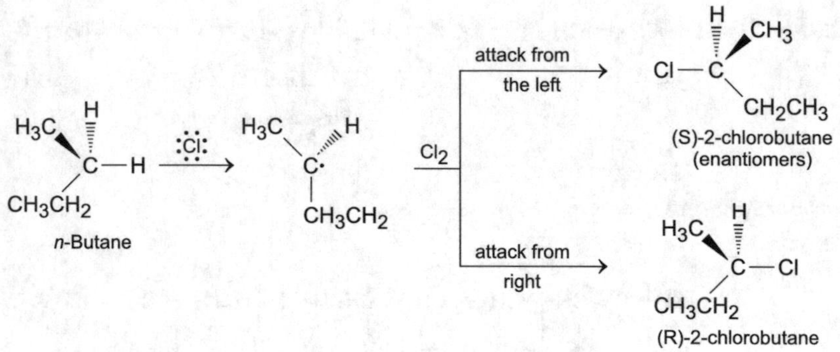

**Fig. 7.84** Formation of the racemic mixture by halogenation of n-butane involving removal of 2°H atom

configuration of the stereogenic centre is retained in the formed product. However, if halogenation occurs at a stereogenic centre, the stereochemistry of the formed product depends on how the mechanism occurs. All these are discussed below: -

(a)  *Halogenation of achiral starting material*: As an example, halogenation of *n*-butane (an achiral starting material) gives two constitutional isomers by the replacement of 1° or 2° hydrogen (Fig. 7.83).

   As seen in Fig. 7.83, the replacement of 1° H gives 1-chlorobutane ($CH_3CH_2CH_2CH_2Cl$), which has no stereogenic centre and is an achiral compound. However, the replacement of 2° H gives 2-chlorobutane ($CH_3CH$ Cl) $CH_2 CH_3$), having a new stereogenic centre and giving a racemic mixture containing an equal amount of the two enantiomers. In the second case, a planar $sp^2$-hybridised radical is generated which reacts with chlorine from either side to form equal amounts of two enantiomers (Fig. 7.84).

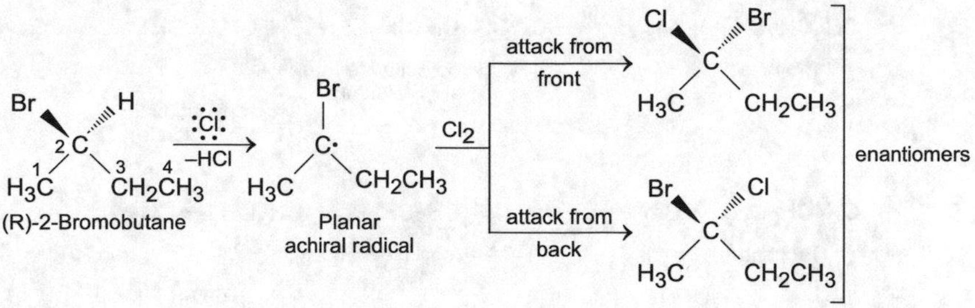

**Fig. 7.85**   Chlorination of (R)-2-bromobutane at C-2 resulting in the formation of a racemic mixture

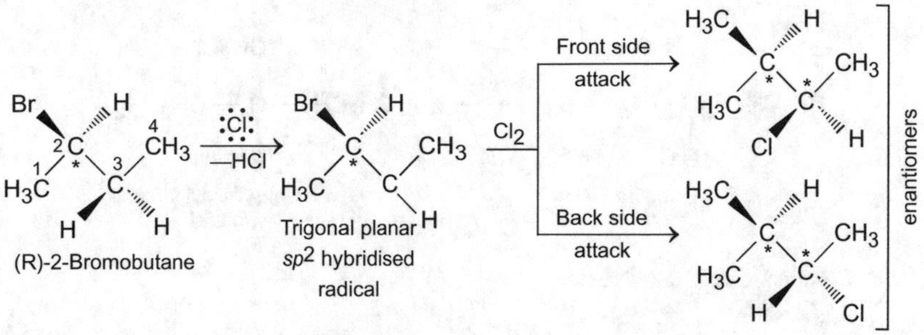

**Fig. 7.86**   Chlorination of (R)-2-bromobutane at C-3 gives a pair of diastereomers

(b)   *Halogenation of a chiral starting material*: As an example in the case of (R)-2-bromobutane chlorination can take place at C-2 and C-3. In case a H from C2 is abstracted, a trigonal planar $sp^2$-hybridised radical (an achiral radical) is formed, which reacts with $Cl_2$ from either side to generate a new stereogenic centre giving an equal amount of two enantiomers (a racemic mixture) (Fig. 7.84).

However, chlorination at C-3 generates a new stereogenic centre. Since no bonds are broken to the stereogenic centre, its configuration is not changed during the reaction. Abstraction of hydrogen from C-3 generates a trigonal planar $sp^2$-hybridised radical, which in turn reacts from either side forming a new stereogenic centre. Thus, the formed products have two stereogenic centres, which have the same configuration at C2 but different configurations at C-3. Thus, the products are diastereomers (Fig. 7.86).

## Chlorination Versus Bromination

Chlorination of alkanes in general is faster than bromination. It is found that chlorination yields a mixture of products and is unselective. However, bromination is generally selective and yields one product in a major amount. As an example, the

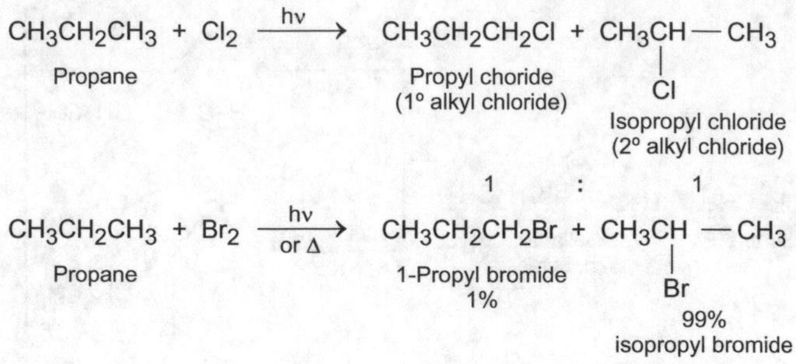

$$CH_3CH_2CH_3 + Cl_2 \xrightarrow{h\nu} CH_3CH_2CH_2Cl + CH_3CH-CH_3$$

Propane                                             Propyl choride        |
                                                   (1° alkyl chloride)    Cl

                                                                          Isopropyl chloride
                                                                          (2° alkyl chloride)

                                                   1          :          1

$$CH_3CH_2CH_3 + Br_2 \xrightarrow[\text{or } \Delta]{h\nu} CH_3CH_2CH_2Br + CH_3CH-CH_3$$

Propane                                             1-Propyl bromide      |
                                                   1%                     Br

                                                                          99%
                                                                   isopropyl bromide

**Fig. 7.87** Comparison of chlorination and bromination of propane

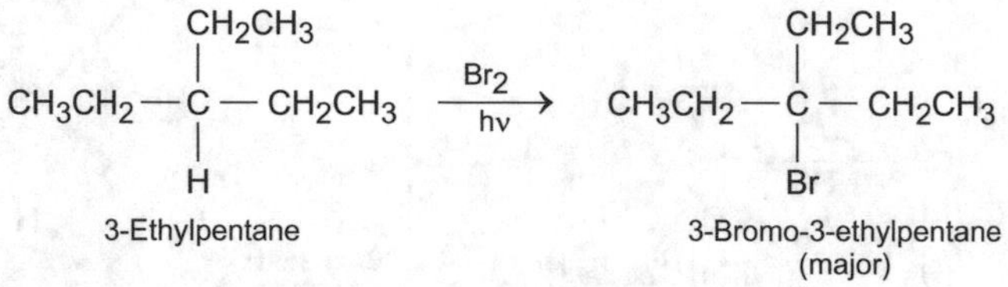

$$
\begin{array}{ccc}
\overset{\displaystyle CH_2CH_3}{|} & & \overset{\displaystyle CH_2CH_3}{|} \\
CH_3CH_2-C-CH_2CH_3 & \xrightarrow[h\nu]{Br_2} & CH_3CH_2-C-CH_2CH_3 \\
| & & | \\
H & & Br
\end{array}
$$

3-Ethylpentane                                        3-Bromo-3-ethylpentane
                                                              (major)

**Fig. 7.88** Bromination of 3-ethylpentane

reaction of propane with chlorine gives a 1:1 mixture of 1° and 2° alkyl chloride. On the other hand, the reaction of propane with bromine reacts much slowly giving 99% 2° bromide (Fig. 7.87).

The above illustration (Fig. 7.87) is in accordance with the principle that less reactive reagent (bromine in this case) is more selective. In bromination, the major product is obtained by the cleavage of the weakest C–H bond.

Another example is the reaction of 3-ethylpentane with bromine which gives 3-bromo-3-ethylpentane as the major product obtained by the cleavage of the weakest 3° C–H bond (Fig. 7.88).

The difference between chlorination and bromination can be easily explained on the basis of the **Hammond Postulate** in order to estimate the relative energy of the transition states of the rate-determining steps in bromination and chlorination. It is known that the rate-determining step is the abstraction of the H atom by the halogen radical. According to the Hammond Postulate,

- The transition state in an endothermic reaction resembles the products. In this case, the more stable product is formed faster.
- The transition state in an exothermic reaction resembles the starting materials. In this case, the relative stability of the products does not greatly affect the relative energy of the transition states and a mixture of products is formed.

Let us first examine the bromination of an alkane like propane. A bromine radical (formed by the action of UV light on bromine) can abstract either a 1° or 2° hydrogen generating either a 1° or a 2° radical. On the basis of calculation of $\Delta H°$ (using bond

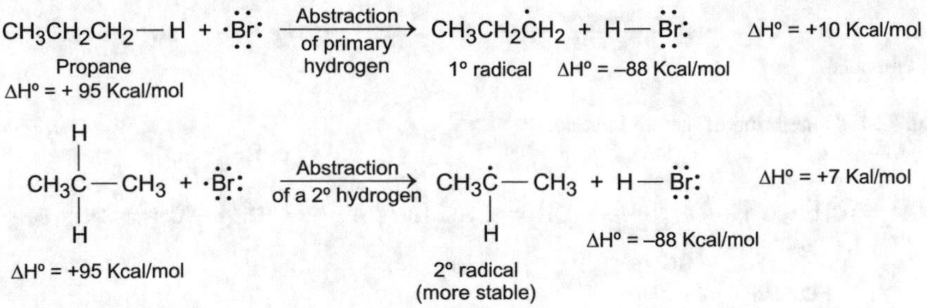

CH$_3$CH$_2$CH$_2$—H + ·Br: $\xrightarrow[\text{of primary}]{\text{Abstraction}}$ CH$_3$CH$_2$ĊH$_2$ + H—Br:    ΔH° = +10 Kcal/mol

Propane                          hydrogen         1° radical   ΔH° = –88 Kcal/mol

ΔH° = + 95 Kcal/mol

$$CH_3\overset{\displaystyle H}{\underset{\displaystyle H}{C}}\!-\!CH_3 + \cdot\ddot{Br}:\ \xrightarrow[\text{of a 2° hydrogen}]{\text{Abstraction}}\ CH_3\overset{\displaystyle\cdot}{\underset{\displaystyle H}{C}}\!-\!CH_3 + H\!-\!\ddot{Br}:$$  ΔH° = +7 Kal/mol

ΔH° = +95 Kcal/mol                              2° radical            ΔH° = –88 Kcal/mol
                                                (more stable)

**Fig. 7.89** In the case of propane, abstraction of 1° or 2° hydrogen both are endothermic reactions

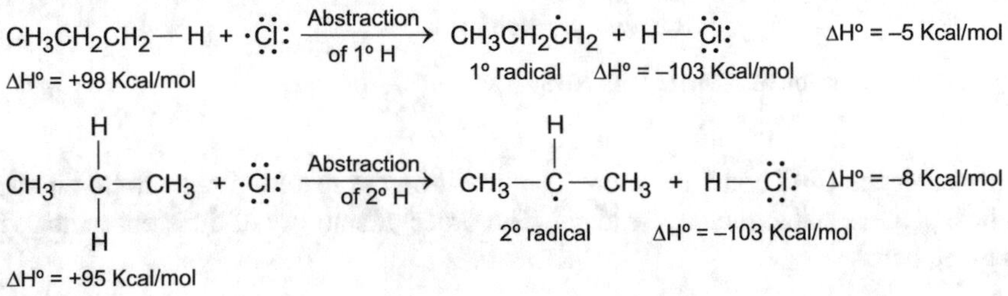

CH$_3$CH$_2$CH$_2$—H + ·Ċl: $\xrightarrow[\text{of 1° H}]{\text{Abstraction}}$ CH$_3$CH$_2$ĊH$_2$ + H—Cl:    ΔH° = –5 Kcal/mol

ΔH° = +98 Kcal/mol                1° radical   ΔH° = –103 Kcal/mol

$$CH_3\!-\!\overset{\displaystyle H}{\underset{\displaystyle H}{C}}\!-\!CH_3 + \cdot\ddot{Cl}:\ \xrightarrow[\text{of 2° H}]{\text{Abstraction}}\ CH_3\!-\!\overset{\displaystyle H}{C}\!-\!CH_3 + H\!-\!\ddot{Cl}:$$  ΔH° = –8 Kcal/mol

ΔH° = +95 Kcal/mol                 2° radical    ΔH° = –103 Kcal/mol

**Fig. 7.90** In the case of propane, abstraction of 1° and 2° hydrogens are both exothermic reactions

dissociation energies), it is found that both reactions (involving abstraction of 1° or 2° hydrogen) are endothermic. However, it takes less energy to form the more stable 2° radical (Fig. 7.89), so the 2° radical is formed faster.

Since a more stable 2° radical is formed faster, the product predominately is CH$_3$CH(Br)CH$_3$.

Let us now examine the chlorination of propane. Like bromine radical, chlorine radical can also abstract a 1° or 2° hydrogen giving either a 1° or 2° radical. On the basis of calculation of ΔH° (using bond dissociation energies), it is found that both reactions (involving abstraction of 1° or 2° hydrogen) are exothermic (Fig. 7.90).

As seen in Fig. 7.90, chlorination has an exothermic rate-determining step, and the transition state to form both radicals (1° or 2 radicals) resembles the same starting material, CH$_3$CH$_2$CH$_3$. Thus, the relative stability of the two radicals is less important and both radicals are formed and so the product obtained is a mixture of propyl chloride (CH$_3$CH$_2$CH$_2$Cl) and isopropyl chloride (CH$_3$CH(Cl)CH$_3$).

Radical Halogenation at Allylic Carbon

The carbon atom adjacent to a double bond is an allylic carbon. Homolysis of the allylic C–H bond of propene generates allyl radical (Fig. 7.91).

As seen in Fig. 7.91, the bond dissociation energy of this process (87 kcal/mol) is less than the bond dissociation for a 3° C–H bond (91 kcal/mol). It is known that

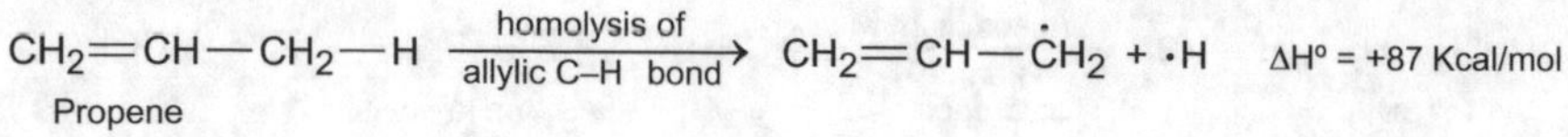

$$CH_2\!=\!CH\!-\!CH_2\!-\!H \xrightarrow[\text{allylic C–H bond}]{\text{homolysis of}} CH_2\!=\!CH\!-\!\overset{\cdot}{C}H_2 \; + \; \cdot H \qquad \Delta H^\circ = +87 \text{ Kcal/mol}$$

Propene

**Fig. 7.91**  Generation of the allyl radical

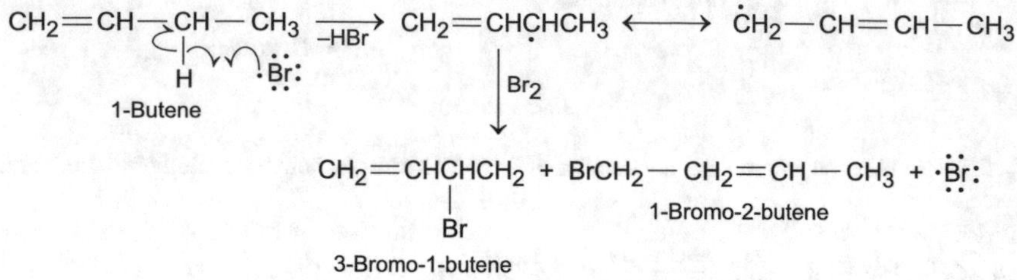

$$CH_2\!=\!CH\!-\!CH\!-\!CH_3 \xrightarrow[-HBr]{} CH_2\!=\!CH\overset{\cdot}{C}HCH_3 \longleftrightarrow \overset{\cdot}{C}H_2\!-\!CH\!=\!CH\!-\!CH_3$$

H   $\cdot\ddot{Br}\!:$

1-Butene

$\Big\downarrow Br_2$

$$CH_2\!=\!CHCHCH_2 \; + \; BrCH_2\!-\!CH_2\!=\!CH\!-\!CH_3 \; + \; \cdot\ddot{Br}\!:$$

|
Br                                        1-Bromo-2-butene

3-Bromo-1-butene

**Fig. 7.92**  Bromination of 1-butane with NBS

the weaker the C–H bond, the more stable will be the resulting radical. So an allyl radical is more stable than a 3° radical. The order of stability of different radicals is as given below:

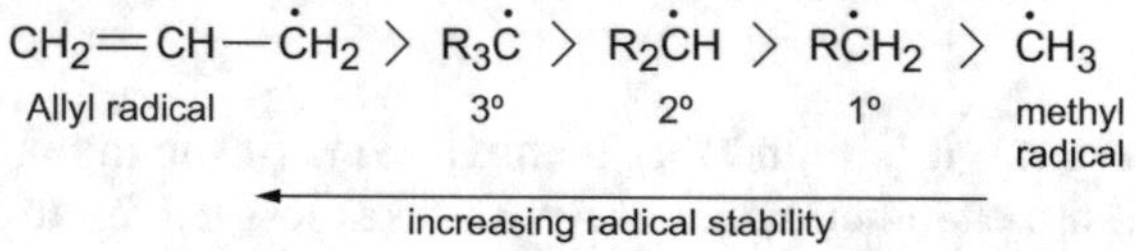

$$CH_2\!=\!CH\!-\!\overset{\cdot}{C}H_2 \; > \; R_3\overset{\cdot}{C} \; > \; R_2\overset{\cdot}{C}H \; > \; R\overset{\cdot}{C}H_2 \; > \; \overset{\cdot}{C}H_3$$

Allyl radical              3°          2°          1°        methyl
radical

$\longleftarrow$ increasing radical stability

The stability of the ally radical is because two resonance structures can be drawn for it

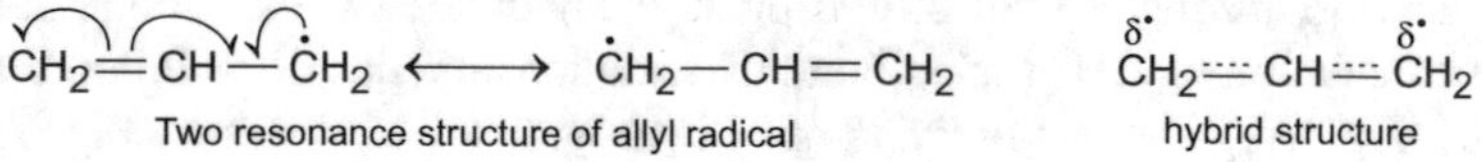

$$\overset{\cdot}{C}H_2\!=\!CH\!-\!\overset{\cdot}{C}H_2 \longleftrightarrow \overset{\cdot}{C}H_2\!-\!CH\!=\!CH_2 \qquad \overset{\delta^{\cdot}}{C}H_2\!\cdots\!CH\!\cdots\!\overset{\delta^{\cdot}}{C}H_2$$

Two resonance structure of allyl radical                    hybrid structure

As seen, the resonance structures differ in the location of the π bonds and non-bonded electrons.

Allyl radical is represented by the hybrid structures (as shown above).

Since allyl radical has two resonance structures, there will be a mixture of two products during halogenation. As an example bromination of 1-butene with N-bromosuccinimide in presence of light or ROOR a mixture of 3-bromo-1-butene and 1-bromo-3-butene is obtained (Fig. 7.92).

The generation of bromine radical from NBS is by the homolysis of the weak N–Br bond in NBS using light energy. This results in the generation of radicals.

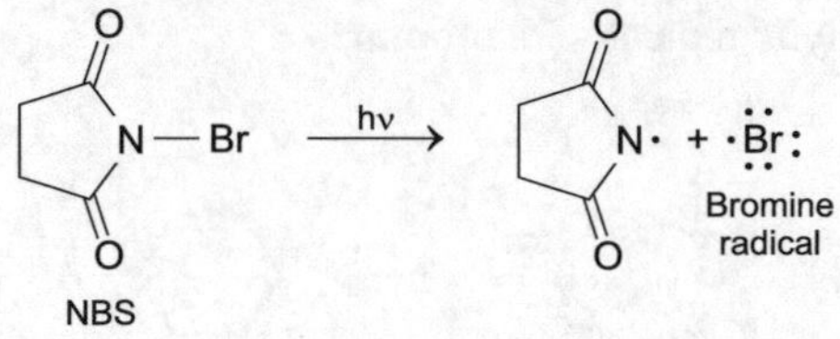

NBS

## Solved Problem 1

What product is obtained by bromination of cyclohexene with NBS in presence of *hv* or peroxide?

The product obtained is 3-bromocyclohexene.

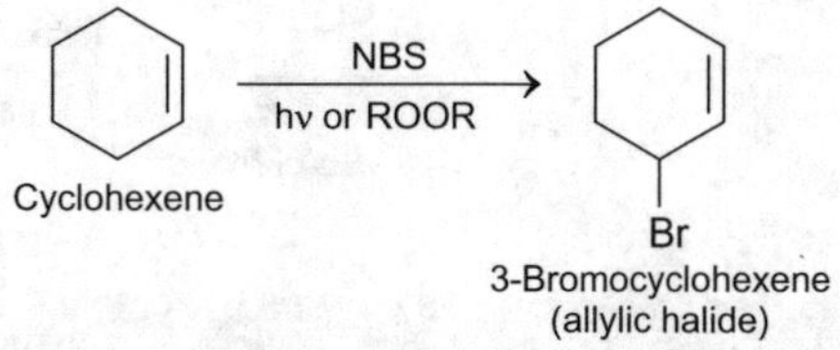

Cyclohexene

3-Bromocyclohexene
(allylic halide)

Since allylic C–H bonds are weaker than $sp^3$-hybridised C–H bonds, the allylic carbon is selectively brominated with NBS in presence of light or peroxide.

The mechanism of allylic bromination is given below:

(i)  Cleavage of N–Br bond of NBS.

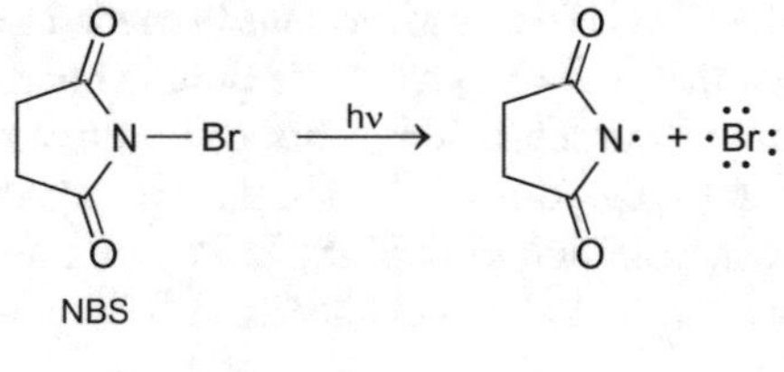

NBS

(ii)  Generation of allylic radical from cyclohexene.

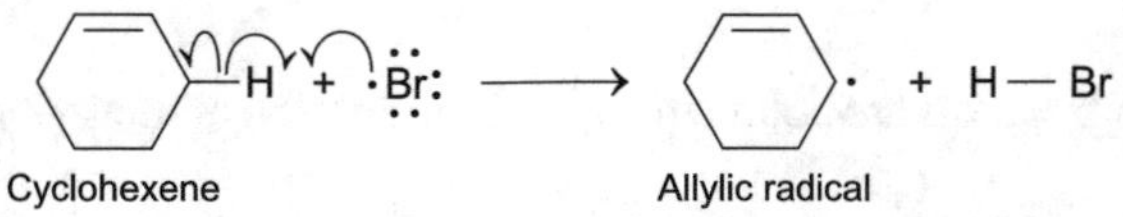

Cyclohexene                              Allylic radical

(iii)   Reaction of allylic radical with bromine.

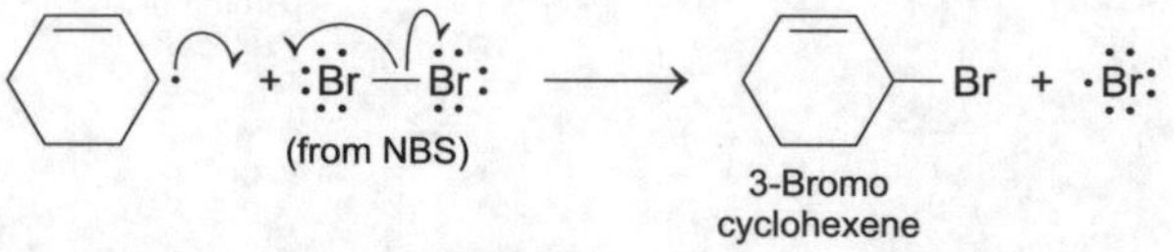

(from NBS)

3-Bromo
cyclohexene

NBS, besides acting as a source of bromine radical [to initiate the reaction, step (*i*)], also generates a low concentration of $Br_2$ which is needed in step (*iii*). The H Br formed in step (*ii*) on reaction with NBS forms $Br_2$, which is used in step (*iii*). Thus, steps (*ii*) and (*iii*) repeatedly occur without the need of step (*i*).

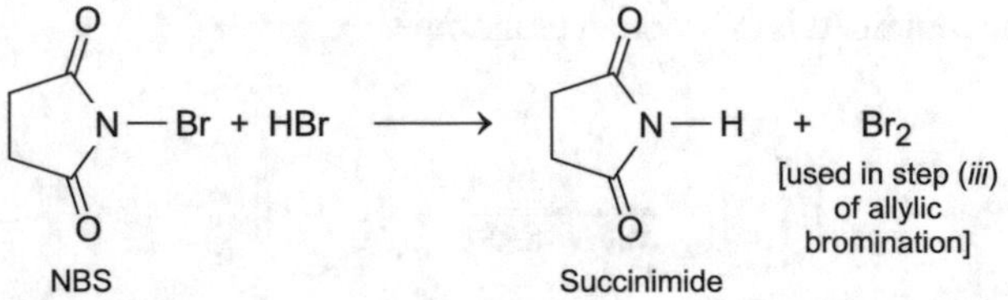

[used in step (*iii*) of allylic bromination]

NBS                          Succinimide

## Solved Problem 2

The reaction of cyclohexene with NBS in presence of light or ROOR gives 3-bromocyclohexene (an allylic bromide) and not 1,2-dibromocyclohexane (an vicinal dibromide). Explain.

For allylic bromination, only a low concentration of $Br_2$ (from NBS) favours allylic substitution over addition. This is because bromine is needed for only one step of the mechanism. It is known that when $Br_2$ adds to a double bond, a low concentration of $Br_2$ will first form a low concentration of bridged brominium ion, which has to react with more bromine (in the form of $Br^-$) in the second step to form a dibromo compound. Since the concentration of both the brominium ion and $Br^-$ is low, the rate of addition is very slow. So the product obtained is only 3-bromocylcohaxene.

## Solved Problem 3

What product or products are obtained by the bromination of methylene cyclohexane with NBS/hv?

A mixture of two products as given below is obtained.

The mechanism of the reaction is

---

## Key Concepts

- **Aldol Condensation**: Reaction of aldehydes containing an $\alpha$-hydrogen atom in presence of dilute alkali to give $\beta$-hydroxyaldehydes called aldols.
- **Beckmann Rearrangement**: Conversion of ketoximes into substituted acid amides by treatment with acidic reagents. Oximes of cyclic ketones give ring enlargements.
- **Baeyer–Villiger Oxidation**: Oxidation of aromatic open chain ketones to esters and cyclic ketones to lactones by reaction with peracids.
- **Bridged Free Radicals**: Radicals having rigid structures with fixed dihedral angles and bond angles.
- **Carbocation**: A carbon atom which is trivalent and has an even number of electrons and has a positive charge. It is planar, with a bond angle of 120°, $sp^2$-hybridised and a reaction intermediates of transitory existence.
- **Carbanion**: A trivalent carbon, negatively charged, $sp^3$-hybridised and has eight electrons in the outer shell of negatively charged carbon. Its configuration is pyramidal.
- **Claisen Condensation**:Base catalysed condensation of an ester containing a $\alpha$-hydrogen with the same or different ester to give $\beta$-ketoester.
- **Claisen Reaction**: The reaction of a carbonyl compound (aldehyde or ketone) without $\alpha$ hydrogen with an ester having active hydrogen in presence of base gives $\alpha, \beta$-unsaturated ester.
- **Cyclooctateraenyl Dianion**: Obtained from cyclooctatetraene by treatment with potassium.

- **Cycloheptyl Carbocation**: A stable carbocation obtained from cycloheptatriene by reaction with triphenylmethyl carbocation.
- **Dieckmann Condensation**: An intramolecular base catalysed reaction of esters of dicarboxylic acids to give β-keto esters.
- **Fenton's Reagent**: An oxidising agent which is a mixture of $H_2O_2$ and $Fe^{2+}$; the effective oxidising agent is the hydroxy radical (HO˙)
- **Free Radicals**: Species possessing an unpaired electron and are produced by homolytic bond fission of a covalent bond. Simple free radicals are $sp^2$-hybridised and are planar.
- **Gomberg Free Radical Reaction**: A reaction useful for the preparation of free radicals by the abstraction of halogen from triarylmethyl halides by reaction with metals.
- **Hammond Postulates**: In endothermic reactions, the transition state is closer in energy to the products, and in exothermic reactions, the transition state is closer in energy to the reactants.
- **Heteroradicals**: Radicals having atoms other than carbons.
- **Hofmann Rearrangement**: Rearrangement of primary amide into primary amine by reaction with sodium hypohalite (NaOH + halogen) via intermediate isocyanate.
- **Hunsdieker Reaction**: Synthesis of organic halides by the thermal decarboxylation of the silver salt of carboxylic acids in presence of halogen.
- **Hydroperoxide Rearrangement**: Acid catalyst rearrangement of hydroperoxides to give phenol. As an example, cumene hydroperoxide gives phenol.
- **Kolbs Electrolytic Reaction**: Formation of hydrocarbons by the electrolysis of alkali salts of carboxylic acid; the reaction involves decarboxylative dimerisation.
- **Michael Addition**: The reaction of an α, β-unsaturated carbonyl compound with a compound containing active methylene group in presence of base gives an adduct.
- **Molecular Rearrangements**: Rearrangements taking place without a change in carbon skeleton or with change in carbon skeleton.
- **Neopentyl Rearrangement**: Solvolysis of neopentyl bromide gives the rearranged product, 2-methylbutan-2-ol and not the expected 2,2-dimethylpropanol (neopentyl alcohol.)
- **Nitrogen Radical**: A radical containing N, e.g., $Ph\dot{N}$ obtained by oxidation of $Ph_2NH$.
- **Perkin Reaction**: Formation of α, β unsaturated carboxylic acid by heating aromatic aldehydes and acid anhydride in presence of sodium or potassium salt of the acid.
- **Photolysis**: Subjecting a compound to UV light (wavelength 320 nm, 3200 Å). It can also cleave strong bonds that do not break readily. It is a decomposition reaction brought about by light. The reaction generally involves the production of free radicals by the breaking of a chemical bond.
- **Redox Reaction**: A reaction in which electrons are transferred from one reactant to another. The reactant which accepts electrons from the other reactant is reduced

and is called the oxidising agent. The reactant which gives out electrons to the other reactant is oxidised and is called a reducing agent.

- **Sandmeyer Reaction**: Replacement of diazonium group in aromatic compounds by halo or cyano groups in presence of cuprous salts.
- **Themolysis**: A reaction involving the formation of free radicals by heating at a suitable temperature.
- **Triphenylmethyl Radical**: Obtained by oxidation of triphenylmethyl anion.
- **Trivalent Carbon**: Carbon in which case one of the orbitals does not form a bond.
- **Thiyl Radical**: A radical containing S, e.g., PhS· obtained by heating diphenyl sulphide.
- **Wurtz–Fittig Reaction**: Formation of alkylated aromatic hydrocarbons by the treatment of a mixture of alkyl halides and an ary halide with sodium.
- **Wurtz Reaction**: Formation of symmetrical hydrocarbons by the treatment of alkyl halides with sodium.
- **2,4,6-tert butyl Phenoxy Radical**: A stable oxygen-containing radical obtained by the $K_3Fe(CN)_6$ oxidation of 2,4,6-tert butyl phenoxide.

## Problems

1. What are carbocations? Discuss their structure and stability.
2. What products are obtained in the following:

(a) $CH_3CH_2\overset{+}{CH_2}$ + $Br^-$ $\longrightarrow$

(b) $H_3C-\overset{\overset{\displaystyle CH_3}{|}}{\underset{\underset{\displaystyle CH_3}{|}}{\overset{+}{C}}}$ + $CH_3-\overset{\overset{\displaystyle CH_3}{|}}{C}=CH_2$

(c) $CH_3-\overset{\overset{\displaystyle CH_3}{|}}{\underset{\underset{\displaystyle CH_3}{|}}{C}}-\overset{+}{C}HCH_3$ $\xrightarrow{\text{rearrangement}}$

3. Why is triphenylmethyl radical stable? Explain.
4. How is cycloheptyl carbocation, tropylium cation, obtained? Discuss their stability.

5.  Write a note on molecular rearrangements involving carbocations. Discuss their stereochemistry.

6.  What products are obtained in the following:

$$
\text{(a)} \quad H_3C - \underset{\underset{CH_3}{|}}{\overset{\overset{CH_3}{|}}{C}} - CH_2Br \xrightarrow{\text{Solvolysis}}
$$

$$
\text{(b)} \quad Ph_2C - \underset{\underset{OH}{|}}{\overset{\overset{Me}{|}}{C}} - Me \;+\; Ph_2C - \underset{\underset{OH}{|}}{\overset{\overset{Et}{|}}{C}} - Et \xrightarrow{H^+}
$$

$$
\text{(c)} \quad H_3C - \underset{\underset{H}{|}}{\overset{\overset{H}{|}}{C}} - \overset{\overset{H}{|}}{\underset{+}{C}} - H \xrightarrow[\text{D}_2\text{O}]{\text{1,2-shift}}
$$

$$
\text{(d)} \quad H_3C - \underset{R^*\text{ is chiral group}}{\overset{\overset{R^*}{|}}{CH}} - CH_2 - NH_2 \xrightarrow[\text{(2) H}_2\text{O}]{\text{(1) Diazotisation}}
$$

7.  What are carbanions? Discuss their structure and stability. On what factors does the stability of carbanions depend?

8.  Arrange the following carbanions in order of their stabilities.
    vinyl, phenyl, ethyl and cyclopropyl.

9.  What products are obtained in the following:

$$
\text{(a)} \quad Ph_3C - CH_2Cl \xrightarrow[\text{(2)}\,H^+]{\text{(1)}\,Na}
$$

$$
\text{(b)} \quad Ph_3 \,\overset{-\;+}{CNa} \xrightarrow{O_2}
$$

$$
\text{(c)} \quad (CH_3CO)_2 \bar{C}H \xrightarrow{I_2}
$$

$$
\text{(d)} \quad CH_3COO\,Et \xrightarrow[\substack{\text{2) C}_6\text{H}_5\text{CHO} \\ \text{3) EtOH}}]{\text{1) NaOEt}}
$$

10. What are free radicals? Discuss their structure and stability.

11. Discuss the comparative stability of allyl free radical, benzylic free radical and triphenylmethyl free radical.

12.    What are heteroradicals? Give examples.
13.    What products are obtained in the following:

$(a)$  $R\!-\!O\!-\!Cl \xrightarrow{h\nu}$

$(b)$  $R\!-\!O\!-\!No \xrightarrow{h\nu}$

$(c)$  $\overset{\bullet}{C}H_3 + CH_2 = CH_2 \longrightarrow$

$(d)$  $Ph_2NH \xrightarrow{MnO_4^-}$

$(e)$  $PhS\!-\!S\!-\!Ph \xrightarrow{\Delta}$

$(f)$  $Ph_2MeCCH_2\!-\!C\!\!\underset{\underset{O}{\parallel}}{\overset{\overset{H}{\mid}}{}} \xrightarrow{MeCO^{\bullet}}$

$(g)$  $(C_6H_5)_3CCl \xrightarrow{Zn}$

$(h)$  $RCOOAg + x_2 \xrightarrow{\Delta}$

# Chapter 8
# Stereochemistry of Fused, Bridged and Caged Rings and Related Compounds

## 8.1 Introduction

Organic compounds containing two (or more) rings may belong to one of the following classes:

- Organic compounds in which the two rings have no common atoms. An example is biphenyl, $C_6H_5$–$C_6H_5$. Such compounds have no stereochemical features except when the two rings are substituted in ortho positions by bulky groups. For details, see Chap. 6.
- Organic compounds in which the two rings have one atom in common. Such compounds are called **spiranes** and present a special type of isomerism. For details, see Sect. 6.4.
- Organic compounds in which the two adjacent rings have two or more atoms in common. In case the two rings have two adjacent atoms in common, the compound is a fused ring compound. However, in bridged compounds, the two rings are linked via non-adjacent atoms. In the former type, i.e., fused ring compounds, the compound can be bicyclic (containing two rings) or polycyclic (containing more than two rings).

Thus, all organic compounds are represented in Chart 8.1.
The present chapter deals with fused ring compounds and bridged compounds.

## 8.2 Fused Rings

As already stated (Chart 8.1), the fused ring compounds may contain two rings (bicyclic compounds) or more than two rings (polycyclic compounds).

© The Author(s), under exclusive license to Springer Nature Switzerland AG 2022
V. K. Ahluwalia, *Stereochemistry of Organic Compounds*,
https://doi.org/10.1007/978-3-030-84961-0_8

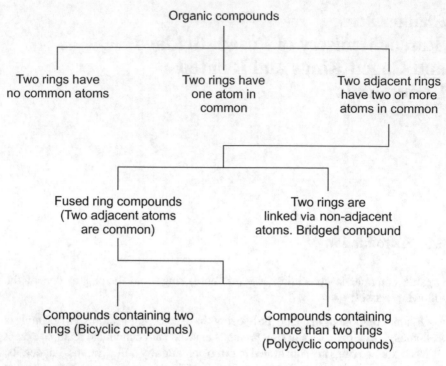

**Chart 8.1**  Different types of organic compounds

## 8.2.1  Fused Ring Compounds Containing Two Rings (Bicyclic Compounds)

The smallest fused ring compound containing two rings isbicyclo[1.1.0] butane (Fig. 8.1a). It is a strained system having a strain of 66.5 kcal mol$^{-1}$ (278 kJ mol$^{-1}$). The strain in this compound (Fig. 8.1a) is more than the sum of the two cyclo-propane moieties (for calculation, refer to Table 2.1 in Chap. 2) by about 48 kJ mol$^{-1}$ (11.5 kcal mol$^{-1}$). However, the strain [240 kJ mol$^{-1}$ (57.3 kcal mol$^{-1}$)] in the next higher fused bicyclo [2.1.0] pentane (Fig. 8.1b) is more than the sum of the strain of cyclopropane and cyclobutane by about 16.7 kJ mol$^{-1}$ (4 kcal mol$^{-1}$) [115 + 109 − 240 = −16].

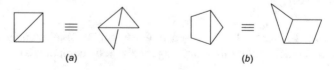

**Fig. 8.1**  Bicyclo [1,1,0] butane (**a**) and bicyclo [2,1,0] pentane (**b**)

The strain in higher bicyclo [*n*,1,0] alkanes is very close to that of the sum of the strains in the two fused rings.

The two hydrocarbons, bicyclo[1,1,0] butane (1a) and bicyclo[2,1,0] pentane (1b) (Fig. 8.1) are *cis* fused. These belong to 3–3 and 3–4 systems, respectively. The 3–5 and 3–6 systems are found in a number of naturally occurring terpenes derived from thujane and carane, respectively (Fig. 8.2).

Among the *four-membered ring system,* the first member (4–3 system), bicyclo[2,1,0] pentane (Fig. 8.1b) has already been mentioned. The next member bicyclo [2, 2, 0] hexane (4–4 system) has been identified (S. Cramer and R. Srinivasan, Tetrahedron lett. 1960, 24) as one of the products in the pyrolysis of bicyclo [3, 2, 0] heptanone (Fig. 8.3).

Bicyclo [3,2,0] heptane (4–5 system), the next higher homologue, is obtained (A.T. Blomquist and J.Kwatek, J Am. Chem., Soc., 1951, **73,** 2098) by the 1,2-addition of ketene to cyclopentadiene (Fig. 8.4).

In the next higher homologues, bicyclo [4,2,0] octane (4–6 system) and bicyclo [5,2,0] nonane (4–7 system) have also been obtained. The latter compound has been obtained in both *cis* and *trans* forms (J. W. Barrett and R. P. Linstead, J. Chem, Soc., 1936, 611) (Fig. 8.5).

The naturally occurring sesquiterpene, caryophyllene, contains the 4–9 system (Fig. 8.6). In this sesquiterpene, there is *trans*-substituted double bond.

Thujane
(3-5 system)

Carane
(3-6 system)

**Fig. 8.2** Bicyclo compounds belonging to 3–5 system (Thujane) and 3–6 system (carane)

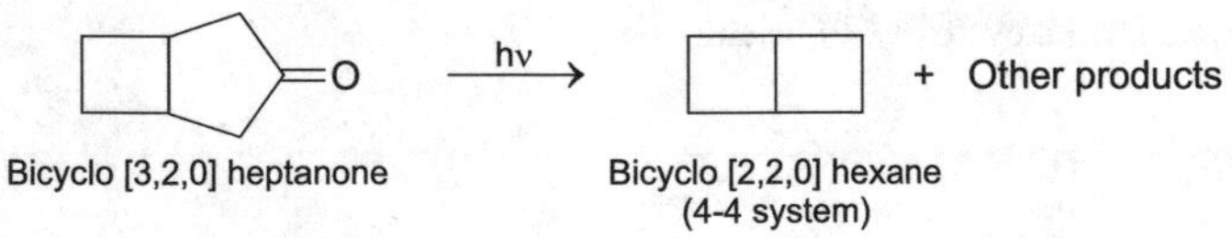

Bicyclo [3,2,0] heptanone

$\xrightarrow{h\nu}$

Bicyclo [2,2,0] hexane
(4-4 system)

+   Other products

**Fig. 8.3** Pyrolysis of bicyclo [3,2,0] heptanone-3

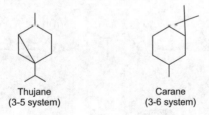

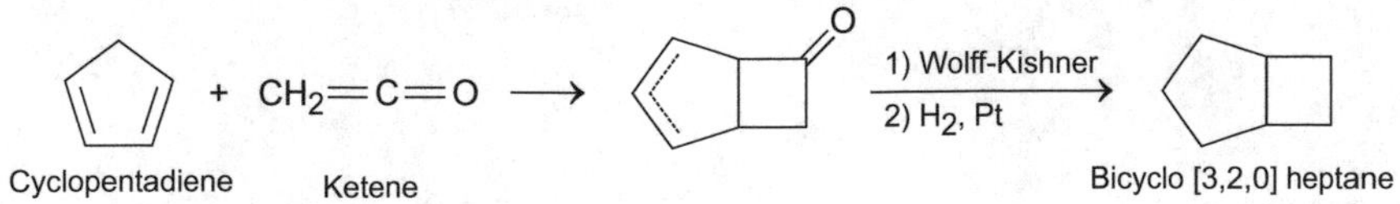

Cyclopentadiene        Ketene

1) Wolff-Kishner
2) $H_2$, Pt

Bicyclo [3,2,0] heptane

**Fig. 8.4** Synthesis of bicyclo [3,2,0] heptane

**Fig. 8.5**  Bicyclo [4,2,0]
octane (**a**) and bicyclo
[5,2,0] nonane (**b**)

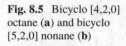

(a)                                (b)

**Fig. 8.6**  Caryophyllene

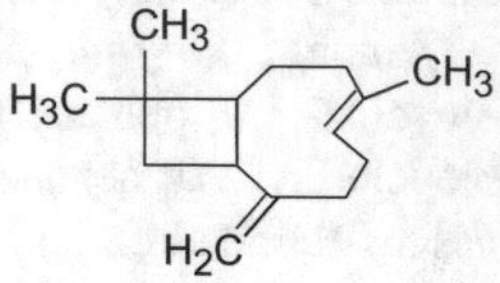

Among the *fused five-membered ring system,* the 5–3 system (thujane, Fig. 8.2)
and 5–4 system (bicyclo [3,2,0] heptane) (Fig. 8.4) have already been described. The
next higher homologue, bicyclo [3,3,0] octane (a 5–5 system), is known in both *cis*
and *trans* forms (J. W. Barrett and R. P. Linstead, J. Chem. Soc., 1936, 611), and the
*trans* isomer is difficult to make (Fig. 8.7).

A derivative of bicyclo [3,3,0] octane (a 5–5 system) is the well-known biotin
(Vitamin H). Its structure is as given in Fig. 8.8. On the basis of chemical work,
biotin was found to be all *cis* isomer, i.e., the ureido ring and the side chain at C-2
are in *cis* position. This has been confirmed on the basis of X-ray studies.

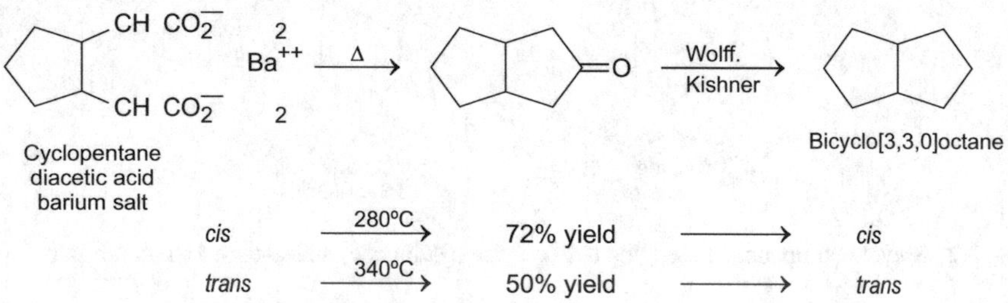

Cyclopentane
diacetic acid
barium salt

                                                                      Bicyclo[3,3,0]octane

|  | | | |
|---|---|---|---|
| *cis* | $\xrightarrow{280°C}$ | 72% yield | $\longrightarrow$ | *cis* |
| *trans* | $\xrightarrow{340°C}$ | 50% yield | $\longrightarrow$ | *trans* |

**Fig. 8.7**  *cis* and *trans* bicyclo [3,3,0] octane

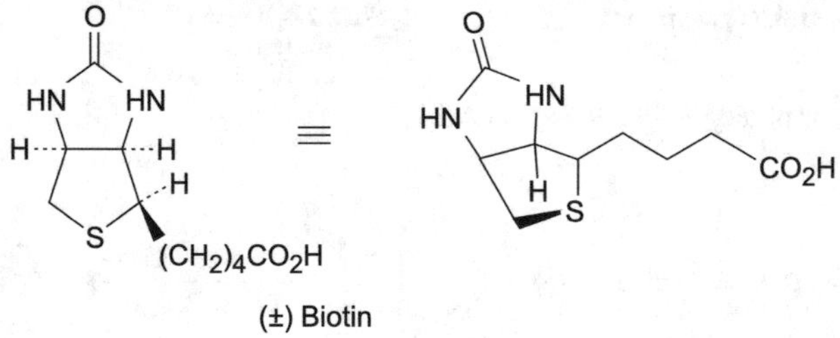

(±) Biotin

**Fig. 8.8**  (±) Biotin

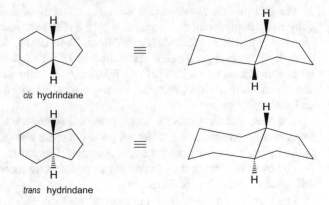

**Fig. 8.9** *cis* and *trans* hydrindane

*cis* hydrindane

*trans* hydrindane

Among the *fused six-membered ring system,* the 6–5 system, the hydrindanes are well known. It exists in both *trans* and *cis* forms (Fig. 8.9).

The well-known compound decalin (which exists in *cis* and *trans* forms) belongs to the 6–6 system [the 6–3, 6–4 and 6–5 systems have already been discussed (see Figs. 8.2 and 8.5)]. The *trans*-decalin (Fig. 8.10) has a centre of symmetry passing midway between C-9 and C-10 and so is optically inactive. It has also a two-fold axis of symmetry passing between C-2, C-3, C-9 and C-10, and C-7 and C-6 and so its symmetry number is 2. The *trans*-decalin is a rigid molecule since the ring fusion can only be through e.e. bonds.

On the other hand, *cis*-decalin (Fig. 8.10) is dissymmetric and unlike *trans*-decalin (which is a rigid molecule), the *cis*-decalin has two interconvertible chair conformations (as in the case of *cis*-1,2-dimethylcyclohexane). By 'flipping', the chair conformation gets converted into its mirror image. Therefore, *cis*-decalin is a non-resolvable dl pair. Also the *cis*-decalin has a two-fold axis of symmetry passing through the C-9–C-10. bond at right angles the bond in a plane which bisects the dihedral angle between the C-9 and C-10 hydrogens. Thus, its symmetry number is also 2.

**Fig. 8.10** *cis*- and *trans*-Decalins

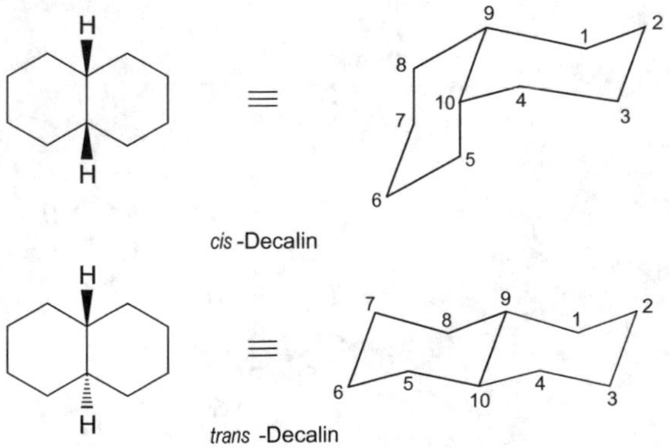

*cis*-Decalin

*trans*-Decalin

The fact that the *trans*-decalin is a rigid molecule and *cis*-decalin has two inter-convertible conformations is supported by the NMR spectra of these two compounds (J. Musher and R.E. Richards, Proc. Chem. Soc., 1958, 230). The *cis*-decalin gives a single absorption peak due to rapid flipping of the molecule from one chair to the other chair form, and all methylene hydrogens become identical. However, *trans*-decalin exhibits a broad, partially resolved band due to non-equivalent equatorial and axial hydrogens.

The *trans*-decalin structure is also present in cadinene and cadinols. Besides the *trans*-decalin structure, the isopropyl group in these compounds is *cis* with respect to the hydrogen atoms nearer the ring Junction (Fig. 8.11).

The *trans*-decalin system is present in pentacyclic triterpenoids amyrin (both β and α-amyrin) and lupeol (Fig. 8.12).

The *fused seven-membered ring system,* the 7–5 system, is bicyclo[5,3,0] decane, which is found to occur in a number of sesquiterpenes like vetivone (Fig. 8.13).

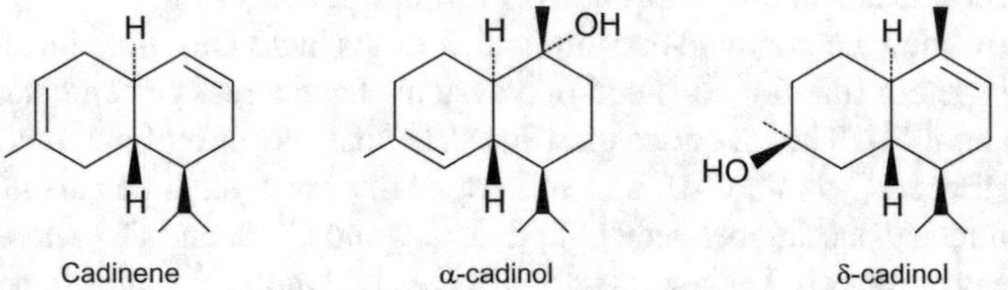

Cadinene          α-cadinol          δ-cadinol

**Fig. 8.11**  Stereochemistry of cadinene and cadinols

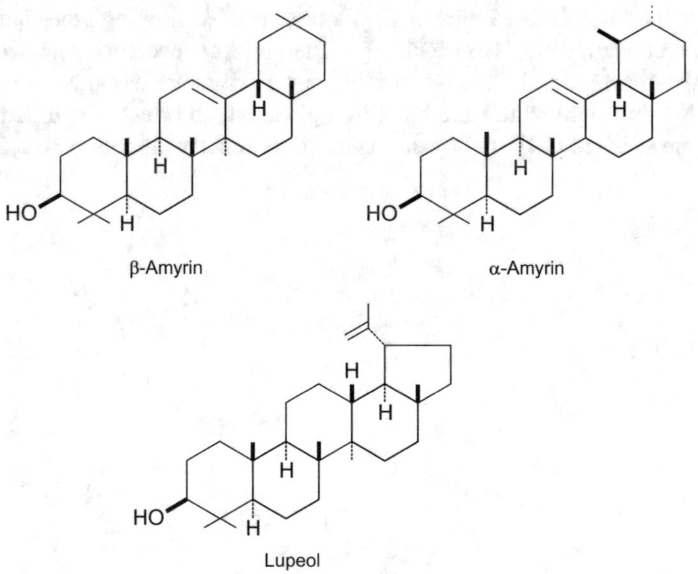

β-Amyrin          α-Amyrin

Lupeol

**Fig. 8.12**  α-Amyrin, β-Amyrin and Lupeol

**Fig. 8.13** Bicyclo [5,3,0]
decane (7–5 system)

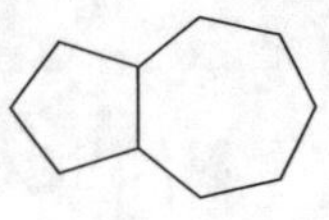

Bicyclo[5,3,0]decane

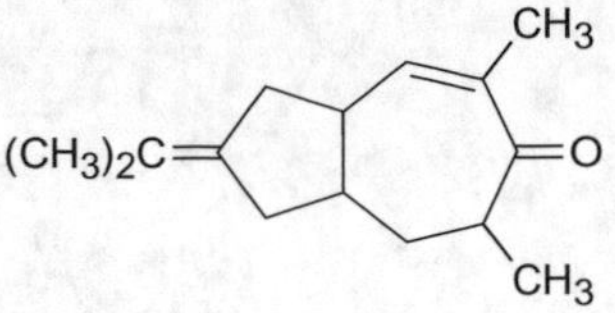

Vetivone

Another fused 7–5 system is azulene (a sesquiterpenoid). One such derivative of azulene is **guaiol**, which occurs in guaiacum wood oil. The fully saturated azulene is named bicyclo [5,3,0] decane.

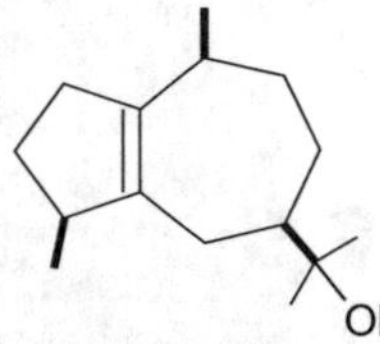

Guaicol

## 8.2.2 Fused Polycyclic Compounds

The simplest polycyclic compound is perhydrophenanthrene, which exists in four dl pairs and two meso forms. The nomenclature of these is given in Fig. 8.14. The prefixes *cis* and *trans* refer to the stereochemistry of fusion of the terminal rings to the central rings. *cisoid* and *transoid* denote the orientation of the terminal rings with respect to each other. The heavy dots indicate a H in front of the plane of the paper and the absence of a heavy dot indicates that the H is behind the plane of the paper.

For more details about the stereochemistry of perhydrophenanthrenes, See Linstead, W. E. Daering, S. B. Davis, P. Levine and R. R. Whetstone, J. Am. Chem. Soc., 1942, **64**, 1985, 1991, 2003, 2006, 2009, 2014.

In the case of perhydroanthracenes, there are five stereoisomers as shown in Fig. 8.15 (W.S. Johnson, E.R. Rugier and J. Ackerman, J. Am. Chem, Soc., 1956, **78**, 6278).

Fused polycyclic compounds are present in the steroid skeleton and the triterpenoids. Two such triterpenoids, amyrin and lupeol, have already been mentioned (Fig. 8.12).

The steroids, as we know, possess a tetracyclic carbon skeleton, the perhydrocyclopentano phenanthrene nucleus (Fig. 8.16), and include a number of natural products like steroids, bile, acids, sex hormones, sapogenins, etc.

In a typical saturated compound like cholesterol, the rings A|B, B|C and C|D are *trans*. This is represented as follows (Fig. 8.17).

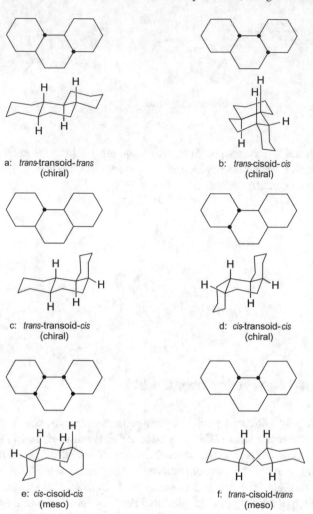

a:  *trans*-transoid-*trans*
(chiral)

b:  *trans*-cisoid-*cis*
(chiral)

c:  *trans*-transoid-*cis*
(chiral)

d:  *cis*-transoid-*cis*
(chiral)

e:  *cis*-cisoid-*cis*
(meso)

f:  *trans*-cisoid-*trans*
(meso)

**Fig. 8.14** Stereoisomers of perhydrophenanthrene dl. pairs-only one enantiomer is shown

In steroids, the A|B ring fusion can be *trans* (as in the case of cholestanol) or *cis* (as in the case of coprostanol). The configuration at the centre of attachment is denoted by 'α' or 'β'. If the substituent is oriented above the plane of the ring (on the same side as the angular methyl groups at C-10 and C-13), the group is in 'β' configuration. This is represented by a bold line. The substituents below the plane of the ring have 'α' configurates and are represented by dotted line bonds (Fig. 8.18).

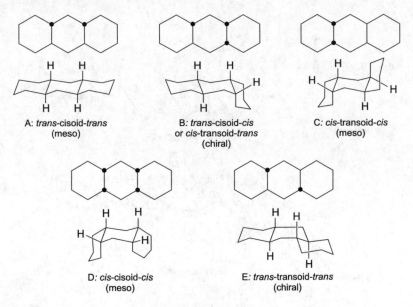

Fig. 8.15  Five stereoisomeric perhydroanthracenes

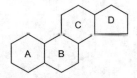

Fig. 8.16  Perhydrocyclopentan ophenanthrene nucleus in steroids

**Fig. 8.17**  Ring Junctions in steroids all junctions are *trans*

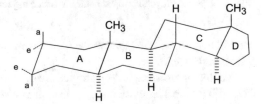

## 8.3  Bridged Compounds

As already mentioned, in bridged compounds, the two rings are linked via non-adjacent atoms. Such compounds are mostly terpenoids. Some of such compounds are given in Fig. 8.19.

In the case of pinane, though two enantiomers are possible, only one pair is known. This is due to fact that the four-membered ring can only be fused to the *six*-membered

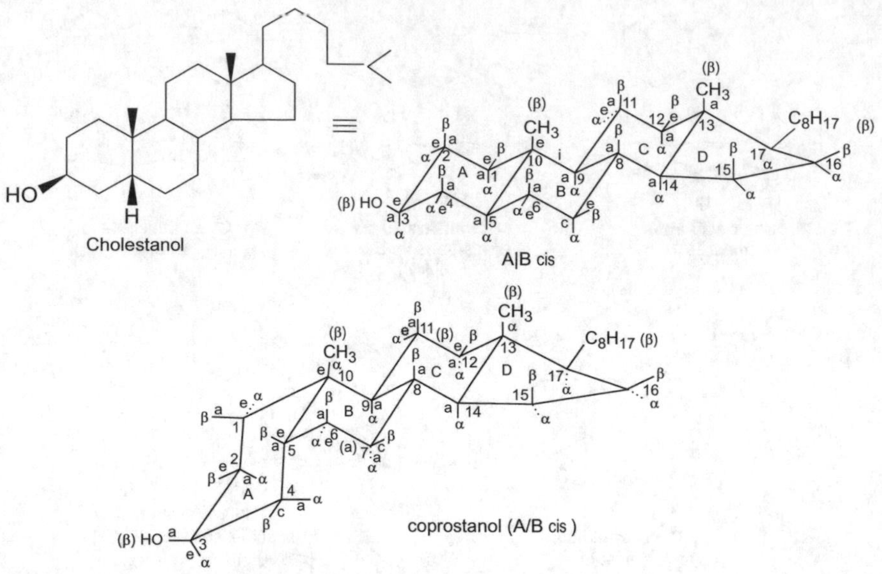

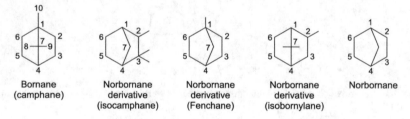

**Fig. 8.18** Cholestanol and coprostanol

**Fig. 8.19** Some bicyclic bridged naturally occurring terpenoids and their derivatives

ring in only the *cis* position; *trans* fusion is impossible. Thus, the enantiomers of the *cis* isomers are known.

Another interesting bicyclic terpene is camphor (structure shown below), a derivative of bornane. Camphor has two dissimilar chiral centres (at positions 1 and 7), but only one pair of enantiomers is known. This is because only the *cis* form is possible; the *trans* fusion of the gem-dimethylmethylene bridge to the cyclohexane ring is not possible. So only the enantiomers of the *cis* isomer are known (as in the case of α-pinene) (Fig. 8.20).

Camphor and its derivatives (borneol and isoborneol) exist in a boat conformation. This is because the gem-dimethyl bridge must be *cis*.

The absolute configuration of aromadendrene, the principal tricyclic sesquiterpenoid hydrocarbon, was elucidated on the basis of ORD studies and also by synthesis. Other tricyclic sesquiterpenoids are cedrene and cedrol. A tricyclic deterpenoids abietic acid is also shown in Fig. 8.21.

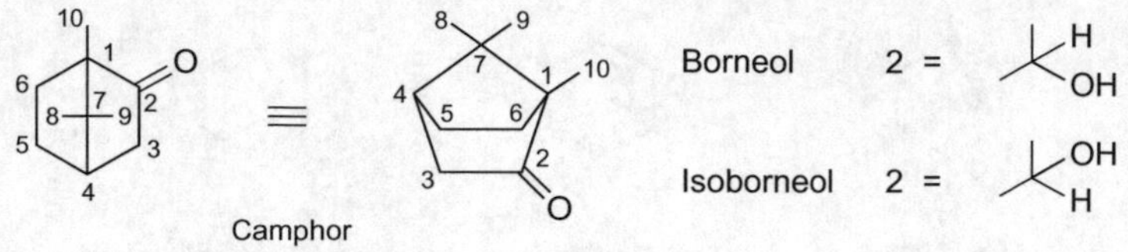

Fig. 8.20 *cis* fusion of gem-dimethyl methylene bridge to cyclohexane ring in campho*r*

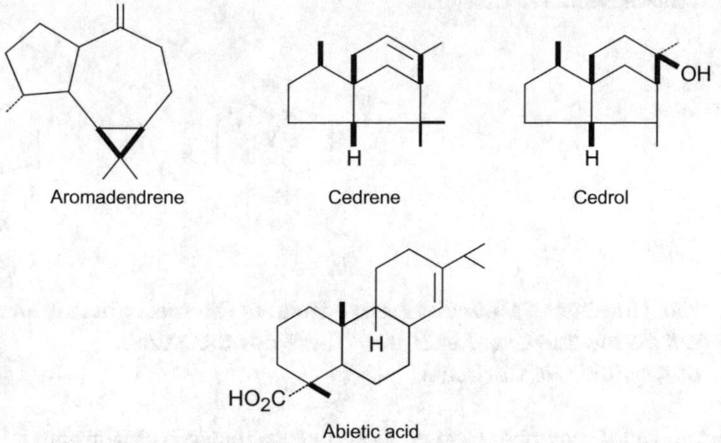

Fig. 8.21 Absolute configuration of some tricyclic terpenoids

Fig. 8.22 Stereochemistry of gibberellic acid

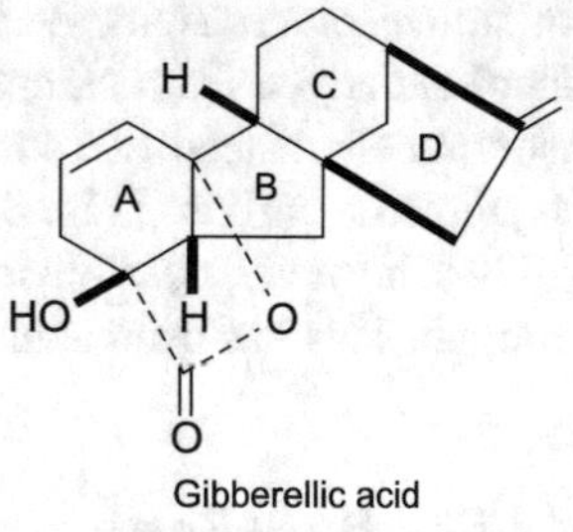

The stereochemistry of the tetracyclic deterpenoid, gibberellic acid, has been assigned as shown in Fig. 8.22.

## 8.3.1 Stereochemical Implications of Bridged Compounds

We come across three stereochemical implications of bridged compounds. These are discussed ahead.

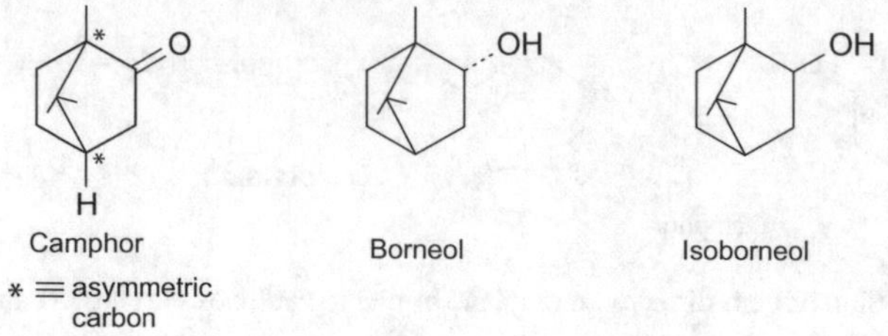

Camphor

* ≡ asymmetric
carbon

Borneol                    Isoborneol

**Fig. 8.23**  Camphor, borneol and isoborneol

**Fig. 8.24**  *cis* and
*trans*-bridged bicyclic ring
systems

cis                               trans

### 8.3.1.1    The Number of Optically Active Isomers (Stereoisomers) Does not Follow the Usual $2^n$ Rule (Where $n$ is the Number of Asymmetric Carbons)

We have known that camphor has two dissimilar asymmetric carbons but it has only one dl pair. This is attributed to the fact that the fusion of the gem-dimethylene bridge to the cyclohexane ring can only be *cis*; the *trans* fusion is not possible. So only the enantiomers of the *cis* isomer are known. This implies that the number of stereoisomers is always half of what will be according to the $2^n$ rule. Thus, the alcohol derived from camphor correspond to two *cis*-bridged dl pairs (borneol and isoborneol) instead of four dl pairs that are possible in other systems with three asymmetric carbons (Fig. 8.22) (see also Sect. 8.3.1) (Fig. 8.23).

It is, however, possible to obtain a *trans*-bridged isomer in case the bridge is large enough. This has been achieved in [4.4.1] system (Fig. 8.24).

### 8.3.1.2    Bredt Rule

According to this rule (J. Bredt, Ann, 1924, **1**, 437; F. S. Fawcett., Chem., Rev., 1950, **47**, 219), in a small bridged system, there can be no double bond at the bridged head position (due to steric reasons). Thus, bromocamphor on treatment with a base does not eliminate to give the unsaturated compound. Also, the unsaturated analogue of camphoric acid does not normally form the anhydride. However, under forcing conditions, the formation of anhydride takes place with the concomitant migration of the double bond to a non-bridge position (Fig. 8.25).

The Bredt rule accounts for the fact that bicyclo [2.2.2] octane-2,6-dione [8.25 (*a*)] is not acidic and lacks the normal acidic properties of a 1,3-diketone. Also,

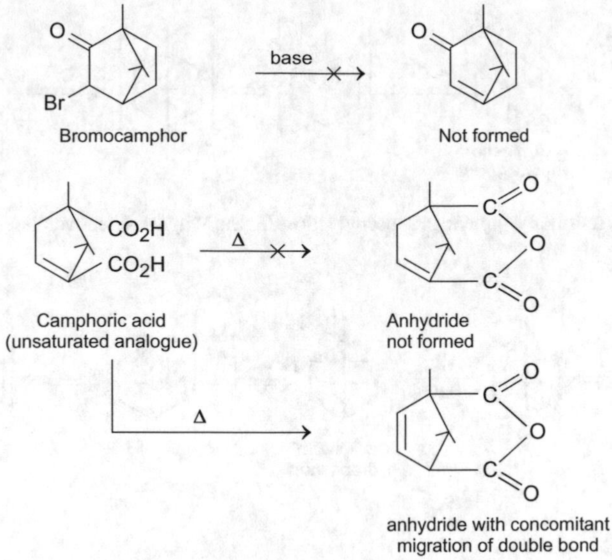

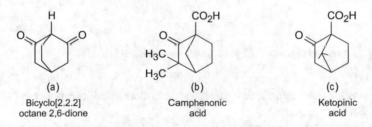

**Fig. 8.25** Examples supporting the Bredt rule

| (a) | (b) | (c) |
|---|---|---|
| Bicyclo[2.2.2] octane 2,6-dione | Camphenonic acid | Ketopinic acid |

**Fig. 8.26** Some other illustrations supporting the Bredt rule

camphenonic acid [8.25 (*b*)] and ketopinic acid [8.25 (*c*)] do not decarboxylate readily although they are β-keto acids (Fig. 8.26) (J. Bredt, J. Prakt. Chem., 1937, (2) **148**, 221).

It has, however, been found that the Bredt rule is not applicable to [5.3.1] system.

### 8.3.1.3 Molecular Rearrangements in Bridged Compounds

In bicyclic monoterpenes, the bicyclo [2.2.1] heptane system is most important. Typical examples are pinane and camphane. Such a system is involved in the molecular rearrangement. An example is the **Wagner–Meerwein rearrangements** in bicyclic monoterpenoids. Thus, treatment of α-pinene in ether with gaseous hydrogen chloride at −15 °C gives an unstable adduct (α-pinene hydrochloride), which on warming undergoes rearrangement to give bornyl chloride (Fig. 8.27).

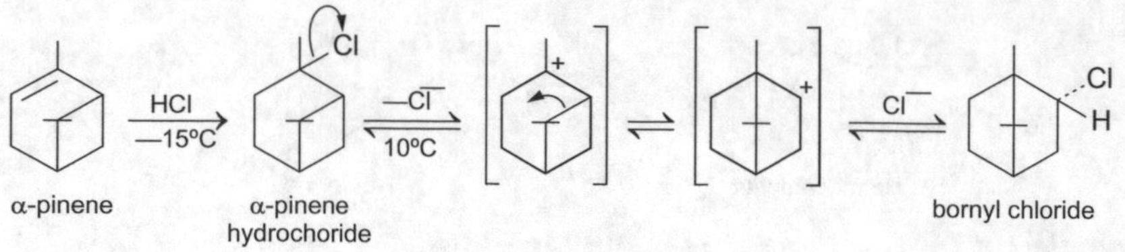

**Fig. 8.27**  Conversion of α-pinene into bornyl chloride (the Wagner–Meerwein rearrangement)

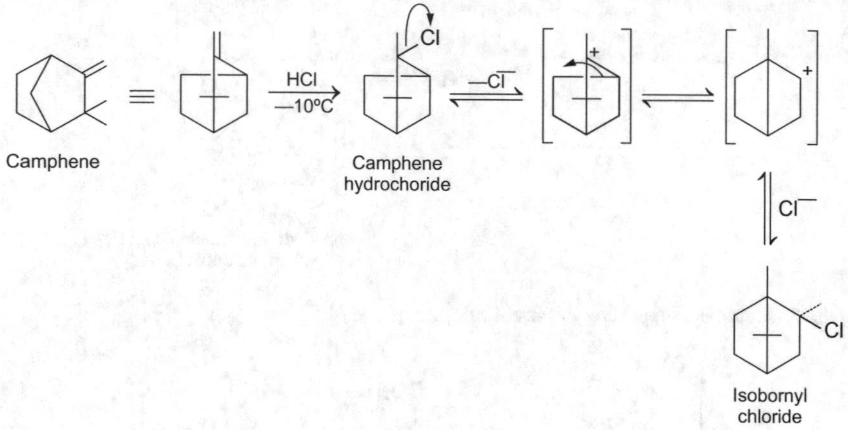

**Fig. 8.28**  Conversion of camphene hydrochloride into isobornyl chloride

In a similar way, camphene hydrochloride is converted into isobornyl chloride (Fig. 8.28).

Both the above rearrangements (Figs. 8.27 and 8.28) proceed through the same carbocation intermediate. However, different epimers are obtained. The reason is not certain.

## 8.4   Catenanes, Rotaxanes and Knots

These differ from all other organic compounds in that the subunits are linked mechanically without involving a chemical bond. The name catenane is derived from the Latin word *catena* meaning chain; molecules containing two interlinked rings. In catenanes, the open chain compounds are made to cyclise. During this process, it may so happen that after a molecule has cyclised, another open chain compound passes through the cyclised ring and then cyclises to form a system of interlocked rings as shown in Fig. 8.29.

The possibility of the formation of the interlocked ring is extremely low.

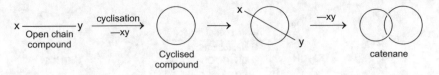

**Fig. 8.29** Steps involved in the formation of interlocked rings (catenanes)

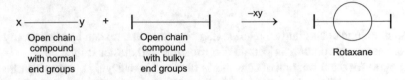

**Fig. 8.30** Formation of a ring having a threaded chain with bulky end groups (rotaxane)

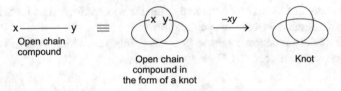

**Fig. 8.31** Formation of knot

In **rotaxanes**, an open chain compound is cyclised in presence of another open chain compound having bulky end groups. There is some possibility that the open chain compound may cyclise in such a way that the other open chain compounds may pass through the cyclised ring. Such compounds are called rotaxanes (In Latin, *rota* means wheel, and axis means the axle) and have a ring like a wheel on an axle. In this case, the bulky end groups prevent the extrusion of the threaded chain from a macromolecule. The formation of a rotaxane is as shown in Fig. 8.30.

In this case also (as in the case of catenanes), the possibility of the formation of threaded compounds like rotaxane is very low.

In knots, a single long chain may cyclise in such a way so as to form a compound with a knot-like look. Its formation may be visualised as shown in Fig. 8.31.

## 8.4.1 *Catenanes*

These are large ring compounds consisting of interlocked rings and are synthesised by using acyloin condensation. The procedure consists in the acyloin condensation of a long chain dicarboxylic ester to give a large ring compound (acyloin) using the high dilution technique. This method is best suited for closing rings of ten carbons or more (Fig. 8.32).

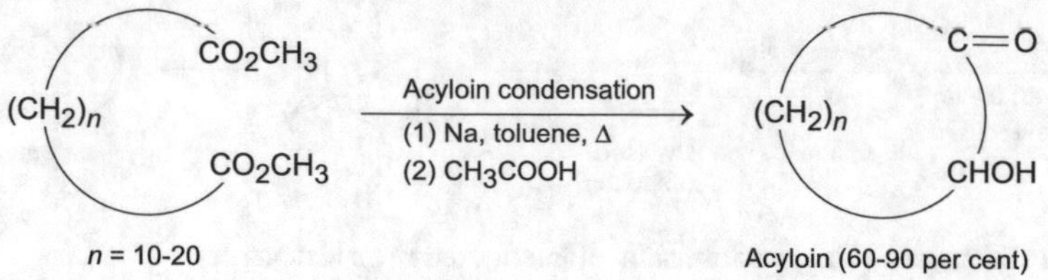

**Fig. 8.32** Acyloin condensation of long chain dicarboxylic ester

A compound containing an interlocking ring (a catenane) can be obtained by the acyloin condensation of a 34-carbon dicarboxylic acid diester (Fig. 8.33).

Support for the formation of catenane is obtained by carrying out the ring closure with the ester of 34-carbon dicarboxylic ester in presence of deuterated 34-carbon cycloalkane. The catenane is obtained with 68 carbons containing one acyloin group and deuterium. Its IR spectra showed $\nu_{C-D}$ bonds (Fig. 8.34).

The structure of 68-carbon deuterated catenane was confirmed by its oxidation to release the deuterated 34-carbon cycloalkane.

The completely reduced catenane (a hydrocarbon) in which both the C=O and CHOH groups are reduced to $CH_2$ is represented as

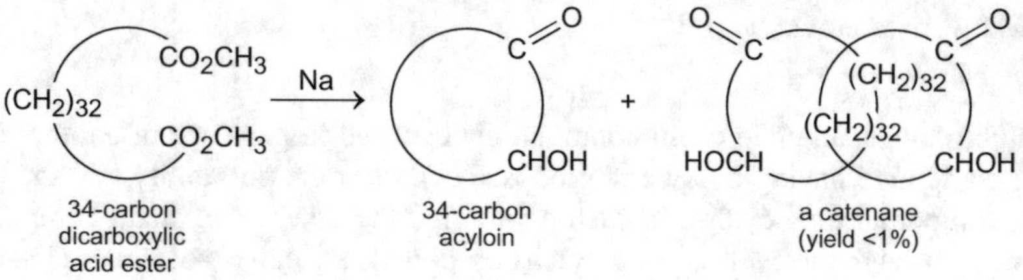

**Fig. 8.33** Synthesis of a catenane

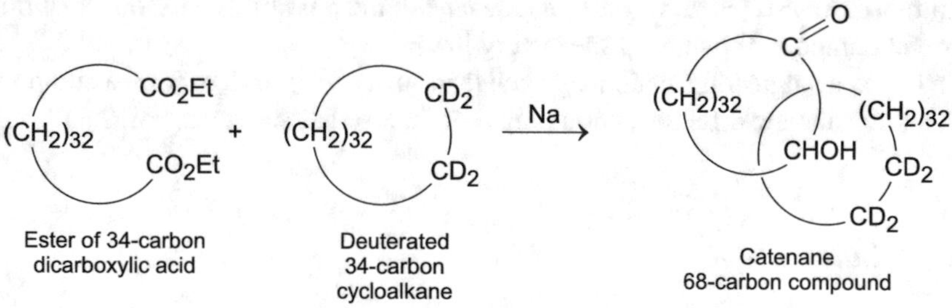

**Fig. 8.34** Synthesis of 68-carbon catenane containing four deuterium atoms

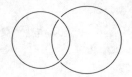

Completely reduced catenane

## 8.4.2  Rotaxanes

A numbered of methods are available for the synthesis of rotaxanes. As already stated, an open chain compound having bulky end groups is cyclised in such a way that the open chain compound may pass through the closed molecule (a macrocycle) as shown in Fig. 8.30.

A synthesis of rotaxane involved the reaction of 2-hydroxycyclotria contanone (1) with succinic anhydride to give the half ester (2). The sodium salt of this half ester (2) on reaction with chloromethylated copolymer from styrene and divinyl benzene gave the resin-bound macrocycle (3), which on reaction with 1, 10-decanediol and triphenyl methyl chloride in presence of pyridine, dimethyl formamide and toluene followed by hydrolysis with sodium bicarbonate in refluxing methanol gives the rotaxane (4) (Figs. 8.33 and 8.35).

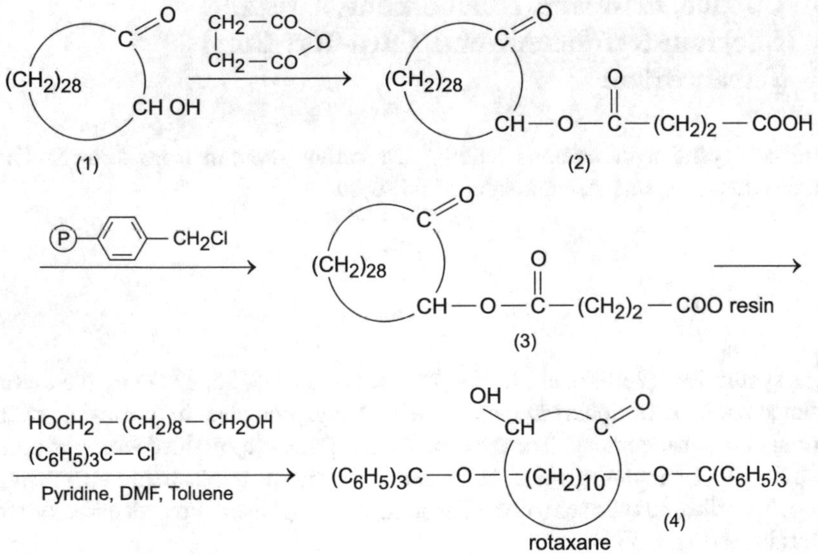

**Fig. 8.35**  Synthesis of a rotaxane

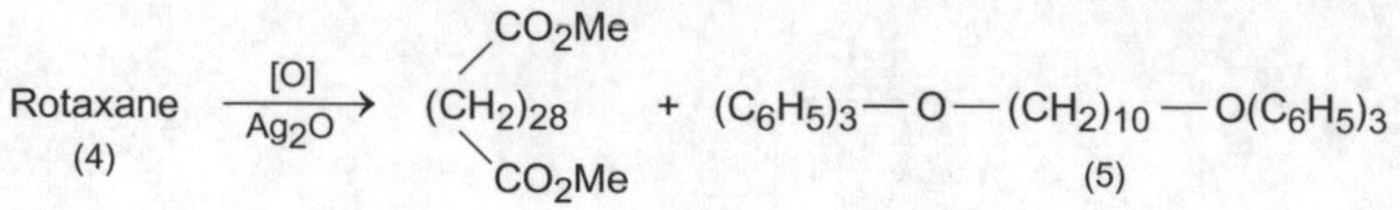

$$\text{Rotaxane} \xrightarrow[\text{Ag}_2\text{O}]{[\text{O}]} \quad (CH_2)_{28} \underset{CO_2Me}{\overset{CO_2Me}{<}} \quad + \quad (C_6H_5)_3-O-(CH_2)_{10}-O(C_6H_5)_3$$

(4)                                                              (5)

**Fig. 8.36**  Oxidation of rotaxane

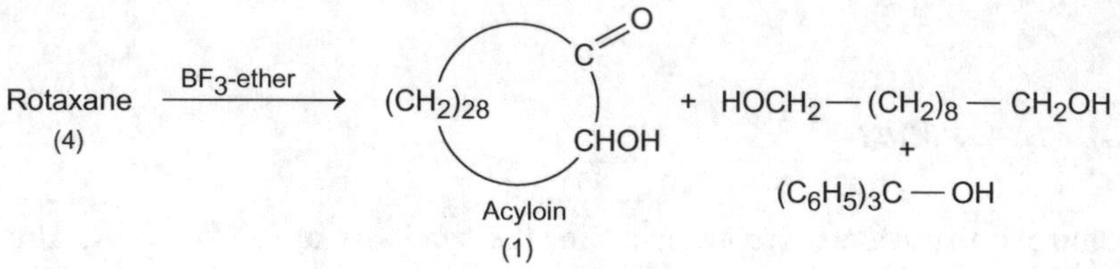

$$\text{Rotaxane} \xrightarrow{\text{BF}_3\text{-ether}} \quad (CH_2)_{28} \quad + \quad HOCH_2-(CH_2)_8-CH_2OH$$

Acyloin
(1)

$$+ \\ (C_6H_5)_3C-OH$$

**Fig. 8.37**  Cleavage of rotaxane with BF$_3$-ether

The structure of rotaxane (4) was confirmed by its IR spectra and chemical degradation. Oxidation with silver oxide gave octacosane-1, 28-dicarboxylic acid as dimethyl ester along with dumshaped molecule (5) (Fig. 8.36).

The cleavage of rotaxane (4) with BF$_3$-etherate in benzene gave decane-1,10-diol triphenyl methanol and the acyloin (1) (Fig. 8.37).

## 8.5  Cubane, Prismane, Adamantane, Twistane, Buckminsterfullerene and Tetra-Tert-Butyl Tetrahedrane

Some polycyclic hydrocarbons having fascinating structures are known. These include cubane, prismane, adamantane and twistane.

### 8.5.1  Cubane

It was synthesised (Pettit et al., J. Am. Chem. Soc., 1966, **88,** 1328) by the thermal decomposition of cyclobutadiene iron tricarbonyl complex in presence of 2,5-dibromo-1,4-benzoquinone. The cyclobutadiene (released by the decomposition of iron tricarbonyl complex) gives an endo complex which on photochemical cyclisation gave cubane dicarboxylic acid by the Favorskii rearrangement. Final decarboxylation gave cubane (Fig. 8.38).

The last step of decarboxylation (Fig. 8.38) is affected by the thermal decomposition of tert butyl ester in isopropyl benzene (Fig. 8.39).

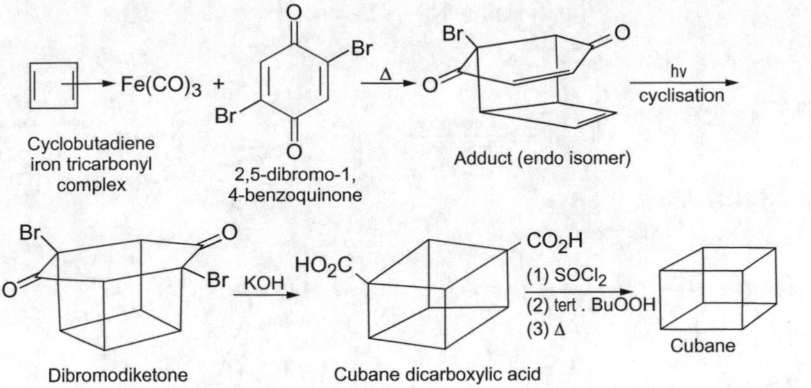

**Fig. 8.38**  Synthesis of cubane

**Fig. 8.39** Mechanism of decarboxylation of cubane dicarboxylic acid

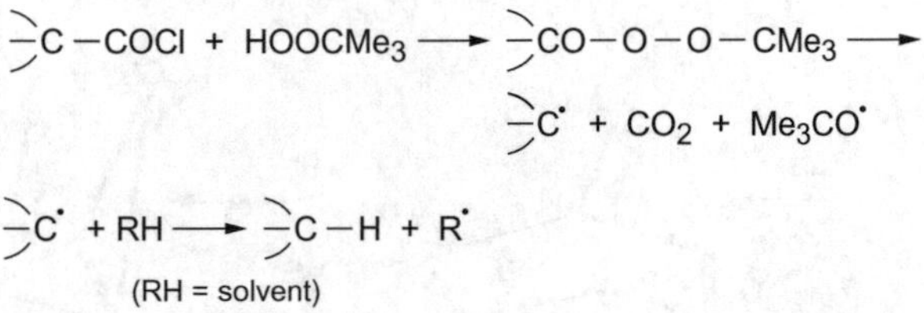

$$\overset{|}{\underset{|}{C}}-COCl + HOOCMe_3 \longrightarrow \overset{|}{\underset{|}{C}}O-O-O-CMe_3 \longrightarrow$$

$$\overset{|}{\underset{|}{C}}{}^{\bullet} + CO_2 + Me_3CO^{\bullet}$$

$$\overset{|}{\underset{|}{C}}{}^{\bullet} + RH \longrightarrow \overset{|}{\underset{|}{C}}-H + R^{\bullet}$$

(RH = solvent)

Cubane shows a single NMR proton signal at δ 4.04 and a single $^{13}C$ resonance at 47.3. The structure of cubane as determined by electron diffraction studies shows a bond length of 157.5 pm (1.57 Å), larger than in cyclobutane. Earlier X-ray data indicated a C–C–H bond angle of 123–127°, suggesting a high degree of s-character of the C–H bond. Surprisingly $J_{13}C–H$ of 60 Hz suggested the high s-character of the C–H bond.

## 8.5.2   Prismane

The structure of prismane was the one as proposed once for benzene by Ladenburg. In spite of a strain energy of about 300 kJ/mole, the prismane molecule is stable (Oth 1968; Woodward et al. 1970). It is synthesised as follows (Fig. 8.40).

A number of alkyl derivatives of prismane are known. These are obtained by photochemical cyclisation of a bicyclo [2.2.0] hexadiene. As an example, hexamethyl prismane is obtained from hexamethyl bicyclo [2.2.0] hexadiene (Schäfer et al. 1967) (Fig. 8.41).

The starting  hexamethylbicyclo [2.0.0] hexadiene is obtained by cyclotrimerisation of but-3-yne.

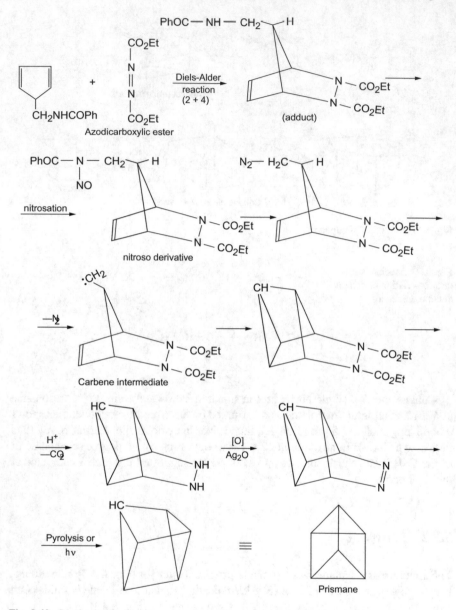

**Fig. 8.40** Synthesis of prismane

## 8.5.3 *Adamantane*

It is a tricyclic system having three cyclohexane rings and was first isolated from a high boiling fraction of petroleum (S. Landa and V. Machecek, Collection Czechoslov, Chem., Commun., 1933, **1,** 5). Its structure is similar to hexamethylene

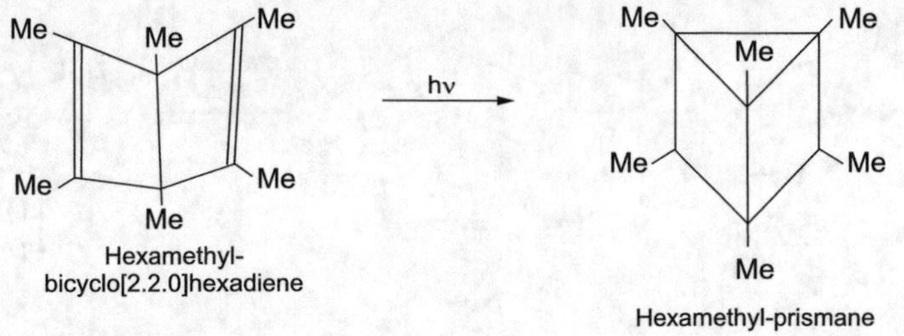

**Fig. 8.41** Synthesis of hexamethyl prismane

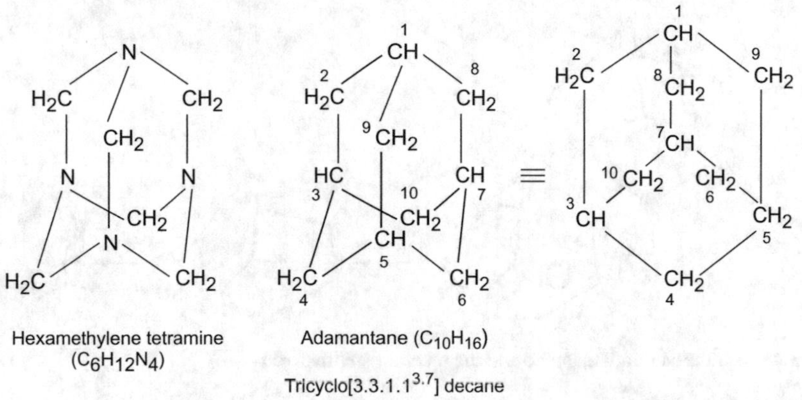

**Fig. 8.42** Adamantane

tetramine (which is obtained by the reaction of formaldehyde and ammonia) with four nitrogens replaced by CH. It is named [3.3.1.1.$^{3,7}$] decane (Fig. 8.42).

Adamantane has a symmetrical and strainless structure. It is obtained from the dimer of dicyclopentadiene by hydrogenation followed by the action of strong acid (J. A.C.S., 1957; **79**, 3292; 1960, **82**, 4645) (Fig. 8.43).

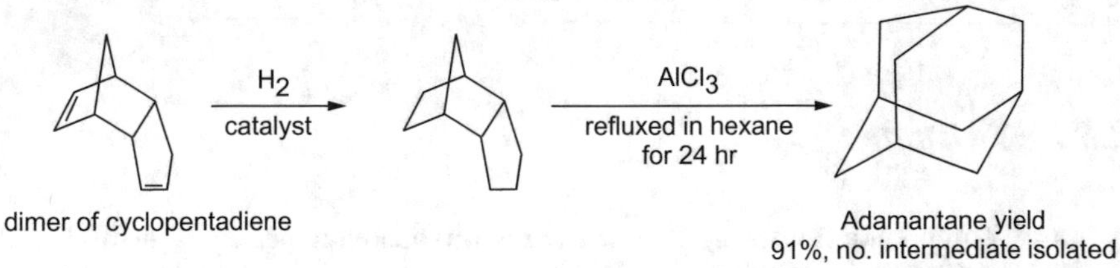

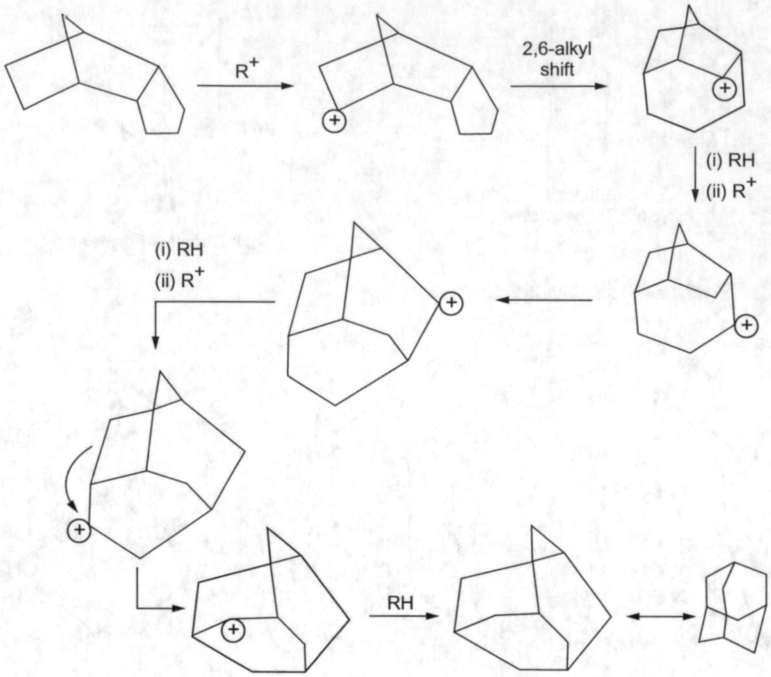

**Fig. 8.43** Synthesis of adamantane and the mechanism involved

## 8.6   Proposed Mechanism

Another synthesis of adamantane involves the reaction of malonic acid with formaldehyde (Fig. 8.44).

Though adamantane appears to be free of strain, it has about 76 kcal/mol strain. This is about four times the strain in cyclohexane. In adamantane, the framework structure has bond angles close to the tetrahedral value. However, the –C–CH$_2$–C– bond angle is not exactly tetrahedral, the optimum value being 112.5°, and is strained cyclohexane, which causes strain in adamantane.

### 8.6.1   Twistane

A novel compound, twistane, is isomeric with adamantane. It is synthesised (Whitelock 1962) as shown in Fig. 8.45.

Twistane is a dissymmetric structure (+). Twistane [[α]$_D$ 41.4] was obtained starting from optically active carboxylic acid (Fig. 8.46).

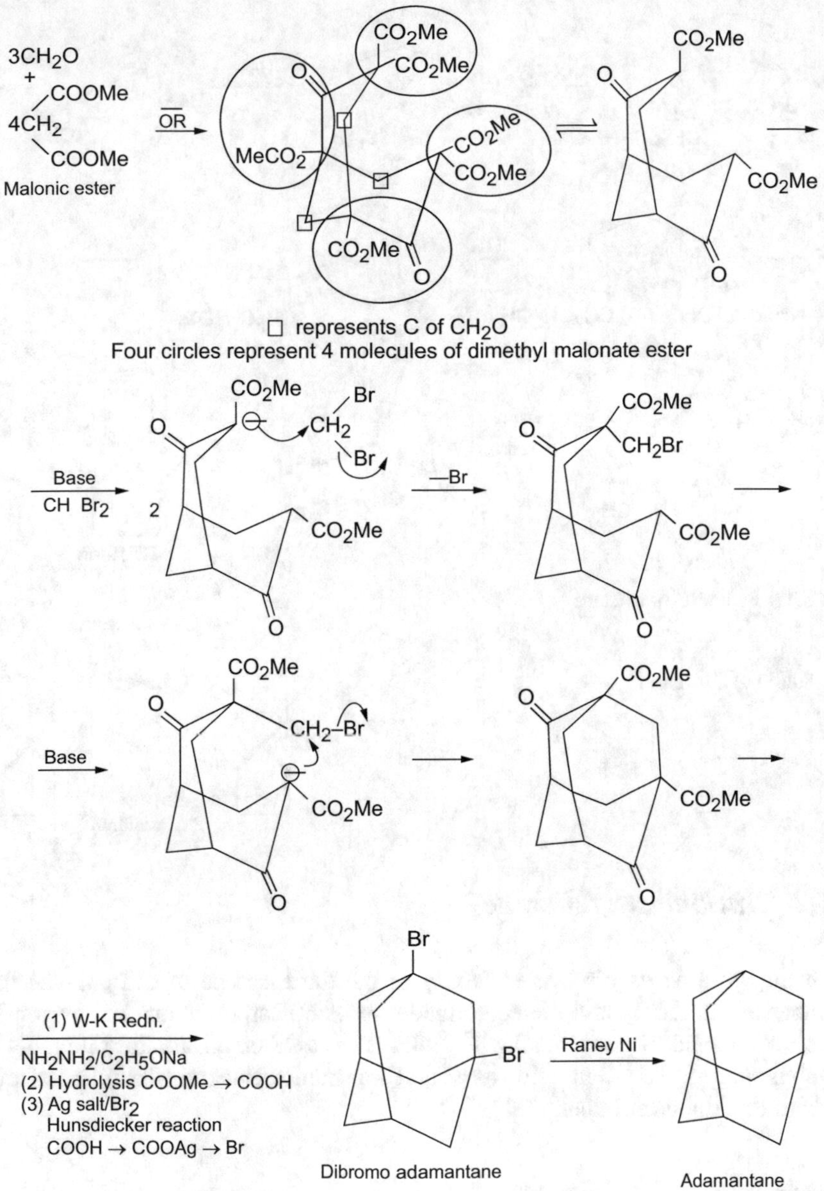

$$3CH_2O$$
$$+$$
$$4CH_2 \begin{array}{c} COOMe \\ \\ COOMe \end{array}$$

Malonic ester

$\overline{OR}$

☐ represents C of CH₂O
Four circles represent 4 molecules of dimethyl malonate ester

$$\xrightarrow[CH\ Br_2]{Base}$$

$$\xrightarrow{-Br}$$

$$\xrightarrow{Base}$$

(1) W-K Redn.
$\xrightarrow{\quad\quad\quad}$
$NH_2NH_2/C_2H_5ONa$
(2) Hydrolysis COOMe → COOH
(3) Ag salt/Br₂
  Hunsdiecker reaction
  COOH → COOAg → Br

$\xrightarrow{Raney\ Ni}$

Dibromo adamantane

Adamantane

**Fig. 8.44** Synthesis of adamantane from malonic ester

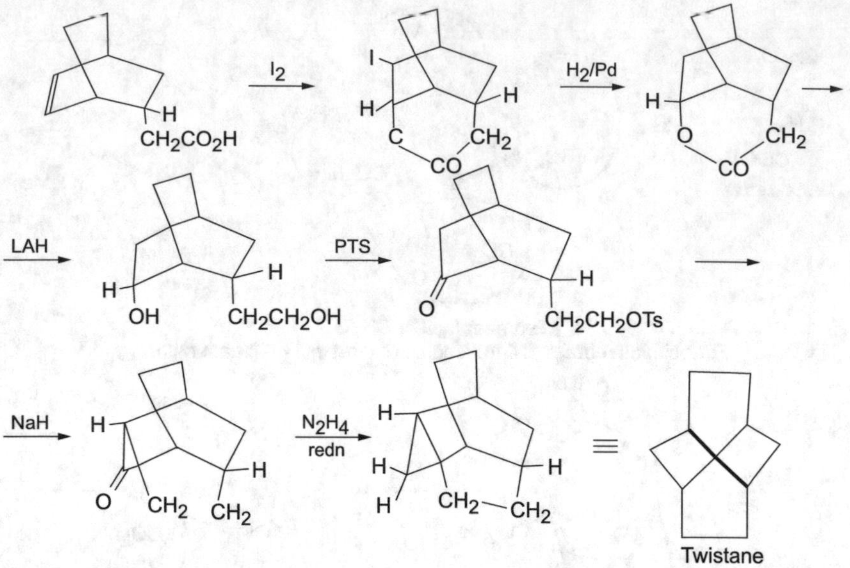

**Fig. 8.45** Synthesis of twistane

**Fig. 8.46** (+) Twistane

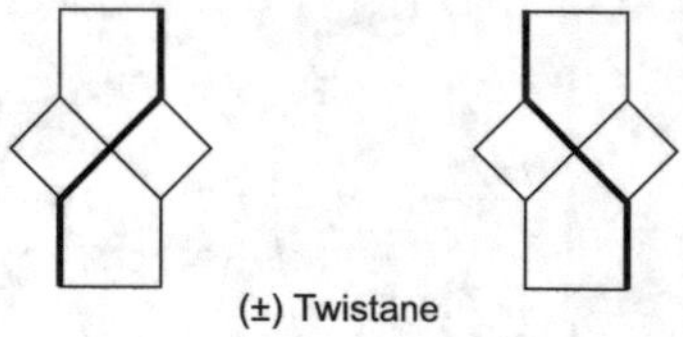

(±) Twistane

## 8.6.2 *Buckminsterfullerene*

It belongs to a newly discovered family of the third isotope of carbon. The first member of the family is a molecule made up of 60 carbon atoms discovered by Harold Kroto and Richard Smalley in 1985. It has a spherical structure (Fig. 8.47), which is shaped like a Geodesic Dome and was named Buckminsterfullerene after Richard Buckminister Fuller.

**Fig. 8.47** Structure of Buckminsterfullerene

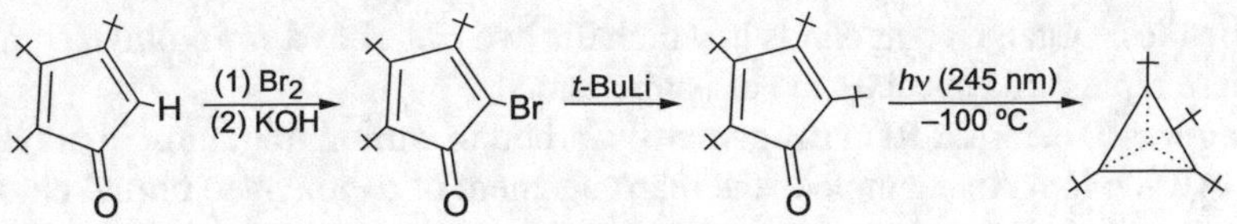

**Fig. 8.48** Synthesis of tetra-tert-butyltetrahedrane

A number of other similar molecules like $C_{28}$, $C_{32}$, $C_{70}$ and $C_{240}$ were discovered and the name fullerene or bulky balls was introduced. These compounds are remarkably stable and have potential use as lubricants. The possibility of their use as superconducting materials at high temperatures when doped with alkali metals has generated interest.

For works on Buckminsterfullerene, its homologues and derivatives see W.E. Billups and M.A. Ciufolinio, Eds, 1993. Buckminsterfullerenes, VCH, New York.

### 8.6.3  Tetra-Tert-Butyltetrahedrane

The parent compound, tetrahedrane, has not been synthesised, but its tetra-tert-butyl derivatives have been synthesised (G. Maier, S. Pfrien, U. Sachäfer, K.–D. Malsch and R. Matusch, Chem. Ber., 1981, **114,** 3965; G. Maier. S. Pfriem, U. Sachäfer and R. Matusch, Angew. Chem. Int. Edn. Engl., 1978, **17,** 520) (Fig. 8.48).

The structure of *tetra-tert*-butyltetrahedrane was established by X-ray crystallography (H. Irngartiner, A. Goldmann, R. Jahn, M. Nirdort, H. Rodewald, G. Maier, K. –D. Malsch and R. Emrich, Angew. Chem. Int. Ed., Engl, 1984, **23**, 993).

**Key Concepts**

- **Adamantane**: A cyclic tricyclic system having three cyclohexane rings.
- **Bicyclic Compounds**: Organic compounds containing two rings.
- **Bredt Rule**: In a small bridged system, there can be no double bond at the bridged head position.
- **Bridged Compound**: Organic compounds in which two rings are linked via non-adjacent atoms.
- **Catananes**: Large ring compounds consisting of interlocked rings.
- **Fused Ring Compound**: Organic compounds in which the two adjacent rings have two or more atoms in common.
- **Polycyclic Compounds**: Organic compounds containing more than two rings.
- **Rotaxanes**: Cyclic organic compounds through which an open chain compound with a bulky end group passes through the ring.

- **Spiranes**: Organic compounds in which the two rings have one atom in common. These represent a special type of isomerism.
- **Wagner–Meerwein Rearrangement**: Carbon to carbon migration of alkyl, aryl or hydride ion. An example is the rearrangement of α-pinene to bornyl chloride.

## Problems

1. Explain what are fused, bridged and caged ring compounds?
2. Give structural formula of the following:

   (a) Bicyclo [1.1.0] butane.
   (b) Bicyclo [2.1.0] pentane.
   (c) Bicyclo [3.2.0] heptane.
   (d) Bicyclo [2.2.0] hexane.
   (e) Bicyclo [3.2.0] heptane.
   (f) Bicyclo [4.2.0] octane.
   (g) Bicyclo [5.2.0] nonane.
   (h) Bicyclo [3.3.0] octane.
   (i) Bicyclo [5.3.0] decane.

3. Discuss the stereochemistry of decalins, perhydrophenanthrenes and perhydroanthracenes.
4. Give the formula of cholestanol and coprostanol giving various ring junctions.
5. Explain why in camphor, the fusion of six-membered ring and gem-dimethyl methylene bridge is *cis*.
6. What product is expected to be obtained in the following:

7. Explain the Wagner–Meerwein rearrangement.
8. What are catenanes, rotaxanes and knots?
9. Give a method for the synthesis of a catenane and a rotaxane.
10. Give a synthesis of cubane, prismane, adamantane and twistane.

# Chapter 9
# Optical Rotatory Dispersion and Circular Dichroism

## 9.1 Introduction

Ordinary light is known to consist of different wavelengths vibrating in many planes (Fig. 9.1a). On passing the ordinary ray of light through a polariser, the emerging light has its vibrations in one plane only (Fig. 9.1b). The emerging beam of light having oscillations in a single plane is said to be plane-polarised light (Fig. 9.1c). This plane is called the plane of polarisation.

It became clear after the discovery of electrons that light is produced due to the oscillation of electrons within the atom. Also, the electromagnetic nature of light was ascertained. The electric and magnetic fields associated with electromagnetic waves are known to oscillate at right angles to each other and also to the direction of propagation of light.

A beam of linearly polarised light (Fig. 9.1b) is, in fact, the vector sum of two components, right-circularised light and a left-circularised component as shown in Fig. 9.2.

A beam of a linearly polarised wave having the electric field vector ($E$) in the $xy$ plane along the direction of propagation ($y$-axis) is the vector sum of a right circularly polarised and a left circularly polarised component, whose projection along $xy$ plane are circles as shown in Fig. 9.2.

The vectors corresponding to the two components at points A to E along the corresponding vector sum which have properties of linearly polarised wave are shown in Fig. 9.3.

When a linearly polarised light is passed through an optically active medium, the two circularly polarised components display different refractive indices and different absorption coefficients. The two chiroptical properties (viz., refractive indices and absorption coefficients) of a chiral medium are known as circular birefringence and circular dichroism, respectively (Fig. 9.4).

Both circular birefringence and circular dichroism are interrelated and are useful in the study of the cotton effect. However, it is best to discuss them separately.

© The Author(s), under exclusive license to Springer Nature Switzerland AG 2022     227
V. K. Ahluwalia, *Stereochemistry of Organic Compounds*,
https://doi.org/10.1007/978-3-030-84961-0_9

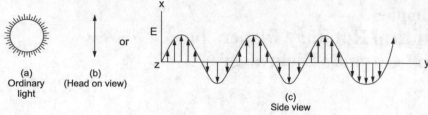

(a)                    (b)
Ordinary        (Head on view)
light

(c)
Side view

Plane polarized light (Linearly polarized light)

**Fig. 9.1**  Ordinary light and plane-polarised light

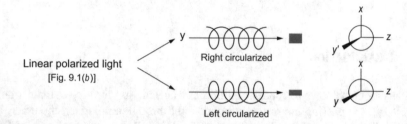

Linear polarized light
[Fig. 9.1(b)]

Right circularized

Left circularized

**Fig. 9.2**  Illustration of right- and left-circularised light

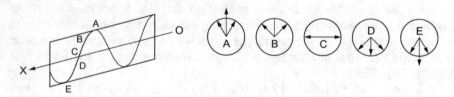

**Fig. 9.3**  Linearly polarised wave shown as the vector sum of two oppositely rotating beams of circularly polarised light

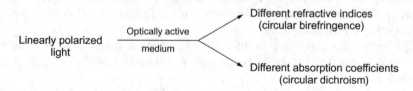

Linearly polarized
light

Optically active
medium

Different refractive indices
(circular birefringence)

Different absorption coefficients
(circular dichroism)

**Fig. 9.4**  Circular birefringence and circular dichroism

## 9.2  Circular Birefringence

It is found that when both the right and left circularly polarised components of a beam of plane-polarised light are passed through a medium with equal speed, there is no rotation of the plane of polarisation. On the other hand, if plane-polarised light is passed through an optically active medium such as I, in which the polarisabilities of the groups or atoms decreases in the order $A > B > X > Y$, then the two components, viz. left- and right-circularised light passes through the medium with unequal speed. In this case, the emerging vector rotates through an angle with respect to the original plane of polarisation. Such a medium is called circularly birefringent. This property of the medium which rotates the plane of polarisation is called *circular birefringence*.

In case the right circularly polarised light moves faster, the medium is dextrorotatory ($\alpha$ is + ve). On the other hand, if the left circularly polarised component moves faster, the medium is laevorotatory ($\alpha$ is – ve).

The angle of rotation is given by the equation:

$$\alpha = \frac{\pi}{\lambda_{vac}} (n_r - n_t) \tag{9.1}$$

where $\lambda_{vac}$ is vacuum wavelength of light employed and $n_r$ and $n_t$ are refractive indices of the medium with respect to right and left circularly polarised components, respectively.

$n = c/v$; $c$ = velocity of light in vacuum.

$v$ = velocity of light in the medium.

$n$ = refractive index.

The system is dextrorotatory if $n_r > n_t$ ($\alpha = +$ ve ) and the system is laevorotary if $n_t > n_r$ ($\alpha + -$ve).

## 9.3 Circular Dichroism

Circular dichroism is the measurement of unequal absorption of right and left circularly polarised compounds of a beam of linearly polarised light. It has already been stated that there are unequal molar absorption coefficients for left and right circularly polarised light in the case of an optically active medium. The difference in the molar extinction coefficients of right and left circularly polarised rays ($\Delta_s$) is called differential dichroic absorption (Eq. 9.2).

$$\Delta_s = \varepsilon_L - \varepsilon_R \neq 0 \tag{9.2}$$

The two circularly polarised components after emerging from an optically active medium are not only out of phase but also have unequal amplitudes. This means that the two components are absorbed to different extents. In case the left circularly polarised component is more strongly absorbed, its electric field vector will be smaller.

In case the two circularly polarised components are in-phase, the resultant ray is elliptically polarised light. As in the case of circularly polarised light, the elliptically polarised ray may be right- or left-handed elliptically polarised.

## 9.4   Cotton Effect

We have now known that a medium that exhibits circular birefringence also exhibits circular dichroism. The combination of these two effects gives rise to the phenomenon called the cotton effect. It can be studied by plotting the change in optical rotation with wavelength. Photoelectric spectropolarimeters are used for such plots easily down to about 220 nm. Compounds absorbing only in far ultraviolet (i.e., below 220 nm) do not show cotton effect since the effect occurs only near an absorption maximum. The optical rotatory dispersion curves are discussed in Sect. 9.5.1.

## 9.5   Optical Rotatory Dispersion (ORD)

Optical Rotatory Dispersion is the measurement of optical rotation as a function of wavelength. The method is applicable only to optically active compounds. ORD measurements are normally made in ultraviolet/visible range, where the functional group of the compound under investigation has an absorption band with a small extinction coefficient. The carbonyl group is ideal for such studies and so studies have been carried out with ketones.

### 9.5.1   Types of Optical Rotatory Dispersion Curves

There are two types of ORD curves. These are simple ORD curves and anomalous ORD curves.

#### 9.5.1.1   Simple ORD Curves

These curves may be plain or normal dispersion curves. In these, the measurement of rotation is restricted to wavelengths away from $\lambda_{max}$ region. The curve resulting shows a steady increase or decrease in optical rotation with the decrease of wavelength. Such a curve is called a plain curve (Fig. 9.5).

The plain curve may or may not cross the zero rotation line. However, it is devoid of inflections or extrema. On the other hand, a normal dispersion curve is devoid of inflections, maxima or minima and crossing of zero rotation axis within measurable range (Djerassi 1990). Two simple ORD curves are illustrated in Fig. 9.5.

**Fig. 9.5** Plain positive curve
(A); Plain negative curve (B)

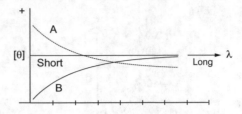

In curve A, the molecular rotation increases in the positive direction on going to a shorter wavelength and is referred to as the plain positive dispersion curve. The curve B is referred to as a plain negative dispersion curve since in this the molecular rotation increases in the negative direction.

The plain dispersion curve is a smooth curve and obeys the Drude equation $[(\alpha) = K/(\lambda^2 - \lambda_0^2)]$ (where $K$ and $\lambda_0$ are empirical constants); $\alpha$ decreases (becoming less positive or negative) with an increase in $\lambda$ which is the wavelength at which specific rotation is measured, $\lambda_0$ corresponds to wavelength due to absorption maxima and $K$ is an empirical constant.

### 9.5.1.2 Anomalous ORD Curves

These curves contain both the peak and trough and do not obey the one-term Drude equation. These curves can be single cotton effect or multiple cotton effect curves.

### Single Cotton Effect Curves

These curves can be positive or negative depending on whether the peak is longer (positive curve) or shorter (negative curve) than the trough. Figure 9.6a shows a positive single cotton effect curve. As seen, the peak and trough are indicated as $P$ and $T$, respectively. The amplitude is the vertical distance, $a$, between the peak and trough and breadth, $b$, is the horizontal distance.

### Multiple Cotton Effect Curve

Such a curve has several peaks and troughs with shoulders and inflections (Fig. 9.6b).

The optical rotatory curves are represented by mathematical Eq. (9.3) known as the Drude Eq. (9.3).

$$[\theta] = \frac{K_1}{\lambda^2 - \lambda_0^2} + \frac{K_2}{\lambda^2 - \lambda_1^2} + \cdots + \frac{K_n}{\lambda^2 - \lambda_{n-1}^2} \tag{9.3}$$

where $K_1, K_2 \ldots K_n$ are constants characteristics of chromophores, which are responsible for the observed rotations, $\lambda_0$ is the wavelength of the measurement and $\lambda_0, \lambda_1, \ldots \lambda_{n-1}$ are the wavelengths of absorption maxima of the compound.

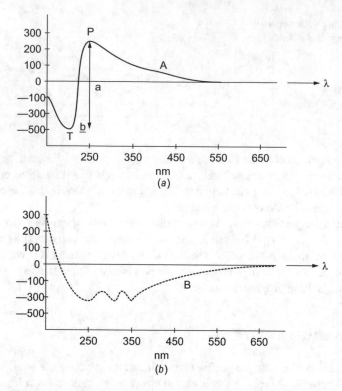

**Fig. 9.6**　Positive **a** and multiple **b** cotton effect curves

## 9.6　Comparison of ORD and CD Curves

It is now known that optical rotatory dispersion involves measurement of rotation and circular dichroism related to the measurement of absorption (see Sects. 9.3 and 9.5). The relationship between ORD (dispersion) and CD (absorption) is expressed by integral transforms known as the Kramers relations, and the corresponding ORD curves can be calculated from experimentally determined CD spectra and vice-versa.

The Kronig–Kramers relations give an expression that is derived from $n \rightarrow \pi^*$ transition of saturated ketones. It relates the amplitude of an ORD cotton effect to $\Delta_s$ of the corresponding CD peak (Eq. 9.4).

$$\alpha = 40.28\Delta_s = \left(1.22 \times 10^{-2}\right)[\theta] \tag{9.4}$$

Both ORD and CD curves give similar information in organic stereochemistry. Like ORD, CD curves can also be positive or negative. Usually, a positive ORD gives a positive CD and a negative ORD gives a negative CD. As per convention, if the CD spectrum is positive, a CD curve invariably encompasses the cotton effect; the area under a CD curve (on integration) gives a measure of the rotational strength of the chromophore undergoing chiral transition.

It is found that a CD curve has an advantage over an ORD curve. This is due to the convenience of the study of the CD as compared to ORD since the ORD curve is normally associated with a background curve due to transition at a distant wavelength which interferes with the determination of the sign of the cotton effect. On the other hand, such a problem does not arise in CD curves. In case a compound has more than one absorption maximum, the CD extrema are well separated. Also, ORD in far UV is much more difficult to measure due to low light transition.

## 9.7 Axial Haloketone Rule

The axial haloketone rule is regarded as a precursor of the more general octant rule (Djerassi 1957). It is known that $\alpha$-halocyclohexanone exists in axial conformation in contrast to other $\alpha$-substituted cyclohexanones which exist in equatorial conformation. This is attributed to lesser unfavourable dipole–dipole interactions resulting in giving larger rotations as compared to other $\alpha$-substituted cyclohexanones which give rise to virtually no perceptible rotation as the substituted resides in the $\beta$-plane.

The axial haloketone rule is applicable to cyclohexanones containing an axial halogen (Cl, Br or I but not F; this is due to the low polarisability of F, and those of Cl, Br or I are higher as compared to hydrogen.

According to the axial haloketone rule, the introduction of an equatorial halogen in either of the $\alpha$-positions does not alter the sign of the cotton effect of the halogen-free cyclohexanone. However, the introduction of an axial halogen (Cl or Br) in the $\alpha$-position leads to alteration of the sign of the cotton effect of the parent ketone.

Looking along the O=C axis (Fig. 9.7), in case the halogen is on the right, there will be a positive cotton effect and if halogen is on the left, there is a negative cotton effect.

In the projected structure, the halogen (X) is in the lower right quadrant, a (+) sector and so the cotton effect is + ve. However, in the enantiomeric structure, in which halogen is in the lower left quadrant a (–) sector, a negative cotton effect is observed.

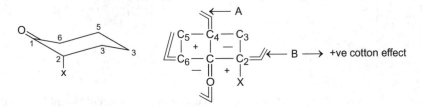

**Fig. 9.7** Positive cotton effect in case of 2-halocyclohexanone (X = Cl or Br)

## 9.8   The Octant Rule

The octant rule (Muffitt et al. 1961) correlates the sign of the cotton effect of chiral cyclohexanone derivatives with their absolute conformations.

According to Muffitt et al (1961), the space around the carbonyl group is divided into eight octants (sectors) about the $X$-, $Y$- and $Z$-axis in the three mutually perpendicular planes A, B and C designated as $XY$, $ZY$ and $XZ$, respectively (Fig. 9.8).

As seen in Fig. 9.8, the vertical plane (A) is intersecting the carbonyl oxygen at carbons 1 and 4 bisecting the cyclohexanone chair form. The plane being perpendicular to plane A contains C=O moiety and the two attached carbons ($C_2$ and $C_6$). The plane being perpendicular to both the planes A and B intersect the carbon–oxygen double bond at midpoint. Thus, the two planes divide the space around the carbonyl group into four quadrants, which are designated as upper left (UF), upper right (UR), lower left (LL) and lower right (LR). The groups in these four quadrants contribute to the sign of the dispersion curve. This is indicated by plus and minus signs (Fig. 9.8).

The four rear quadrants constituting the rear octants are away from the observer. There is another set of four quadrants (which are mirror images of the former) on the left side of plane C (not shown in Fig. 9.8).

It is found that in most of the substituted cyclohexanones, the substituents appear only in the four rear quadrants. However, in rare cases, a part of some molecules appears in the front quadrants. Thus, for simple cyclohexanones, plane C is not considered unless there are substituents to the left of it. Only the four rear quadrants due to planes A and B are to be considered.

Let us consider a molecule possessing groups extending into four quadrants. A typical example is that of 7-keto-steroids. The octant representation is shown in Fig. 9.9.

In the four rear octants, the contribution of substituents towards the sign of the cotton effect is according to the following rules:

(i)    Substituents that lie in the coordinate planes A and B do not make any contribution to the cotton effect. This also includes $2e$, $4a$, $4e$ and $6e$ bonds.

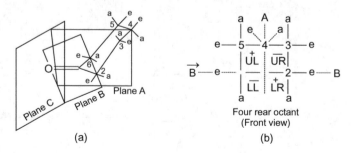

(a)                                        (b)

**Fig. 9.8   a** Front view of the orientation of cyclohexanone framework and **b** Signs of four octants along the carbonyl carbon

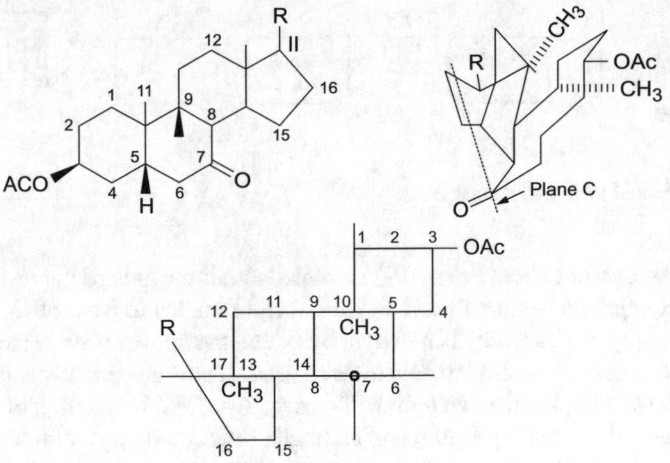

**Fig. 9.9**  Octant representation of four quadrants of 7-ketosteroid with substituting group

(ii)   Positive contribution is made by substituents lying in the (+) sectors (UL and
        LR) and negative contribution is made by substituents lying in the (–) sectors
        (UR and LL). Thus, the groups in the upper left (UL) quadrant and axial groups
        in the lower right (LR) contribute to a positive cotton effect, and the group in
        the upper right (UR) and axial substituents in the lower left (LL) contribute to
        the negative cotton effect. The octant rule is summarised as follows:

| Axial and equatorial groups positive cotton effect + | Axial and equatorial groups negative cotton effect — |
|---|---|
| Only axial groups negative cotton effect — | Only axial groups positive cotton effect + |

# 9.9   Instrumentation for ORD and CD Measurements

## 9.9.1   Instruments for ORD Measurements

Two types of spectropolarimeters, viz., visual and photoelectric spectropolarimeters.
     **Visual Polarimeter**: A visual polarimeter (Fig. 9.10) consists of a light source,
which is usually a sodium vapour lamp. Its light is made parallel by a lens ($L_1$) and
polarised by a Nicol Prism ($P_1$). In the absence of any optically active sample in the

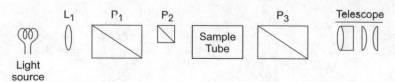

**Fig. 9.10**  Optical system of a polarimeter

sample tube, another Nicol Prism (P₃) is rotated about the axis of the instrument so that the principal planes (or P₁ and P₃) are at right angles to each other. In such a situation, there is no transmittance of light by P₃ and the field of view in the telescope is dark. In this case, P₁ and P₂ are said to be crossed. If now an optically active sample is placed in the sample tube, the light will emerge from P₃. Restoration of extinction can be achieved by rotating P₃ through an angle, (measured on a calibrated circular scale) which will be equal in magnitude and sign to the rotation of the sample. Determining precisely at which angular positions of P₃, the illumination of the field is of a minimum is difficult. This is because the transmitted intensity change per unit angular rotation is small near the extinction point. Therefore, a third prism (P₂) is inserted in order to cover half the field. This enables the two plane-polarised rays which differ in phase by half a wavelength to pass through the tube. In case the analysing prism is crossed with an unobstructed part of the field, the part is darkened as shown in Fig. 9.11a. Rotating the analyser in the correct direction gives a field as shown in Fig. 9.11b. Further rotation gives a situation as in Fig. 9.11c, in which light from the obstructed half of the field is emitted but some light from the unobstructed half passes through. By this procedure, the balance point (as shown in Fig. 9.11b) is obtained by judging when the two areas are illuminated by equal intensity.

### *Photoelectric Spectropolarimeter*

The light between 190 and 700 nm from a light source (like a xenon arc lamp) is passed through a monochromator. The monochromatic light, thus, obtained is passed through a polariser, sample cell and an analyser (Fig. 9.12).

In the spectropolarimeter, the polariser is mechanically caused to oscillate through an angle of about 2°. The plane of polarisation of the beam which passes through the sample is modulated through the same angle. The analyser is positioned at its extinction angle by rotation till the intensity of light reaching the photomultiplier is the same for both the extreme positions of the polariser.

Alternatively, modulation of the plane of polarisation is achieved electrically by interposing a Faraday cell in the light path between the polariser and analyser. The

**Fig. 9.11**  Fields of view of a polarimeter [see (a), (b) and (c)

         (a)                    (b)                    (c)

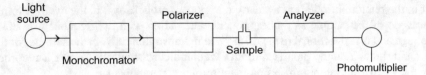

**Fig. 9.12** Optical components of a spectropolarimeter

Faraday cell consists of a cylinder surrounded by a coil through which an alternating current (60 Hz) is passed. The magnetic field of the coil induces silica optical rotation; the sign of optical rotation depends on the direction of the current passed through the coil. Thus, the plane of polarisation of the beam which leaves the Faraday cell is modulated. The ORD spectra are recorded automatically.

### 9.9.2 Instrumentation for CD Measurements

An ordinary spectrophotometer is used to measure circular dichroism (CD). However, in this, there should be a procedure for producing $d$- and $l$-circularly polarised radiation. This is achieved by passing a plane-polarised light through a quartz wave plate. Rotation of the plane from $+45°$ to $-45°$ produces first $d$- and then $l$-circularly polarised light. A number of quartz-wavelength plates are necessary to cover a large wavelength region. Other devices are also available to produce a circularly polarised beam in order to measure the relative absorption by the spectrophotometer.

A typical instrument for the measurement of circular dichroism is diagrammatically shown in Fig. 9.13.

The method consists in passing a beam of plane-polarised light through an electro-optic modulator (EOM) or pockets cell. This is a $0°$–$Z$ cut plate made of potassium dideuterium phosphate oriented in such a way so that the incident light is parallel to its $x$ or crystallographic axes. An alternating electric field (325 Hz) is applied to transparent conducting films on the opposing Z-surfaces of the plate, it alternately transmits the left and right circularly polarised components of the plane-polarised beam. On passing these through a circularly dichromic sample, the photo-current

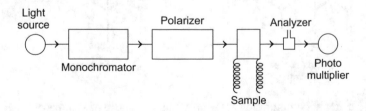

**Fig. 9.13** Optical components of a circular dichrometer

from the photomultiplier has a steady component proportional to the average transmittance of the sample and a component, alternating with the frequency of EOM voltage and is proportional to the difference in transmittance from the sample for the left and right circularly polarised light. Circular dichroism of the sample is recorded by the electrical system processing the alternating component.

## 9.10   Applications of Optical Rotatory Dispersion and Circular Dichroism

Both optical rotatory dispersion and circular dichroism have been used in the determination of structures of a number of optically active components like amino acids, proteins, steroids, antibiotics, terpenes, etc. These have been helpful in determining the stereochemistry of various components. On the basis of a number of studies, some general rules have been formulated.

### Stereochemistry of Aliphatic Amino Acids

It has been found that amino acids show a characteristic cotton effect, whose sign is helpful in the determination of the stereochemistry at the asymmetric centre. In the case of amino acids, a positive cotton effect around 215 nm indicates laevo configuration and a negative cotton effect indicates dextro enantiomer. In the case of polypeptides, measurements of optical rotatory dispersion can be used to estimate the per cent of $\alpha$-helix structure.

### Stereochemistry of Steroids

It has been found that in steroids, *cis* and *trans* ring junctions lead to different kinds of optical rotatory dispersion curves. Thus, as an illustration, it is easy to decide whether a 3-kelkosteroid belongs to the cholestane system (*A/B* ring *trans* fused) or the coprostane system (*A/B* rings *cis* fused).

Cholestane
system

Coprostane
system

These differ in configuration at $C_5$ only but this centre of symmetry is near to the keto group to have an important effect on the asymmetric environment of the latter and so the systems can be identified.

## Location of Hydroxyl Group in Steroidal Alkaloid Rubijervine

A naturally occurring steroidal alkaloid rubijervine was found to contain two hydroxy groups one of which was present at position 3. The location of the other OH group was found by partial oxidation of the alkaloid, keeping the C-3 hydroxyl group intact. The product obtained was a ketone (CHOH → CO). The ORD curve of this ketone was found to be typical of 12-ketosteroid type. So the second hydroxyl group in rubijervine is located at the C-12 position.

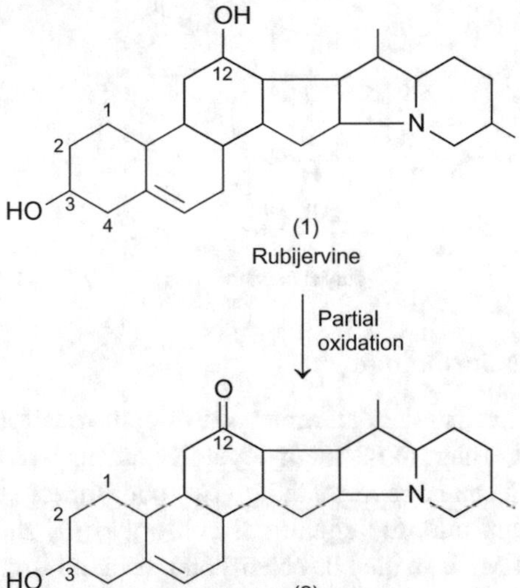

(1)
Rubijervine

Partial oxidation

(2)
Rubijervone

## Determination of Absolute Configuration

It has already been stated that for the determination of the absolute configuration of a compound, the presence of the carbonyl group is best suited for ORD studies. The carbonyl group is present frequently in a number of natural products. However, even in the absence of the carbonyl group, it may still be possible to convert other functional groups into the carbonyl group. One such example is given above. In another illustration, Cafestol (3), which is found in coffee beans, can be degraded to the corresponding ketone (4). A comparison of the ORD curve of (4) with that of $4\alpha$-elhylcholestan-3-one (5) revealed that these two molecules must have opposite configurations. As the absolute configuration of (5) is known, configuration (3) can be deduced.

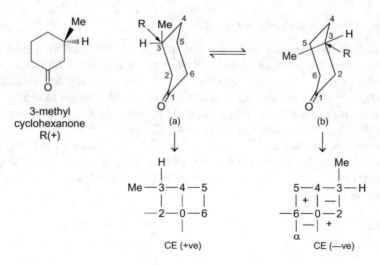

(3)
Cafestol

(4)

(5)
4 α-ethylcholestan-3-one

## Determination of Conformation

(i)  A typical example for the determination of conformation is that of 3-methyl
cyclohexanone. Thus, (+)-3-methylcyclohexanone is known to have the R
configuration. It can be represented by two conformers (shown in Fig. 9.14).
As per the octant rule, the equatorial conformer *(a)* should have a positive
cotton effect (3-Me is in the UL octant) and the axial isomer *(b)* will display a
negative cotton effect (3-Me is in UR octant). As the ketone shows a positive

**Fig. 9.14** Conformation of (+) 3-Methylcyclohexanone

cotton effect, the equatorial conformer is preferred. This is inconsistent with the principle of conformational analysis.

(ii)   Another example is determination of conformation of (+) *cis*-10-methyl-2-decalone. It can exist in any of the two conformations (I or II). These exhibit different contributions (positive or negative) to the cotton effect and so can be identified.

On the basis of the application of the cotton effect, conformation I should give a negative cotton effect (ring *B* is in the negative sector) and structure II should display a positive cotton effect (ring *B* is in the positive sector). It, however, has been found that the compound gives a negative cotton effect and hence should have conformation I (Fig. 9.15).

### Determination of Configuration

*Trans*-10-methyl-2-decalone exists in a rigid conformation (Fig. 9.16). Thus, the absolute conformation of (+) enantiomer (which shows a positive cotton effect) is

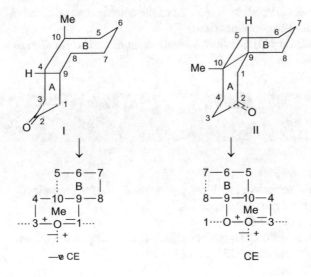

**Fig. 9.15**  Conformation of (+) *cis*-10-methyl-2-decalone

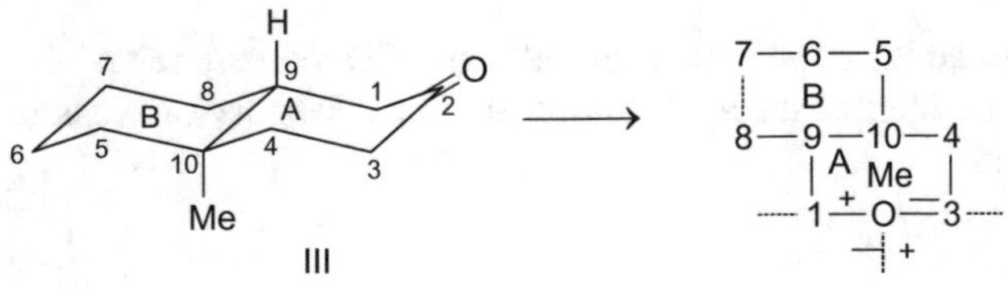

**Fig. 9.16**  Configuration of *trans*-10-methyl-2-decalone

III. This is inconsistent with the octant rule (ring *B* in the projection is in the positive sector). The mirror image configuration will project a negative cotton effect.

## Key Concepts

- **Axial Haloketone Rule**: Precursor of octant rule *see* octant Rule.
- **Circular Bifriengence**: When a linearly polarised light is passed through an optically active medium, the two circularly polarised components exhibit different refractive indices and different absorption coefficients. The two chiral properties, viz., refractive indices and absorption coefficients of a chiral medium are known as circular birefringence and circular dichroism, respectively.
- **Circular Dichroism**: *See* circular birefringence.
- **Cotton Effect**: The combination of circular birefringence and circular dichroism is known as the cotton effect.
- **Octant Rule**: Correlation of the sign of the cotton effect of chiral cyclohexanone with their absolute confirmation.
- **Optical Rotatory Dispersion**: Measurement of optical rotation as a function of wavelength.

## Problems

1. What is plane-polarised light?
2. Explain the terms circular birefringence and circular dichroism.
3. Define optical rotatory dispersion.
4. What is the cotton effect?
5. Describe the different types of optical rotatory dispersion curves.
6. Give a comparison of ORD and CD curves.
7. Write notes on:

   (a)  Axial haloketone rule.
   (b)  Octant Rule.

8. Describe the instrumentation for ORD and CD measurements.
9. Give an account of the applications of optical rotatory dispersion and circular dichroism.

# Part III
# Stereochemistry of Reactions

# Chapter 10
# Stereochemistry of Addition Reactions

## 10.1 Introduction

Addition reactions are of three types, viz., *electrophilic addition* reactions, *nucleophilic addition* reactions and free radical addition reactions. The electrophilic *addition reactions* take place in unsaturated compounds like alkenes and alkynes. However, nucleophilic addition reactions take place in compounds containing a carbonyl group.

## 10.2 Electrophilic Addition Reactions

These reactions are typical of alkenes and alkynes.

### 10.2.1 Electrophilic Addition Reactions of Alkenes

The common electrophilic addition reactions involve addition of hydrogen halide, water, halogen, *halohydrin* (halogen and water). Symmetrical alkenes give only one product. However, unsymmetrical alkene may give two products. We are concerned only with unsymmetrical alkenes.

#### 10.2.1.1 Electrophilic Addition of Hydrogen Halides

In case of unsymmetrical alkenes (like propene), in the *addition of H–Br formation of two products are possible*, viz., isopropyl bromide and *n*-propyl bormide (Fig. 10.1). It is found that in case of propene only *isopropyl bromide* is obtained

© The Author(s), under exclusive license to Springer Nature Switzerland AG 2022    245
V. K. Ahluwalia, *Stereochemistry of Organic Compounds*,
https://doi.org/10.1007/978-3-030-84961-0_10

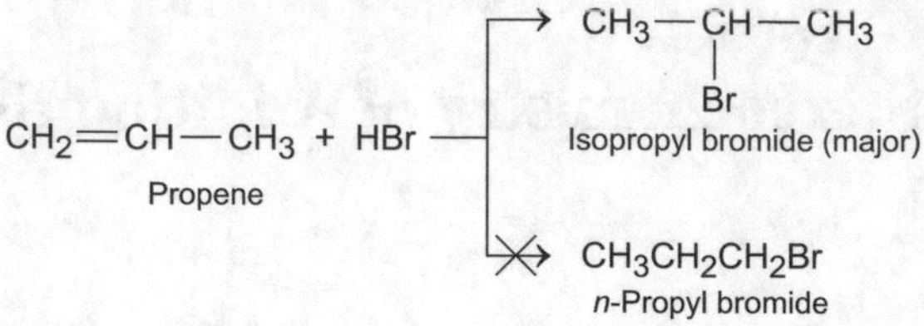

**Fig. 10.1** Reaction of propene with HBr

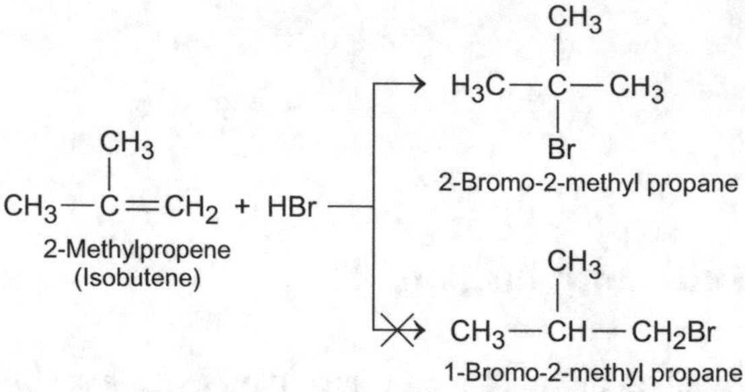

$$CH_2=CH-CH_3 + HBr \longrightarrow$$

Propene

$$\longrightarrow CH_3-CH-CH_3$$
$$\quad\quad\quad\quad\quad\quad |$$
$$\quad\quad\quad\quad\quad\quad Br$$
Isopropyl bromide (major)

$$\longrightarrow\!\!\!\times CH_3CH_2CH_2Br$$
n-Propyl bromide

**Fig. 10.2** Reaction of 2-methylpropene with HBr

$$\quad\quad CH_3$$
$$\quad\quad\ |$$
$$CH_3-C=CH_2 + HBr \longrightarrow$$

2-Methylpropene
(Isobutene)

$$\quad\quad\quad\quad\quad CH_3$$
$$\quad\quad\quad\quad\quad\ |$$
$$\longrightarrow H_3C-C-CH_3$$
$$\quad\quad\quad\quad\quad\ |$$
$$\quad\quad\quad\quad\quad Br$$
2-Bromo-2-methyl propane

$$\quad\quad\quad\quad\quad CH_3$$
$$\quad\quad\quad\quad\quad\ |$$
$$\longrightarrow\!\!\!\times CH_3-CH-CH_2Br$$
1-Bromo-2-methyl propane

as the major product. Similarly, *2-methyl propene* on reaction with HBr gives 2-bromo-2-methylpropane (Fig. 10.2).

The addition of unsymmetrical reagents like HBr to unsymmetrical alkenes is decided on the basis of **Markovnikov's rule**. According to this rule, the positive end of the reagent gets attached to the carbon atom of the double bond bearing larger number of H atoms. It can be said that the halogen atom in case of H–X gets attached to the carbon atom with fewer H atoms.

The formation of isopropyl bromide (Fig. 10.1) can be explained on the basis of the following:

*Propene* reacts with HBr to give a π complex, which gives rise to the formation at a cyclic intermediate and bromide ion. The cyclic intermediate can form two carbocations, i.e., primary and secondary. The secondary carbocation is more stable and reacts with the nucleophile (Br⁻) to give isopropyl bromide as the major product (Fig. 10.3).

However, if the above reaction is carried out in presence of peroxide, the product obtained is *n*-propyl bromide and the addition takes place in anti-Markovnikov fashion. This is known as **peroxide effect**. In this case, the addition takes place by free radical mechanism involving the following three steps (Fig. 10.4).

In case of 1-methylcyclohexene (an unsymmetrical alkene), the addition of HBr gives 1-bromo-1-methylcyclohaxane as shown below (Fig. 10.5).

The addition of HX to an unsymmetrical alkene is **regioselective**, which implies that some of the product with anti-Markonikov orientation is also obtained as a minor product.

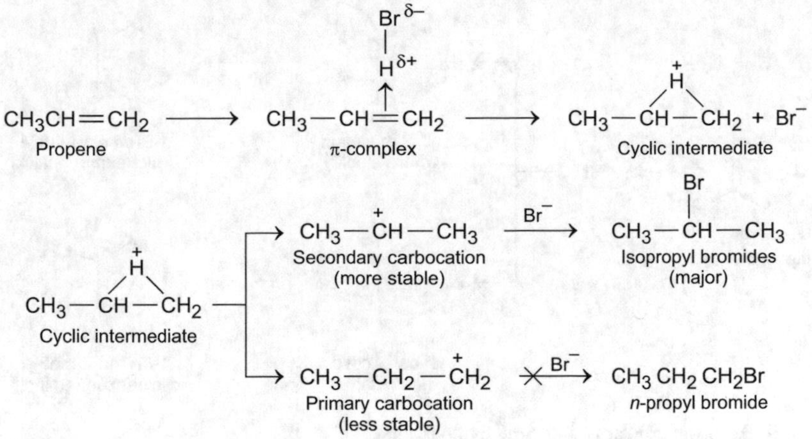

$$\underset{\text{Propene}}{CH_3CH=CH_2} \longrightarrow \underset{\pi\text{-complex}}{CH_3-CH\overset{\overset{\displaystyle Br^{\delta-}}{\overset{|}{\underset{\uparrow}{H^{\delta+}}}}}{=}CH_2} \longrightarrow \underset{\text{Cyclic intermediate}}{CH_3-\overset{\overset{+}{\overset{H}{\diagup\diagdown}}}{CH}-CH_2 + Br^-}$$

$$\underset{\text{Cyclic intermediate}}{CH_3-\overset{\overset{+}{\overset{H}{\diagup\diagdown}}}{CH}-CH_2}$$

$$\rightarrow \underset{\substack{\text{Secondary carbocation} \\ \text{(more stable)}}}{CH_3-\overset{+}{CH}-CH_3} \xrightarrow{Br^-} \underset{\substack{\text{Isopropyl bromides} \\ \text{(major)}}}{CH_3-\overset{\overset{\displaystyle Br}{|}}{CH}-CH_3}$$

$$\rightarrow \underset{\substack{\text{Primary carbocation} \\ \text{(less stable)}}}{CH_3-CH_2-\overset{+}{CH_2}} \overset{Br^-}{\times\!\!\longrightarrow} \underset{n\text{-propyl bromide}}{CH_3\,CH_2\,CH_2Br}$$

**Fig. 10.3** Mechanism of the formation of isopropyl bromide by the reaction of propene with HBr

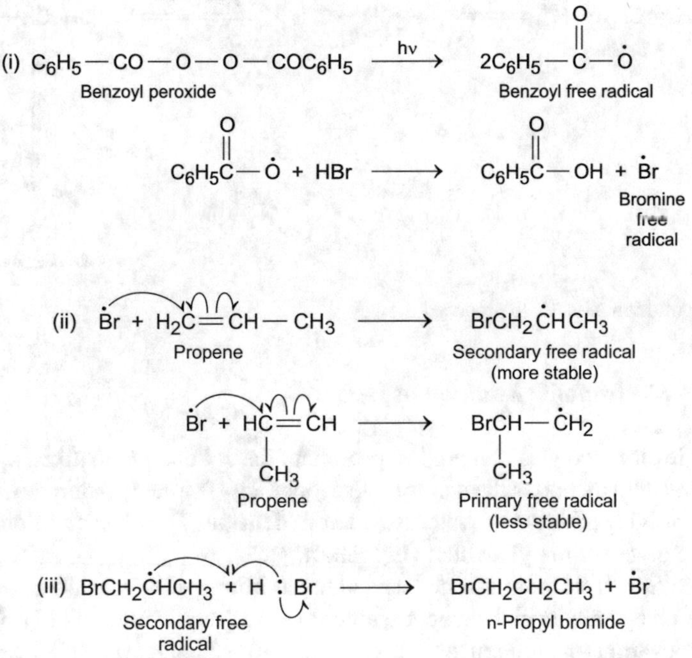

$$\text{(i) } \underset{\text{Benzoyl peroxide}}{C_6H_5-CO-O-O-COC_6H_5} \xrightarrow{h\nu} \underset{\text{Benzoyl free radical}}{2C_6H_5-\overset{\overset{\displaystyle O}{\|}}{C}-\overset{\displaystyle \cdot}{O}}$$

$$\underset{}{C_6H_5\overset{\overset{\displaystyle O}{\|}}{C}-\overset{\displaystyle \cdot}{O} + HBr} \longrightarrow \underset{}{C_6H_5\overset{\overset{\displaystyle O}{\|}}{C}-OH} + \underset{\substack{\text{Bromine} \\ \text{free} \\ \text{radical}}}{\overset{\displaystyle \cdot}{Br}}$$

$$\text{(ii) } \overset{\displaystyle \cdot}{Br} + \underset{\text{Propene}}{H_2C=CH-CH_3} \longrightarrow \underset{\substack{\text{Secondary free radical} \\ \text{(more stable)}}}{BrCH_2\overset{\displaystyle \cdot}{C}HCH_3}$$

$$\overset{\displaystyle \cdot}{Br} + \underset{\substack{| \\ CH_3 \\ \text{Propene}}}{HC=CH} \longrightarrow \underset{\substack{| \\ CH_3 \\ \text{Primary free radical} \\ \text{(less stable)}}}{BrCH-\overset{\displaystyle \cdot}{C}H_2}$$

$$\text{(iii) } \underset{\substack{\text{Secondary free} \\ \text{radical}}}{BrCH_2\overset{\displaystyle \cdot}{C}HCH_3} + H:Br \longrightarrow \underset{n\text{-Propyl bromide}}{BrCH_2CH_2CH_3} + \overset{\displaystyle \cdot}{Br}$$

**Fig. 10.4** Formation of n-propyl bromide by the addition of HBr to propene in presence of peroxide

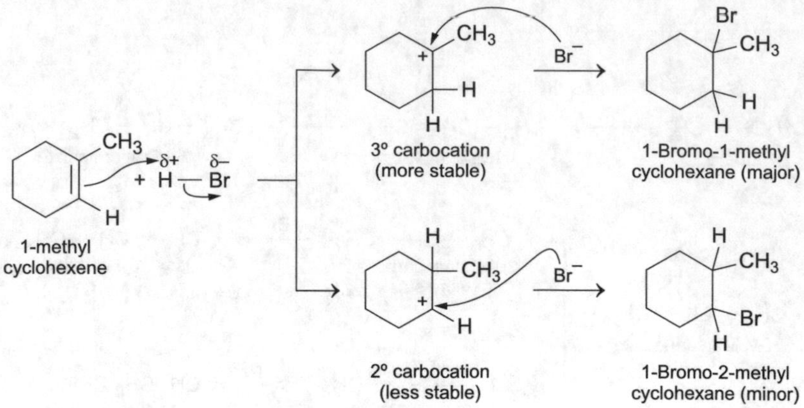

**Fig. 10.5** Addition of HBr to 1-methyl cyclohexene

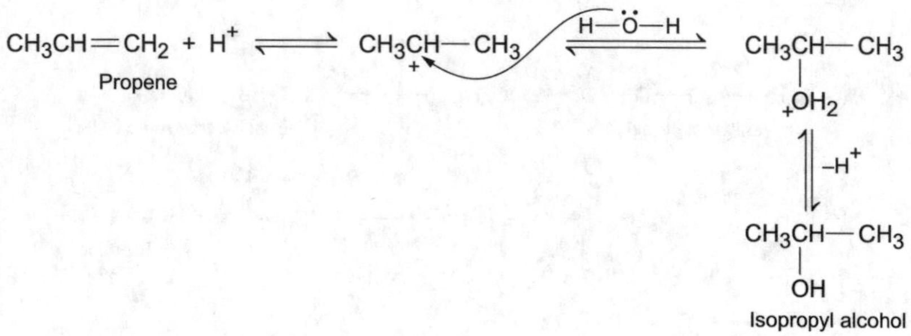

**Fig. 10.6** Addition of water to propene

### 10.2.1.2 Electrophilic Addition of H₂O

Water by itself is too weak an acid to protonate the π bond on an alkene. Addition of a mineral acid increases the acidity (electrophilicity) of the reaction medium. The process involving addition of water is called **hydration**. Thus, the addition of water to propene gives isopropyl alcohol (Fig. 10.6).

The addition of water to alkene takes place in Morkovnikov fashion.

In case of 1-methylcyclohexene, treatment with aqueous solution of $H_2SO_4$ gives the 3° carbocation (in preference to 2° carbocation), which reacts with water to give alcohol as shown in Fig. 10.7.

### 10.2.1.3 Electrophilic Addition of Halogen

Addition of halogen to an alkene in an inert solvent gives vicinal dihalides (Fig. 10.8).

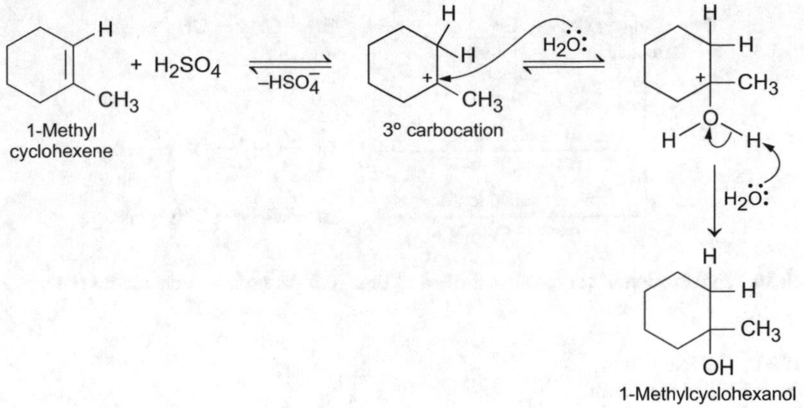

**Fig. 10.7** Addition of water to 1-methylcyclohexene

**Fig. 10.8** Addition of halogen to an alkene

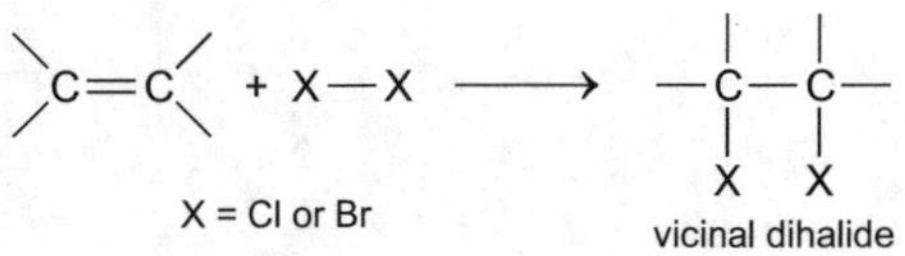

$X = Cl$ or $Br$

vicinal dihalide

Since the halogen (chlorine or bromine) are symmetrical, regiochemistry is not of importance. In view of this only symmetrical alkenes are studied.

It was earlier believed that when an alkene comes in close proximity of a halogen molecule, it gets polarised due to $\pi$ electrons of the alkene resulting in the formation of dipole in the halogen molecule. The positive end of this polarised halogen gets loosely attached to the $\pi$ electron cloud of the alkene forming a $\pi$ complex, which breaks down to give a carbocation. The formed carbocation combines readily with the nucleophilic halogen giving the adduct as shown below by using bromine as the halogen (Fig. 10.9).

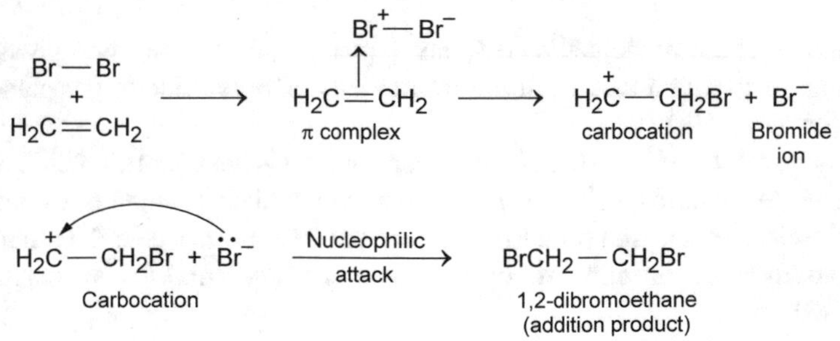

**Fig. 10.9** Addition of bromine to ethene via the formation of $\pi$ complex

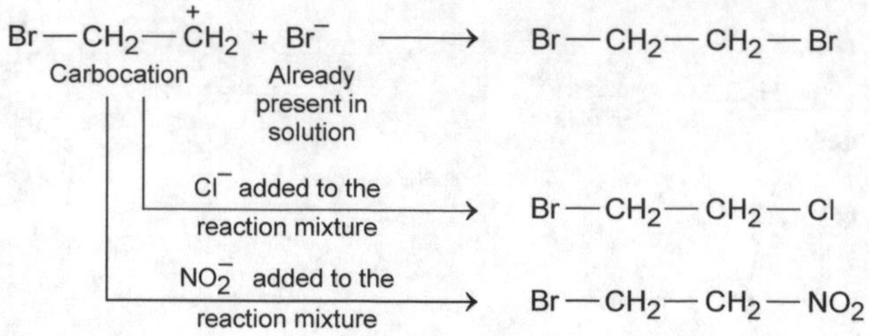

$$\underset{\text{Carbocation}}{Br-CH_2-\overset{+}{C}H_2} + \underset{\substack{\text{Already}\\\text{present in}\\\text{solution}}}{Br^-} \longrightarrow Br-CH_2-CH_2-Br$$

$$\underset{\text{reaction mixture}}{Cl^- \text{ added to the}} \longrightarrow Br-CH_2-CH_2-Cl$$

$$\underset{\text{reaction mixture}}{NO_2^- \text{ added to the}} \longrightarrow Br-CH_2-CH_2-NO_2$$

**Fig. 10.10**  The reaction of carbocation formed in Fig. 10.9 with other nucleophiles ($Cl^-$, $NO_2^-$)

**Fig. 10.11**  The reaction of bromine with alkene to form *trans* addition product via the formation of a cycle bromonium ion

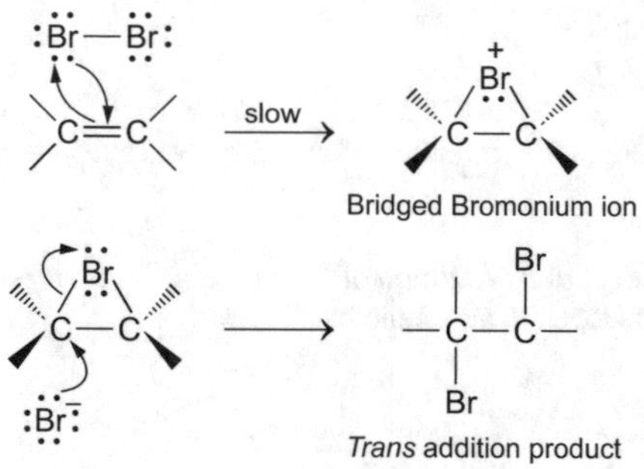

Bridged Bromonium ion

*Trans* addition product

The above mechanism (Fig. 10.9) is supported by the observation that mixed products are obtained when an alkene is reacted with bromine in an inert solvent in presence of other nucleophiles such as chloride or nitrate ions. In this case, a stepwise addition takes place (Fig. 10.10).

It has now been conclusively proved that the above mechanism (Fig. 10.9) does not hold good. This is because the addition $X_2$ gives *trans* addtion products. It is now believed that the reaction takes place via a cyclic intermediate bridged bromonium ion). Subsequently, the ring opening of the cyclic intermediate takes place to give *trans* addition product (Fig. 10.11).

The formation of the cyclic bromonium ion has been supported by spectroscopic data.

The attack of the nucleophile ($Br^-$) takes place from the backside giving *trans* addition product. The attack from the frontside is hindered due to the presence of bromo atom (Fig. 10.12).

The above reaction (Fig. 10.12) is stereospecefic; it is not enantiospecific or enantioselective. An enantiospecific reaction is one which gives a single enantiomer and an enantioselective reaction produces more of one enantiomer than the other.

The addition of bromine to cyclohexene gives a mixture of enantiomers (Fig. 10.13).

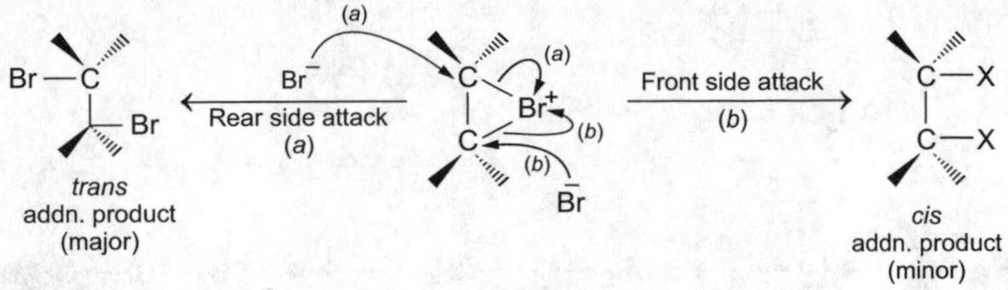

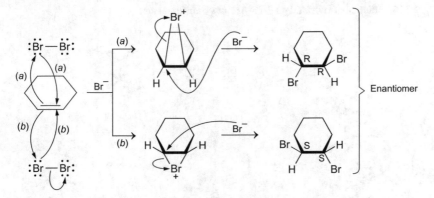

**Fig. 10.12** Formation *trans* addition product as the major component

**Fig. 10.13** Addition of Br$_2$ to cyclohexane giving a mixture of enantiomers

### 10.2.1.4 Electrophilic Addition of HOBr (Br$_2$ + H$_2$O)

A solution of bromine in water gives HOBr. Thus, the first step in the reaction of an alkene is with bromine to give the bridged bromonium ion (as was shown in Figure 10.11). In the second step, water reacts with bridged bromonium ion (and not the Br$^-$); this is because the concentration of Br$^-$ is much less than that of H$_2$O. The stereochemistry of addition in this case also is *trans*. The final step is loss of a proton to yield a vicinal halohydrin (Fig. 10.14).

In case of 1-methylcyclohexene, the addition of HOBr (Br$_2$ + H$_2$O) gives a vicinal halohydrin (Fig. 10.15).

In the above case (Fig. 10.15), the two groups which add to the π bond are different and so the regiochemistry of the addition is rationalised. The major product in this reaction has the bromine atom attached to carbon atom with more H atoms.

Thus, in the above reaction the major product is a 3° alcohol and so the reaction is regioselective and that the orientation of OH to Br is *trans* and so the reaction is stereospecific. Since a racemic mixture is obtained, the reaction is not enantioselective.

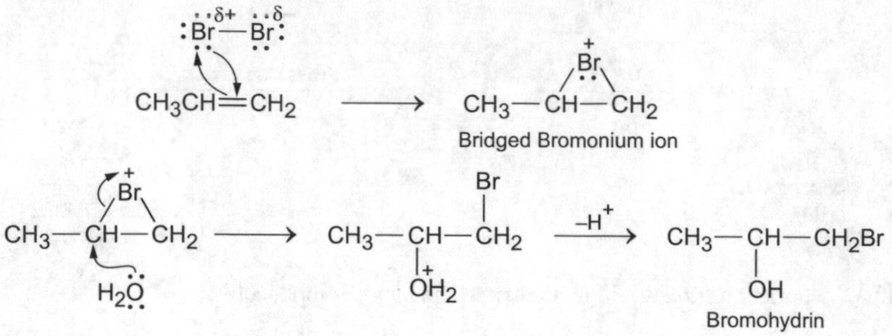

**Fig. 10.14**  Addition of HOBr (H$_2$O + Br$_2$) to unsymmetrical alkene

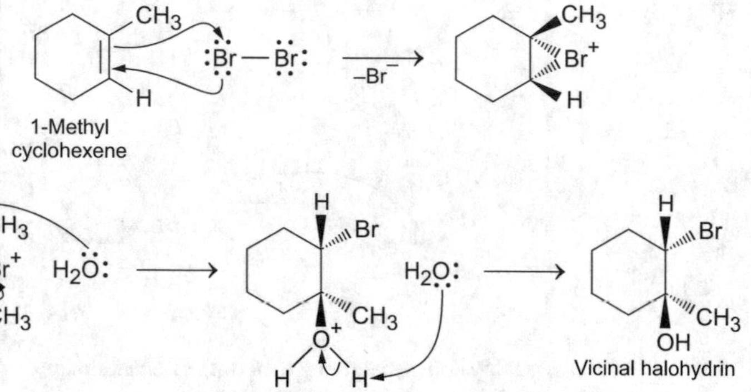

**Fig. 10.15**  The reaction of 1-methylcyclohexene with HOBr (Br$_2$ + H$_2$O) gives the vicinal halohydrin

## Solved Problem

What are the major products obtained by the reaction of 1-pentene and bromine in water? Discuss the stereochemistry of the products and assign the configurations of the stereocentres.

**Ans**. The reaction of bromine with 1-pentene gives the bromonium ion. In this case, the bromine atom can be attached on either side of the plane of the carbon double bond of 1-pentene. Reaction of bromine with the front face (path *a*) gives the intermediate (I) and reaction at the back face of the double bond via path (*b*) gives the intermediate (II). In the halonium ion (I or II), the secondary carbon of the three-membered ring carries greater share of the charge than the primary carbon.

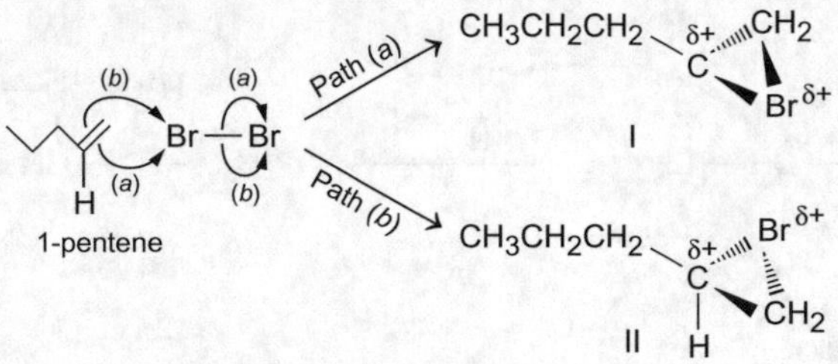

In the second step, the intermediates I and II react with water. In this case, the reaction occurs on the side of the three-membered ring away from the bromine atom. Final deprotonation gives vicinal bromohydrin products.

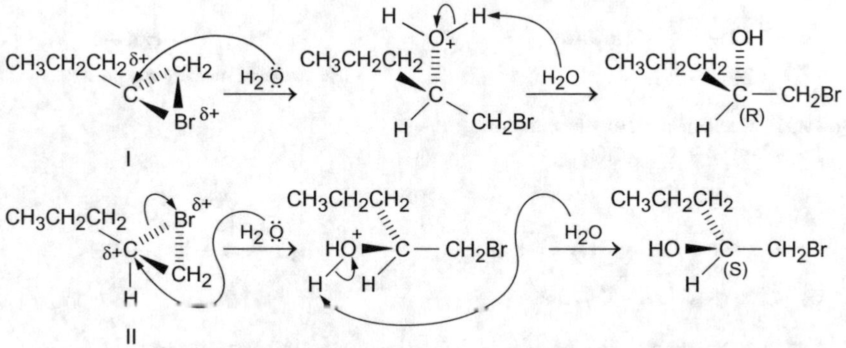

The product obtained is a racemic mixture since both the starting reactants are achiral and a new chiral centre is produced.

### 10.2.1.5 Oxymercuration-Demercuration: An Electrophilic Addition

Metal ions are considered to be electrophiles. These react with alkenes. The reaction is called **oxymercuration**. In this reaction, the addition of OH (from water) and HgOAc takes place to a double bond. A reagent commonly used is mercury (II) acetate, which ionises in water to form (aceto) mercury (II) cation and the acetate ion.

$$Hg(OAc)_2 \underset{}{\overset{H_2O}{\rightleftharpoons}} {}^+HgOAc + AcO^-$$

In oxymercuration, the mercury species, $^+HgOAc$ attacks the less substituted carbon of the double bond (i.e., at the carbon atom that bears greater number of H atoms. Subsequent steps are given below (Fig. 10.16).

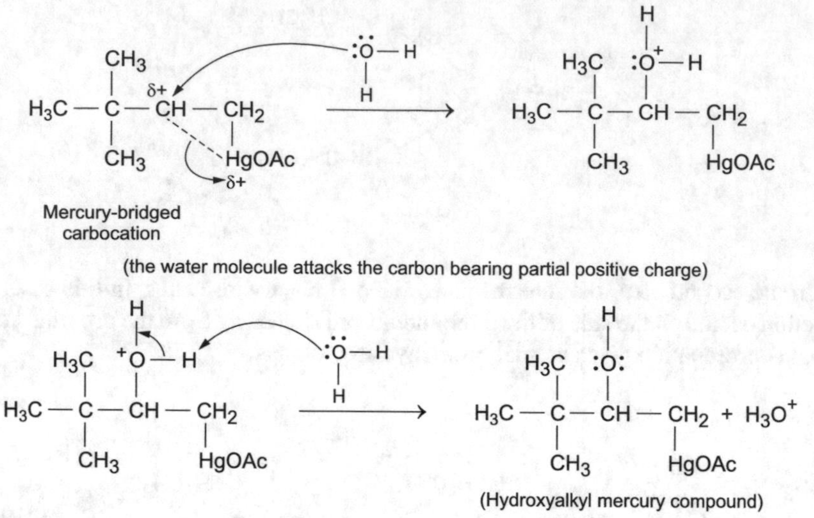

Fig. 10.16 Mechanism of oxymercuration

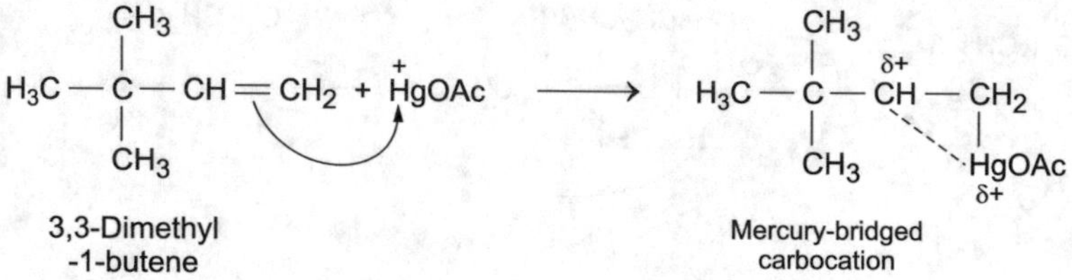

3,3-Dimethyl
-1-butene

Mercury-bridged
carbocation

In the mercury-bridged carbocation, the positive charge is shared between 2° carbon atom and the mercury atom. The charge on the carbon atom is sufficiently large in order to account for the Markovnikov orientation of the addition.

Finally in the **demercuration**, the formed (hydroxyalkyl) mercury compound on treatment with NaBH₄ gives the tertiary alcohol (Fig. 10.17).

The reaction of 1-methyl cyclohexene with ⁺HgOAc gives mercury-bridged carbocation. In this case, the greater stability of a 3° carbocation makes the bridged mercuric ions unsymmetrical. Subsequently water reacts as a nucleophile at the more highly substituted carbon giving a *trans* product. Since HgOAc group (the electrophile) in the product is attached to the carbon atom bearing more H atoms,

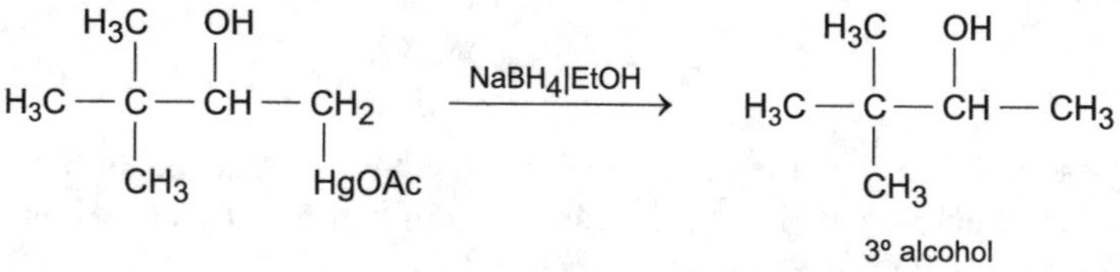

3° alcohol

Fig. 10.17 Demercuration

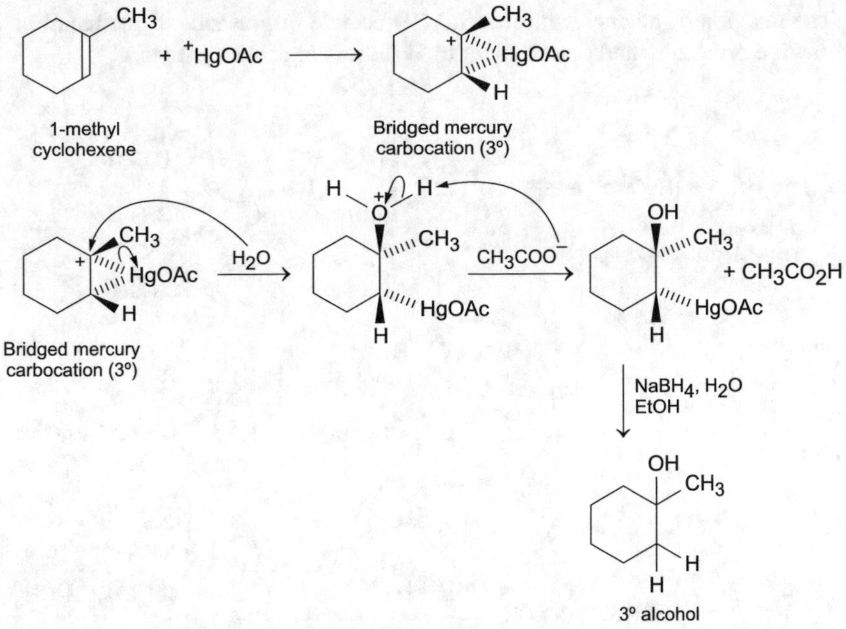

**Fig. 10.18** Mercuration-Demercuration of 1-methyl cyclohexene

it can be said that Markovnikov addition has occurred. Thus, oxymercuration is stereospecific (anti) and regioselective. Final treatment of the product with $NaBH_4$ (demercuration) gives a tertiary alohol. Various steps involved are given in Fig. 10.18.

In the above case, the final product 3° alcohol is achiral. The same alcohol is also obtained with the acid catalysed addition of $H_2O$ to the alkene double bond.

---

### Solved Problem

---

What products are expected to be obtained from the reaction of 1-hexene and mercury(II) acetate in methanol? Discuss the stereochemistry of the products obtained by demercuration.

**Ans**. 1-Hexene on reaction with $^+HgOAc$ gives two bridged mercury carbocation since the reaction can take place at each face of the double bond. Subsequently methanol reacts with the cationic species and gives enantiomeric ethers after deprotonation (see Fig. on next page).

In the above case, the stereo centre at $C_2$ in each molecule remains the same, but the configuration changes as the group priorities are altered in the demercuration product.

The reaction of alkenes with mercury(II) acetate in presence of methanol is an alternative procedure and complement to **Williamson ether synthesis**.

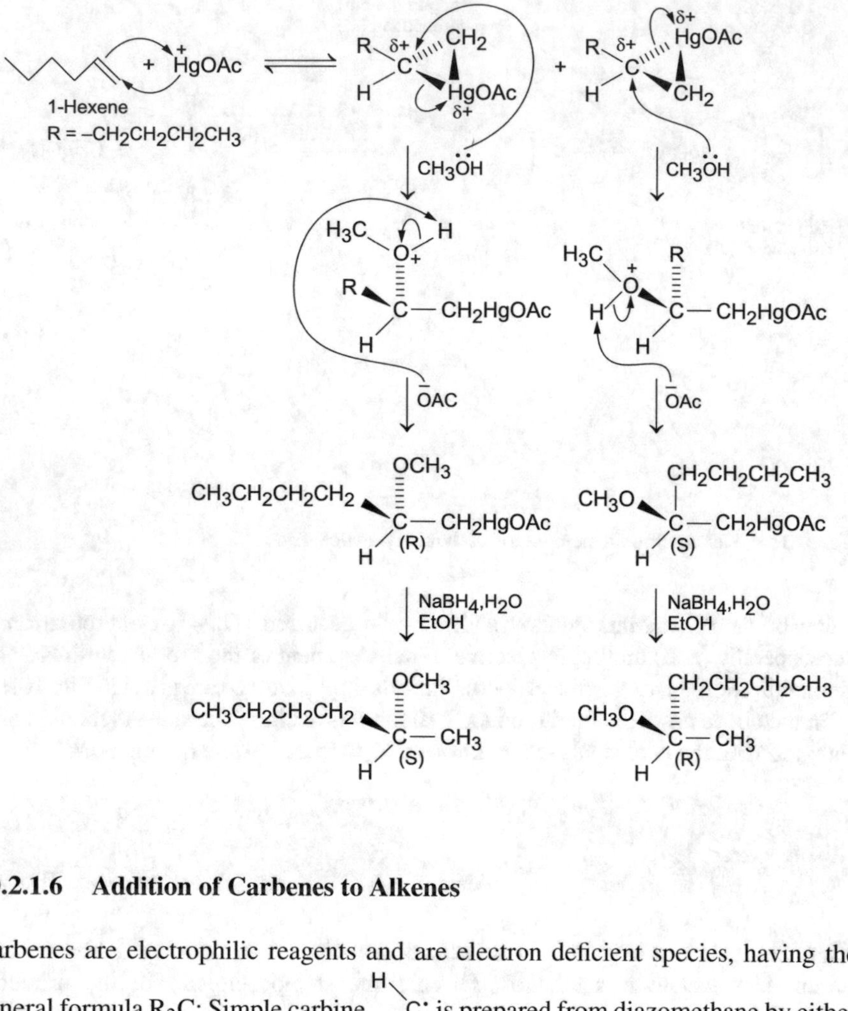

### 10.2.1.6 Addition of Carbenes to Alkenes

Carbenes are electrophilic reagents and are electron deficient species, having the general formula $R_2C:$ Simple carbine $\overset{H}{\underset{H}{\diagdown}}C:$ is prepared from diazomethane by either heating or irradiation by light (photolysis) (Fig. 10.19).

**Fig. 10.19** Preparation of methylene

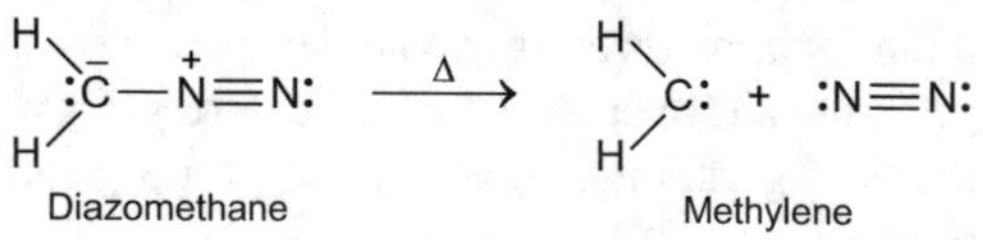

Diazomethane                           Methylene

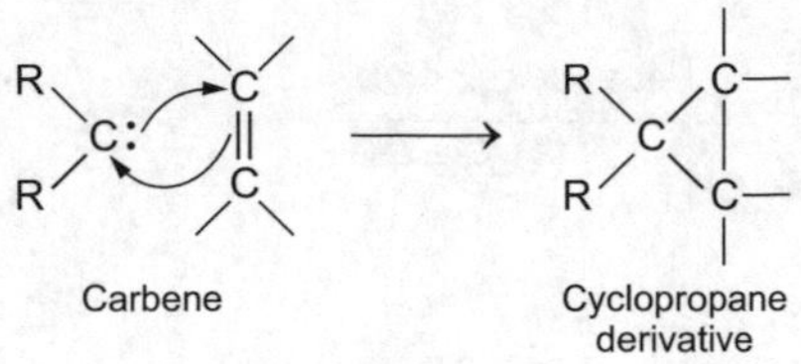

Fig. 10.20 Reaction of alkene with carbene

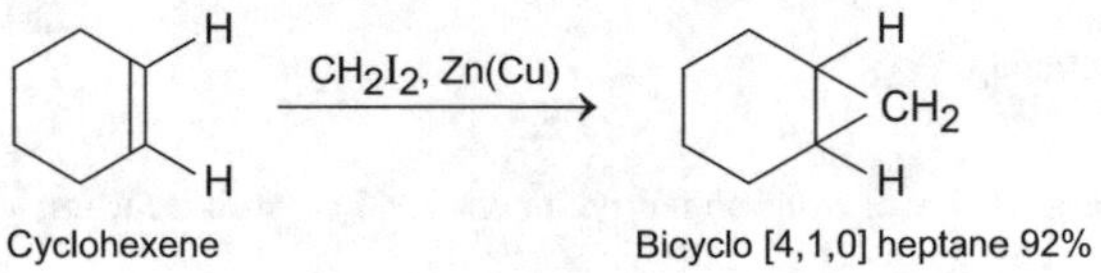

Cyclohexene                Bicyclo [4,1,0] heptane 92%

**Fig. 10.21** Reaction of cyclohexene with Simmon-Smith reagent

The carbene carbon has only six electrons and so it is a potent electrophile and reacts with the nucleophilic $\pi$ bond of an alkene to form a three-membered ring (cyclopropane). The process is called cyclopropanation and is a concerted process (Fig. 10.20).

In place of carbenes it is convenient to use a **carbenoid reagent**, which reacts as if it was an carbene. A common procedure is **Simmon–Smith reaction**, which uses **Simmon–Smith reagent** [$CH_2I_2$, $Zn(Cu)$], which is prepared by the treatment of $CH_2I_2$ with Zn that has been coated with copper. Thus, cyclohexene gives bicyclo [4.1.0] heptane (Fig. 10.21).

It is found that the Simmon–Smith reaciton is not universally applicable. Another reagent used for cyclopropanation is iodomethylene zinc trifluoroacetate, which is prepared by reacting diethyl zinc with trifluoroacetic acid and methylene iodide (Fig. 10.22).

This reagent reacts with alkenes to give the corresponding cyclopropane derivative. In this reaction, the stereochemistry of the starting alkene is retained in the cyclopropane product showing that the reaction is concerted. Two examples are given below (Fig. 10.23).

The function of Zn in this reaction also in Simmon–Smith reaction is to capture the iodide ion as shown below:

$$(CH_3CH_2)_2Zn + CF_3COOH + CH_2I_2 \xrightarrow[0°C]{CH_2Cl_2} (CF_3COO)Zn - CH_2 - I + CH_3CH_2I + CH_3CH_3$$

Diethyl zinc    Trifluoro    Methylene     Iodomethylene zinc
             acetic acid    iodide       trifluoroacetate

**Fig. 10.22** Preparation of idomethylene zinc trifluoroacetate

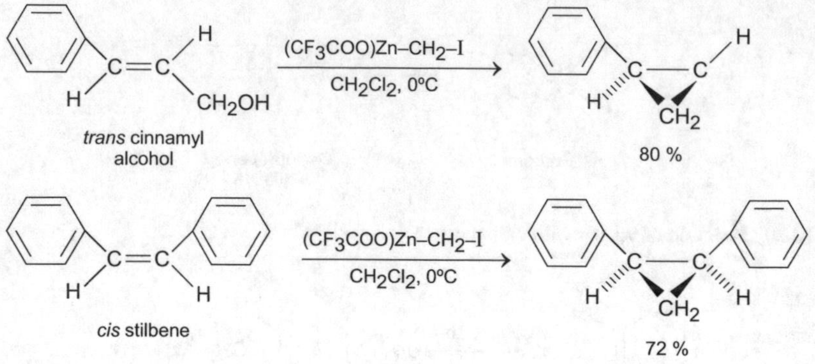

**Fig. 10.23** Cyclopropanation of *trans* cinnamyl alcohol and *cis* stilbene using iodomethylene zinc trifluoroacetate

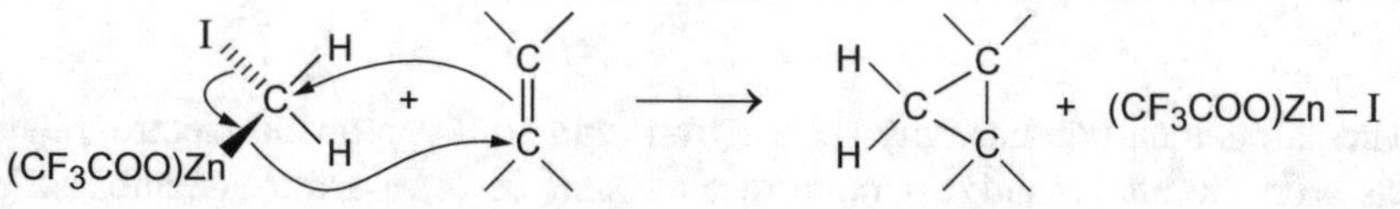

### 10.2.1.7   Addition of Borane to Alkenes

Borane ($BH_3$) has an electron-deficient boran atom and is an electrophile. It reacts with alkenes to give organoboranes and the process is called **hydroboration**.

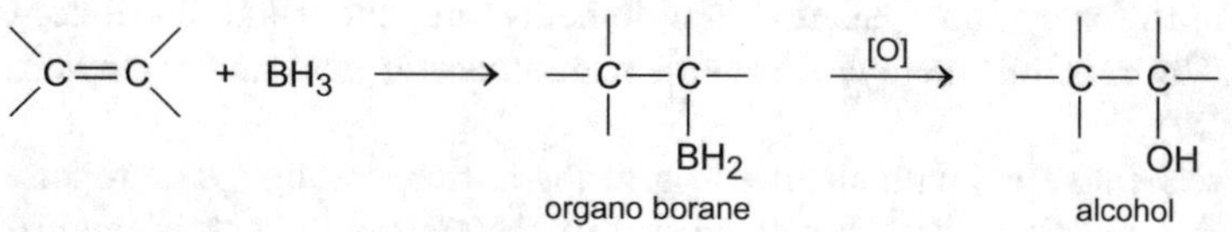

In case of cyclohexene, the following reaction takes place.

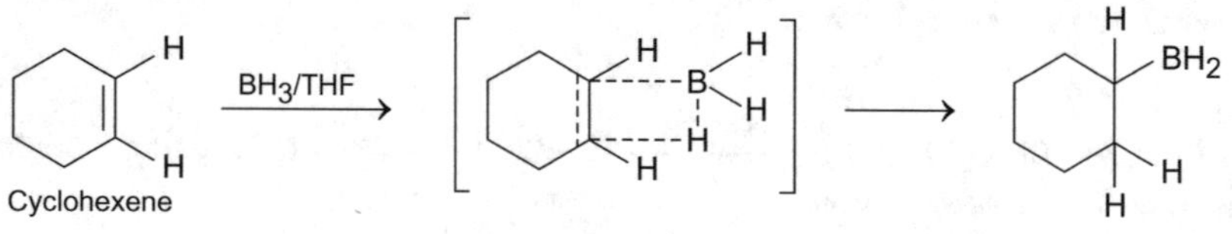

The organoboranes have tremendous synthetic applications. For details see Sect. 19.3.7, Chap. 19.

## 10.2.2   Electrophilic Addition Reactions of Alkynes

The common electrophilic addition reactions of alkyne involve addition of hydrogen halides, halogen, water and carbenes.

### 10.2.2.1   Electrophilic Addition of Hydrogen Halides

The addition of H.X to an alkyne is an example of electrophilic addition, since the electrophilic ($H^+$) end of the reagent is attracted to the electron-rich triple bond.

   In case of symmetrical alkyne (like 2-butyne) addition of one mole HX (HBr or HCl) gives E or Z alkene, which on reaction with the second equivalent gives the dihalide, the two halogen atoms are attached to the same carbon (Fig. 10.24).

   In case of terminal alkynes, both H atoms are bonded to the terminal carbon, i.e., the hydrohalogeneration follows Markovnikov's rule (Fig. 10.25).

   The mechanism of the reaction in the above case (Fig. 10.25) involves the attack of H atom of H–Br to form a new bond generating a vinyl carbocation. In this case, the addition follows Markovnikov's rule and forms the more substituted (more stable) carbocation (vinyl carbocation). Subsequent nucleophilic attack of $Br^-$ gives vinyl bromide. Addition of second molecule of H–Br also occurs in similar two-step process. Thus, addition of $H^+$ to π bond of vinyl bromide gives a carbocation. This is followed by attack of $Br^-$ to form a geminal dibromide (2,2-dibromobutane) (Fig. 10.26).

   In the second step (addition of second molecule of HBr (Fig. 10.26), the carbo-cation (A) is formed. The alternative carbocation (B) is not formed. This is because Markovnikov addition of HBr to the viniyl bromo compound places H on the terminal carbon (C-1) to form more substituted carbocation (A) and not the less substituted carbocation (B).

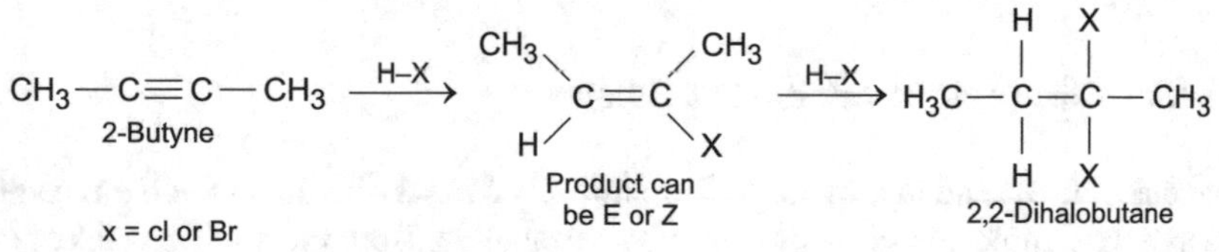

**Fig. 10.24** Addition of H–X to 2-butyne

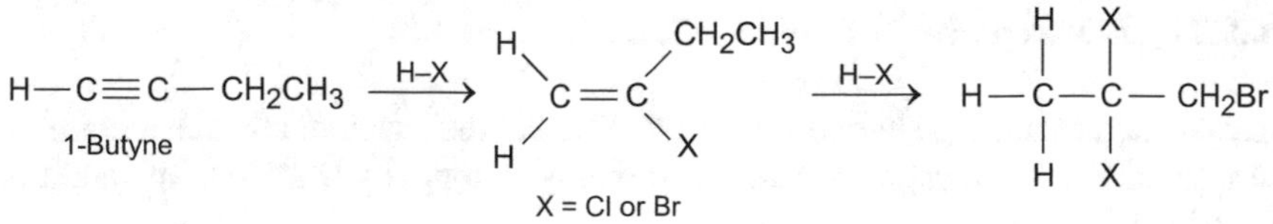

**Fig. 10.25** Addition of H–X to 1-butyne

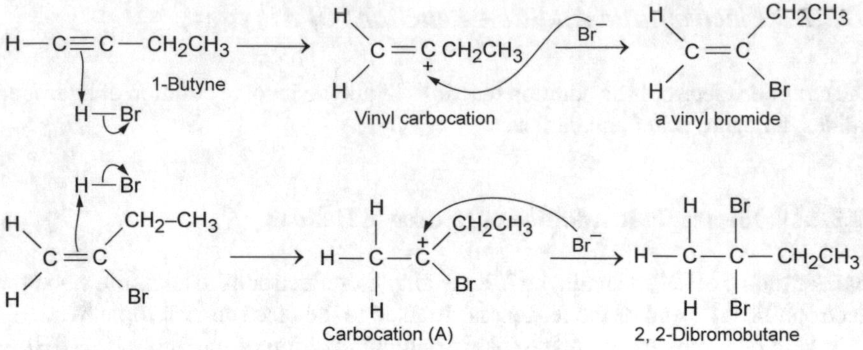

Fig. 10.26  Mechanism of the reaction of 1-butyne with HBr

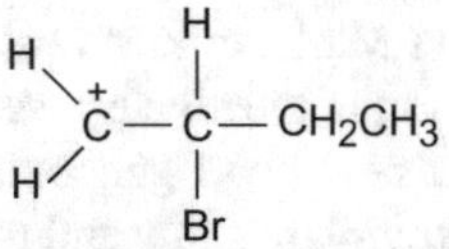

B (less substituted carbocation)
Not formed.

With the addition of only one equivalent of H Br to propyne, the reaction stops forming only vinyl bromide.

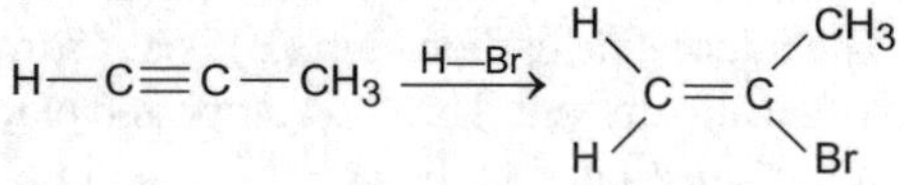

vinyl bromide
(2-Bromopropene)

### 10.2.2.2  Electrophilic Addition of Halogen

Halogens (Cl, Br) add to alkynes to form initially a *trans* dihalide, which on reaction with a second molecule of halogen gives a tetrahalide. Both the additions take place via the formation of bridged halonium ion (Fig. 10.27).

### 10.2.2.3  Electrophilic Addition of Water

Internal alkynes undergo hydration with $H_2SO_4$ but the terminal alkynes require the presence of $Hg^{2+}$ catalyst in order to yield methyl ketones by Markovnikov addition of $H_2O$ (Fig. 10.28).

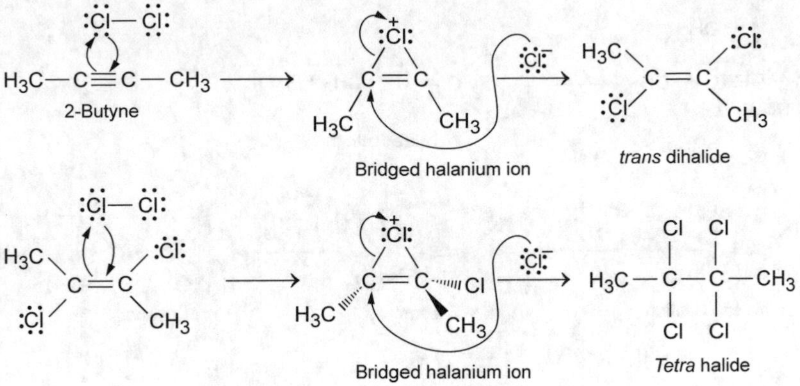

Fig. 10.27 Addition of halogen to 2-butyne

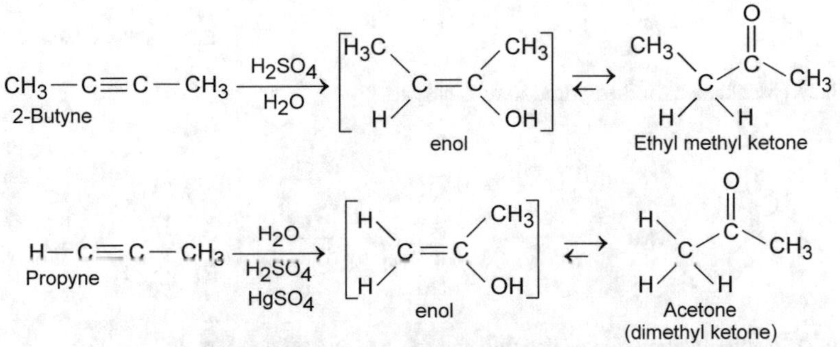

Fig. 10.28 Hydration of alkynes

The mechanism of hydration of an alkyne involves addition of $H^+$ (from $H_3O^+$) forming an $sp$ hybridised vinyl carbocation. This is followed by nucleophilic attack on carbocation and loss of a proton to give an enol, which tautomerises to the ketone (Fig. 10.29).

## 10.3  Nucleophilic Addition Reactions

Most of the nucleophilic addition reactions are encountered in compounds containing a carbonyl group, particularly aldehydes and ketones. The carbon atom of a carbonyl group is known to bear a partial positive charge which is due to the polarity of the $C{=}O$ double bond. Thus, the C atom of the $C{=}O$ group is electrophilic. A nucleophile reacts with the electrophilic carbon of the carbonyl group. In this case, the electrons in the $\pi$ bond move to the oxygen to generate a nucleophilic site. The formed alkoxide on treatment with $H_3O^+$ gives the alcohol (Fig. 10.30).

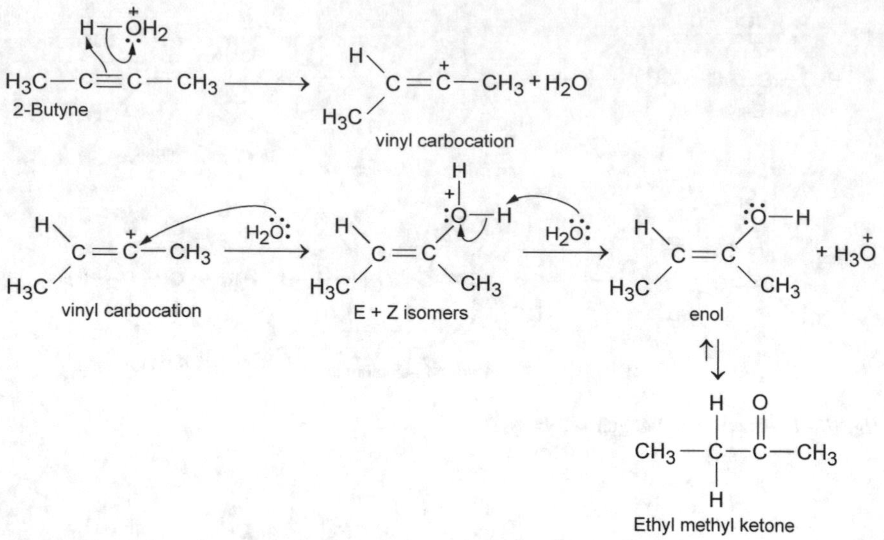

**Fig. 10.29**  Mechanism of the hydration of 2-butyne

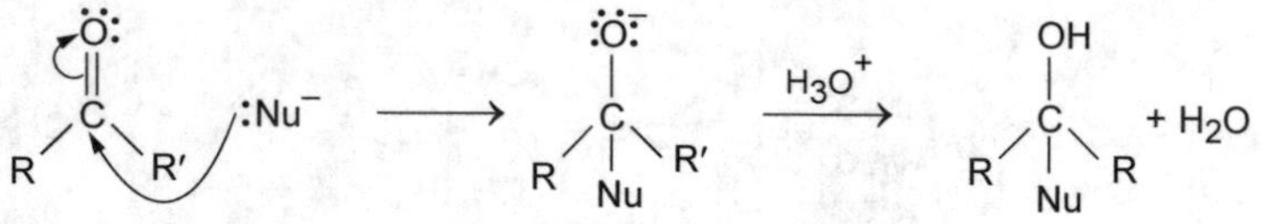

**Fig. 10.30**  Representation of a nucleophilic addition to carbonyl group

The nucleophiles that are mostly used include hydrogen cyanide, water, halogen halides, organometallic compounds, hydride and the well-known Wittig reagent.

## 10.3.1  Nucleophilic Addition of $^-CN$

Hydrogen cyanide does not add to carbonyl group. However, in presence a base, $^-CN$ is generated which adds on to C=O group to give an alkoxide, which reacts with HCN giving a cynohydrin and regenerating $^-CN$ (Fig. 10.31).

## 10.3.2  Nucleophilic Addition of Water

Water undergoes addition to form an unstable geminal diol under basic conditions. At higher pH, water contains the hydroxide ion, which acts as a base and also a

$$HCN + OH^- \rightleftharpoons CN^- + H_2O$$

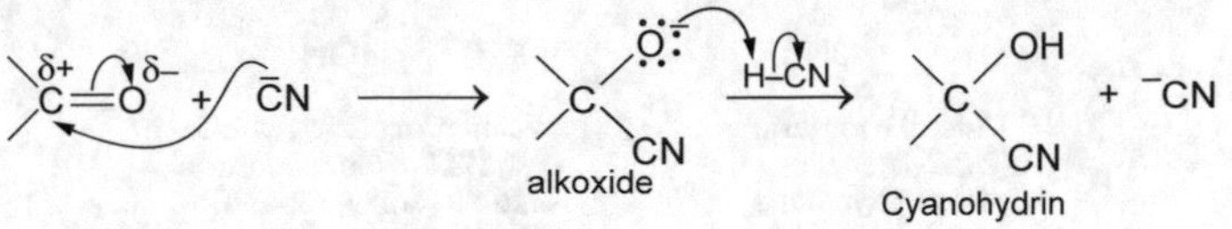

**Fig. 10.31**  Nucleophilic addition of $^-CN$ to carbonyl group

nucleophile. It reacts with C=O double bond giving an alkoxide intermediate, which being a strong base removes a proton from water from the geminal diol, also called a hydrate (Fig. 10.32).

The above reaction (Fig. 10.32) is called hydration and can also be conducted in acidic conditions (in presence of mineral acids) (Fig. 10.23).

The geminal diols (as shown in Figs. 10.32 and 10.33) are in general unstable. However, some hydrates are reasonably stable. As an example, formaldehyde, the smallest aldehyde exists in aqueous solution as its hydrate, known as formalin (Fig. 10.34).

In case an electron-withdrawing group is attached to the carbonyl group, the formed gem diol is stabilised. An example is trichloroacetaldehyde (exists as chloral hydrate) and hexafluoroacetone (forms stable gem diol on exposure to water.

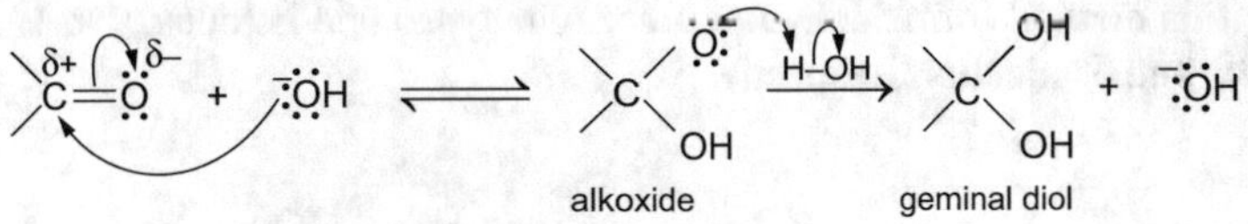

**Fig. 10.32**  Addition water (pH > 8) to C=O group

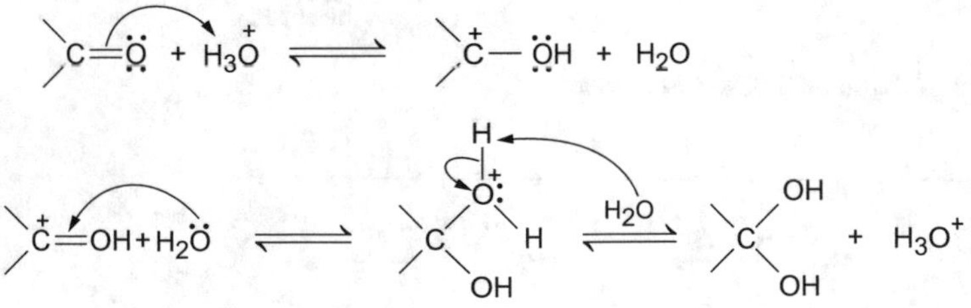

**Fig. 10.33**  Addition of water to C=O under acidic conditions

**Fig. 10.34**  The reaction of formaldehyde with water to give the diol (Formalin)

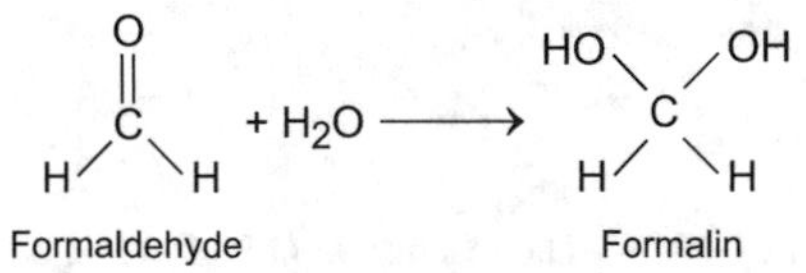

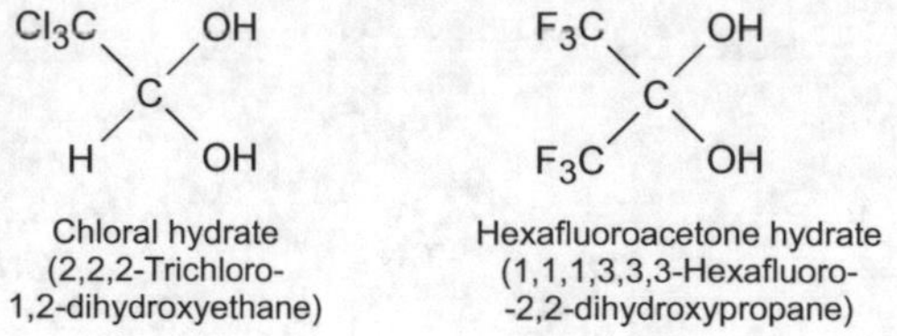
Cl3C
OH
C
H
OH
Chloral hydrate
(2,2,2-Trichloro-
1,2-dihydroxyethane)
F3C
OH
C
F3C
OH
Hexafluoroacetone hydrate
(1,1,1,3,3,3-Hexafluoro-
-2,2-dihydroxypropane)

## *10.3.3 Nucleophilic Addition of Hydrogen Halides*

Like water under acidic or basic conditions, the hydrogen halides (HCl, HBr, HI or HF) add to the carbonyl group under acidic conditions which form geminal halohydrin (Fig. 10.35).

### 10.3.3.1 Addition of Organometallic Compounds to Aldehydes and Ketones

In Grignard reagent, RMgX, the nucleophilic portion is R, which reacts with the electrophilic carbon atom of the carbonyl group after association of the O with Mg. The adduct after hydrolysis yields the corresponding alcohol (Fig. 10.36).

Using this method formaldehydes, other aldehydes and ketones give 1° alcohol, 2° alcohols and 3° alcohols, respectively.

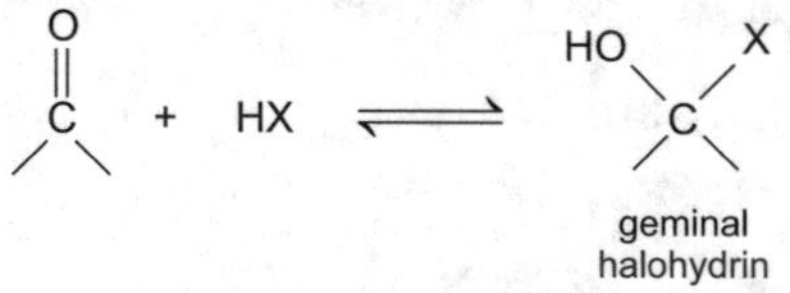
O
C
+
HX
HO
X
C
geminal
halohydrin

**Fig. 10.35** Addition of HX to C=O group

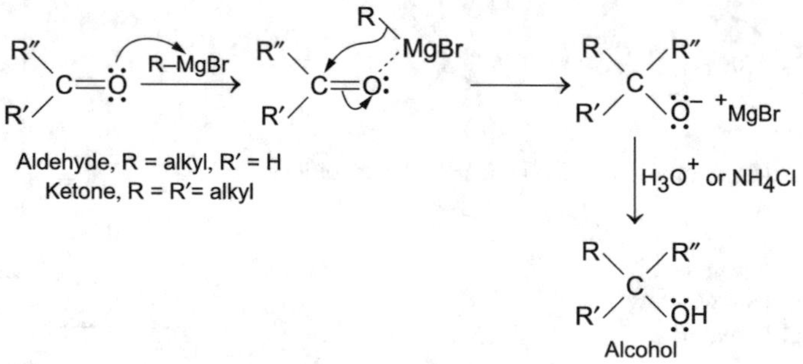
R″
C=O
R′
R–MgBr
Aldehyde, R = alkyl, R′ = H
Ketone, R = R′= alkyl
R
MgBr
R″
C=O:
R′
R
R″
C
R′
O⁻ ⁺MgBr
H3O⁺ or NH4Cl
R
R″
C
R′
OH
Alcohol

**Fig. 10.36** The reaction of Grignard reagent with carbonyl group

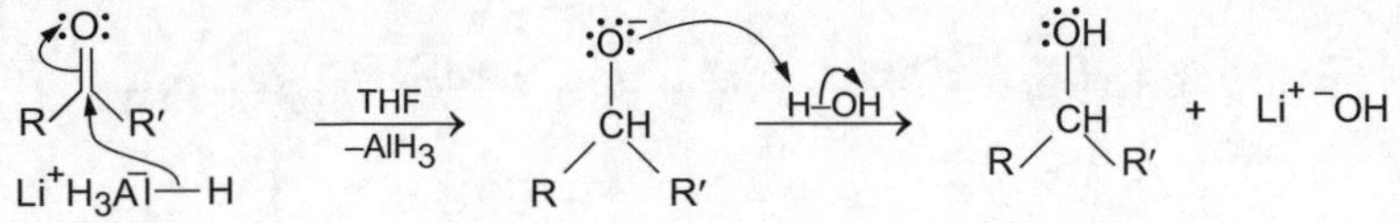

**Fig. 10.37** The reaction of C==O group with Li Al H$_4$

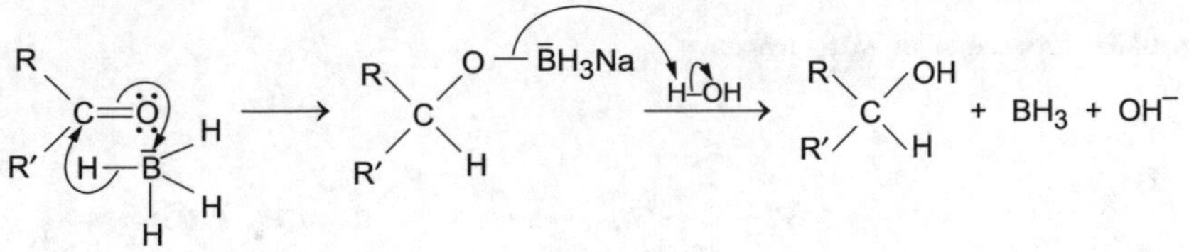

**Fig. 10.38** The reaction of C==O group with NaBH$_4$

In case an unsymmetrical ketone is used a new chiral centre is formed. See also Sect. 19.3.8.1.2 in Chap. 19.

#### 10.3.3.2 Addition of Hydride to C=O Group

The common source of hydride ion is sodium borohydride and lithium aluminium hydride. As in the case of Grignard reagent, RMgX, R is the nucleophilic portion, the nucleophilic hydride ion from NaBH$_4$ or LiAlH$_4$ can add to aldehyde or ketone carbonyl group giving alcohol after hydrolysis. In fact both NaBH$_4$ and LiAlH$_4$ serve as a source of hydride, H$^-$: the nucleophile. This process is commonly known as reduction of C==O group. Thus, an aldehyde gives 1° alcohol and ketone gives 2° alcohol (Fig. 10.37).

In case of LiAlH$_4$ (Fig. 10.37), all the four hydride ions have nucleophilic properties. This means that in this case 4 equivalent of the ketone or aldehyde can be reduced with only 1 equivalent of LiAlH$_4$.

NaBH$_4$ is comparatively less reactive and its reactions are carried out in alcohol and water (Fig. 10.38).

Unsymmetrical ketones can be enantioselectively reduced.

See also Sect. 19.3.8.2.2 in enantioselective synthesis.

### 10.3.4 Addition of Wittig Reagent

Wittig reagent is an organophosphorus reagent (containing a carbon-phosphorus bond). A typical Wittig reagent is prepared as shown below (Fig. 10.39).

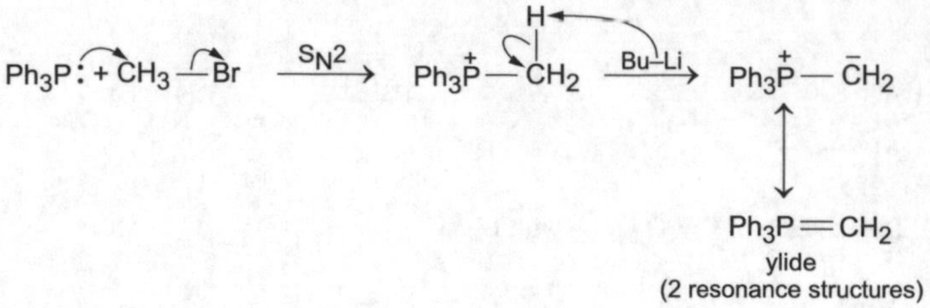

$$Ph_3P: + CH_3 \overset{\frown}{-} Br \xrightarrow{S_N2} Ph_3\overset{+}{P} - \overset{H}{\underset{}{C}}H_2 \xrightarrow{Bu-Li} Ph_3\overset{+}{P} - \overset{-}{C}H_2$$

$$Ph_3P = CH_2$$
ylide
(2 resonance structures)

**Fig. 10.39** Preparation of Wittig reagent

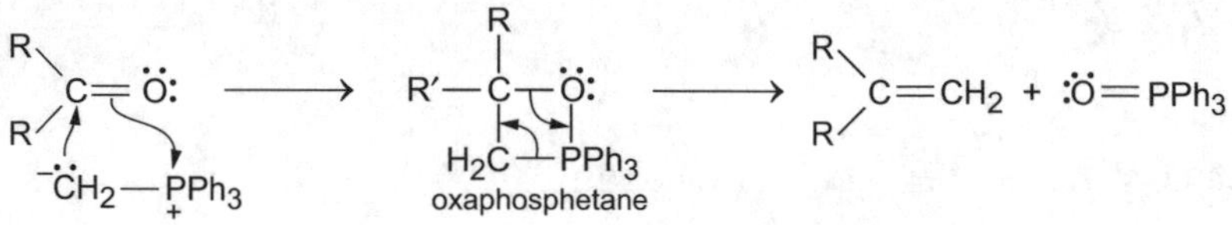

$$\underset{R'}{\overset{R}{\diagdown}}C = \overset{..}{O}: \longrightarrow R' - \overset{R}{\underset{H_2C \overset{}{\mid} PPh_3}{\overset{\mid}{C}}} - \overset{..}{O}: \longrightarrow \underset{R'}{\overset{R}{\diagdown}}C = CH_2 + :\overset{..}{O} = PPh_3$$
oxaphosphetane

**Fig. 10.40** The reaction of Wittig reagent with ketones

In one of the resonance structures of Wittig reagent, the carbon bears a negative charge and so it is nucleophilic. The Wittig reagent reacts with ketones to give an alkene (Fig. 10.40).

For more details about Wittig reagent see Wittig Reaction, Sect. 19.10 in Chap. 19.

## 10.4    Addition Reactions of Conjugated Dienes

Dienes are compounds containing two double bonds. The double bonds can be isolated (as in the case of 1,4-hexadiene, $CH_2=CH-CH_2-CH=CH_2$) or conjugated (as in the case of 1,3-butadiene, $CH_2=CH-CH=CH_2$); the two double bonds can also be cummulated (as in the case of 2,3-pentadiene, $CH_3CH=C=CHCH_3$). The isolated dienes undergo all the addition reactions as in the case of alkenes.

However, conjugated diens undergo addition reactions in a different fashion.

### 10.4.1    Addition of HBr

Conjugated diene add on to HBr to give a mixture of products. As an example, 1,3-butadiene on reaction with 1 equivalent HBr gives a mixture of 1,2- and 1, 4-addition products (Fig. 10.41).

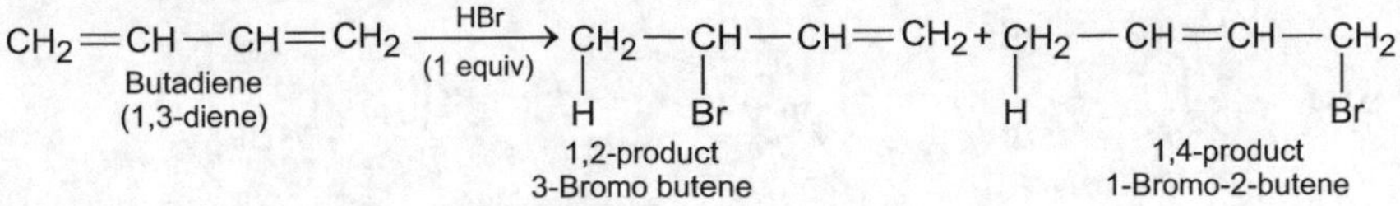

$CH_2=CH-CH=CH_2$ $\xrightarrow[\text{(1 equiv)}]{\text{HBr}}$ $CH_2-CH-CH=CH_2$ + $CH_2-CH=CH-CH_2$

Butadiene
(1,3-diene)

H    Br                    H                    Br

1,2-product              1,4-product
3-Bromo butene          1-Bromo-2-butene

**Fig. 10.41**  Addition of HBr to 1,3-butadiene

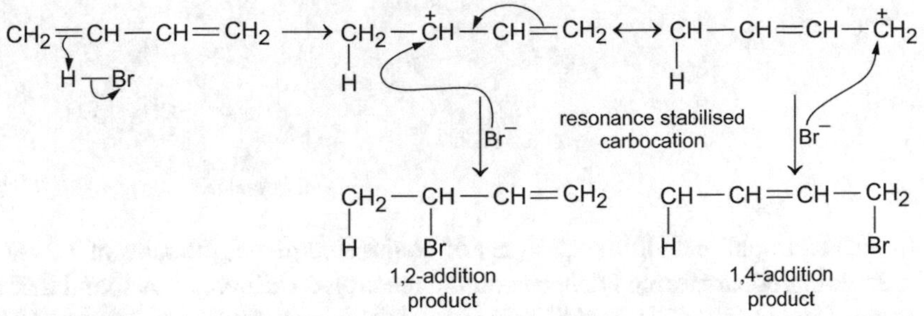

$CH_2=CH-CH=CH_2$ $\longrightarrow$ $CH_2-\overset{+}{C}H-CH=CH_2$ $\longleftrightarrow$ $CH-CH=CH-\overset{+}{C}H_2$

H—Br                              H                              H

Br⁻

resonance stabilised
carbocation

Br⁻

$CH_2-CH-CH=CH_2$        $CH-CH=CH-CH_2$

H    Br                      H              Br

1,2-addition                  1,4-addition
product                        product

**Fig. 10.42**  Mechanism to the formation of 1,2- and 1,4-adducts by the reaction of 1, 3-butadiene with HBr

The 1,2-addition product results from the Markovnikov addition of HBr to adjacent carbons of the diene. On the other hand, 1,4-addition of HBr to the two end carbons of the diene is also called conjugate addition.

The mechanism of the addition of the electrophile ($H^+$) from HBr results in the formation of a new C–H bond. $H^+$ invariably adds to the terminal carbon of the 1,3-diene resulting in the formation of a resonance-stabilised carbocation. Subsequently, nucleophilic attack occurs at either site of the resonance-stabilised carbocation that bears a (+) charge resulting in the formation is 1,2 or 1,4-addition products (Fig. 10.42).

**Solved Problem**

What product or products are obtained by the reaction of 1,2-dimethyl-3, 5-cyclohexadiene?

**Solution**

1,2-Dimethyl-3,5-cyclohexadiene is a congugated diene and so gives a mixture of 1,2 and 1,4-addition products as shown below:

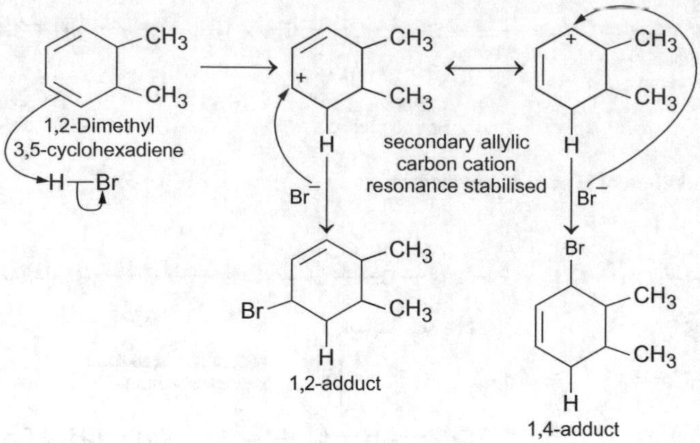

In the electrophilic addition reactions of conjugated diene, the ratio of 1,2- and 1,4-addition products formed depends on the reaction conditions. It is found that at low temperature (–80 °C), 1,2-addition product is obtained in 80% yield compared to 20% yield for the 1,4-adduct. However, at higher temperature (40°), the 1, 2-adduct is obtained in 20% yield compared to 80% yield of 1,4-adduct.

## 10.4.2   Cycloaddition Reactions

The conjugated dienes readily undergo cycloaddition reactions. A typical example is the Diels–Alder reaction. It involves the reaction of a conjugated diene and a dienophile to give a cyclohexene derivative.

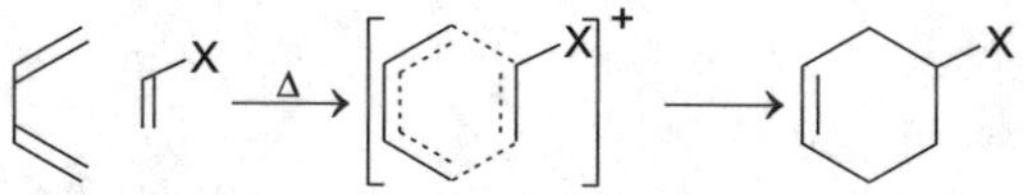

For more details, see Diels–Alder reactions (Sect. 20.3) in Chap. 20.

## 10.4.3   Addition of Halogens

Halogens like chlorine and bromine also undergo 1,2- and 1,4-additions. A typical example is given below

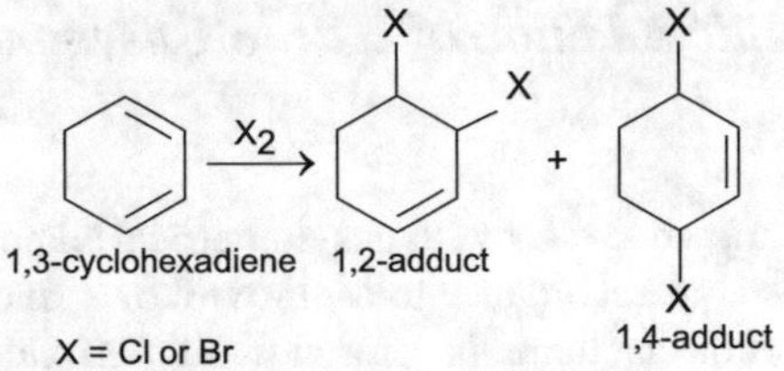

1,3-cyclohexadiene   1,2-adduct

1,4-adduct

X = Cl or Br

## 10.5   Free Radical Addition Reactions

### 10.5.1   *Introduction*

The term free radical is used for any species which possess unpaired electron. These are very important neutral reaction intermediates and are of transitory existence. These radicals react with other radicals or molecules by gaining one more electron to restore stable bonding pair.

Free radicals are obtained by homolytic fission of a covalent bond, the two departing atoms take one electron each and are called free radicals.

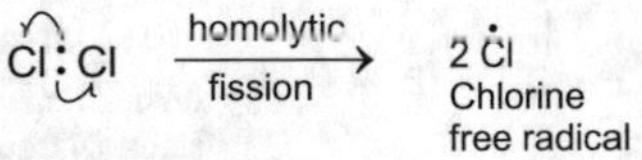

These are obtained by photolysis of compounds like acetone or peroxides (Fig. 10.43).

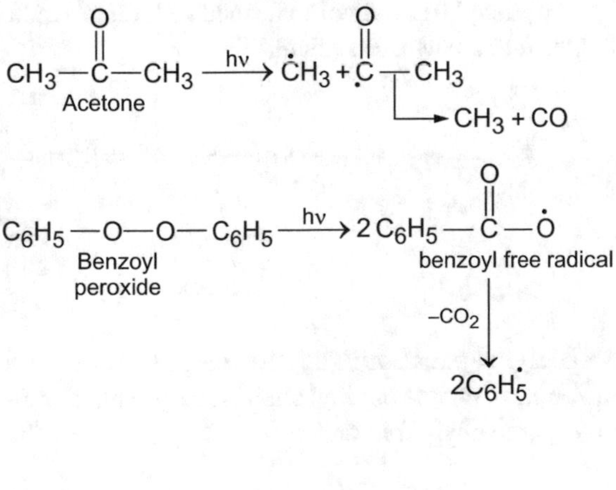

**Fig. 10.43** Generation of free radicals

## 10.5.2  Electrophilic Addition of HBr to Unsymmetrical Alkenes

Hydrogen bromide is known to react with unsymmetrical alkene like propene to give 2-bromopropane. This is in accordance to Morkovnikov's rule (see Sect. 10.2.1.1). However, if the reaction is done in presence of peroxide the product is 1-bromopropane. This addition is as per anti-Markovnikov addition and takes place via the formation of free radical. The mechanism of the generation of free radical (B$^{\bullet}$r) and its subsequent reaction with propene has been discussed earlier (see Fig. 10.4).

## 10.6  Free Radical Polymerisation

The reaction of free radical (R$^{\bullet}$) to ethene or vinyl chloride gives the polymers polyethylene and polyvinyl chloride (PVC), respectively. This polymerisation involves three steps, viz., initiation, propagation and termination.

**Step (*i*) Initiation**: The polymerisation is initiated by a free radical initiator like benzoyl peroxide. The free radical is generated as follows:

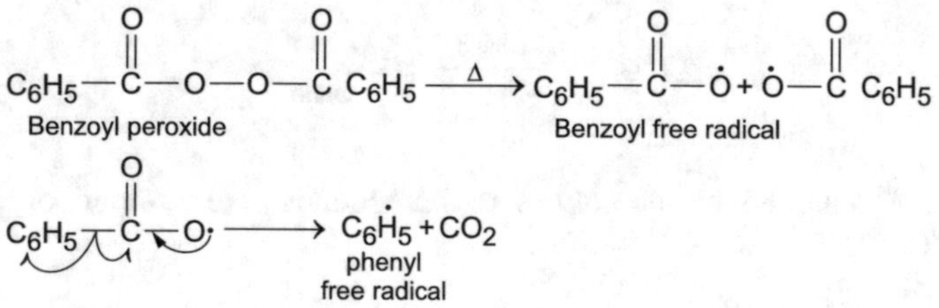

The phenyl free radical so formed adds on to a molecule of ethene or vinyl chloride to generate a new free radical.

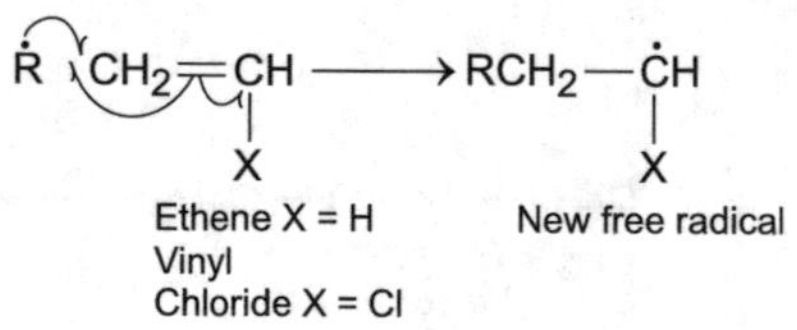

**Step (*ii*) Propagation**: The new free radical generated in step (*i*) adds on to another molecule of ethene or vinyl chloride to give another free radical. Successive additions give a long chain free radical.

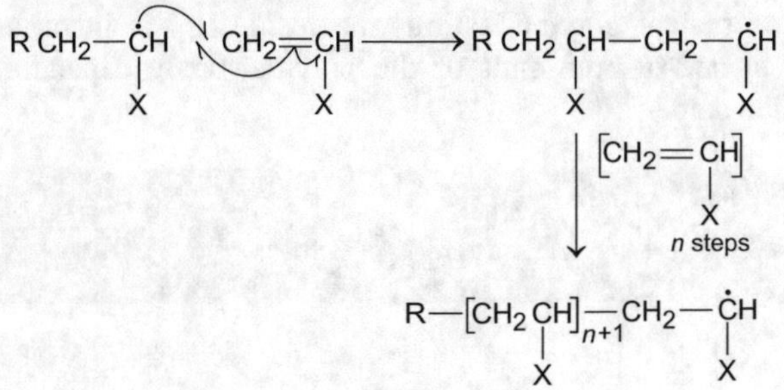

$$R\ CH_2-\overset{\bullet}{C}H \quad CH_2=CH \longrightarrow R\ CH_2\ CH-CH_2-\overset{\bullet}{C}H$$

with X substituents, leading through $[CH_2=CH]$ with X, $n$ steps

$$R-[CH_2\ CH]_{n+1}-CH_2-\overset{\bullet}{C}H$$

with X substituents

**Step (iii) Termination**: The termination of the long chain per radical formed in step (ii) is achieved by radical coupling or disproportion as given below:

### Radical Coupling

Two radical couple to give a long chain polymer.

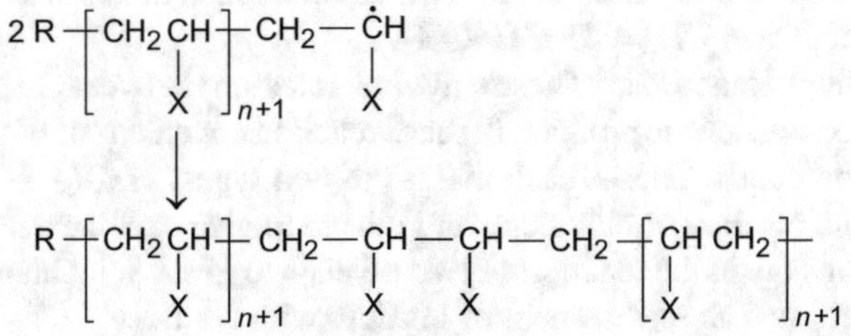

$$2\ R-[CH_2\ CH]_{n+1}-CH_2-\overset{\bullet}{C}H$$

$$R-[CH_2\ CH]_{n+1}-CH_2-CH-CH-CH_2-[CH\ CH_2]_{n+1}$$

Polyethylene  R = H
PVC       R = Cl

### Disproportionation

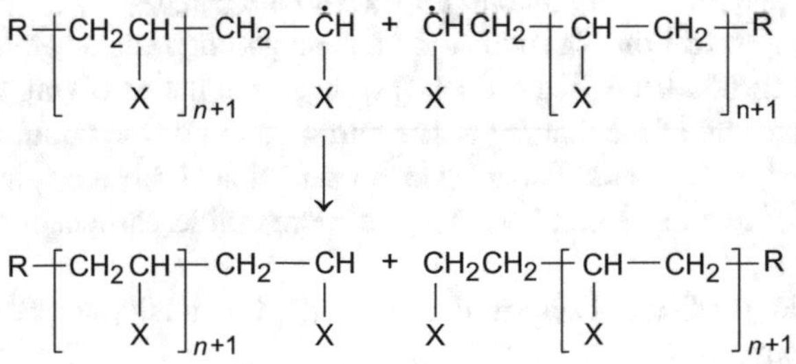

$$R-[CH_2\ CH]_{n+1}-CH_2-\overset{\bullet}{C}H \ +\ \overset{\bullet}{C}H\ CH_2-[CH-CH_2]_{n+1}-R$$

$$R-[CH_2\ CH]_{n+1}-CH_2-CH \ +\ CH_2\ CH_2-[CH-CH_2]_{n+1}-R$$

Polyethylene  X = H
PVC       X = Cl

As seen, in free radical polymerisation, the reside (R) is incorporated into the polymer chain at one or both ends of the polymer chain depending on the type of the termination step.

## Key Concepts

- **Bromonium Ion**: An ion formed by the reaction of bromine molecule to an alkene.

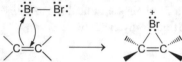

- **Carbenes**: These are electron-deficient species of the formula $R_2C:$ and are obtained by the irradiation of diazomethane.
- **Conjugated Dienes**: Dienes contain two double bonds in alternate positions, for example, butadiene, $CH_2=CH–CH=CH_2$.
- **Cycloaddition Reactions**: These involve reaction between two unsaturated molecules to give a cyclic product. In these reactions, π electrons of two molecules form 2 sigma bonds. These reactions are to two types, viz., (4 + 2) cycloaddition (as in the reaction of butadiene and ethene to give cyclohexene) and (2 + 2) cycloadditions (as in the reaction of two alkenes to give cyclobutane derivatives)
- **Demercuration**: The replacement of HgOAc with Na $BH_4$.
- **Electrophilic Addition Reaction**: The addition of an electrophilic reagent to unsaturated compounds (alkenes and alkynes).
- **Free Radicals**: Any species which possess unpaired electron. These are obtained by homolytic fission of a covalent bond.
- **Hydration**: The process involves addition of water to an alkene.
- **Hydroboration**: The process of treatment of an alkene with $BH_3$ followed by oxidation of the formed organobarane with alkaline $H_2O_2$.
- **Markovnikov Addition**: The addition of an unsymmetrical reagent (like H Br) to an unsymmetrical alkene (like propene) gives an adduct involving the attachment of the positive end of the reagent to the carbon atom of the double bond bearing large number of H atoms. It can also be said that the electrophilic portion of the reagent adds to a π bond so that the more stable carbocation intermediate predominates.
- **Nucleophilic Addition Reaction**: Addition of a nucleophile (like ⁻CN) to carbonyl group.
- **Oxymercuration**: The reaction of metallic ions to alkenes.
- **Peroxide Effect**: The addition of an unsymmetrical reagent (like HBr) to an unsymmetrical alkene (like propene) in presence of peroxide gives an adduct, which is reverse to Markovnikov addition. In this case, the halogen gets attached to the carbon atoms of the double bond bearing maximum number of H atoms.

- **Regioselective Reaction**: Also known as enantioslective reaction, it is a reaction in which one enantiomer is obtained in major amount.
- **Regiospecific Reaction**: Also known as enantroselective reaction, it is a reaction in which product with only a single orentiation is obtained.
- **Simmon-Smith Reagent**: [$CH_2I_2$, Zn(Cu)] obtained by the treatment of $CH_2I_2$ with Zn that has been coated with copper. It is used for epoxidation of alkenes.
- **Vicinal Dihalide**: A compound in which two halogens are attached to adjacent carbons.
- **Vicinal diol**: An organic compound in which two OH groups are attached to the same carbon.
- **Wittig Reagent**: An organo phosphorus, reagent containing a carbon-phosphorus bond. A typical Wittig reagent is prepared by the $S_N2$ reaction of triphenyl phosphine with an alkyl halide

$$Ph_3P: + CH_3-Br \xrightarrow{S_N2} Ph_3\overset{+}{P}-CH_2 \xrightarrow{Bu-Li} Ph_3\overset{+}{P}-\bar{C}H_2 \longleftrightarrow Ph_3P=CH_2$$

## Problems

1. 2-Methyl propene on reaction with HBr gives 2-bromo-2-methylpropane and not the alternative 1-bromo-2-methyl propane. Explain.
2. What product is expected to be obtained by the addition of HBr to 3, 3-dimethyl-1-butene?
3. What product is expected to be obtained by the reaction of 1-pentene with HBr.
4. What product is obtained by the treatment of 2-methyl propene with methanol in presence of $H_2SO_4$?

   **Ans**. Tert.butyl methyl ether
5. What product is expected to be obtained in the oxymercuration-demercuration of 3-methyl butene?

   **Ans**. A 2° alcohol is obtained.
6. What product (or products) is expected to be obtained by the reaction of 2, 4-hexadiene with one equiv. HBr?

   **Ans**. 4-Bromo-2-hexene (1, 2-adduct) and 5-Bromo-3-hexene (1, 4-adduct)
7. What products are obtained?

   (a) by treatment of 2, 5-dimethyl-hexa-2, 3-diene with (*i*) HBr and (*ii*) $Cl_2/CH_2Cl_2$
   (b) By treatment of 2, 3-dimethyl-buta-1, 3-diene with (*i*) HBr and (*ii*) $Br_2/CH_2Cl_2$.

8.   Starting with 2-methyl propene, how will you obtain 2-bromo-2-methyl propane and 1-bromo-2-methyl propane.

9.   The addition of HBr to propene is regioselective. What do you understand by this?

10.  What products are obtained in the following?

(a)   [cyclohexene with CH₃] + $\xrightarrow{\text{Dilute } H_2SO_4}$

(b)   $\text{C}={=}\text{C}$ + $Br_2 \longrightarrow$

(c)   [cyclohexane] + $Br_2 \longrightarrow$

(d)   [cyclohexene with CH₃] + HOBr $\longrightarrow$

(e)   $CH_3\,CH_2\,CH{=}CH_2 + Br_2 + H_2O \longrightarrow$

(f)   [cyclohexene with CH₃] + $HgOAc + H_2O \longrightarrow$

(g)   $CH_3\,CH_2\,CH_2\,CH{=}CH_2 + HgOAc + H_2O \longrightarrow$

(h)   $CH_3\,C{\equiv}C{-}CH_3 \xrightarrow{H_2SO_4/HgSO_4/H_2O}$

(i)   $\underset{R'}{\overset{R}{>}}C{=}O + RMgX \longrightarrow$

(j)   $CH_2{=}CH{-}CH{=}CH_2 + 1 \text{ equiv. HBr} \longrightarrow$

11.  What is the best method for obtaining cyclopropane derivatives from alkanes?

12.  Write notes on

(a)   Free radical addition reactions

(b)   Manufacture of PVC by free radical reaction.

# Chapter 11
# Stereochemistry of Elimination Reactions

## 11.1 Introduction

Elimination reactions are the reverse of addition reactions. These involve elimination of two or four groups attached to adjacent carbon atoms in a substrate forming a multiple bond. An example is elimination of HX from an alkyl halide to form an alkene (Fig. 11.1).

Elimination reactions of different types are known. These include

- Elimination reactions of alkyl halides.
- Elimination reactions of alcohols.
- Eliminations involving ammonium compounds.

In general, elimination reactions are of two types, viz., bimolecular elimination reactions (E2) and unimolecular reactions (E1). As already stated in elimination reaction, two groups are eliminated from adjacent atoms. In fact, two σ bonds are broken and one π bond is formed (Fig. 11.2).

In these reactions, if one of the atoms to be eliminated is H, then the other is a heteroatom, more electronegative than carbon.

## 11.2 Bimolecular Elimination Reactions (E2)

A common example of bimolecular elimination reaction is the reaction of propyl bromide with a base to form an alkene (Fig. 11.3).

The E2 reaction, like $S_N2$ reaction, depends on the concentration of both the substrates and the nucleophile and the reaction is of second order. It is represented as E2.

The removal of HX from an alkyl halide is called **dehydrohalogenation** and is a commonly used method for the introduction of a π bond and preparation of alkene. In fact, dehydrohalogenation is an example of β-elimination since it involves loss

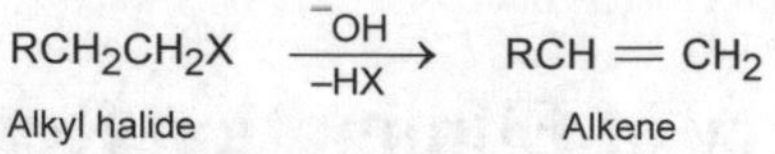

$$RCH_2CH_2X \xrightarrow[-HX]{^-OH} RCH = CH_2$$

Alkyl halide                    Alkene

**Fig. 11.1** The elimination of H–X from alkyl halide

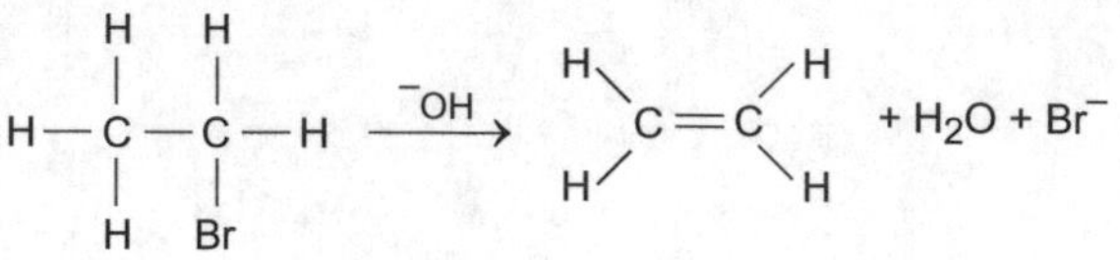

$$H-\underset{\underset{H}{|}}{\overset{\overset{H}{|}}{C}}-\underset{\underset{Br}{|}}{\overset{\overset{H}{|}}{C}}-H \xrightarrow{^-OH} \underset{H}{\overset{H}{\diagdown}}C=C\underset{\diagdown H}{\diagup H} + H_2O + Br^-$$

**Fig. 11.2** In elimination reactions two σ bonds are broken and one π bond is formed

$$H_3C - \overset{\overset{H}{|}}{\underset{\underset{Br}{|}}{CH}} - CH_2 \xrightarrow{base} H_3C - CH = CH_2 + BH^+ + Br^-$$

:B

Propyl bromide

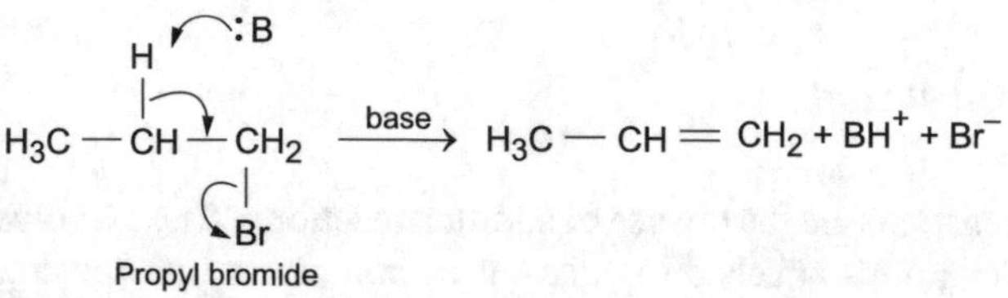

**Fig. 11.3** Reaction of propyl bromide with base

of elements from two adjacent atoms, the α-carbon bonded to the leaving group X, and the β-carbon adjacent to it. The three curved arrows show how two sigma bonds are broken and one π bond is formed (see Figs. 11.2 and 11.3). Thus as seen, the base (B:) eliminates a proton on the β-carbon forming the H–B. This is followed by formation of a new π bond between the α and β-carbon by the electron pair in the β carbon. The electron pair in the C—X bond ends up on halogen resulting in the formation of the leaving group X$^-$.

Like the $S_N2$ reaction, the E2 reaction is also a concerted reaction. All bonds are broken and formed in a single step. The reaction occurs via a transition state and is known as 1,2-elimination or β-elimination. In these reactions, the two groups to be eliminated (i.e., H and X) are *trans* to each other and so the E2 reactions are generally *trans* eliminations (Fig. 11.4).

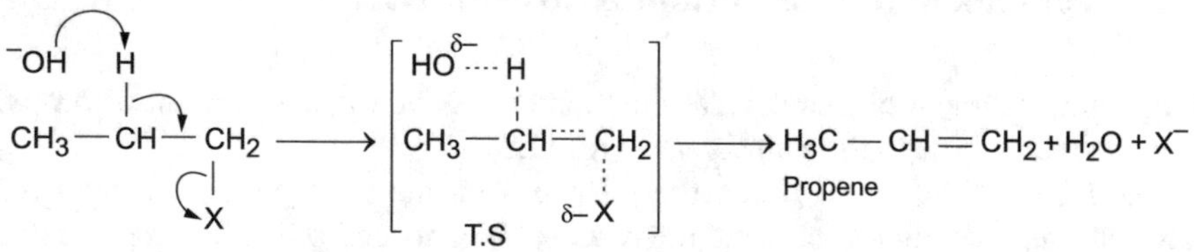

$$^-OH \quad H$$
$$CH_3 - \overset{|}{\underset{\underset{X}{|}}{CH}} - CH_2 \longrightarrow \begin{bmatrix} HO \overset{\delta-}{\cdots} H \\ CH_3 - CH \overset{\vdots}{=\!=\!=} CH_2 \\ \underset{T.S}{\overset{}{}} \quad \delta- X \end{bmatrix} \longrightarrow H_3C - CH = CH_2 + H_2O + X^-$$

Propene

**Fig. 11.4** E-2 reactions are *trans* elimination reactions

In the transition state (Fig. 11.4), the C–H and C–X bonds are partially broken and the O–H and π bonds are partially formed and both the base and the departing leaving group bear a partial negative charge.

In E2 reactions, as the number of alkyl group on the carbon atom having the leaving group increases, the rate of reaction increases. Thus, 3° alkyl halides are more reactive than 2° alkyl halides, which in turn are more reactive than the 1° alkyl halide.

$$R_3C - X \quad > \quad R_2CH\,X \quad > \quad RCH_2X$$

| 3° alkyl halide | 2° alkyl halide | 1° alkyl halide |

This trend is just the opposite to the reactivity of the alkyl halides in $S_N2$ reaction in which increasing alkyl substitution decreases the rate of the reaction (see Sect. 12.5.1.1).

In E2 reaction, since the bond to the leaving group is partially broken in the transition state, the reaction is faster if the leaving group is better. The order of reactivity of the alkyl halides is

$$R—I > R—Br > R—Cl > R—F$$

The rate of E2 reaction also depends on the strength of the base. The rate increases as the strength of the base increases. Normally in E2 reactions, the base used are ⁻OH and ⁻OR. However, two other string bases like DBN (1,5-diazabicyclo [4.3.0] non-5-ene) and DBU (1,8-diazabicyclo [5.4.0] undec-7-ene are also used.

DBN

DBU

An example of the use of DBN in E2 reaction is in the synthesis of a prostaglandin as shown in Fig. 11.5.

The E2 reaction is regioselective. A regioselective reaction yields one constitutional isomer exclusively or predominantly. As an example, the reaction of 2-bromo-2-methylbutane with base can yield either 2-methyl 2-butene or 3-methylbutene. The alkene 2-methyl-2-butene being trisubstituted is obtained as the major product. Thus, if an alkylhalide has more than one β-carbons (as in the case 2-bromo-2-methylbutane), the course of the reaction is governed by **Saytzeff rule**, according to which the major product of β-elimination has more substituted double bond (Fig. 11.6).

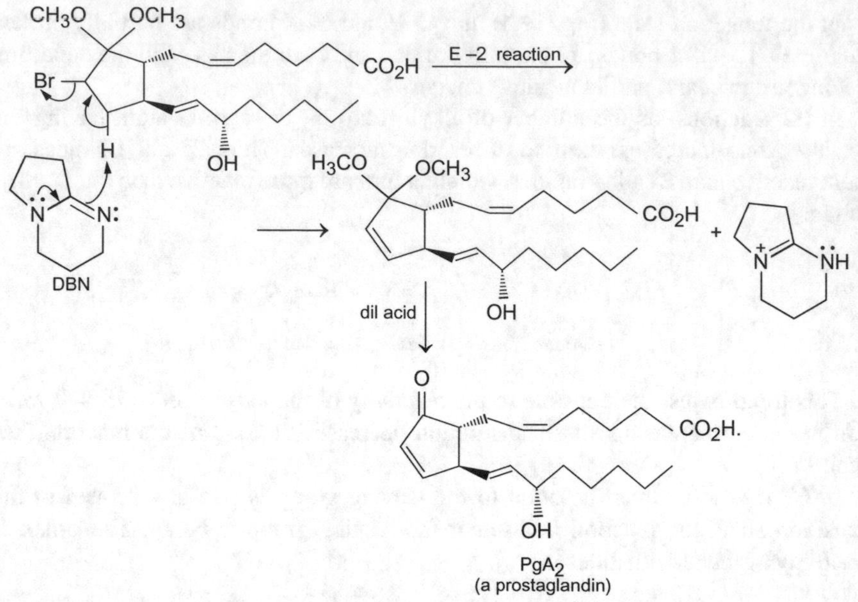

**Fig. 11.5** E2 reaction using the base DBN (Synthesis of PgA$_2$)

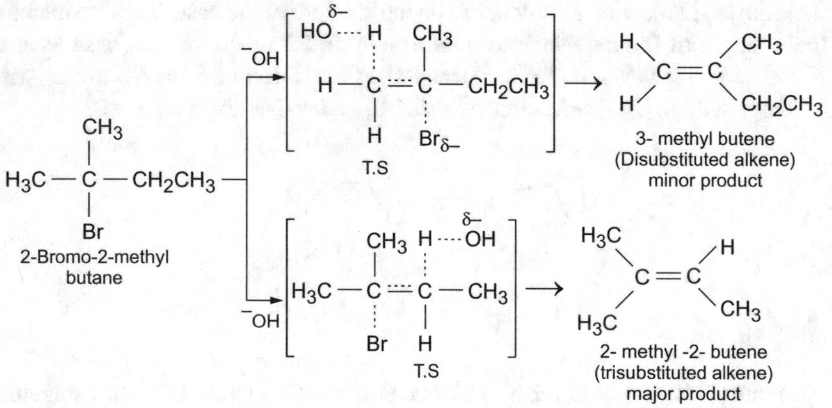

**Fig. 11.6** E2 elimination 2-bromo-2-methyl butane

In case, in a dehydrohalogenation a mixture of stereoisomers is possible, the major product is the more stable stereoisomer. As an example, the alkyl halide (A) yields a mixture of *trans* and *cis* alkenes. In this case, the *trans* alkene is the major product since it is more stable (Fig. 11.7).

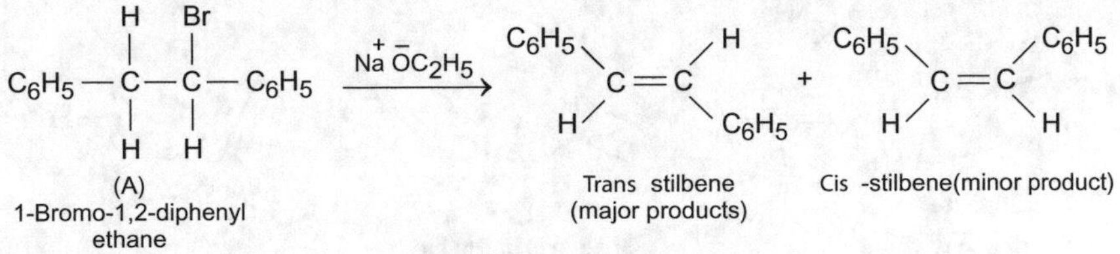

$$C_6H_5 - \overset{\overset{\displaystyle H}{|}}{\underset{\underset{\displaystyle H}{|}}{C}} - \overset{\overset{\displaystyle Br}{|}}{\underset{\underset{\displaystyle H}{|}}{C}} - C_6H_5 \xrightarrow{\text{Na } \overset{+}{O}\overset{-}{C_2H_5}}$$

(A)
1-Bromo-1,2-diphenyl
ethane

$$\underset{H}{\overset{C_6H_5}{>}}C=C\underset{C_6H_5}{\overset{H}{<}}$$

Trans stilbene
(major products)

$+$

$$\underset{H}{\overset{C_6H_5}{>}}C=C\underset{H}{\overset{C_6H_5}{<}}$$

Cis -stilbene(minor product)

**Fig. 11.7** E2 elimination of 1-bromo-1,2-diphenyl ethane

## 11.2.1 Stereochemistry of E2 Reaction

We have seen that though the E2 reaction does not yield products with stereogenic centres, its transition state consists of four atoms that react at the same time. In fact, these react only if they possess a particular stereochemical arrangement of atoms or groups.

As already stated, the transition state of an E2 reaction consists of four atoms from the alkyl halide—one H atom, two C atoms and the leaving group (X)-all alligned in a plane. In case the H and X atoms are oriented on the same side of the molecules the geometry is referred to as syn periplanar. Alternatively, the H and X atoms are oriented on opposite sides of the molecule, the geometry is referred to as anti-periplanar (as shown in Fig. 11.8).

It has been shown by all evidences that E2 elimination occurs in the anti-periplanar geometry. This arrangement permits the molecule to react in the lower energy staggered conformation. In fact, anti-periplanar geometry is the preferred arrangement for aryl alkyl halide (acyclic or cyclic) undergoing E2 elimination.

For cyclic compounds (e.g., cyclohexyl halide), the geometrical constraints of the E2 reaction are more crucial. In these cases, the proton and the leaving groups must be *trans* and diaxial. As an example, consider the reaction between bromocyclohexane and a strong base to give cyclohexene (Fig. 11.9).

**Fig. 11.8** Two conformations of transitate state of E2 reaction

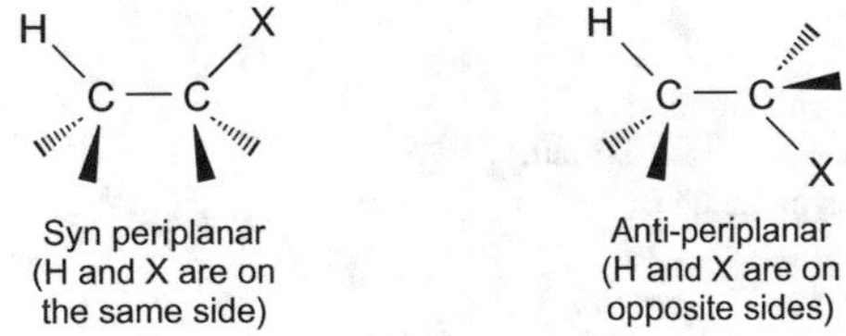

Syn periplanar
(H and X are on
the same side)

Anti-periplanar
(H and X are on
opposite sides)

**Fig. 11.9** Reaction of bromocyclohexane with base

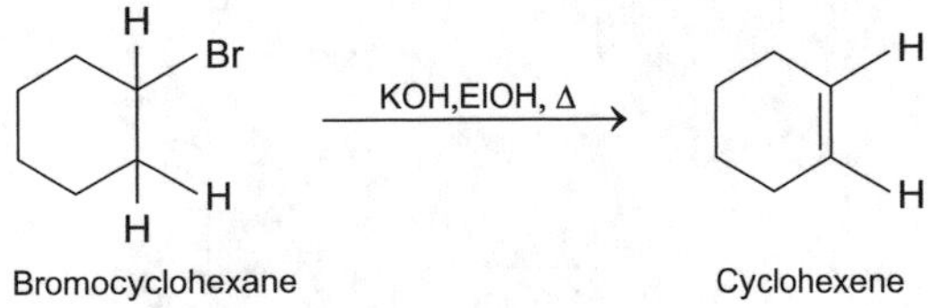

$$\xrightarrow{\text{KOH,EtOH, } \Delta}$$

Bromocyclohexane

Cyclohexene

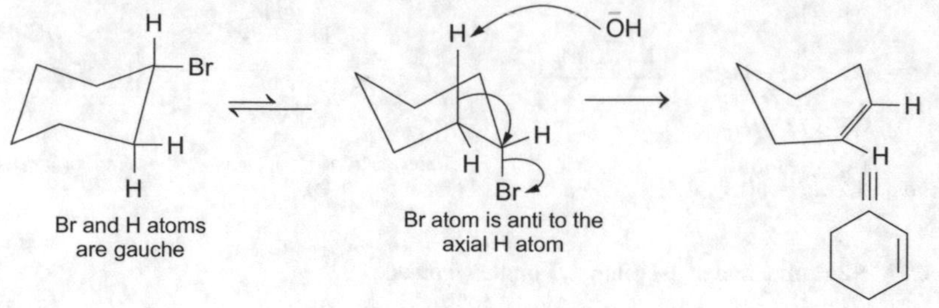

Br and H atoms          Br atom is anti to the
are gauche              axial H atom

**Fig. 11.10** Two conformation of bromocyclohexane

For dehydrohalogenation both the H atoms and the leaving group can assume an anti-conformation only if both group are *trans* and diaxial. In case, they are not diaxial, then the cyclohexane ring must undergo a ring-flip into the other chair conformation. (Fig. 11.10).

In case, there is more than one pathway by which elimination can occur, the Saytzeff product (the more highly substituted alkene) will be preferentially obtained. As an example consider the dehydrohalogenation of 1-chloro-2-methylcyclohexane, which exists in *cis* and *trans* forms.

*Cis*-1-chloro-2-methyl cyclohexane can exist in two conformations (A and B); each of these conformations has one group axial and one group equatorial. E2 reaction occurs from conformation B, which has an axial Cl atom (Fig. 11.11).

Since the conformation B has two different axial β-H atoms (labelled $H_a$ and $H_b$), the E2 reaction can occur from both giving two alkenes. The major product has the more stable trisubstituted double bond as per Saytzeff rule (Fig. 11.12).

The *trans*-1-chloro-2-methylcyclohexane also exists in two conformations, C (having two equatorial substituents) and D (having two axial substituents. The E2 reaction occurs from conformation D, which has an axial Cl atom (Fig. 11.13).

Since the conformation D has only one β-H, E2 reaction occurs in one direction only to give a single product having the disubstituted double bond (Fig. 11.14).

**Fig. 11.11** Two conformations of *cis*-1-chloro-2-methyl cyclohexane

(A)                                      (B)

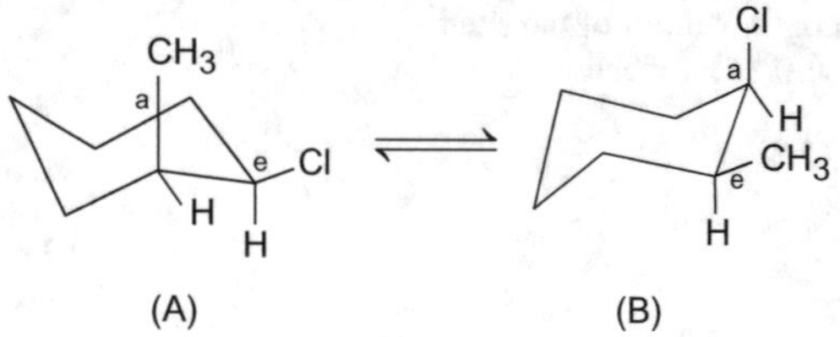

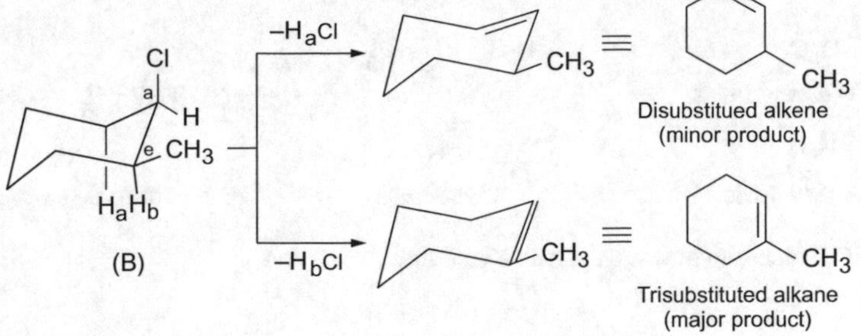

**Fig. 11.12** E2-reaction of *cis*-1-chloro-2-methyl cyclohexane in conformation B

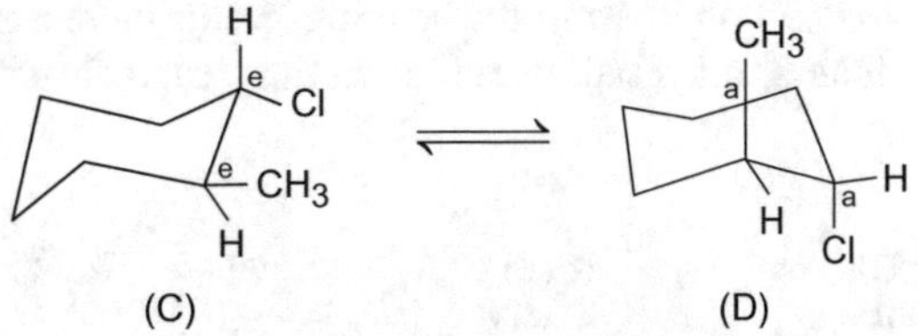

**Fig. 11.13** Two conformations of *trans*-1-chloro-2-methyl cyclohexane

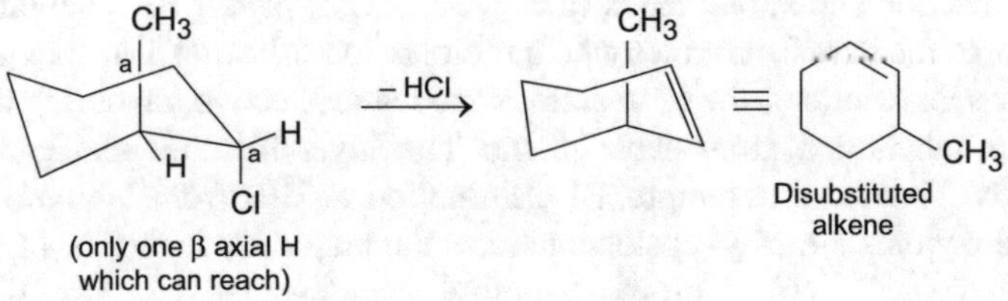

**Fig. 11.14** E2 reaction of *trans*-1-chloro-2-methyl cyclohexane in conformation D

## 11.3 Unimolecular Elimination Reaction (E1)

The rate of unimolecular elimination reaction is dependent on the concentration of the substrate only and is independent of the concentration of the nucleophile. The reaction is of first order and is designated as E1. As in the case of $S_N1$ reaction, the E1 reaction is also a two-step reaction. The first step is slow ionisation of the alkyl halide to give a carbocation. This is followed by fast abstraction of a proton from the β-carbon by a base giving an alkene (Fig. 11.15).

Both E1 and E2 mechanisms involve the same number of bonds broken and formed. The difference between the two is that in an E1 reaction, the leaving group comes off to form a carbocation before the β proton is removed and the reaction is a two-step reaction. However, in an E2 reaction, both the leaving group and the β

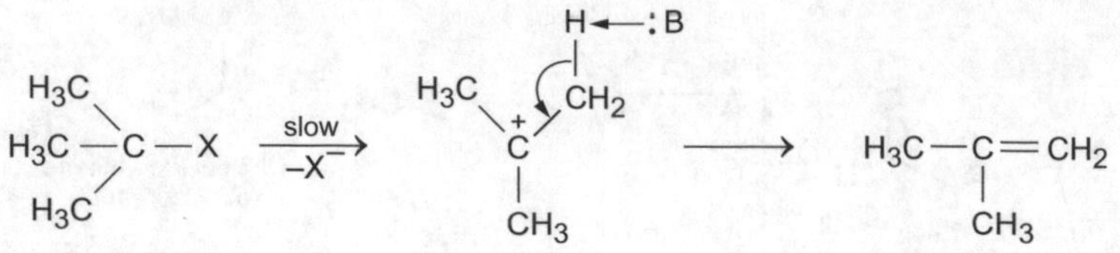

**Fig. 11.15** E1 reaction of tert. butyl halide with base

protons are removed simultaneously in a single step and the reaction occurs via the formation of a transition state.

As in the case of E2 reaction, the rate of E1 reaction also increases as the number of alkyl group on the carbon bearing the leaving group increases. Thus, 3° alkyl halide reacts faster than 2° alkyl halide, which in turn reacts faster than the 1° alkyl halide.

$$R_3CX \quad > \quad R_2CHX \quad > \quad RCH_2X$$

|  |  |  |
|---|---|---|
| 3° alkyl | 2° alkyl | 1° alkyl halide |
| halide | halide |  |

The rate of an E1 reaction is not affected by base, weak base (like $H_2O$ and ROH) favour E1 reaction and strong bases (like $^-OH$, $^-OR$) favour E2 reactions. In fact, the strength of the base determines whether a reaction follows E1 or E2 mechanism.

Also, like E2 reactions, the E1 reactions are regeoselective, favouring the formation of more substituted, more stable alkene. The Saytzeff rule is also applicable for E1 reactions. Thus, as an example, E1 elimination of HBr from 1-bromo-1-methyl cyclopentane yields 1-methyl cyclopentene as the major product (Fig. 11.16).

As seen (Fig. 11.16), 1-bromo-1-methyl cyclopentane has two different β carbons (at positions 2 or 5 and H of $CH_3$). Elimination of H from position 2 gives 1-methylcyclopentene, while elimination of H from $CH_3$ gives methylene cyclopentane.

## Comparison of E1 and $S_N1$ reactions

Both E1 and $S_N1$ reactions have the same first step, *i.e.*, formation of a carbocation. They differ in the second step, *i.e.*, what happens to the carbocation. In an E1 reaction,

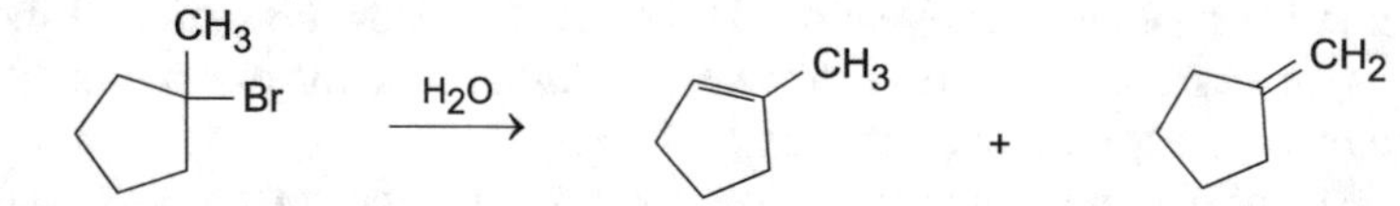

**Fig. 11.16** E1 elimination of 1-bromo-1-methyl cyclopentane

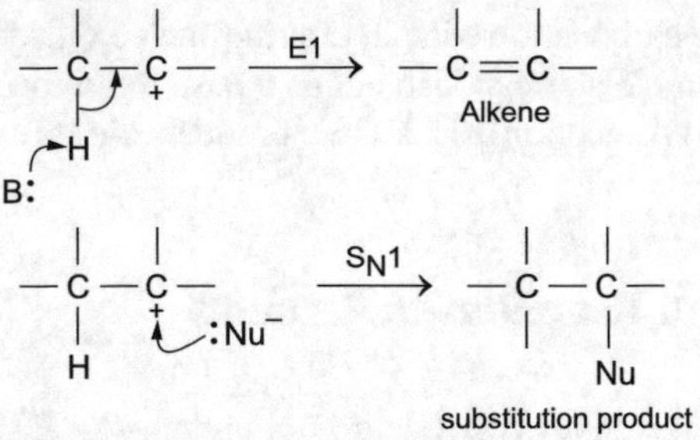

**Fig. 11.17** Difference between E-1 and $S_N1$ reactions

a base removes a proton forming a new $\pi$ bond. However, in an $S_N1$ reaction, a nucleophile attacks the formed carbocation giving a substitution product Fig. 11.17.

In fact, same conditions favour substitution by $S_N1$ mechanism and elimination by E1 mechanism. Thus, using a tertiary alkyl halide as a substrate and a weak base or a weak nucleophile as a reagent, both reactions (E1 and $S_N1$) occur simultaneously giving as mixture of products. This is illustrated by the reaction of $(CH_3)_3CBr$ with $H_2O$ (Fig. 11.18).

The first step in both reactions is the formation of a carbocation:

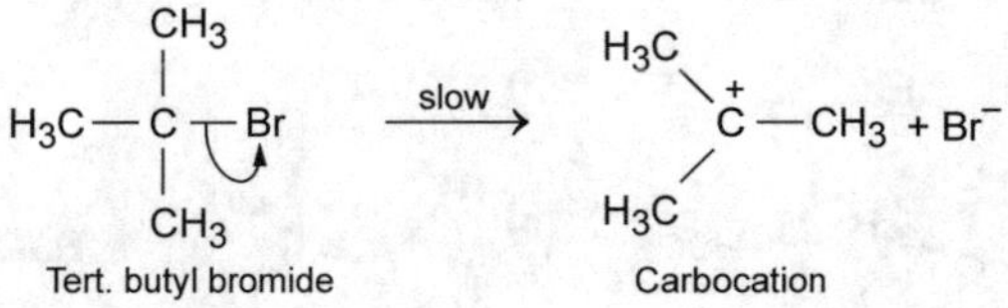

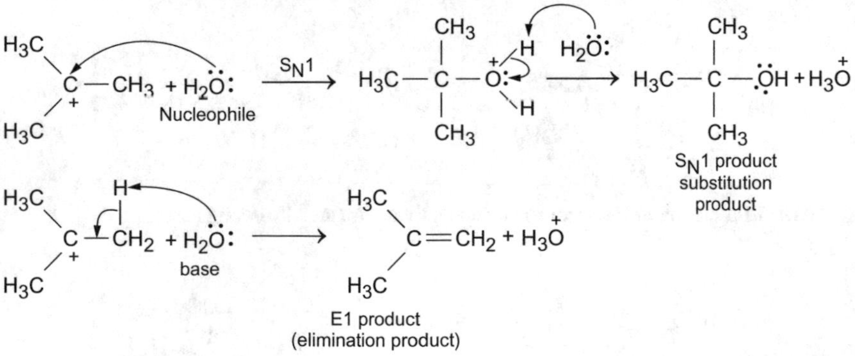

**Fig. 11.18** Simultaneous occurring of E1 and $S_N2$ reaction by thereaction of tert. butyl bromide with water

The reaction of the carbocation with $H_2O$ as the nucleophile yields the substitution product ($S_N1$ reaction). Water can also act as a base to remove a proton giving the elimination product (E1 reaction). Thus, two products are formed (Fig. 11.18).

## 11.4   Elimination Reactions in Alcohols

Primary alcohols unlike an alkyl halide do not undergo an E2 reaction with strong base since under such conditions OH is a poor leaving group. However, as we know, primary alcohols can undergo elimination on treatment with a strong acid, which generate a good leaving group, $H_2O$ (Fig. 11.19).

Since in primary alcohols, there is no steric hindrance at the carbon atom bearing the leaving group ($H_2O^+$), substitution can also take place with nucleophile, even if it is weak. Thus, with $H_2SO_4$ as the reagent, an alkyl sulphate ester can also be obtained (Fig. 11.20).

Any alcohol molecule not protonated (in Fig. 11.19) can act as a nucleophile to yield the corresponding symmetrical ether (Fig. 11.21).

In case of 2° or 3° alcohol dehydration occurs by the E1 pathway via. the formation of a carbocation intermediate. Thus, 2-pentanol on heating with conc. $H_2SO_4$ gives a mixture of 2-pentene and 1-pentene in the ratio 70: 30. The formation of more substituted alkene (2-pentene) is as per Saytzeff rule (Fig. 11.22).

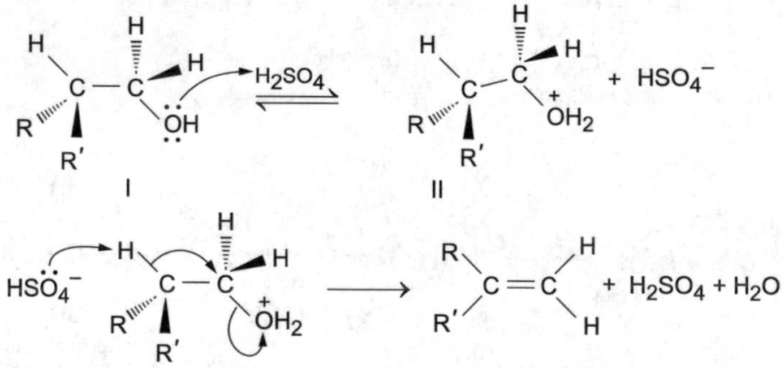

Fig. 11.19  Elimination of $H_2O$ from 1° alcohols by treatment with strong acids

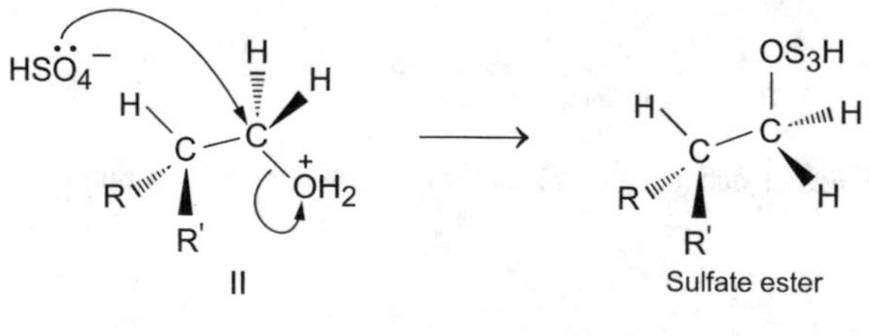

Fig. 11.20  Formation of sulfate ester in the reaction of 1° alcohol with conc. $H_2SO_4$

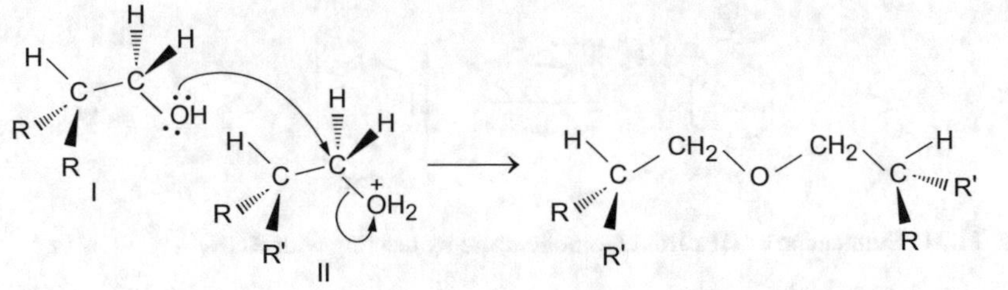

**Fig. 11.21** Formation of symmetrical ether in the reaction of 1° alcohol with the protonated alcohol

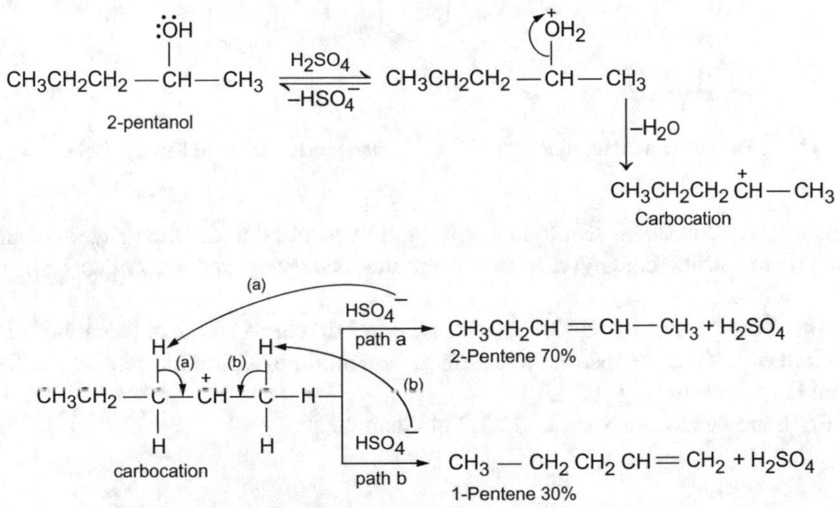

**Fig. 11.22** The reaction of 2-pentanol with conc. $H_2SO_4$

## 11.5 Elimination of HBr from Bromobenzene (Formation of Benzyne as an Intermediate)

The reaction of bromobenzene with $KNH_2$/liquid ammonia is known to given aniline (Fig. 11.23).

The above reaction was originally believed to occur by a nucleophilic pathway (see also nucleophilic aromatic substitution via an elimination—addition reaction,

**Fig. 11.23** Conversion of bromobenzene into aniline

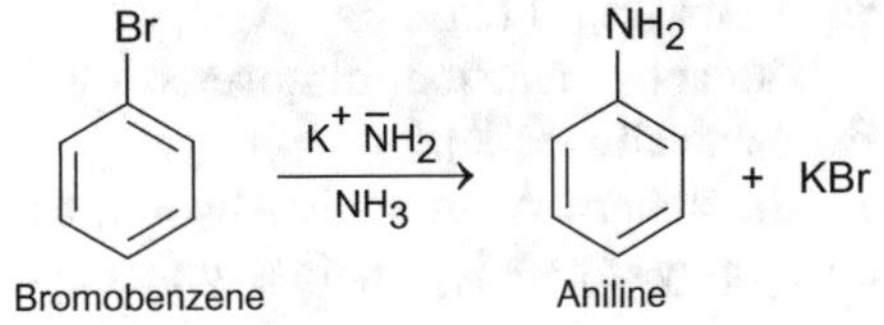

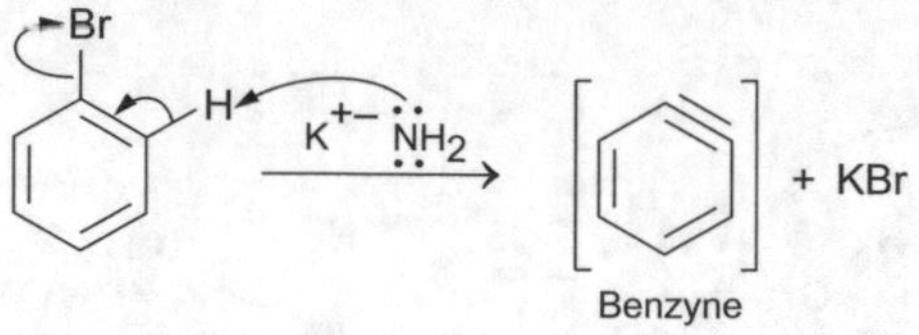

**Fig. 11.24** Elimination of HBr from bromobenzene by reacting with KNH$_2$

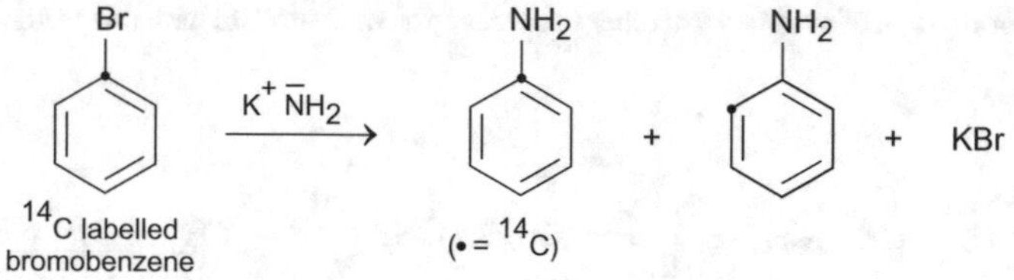

**Fig. 11.25** The elimination reaction with $^{14}$C labelledbromobenzene with K NH$_2$ | NH$_3$

Sect. 12.11.1.2 in Chap. 12). In fact, elimination occurred under the reaction conditions (using strong base) giving an unsaturated benzene derivative called benzyne (Fig. 11.24).

The above mechanism finds support in the fact that the reaction of bromobenzene labelled with $^{14}$C at the ipso carbon atom gives a mixture of products having labelled atom in two places (Fig. 11.25).

For more details see Sect. 12.11.1.2 in Chap. 12.

## 11.6 Eliminations Involving Ammonium Compounds

### 11.6.1 Hofmann Elimination

Besides the elimination reaction so far studied, there is another type of elimination involving ammonium compounds. In these eliminations, the substrate bears a positive charge. Important reaction of this type is the E2 type elimination which takes place when a quaternary ammonium hydroxide is heated to give an alkene, water and a tertiary amine as products. Thus, quaternary ammonium hydroxide (viz., trimethylpropyl ammonium hydroxide), having alkyl groups that are larger than methyl group, on heating gives an alkene (propene, tertiary amine (trimethyl amine) and water (Fig. 11.26).

The above reaction, discovered by August W. von Hofmann (1851), is called Hofmann elimination.

The quaternary ammonium hydroxides are obtained by treating the corresponding quaternary ammonium halides with moist silver oxide. The hydroxide ion acting as a

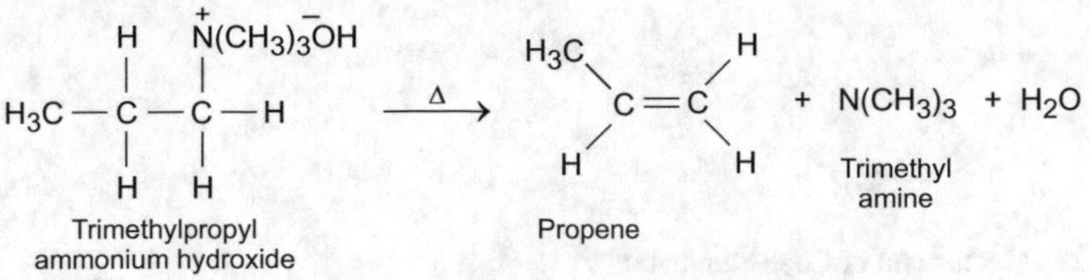

**Fig. 11.26** Products obtained by heating quaternary ammonium hydroxide

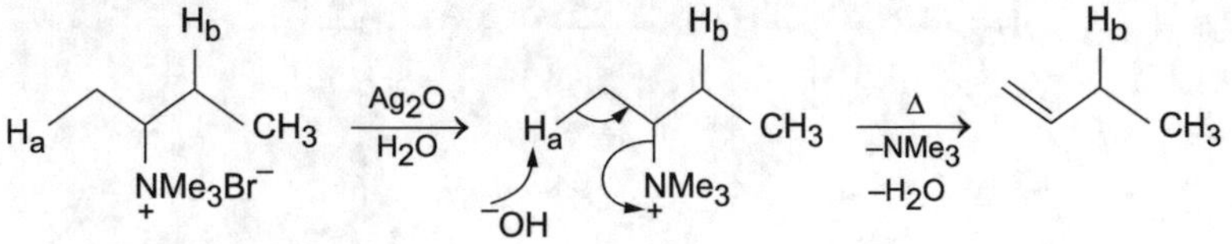

**Fig. 11.27** Mechanism of Hofmann elimination

strong base abstracts the β-hydrogen of the larger alkyl group and a π-bond is formed via the expulsion of triethylamine (E2 elimination). When there are two acidic β-hydrogens, $H_a$ and $H_b$, it is the $H_a$ which is removed by base so that least substituted alkene is the predominant product (Fig. 11.27).

The Hofmann elimination yield mainly the least substituted alkene. This is called the **Hofmann rule** and contrasts the usual Saytzeff elimination in which more substituted alkene is obtained (see Sect. 16.11.2).

## 11.6.2 Cope Elimination

It involves heating a tertiary amine oxide (prepared by treating tertiary amine with $H_2O_2$) to give an alkene and N,N-dimethylhydroxyl amine (Fig. 11.28). The Cope elimination is a syn elimination and proceeds through a cyclic transition state as shown in Fig. 11.29.

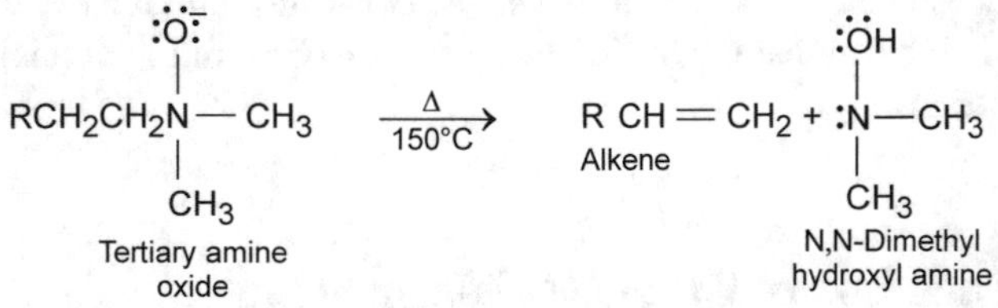

**Fig. 11.28** Cope Elimination

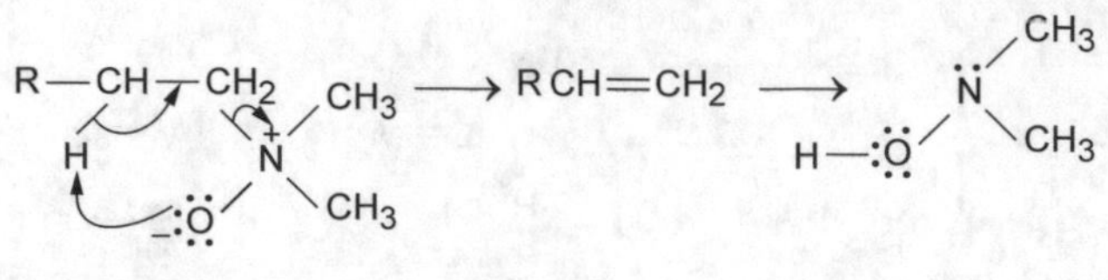

**Fig. 11.29** Mechanism of Cope elimination

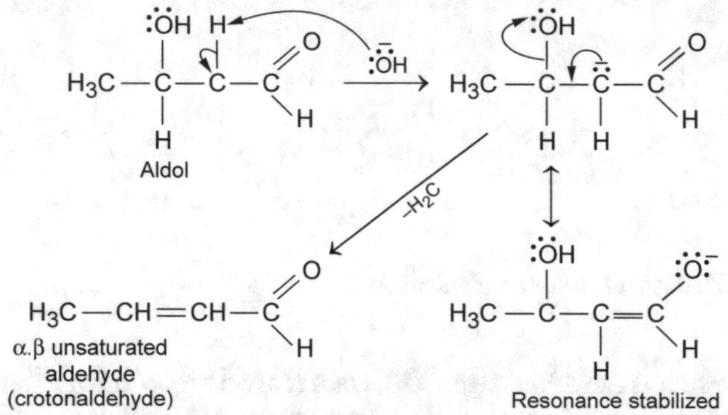

**Fig. 11.30** Dehydration of β-hydroxycarbonyl compound (aldol) with base

## 11.7   Elimination Reaction of β-Hydroxycarbonyl Compounds

We have known dehydrohalogenation of alkyl halides by E2 or E1 mechanism and alcohols on dehydration by heating with conc. $H_2SO_4$ give alkenes. Alcohols (primary) dehydrate only in the presence acid but not base, since hydroxide is a poor leaving group. However, in case hydroxyl group is β to a carbonyl group (as in the case of aldol obtained in aldol condensation), loss of H and OH from α and β carbon occurs giving a conjugated double bond. The dehydration of β hydroxy carbonyl compounds takes place in the following steps (Fig. 11.30).

The above mechanism is called E1CB mechanism (E1CB stands for elimination, unimolecular, conjugate base). It differs from E2 and E1 eliminations. Like E1 elimination E1CB mechanism occurs in two steps. However, unlike E1 elimination, the intermediate in E1 elimination, the intermediate is E1CB elimination is a carbanion, not a carbocation.

## 11.8   Comparison of E1 and E2 Mechanisms

The following Table gives a comparison between E1 and E2 Mechanisms:

|  | E2 mechanism | E1 mechanism |
|---|---|---|
| 1. Usefulness | Commonly used | Less useful since a mixture of $S_N1$ and E1 products are usually obtained (see Fig. 11.18) |
| 2. Base used | Strong, negatively charged base like $^-$OH and $^-$OR favour E2 mechanism | Favoured by weaker and neutral bases like $H_2O$ and ROH |
| 3. Alkyl halids | E2 mechanism occurs with 1°, 2° and 3°-alkyl halides ($R_3CX > R_2CHX >$ $RCH_2X$ | 1° alkyl halides do not undergo E1 mechanism since these form unstable 1°, carbocations |

## 11.9   Synthesis of Alkynes

Alkynes can be synthesised by two successive dehydrogenation reactions (E2 reactions) using a vicinal dihalide (in which two X atoms are on adjacent carbons) or geminal dihalide (in which two X atoms are on the same carbon) (Fig. 11.31).

As seen (Fig. 11.31) stronger base is needed to synthesise alkynes by dehydrohalogenation of dihalides. This is because in the second stage of elimination the carbon atom bearing X group is $sp^2$-hybridised and $sp^2$-hybridised C–H bonds and stronger than $sp^3$-hybridised C–H bonds.

### Solved Problem

Alkyl halides are known to undergo substitution ($S_N1$, $S_N2$) or elimination (E1, E2) reactions. How can you know that a particular alkyl halide will undergo substitution or elimination with a given base or nucleophile and by what mechanism?

**Ans.** Normally the answer is not easy and a mixture of products results. However, the following two procedures can help to determine whether substitution or elimination occurs.

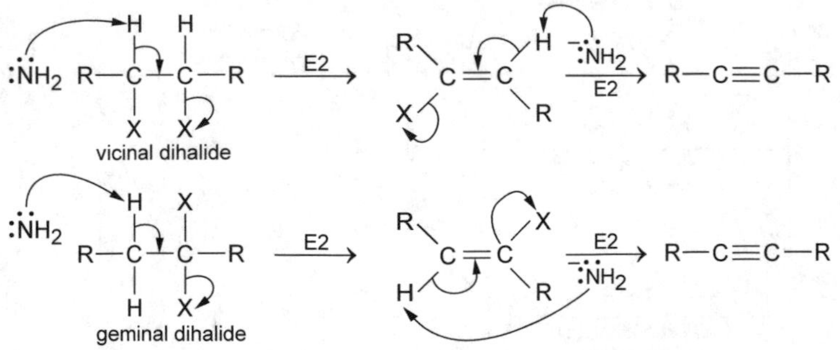

**Fig. 11.31** Synthesis of alkynes from vicinal and geminal dihalides

(i)  Substitution is favoured over elimination if good nucleophiles that are weak bases are used. Some nucleophiles like $^-I$, $Br^-$, $HS^-$, $^-CN$, $CH_3COO^-$ give substitutions.

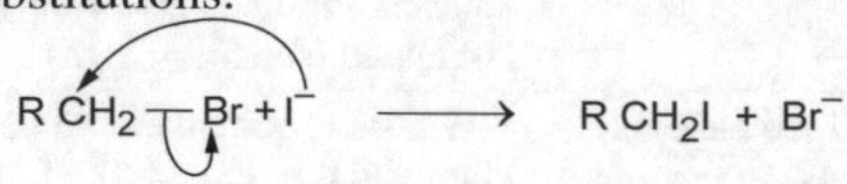

$$R\ CH_2 - Br + I^- \longrightarrow R\ CH_2I + Br^-$$

(ii)  Elimination is favoured over substitution by bulky non-nucleophilic bases like $KOC(CH_3)_3$, DBU and DBN.

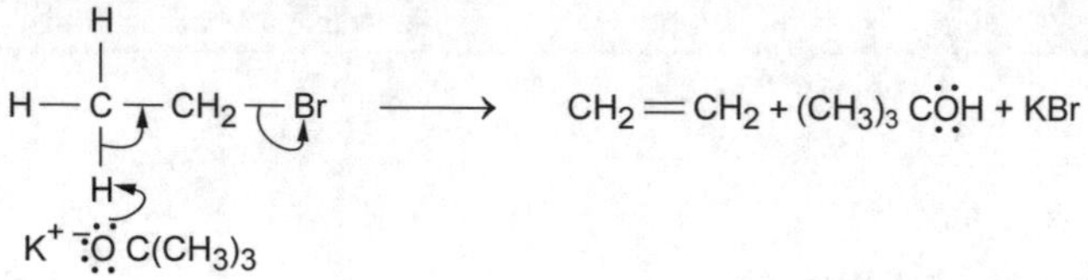

$$H-C-CH_2-Br \longrightarrow CH_2 = CH_2 + (CH_3)_3\ C\ddot{O}H + KBr$$

$$K^+\ \ddot{:}\ddot{O}\ C(CH_3)_3$$

## Summary

Summary of the reaction of alkyl halides in substitution ($S_N1$, $S_N2$) or elimination (E1, E2) reactions.

1.   Primary alkylhalides react by both $S_N2$ and E2 mechanism.

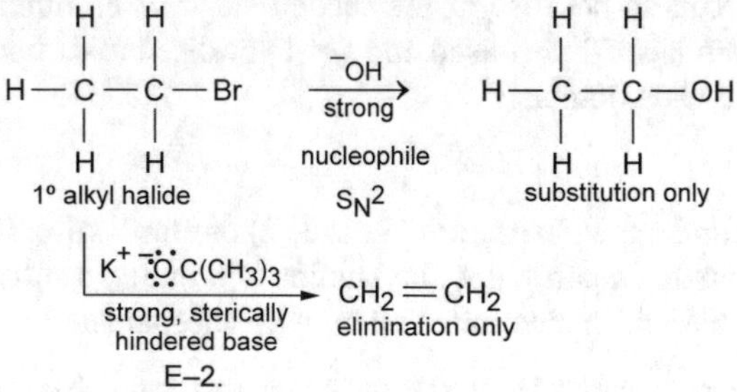

$$H-C-C-Br \xrightarrow[\text{strong}]{^-OH} H-C-C-OH$$

1° alkyl halide        strong nucleophile        substitution only
                       $S_N2$

$$\xrightarrow[\substack{\text{strong, sterically} \\ \text{hindered base} \\ \text{E-2.}}]{K^+\ \ddot{O}\ C(CH_3)_3} CH_2 = CH_2$$
elimination only

2.   Secondary alkyl halides react with all the mechanisms.

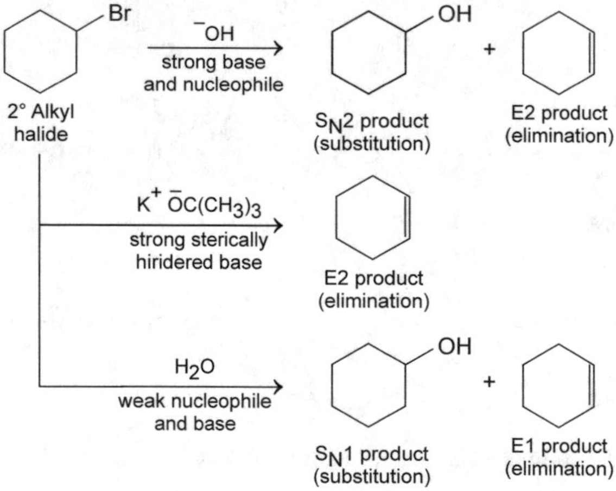

2° Alkyl halide  $\xrightarrow[\substack{\text{strong base} \\ \text{and nucleophile}}]{^-OH}$  $S_N2$ product (substitution)  +  E2 product (elimination)

$\xrightarrow[\substack{\text{strong sterically} \\ \text{hiridered base}}]{K^+\ \bar{O}C(CH_3)_3}$  E2 product (elimination)

$\xrightarrow[\substack{\text{weak nucleophile} \\ \text{and base}}]{H_2O}$  $S_N1$ product (substitution)  +  E1 product (elimination)

3.   Tertiary alkyl halides react by all mechanism except $S_N2$ mechanism.

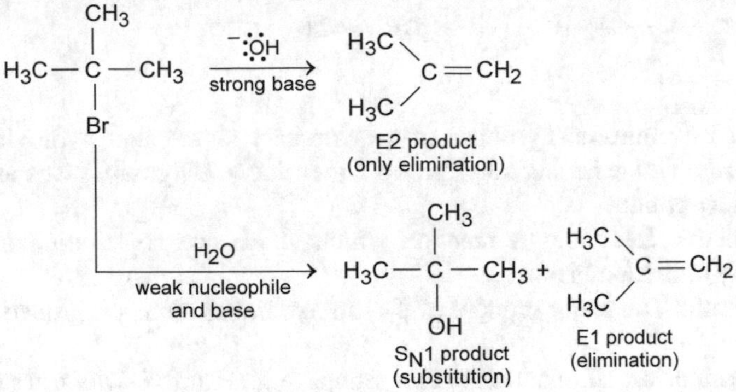

E2 product
(only elimination)

$S_N1$ product
(substitution)

E1 product
(elimination)

**Key Concepts**

- **Benzyne**: Dihydrobenzene, an unstable reaction intermediate obtained by the action of strong base on bromobenzene.

- **Bimolecular Elimination Reaction (E2)**: A second-order concerted stereoselective reaction depending on the concentration of both the substrate and the nucleophile. The reaction involves *trans* elimination (β-elimination).

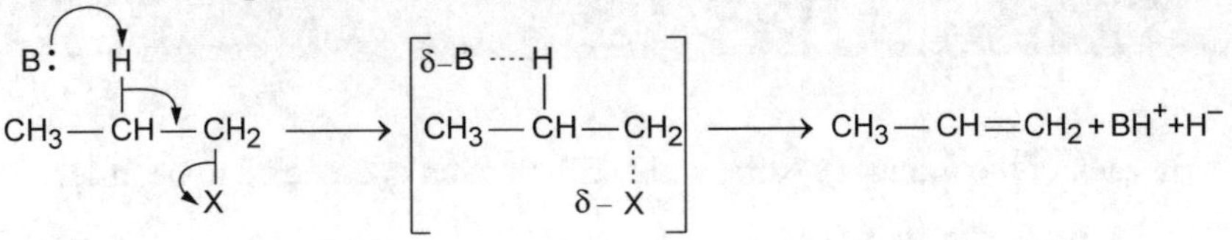

- **Concerted Reaction**: Reaction in which all bonds are broken and formed in a single step.
- **Cope elimination**: Formation of an olefin and a hydroxylamine by thermal decomposition of a tertiary amine oxide.
- **DBN**: A storng base, 1,5-diazabicyclo [4.3.0] non-5-ene.
- **DBU**: A strong base, 1,8-diazabicyclo [5.4.0] undec-7-ene.
- **Dehydrohalogenation**: Removal of HX from an alkylhalide.
- **Elimination Reaction**: Involving elimination of two or four groups attached to adjacent carbon atoms in a substrate forming a multiple bond.

- **E1CB Elimination**: In case the OH group is β to α carbonyl group (as in the case of aldol), loss of H and OH from α-and β-carbon atoms form a conjugated double bond

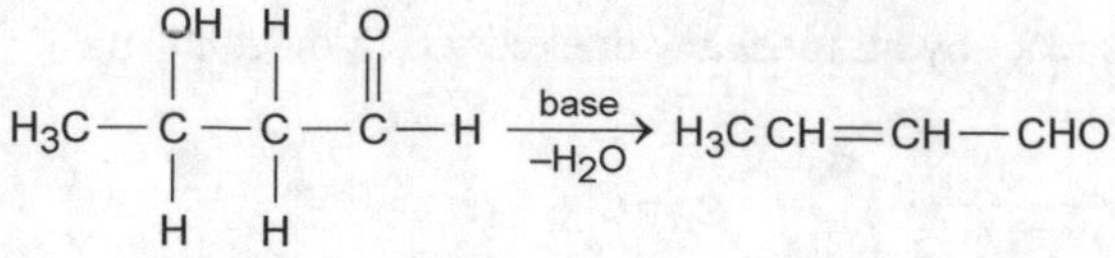

$$H_3C-\underset{\underset{H}{|}}{\overset{\overset{OH}{|}}{C}}-\underset{\underset{H}{|}}{\overset{\overset{H}{|}}{C}}-\overset{\overset{O}{||}}{C}-H \xrightarrow[-H_2O]{base} H_3C\ CH=CH-CHO$$

Aldol

- **Hofmann Elimination**: Pyrolysis of a quaternary ammonium hydroxide salts (derived from amine having alkyl group larger than $CH_3$ group) gives an olefin and a tertiary amine.
- **Regeoselective Reaction**: A reaction which yields one constitutional isomer exclusively or predominantly.
- **Saytzeff Rule**: The major product of β-elimination has more substituted double bond.
- *Trans* **Elimination**: Elimination of two groups (e.g., H and X) which are *trans* to each other.

- **Unimolecular Elimination Reaction (E1)**: A first-order elimination reaction depending on the concentration of only the substrate and is a two-step process.

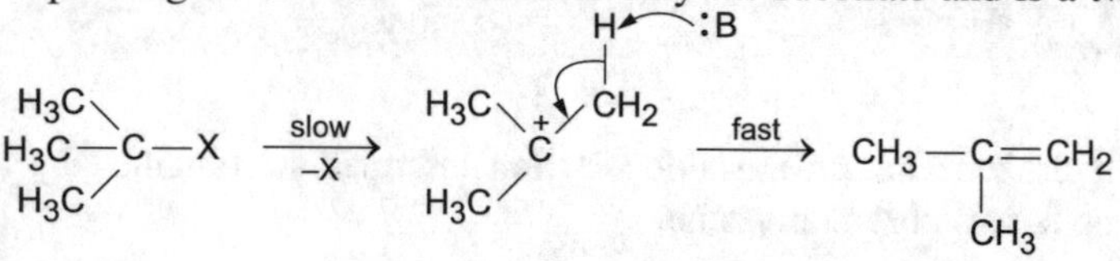

$$\underset{\underset{H_3C}{\overset{H_3C}{}}}{\overset{H_3C}{}}C-X \xrightarrow[-X]{slow} \underset{\underset{H_3C}{\overset{H_3C}{}}}{\overset{H_3C}{}}\overset{+}{C}\overset{\overset{H\frown :B}{CH_2}}{} \xrightarrow{fast} CH_3-\underset{\underset{CH_3}{|}}{C}=CH_2$$

## Problems

1. In each of the following pairs, which E2 reaction is expected to be faster?

    (a)  $RCH_2CH_2Br + {}^-OH \longrightarrow$
         $RCH_2CH_2Br + {}^-OC(CH_3)_3 \longrightarrow$

    (b)  $RCH_2CH_2Br + {}^-OC(CH_3)_3 \longrightarrow$
         $RCH_2CH_2Cl + {}^-OC(CH_3)_3 \longrightarrow$

2. What product is expected to be obtained in the following reaction?
   $CH_3CH_2-CH_2-CH_2Br + RO^- \longrightarrow$

   **Ans.** 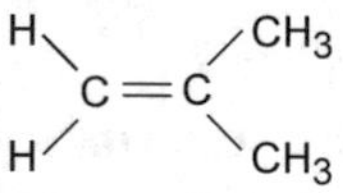  $\underset{\underset{H}{H}}{\overset{H}{}}C=C\underset{\underset{CH_3}{}}{\overset{CH_3}{}}$

3. What product is expected to be obtained by the β-elimination of α-chloro ethylcyclope?

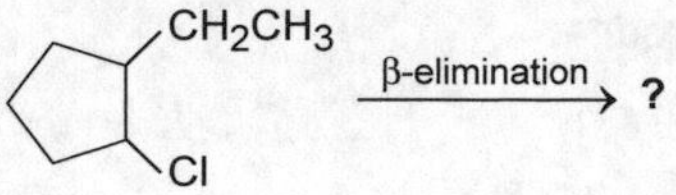

**Ans.** Ethylcyclopentene

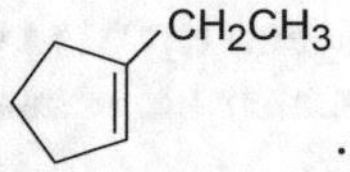

4. What alkenes are expected to be obtained from each of the following alkyl halides by an E2 reaction?

(a)   (b)

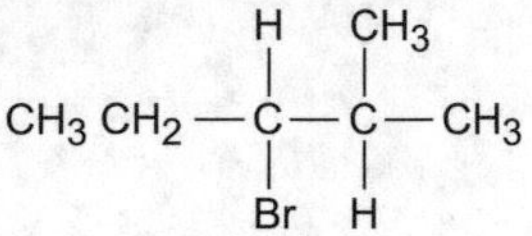

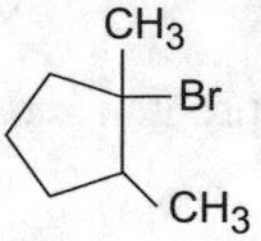

(c) 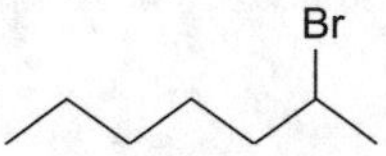   (d)

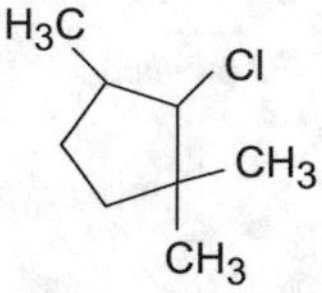

5. What alkenes are expected to be obtained as major product from each of the following alkyl halide by E1 reaction?

(a)   (b)

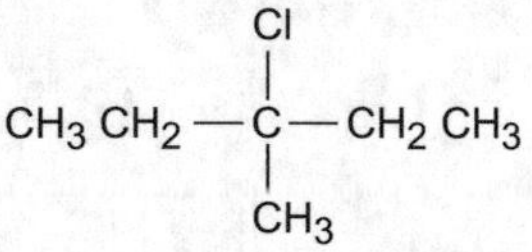

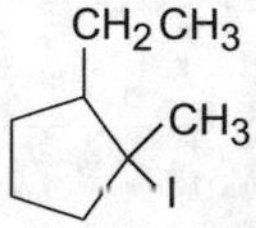

**Hint.** Use Saytzeff's rule to predict the major product.

6. Which alkene is expected to be obtained as major product in the following E2 reaction?

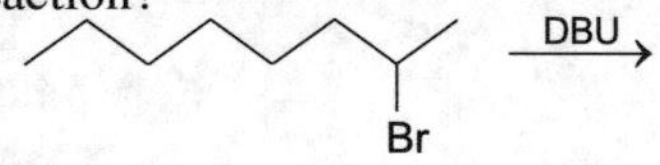

**Ans.** 2-octene and 1-octene in the ratio 4:1.

7. How will you decide whether an elimination reaction follows E1 or E2 mechanism?

**Hint.** Strong base favour E2 mechanism and weak base favour E1 mechanism.

8. Explain why stronger base is needed in the following dehydrohalogenation?

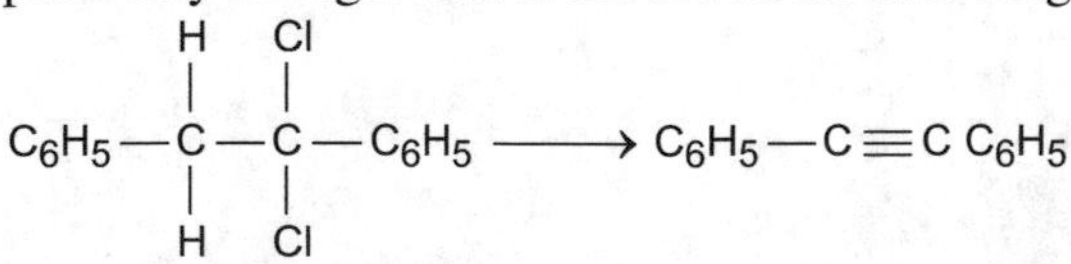

9. What product or products are expected to be obtained in the following reaction?

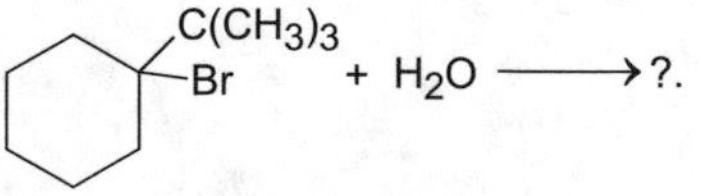

**Ans**. A mixture of $S_N1$ product and E1 product

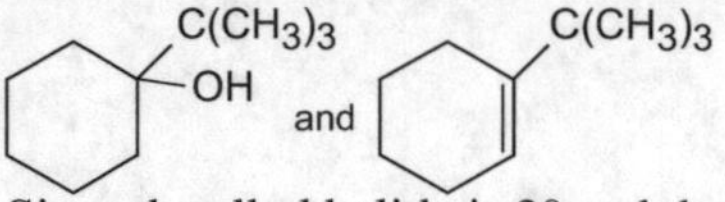

C(CH₃)₃ structure with OH and C(CH₃)₃ structure with double bond, "and" between them

Since the alkyl halide is 3° and the reagent ($H_2O$) is a weak base and nucleophile, so products from both $S_N1$ and E1 mechanism are obtained.

10.    What product or products are expected to be obtained in the following reaction?

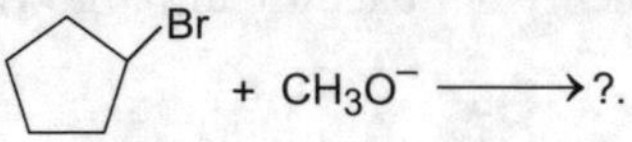

cyclopentyl Br structure + $CH_3O^-$ ⟶ ?.

**Ans**. A mixture of $S_N2$ product and E2 product

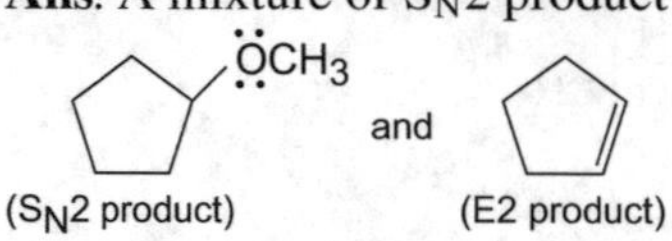

cyclopentyl ÖCH₃ structure and cyclopentene structure

(S$_N$2 product)            (E2 product)

# Chapter 12
# Stereochemistry of Substitution Reactions

## 12.1 Introduction

The reaction in which one or more atoms or groups in a compound are replaced or substituted by other atoms or groups is known as substitution reaction. The products that are obtained in substitution reactions are called substitution products. An example is

$$CH_3 - I + Cl^- \rightarrow CH_3Cl + I^-$$

In these reactions, one sigma bond breaks and another forms at the same carbon atom.

Different types of substitution reactions are known:

- Free radical substitution reaction
- Electrophilic substitution reactions
- Nucleophilic substitution reactions.

Some of these include

- Substitution reactions of alcohols
- Substitution reactions of ethers
- Substitution reactions of epoxides
- Substitution reactions of thiols
- Substitution reactions of thioethers
- Aromatic substitutions
- Substitution reactions of aryl diazonism salts.
- Substitution of sulfonic acid group in benzene sulfonic acid.
- Substitution of active hydrogen by alkyl and acyl groups
- Substitution of OH groups of alcohols by Cl or Br.

© The Author(s), under exclusive license to Springer Nature Switzerland AG 2022
V. K. Ahluwalia, *Stereochemistry of Organic Compounds*,
https://doi.org/10.1007/978-3-030-84961-0_12

## 12.2   Free Radical Substitution Reactions

These reactions are initiated by free radicals. As has already been stated (Sect. 7.4), the term free radical is used for any species which possess unpaired electron. These are obtained by the homolytic fission of a covalent bond, the two departing atoms take one electron each and are called free radicals (Fig. 12.1).

The most common substitution reactions of free radicals are halogenation. Some of such reactions are discussed below.

### 12.2.1   Conversion of Methane into Carbon Tetrachloride

Methane on treatment with chlorine gives carbon tetrachloride. The reaction proceeds as shown below (Figs. 12.1, 12.2 and 12.3).

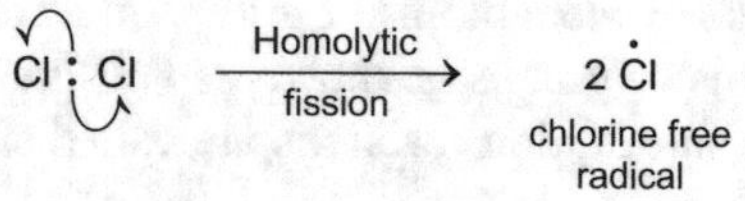

**Fig. 12.1**   Formation of free radicals by homolytic fission

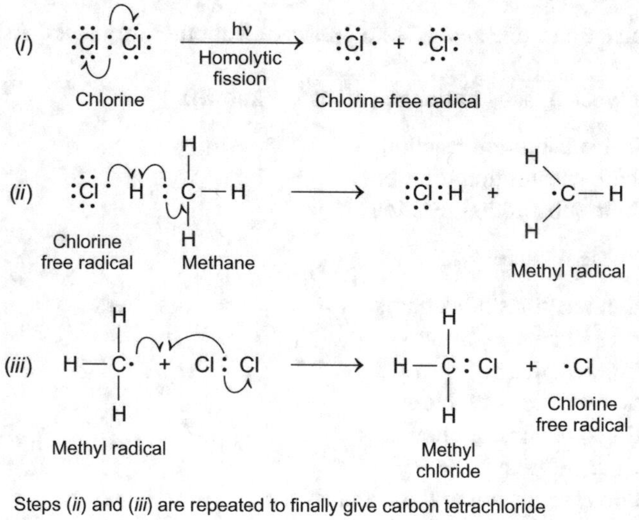

Steps (*ii*) and (*iii*) are repeated to finally give carbon tetrachloride

**Fig. 12.2**   Mechanism of free radical reaction of the reaction of methane and chlorine

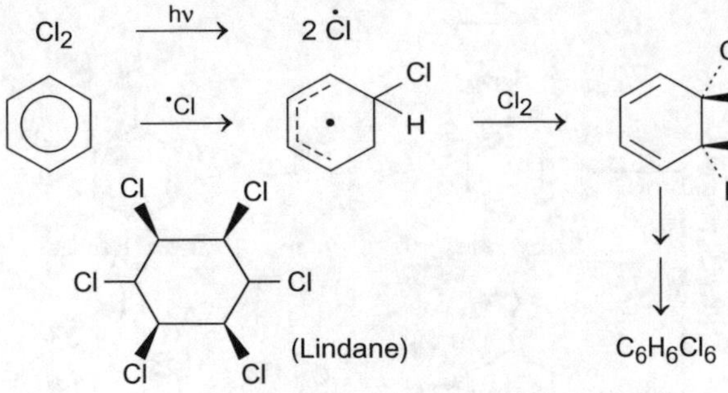

**Fig. 12.3**   Reaction of benzene with chlorine

## 12.2.2   Conversion of Benzene into Benzene Hexachloride

The reaction of benzene with chlorine takes place via free radical pathway. The reaction is catalysed by light (Fig. 12.3).

The reaction gives a mixture of eight non-convertible stereoisomers of hexachlorobenzene. The gamma isomer is known as lindane (gammexame) and is a useful insecticide.

## 12.2.3   Conversion of Toluene into Benzyl Chloride

The reaction of toluene with chlorine in presence of sunlight gives benzyl chloride. The Cl free radical abstracts a proton from $CH_3$ group to give delocalized benzyl radical, $C_6H_5 - \overset{\bullet}{C}H_2$ (which is reasonance stabilised) rather than hexadienyl radical, in which the aromatic stablisation of the starting material is lost (Fig. 12.4).

## 12.2.4   Conversion of Propene into n-Propyl Bromide

In the reaction of propene with HBr, in presence of a peroxide (which generates a free radical, $\overset{\bullet}{B}r$), the addition takes places as per anti-Markovnikov rule to give $n$-propyl bromide.

For details see Fig. 10.5 in Chap. 10.

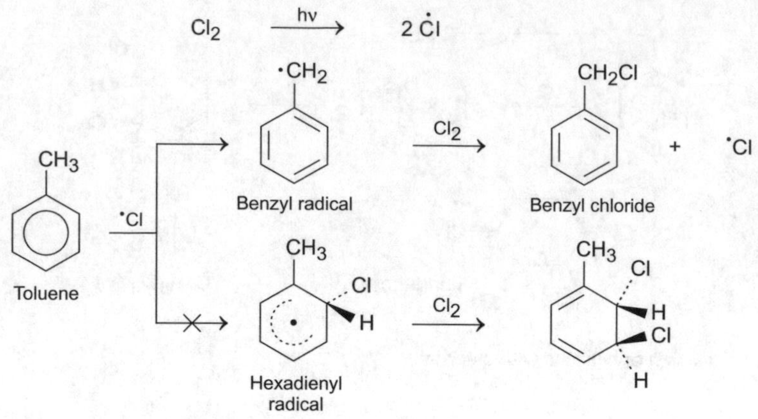

Fig. 12.4  Conversion of toluene into benzyl chloride

## 12.2.5  Allylic Substitution

The carbon atom adjacent to a double bond is known as allylic carbon atom. As an example, propene on reaction with chlorine at 500–600 °C gives allylic substitution product (Fig. 12.5).

The mechanism of the reaction is given below (Fig. 12.6).

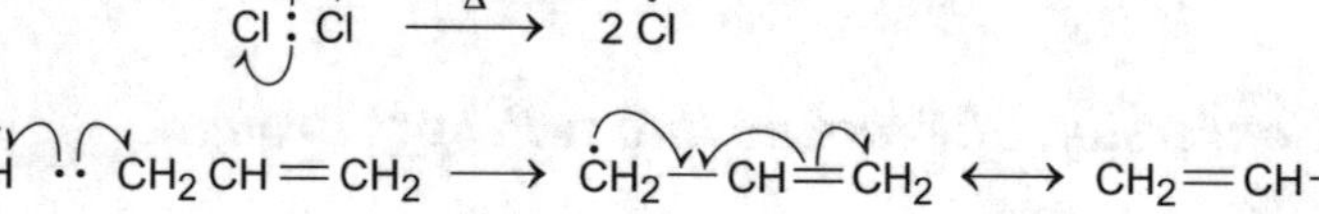

Fig. 12.5  Allylic substitution

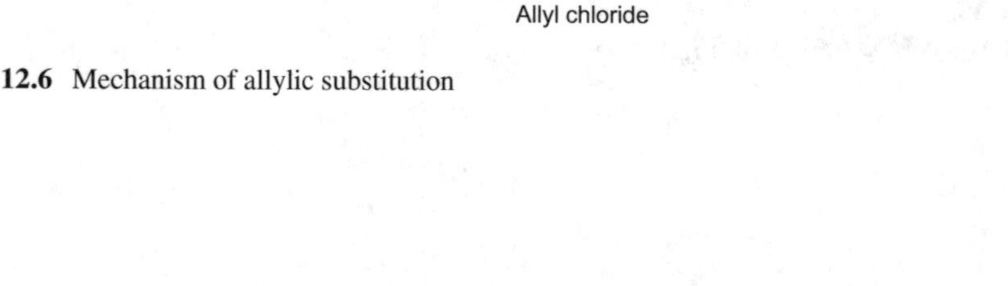

Fig. 12.6  Mechanism of allylic substitution

**Fig. 12.7** The reaction of ethene with chlorine on heating

$$CH_2 = CH_2 + Cl_2 \xrightarrow{500°C \text{ or } hv} CH_2 = CHCl + HCl$$

Ethene           vinyl chloride

## *12.2.6 Vinylic Substitution*

The carbon atom attached to a double bond is called vinylic carbon. Alternatively, compounds in which halogen is attached to $sp^2$ hybridised carbon are called vinylic halide. The vinyl group is $CH_2 = = CH_-$ (Fig. 12.7).

## *12.2.7 Benzylic Bromination*

The reaction of ethylbenzene with bromine in presence of UV light gives 2-bromo-2-phenyl ethane. The steps involved are given below (Fig. 12.8).

## 12.3 Electrophilic Substitution Reaction of Monosubstituted Benzenes

These types of reactions involve attack of an electrophile to a substrate (aromatic compounds) and the reaction is represented as SE (S stands for substitution and E for electrophilic). The most common examples of electrophilic substitution reactions of monosubstituted benzenes are nitration, halogenation, sulphonation, alkylation and acylation of benzene (See, Sects. 12.3.1, 12.3.2, 12.3.3 and 12.3.4).

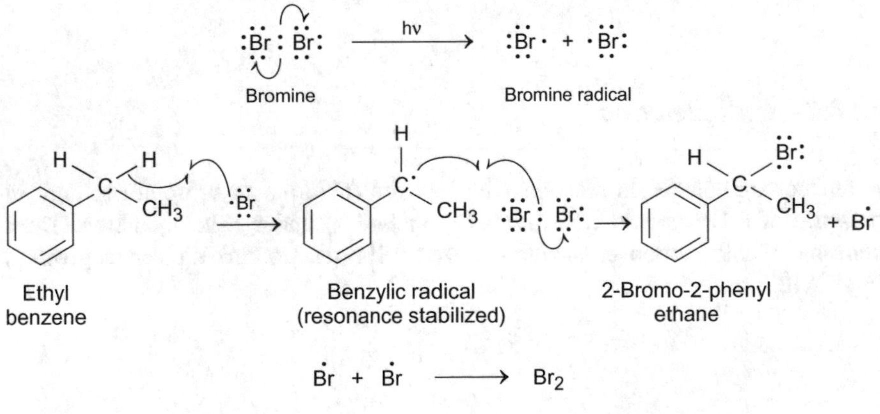

**Fig. 12.8** Benzylic bromination

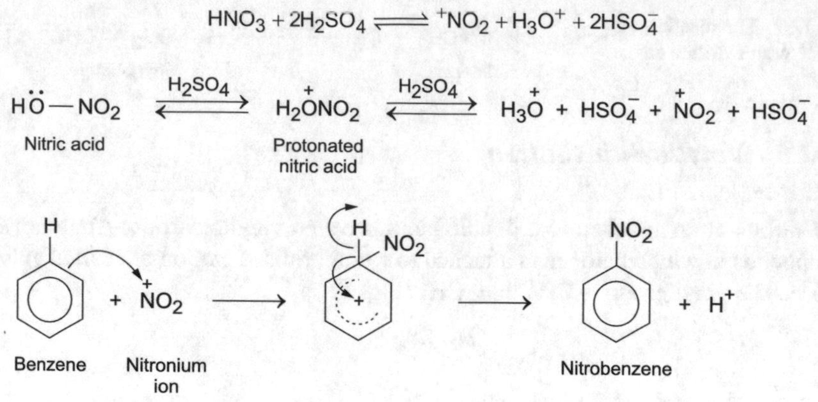

$$HNO_3 + 2H_2SO_4 \rightleftharpoons {}^+NO_2 + H_3O^+ + 2HSO_4^-$$

Fig. 12.9  Nitration of benzene to give nitrobenzene

## 12.3.1  Nitration

Aromatic compounds (e.g., benzene) on reaction with nitration mixture (a mixture of conc. $HNO_3$ and conc. $H_2SO_4$) gives nitro derivative. The process is called nitration.

$$C_6H_6 \xrightarrow[\text{Nitration}]{\text{Conc. } HNO_3 | \text{Conc. } H_2SO_4} C_6H_5NO_2 + H_2O$$
$$\text{Nitrobenzene}$$

The reaction is believed to take place by the formation nitronium ion ($^+NO_2$) (Fig. 12.9).

$$HNO_3 + 2H_2SO_4 \rightleftharpoons {}^+ NO_2 + H_3O^+ + 2HSO_4^-$$

## 12.3.2  Halogenation

Aromatic compounds on reaction with halogen (chlorine or bromine) in dark in presence of a Lewis acid catalyst ($FeCl_3$ or $FeBr_3$) undergo halogenation. Thus, benzene on chlorination gives chlorobenzene. The reaction takes place as given in Fig. 12.10.

**Fig. 12.10** Mechanism of chlorination of benzene

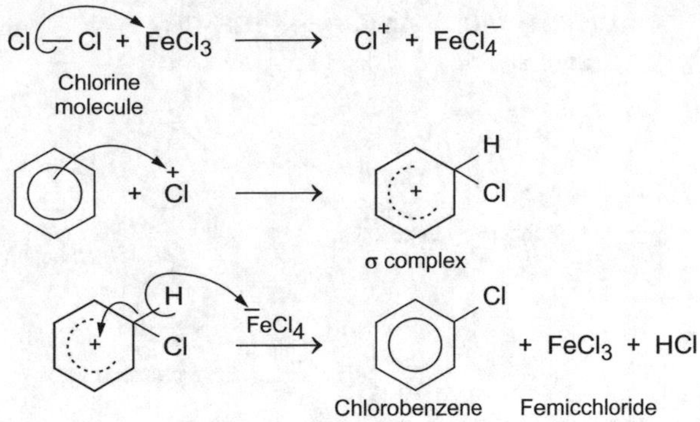
Cl — Cl + FeCl3 → Cl+ + FeCl4−
Chlorine molecule
+ Cl+
H
Cl
+
σ complex
H
Cl
FeCl4−
Cl
+ FeCl3 + HCl
Chlorobenzene
Femicchloride

## 12.3.3 Sulfonation

Aromatic compounds like benzene on treatment with conc. $H_2SO_4$ or fuming $H_2SO_4$ gives benzene sulphonic acid. In sulfonation, the electrophilic reagent is $SO_3$. The reaction takes place as given below (Fig. 12.11).

$$2H_2SO_4 \rightleftarrows H_3O^+ + HSO_4^- + SO_3$$

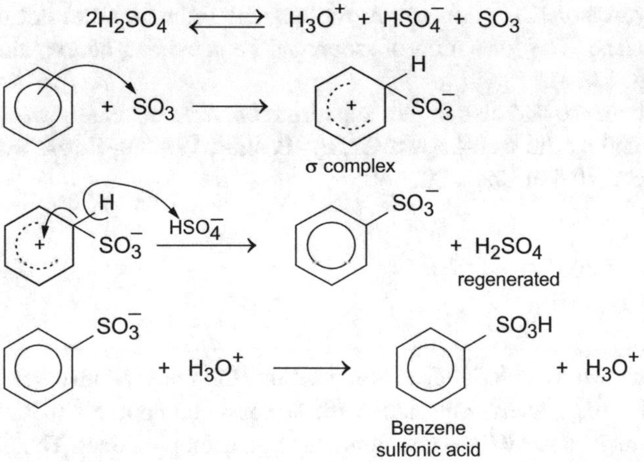
2H2SO4 ⇌ H3O+ + HSO4− + SO3
+ SO3
H
SO3−
σ complex
H
SO3−
HSO4−
SO3−
+ H2SO4
regenerated
SO3−
+ H3O+
SO3H
+ H3O+
Benzene sulfonic acid

**Fig. 12.11** Mechanism of sulfonation of benzene

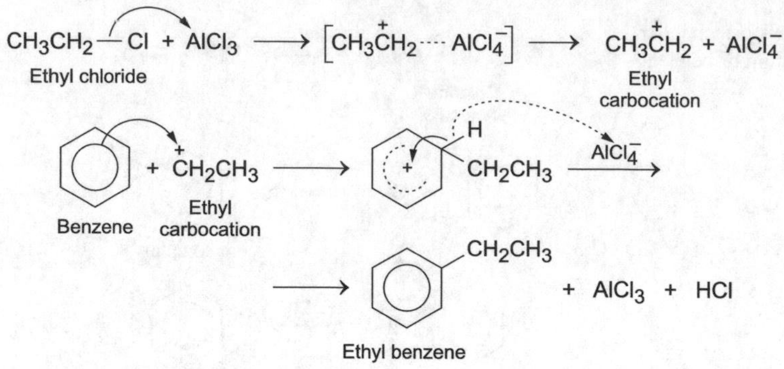

$$CH_3CH_2-Cl + AlCl_3 \longrightarrow \left[CH_3\overset{+}{C}H_2 \cdots AlCl_4^-\right] \longrightarrow CH_3\overset{+}{C}H_2 + AlCl_4^-$$

Ethyl chloride                                     Ethyl
carbocation

Benzene   Ethyl carbocation

Ethyl benzene

**Fig. 12.12** Mechanism of Friedel Crafts alkylation

## 12.3.4 Alkylation

Commonly known as **Friedel Crafts alkylation**, the reaction involves treatment of aromatic compounds (like benzene) with alkyl halides in presence of anhydrous Lewis acid catalysts like $AlCl_3$, $FeCl_3$, $BF_3$, etc. to give alkyl-substituted benzene. Following steps are involved in the alkylation reaction (Fig. 12.12).

Friedel Crafts alkylation is not useful to introduce alkyl group higher than $CH_3CH_2$. This is due to skeletal rearrangement taking place in case of higher alkenes. As an example, the reaction of benzene with *n*-propyl chloride in presence of anhydrous $AlCl_3$ gives isopropyl benzene (cumene) and not the expected *n*-propyl benzene. The formation of isopropyl benzene can be explained as given below (in Fig. 12.13).

*n*-Propyl benzene and also higher substituted benzenene can however be conveniently obtained by the well-known **Corey–Posner, Whiles–House synthesis**. For details see Sect. 20.8 in Chap. 20.

## 12.3.5 Acylation

Commonly known as **Friedel Crafts acylation**, the reaction involves the reaction of acyl halides (e.g., acetyl chloride) with aromatic compounds (e.g., benzene) in presence of anhydrous $AlCl_3$ to give ketones (e.g., acetophenone). Thus, the reaction of benzene with acetyl chloride in presence of anhydrous $AlCl_3$ gives acetophenone. The reaction involves the following steps (Fig. 12.14).

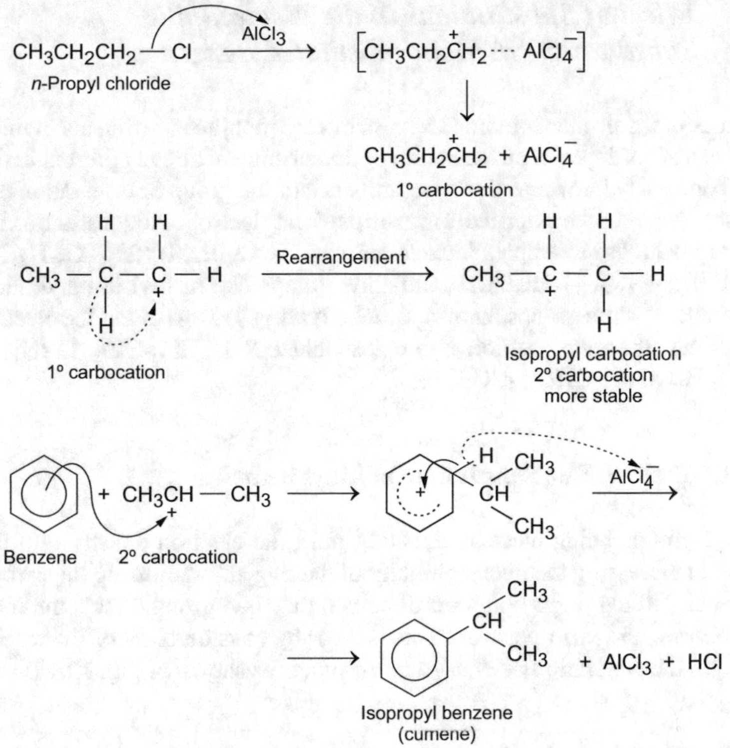

$$CH_3CH_2CH_2-Cl \xrightarrow{AlCl_3} \left[ CH_3CH_2\overset{+}{C}H_2 \cdots AlCl_4^- \right]$$

n-Propyl chloride

$$CH_3CH_2\overset{+}{C}H_2 + AlCl_4^-$$

1° carbocation

$$CH_3-\underset{\underset{H}{|}}{\overset{\overset{H}{|}}{C}}-\overset{\overset{H}{|}}{\underset{+}{C}}-H \xrightarrow{Rearrangement} CH_3-\underset{+}{\overset{\overset{H}{|}}{C}}-\overset{\overset{H}{|}}{\underset{\underset{H}{|}}{C}}-H$$

1° carbocation

Isopropyl carbocation
2° carbocation
more stable

Benzene + $CH_3CH-CH_3$ (2° carbocation) →

$\overset{H}{\underset{CH}{\diagdown}}\;CH_3$, $CH_3$, $AlCl_4^-$ →

$$\overset{CH_3}{\underset{CH}{|}}\;CH_3 + AlCl_3 + HCl$$

Isopropyl benzene
(cumene)

**Fig. 12.13** Formation of cumene by the reaction of benzene with propylbromide

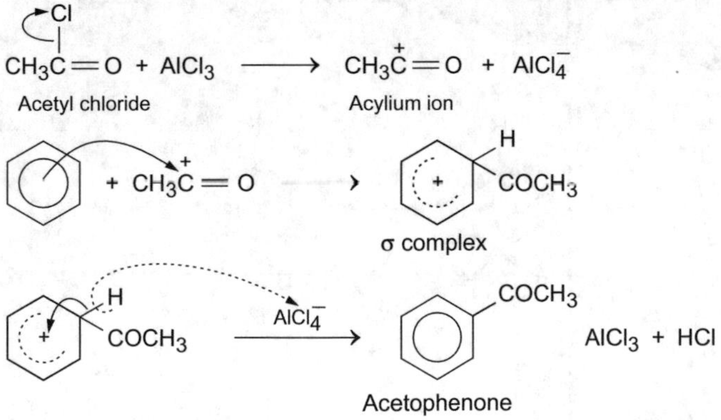

$$CH_3\overset{\frown}{C}=O + AlCl_3 \longrightarrow CH_3\overset{+}{C}=O + AlCl_4^-$$

Acetyl chloride

Acylium ion

+ $CH_3\overset{+}{C}=O$ ＞

$\overset{H}{\diagdown}COCH_3$

σ complex

$\overset{H}{\diagdown}COCH_3 \xrightarrow{AlCl_4^-}$

$\overset{COCH_3}{\diagup}$ AlCl_3 + HCl

Acetophenone

**Fig. 12.14** Mechanism of Friedel Crafts acylation

## 12.3.6   Effect of Substitutions in the Electrophilic Substitutions in Monosubstituted Benzenes

The nature of the product obtained in case of electrophilic substituents in monosubstituted benzenes depends on the nature of the substituent already present in the ring. This is commonly known as directive influence of the group or orientation effect.

Some groups known as **activating groups** direct electrophilic substitution in *ortho* and *para* positions. Examples include halogen (F, Cl, Br, I), OH, $-OCH_3$, $-OR$, $-NH_2$, $-NHR$, $-NR_2$, $-NHCOCH_3$ and alkyl groups like methyl or ethyl. However, the presence of some groups, known as **deactivating groups** direct the electrophilic substitution in the *meta* position. Examples include $NO_2$, $CF_3$, $CCl_3$, $C\equiv N$, $-SO_3H$, $-COOH$, $-COOR$, $-CHO$, $-COR$, etc.

### 12.3.6.1   Electrophilic Substitution in Alkyl Benzenes

The alkyl groups being electron donating push the electron density into the ring resulting in increasing the nucleophilicity of the ring and stabilising the carbocation intermediate. Being *ortho* and *para* directive, the alkyl group directs the incoming group to *ortho* and *para* position. This is clear by consideration of the mechanism involved in the generation of *o-*, *m-* and *p*-isomers as shown in Fig. 12.15.

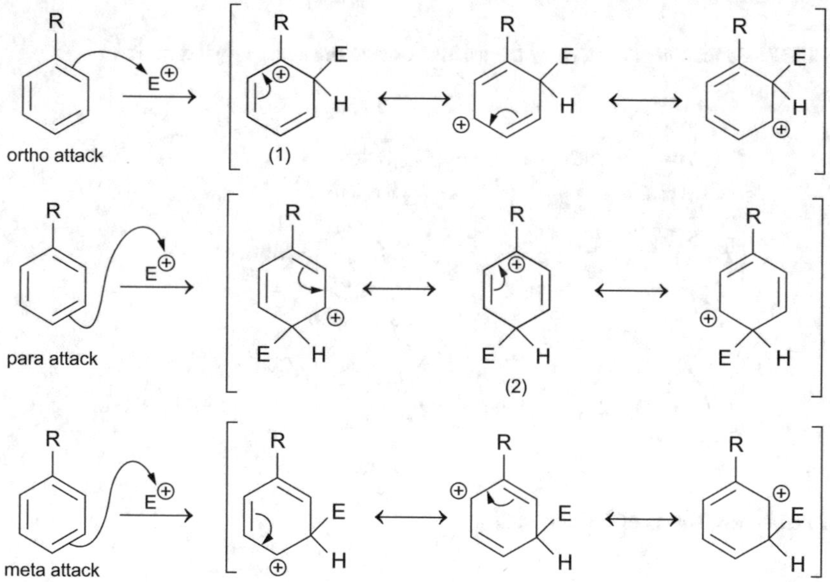

**Fig. 12.15**  Canonical structures arising from *o-*, *m-* and *p*-attacks in case of alkyl benzenes

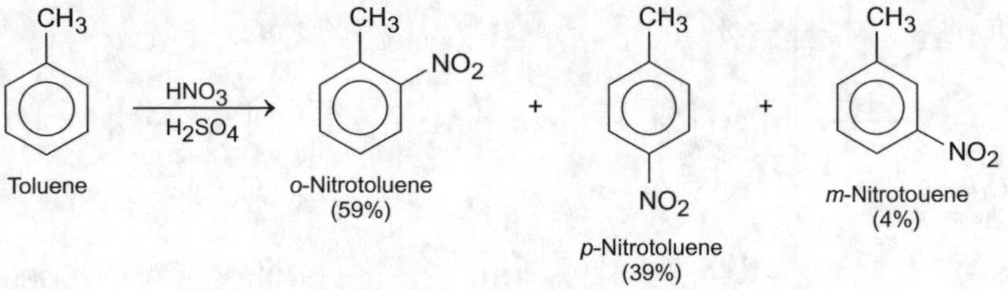

**Fig. 12.16**  Nitration of toluene gives mainly a mixture of *o-* and *p*-nitrotoluenes

As seen, in case attack occurs at *ortho* position the structure I (out of the three canonical structures) is most stable as in this case positive charge is on the carbon atom immediately next to the alkyl group. In turn, the alkyl group stabilises a neighbouring +ve charge more strongly due to the inductive effect (–I). In case of para attack also, one of the canonical structure (2) has positive charge on carbon atom immediately next to the alkyl group. However, in case of *meta* attack, there is no such structure. Therefore, the reaction pathway will preferably be on the one which goes through the most stable intermediate. Thus, the alkyl group is *ortho* and *para* directing.

As an example, toluene on nitration gives a mixture of *o-* and *p*-nitrotoluene in 59% and 39% yields, respectively. A small amount (4%) of the *m*-isomer is also obtained (Fig. 12.16).

### 12.3.6.2   Electrophilic Substitution in Phenol

In a similar way other activating groups like –OH (in phenols), –OR (in ethers), –NHCOR (in amides) or –NR$_2$ (in amines) direct the substitution to *o-* and *p*-positions. In all these cases, the aromatic ring is activated through + M (mesomeric) effect.

This is clearly understood by the comparison of the intermediate canonical structures arising out of *o-*, *m-*, and *p*-attacks in case of phenol (Fig. 12.17).

In the above case (Fig. 12.17), the structures (1) and (2) arising from *o-* and *p*-attacks by the electrophile have positive charge on carbon atom next to the OH group. Oxygen being nucleophilic can share its lone pair of electrons to form a new π-bond to the neighbouring electrophilic centre; this results in a fourth resonance structure (3) and (4) in which the positive charge is on the heteroatom (oxygen). In this case, the resonance effect is more prominent than the inductive effect of electronegative hetero atom. In view of this, the delocalisation of the positive charge stabilises the intermediate carbocation. However, in case of *m*-attack none of the canonical structures has positive charge on C next to the heteroatom and the 4th structure is not possible. So *m*-attack is not favoured.

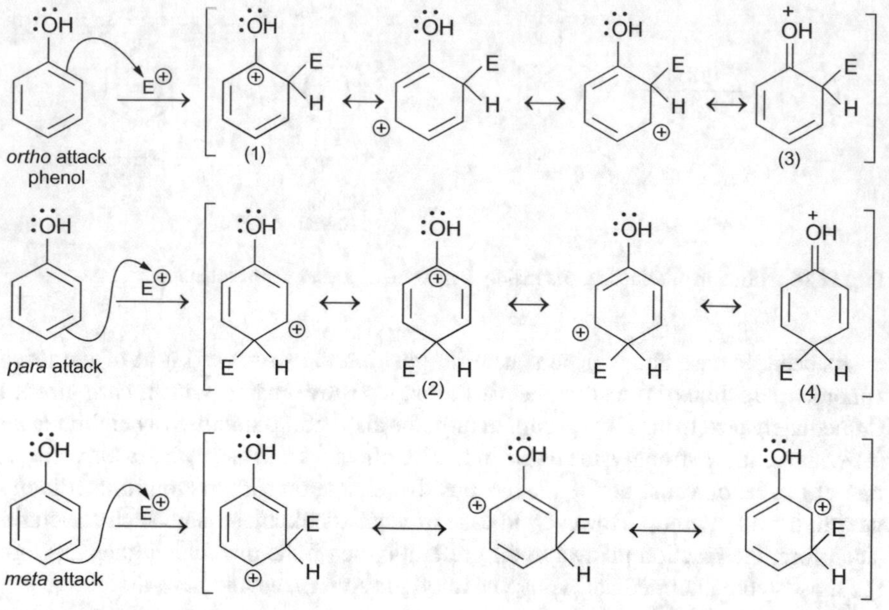

**Fig. 12.17** Canonical structures arising from *o-*, *m-* and *p-*attacks in case of phenol

### 12.3.6.3  Electrophilic Substitution in Aniline

In case of $NH_2$ group *o-* and *p-*attacks are observed. This is due to the contributions of the structures (1) and (2) (in Fig. 12.18). So $-NH_2$ group directs the substituents to *o-* and *p-*positions (Fig. 12.18).

### 12.3.6.4  Electrophilic Substitution in Nitrobenzene

Nitro group is a deactivating group and is *meta* directing. The deactivating groups increase positive charge and destablize the understood carbocation. This destabilisation is quite pronounced in the intermediates arising from *o-* and *p-*attacks. So in these cases, *m-*attack is favoured. This is understood by looking at the canonical structure arising from *o-*, *m-* and *p-*attacks in case of nitrobenzene (Fig. 12.19).

### 12.3.6.5  Electrophilic Substitution in Haloarenes

In case of haloarenes, the results are quite different. The halo group is deactivating and so is *m-*directing. However, in chlorobenzene, the chlorine atom being highly electronegative is expected to withdraw electrons from the benzene ring resuare called nucleophilic substitution ring.

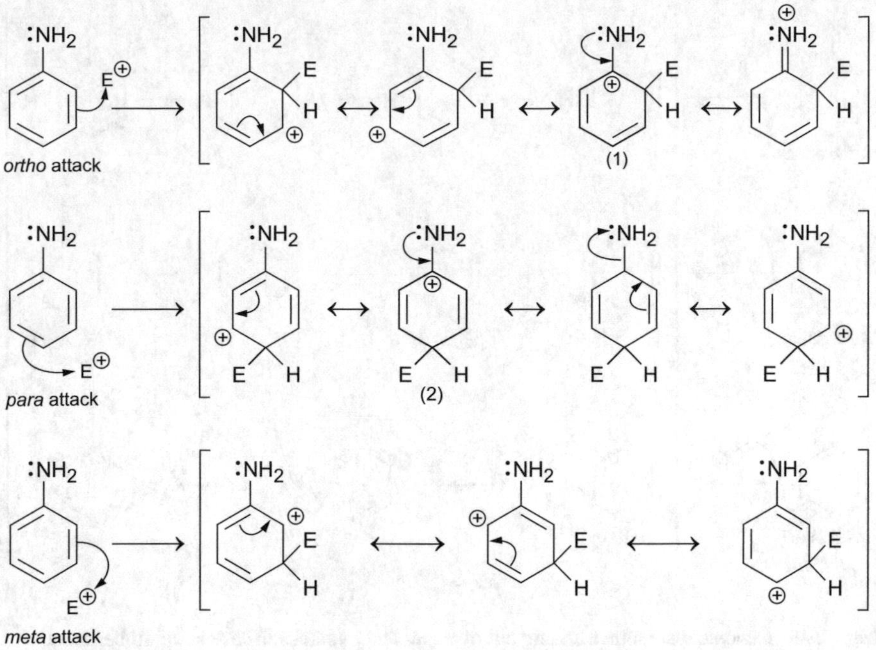

**Fig. 12.18** Canonical structures arising from *o*-, *m*- and *p*-attacks in case of aniline

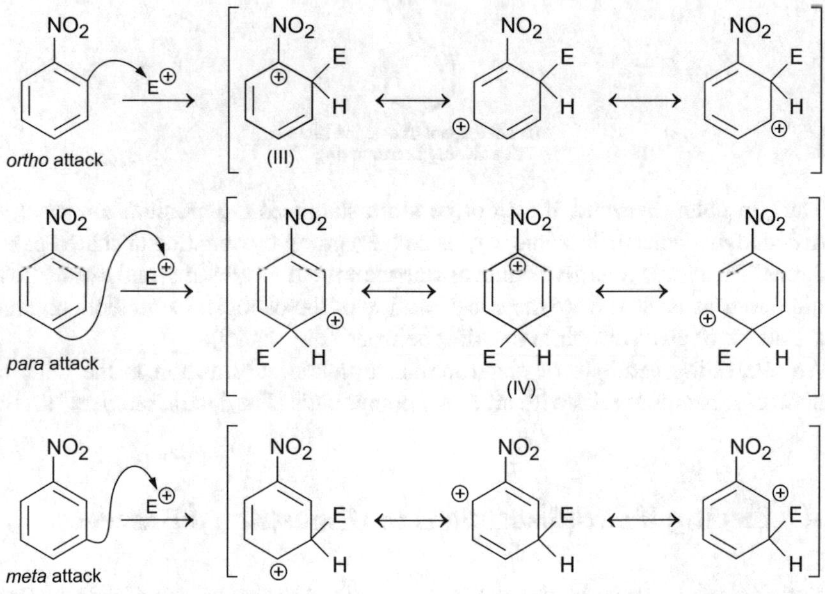

**Fig. 12.19** Canonical structures arising due to *o*-, *m*- and *p*-attack in case of nitrobenzene

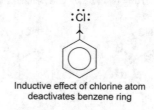

**Fig. 12.20** Canonical structures arising out of *o*-, *m*- and *p*-attacks in case of chlorobenzene

Inductive effect of chlorine atom
deactivates benzene ring

Thus, in chlorobenzene, the chlorine atom stabilises the arenium ion resulting from *o*- and *p*-attacks in the same way as an $NH_2$ group by donating unshared pair of electrons. These electrons give stable resonance structures giving *o*- and *p*-substituted arenium ions as is clear from the consideration of the canonical structures obtained by the attack of an electrophile to chlorobenzene (Fig. 12.20).

An interesting example of electrophilic aromatic substitution is the coupling reaction of diazonium salts with aromatic compounds (For details see Sect. 12.12).

## 12.4 Electrophilic Substitutions in Disubstituted Benzenes

In case of electrophilic substitutions in disubstituted benzenes, the directing effects of both the substitutents have to be considered. There can be three situations.

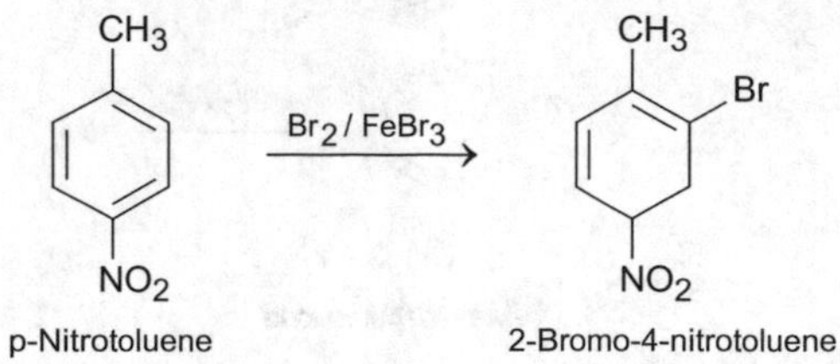

**Fig. 12.21** Bromination of p- nitrotoluene gives 2-bromo-4-nitrotoluene

Br$_2$/ FeBr$_3$

p-Nitrotoluene

2-Bromo-4-nitrotoluene

(a) **The directing effects of both the substituents reinforce**: In such a situation, the new substituent enters the position as directed by both the groups. For example, in p-nitrotoluene, the CH$_3$ group is o-, p-directing and the NO$_2$ group is m-directing. Since these two substitutents reinforce each other, only one product is formed by treatment with Br$_2$ in presence of FeBr$_3$. In this case, the position *para* to CH$_3$ is already blocked (by NO$_2$ group) and the position m- to NO$_2$ group is the same as o- to CH$_3$ group. So the product formed is *ortho* to CH$_3$ group and *meta* to NO$_2$ group (Fig. 12.21).

(b) **The directing effects of both the substituents oppose each other**: In case of **disubstitued** benzenes, if one group is o- and p- directing and the other is m-directing, the group that is strongly activating and o-, p-directing will determine the orientation of the incoming electrophile. As an example, in case of 2-fluoromethoxy benzene (F is m-directing and OCH$_3$ is o-, p-directing), nitration will give a mixture of o- and p-nitroproducts (Fig. 12.22).

Another example is electrophilic substitution of p-methyl acetanilide. In this case both NHCOCH$_3$ group and CH$_3$ group are o-, p-directing since NHCOCH$_3$ is a stronger activator, substitution occurs *ortho* to it (Fig. 12.23).

The order of directing influence of the following o- and p-directing groups is

$$-NH_2 > -NR_2 > OMe > -N\overset{\overset{O}{\|}}{H}C\,CH_3 > -Me - X.$$

(c) **No substitution occurs in between two *meta* substituted (1,3-disubstituted) substituents**: As an example in case of m-xylene (1,3-disubstituted benzene)

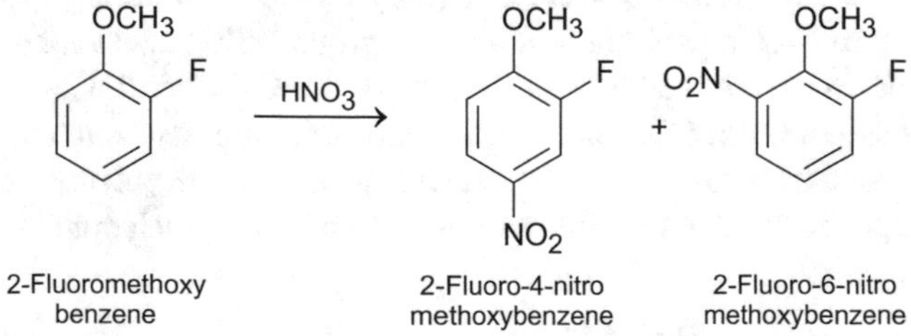

HNO$_3$

2-Fluoromethoxy benzene

2-Fluoro-4-nitro methoxybenzene

2-Fluoro-6-nitro methoxybenzene

**Fig. 12.22** Nitration of 2-fluoromethoxybenzene

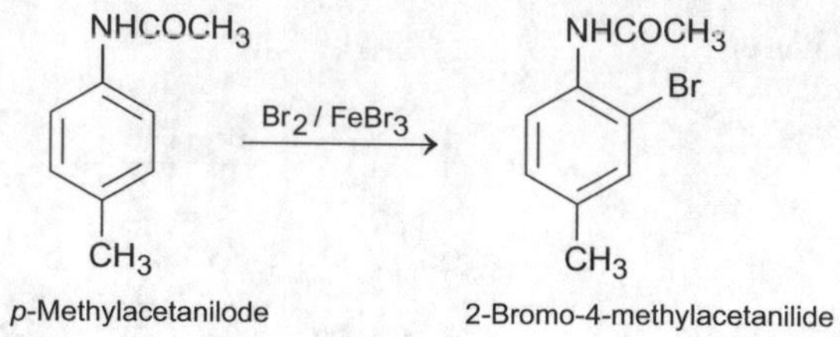

Fig. 12.23  Bromination of *p*-methylacetanilide

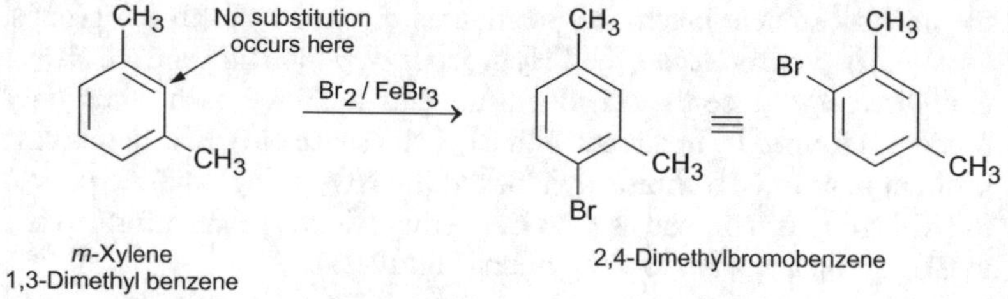

Fig. 12.24  Bromination of *m*-xylene

no substitution will occur in between the two substitutions, even though both the methyl groups activate this position. So the product of bromination is 2,4-dimethyl-bromobenzene (Fig. 12.24).

The position occupied by Br is *ortho* to one CH$_3$ group and *para* to the other CH$_3$ group.

## 12.5  Nucleophilic Substitution Reactions

The substitution brought about by a nucleophile is called nucleophilic substitution and is denoted by S$_N$ (S stands for substitution and N for nucleophilic). In general, in a substitution reaction, the substrate (most commonly an alkyl halide) reacts with a nucleophile to give a product and the leaving group is formed (Fig. 12.25).

The substrate, RX, must contain an R (alkyl) group containing *sp*$^3$ hybridised carbon bonded to X. X represents an atom or a group called a leaving group, which should be able to accept the electron density in the C–X bond. Common leaving groups are halogen atom (–X), but –OH$_2$$^+$ and –N$_2$$^+$ are also known. The nucleophile must contain a lone pair of electrons or a π-bond but not necessarily a negative charge. Since the substitution reactions involve nucleophiles (which are

$$R—X + :Nu^- \longrightarrow R—Nu + X:^-$$

Fig. 12.25  A substitution reaction

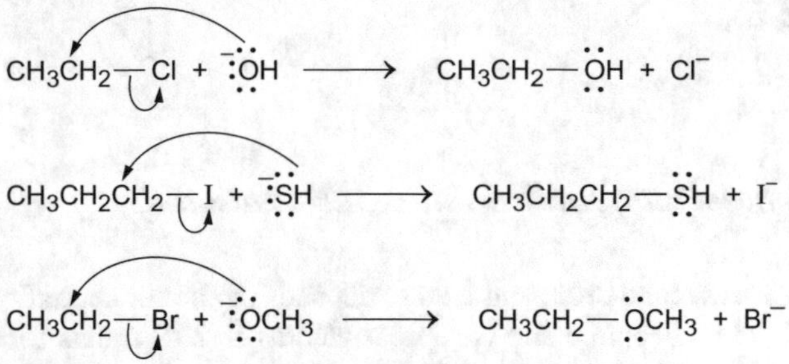

$$CH_3CH_2{-}Cl + \text{:}\ddot{O}H \longrightarrow CH_3CH_2{-}\ddot{O}H + Cl^-$$

$$CH_3CH_2CH_2{-}I + \text{:}\ddot{S}H \longrightarrow CH_3CH_2CH_2{-}\ddot{S}H + I^-$$

$$CH_3CH_2{-}Br + \text{:}\ddot{O}CH_3 \longrightarrow CH_3CH_2{-}\ddot{O}CH_3 + Br^-$$

**Fig. 12.26** Some typical nucleophilic substitution

electron rich), they are called nucleophilic substitution reactions. Some examples of nucleophilic substitutions are given below (Fig. 12.26).

As seen in all the above case, the alkyl group contains an $sp^3$ hybridised carbon bonded to an halogen (Cl, I, Br), which form the leaving group. The nucleophile in all these cases contain a lone pair and also negative charge ($\text{:}\ddot{O}H$, $\text{:}\ddot{S}H$, $\text{:}\ddot{O}CH_3$). The product of the reaction is neutral.

However, when a neutral nucleophile (like:$N(CH_3)_3$, the substitution product bears a positive charge, and all the groups ($CH_3$ in this case) bonded to N atom in the nucleophile stay bonded in the product (Fig. 12.27).

In nucleophilic substitution reaction, we come across two situations.

(i)  **Bond breaking occur simultaneously**: This is a one-step process and is represented as (Fig. 12.28).

The reaction by which the above nucleophilic substitution occurs is called bimolecular nucleophilic substitution ($S_N2$).

(ii)  Nucleophilic substitutions in which bond breaking occurs before bond making. This is a two-step process and is represented as (Fig. 12.29).

The reaction by which the two-step substitution takes place is called unimolecular nucleophilic substitution ($S_N1$).

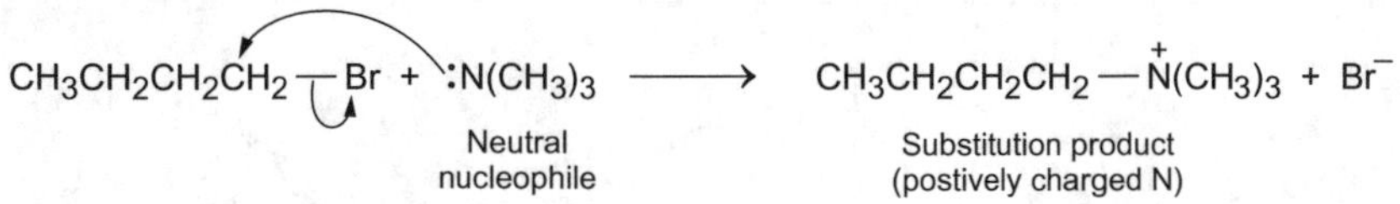

$$CH_3CH_2CH_2CH_2{-}Br + \text{:}N(CH_3)_3 \longrightarrow CH_3CH_2CH_2CH_2{-}\overset{+}{N}(CH_3)_3 + Br^-$$

Neutral                          Substitution product
nucleophile              (postively charged N)

**Fig. 12.27** Substitution reaction involving a neutral nucleophile

**Fig. 12.28** One-step nucleophilic substitituon

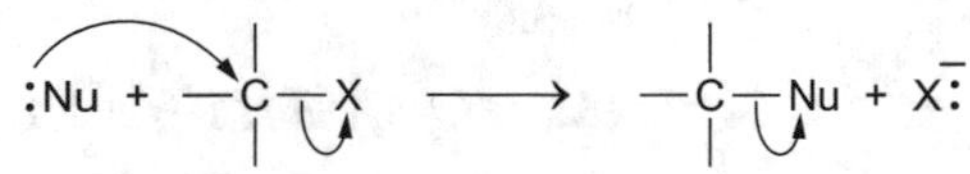

$$\text{:}Nu + {-}\overset{|}{\underset{|}{C}}{-}X \longrightarrow {-}\overset{|}{\underset{|}{C}}{-}Nu + X\text{:}^-$$

**Fig. 12.29** Two-step
nucleophilic substitution

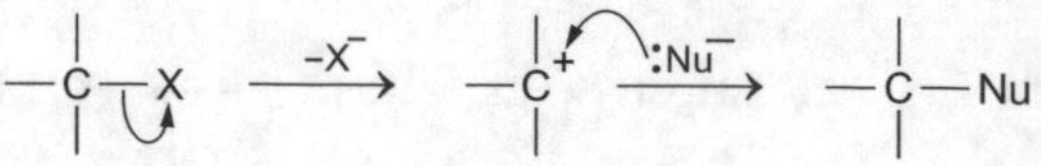

## 12.5.1  *Bimolecular Nucleophilic Substitution ($S_N2$)*

This reaction is of second order and its rate depends on the concentration of both the substrate and the nucleophile and is represended as $S_N2$ (S stands for substitution, N for nucleophilic and 2 for bimolecular).

$$\text{Rate} \propto [\text{substrate}]\,[\text{Nucleophile}]$$

An example of a bimolecular nucleophilic substitution is the hydrolysis of a primary alkyl halide (e.g. methyl bromide) with aqueous alkali. According to Ingold, there is participation of both alkyl halide and the hydroxyl ions in the rate determining step (i.e., slowest step). It was suggested that the reaction proceeds via a transition state, in which the attacking hydroxyl ion gets partially bonded to the reacting carbon of methyl bromide before the departure of the bromide ion (Fig. 12.30).

As seen, in the transition state both Br and OH are partially bonded (shown by dotted line) to the C atom of the $CH_3$ group. The C–Br bond is not completely cleaved in the TS, and the C–OH bond is not completely formed in the T.S. Also, the hydroxide has a partial negative charge ($\delta-$) since it has began to share its electrons with the C atom and the bromine also carries a partial negative charge ($\delta-$) since it has started moving its shared pair of electrons away from the carbon atom. According to Ingold, part of the energy, necessary for the cleavage of C–Br bond comes from the energy released in forming HO–C bond.

The $S_N2$ reaction is a concerted reaction. Bond forming and bond breaking occur simultaneously. Also, in $S_N2$ reaction, changing the concentration of either nucleophile or alkyl halide affects the rate of the reaction. Thus, doubling the concentration of either of the reactants doubles the rate of the reaction. Doubling the concentration of both the reactants increases the rate of the reaction four times.

Another example of a $S_N2$ reaction is the reaction of methyl bromide with acetate ion (Fig. 12.31).

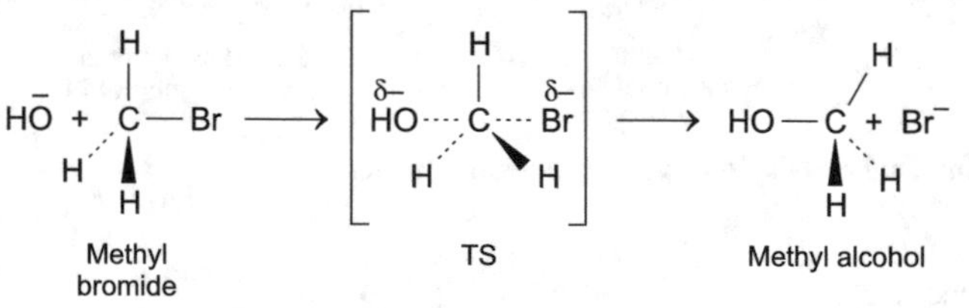

**Fig. 12.30**  $S_N2$ reaction of methyl bromide with alkali

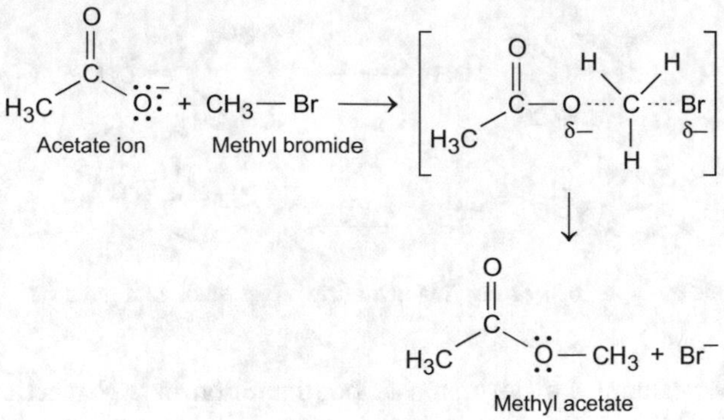

$$H_3C-\overset{\overset{O}{\parallel}}{C}-\overset{..}{\underset{..}{O}}{:}^- + CH_3-Br \longrightarrow \left[ H_3C-\overset{\overset{O}{\parallel}}{C}-\overset{..}{O}\cdots\overset{\overset{H}{|}}{\underset{\underset{H}{|}}{\underset{\delta-}{C}}}\cdots\overset{H}{\underset{\delta-}{Br}} \right]$$

Acetate ion    Methyl bromide

$$\downarrow$$

$$H_3C-\overset{\overset{O}{\parallel}}{C}-\overset{..}{\underset{..}{O}}-CH_3 + Br^-$$

Methyl acetate

**Fig. 12.31** $S_N2$ reaction of methyl bromide with acetate ion

### 12.5.1.1 Stereochemistry of $S_N2$ Reaction

The stereochemical implication of an $S_N2$ reaction is that there is inversion of configuration (Fig. 12.32).

As seen, the spatial arrangement of the three groups (R, R′ and R″) attached to carbon atom is effectively turned inside out and we say the carbon atom has undergone inversion of its configuration.

In case there is frontside attack of the nucleophile, the product obtained will be an enantiomer of the product obtained by backside attack (Fig. 12.33).

It has been found that all $S_N2$ reactions proceed with backside attack of the nucleophile (Fig. 12.32) resulting in inversion of configuration at the stereogenic centre. It is believed that the backside attack is based on electronic argument. Both the nucleophile and the leaving groups being electron rich and like charges repel each other. In view of this, the backside attack keeps the two groups as far away as possible. In the transition state (Fig. 12.30), the nucleophile and the leaving groups are 180° away from each other and the other three groups around the carbon occupy a plane.

**Fig. 12.32** Backside approach of the nucleophile in $S_N2$ reaction

$$HO\overset{..}{:}^- + \overset{R'}{\underset{R''}{\overset{|}{C}}}-Br \xrightarrow[\text{attack}]{\text{Back side}} HO-\overset{R'}{\underset{R''}{\overset{|}{C}}}\diagdown R$$

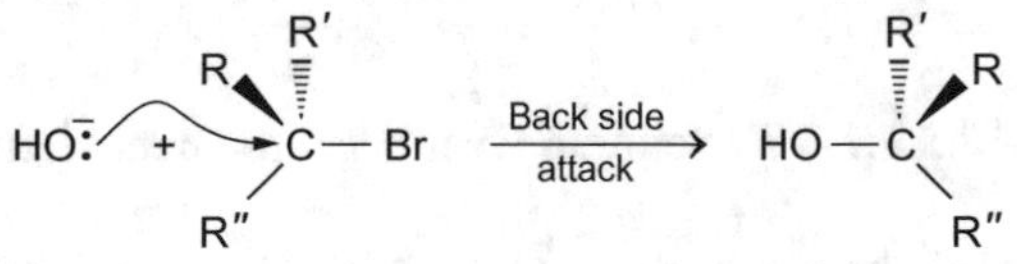

**Fig. 12.33** Frontside attack of the nucleophile in $S_N2$ reaction

$$\overset{R'}{\underset{R''}{\overset{|}{C}}}-Br \quad \overset{..}{:}\overline{O}H \xrightarrow[\text{attack}]{\text{Front side}} \overset{R'}{\underset{R''}{\overset{|}{C}}}-HO + Br^-$$

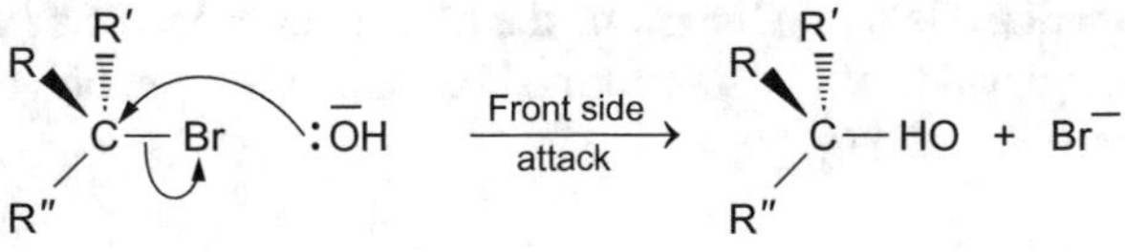

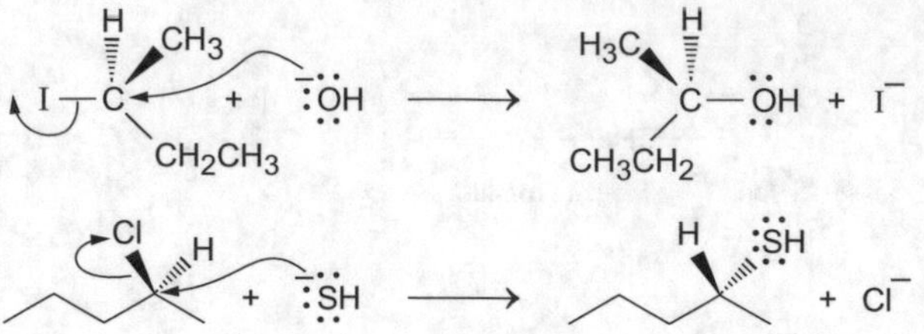

**Fig. 12.34** Some examples of $S_N2$ reaction involving inversion of configuration

Some other examples of inversion of configuration in $S_N2$ reactions are given below (Fig. 12.34).

The rate of $S_N2$ reaction decreases as the number of alkyl groups on the carbon attached to the leaving group increases.

$$CH_3-X \quad R\,CH_2-X \quad R_2CH-X \quad R_3C-X$$

Methyl          1°              2°              3°

$\longrightarrow$

decreasing rate of $S_N2$ reaction

Thus, methyl and 1° alkyl halides undergo $S_N2$ reactions easily, 2° alkyl halides react more slowly and 3° alkyl halides do not undergo $S_N2$ reaction. This order can be explained due to steric effects.

To sum up, bimolecular nucleophilic substitution reaction is a concerted bimolecular one-step reaction in which the attack of the nucleophile is from backside and the product obtained is by the inversion of configuration.

In case of aliphatic halides, there is no way to prove that the attack by a nucleophile proceeds by inversion of configuration. However, with a cyclic molecular like *cis*-1-chloro-3-methyl cyclopentane, one can observe the result of configuration inversion. The reaction of *cis*-1-chloro-3-methyl cyclopentane with ⁻OH in an $S_N2$ reaction gives *trans* 3-methylcyclopentanol. The hydroxide ion gets bonded on the opposite side of the ring from the chlorine it replaces as shown below (Fig. 12.35).

### 12.5.1.2  Functional Group Transformations Using $S_N2$ Reactions

$S_N2$ reactions are very useful in organic synthesis since these enable us to convert one functional group into another—a process known as functional group transformation. Using $S_N2$ reaction, the functional group of a methyl, primary or secondary alkyl halide can be transformed into that of an alcohol, ether, thioether, nitrile, ester, etc.(Fig. 12.36).

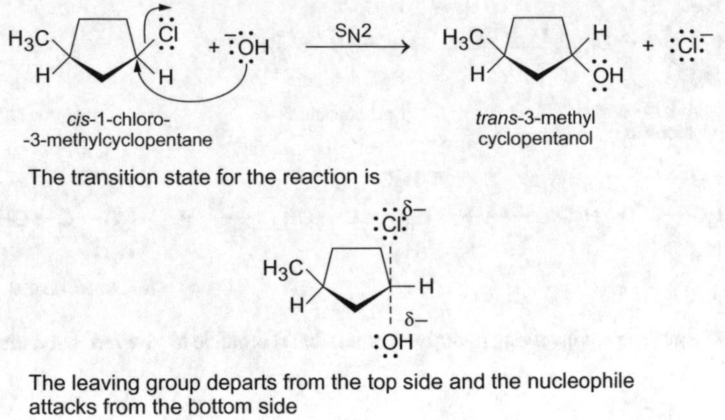

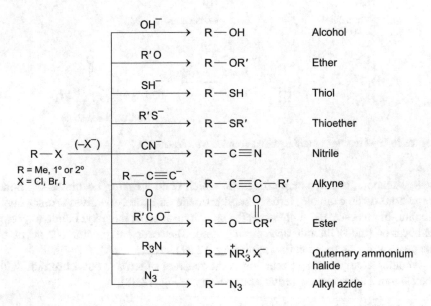

The transition state for the reaction is

The leaving group departs from the top side and the nucleophile attacks from the bottom side

**Fig. 12.35** Inversion in $S_N^2$ reaction

**Fig. 12.36** Functional group interconversions of methyl, primary and secondary alkyl halides using $S_N2$ reactions

## 12.5.2 *Unimolecular Nucleophilic Substitution ($S_N1$)*

Unlike the $S_N2$ reaction, the unimolecular reaction is of first order and its rate depends on the concentration only of the substrate (alkyl halide) and is independent of the concentration of the nucleophile ($^-OH$). It is designated as $S_N1$ (S, N and 1 stand for substitution, nucleophilic and unimolecular, respectively).

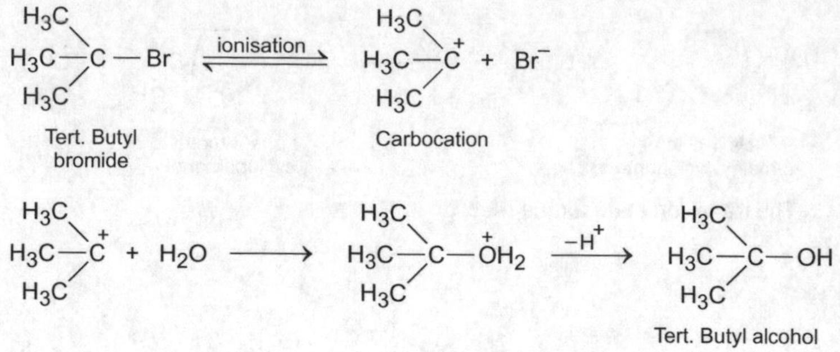

$$H_3C-\underset{\underset{H_3C}{|}}{\overset{\overset{H_3C}{|}}{C}}-Br \underset{}{\overset{\text{ionisation}}{\rightleftharpoons}} H_3C-\underset{\underset{H_3C}{|}}{\overset{\overset{H_3C}{|}}{\overset{+}{C}} + Br^-$$

Tert. Butyl
bromide

Carbocation

$$H_3C-\underset{\underset{H_3C}{|}}{\overset{\overset{H_3C}{|}}{\overset{+}{C}}} + H_2O \longrightarrow H_3C-\underset{\underset{H_3C}{|}}{\overset{\overset{H_3C}{|}}{C}}-\overset{+}{O}H_2 \overset{-H^+}{\longrightarrow} H_3C-\underset{\underset{H_3C}{|}}{\overset{\overset{H_3C}{|}}{C}}-OH$$

Tert. Butyl alcohol

**Fig. 12.37** $S_N1$ reaction involving hydrolysis of tert. butyl bromide to give tert. butyl alcohol

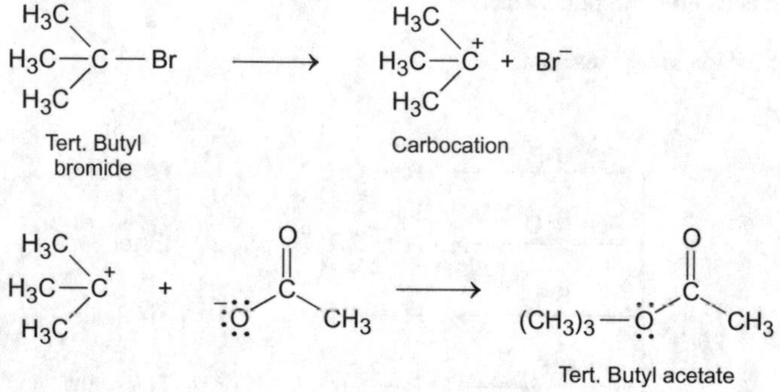

$$H_3C-\underset{\underset{H_3C}{|}}{\overset{\overset{H_3C}{|}}{C}}-Br \longrightarrow H_3C-\underset{\underset{H_3C}{|}}{\overset{\overset{H_3C}{|}}{\overset{+}{C}} + Br^-$$

Tert. Butyl
bromide

Carbocation

$$H_3C-\underset{\underset{H_3C}{|}}{\overset{\overset{H_3C}{|}}{\overset{+}{C}}} + \overset{..}{\underset{..}{O}}\overset{\overset{O}{||}}{\underset{}{C}}_{CH_3} \longrightarrow (CH_3)_3-\overset{..}{O}\overset{\overset{O}{||}}{\underset{}{C}}_{CH_3}$$

Tert. Butyl acetate

**Fig. 12.38** $S_N2$ reaction of tertiary butyl bromide with acetate ion

$S_N1$ reaction is a two-step process in which bond breaking occurs before bond formation. As an example, tertiary butyl bromide on hydrolysis gives tertiary butyl alcohol. In this case, the first step is the slow ionisation of the alkyl halide to give a carbocation (this is the rate determining step). The second step is the fast attack of the nucleophile on to the carbocation (Fig. 12.37).

Another example of $S_N1$ reaction is the reaction of tertiary butyl bromide with acetate ion. The product is tertiary butyl acetate (Fig. 12.38).

#### 12.5.2.1   Stereochemistry of $S_N1$ Reaction

In $S_N1$ reaction, carbocations are intermediates. The carbocation having three groups around carbon is $sp^2$ hybridised and is trigonal planar and contains a vacant $p$-orbital extending above and below the plane.

As an illustration, let us consider the $S_N1$ reaction of a 3° alkyl halide having a leaving group attached to the stereogenic carbon. On ionisation, the 3° alkyl halide looses the leaving group and gives a planar carbocation, which is chiral. The nucleophile can attack the carbocation from either side to give two products, which are

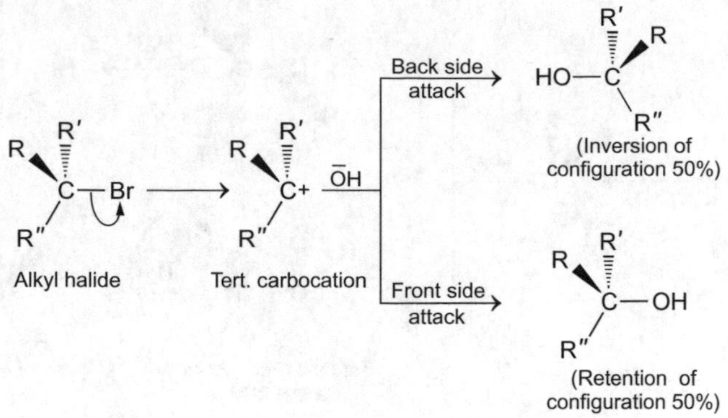

**Fig. 12.39** $S_N1$ reaction of a 3° alkyl halide (having three different groups) with a nucleophile ($^-OH$)

different compounds containing one stereogenic centre. Both these compounds are enantiomers. Thus, the attack of a nucleophile on the carbocation can take place with equal ease from either side of the planar carbocation giving a 1: 1 mixture of products having the same but opposite configuration of the starting material. Thus, racemisation occurs yielding an optically inactive mixture of products. This is shown in Fig. 12.39.

In the above $S_N1$ reaction (Fig. 12.39) two products are obtained; these are different compounds having one stereogenic centre. Both and stereoisomers are not superimposable and are enantiomers. As there is no preference for the attack of nucleophile from either side, an equal amount of the enantiomer is formed, i.e., we get a racemic mixture. Thus, we say that racemisation has occurred (**Racemisation** is the formation of equal amounts of two enantiomers from a single starting material).

We have seen the $S_N1$ reaction of a 3° alkyl halide with a negatively charged nucleophile ($^-OH$ in the above case, Fig. 12.39). Let us now consider the reaction of an alkyl halide (having three different groups attached to C) with a neutral nucleophile (e.g., water). In this case, the formed carbocation reacts with $H_2O$ to give two products bearing a positive charge. These products lose a proton readily to form neutral products (Fig. 12.40).

Thus, we see that the $S_N1$ reaction of a 3° alkyl halide with a negative nucleophile is a two-step process and the reaction with a neutral nucleophile ($H_2O$) is a three-step process.

As in the case of $S_N2$ reaction, in case of $S_N1$ reaction also, the reaction rate depends on the number of alkyl groups on the carbon atom attached to the leaving group. More the number of alkyl group, better is the rate of the $S_N2$ reaction.

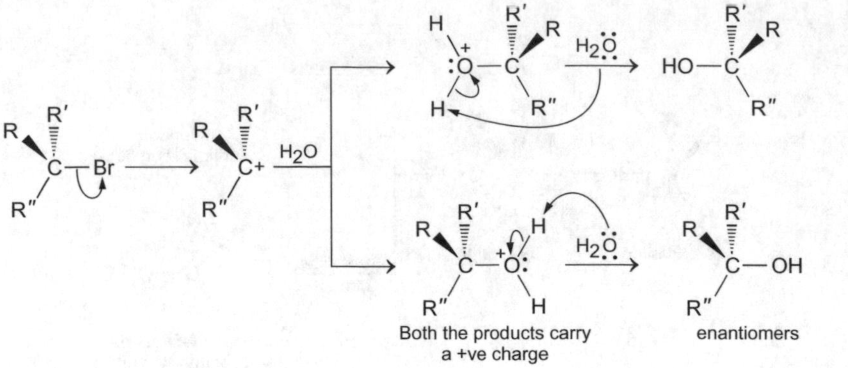

**Fig. 12.40** $S_N1$ reaction of a 3° alkyl halide (having three different groups) with a neutral nucleophile ($H_2O$)

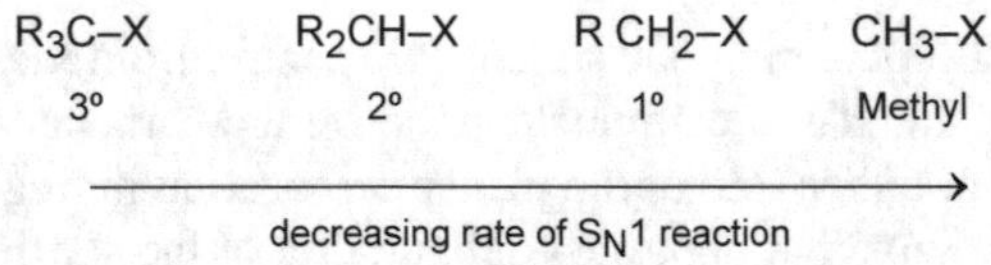

|  |  |  |  |
|---|---|---|---|
| $R_3C–X$ | $R_2CH–X$ | $R\,CH_2–X$ | $CH_3–X$ |
| 3° | 2° | 1° | Methyl |

$$\longrightarrow$$

decreasing rate of $S_N1$ reaction

The special features of $S_N1$ reaction of an alkyl halide with a nucleophile include

- 3° alkyl halides undergo $S_N1$ reaction rapidly.
- 2° alkyl halides react slowly.
- 1° alkyl halides and methyl halides do not undergo $S_N1$ reaction. This trend is exactly opposite to that seen for $S_N2$ reaction.

The characteristics of an $S_N1$ reaction are

- It is a unimolecular reaction; the rate of reaction is proportional to the concentration of the substrate only (alkyl halide).
- It involves two-step mechanism (However, in case of neutral nucleophiles, there is three-step mechanism).
- The intermediate carbocation is trigonal, planar.
- More substituted halides react faster.

In nucleophilic substitution reaction of an alkyl halide, the 1° halide reacts with $S_N2$ mechanism, a tertiary halide reacts with $S_N1$ mechanism but a 2° halide can react with both $S_N1$ and $S_N2$ mechanism.

As an example, let us consider nucleophilic substitution of a secondary halide, *cis*-1-bromo-4-methyl cyclohexane. A strong nucleophile, $^-OH$ favours $S_N2$ reaction, which occurs from backside attack of the nucleophile; this results in inversion of configuration. Since the leaving group $Br^-$ is above the plane of the ring, the nucleophile attack from below resulting in the formation of a single product (Fig. 12.41).

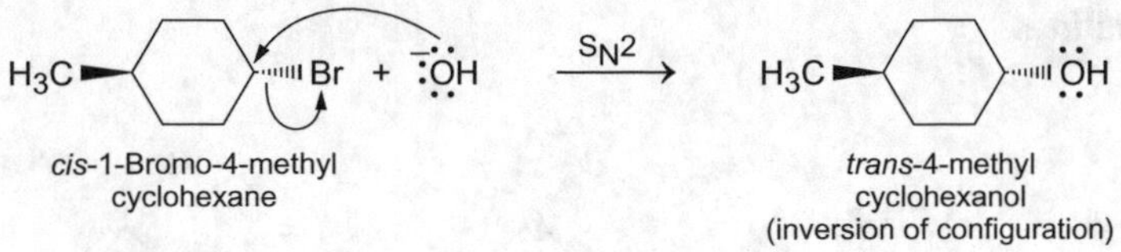

**Fig. 12.41**  S$_N$2 reaction of *cis*-1-bromo-4-methylcyclohexane with a strong nucleophile

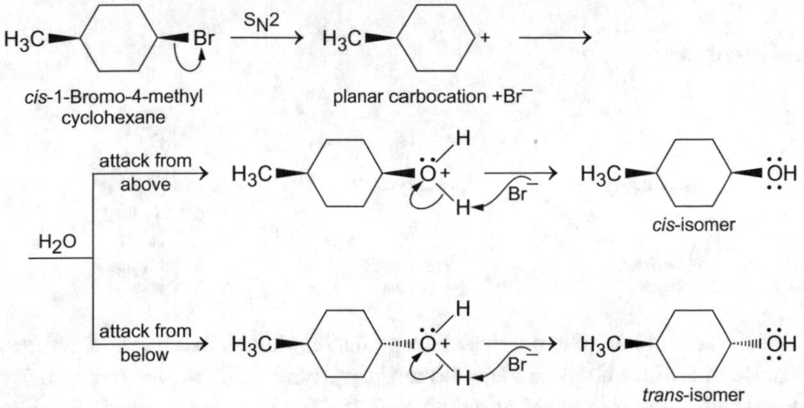

**Fig. 12.42**  The S$_N$2 reaction of *cis*-1-bromo-4-methyl cyclohexane with a weak nucleophile (H$_2$O)

In S$_N$1 reaction, a weak nucleophile (H$_2$O) is used. The reaction occurs via the formation of a carbocation. The nucleophilic attack can take place from both above and below the plane of the ring to produce two products which are diastereomers (one is *cis* and the other is *trans*) (Fig. 12.42).

Thus, the stereochemistry of the substitution product depends on the mechanism of the reaction (S$_N$1 or S$_N$2).

See also Nucleophilic Aromatic substitution (Chap. 11).

## 12.5.3  *Nucleophilic Substitutions in Allylic and Benzylic Halides*

The allylic and benzylic halides are classified in the same way as organic halides as shown below

## Allylic halides

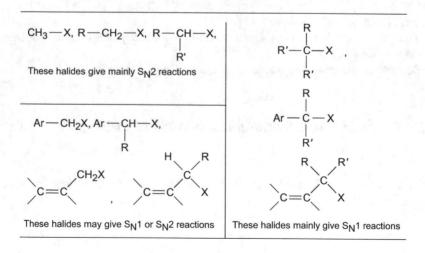

Both allylic and benzylic halides undergo nucleophilic substitution reactions. As in the case of tertiary alkyl halides, the tertiary allylic halides and tertiary benzylic halides do not undergo substitution by $S_N2$ mechanism due to steric hindrance associated with three bulky groups on the carbon bearing halogen.

However, primary and secondary allylic and benzylic halides can react either by $S_N2$ mechanism or $S_N1$ mechanism. The following table summarises alkyl, allylic and benzylic halides in $S_N$ reactions.

## 12.6  Substitution Reaction of Alcohols

Substitution reactions of alcohols are not possible since the OH group of alcohols is not a good leaving group. The only alternative is to convert the OH group into a better leaving group. A number of ways are available, some of which are discussed below.

### 12.6.1  Conversion of OH of Alcohols into $^+OH_2$

Some alcohols, particularly tertiary alcohols on treatment with hydrohalic acid (HCl, HBr or HI), protonates the OH group and simultaneously delivers a good nucleophile (Br$^-$). Thus, tertiary butyl alcohol reacts with HBr to give protonated alcohol, which ionises to give a carbocation (as in $S_N1$ reaction). Finally, the carbocation reacts with the nucleophile (Br$^-$) to give substitution product (Fig. 12.43).

This process is only of limited applicability since 1° alcohols do not react with HCl or HBr. However, the reaction can take place if the 1° alcohol is heated with HBr in presence of $H_2SO_4$.

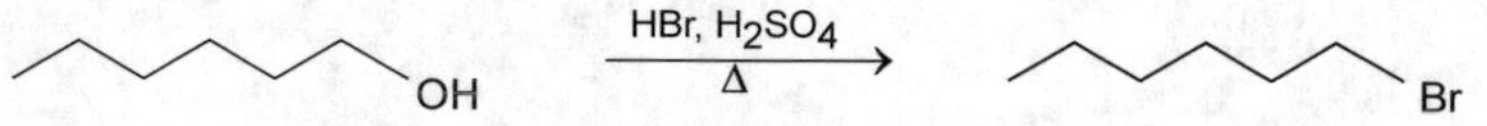

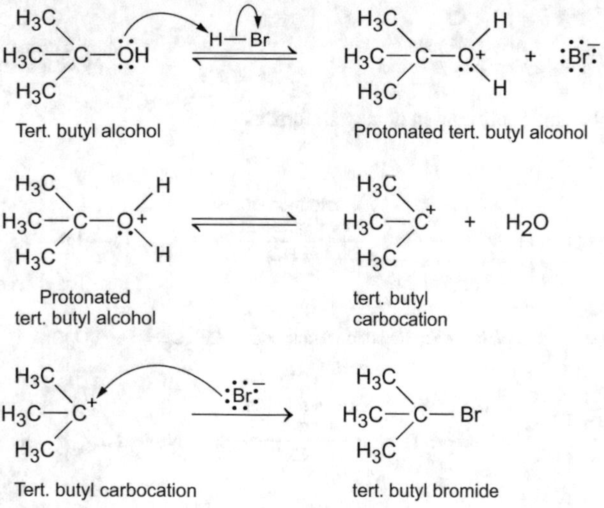

**Fig. 12.43**  Substitution reaction of tert. butyl alcohol with HBr

## 12.6.2   Conversion of Alcohols into Tosylates or Mesylates

Alkyl sulfonates are commonly used as substrates for nucleophilic substitution reactions since sulfonate ions are good leaving groups (Fig. 12.44).

The above reaction of alkyl sulfonate with nucleophile (Fig. 12.44) gives an indirect method for the nucleophilic substitution reactions of alcohols. The method consists in converting an alcohol to an alkyl sulfonate and then reaction of the formed alkyl sulfonate with a nucleophile. In case the carbon bearing the OH group is a stereocentre, the sulfonate ester formation proceeds with retention of configuration. Finally, if the reaction is $S_N2$ there is inversion of configuration (Fig. 12.45).

Alkyl sulfonates (tosylates, etc.) undergo all nucleophilic substitution reactions that an alkyl halide do.

The mechanism of the formation of tosyl ester involves the displacement of chloride ion from p-toluene sulphonyl chloride by the alcohol followed by treatment with base (Fig. 12.46).

In place of alkyl tosylate, alkyl methanesulfonate can also be used. These are prepared by the reaction of alcohol with methane sulfonyl chloride followed by treatment of the formed product with a base (Fig. 12.47).

The reaction of alkyl sulfonate with a nucleophile can be either by $S_N1$ or $S_N2$ process depending on the structure of the substrate. Some examples are given below (Fig. 12.48).

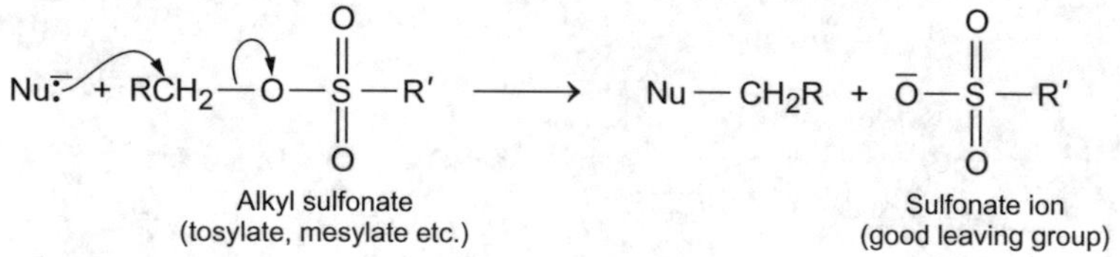

**Fig. 12.44**  Nucleophilic substitution of alkylsulfonates

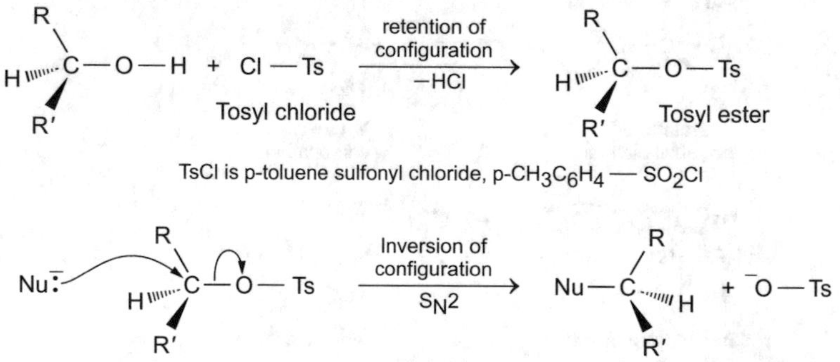

**Fig. 12.45**  $S_N2$ reaction of alcohol via the formation of the corresponding tosyl ester

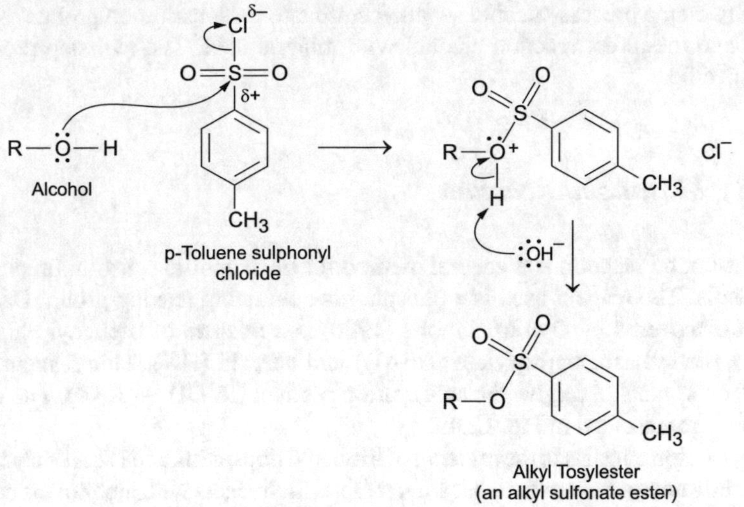

Fig. 12.46  Mechanism of the formation of alkyl tosylate

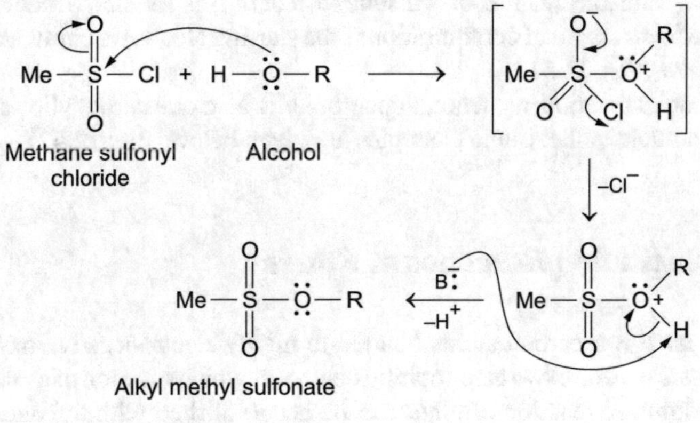

Fig. 12.47  Formation of alkyl methyl sulfonate

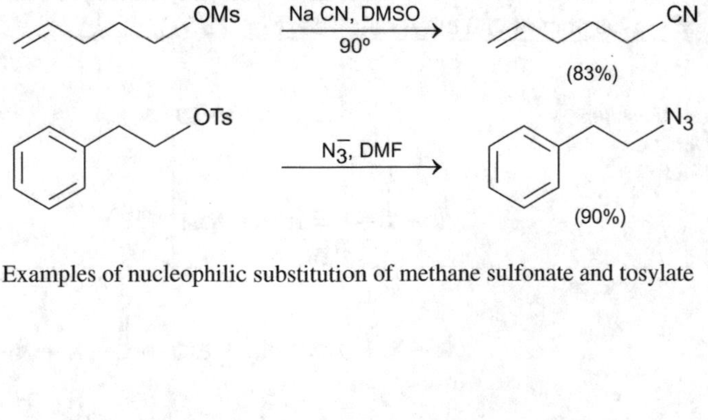

Fig. 12.48  Examples of nucleophilic substitution of methane sulfonate and tosylate

The two-step process, alcohol → alkyl sulfonate → substitution product is much superior to the direct reaction alcohol with mineral acid. The two-step process is stereospecific.

### 12.6.3   Mitsunobu Reaction

The Mitsunobu reaction is a general method for nucleophilic substitution reactions of alcohols. The reagent used is a phosphorus-containing leaving group. One such reagent discovered by Oyo Mitsunobu (1980) is a mixture of triphenyl phosphine (Ph$_3$ P), diethyl azodicarboxylate (DEAD) and an acid (HX). This reagent reacts with 1° or 2° alcohol to give the substitution product (R–OH → R–X). The overall reaction is represented in Fig. 12.49.

The reaction consists in the reaction of triphenyl phosphine and DEAD in presence of an acidic component to form an adduct (1), which reacts with alcohol (as a nucleophile) to yield the corresponding (alkoxy) triphenyl prosphonium ion, [RO–PPh$_3$]$^+$ (2). Subsequent steps are shown in (Fig. 12.50).

A special valuable feature of Mitsunobu reaction is its stereochemical course which leads to inversion of configuration if the starting alcohol is chiral. An example is given below (Fig. 12.51).

In Mitsunobu reaction, a phenol, thiocarboxylic acid or a carboxylic acid can also be used as a nucleophile. Some examples are given below (Fig. 12.52).

## 12.7   Substitution Reactions of Ethers

The ethers have two carbon atoms bonded to highly electronegative oxygen atom. Thus, in ethers there are two electrophilic centres at which reaction can occur. Nucleophilic substitution reaction of ether can be accomplished with a reagent that can react with oxygen atom to form a good leaving group that results from protonation of the oxygen atom (as in the case of alcohol). Thus, the reaction of ethyl phenyl ether with HBr gives phenol and ethyl chloride (Fig. 12.51).

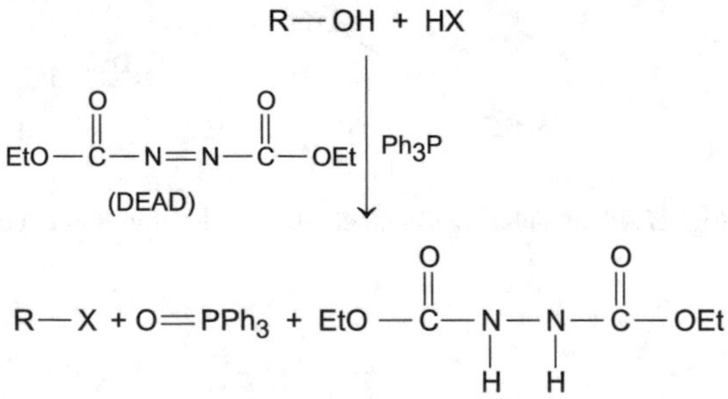

Fig. 12.49  The Mitsunobu reaction (conversion of R–OH into R–X)

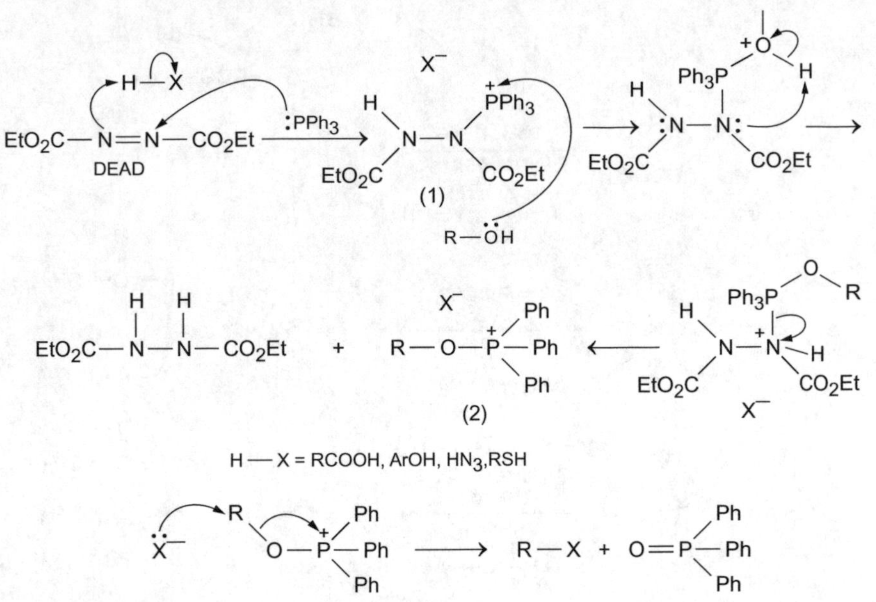

**Fig. 12.50** Mechanism of Mitsunobu reaction

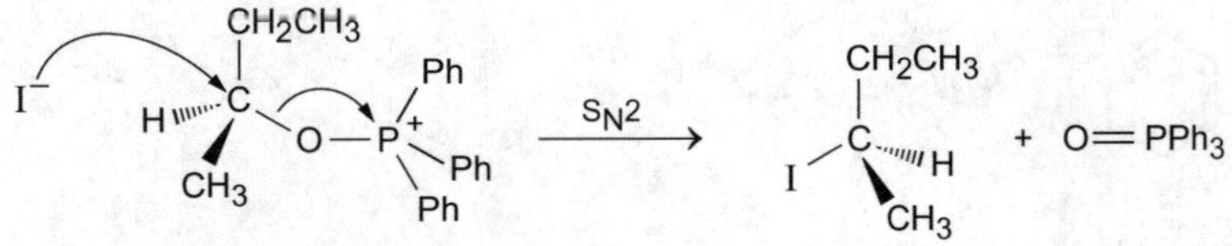

**Fig. 12.51** Inversion of configuration in Mitsunobu reaction if the starting alcohol in chiral

The phenol formed in the above reaction (Fig. 12.53) do not react since the OH group is attached to $sp^2$ hybridised carbon atom just as aryl halides do not undergo $S_N1$ or $S_N2$ reactions.

However, in case of dialkyl ethers, two molecules of alkyl halide are formed; in this case, the alcohol formed in the first step reacts further to give another alkyl halide (Fig. 12.54).

The use of HBr or HI in the cleavage of ethers is limited (as both HBr and HI are strong acids) to ethers that have few other substituents. In most of the reactions of ethers iodotrimethyl silane, $(CH_3)_3SiI$, being a milder reagent is used to cleave the ethers. In the later case, the oxygen atom of ether reacts with electrophilic silicon atom with simultaneous displacement of iodine ion. By this a good leaving group is generated. Subsequently, the iodide ion being an excellent nucleophile reacts with the electrophilic carbon atom displacing the leaving group (alkyl trimethyl silyl ether) (Fig. 12.55).

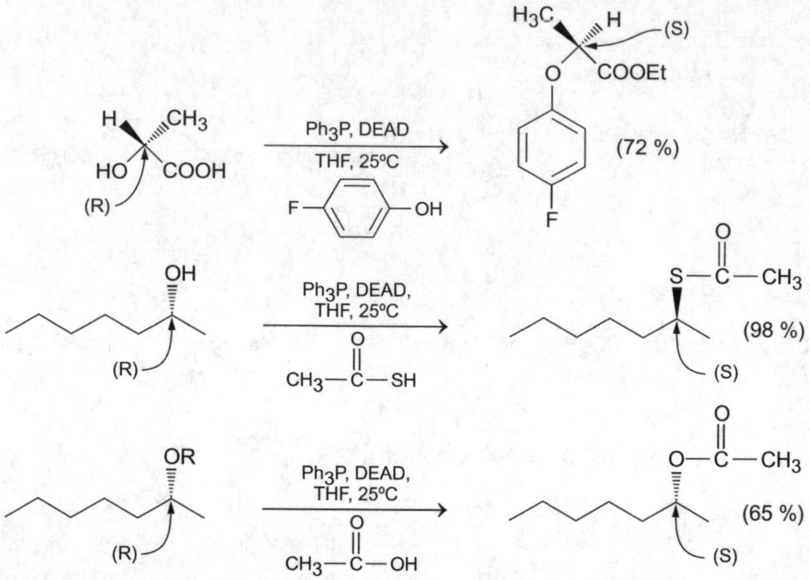

**Fig. 12.52** Some examples of Mitsunobu reactions using phenol, thiocarboxylic acid or a carboxylic acid as a nucleophile

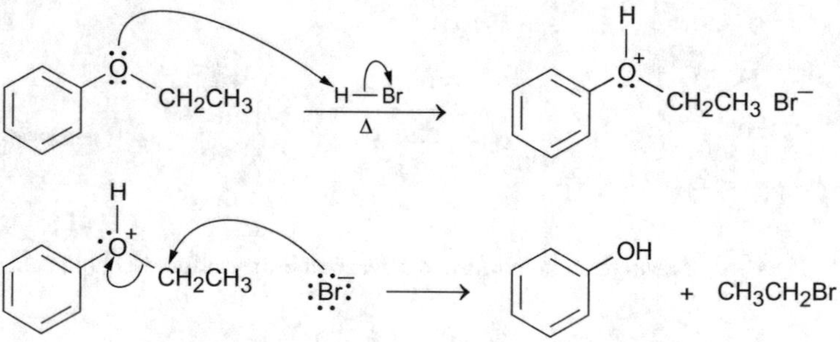

**Fig. 12.53** Nucleophilic substitution of ether with HBr

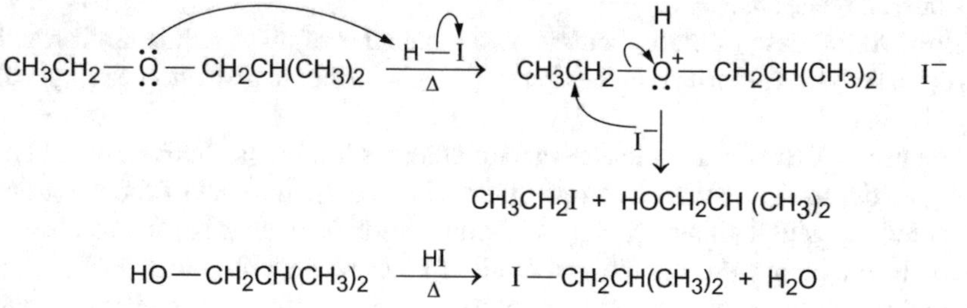

**Fig. 12.54** Reaction of dialkyl ether with HI

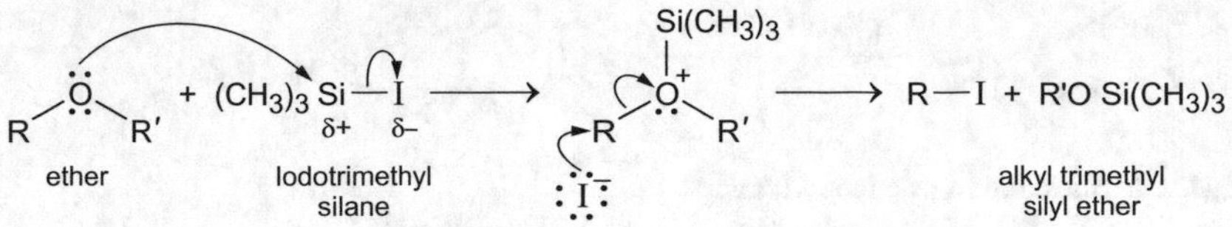

**Fig. 12.55** Cleavage of ether with iodotrimethyl silane

## 12.8 Substitution Reactions of Epoxides

Epoxides, also called oxiranes or oxacyclopropanes readily react with nucleophiles. In fact, epoxides are more reactive than most of the other ethers (Fig. 12.56).

An interesting example of epoxide ring opening is by reaction with azide ion, $N_3^-$, which gives a racemic mixture of azido alcohols (Fig. 12.57).

The above reaction is believed to take place as shown below (Fig. 12.58).

For more details about epoxide ring opening see Sect. 19.3.3.

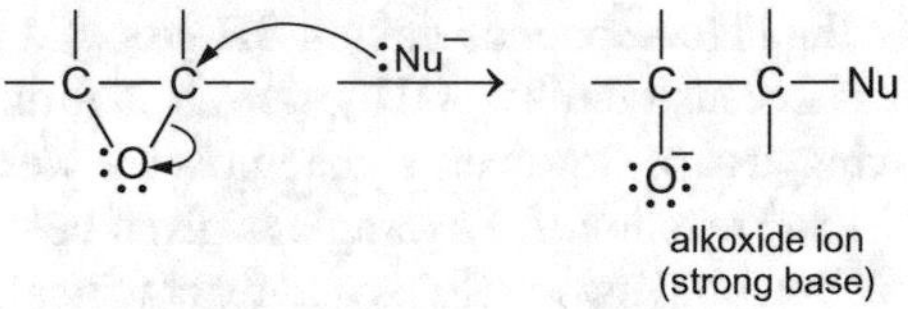

alkoxide ion
(strong base)

**Fig. 12.56** Ring opening of epoxide with nucleophiles

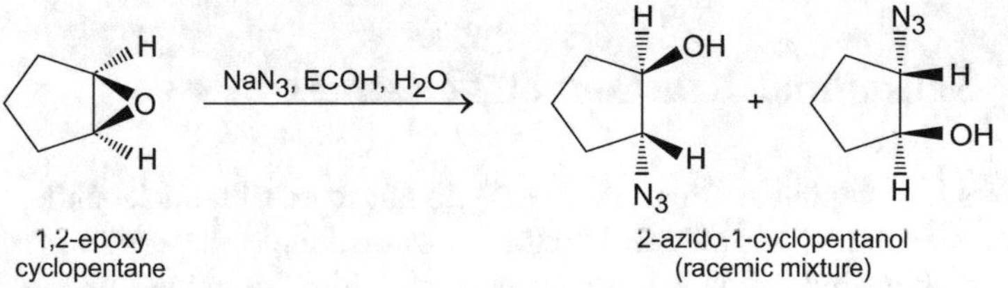

1,2-epoxy
cyclopentane

2-azido-1-cyclopentanol
(racemic mixture)

**Fig. 12.57** Epoxide ring opening by azido ion

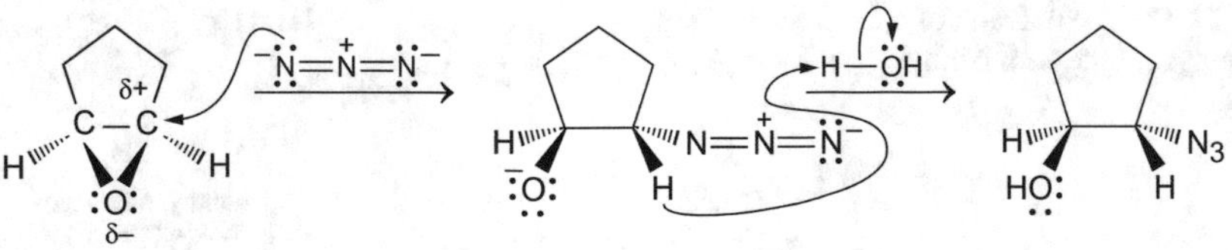

**Fig. 12.58** Mechanism of ring opening of epoxide by reaction with azide ion

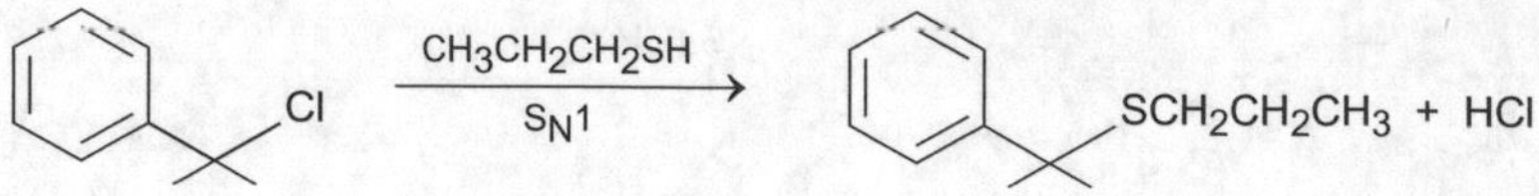

**Fig. 12.59**  Substitution reaction of thiols

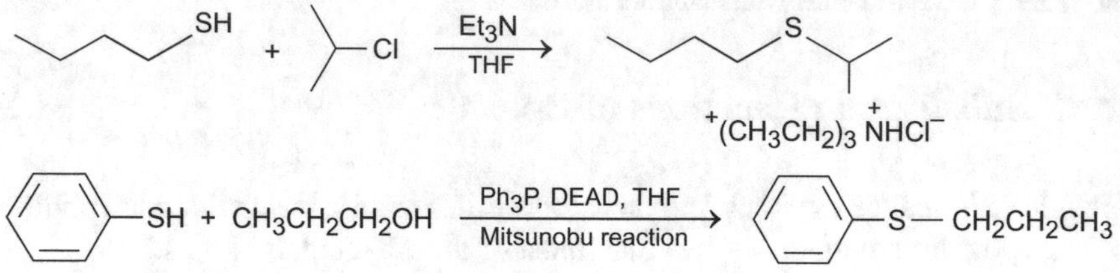

**Fig. 12.60**  Reaction of thiolate ion with alkyl halides and alcohols

## 12.9    Substitution Reaction of Thiols

Compared to alcohols, thiols (compounds having SH group) do not undergo substitution reactions. This is because unlike –OH group, it is difficult to make the SH group into a good leaving group. However, since thiols are weak bases, they can be used as nucleophiles in $S_N{}^1$ reaction. An example is given below (Fig. 12.59).

The thiolate ions are potential nucleophiles and react with alkyl halides and alcohols [via Mitsunobu reaction, Sect. 12.6.3]. Reaction with alkyl halides is conducted in presence of a base, when thiols are converted into thiolate ion (Fig. 12.60).

## 12.10    Substitution Reactions of Thioethers

Thioethers (the sulphur analogue of ethers) do not react with nucleophiles, but can react as nucleophiles (as in the case of thiols). As an example, the reaction of dimethyl sulphide with methyl iodide in ether gives a crystalline trimethyl sulfonium iodide in 100% yield (Fig. 12.61).

**Fig. 12.61**  The reaction of dimethyl sulfide with methyl iodide

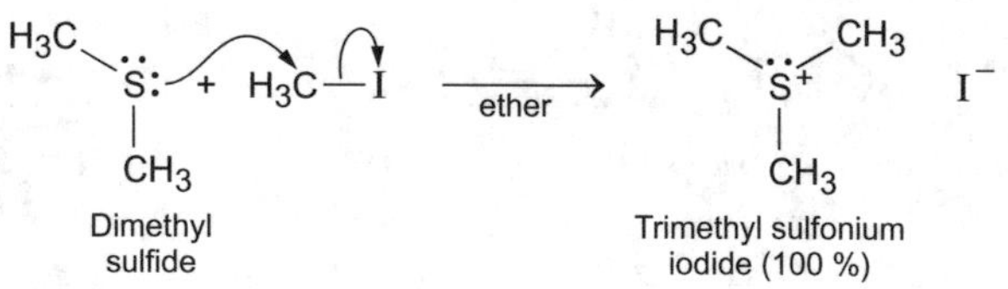

Dimethyl
sulfide

Trimethyl sulfonium
iodide (100 %)

# 12.11 Aromatic Substitution

We have come across only one type of aromatic substitution and that is chloronation of benzene to give hexachlorobenzene. This involves free radical substitution (for details see Sect. 12.2.2).

Besides free radical substitution, there are some cases of nucleophic aromatic substitution and electrophilic aromatic substitution.

## 12.11.1 Nucleophilic Aromatic Substitution

This can take place either by addition elimination pathway or elimination–addition pathway.

### 12.11.1.1 Nucleophilic Aromatic Substitution via Addition–Elimination Reaction

In aryl halides (like chlorobenzene) nucleophilic substitutions do not occur. However, such nucleophilic substitutions occur readily in case the aryl halide has strong electron-withdrawing groups in *ortho* or *para* position to the halogen atom. Some examples are given below (Fig. 12.62).

*m*- Chloronitrobenzene does not undergo nucleophilic substitution.

The mechanism of the above aromatic aromatic substitution (Fig. 12.62) is believed to be a **addition–elimination mechanism** involving formation of carbanion (with delocalised electrons) known as **Meisenheimer complex**; its correct structure was proposed by a German chemist, Jacob Meisenheimer. Thus, in the reaction of *p*-choronitrobenzene with OH⁻, the first step is the addition of OH⁻ to give the carbanion. This is followed by elimination of a chloride ion to give a substitution product since the aromaticity of the ring is restored. This mechanism is known as the $S_N$ Ar mechanism (Fig. 12.63).

p-Nitro chlorobenzene          carbanion
(Meisenheimer complex)
(only the stable carbonion
is shown)

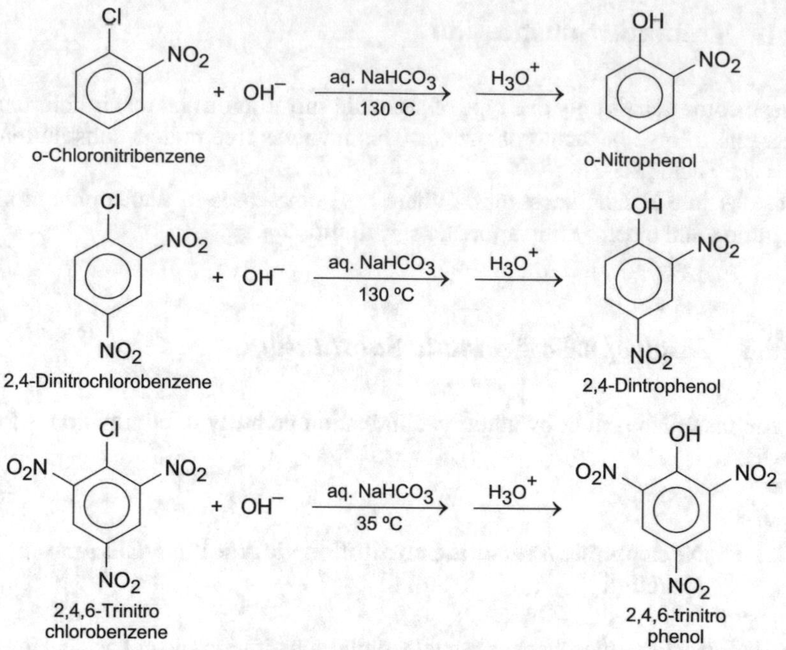

**Fig. 12.62** Some nucleophilic aromatic substitutions

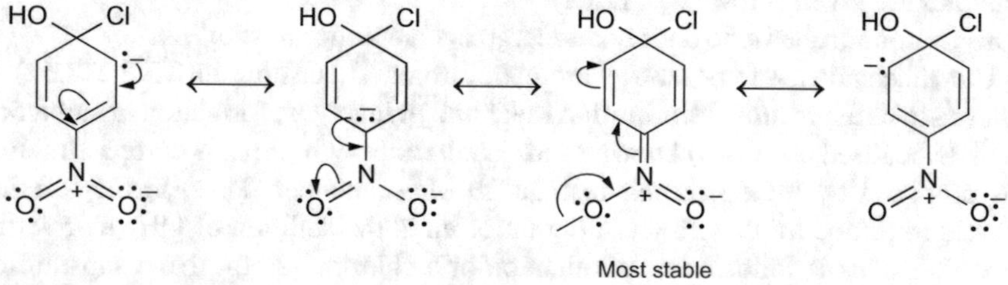

**Fig. 12.63** SN Ar Mechanism

In the above case, the carbonion is stabilised by electron-withdrawing groups in positions *ortho* and *para* to the halogen atom.

### 12.11.1.2   Nucleophilic Aromatic Substitution via an Elimination–Addition Reaction

It is known that aryl halides such as chlorobenzene and bromobenzene do not undergo nucleophilic substitution under ordinary conditions. This is, however, possible under forcing conditions. As an example, chlorobenzene can be converted to phenol by heating with aqueous sodium hydroxide under pressure at 350 °C (Fig. 12.64).

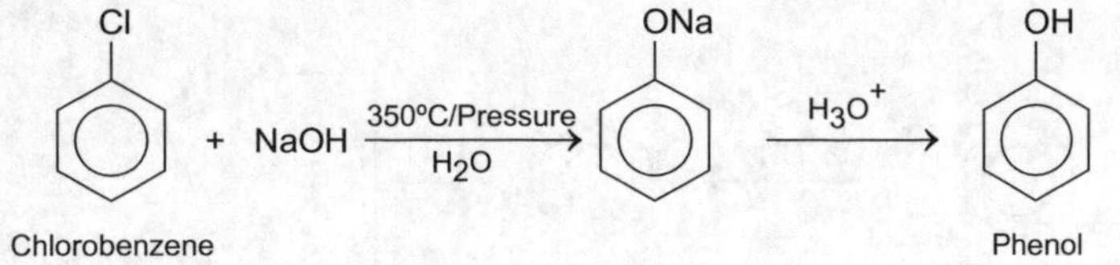

**Fig. 12.64**  Conversion of chlorobenzene into phenol

**Fig. 12.65**  Conversion of bromobenzene into aniline

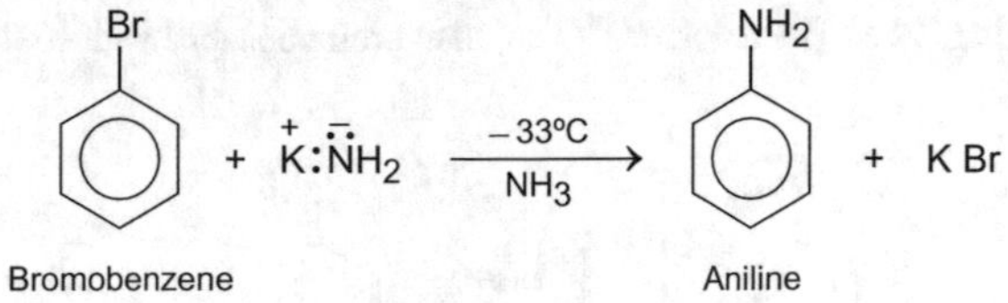

Bromobenzene on reaction with a powerful base ($KNH_2$ | liquid ammonia) gives aniline (Fig. 12.65).

The above reaction (Fig. 12.65) takes place via an **elimination–addition mechanism** as shown in Fig. 12.66.

It was shown (J.D. Roberts, 1953) that $^{14}$C-labelled (C*) bromobenzene on treatment with $KNH_2$ in liquid ammonia, the produced aniline had the label equally divided between 1 and 2 positions. This result is in agreement with elimination–addition mechanism (Fig. 12.67).

However, *o*-(trifluoromethyl) chlorobenzene on reaction with $NaNH_2$ | $NH_3$ gave only one product, *m*-(trifluoromethyl) aniline. The reaction proceeds via the formation of benzyne, which adds on amide ion in a way that produces more stable carbanion (in which the negative charge is closer to electronegative $CF_3$ group), which in turn accepts a proton from ammonia to form *m*-(trifluoromethyl) aniline (Fig. 12.68).

Generation of benzyne in presence of a diene (e.g., furam) gives Diel's Alder adduct (Fig. 12.69).

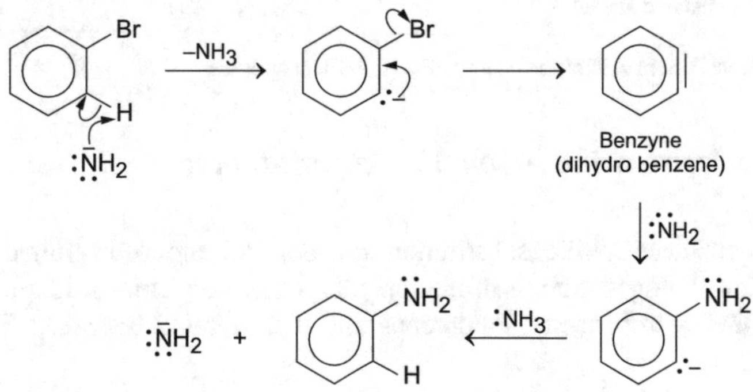

**Fig. 12.66**  Mechanism of elimination–addition reaction

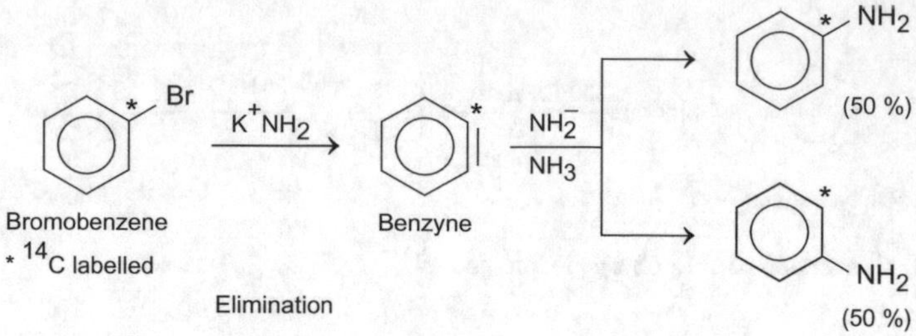

**Fig. 12.67**   Reaction of $^{14}$C-labelled bromobenzene with KNH$_2$|liquid NH$_3$

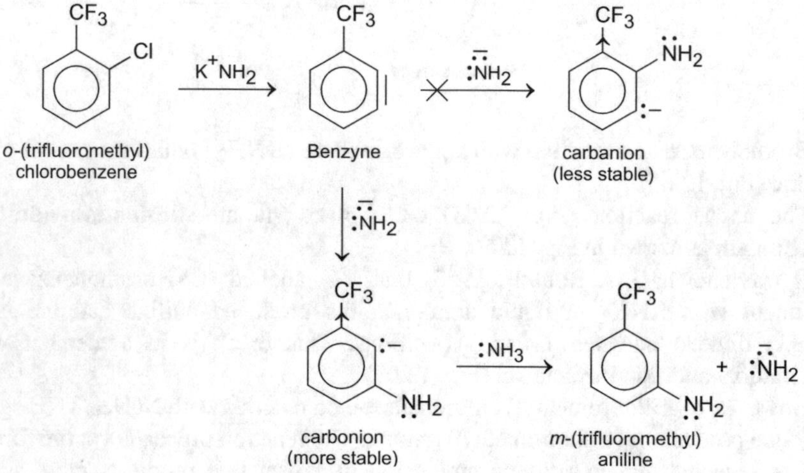

**Fig. 12.68**   Reaction of o-(trifluoromethyl) chlorobenzene with KNH$_2$ | NH$_3$ gives only one product

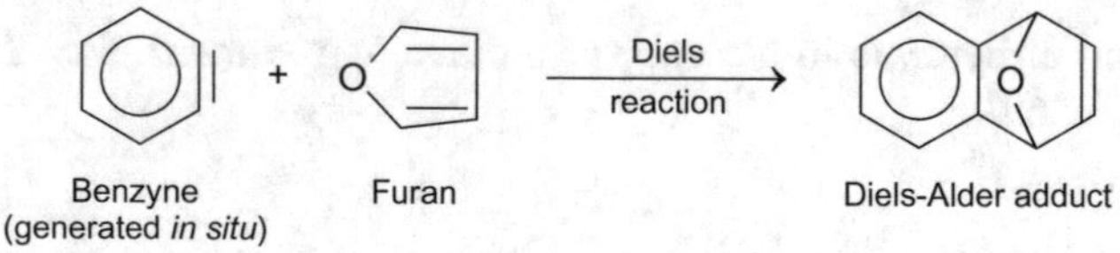

**Fig. 12.69**   Diels Alder reaction of benzynePositive Electrophiles

## 12.11.2   *Electrophilic Aromatic Substitution*

We have seen electrophilic substitution reactions of monosubstituted benzenes (like nitration, halogenation, sulfonation, alkylation, acylation), alkyl benzenes, phenol, aniline, nitrobenzene, haloarenes and disubstituted benzenes (For details see Sect. 12.3).

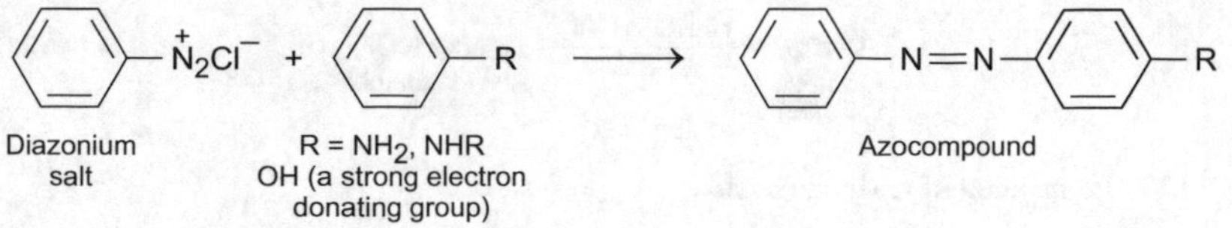

Diazonium salt

R = NH₂, NHR
OH (a strong electron donating group)

Azocompound

**Fig. 12.70** Coupling reaction of diazonium salt

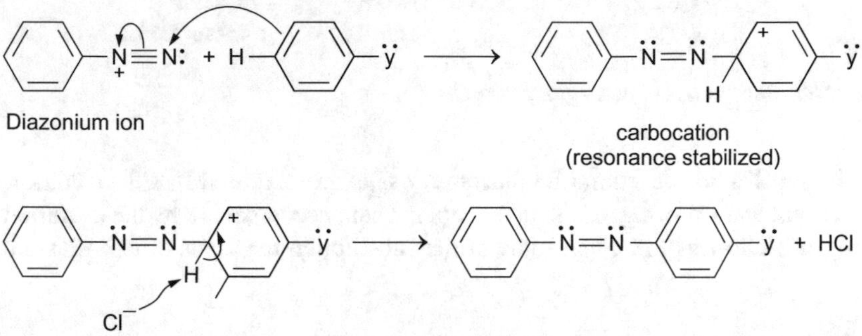

Diazonium ion

carbocation
(resonance stabilized)

**Fig. 12.71** Mechanism of azo coupling

Besides these, we come across another type of electrophilic aromatic substitution. This is the coupling reaction of aryl diazonium salts with an aromatic compound that contains a strong electron-donating group. The two rings couple to form an azocompound (containing nitrogen-nitrogen double bond) in Fig. 12.70.

This reaction is an example of electrophilic aromatic substitution; in this case the diazonium salt acts as the electrophile. The azo coupling occurs by the reaction of the electrophilic diazonium ion with electron rich benzene ring to give a carbocation (which is resonance stabilised). Finally, loss of a proton generates the aromatic ring in Fig. 12.71.

Since a diazonium salt is weakly electrophilic, it is necessary to couple with aromatic compound having strong electron-donating group (y=NH₂, NHR, NR₂ or OH).

Using the azo coupling reaction, a large number of azo-dyes have been prepared.

## 12.12 Substitution Reactions of Aryl Diazonium Salts

Aryl diazonium salts are prepared by the diazotiation of aryl primary amines by treatment with NaNO₂ | HCl at 0–5 °C (Fig. 12.72).

Aryl diazonium salts react with a variety of reagents to form substitution products; the diazo group can be substituted by a number of groups inducting –F, –Cl, –Br, –I, –CN, –OH and –H.

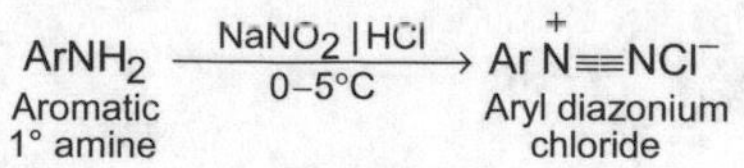

$$\text{ArNH}_2 \xrightarrow[\text{0–5°C}]{\text{NaNO}_2 \mid \text{HCl}} \text{Ar } \overset{+}{\text{N}} \equiv \text{NCl}^-$$

Aromatic                                    Aryl diazonium
1° amine                                     chloride

**Fig. 12.72**  Preparation of diazonium salts

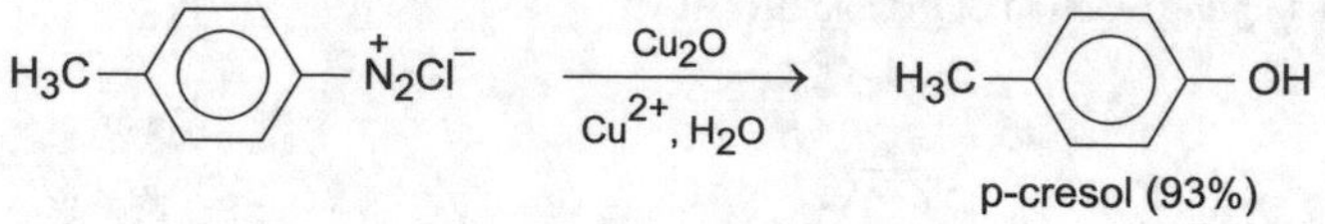

$$\text{H}_3\text{C}-\underset{}{\bigcirc}-\overset{+}{\text{N}}_2\text{Cl}^- \xrightarrow[\text{Cu}^{2+}, \text{H}_2\text{O}]{\text{Cu}_2\text{O}} \text{H}_3\text{C}-\underset{}{\bigcirc}-\text{OH}$$

p-cresol (93%)

**Fig. 12.73**  Substitution of diazo group with OH

For substitution reaction, the diazonium salts are prepared in situ. Addition of the reagent and warming results in the replacement accompanied by the evolution of nitrogen. Following are given some of the substitution reactions of aryl diazonium salts.

## 12.12.1  Substitution by OH

The diazonium group can be replaced by a hydroxyl group by adding cuprous oxide to a dilute solution of the diazonium salt containing an excess of cupric nitrate (Fig. 12.73).

This method developed by T. Cohen (University of Pittsburg) is a much simpler method than the older methods (involving heating the salt with concentrated aqueous acid) and has been used for the synthesis of a variety of phenols.

## 12.12.2  Substitution by Cl, Br and CN

Aryl diazonium salts on treatment with cuprous chloride, cuprous bromide and cuprous cyanide give products in which diazonium group is replaced or substituted by –Cl, –Br and –CN, respectively. These reactions are known as **Sandmeyer reactions** (Fig. 12.74).

This method is a useful alternative to the direct chlorination and bromination of an aromatic ring using $Cl_2$ or $Br_2$ in presence of a Lewis acid catalylst.

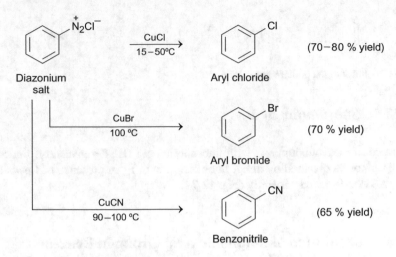

**Fig. 12.74**  Substitution of diazo group with Cl, Br or CN

**Fig. 12.75**  Substitution of diazo group with iodo group

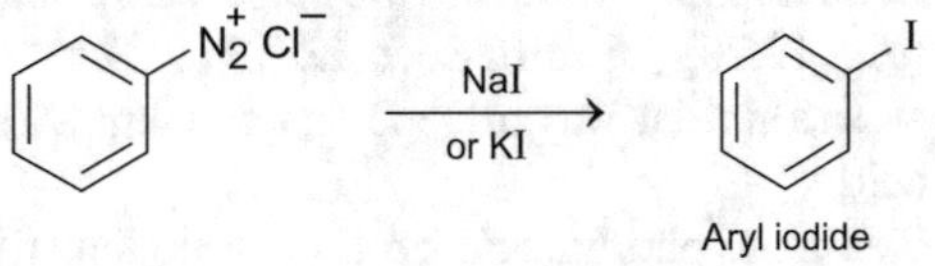

### 12.12.3  Substitution by I

Diazonium salts react with sodium or potassium iodide to give aryl iodides. This is a useful reaction, since direct iodination of aromatic compounds is not possible (Fig. 12.75).

### 12.12.4  Substitution by H

Aryl diazonium salts on reaction with hypophosphorus acid ($H_3PO_2$) give product in which diazo group is replaced by H (Fig. 12.76).

**Fig. 12.76**  Substitution of diazo group with H

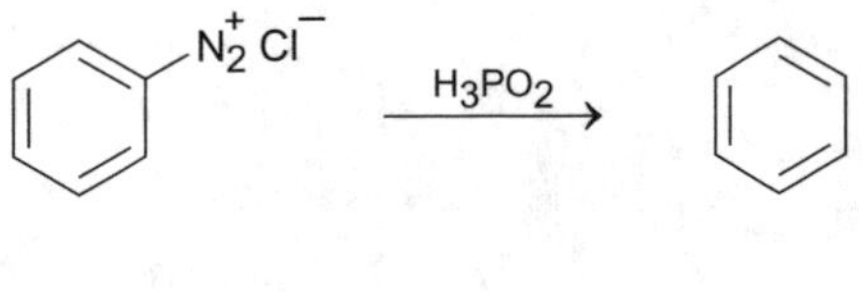

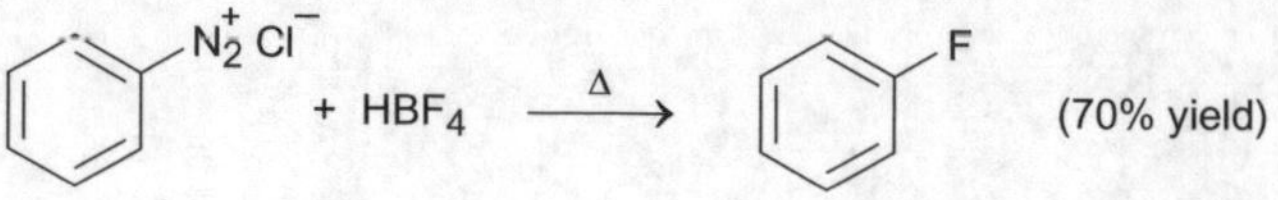

**Fig. 12.77** Substitution of diazo group by F

## 12.12.5  Substitution by F

The reaction of diazonium salt with fluoroboric acid (HBF$_4$) gives aryl fluorides, which cannot be obtained by direct fluorination with F$_2$ in presence of Lewis acid catalyst, since F$_2$ reacts violently (Fig. 12.77).

## 12.13  Substitution of Sulphonic Acid Group in Benzene Sulphonic Acid

Aromatic sulphonic acid are prepared by heating benzene with fuming sulphuric acid (Fig. 12.78) (see also Sect. 12.3.3).

In a similar way toluene on treatment with H$_2$SO$_4$–SO$_3$ gives p-toluene sulphonic acid.

The sulphonic acid group can be substituted (or replaced) by groups like OH, NH$_2$, CN, SH, NO$_2$ and Br.

### 12.13.1  Substitution by OH

Sodium salt of aryl sulphonic acid on fusion with NaOH followed by acidification gives phenol (Fig. 12.79).

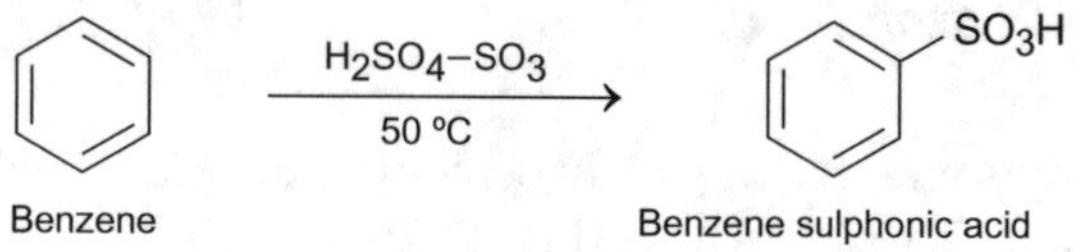

**Fig. 12.78** Preparation of benzene sulphonic acid

$$C_6H_5SO_3H \xrightarrow{\text{NaOH}} C_6H_5SO_3^-Na^+ \xrightarrow[\text{Fuse}]{\text{NaOH}} C_6H_5O^- Na^+ + Na_2SO_3 + H_2O$$

Benzene sulphonic
acid

Sodium salt of
benzene sulphonic
acid

$$\xrightarrow{H^+} C_6H_5OH$$

Phenol

**Fig. 12.79**  Substitution of $SO_3H$ group by OH

**Fig. 12.80**  Substitution of
$SO_3H$ by $NH_2$

$$ArSO_3^-Na + \overset{+}{N}aNH_2 \xrightarrow{\text{Fuse}} ArNH_2 + Na_2SO_3$$

1° Amine

**Fig. 12.81**  Substitution of
$SO_3H$ groups by CN

$$ArSO_3^-Na^+ + NaCN \xrightarrow{\text{Fusion}} ArCN + Na_2SO_3$$

Aryl
cyanide

## 12.13.2   Substitution by $NH_2$

Sodium salt of aryl sulphonic acid on fusion with sodamide gives 1° amines (Fig. 12.80).

## 12.13.3   Substitution of CN

Sodium salt of aryl sulfonic acids on fusion with NaCN gives the corresponding cyanides (Fig. 12.81).

## 12.13.4   Substitution by SH

Sodium salt of aryl sulfonic acids on fusion with potassium hydrosulphide give thiophenols (Fig. 12.82).

**Fig. 12.82**  Substitution of
$SO_3H$ group by SH

$$ArSO_3^- \overset{+}{K} + KSH \xrightarrow{\text{Fusion}} ArSH + K_2SO_3$$

Thiophenol

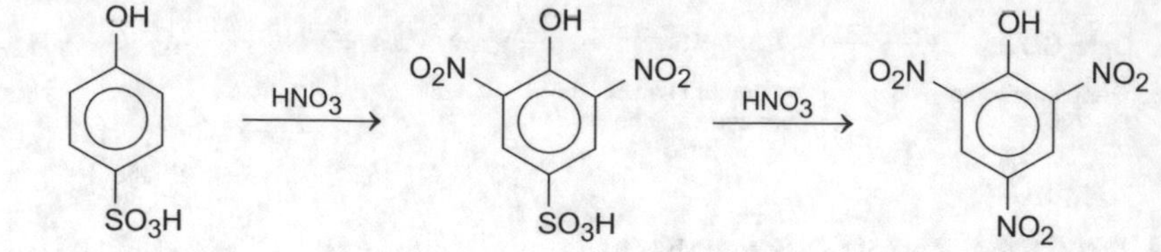

*p*-Hydroxy benzene sulphonic

2,4,6-Trinitrophenol

**Fig. 12.83** Conversion of *p*-hydroxybenzene sulphonic acid into 2,4,6-trinitrophenol

### *12.13.5 Substitution by $NO_2$ Group*

There is no method to substitute $SO_3H$ group by $NO_2$ group. However, 2,4,6-trinitrophenol can be obtained by treatment of *p*-hydroxybenzene sulphonic acid with $HNO_3$ (Fig. 12.83).

### *12.13.6 Substitution by Br*

There is no method to substitute $SO_3H$ group by Br. However 2, 4, 6-tribromoaniline can be obtained by the treatment of *p*-aminobenzene sulphonic acid with $Br_2$ | $H_2O$ (Fig. 12.84).

## 12.14 Substitution of Active Hydrogen by Alkyl and Acyl Groups

The active hydrogen in terminal alkynes, alcohols, phenols and amines can be substituted by alkyl groups as shown below (Fig. 12.85).

$$R-C\equiv CH \xrightarrow{NaNH_2} R-C\equiv C^- \xrightarrow{RX} R-C\equiv C-R'$$

**Fig. 12.84** Conversion of *p*-aminobenzene sulphonic acid into 2,4,6-tribromoaniline

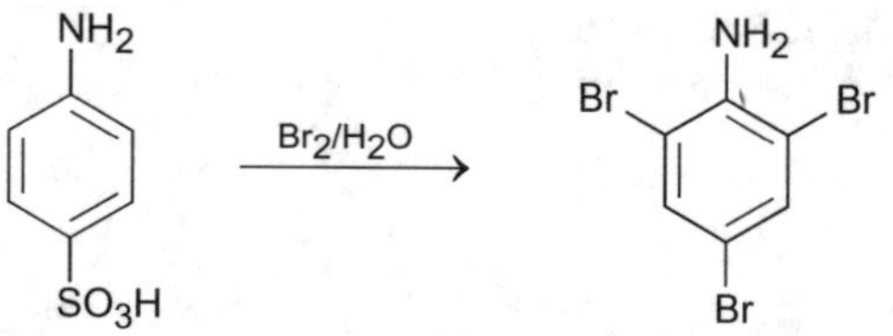

*p*-Amino benzene sulphonic acid

2,4,6-Tribromoaniline

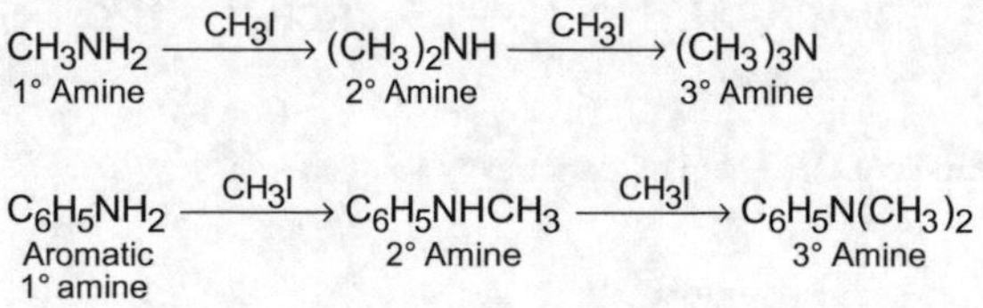

$$CH_3NH_2 \xrightarrow{CH_3I} (CH_3)_2NH \xrightarrow{CH_3I} (CH_3)_3N$$

1° Amine         2° Amine         3° Amine

$$C_6H_5NH_2 \xrightarrow{CH_3I} C_6H_5NHCH_3 \xrightarrow{CH_3I} C_6H_5N(CH_3)_2$$

Aromatic         2° Amine         3° Amine
1° amine

**Fig. 12.85** Substitution of active hydrogen in terminal alkyne, alcohols, phenols and amines with alkyl groups

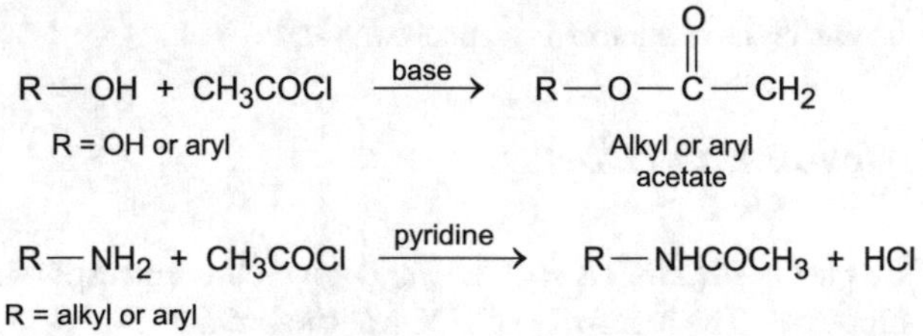

$$R-OH + CH_3COCl \xrightarrow{base} R-O-\overset{\overset{\displaystyle O}{\|}}{C}-CH_2$$

R = OH or aryl         Alkyl or aryl
                 acetate

$$R-NH_2 + CH_3COCl \xrightarrow{pyridine} R-NHCOCH_3 + HCl$$

R = alkyl or aryl

**Fig. 12.86** Substitution of active H in alcohols, phenols and amines by acyl group

$$R-\overset{-}{O}-\overset{+}{Na} + R'-X \xrightarrow{S_N2} R-O-R'$$

Sodium                     ether
alkoxide

Phenoxide               Ethers
ion

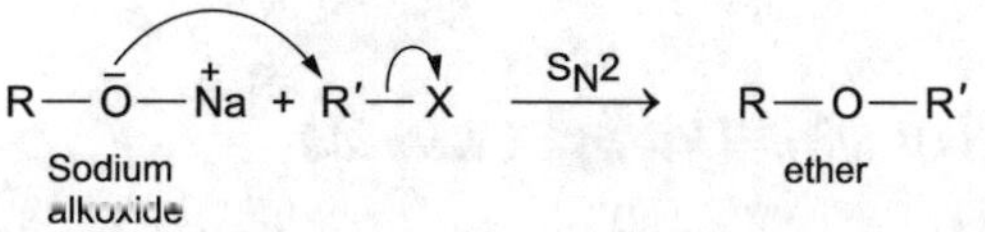

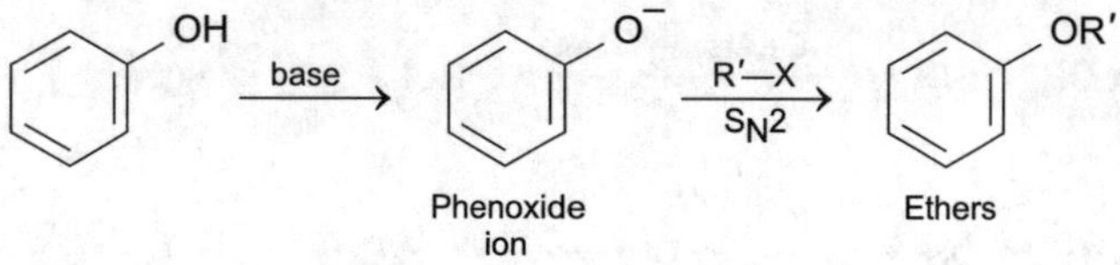

The above reaction is known as Williamson ether synthesis.

Substitution of active H in alcohol, phenols and amines by acyl group can be affected by treatment of alcohols and amines with acetyl chloride in presence of base (Fig. 12.86).

## 12.15 Substitution of Hydroxyl Group of Alcohols by Cl or Br

The OH group in alcohols can be replaced by halogen by treatment with HX, $SOCl_2$ or $PBr_3$.

$$R-CH_2OH + HX \longrightarrow RCH_2X + H_2O$$
Alcohol

**Fig. 12.87**  Substitution of OH group of alcohols by halogen

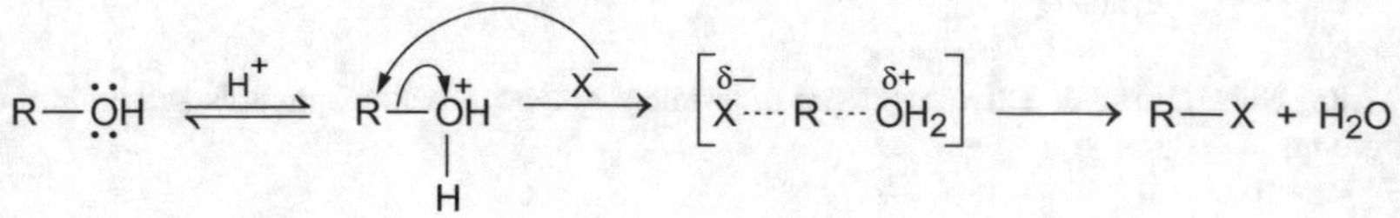

$$R-\ddot{O}H \rightleftharpoons R\overset{+}{-}\underset{\underset{H}{|}}{O}H \xrightarrow{X^-} \left[ \overset{\delta-}{X}\cdots R\cdots \overset{\delta+}{O}H_2 \right] \longrightarrow R-X + H_2O$$

**Fig. 12.88**  Mechanism of the reaction of 1° alcohols with HX

## 12.15.1  *Reaction with HX*

Treatment of alcohol with HX (X = Cl, Br, I) gives the corresponding halogen-substituted compounds. The reactivity of HX is of the order 3° > 2° > 1° (Fig. 12.87).
  Primary alcohols react by $S_N2$ mechanism (Fig. 12.88).
  Secondary alcohols react by $S_N1$ or $S_N2$ mechanism.

## 12.15.2  *Reaction with Thionyl Chloride*

See Fig. 12.89.

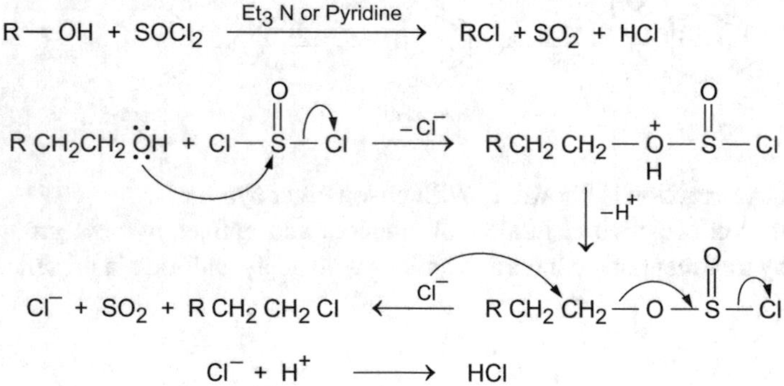

$$R-OH + SOCl_2 \xrightarrow{Et_3 N \text{ or Pyridine}} RCl + SO_2 + HCl$$

**Fig. 12.89**  Reaction of alcohols with thionyl chloride to give alkyl chlorides

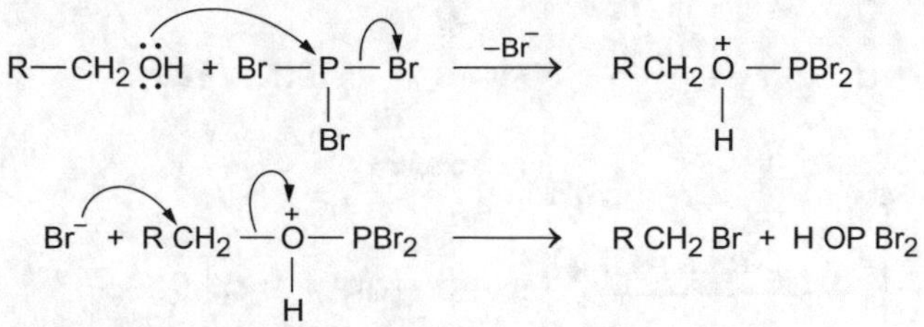

$$R-CH_2\ddot{O}H + Br-P-Br \xrightarrow{-Br^-} R\,CH_2\overset{+}{O}-PBr_2$$
$$\underset{Br}{|} \qquad\qquad \underset{H}{|}$$

$$Br^- + R\,CH_2-\overset{+}{\underset{H}{O}}-PBr_2 \longrightarrow R\,CH_2\,Br + H\,OP\,Br_2$$

**Fig. 12.90**  Reaction of alcohols with PBr$_3$ to give alkyl bromides

### 12.15.3  Reaction with Phosphorus Halide

See Fig. 12.90.

## 12.16  Substitutions in Carboxylic Acids

In case of carboxylic acids, the OH of carboxyl group can be substituted by Cl, OR'
or NH$_2$ to give the corresponding acid chlorides, esters or amides. Besides, this, the
COOH group can be replaced by Br and the α-hydrogen in carboxylic acid can be
replaced by Br.

### 12.16.1  Substitution of OH of Carboxyl Group by Cl, OR
or NH$_2$

The reaction of carboxylic acid with SOCl$_2$ gives the corresponding acid chloride and
reaction with alcohol in presence of acid gives esters. Also treatment of carboxylic
acid with ammonia gives ammonium salt, which on heating gives the corresponding
amide (Fig. 12.91).

### 12.16.2  Substitution of COOH Group by Bromine

The reaction is known as **Hunsdiecker reaction** and consists in the thermal decomposition of the silver salt of the carboxylic acid in presence of halogen. The reaction
involves decarboxylative halogenation and proceeds via the formation of free radical
(Fig. 12.92).

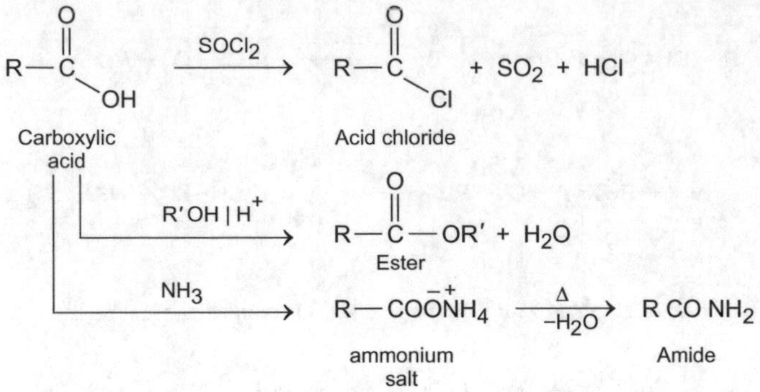

$$R - \underset{\underset{OH}{}}{\overset{\overset{O}{\|}}{C}} \xrightarrow{SOCl_2} R - \underset{\underset{Cl}{}}{\overset{\overset{O}{\|}}{C}} + SO_2 + HCl$$

Carboxylic acid          Acid chloride

$$\xrightarrow{R'OH\,|\,H^+} R - \underset{}{\overset{\overset{O}{\|}}{C}} - OR' + H_2O$$

Ester

$$\xrightarrow{NH_3} R - \overset{-+}{COONH_4} \xrightarrow[-H_2O]{\Delta} R\,CO\,NH_2$$

ammonium          Amide
salt

**Fig. 12.91** Substitution of OH of COOH group by Cl, OR or NH$_2$

**Fig. 12.92** Substitution of COOH group by bromine

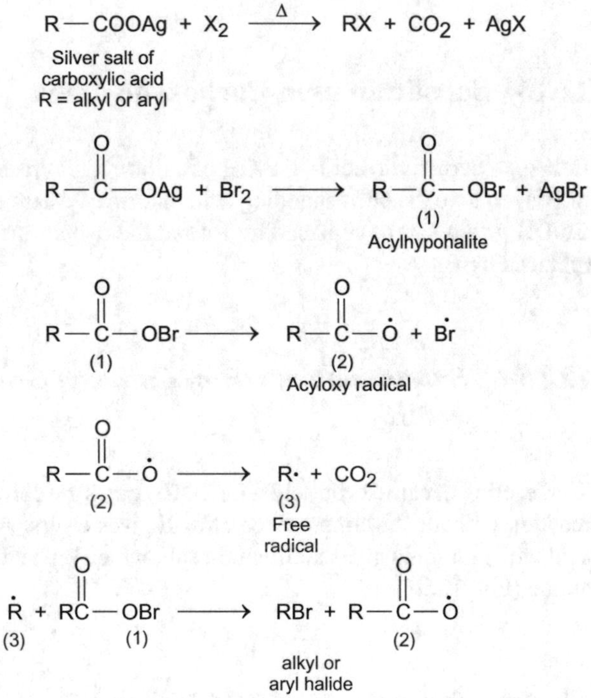

$$R - COOAg + X_2 \xrightarrow{\Delta} RX + CO_2 + AgX$$

Silver salt of
carboxylic acid
R = alkyl or aryl

$$R - \underset{}{\overset{\overset{O}{\|}}{C}} - OAg + Br_2 \longrightarrow R - \underset{}{\overset{\overset{O}{\|}}{C}} - OBr + AgBr$$
(1)
Acylhypohalite

$$R - \underset{(1)}{\overset{\overset{O}{\|}}{C}} - OBr \longrightarrow R - \underset{(2)}{\overset{\overset{O}{\|}}{C}} - \dot{O} + \dot{B}r$$
Acyloxy radical

$$R - \underset{(2)}{\overset{\overset{O}{\|}}{C}} - \dot{O} \longrightarrow \underset{(3)}{R\cdot} + CO_2$$
Free
radical

$$\underset{(3)}{\dot{R}} + \underset{(1)}{R\overset{\overset{O}{\|}}{C} - OBr} \longrightarrow RBr + R - \underset{(2)}{\overset{\overset{O}{\|}}{C}} - \dot{O}$$
alkyl or
aryl halide

## 12.16.3    *Substitution of α-hydrogen in Carboxylic Acids by Br*

This is achieved by **Hell–Volhard–Zelinsky reaction** and consists in the reaction of carboxylic acid with halogen (Br$_2$ or Cl$_2$) in presence of small amount of phosphorus to give α-halogen-substituted carboxylic acid (Fig. 12.93).

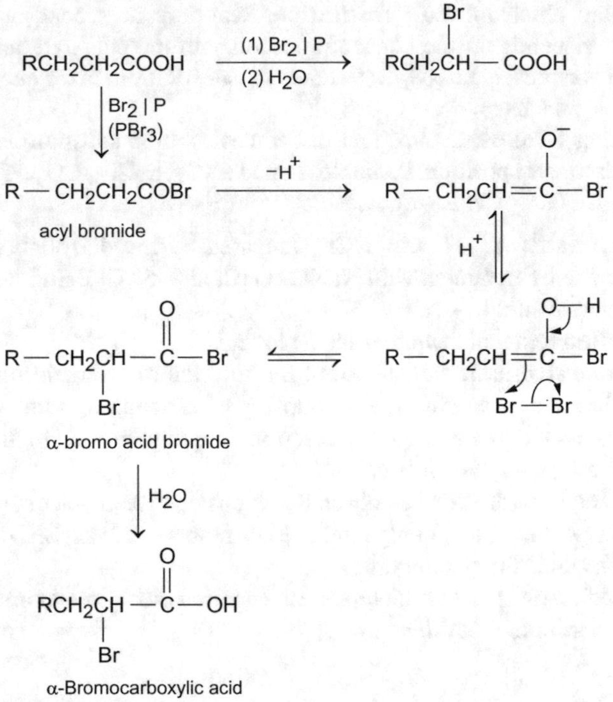

**Fig. 12.93** Substitution of α-hydrogen in carboxylic acid by Br

## Key Concepts

- **Activating Group**: Group directs electrophilic substitution in aromatic compounds to *ortho* and *para* positions. Examples include halogen (Cl, Br, I, F), $^-$OH, –OR, –NH$_2$ –NHR, –NR$_2$, –NHCOCH$_3$ and alkyl groups like methyl or ethyl.
- **Allylic Carbon**: The carbon atom adjacent to a double bond.
- **Allylic Substitution**: Substitution of allylic carbon (the carbon atom adjacent to a double bond).
- **Benzyne**: An unstable reaction intermediate having a triple bond in benzene ring. There are aromatic counterpart of acetylene and are also called dehydro benzene. These are obtained in situ by the reaction halogen-containing aromatic compounds with sodamide and liquid ammonia.

- **Bimolecular Nucleophilic Substitution Reaction**: A second-order reaction, whose rate depends on the concentration of both the substrate and the nucleophile and is represented as $S_N2$ (S stands for substitution, N for nucleophilic and 2 for bimolecular).
- **Deactivating Groups**: Groups that direct electrophilic substitution in aromatic compounds to *meta* position. Examples include $NO_2$, in $CF_3$, $CCl_3$, $C\equiv N$, $-SO_3H$, $-COOH$, $-COOR$, $-CHO$, $-COR$, etc.
- **Diazonium Salts**: $Ar\ \overset{+}{N} \equiv\equiv NCl^-$ Prepared by the diazotiation of an aryl primary amine by treatment with $NaNO_2$ l HCl at 0–5 °C. Being unstable, these are generated in situ.
- **Electrophiles**: Reagent that are electron loving.
- **Electrophilic Aromatic Substitution**: Substitution of aromatic substrates with electrophiles. An example is the coupling of diazonium salts with organic compounds that contain a strong electron-donating group. In this case, the diazonium salt acts as an electrophile.
- **Electrophilic Substitution Reaction**: Reactions involve attack of an electrophile to a substrate (aromatic compound). It is represented as SE (S stands for substitution and E for electrophilic).
- **Free radical**: Any species which possess unpaired electron and are obtained by homolytic fission of a covalent bond.

$$Cl : Cl \longrightarrow 2\ \overset{\cdot}{Cl}$$

- **Free Radical Substitution**: Reactions initiated by free radicals.
- **Friedel Crafts reaction**: The reaction of aromatic compounds (benzene) with alkyl halides or acyl halides in presence of Lewis acid catalysts to give alkyl-substituted benzenes (Friedel Crafts alkylation) or acyl substituted benzenes (ketones) (Friedel Crafts acylation).
- **Halogenation**: The substitution of H in an organic compound by halogen. It is carried out by reaction with halogen and Lewis acid catalyst ($FeCl_3$).
- **Hell-Volhard-Zelinsky Reaction**: The reaction of a carboxylic acid with halogen ($Br_2$ or $Cl_2$) in presence of small amount of phosphorus to give α-halogen-substituted carboxylic acid.
- **Hunsdiecker Reaction**: Thermal decomposition of silver salt of carboxylic acid in presence of halogen gives alkyl halides. The reaction involves decarboxylative halogenation and proceeds via the formation of free radical.
- **Mitsunobu Reaction**: A general method for the nucleophilic substitution reactions of alcohols using a reagent containing phosphorus containing leaving group. One such reagent is a mixture of triphenyl phosphine, diethylazodicarboxylate and an acid (H–X).
- **Negative Nucleophiles** ($\overset{..}{N}\bar{u}$): Have a pair of electrons and carry negative charge (Examples, $R\bar{C}H_2$, $\bar{Cl}$, $\bar{O}H$, $\bar{C}N$).

- **Neutral Electrophiles (E)**: Neutral species, the central atom having six electrons (Examples, $BF_3$, $AlCl_3$, $SO_3$, $FeCl_3$)
- **Neutral Nucleophiles (Nu)**: Have a pair of electrons and carry no charge (Examples, $H_3N :$, $H - \ddot{O} - H$, $\ddot{R} - SH$, $\ddot{R} - S - R$)
- **Nitration**: The procedure of introducing $NO_2$ group in an aromatic compound by reaction with nitrating mixture (Conc. $HNO_3$ + Conc. $H_2SO_4$). The reaction takes place via the formation of nitronium ion.
- **Nucleophilic Substitution Reaction**: Substitution brought about by a nucleophile and is denoted by $S_N$ (S stands for substitution and N for nucleophilic).
- **Nucleophiles**: Reagents that are electron rich.
- **Nitronium Ion**: A species formed by the reaction of conc. $HNO_3$ and conc. $H_2SO_4$

$$HNO_3 + 2H_2SO_4 \rightleftharpoons \overset{+}{N}O_2 + H_3O^+ + 2HSO_4^-$$

It is responsible for the nitration of aromatic compounds.

- **Positive Electrophiles**: ($E^+$) carry positive charge and are deficient of two electrons (Examples $R\overset{+}{C}H_2$, $\overset{+}{B}r$, $H\overset{+}{N}\overset{+}{O_2}$).
- **Racemisation**: Formation of equal amounts of two enantiomers from a single starting material.
- **Substitution Reaction**: A reaction in which one or more atoms or groups in a compound are replaced or substituted by other atoms or groups.
- **Sulfonation**: The substitution or replacement of H in an organic compound by $SO_3H$ group. It is carried out by reaction of organic, compound with conc. $H_2SO_4$ or fuming $H_2SO_4$. The electrophilic reagent is $SO_3$.

$$2H_2SO_4 \rightleftharpoons H_3O^+ + HSO_3^- + SO_3$$

- **Sandmeyer Reaction**: The replacement of diazonium salt by halogen or cyano group by reacting with cuprous salt or $^-CN$. The reaction proceeds via the formation of free radical.
- **Unimolecular Nucleophilic Substitution Reaction**: A first-order reaction, whose rate depends on the concentration only of the substrate (alkyl halide) and is independent of the concentration of the nucleophile ($^-OH$). It is designated as $S_N1$ (S, N and 1 stand for substitution, nucleophilic and unimolecular.
- **Vinylic Carbon**: The carbon atom is attached to a double bond. The vinylic carbon is attached to a $sp^2$ hybridised carbon.
- **Williamson Ether Synthesis**: The reaction of an alkoxide with an alkyl halide ($S_N2$ reaction) gives ethers.

## Problems

1. What product is expected to be obtained by nitration of (*a*) *p*-cresol and (*b*) *m*-cresol?

**Ans.** (a) 2-Nitro-4-methyl phenol.

(b) A mixture of 2-nitro-5-methyl phenol and 4-nitro 5-methyl phenol.

2.   What products or products are expected to be obtained by the nitration of

    (a)   methyl *p*-methoxybenzoic acid
    (b)   2-Bromo-anisole
    (c)   *o*-Nitrololuene
    (d)   *m*-Bromochlorobenzane

3.   How will you synthesise the following:

    (a)   *o*-Bromo-nitrobenzene from benzene
    (b)   *m*-Bromo-nitrobenzene from benzene
    (c)   *o*-Nitrotoluene from benzene

4.   What producted is expected to be obtained (giving the stereochemistry) in the following $S_N2$ reaction

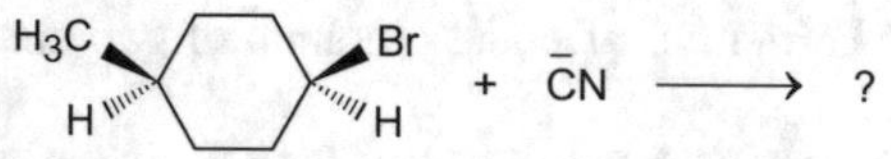

**Ans.** The product is

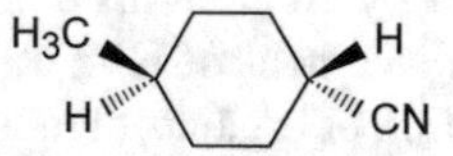

5.   What product is expected to be obtained in the $S_N1$ reaction (giving the stereochemistry in the following halides:

    (a)

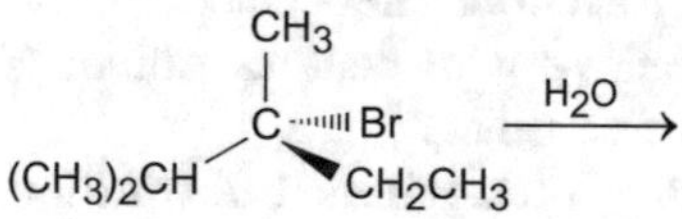

    (b)

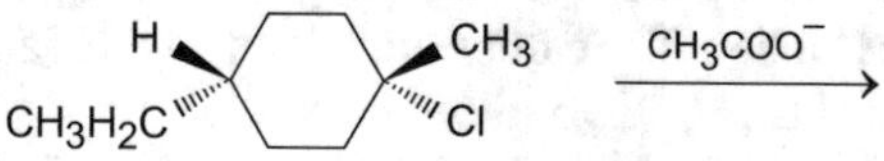

6.   Indicate whether electrophilic substitution with following halides takes place by $S_N1$ or $S_N2$ mechanism

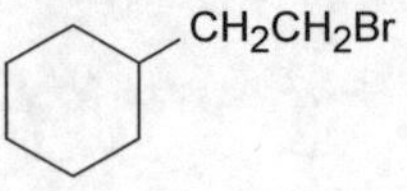

I 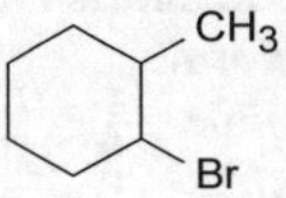                           II 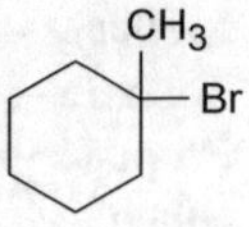                          III

**Ans**. I being a primary halide reacts by $S_N2$ mechanism.
II being a secondary halide can react both by $S_N1$ and $S_N2$ mechanism.
III being a tertiary halide reacts by $S_N1$ mechanism.

7. Give the mechanism of nucleophilic substitution for the following reaction giving the product formed:

(i)                    $CH_3\, CH_2\, CH_2\, CH_2{-}Br\, {-}1 + :\, C{\equiv}CH \longrightarrow$

(ii)

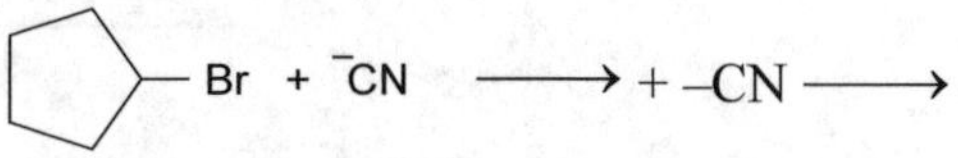

(iii)

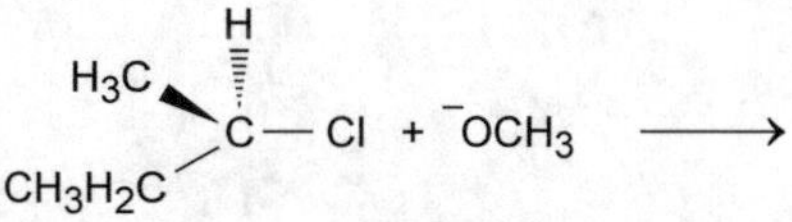

**Ans**. (i) $S_N2$ mechanism. Product is $CH_3\, CH_2\, CH_2\, CH_2C{\equiv}CH$.
(ii) Can react with $S_N1$ or $S_N2$ reaction strong nucleophile favours $S_N2$ reaction giving

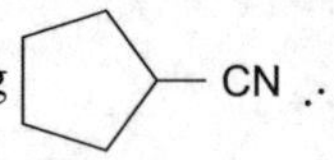

(iii) The $2°$ alkyl halide can react by both $S_N{}^1$ or $S_N{}^2$ mechanisms. Using $^-OCH_3$, the $S_N{}^2$ reaction there is inversion of configuration giving

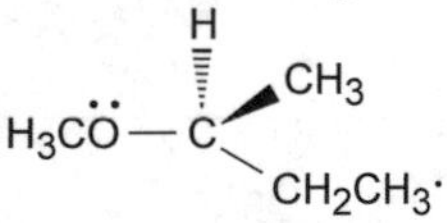

8. How will you affect the following conversion:

   (i)     Bromobenzento → Aniline
   (ii)    Aniline → Fluorobenzene
   (iii)   Phenol → 2, 4, 6-Trinitrophenol
   (iv)    Benzene → Benzonitrile
   (v)     Phenol → Anisole
   (vi)    Benzoic acid → Bromobenzene
   (vii)   Phenyl acetic acid → α-promophenyl acetic acid

(viii)   Methane → Carbontetrachloride
(ix)    Toluene → Benzyl chloride
(x)     Propene → Allyl chloride
(xi)    Ethene → Vinyl chloride

9.   What product is expected to be obtained in the following:

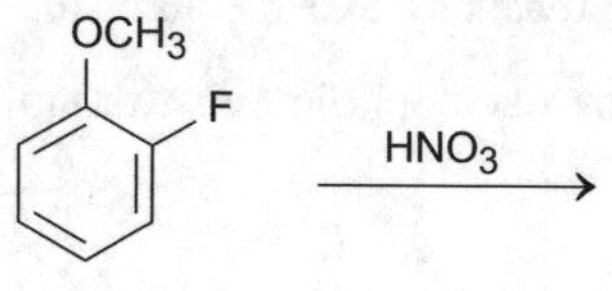

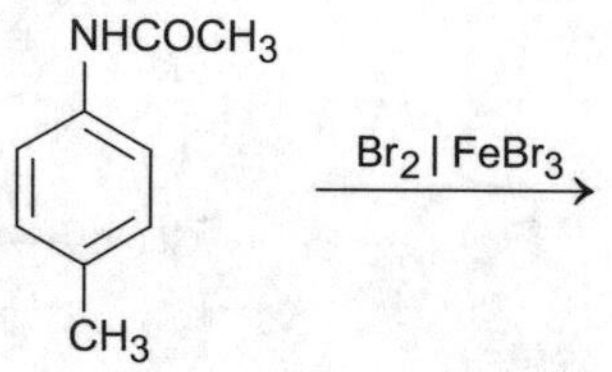

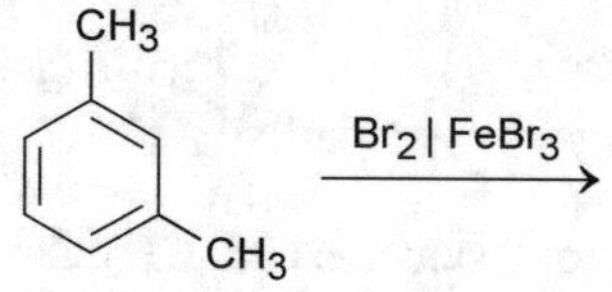

10.   Discuss the stereochemistry of $S_N1$ and $S_N2$ reactions.
11.   Discuss substitution reactions of alcohols, ethers, epoxides.
12.   Write a note on aromatic substitutions.

# Chapter 13
# Stereochemistry of Rearrangement Reactions

## 13.1 Introduction

All organic reactions belong to one of the four categories, viz., addition, elimination, substitution or rearrangements. As has already been discussed (Chap. 10), addition reactions are characteristics of compounds containing multiple bonds. In such reactions, all parts of the adding reagents appear in the product. An example of an addition reaction is given below (Fig. 13.1).

The substitution reactions, as have already been discussed (Chap. 12), are characteristic reactions of saturated compounds like alkanes, alkyl halides and aromatic compounds etc. In such reactions, one group is replaced by another. An example of substitution reaction is given below (Fig. 13.2).

The elimination reaction, which are opposite to addition reactions, (*see* Chap. 11) a molecule loses the elements of another molecule to give a product containing a compound containing a multiple bond. A typical example of an elimination reaction is given below (Fig. 13.3).

In a rearrangement reaction, a molecule undergoes reorganisation of its parts. In other words, the molecular skeleton is altered. This means that the sequence in which atoms are arranged (in a molecule) is changed. An example is given below (Fig. 13.4).

In the above example of rearrangement reaction, not only the position of the double bond and a hydrogen atom has changed but also a methyl group has moved from one carbon to another carbon. As seen, the rearrangement reactions result in the skeletal rearrangement.

V. K. Ahluwalia, *Stereochemistry of Organic Compounds*,
https://doi.org/10.1007/978-3-030-84961-0_13

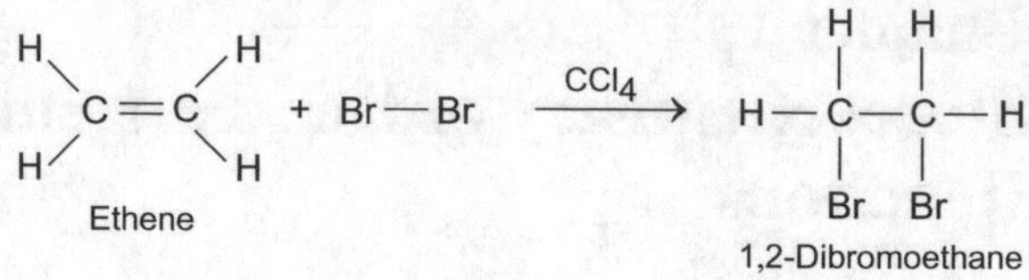

**Fig. 13.1**  An addition
reaction

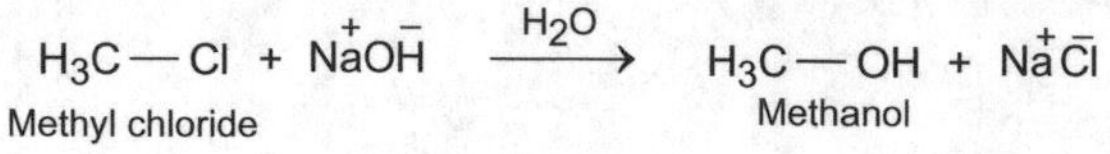

Ethene                    1,2-Dibromoethane

**Fig. 13.2**  A substitution
reaction

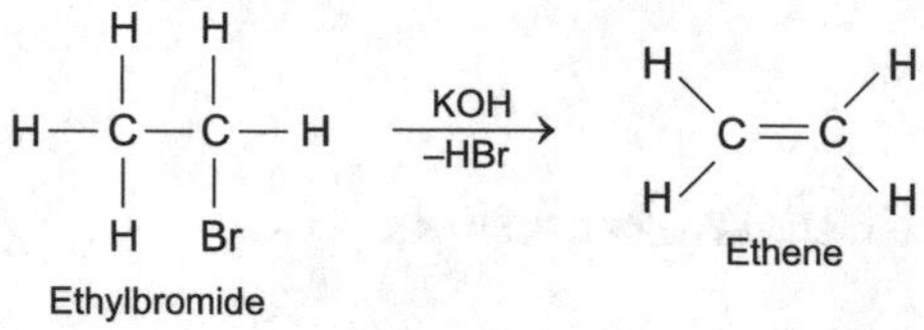

Methyl chloride              Methanol

**Fig. 13.3**  An elimination
reaction

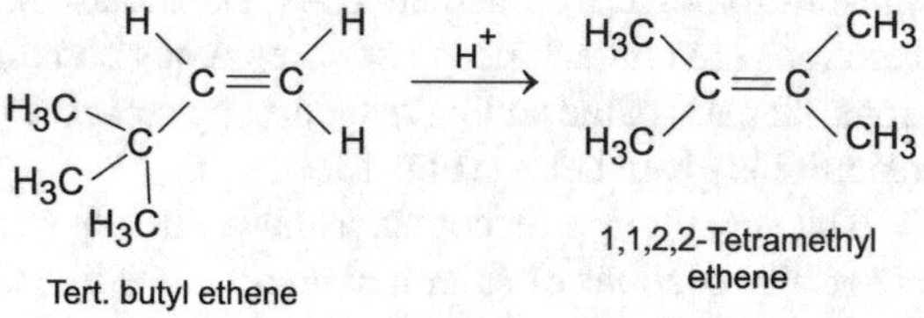

Ethylbromide                    Ethene

**Fig. 13.4**  A rearrangement
reaction

Tert. butyl ethene                1,1,2,2-Tetramethyl ethene

## 13.2   Classification of Rearrangement Reactions

The rearrangement reactions can be classified as intramolecular or intermolecular. Alternatively, these can be classified as belonging to one of the three groups, viz., rearrangements involving carbon–carbon rearrangement, carbon–nitrogen rearrangement or carbon–oxygen rearrangements.

### 13.2.1   Classification as Intramolecular or Intermolecular Rearrangements

#### 13.2.1.1   Intramolecular Rearrangements

In these rearrangements, the migrating group is not completely detached from the system during migration. Some examples include Claisen rearrangement (*see* Sect. 13.2.4.2), Beckmann rearrangement (*see* Sect. 13.2.3.1) and *see* also Fries rearrangement (Sect. 13.2.1.2.3).

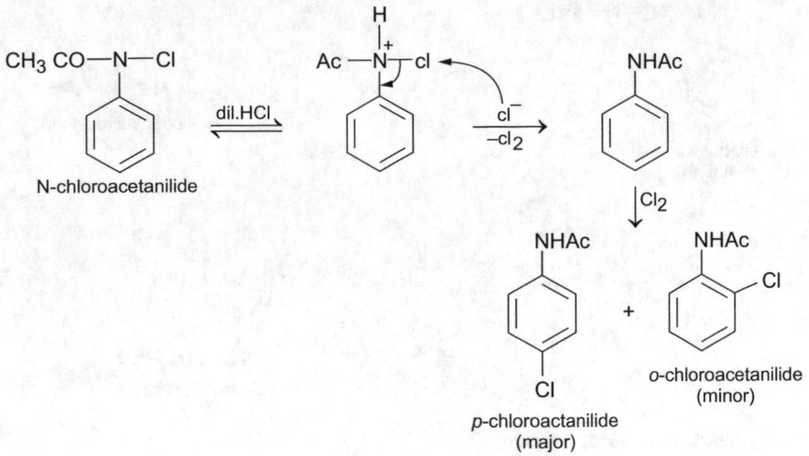

**Fig. 13.5** Orton rearrangement

### 13.2.1.2 Intermolecular Rearrangements

In these rearrangements, the atoms or groups that undergo migration are completely detached from the substrate and subsequently get attached at some other reactive sites of the molecule. Some examples include Orton rearrangement, diazoamino rearrangement and Fries rearrangement.

Orton Rearrangement

Treatment of N-chloroacetanilide with dilute hydrochloric acid gives[1,2] a mixture of o-and p-chlorotoluene (Fig. 13.5).

### References

1. K. J. Orton. J. Chem. Soc., 1905, **95**, 1456.
2. M. Richardson, J. Chem. Soc., 1929, 1873.

Diazoamino Rearrangement

Treatment of diazoamino benzene with hydrochloric acid at 40 °C gives p-aminoazobenzene. The reaction involves the generation of $ArN_2^+$ followed by coupling of diazonium salt with aniline (Fig. 13.6).

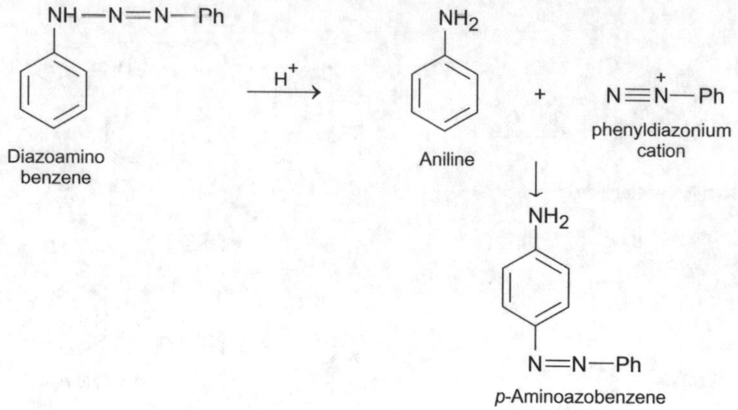

**Fig. 13.6**  Diazoamino rearrangement

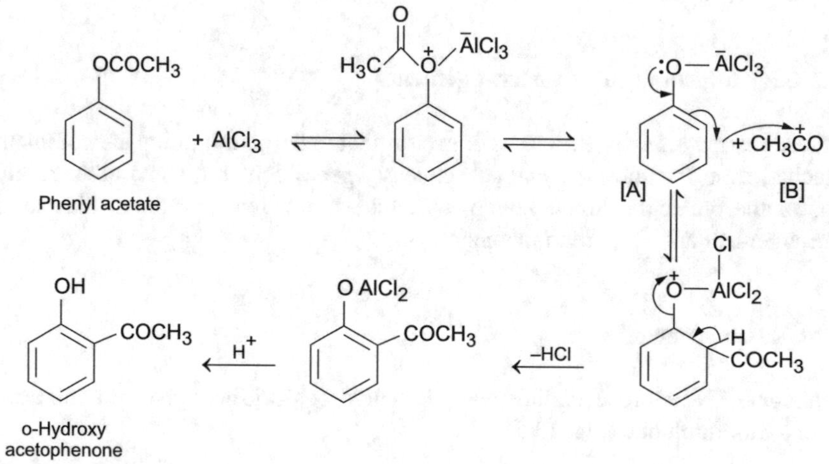

**Fig. 13.7**  Intermolecular Fries rearrangement

## Fries Rearrangement [1]

Treatment of phenylacetate with anhydrous $AlCl_3$ gives a mixture of *o*- and *p*-hydroxyacetophenone. The reaction proceeds via intermolecular rearrangement[2] as shown below (Fig. 13.7).

The *p*-isomer is formed as shown below (Fig. 13.8).

Fries rearrangement can also take place via intermolecular rearrangement[2] as shown below (Fig. 13.9).

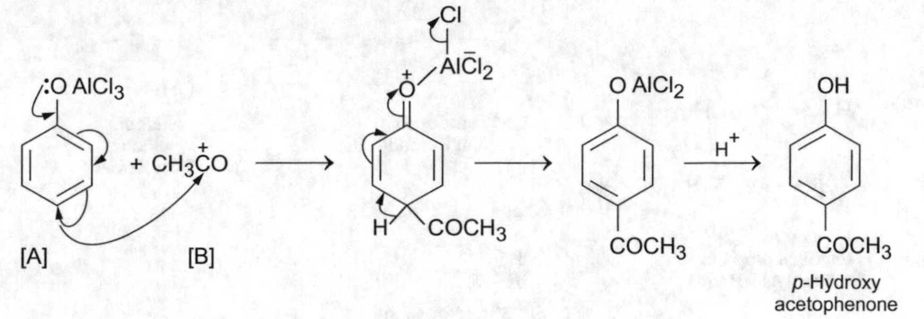

**Fig. 13.8** Formation of *p*-isomer

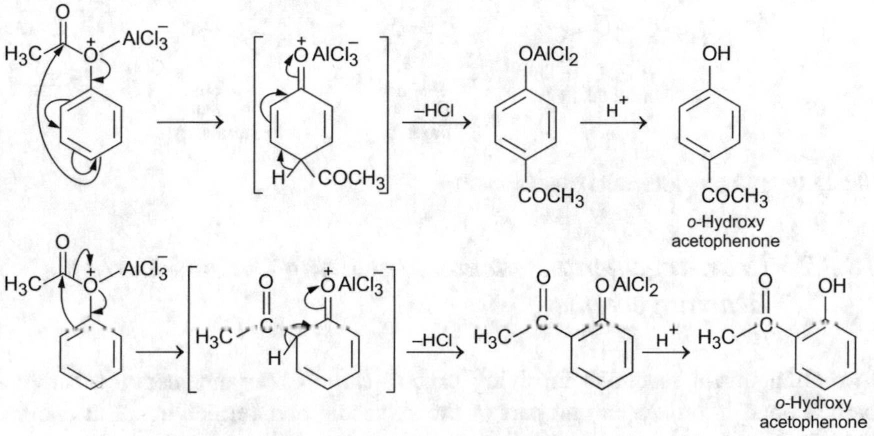

**Fig. 13.9** Intermolecular Fries rearrangement

### References

1. K. Fries and G. Fink, Ber., 1908, **41**, 427; K. Fries and W. Plaftendorf, Ber., 1910, **43**, 212. A. H. Bhatt., Chem. Rev., 1940, **27**, 429; Eftenberg, Angew Chem. Int. Ed., 1973, **12**, 776; H. H. Bhatt., Org. Rect., 1942, **1**, 342.
2. N. M. Cullinane, A. G. Evans and E. T. Lloyd., J. Chem. Soc., 1956, 2222.

### 13.2.1.3 Classification of Rearrangement Reactions Involving Carbon–Carbon, Carbon–Nitrogen and Carbon–Oxygen Rearrangements

As already stated, the rearrangement reactions can belong to one of the three groups, viz., rearrangements involving carbon–carbon rearrangements, carbon–nitrogen rearrangement or carbon–oxygen rearrangements.

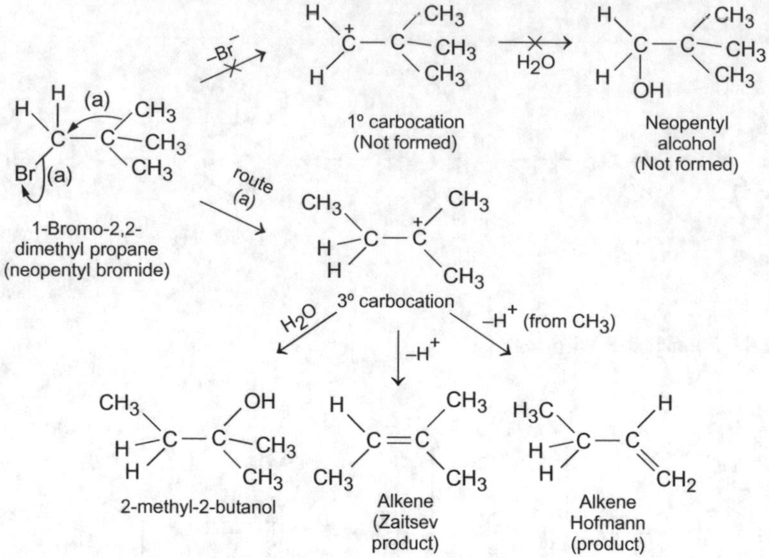

**Fig. 13.10**    Wagner–Meerwein rearrangement

## 13.2.2    Rearrangements Reactions Involving Carbon–Carbon Rearrangements

In rearrangement reactions involving carbon–carbon rearrangements, a carbon–carbon bond is broken in one part of the molecule and formed again at another part. Such rearrangements are of three types, viz., cationic rearrangements, anionic rearrangements and pericyclic rearrangements.

### 13.2.2.1    Cationic Rearrangements

Wagner–Meerwein Rearrangement

Such rearrangements involve a change in carbon skeleton via rearrangement of carbocations as intermediates and are known as Wagner–Meerwein rearrangements[1]. As an example, consider the solvolysis[2] of neopentyl bromide (1-bromo-2,2-dimethylpropane). The heterolytic cleavage of C–Br bond of 1-bromo-2,2-dimethylpropane does not give primary carbocation but gives a more stable tertiary carbocation, which is produced when a $CH_3$ group migrates from C–2 to C–1 as the C–Br bond breaks. The tertiary carbocation on reaction with water gives[3] an alcohol. Alternatively, the 3° carbocation can lose a proton to form alkene (Zaitsev product or Hofmann product depending on which proton is in loss (Fig. 13.10).

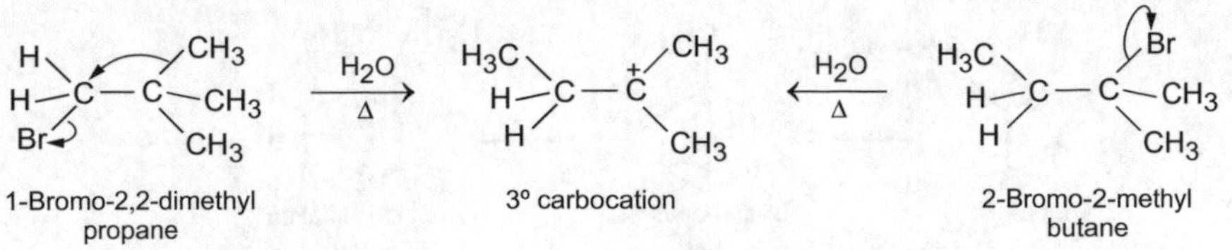


**Fig. 13.11** Solvolysis of 1-bromo-2,2-dimethyl propane and 2-bromo-2-methylbutane

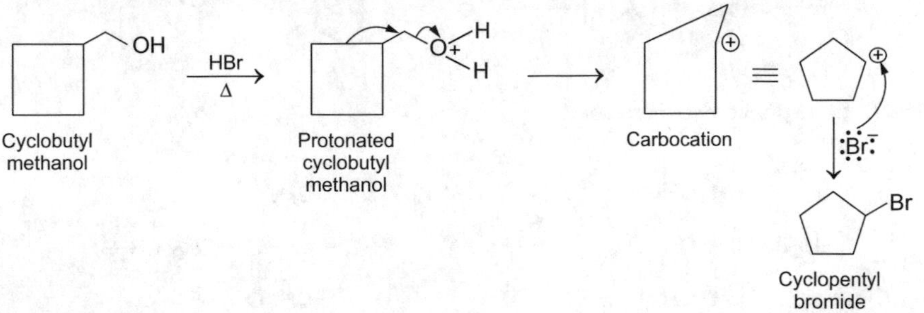


**Fig. 13.12** Ring expansion in cationic rearrangement

It is interesting to note that both 1-bromo-2, 2-dimethyl propane and 2-bromo-2-methylbutane gavc thc same products on solvolysis via the formation of the same 3° carbocation (Fig. 13.11).

Cationic rearrangements in cyclic compounds can also involve ring expansion. Thus, cyclobutylmethanol on heating with HBr gives cyclopentyl bromide. In such rearrangements, ring strain is also important. Since cyclopentane is more stable than cyclobutane, the former (cyclopentane) ring is formed (Fig. 13.12).

## Solved Problem

Acid-catalysed solvolysis of α-pinene gives borneol. Give the steps involved. Why in this case the reaction occurs via a 2° carbocation and not a 3° carbocation?

**Ans**: Various steps involved are in Fig. 13.13.

In the above sequence of reactions, the formation of a 3° carbocation is favoured by protonation of α-pinene. However, migration of a carbon group from the adjacent position gives a 2° carbocation, which is more stable than the initially formed 3° carbocation since the ring strain in the four-membered cyclobutane ring of α-pinene is relieved. Thus, the 2° carbocation reacts with water giving alcohol having carbon skeleton different from the starting α-pinene.

Other examples of cationic rearrangements are pinacol–pinacolone rearrangement and Wolf rearrangement. These are discussed below.

## References


1. G. Wagner, J. Russ. Phys. Chem. Soc., 1899, **31**, 680; Ber., 1899, **32**, 2302; H. Meerwein, Ann., 1914, **405**, 129.

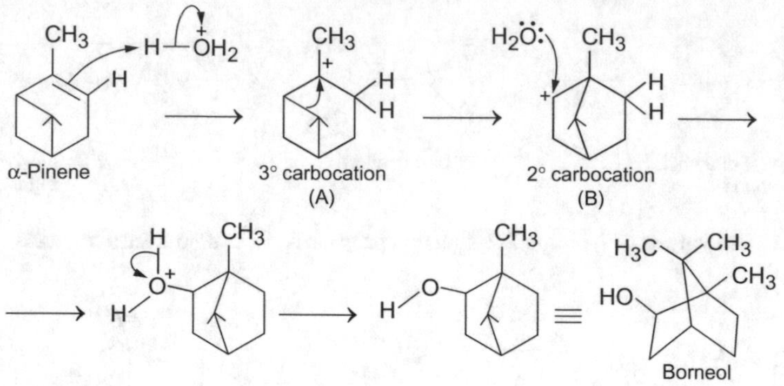

**Fig. 13.13**  α-pinene gives borneaol

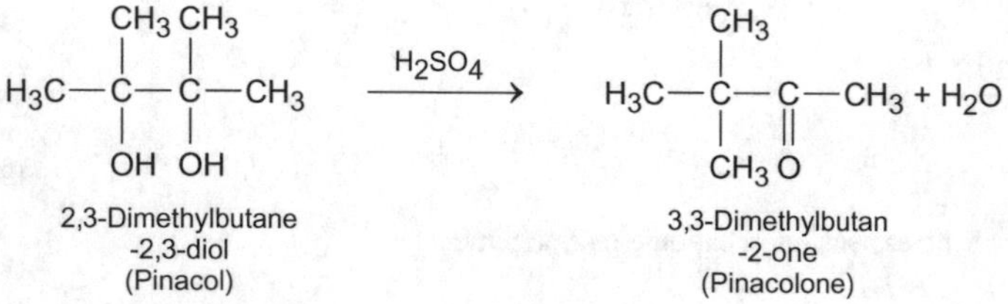

**Fig. 13.14**   Pinacol–Pinacolone rearrangement

2.   F. C. Whitmore, J. Chem. Soc., 1948, 1090; F. C. Whitmore and H. S. Rothrock, J. Am Chem. Soc., 1933, **55**, 1100; 1932, **54**, 3431.
3.   Cited in Organic Chemistry, M. A. Fox and J. K. Whilesell, Jones and Barlett., 2004, page 680.

Pinacol–Pinacolone Rearrangement

Acid-catalysed dehydration of 1, 2-diols leads to rearrangement leading to the forma-tion of a ketone. Thus, 2,3-dimethylbutane-2, 3-diol (pinacol) on treatment with hot 30% $H_2SO_4$ gives 3,3-dimethyl-2-butanone (commonly known as pinacolone). This type of acid-catalysed dehydration reaction involving rearrangement of vicinal diols (1,2-diols) is called pinacol–pinacolone rearrangement[1] (Fig. 13.14).

The reaction proceeds by the addition of a proton to one of the hydroxyl groups to give oxanium ion (1), which loses a molecule of water to give the carbocation (2). The carbocation (2), in turn, undergoes rearrangement involving the migration of a methyl group from carbon atom adjacent to the carbon bearing positive charge. The carbocation (3) so formed is stabilised by a shift of charge from carbon to oxygen to form oxonium ion (4). The driving force for 1, 2-methyl shift comes from the tendency of the initially formed carbocation intermediate (2) to form stable oxonium ion (4) (Fig. 13.15).

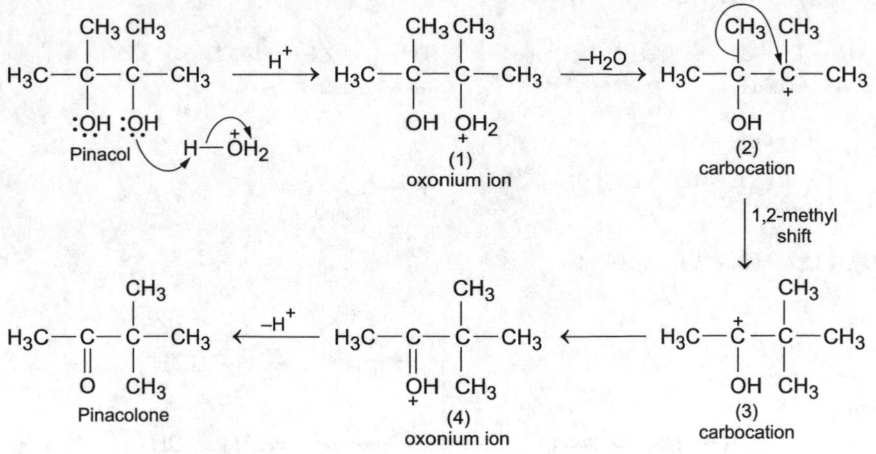

CH$_3$ CH$_3$     H$^+$     CH$_3$ CH$_3$     −H$_2$O     CH$_3$ CH$_3$

H$_3$C—C—C—CH$_3$ ⟶ H$_3$C—C—C—CH$_3$ ⟶ H$_3$C—C—C—CH$_3$

:OH :OH     OH $\overset{+}{O}$H$_2$     OH $\overset{+}{}$

Pinacol    H—$\overset{+}{O}$H$_2$     (1) oxonium ion     (2) carbocation

1,2-methyl shift ↓

CH$_3$     −H$^+$     CH$_3$     CH$_3$

H$_3$C—C—C—CH$_3$ ⟵ H$_3$C—C—C—CH$_3$ ⟵ H$_3$C—$\overset{+}{C}$—C—CH$_3$

O   CH$_3$     $\overset{+}{O}$H   CH$_3$     OH   CH$_3$

Pinacolone     (4) oxonium ion     (3) carbocation

**Fig. 13.15** Mechanism of Pinacol–Pinacolone rearrangement

OH OH     1) H$^+$     OH     1) 1,2-methyl shift     CH$_3$ O

Ph—C—C—CH$_3$ $\xrightarrow[\text{2) −H}_2\text{O}]{}$ Ph—$\overset{+}{C}$—C—CH$_3$ $\xrightarrow[\text{2) −H}^+]{}$ Ph—C—C—CH$_3$

Ph   CH$_3$       Ph   CH$_3$       Ph

2-Methyl-1,1-diphenyl propane-1,2-diol     (5) carbocation     3,3-Diphenyl propane-2-one

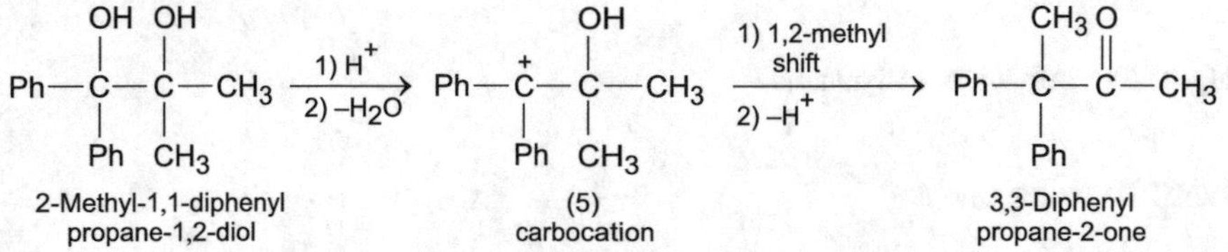

**Fig. 13.16** Pinacol–Pinacolone rearrangement

With symmetric diols (as in the case of pinacol), it does not matter which hydroxyl group gets protonated. However, in unsymmetrical diols, there is a choice as to which hydroxyl group is removed preferentially. The hydroxyl group that is lost is the one that gives a more stable carbocation. Thus, in case of unsymmetrical diols, the stability of the carbocation, relative migratory aptitude of the substituents are the reaction conditions that play an important role in deciding which product is formed. It has been found that aryl group migrates preferentially in competition with alkyl groups. The order of migratory aptitude is Ar > H > CH$_3$. Thus, as an example, in 2-methyl-1,1-diphenylpropane-1,2-diol, the OH group removed is the one on the phenyl side, so that the carbocation (5) is formed. Shift of CH$_3$ group from adjacent carbon atom in (5) followed by deprotation of the formed oxonium ion gives 3, 3-diphenylpropane-2-one. The carbocation (5) is resonance stabilised and so is preferred (Fig. 13.16).

*See also* (Sect. 7.2.4).

## Reference

1. R. Fittig, Ann., 1859, **110**, 17; 1860, **114**, 54; C.J. Collins, Quart Rev., 1960, **14**, 357.

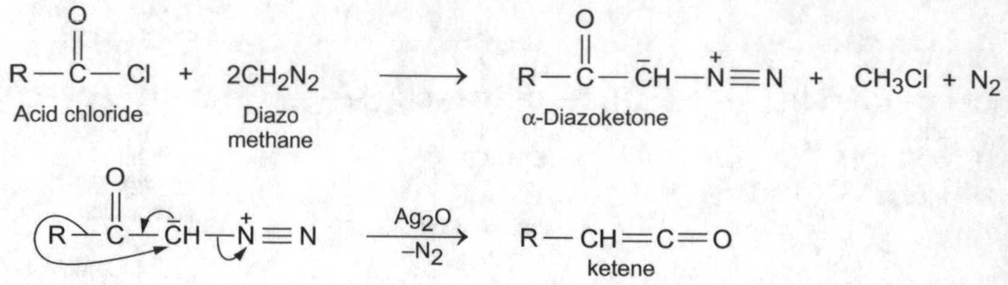

$$R-\overset{\overset{\displaystyle O}{\|}}{C}-Cl \ + \ 2CH_2N_2 \ \longrightarrow \ R-\overset{\overset{\displaystyle O}{\|}}{C}-\overset{-}{C}H-\overset{+}{N}\equiv N \ + \ CH_3Cl \ + \ N_2$$

Acid chloride        Diazo         α-Diazoketone
                    methane

$$\widehat{R\!-\!\overset{\overset{\displaystyle O}{\|}}{C}\!-\!\overset{-}{C}H\!-\!\overset{+}{N}}\equiv N \ \xrightarrow[-N_2]{Ag_2O} \ R-CH=C=O$$

ketene

**Fig. 13.17**  Wolff rearrangement

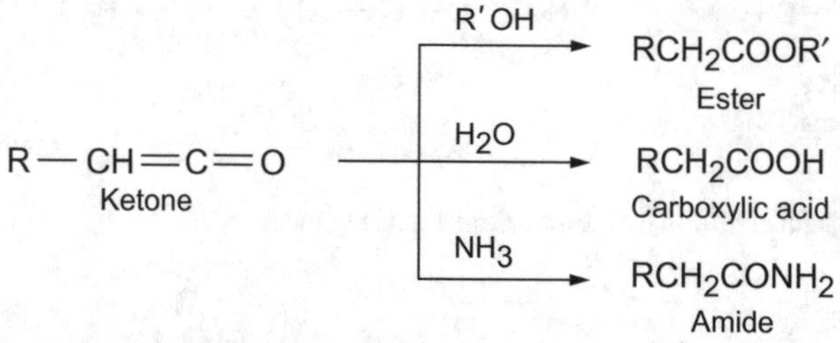

$$R-CH=C=O$$
Ketone

R′OH → $RCH_2COOR'$  Ester

$H_2O$ → $RCH_2COOH$  Carboxylic acid

$NH_3$ → $RCH_2CONH_2$  Amide

**Fig. 13.18**  Arndt–Eistert Synthesis

## Wolff Rearrangement

The α-diazoketones on reaction with silver oxide undergo rearrangement[1] with the elimination of nitrogen and form a ketene. The reaction is known as Wolff rearrangement. The α-diazoketones are obtained from acid chlorides by treatment with diazomethane (Fig. 13.17).

When the Wolff rearrangement is carried out in the presence of a nucleophile like water, alcohol or an amine, the formed ketene is converted into carboxylic acid, ester or amide, respectively. The overall reaction is called **Arndt–Eistert Synthesis**[2] (Fig. 13.18).

### References

1.   L. Wolff, Ann., 1912, **394**, 23; H. Meier, Angew. Chem. International Ed., 1975, **14, 32**.
2.   B. Eistert and F. Arndt. Chem. Ber., 1927, **60**, 1364.

### 13.2.2.2  Anionic Rearrangements

These rearrangements involving rearrangements of carbanions are much less common than those of cationic rearrangements. Such rearrangements occur when a more stable anion is formed in a reaction.

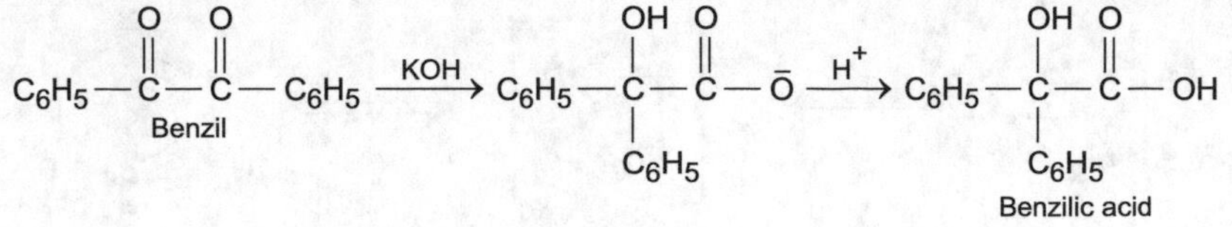

$$C_6H_5-\overset{\overset{O}{\|}}{C}-\overset{\overset{O}{\|}}{C}-C_6H_5 \xrightarrow{KOH} C_6H_5-\underset{\underset{C_6H_5}{|}}{\overset{\overset{OH}{|}}{C}}-\overset{\overset{O}{\|}}{C}-\bar{O} \xrightarrow{H^+} C_6H_5-\underset{\underset{C_6H_5}{|}}{\overset{\overset{OH}{|}}{C}}-\overset{\overset{O}{\|}}{C}-OH$$

Benzil                                                   Benzilic acid

**Fig. 13.19** Benzil–Benzilic acid rearrangement

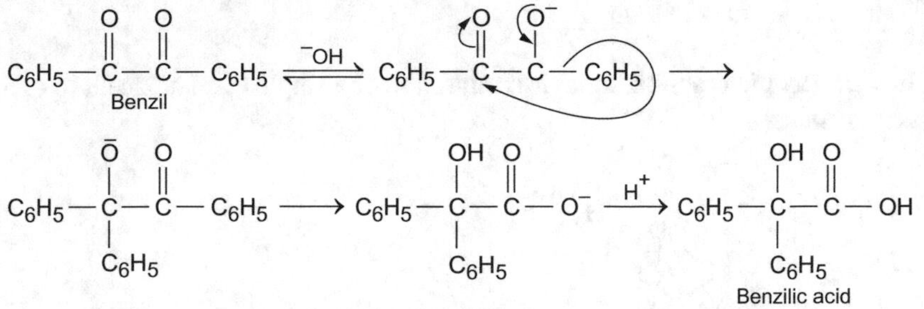

**Fig. 13.20** Mechanism of Benzil–Benzilic and rearrangement

## Benzil–Benzilic Acid Rearrangement

It consists[1] of the treatment of α-diketones (benzils) with a base. The formed salt of α-hydroxy carboxylic acid on acidification gives benzilic acid (Fig. 13.19.).

The mechanism of the reaction is as given below (Fig. 13.20).

**Reference**

1. A. W. Hoffmann, Proc. Rog. Soc. London, 1963, **12**, 576; P. Jacobson, F. Henrich and J. Klein, Ber, 1893, **26**, 688.

## 13.2.2.3 Pericyclic Rearrangements

Reactions in which the movement of electrons occurs in a concerted process are called pericyclic reactions. These reactions proceed via a cyclic transition state and bonds are broken and made at the same time; no intermediate radicals, carbocations or carbanions are formed. Common types of pericyclic reactions are electrocyclic reactions, sigmatropic reaction and cycloaddition reactions.

In **electrocyclic reaction**, a single bond is formed between the termini of a conjugated polyene system.

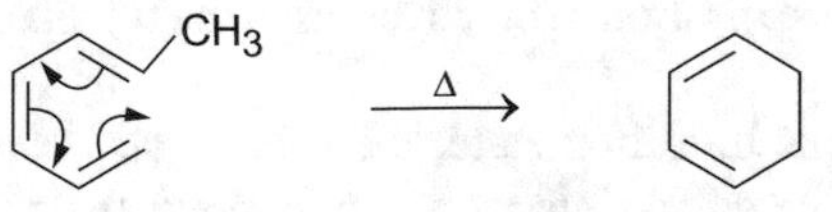

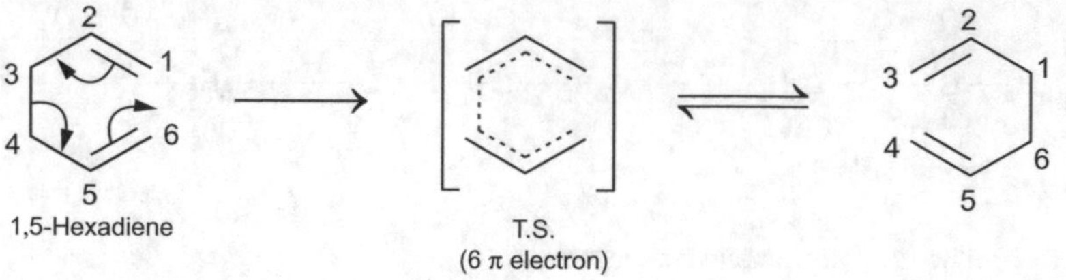

**Fig. 13.21**   Cope rearrangement

In **sigmatropic reaction**, there is a migration of a single bond adjacent to one or more π systems.

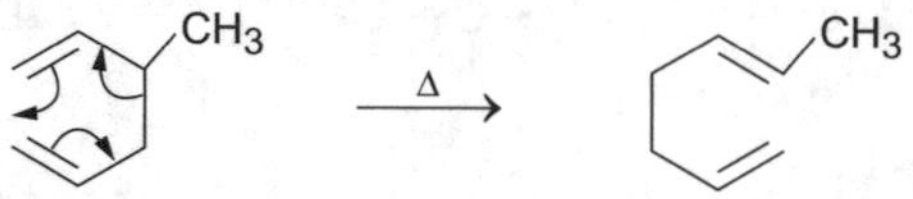

In **cycloaddition reaction**, there is a formation of a ring from the reaction of two π systems.

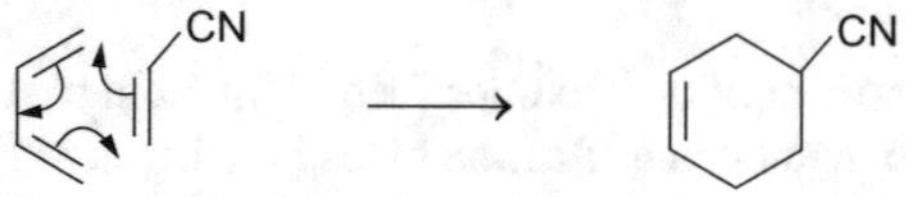

*See* also stereochemistry of Pericyclic reactions (Chap. 15).

Diels Alder Reaction

It proceeds via a transition state involving six electrons. Since Diels Alder reaction involves the combination of two starting materials, it is also referred as a cycloaddition reaction. For more details about Diels Alder reaction, see Sect. 19.3 in Chap. 19

Cope Rearrangement

This rearrangement[1] (discovered by Arthur cope of Massachusetts its Institute of Technology) proceeds via a cyclic transition state (having an array of six electrons) as in the case of Diels Alder reaction. As an example, the rearrangement of 1,5-hexadiene involves delocalisation of electrons as in the transition state of Diels–Alder reaction (Fig. 13.21).

As seen, in the cyclic transition state, the σ bond between C-3 and C-4 (in 1,5-hexadiene) breaks at the same time as a new σ bond formation between C-1 and

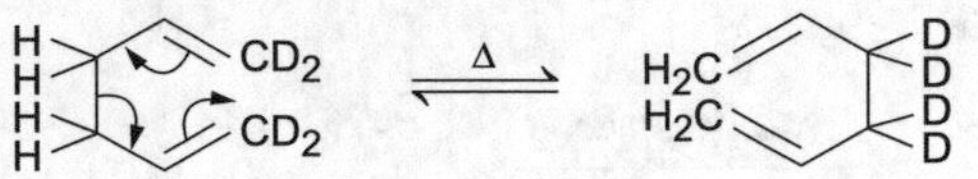

**Fig. 13.22** Degenerate Cope rearrangement

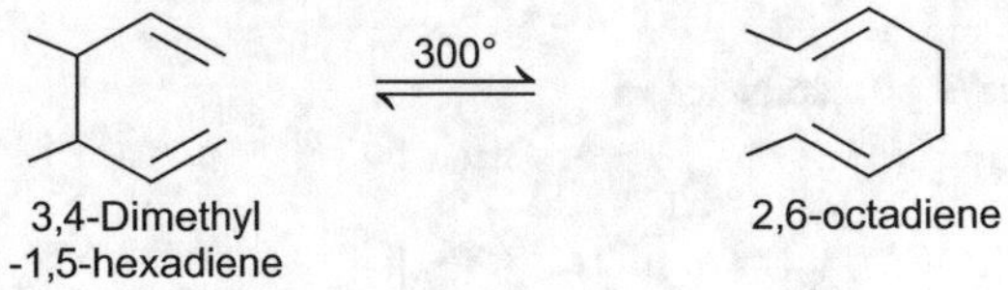

3,4-Dimethyl
-1,5-hexadiene

2,6-octadiene

**Fig. 13.23** Cope rearrangement of 3,4-dimethyl-1,5-hexadiene

C-6 takes place. Both the $\pi$ bonds in 1,5-hexadiene shift their positions (in the formed product) to form a degenerate 1,5-hexadiene. This type of reaction is called a **sigmatropic rearrangement**, and the migration of bond is called a **sigmatropic shift**. In 1,5-hexadiene, the end of the $\sigma$ bond has shifted, one end by three carbons in one direction from C-4 to C-6 and the other end by three carbons in the other direction from C-3 to C-1. So this migration is called [3, 3]-sigmatropic shift.

In the cope rearrangement of 1,5-hexadiene, the product is identical with the starting material except with respect to the identity of the individual carbon atoms. Such a process is said to be degenerate and so the Cope arrangement of 1,5-hexadiene is a degenerate rearrangement. This finds support if, in starting material, the two hydrogens on each of the terminal carbons are replaced by deuterium atoms (Fig. 13.22).

As seen, the reactant has deuterium atoms bonded to vinyl carbons, and the product has deuterium atoms located at the allylic positions. This can be confirmed by the NMR spectra of the starting and the formed product. The above reaction has demonstrated that a degenerate rearrangement has taken place.

Cope rearrangement of 3,4-dimethyl-1,5-hexadine by a 3,3-sigmatropic shift gives 2,6-octadiene (Fig. 13.23).

In the above cope rearrangement, heating at high temperature (300 °C) is necessary. This has limitations in preperative work. It has been found that the reaction can take place[2] at room temperature in presence of catalytic amount of $PdCl_2(PhCN)_2$ in THF. As an example cope, rearrangement of 2-methyl-3-phenyl-1,5-hexadiene is the presence of the catalyst that gives the dienes in 97% yield in the ratio (97: 3) (Fig. 13.24).

Thermal rearrangement of 1,5-hexadiene with a hydroxyl group at C-3 gives unsaturated aldehydes (Fig. 13.25).

In case there are hydroxy substituents at C-3 and C-4 of 1,5-hexadiene, 1,6-dicarbonyl compounds result (Fig. 13.26).

Such rearrangements of hydroxy-substituted 1,5-hexadienes are called **oxy-Cope rearrangements**[3].

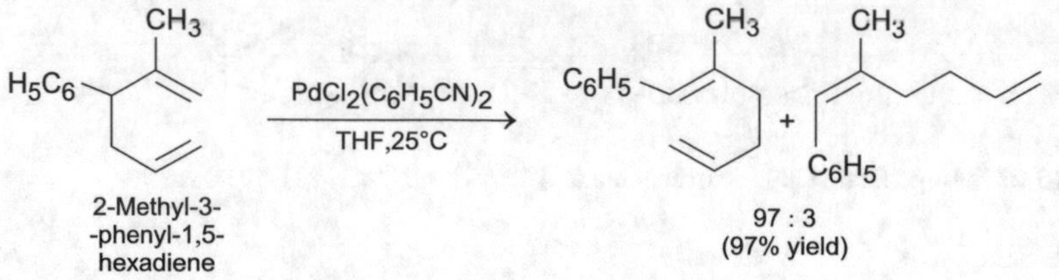

**Fig. 13.24**  Catalytic Cope rearrangement

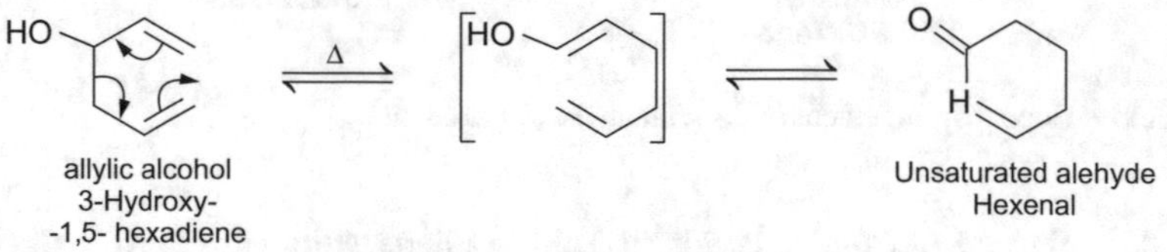

**Fig. 13.25**  Oxy cope rearrangement

**Fig. 13.26**  Oxy-Cope rearrangement

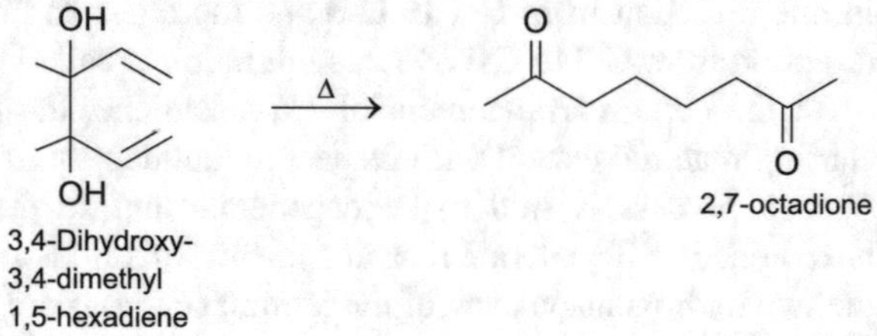

**Fig. 13.27**  Aza Cope rearrangement

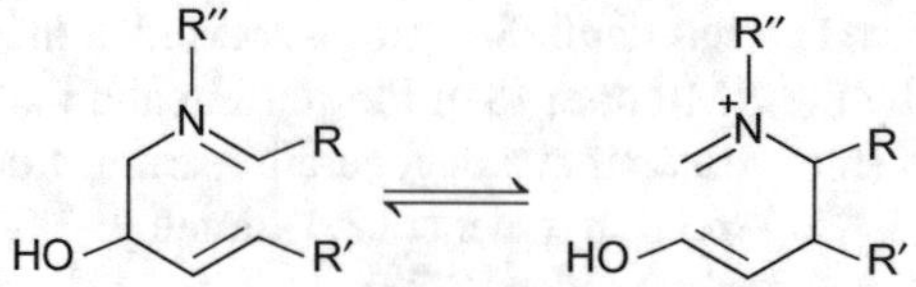

One comes across another type of Cope rearrangement called **aza-Cope rearrangement**[4]. It occurs under mild conditions. One example is given below (Fig. 13.27).

## References

1. A. C. Cope et al., J. Am. Chem. Soc., 1940, **62**, 441; S. J. Roth, Angew. Chem. Int. Ed., 1963, **2**, 115.
2. L. E. Overman and A. F. Renaldo, Tet. Lett., 1980, **24**, 3757.
3. N. Bluthe, M. Malacria and J. Gore, Tet. Lett., 1983, **27**, 1157.

**Electrocyclic Reactions**

As already mentioned, in electrocyclic reactions, a single bond is formed between the termini of a conjugated polyene system (Sect. 13.2.2.3). Thus, in an electrocyclic reaction, intramolecular interaction of both ends of a π system leads to intramolecular cyclisation. Thus, 1,3,5-hexatriene gives 1,3-cyclohexadiene.

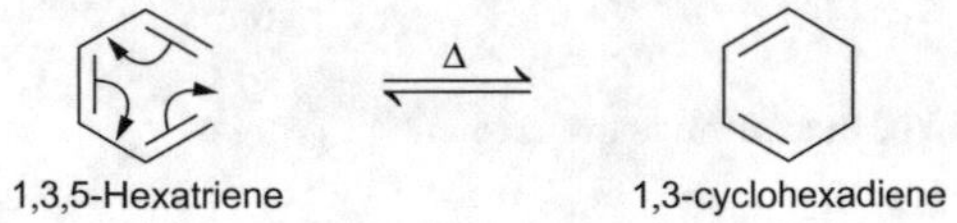

1,3,5-Hexatriene                1,3-cyclohexadiene

The electrocyclic reaction of 1,3,5-hexatriene, like Diels–Alder reaction and Cope rearrangement, also proceeds via a six-electron transition state.

## 13.2.3 Rearrangement Reactions Involving Carbon–Nitrogen Rearrangements

In rearrangement reactions involving carbon–nitrogen rearrangements, a carbon–carbon bond is broken in one part of the molecule, and a carbon–nitrogen bond is formed in another part. Some examples of this type of rearrangement include Beckmann rearrangement and Hofmann rearrangement.

### 13.2.3.1 Beckmann Rearrangement

The oximes of ketones on treatment with acidic reagent undergo rearrangement to give N-substituted amides. This rearrangement is known as Beckmann rearrangement[1].

The acidic reagents used include phosphorus pentachloride, concentrated sulphuric acid, phosphoric acid, thionyl chloride etc. A typical example is the rearrangement of benzophenone oxime to benzanilide (Fig. 13.28).

The mechanism of the reaction is as given below (Fig. 13.29).

As seen, the treatment of ketoxime with acid generates a good leaving group (–OH is converted into $H_2O^+$ with $H_2SO_4$, $–OPCl_4$ with $PCl_5$ and $–OSOC_6H_5$ with $C_6H_5SO_2Cl$) on the N atom. Loss of the leaving group generates an electron-deficient species; this is accompanied by the migration of a group from the adjacent carbon

**Fig. 13.28** Beckmann Rearrangement

$$\underset{Ph}{\overset{Ph}{>}}C=NOH \xrightarrow{H^+} PhCONHPh$$

Benzophenone oxime                Benzanilide

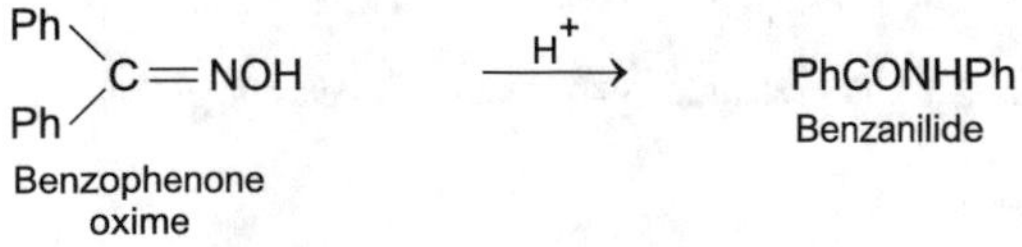

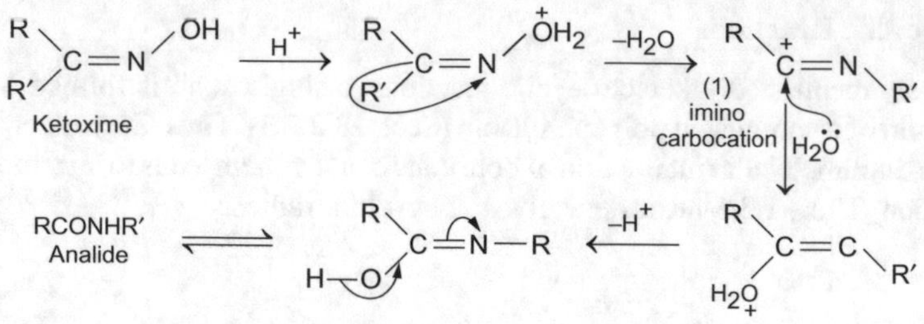

**Fig. 13.29**   Mechanism of Beckmann rearrangement

to the electron-deficient N to give iminocarbocation (1), which reacts with water to give an anilide as the final product. The driving force for the 1,2-alkyl shift is the formation of more stable carbocation (1). It has been found that the rate of the reaction is retarded if an electron withdrawing group (e.g., *p*-nitro, *p*-chloro) is present on the migrating group, and the rate is accelerated if electron-donating groups (e.g., *p*-CH₃ and *p*-OCH₃) are present on the migrating group. The rearrangement is highly stereospecific. It is found that the group anti to the oxime hydroxyl group always migrates irrespective of the relative migratory aptitude of the two groups. Some examples are given below (Fig. 13.30).

The oximes as we know exhibit geometrical isomerism. The product obtained depends on whether syn or anti-oxime is used.

The reaction of oxime with $PCl_5$ is represented below (Fig. 13.31).

The chiral groups migrate intramolecularly with retention of configuration. So it is most likely that the migrating group is never completely detached from the remainder of the molecule. As an example, (+)-methyl-3-heptyl ketoxime is converted by Beckmann rearrangement into 3-acetamido heptane with retention of configuration[3] (Fig. 13.32).

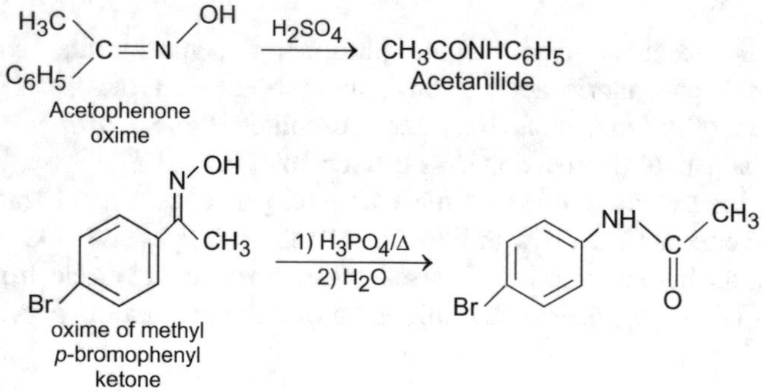

**Fig. 13.30**   Examples of Beckmann rearrangement

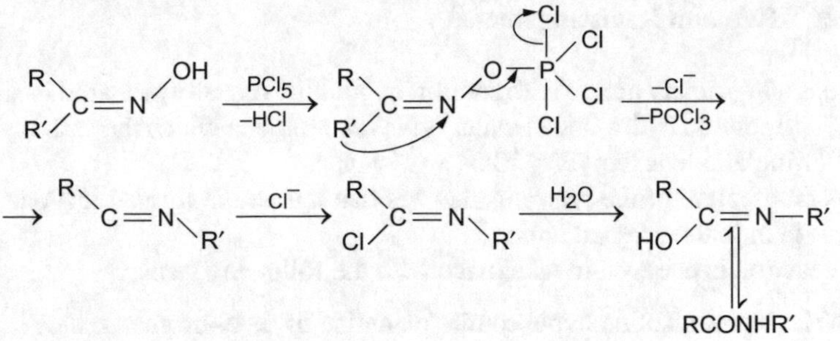

**Fig. 13.31**  Beckmann rearrangement using PCl$_5$

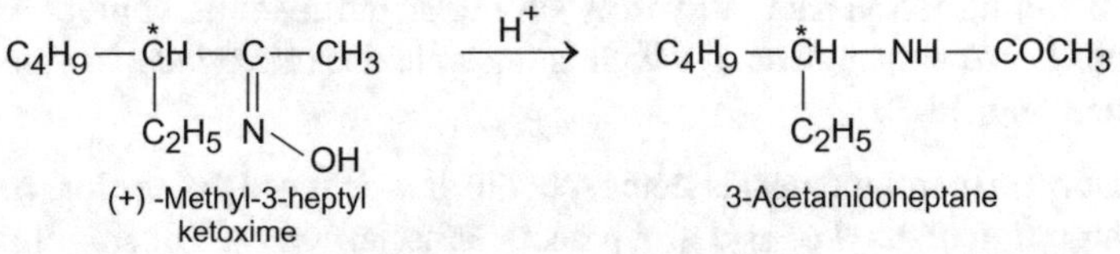

$$C_4H_9—\overset{*}{C}H—\underset{\underset{OH}{\overset{\parallel}{N}}}{\overset{C_2H_5}{\underset{|}{C}}}—CH_3 \xrightarrow{H^+} C_4H_9—\overset{*}{C}H—NH—COCH_3$$

(+) -Methyl-3-heptyl
ketoxime

3-Acetamidoheptane

**Fig. 13.32**  Beckmann reagent proceeds with retention of configuration

It is well known that the Beckmann rearrangement of cyclohexanone oxime gives caprolactam[4], a percussor of nylon-6. This method is used on a commerical scale (Fig. 13.33).

In the above reaction, ammonia is used to neutralise H$_2$SO$_4$ and (NH$_4$)$_2$SO$_4$, a fertiliser is obtained as a byproduct.

## References

1.  W.Z. Heldt and L.G. Donaruma, Org. React, 1960, XI, Chapt 1.
2.  E. Beckmann, Ber, 1886, **19**, 988 M.J.S. Dewar. The Electronic Theory of Organic Chemistry, Clarendon Press., 1949, 219, 1.
3.  J. Kenyon and D.P. Young, J. Chem. Soc. (London), 1941, 263.
4.  P.E. Eaton, J. Org. Chem., 1973, **38**, 4071.

**Fig. 13.33**  Synthesis of caprolactam

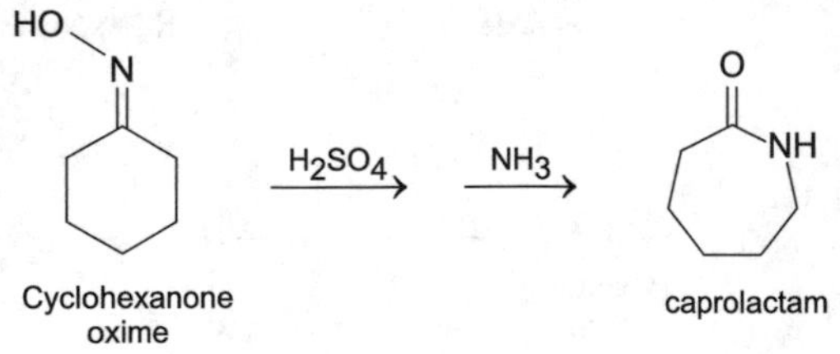

Cyclohexanone
oxime

caprolactam

### 13.2.3.2    Hofmann Rearrangement[1]

Treatment of a primary amide with sodium hypohalite (usually prepared in situ from halogen and sodium hydroxide in water) gives an amine in which the carbonyl group of the starting amide is lost (Fig. 13.34).

A special feature of the rearrangement is that the amine formed has one carbon atom less than in the original amide.

The reaction is believed to take place with the following steps:

1.  The action of alkaline hypobromite on amide gives N-bromoamide.
2.  Loss of acidic hydrogen from N-bromamide by basic hydroxide ion gives the nitrogen anionic species, which being unstable loses bromide ion with simultaneous migration of arylor alkyl group from the adjacent C atom to the N.
3.  The resulting isocyanate is hydrolysed under the reaction conditions to give N-substituted carbamic acid, which is unstable decarboxylates to give primary amines (Fig. 13.35).

The Hofmann rearrangement is stereospecific. It is believed that the loss of halogen ion and migration of alkyl or aryl group occur simultaneously. The migrating group does not break away from the carbon till it has started to attach itself to nitrogen. So, it is believed that the reaction proceeds via a transition state in which the migratory group is partially bonded to both the migrating origin and the migration terminus. In fact optically active amide undergoes Hofmann rearrangement with complete

$$RCNH_2 + Br_2 + 4KOH \longrightarrow RNH_2 + 2KBr + K_2CO_3 + H_2O$$

Primary amide                                    Primary amine

**Fig. 13.34**    Hofmann rearrangement

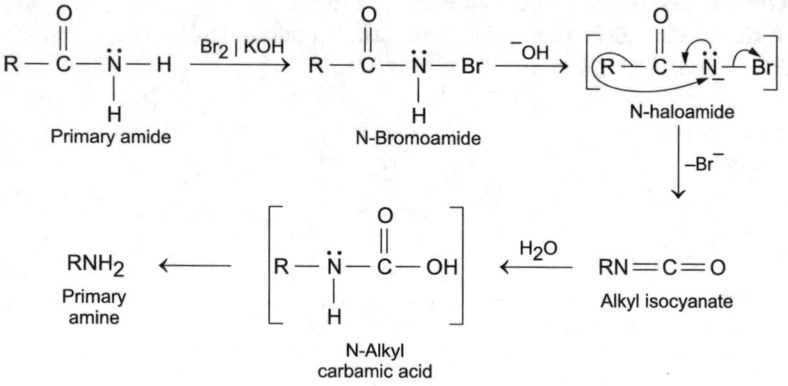

**Fig. 13.35**    Mechanism of Hofmann rearrangement

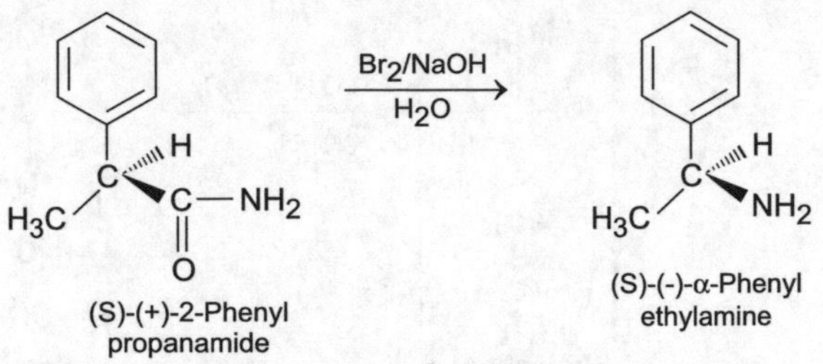

**Fig. 13.36**  Stereospecific Hofmann rearrangement

retention of configuration. Thus, the rearrangement is stereospecific. As an example, (S)-(+)-2 phenyl propanamide gives 96% optically pure (S)-(−)-α-phenyl ethylamine (Fig. 13.36).

**Reference**

1.  E. S. Wallis and J. F. Lane, Organic Reactions, 1949, **3**, 267; A. W. Hofmann, Ber., 1881, **14**, 2725.

### 13.2.4  Rearrangement Reactions Involving Carbon–Oxygen Rearrangements

In these rearrangements, a carbon–carbon bond is broken in one part of the molecule, and a carbon–oxygen bond is formed in another part. Some examples of this type of rearrangement include Baeyer–Villiger oxidation and Claisen rearrangement.

#### 13.2.4.1  Baeyer–Villiger Oxidation[1,2]

It involves the oxidation of ketones with hydrogen peroxide or with peracids (RCO$_3$H) to give esters. An example is the Baeyer–Villiger oxidation of acetophenone with perbenzoic acid to give phenyl acetate (Fig. 13.37).

The peracids used include trifluoroacetic acid, per benzoic acid, performic acid, peracetic acid, *m*-chloroperbenzoic acid (*m*-CPBA).

The reaction takes place by protonation of the carbonyl oxygen of the ketone. The addition of peracid to the protonated ketone gives a tetrahedral intermediate

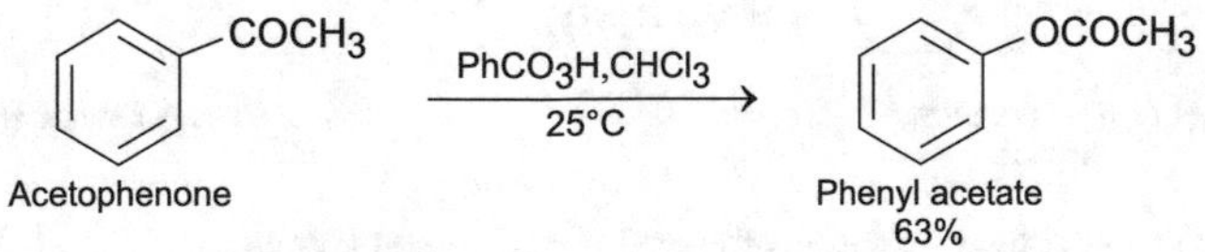

**Fig. 13.37**  Baeyer–Villiger oxidation

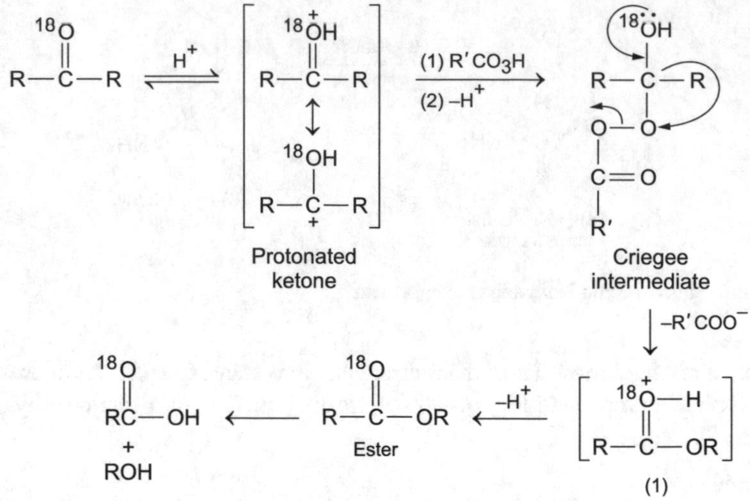

**Fig. 13.38**  Mechanism of Baeyer–Villiger Protonated ester oxidation

(Criegee intermediate). This is followed by the elimination of carboxylate anion and migration of R to the electron-deficient oxygen occurring simultaneously. The resulting protonated form of ester (1) loses a proton to give ester (Fig. 13.38).

As seen, the reaction involves migration of aryl or alkyl group from adjacent carbon to oxygen. As the leaving group $(R'COO^-)$ departs, partially, the oxygen develops a positive charge and 1,2-alkyl shift from adjacent carbon to oxygen takes place. The loss of $RCOO^-$ and migration of R are concerted. This finds support in using $^{18}O$-labelled ketone. It is found that the carbonyl oxygen of the ketone becomes the carbonyl oxygen of the ester, and the ester has the same $^{18}O$ content as the starting ketone.

The order of preference for the migration among alkyl groups follows the order $3°$ $> 2° > 1° > CH_3$. The aryl group migrates in preference to $1°$ alkyl groups and methyl. Among the aryl groups, migration is favoured by electron-releasing substituents. The migratory aptitude in aryl group is of the order $p\text{-}CH_3OC_6H_4 > C_6H_5 > p\text{-}NO_2C_6H_4$. As an example, Baeyer–Villiger oxidation of phenyl $p$-nitrophenyl ketone gives only phenyl $p$-nitrobenzoate (Fig. 13.39).

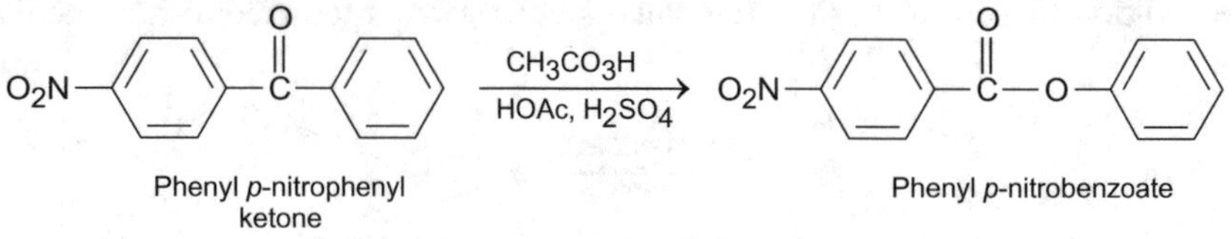

**Fig. 13.39**  Baeyer–Villiger oxidation of phenyl $p$-nitrophenyl ketone

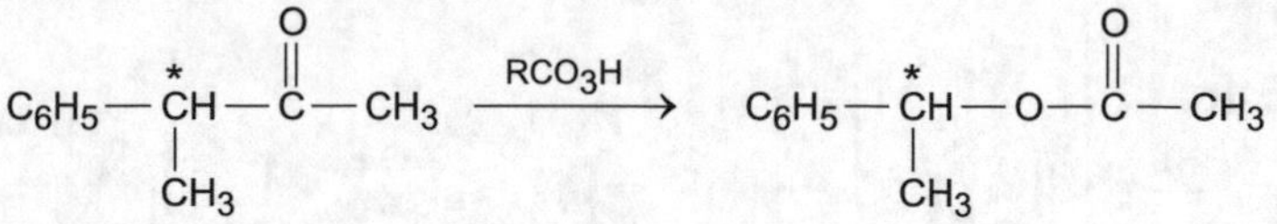

**Fig. 13.40**  Stereospecific Baeyer–Villiger oxidation

The Baeyer–Villiger oxidation is stereospecific. In case of chiral ketones, the reaction proceeds with complete retention of configuration. One example is given below (Fig. 13.40).

As seen, in Baeyer–Villiger oxidation, an oxygen atom is inserted between the carbonyl group and the $\alpha$-carbon of the ketone to give an ester.

## References

1.  A.V. Baeyer and V. Villiger., Ber., 1899, **32**, 13625; 1900, **33**, 858.
2.  H.O. House, Modern synthetic Readions, 2nd Edn., Benjamin, Menlo Park, New York, 1972, P, 321) W.D. Emmons and G.B. Lucas, J. Am. Chem. Soc., 1955, **77**, 2287.

### 13.2.4.2  Claisen Rearrangement

Allyl–aryl ethers on heating give o- or p-allylphenols. This is an example of sigmatropic rearrangement and is called Claisen rearrangement[1]. The alkyl group migration takes place from oxygen to the ring at *ortho* position (Fig. 13.41).

Claisen rearrangement is a [3,3]-sigmatropic rearrangement and proceeds via a cyclic transition state. The formed cyclohexadienone intermediate regains aromatic nucleus by tautomerisation yielding the rearranged product. The Claisen rearrangement is, thus, a cyclic intramolecular rearrangement (Fig. 13.42).

On the basis of [14]C labelling studies, it has been shown that during migration to o-position, the position of [14]C atom in the allyl group is inverted (Fig. 13.43).

In case, both the o-positions are occupied, p-substituted phenol results via two successive shifts of the allyl group. In this case, there is double inversion of the position of the [14]C label in the allyl group (Fig. 13.44).

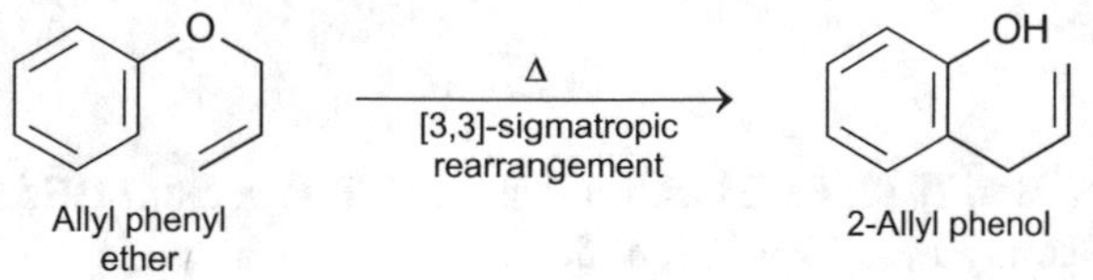

Allyl phenyl
ether

2-Allyl phenol

**Fig. 13.41**  Claisen rearrangement

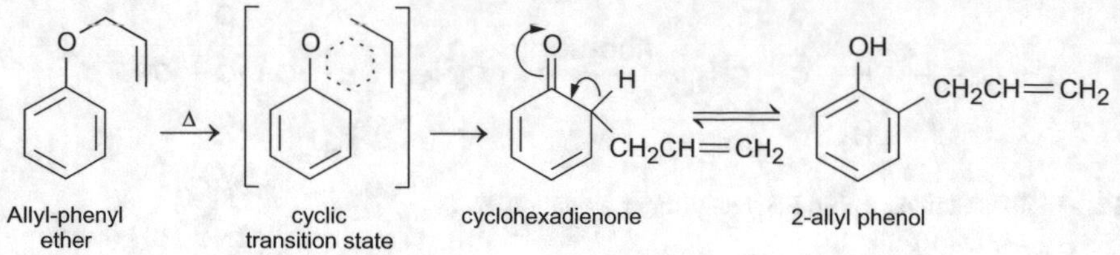

**Fig. 13.42**   Mechanism of Claisen rearrangement

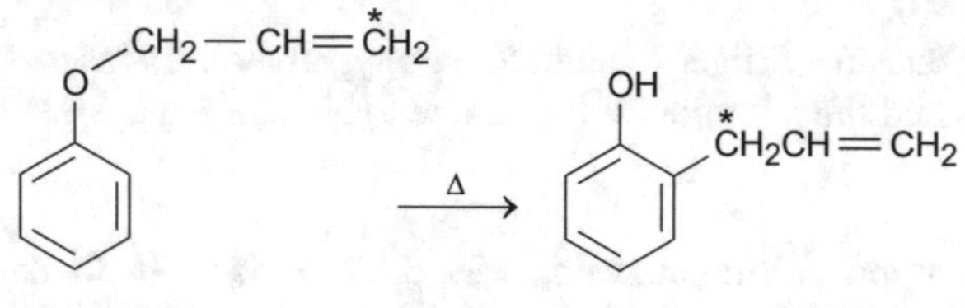

**Fig. 13.43**   Inversion of allyl group during Claisen rearrangement

4-Allyl-2,6-dimethyl phenol

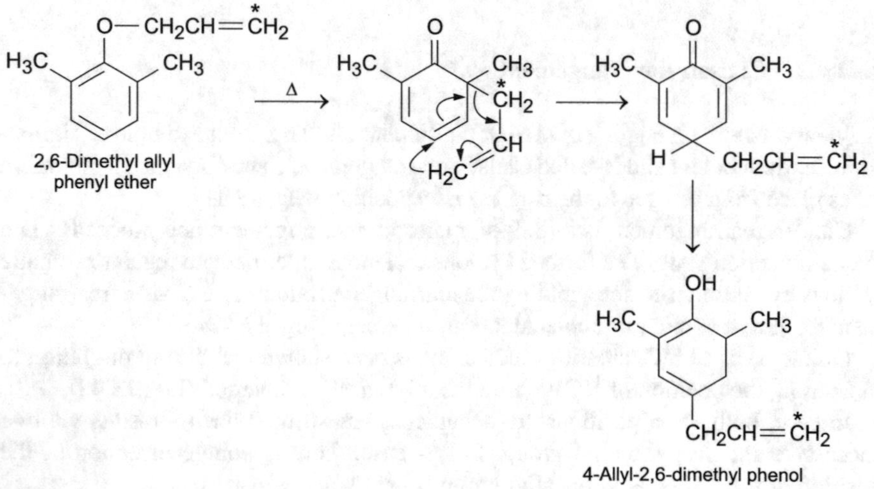

**Fig. 13.44**   Claisen rearrangement of 2, 6-dimethylallyl phenyl ether

## Reference

1.   L. Claisen, Ber, 1912, **45**, 3175; L.Claisen, E. Tietze, Ber., 1925, **58**, 275. D.S.
     Tabell., Chem. Revs., 1940, **27**, 495.

## 13.2.5 *Rearrangement Reactions*

A number of rearrangements belonging to various types have, so far, been discussed. Following are given some other rearrangement reactions.

### 13.2.5.1 Baker–Venkataraman Rearrangement[1]

The base-catalysed rearrangement of o-acyloxyketones to β-diketones is known as Baker–Venkataraman rearrangement (Fig. 13.45).

The reaction takes place by abstraction of a proton from acyl group of o-acyloxyketone. The resulting carbanion undergoes cyclisation followed by ring opening to give β-diketones (Fig. 13.46).

β-diketones are important intermediates for the synthesis of 2-methyl-chromones and flavones.

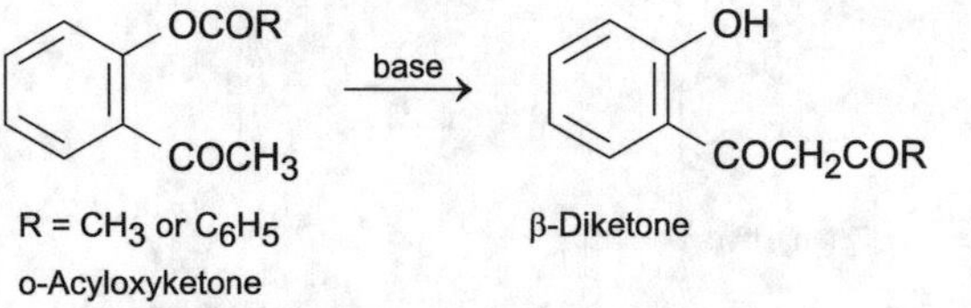

R = CH$_3$ or C$_6$H$_5$

o-Acyloxyketone

β-Diketone

**Fig. 13.45** Baker–Venkataraman rearrangement

Carbanion

β-Diketone

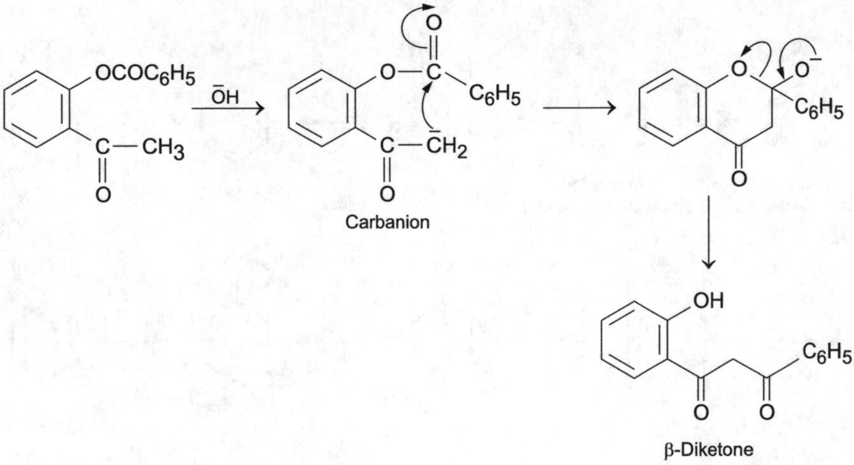

**Fig. 13.46** Mechanism of Baker–Venkataraman rearrangement

**Reference**

1.   W. Baker, J. Chem. Soc., 1933, 1381; H.S. Mahl, K. Venkataraman, J. Chem. Soc., 1934, 1767.

### 13.2.5.2   Bamberger Rearrangement

N-Phenyl hydroxylamine on treatment with strong acid undergoes rearrangement[1] to give *p*-aminophenol. This is known as Bamberger rearrangement[1] (Fig. 13.47).

The reaction is found to be intermolecular as shown by rearrangement in the presence of $^{18}O$-labelled[2] water (solvent) as shown below (Fig. 13.48).

**References**

1.   H. J. Shine, Aromatic Rearrangement, Elsevier, New York, 1967, p.182-190.
2.   H.F. Heller, E.D. Hughes and C.K. Ingold, Nature, 1951, 168, 909.

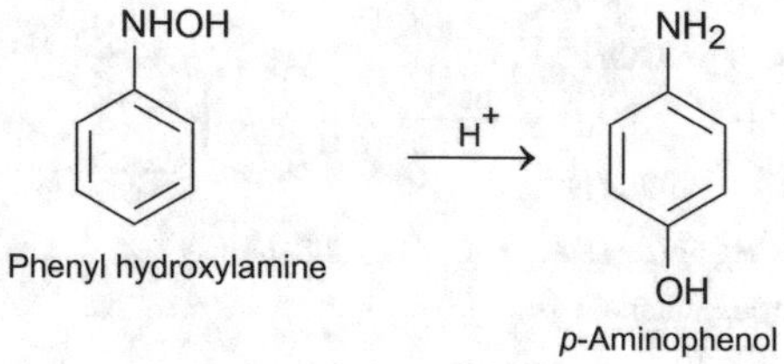

**Fig. 13.47**   Bamberger rearrangement

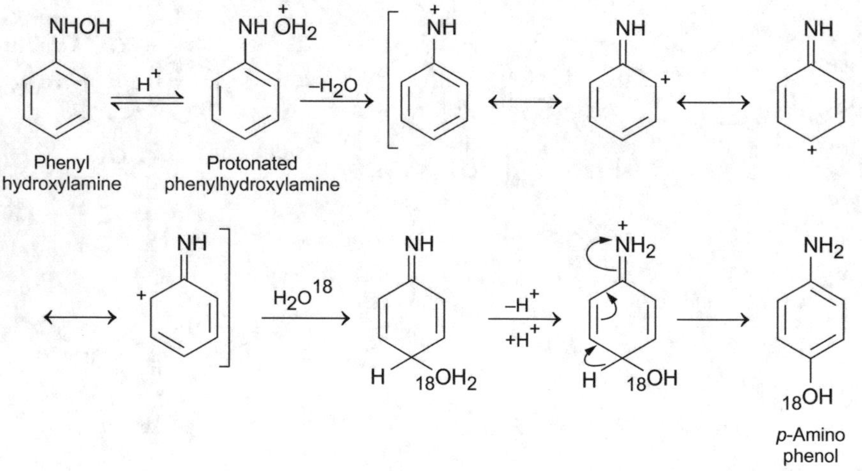

**Fig. 13.48**   Mechanism of Bamberger rearrangement

### 13.2.5.3  Benzidine Rearrangement

Hydrazobenzene on treatment with acid undergoes rearrangement to give 4, 4'-diamino biphenyl (p-benzidine). This is known as benzidine rearrangement[1] (Fig. 13.49).

The reaction involves intermolecular rearrangement as given below (Fig. 13.50).

The above mechanism finds support in the observation that a mixture of two different hydrazo benzenes on subjecting to benzidine rearrangement gives a mixture of only two benzidines and no cross-coupling product is obtained.

In case the hydrazobenzene contains a para substituent, the product is p-amino diphenylamine (semidene rearrangement).

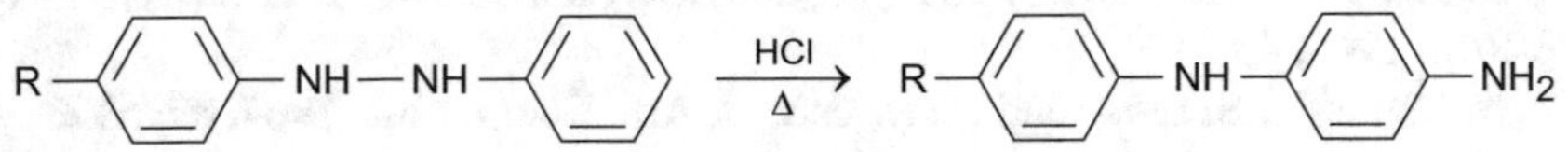

### Reference

1.  A.W. Hoffmann, Proc. Roy. Soc. London, 1863, **12**, 576; T. Shlradsky and S. Auramovki Grisaru, J. Het. Chem, 1980, **17**, 189.

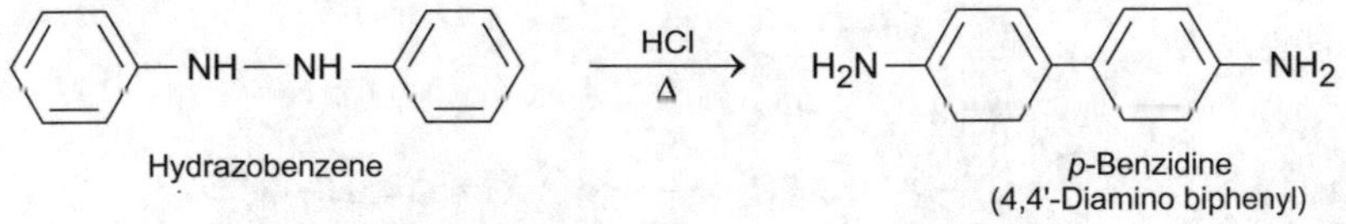

Fig. 13.49  Benzidine rearrangement

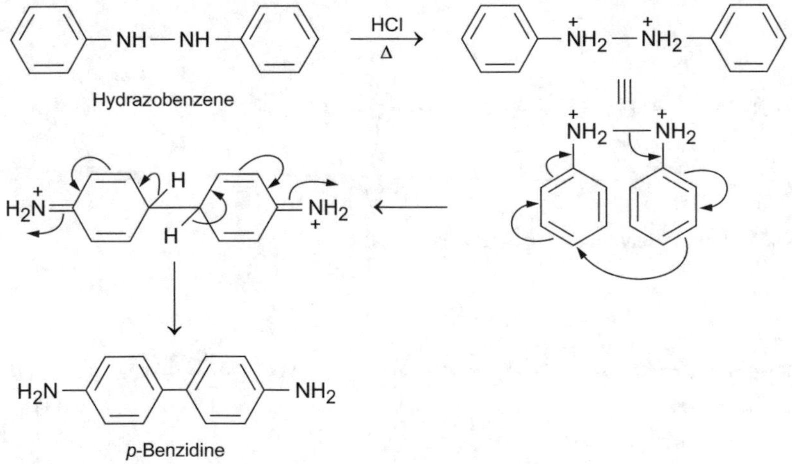

Fig. 13.50  Mechanism of benzidine rearrangement

### 13.2.5.4  Curtius Rearrangement

Rearrangement of acid azides (obtained by the action of sodium azide on an acid chloride) by heating in non-aqueous solvents such as chloroform gives isocyanates. This reaction is known as Curtius rearrangement[1] (Fig. 13.51).

The reaction is believed to proceed via a concerted rearrangement (Fig. 13.52).

However, in case tertiary alkyl azide is used, the Curtius rearrangement goes via a nitrene intermediate[2] to give imines (Fig. 13.53).

The isocyanides are important intermediates for the synthesis of 1° amines (Fig. 13.54).

### References

1.  T. Curtius, J. Prakt. Chem., 1894, **50**, 275; J.H. Saunders, R. Slocombe, J. Chem. Rev., 1948, **43**, 203.
2.  W. Lwowski, S. Linke and G. T. Tisue, J. Am. Chem. Soc., 1967, **89**, 6308.

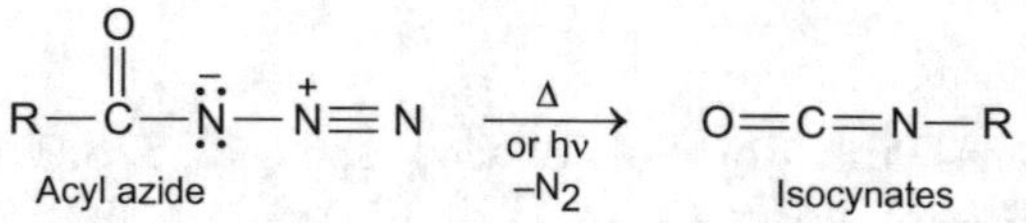

**Fig. 13.51**  Curtius rearrangement

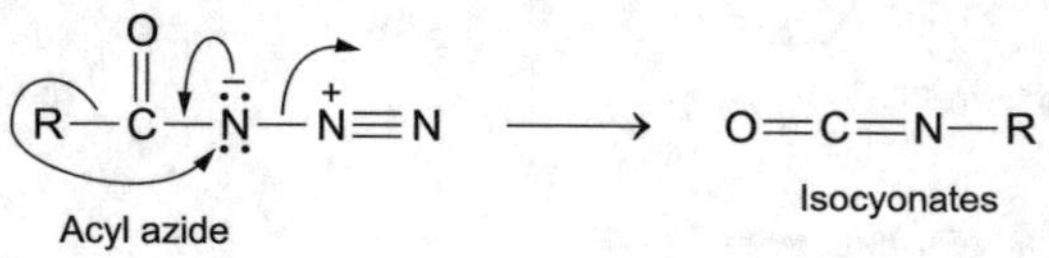

**Fig. 13.52**  Mechanism of Curtius rearrangement

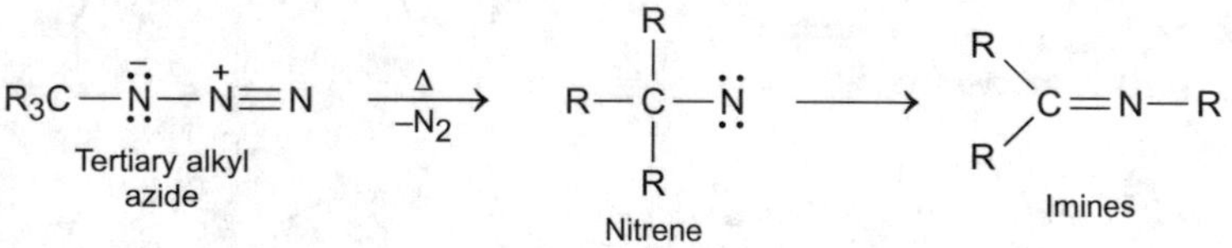

**Fig. 13.53**  Curtius rearrangement of tert. alkyl azides

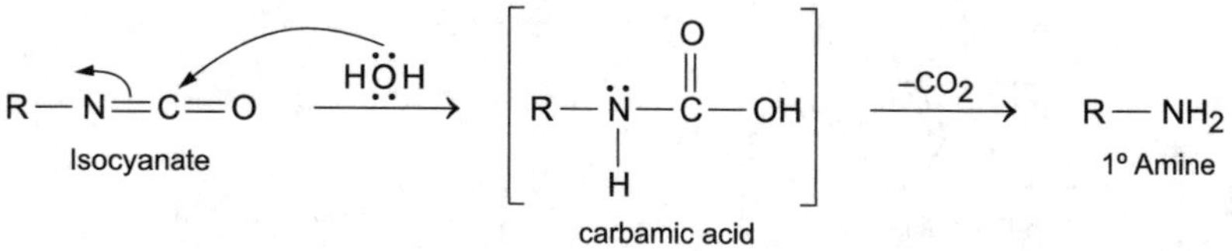

**Fig. 13.54**  Synthesis of 1° amines from isocyanates

### 13.2.5.5 Fischer–Hepp Rearrangement

N-Alkyl-N-nitroso aryl amines (obtained by nitrosation of N-alkylamines in the presence of $NaNO_2$ | HCl at 0 °C) on treatment with HCl undergo rearrangement to give the *p*-nitroso product (Fig. 13.55).

On the basis of [15] N labelled studies, it has been shown that the mechanism of the reaction is intermolecular[2].

This rearrangement has been used for the synthesis of 4-nitroso-N-alkylanilines (Fig. 13.56).

### References

1. O. Fischer and E. Hepp, Chem. Ber, 1886, **19**, 2991.
2. Williams, Tetrahedron, 1975, **31**, 1343; J. Chem. Soc. Perkin Trans II, 1975, 655; 1982, 801.

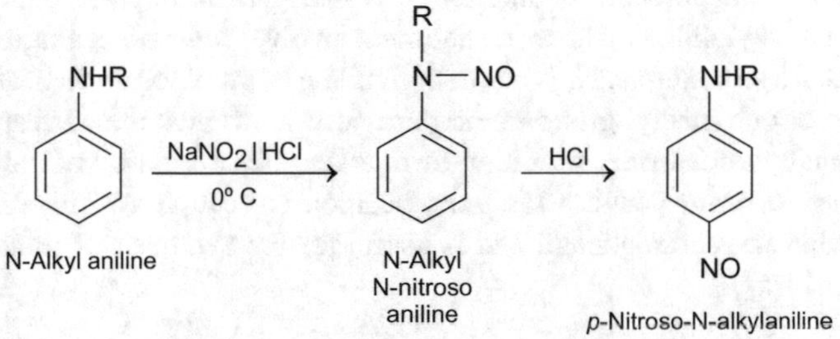

**Fig. 13.55** Fischer–Hepp rearrangement

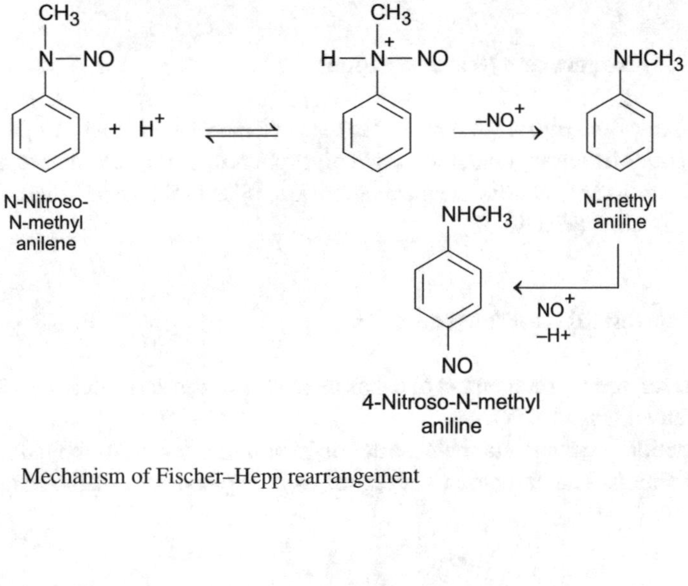

**Fig. 13.56** Mechanism of Fischer–Hepp rearrangement

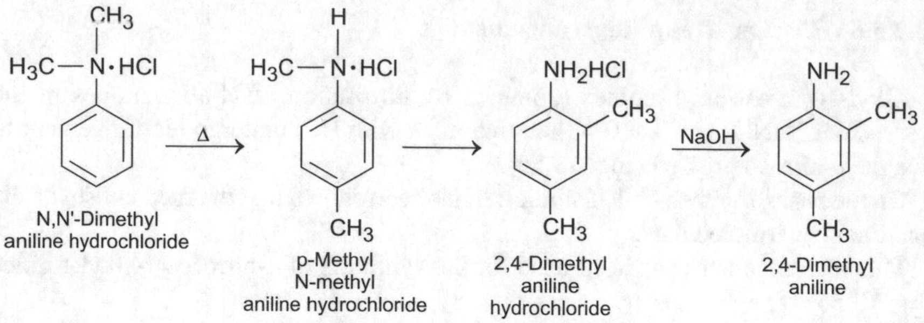

**Fig. 13.57**   Hofmann–Martius rearrangement

### 13.2.5.6   Hofmann–Martius Rearrangement[1]

The thermal rearrangement of N-alkyl or N, N'-dialkyl aniline hydrochlorides gives ortho and para alkylanilines; the rearrangement involves intermolecular migration of alkyl groups. As an example, N, N'-dimethylaniline hydrochloride on heating results in migration of one methyl group to the para position to give the corresponding N-methylaniline hydrochloride, which, in turn, undergoes migration of the remaining methyl group to *ortho* position (as *para* position is occupied). This is known as Hofmann–Martius rearrangement and is useful for the synthesis of homologues of aniline (Fig. 13.57).

### Reference

1.   A.W. Hofmann and C.A. Martius, Ber., 1971, **4**, 742; A.W. Hofmann, Ber., 1972, **5**, 720.

### 13.2.5.7   Hydroperoxide Rearrangement

Hydroperoxides undergo rearrangement in the presence of acid. For example, cumene hydroperoxide on rearrangement with acid gives phenol. This is a commercial method for the preparation of phenol. The mechanism is similar to Baeyer–Villiger oxidation and is given below (Fig. 13.58).

### 13.2.5.8   Favorskii Rearrangement[1, 2]

It is base-catalysed rearrangement of α-haloketones (bromo and chloro) to carboxylic acid derivatives (Fig. 13.59).

   The reaction proceeds via the abstraction of α-hydrogen from the haloketone to give a carbanion. The carbanion undergoes S$_N$2 displacement of the halide ion to

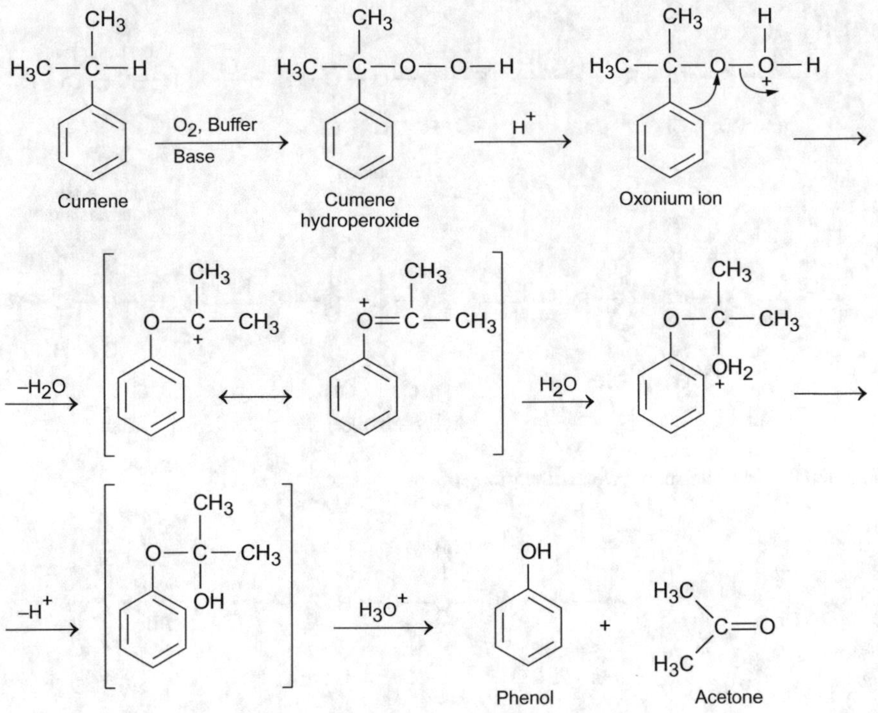

**Fig. 13.58** Mechanism of hydroperoxide rearrangement

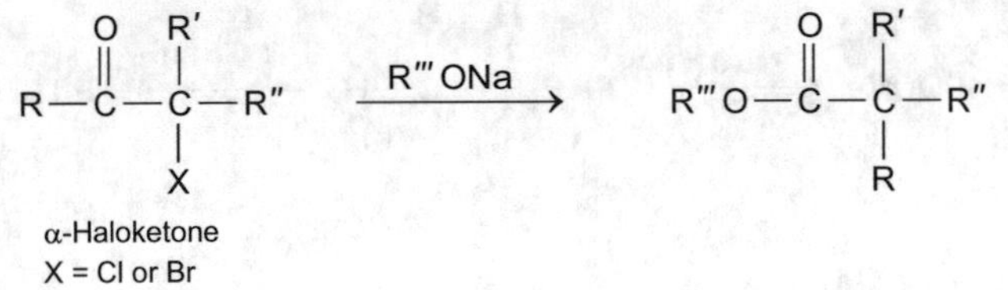

**Fig. 13.59** Favorskii rearrangement

give a cyclopropanone intermediate. Its ring opening under the reaction conditions gives a more stable carbanion, which accepts a proton from the solvent to give the ester (Fig. 13.60).

The above mechanism finds support in the observation that the isomoric haloketones A and B on Favorskii rearrangement give the same ester.

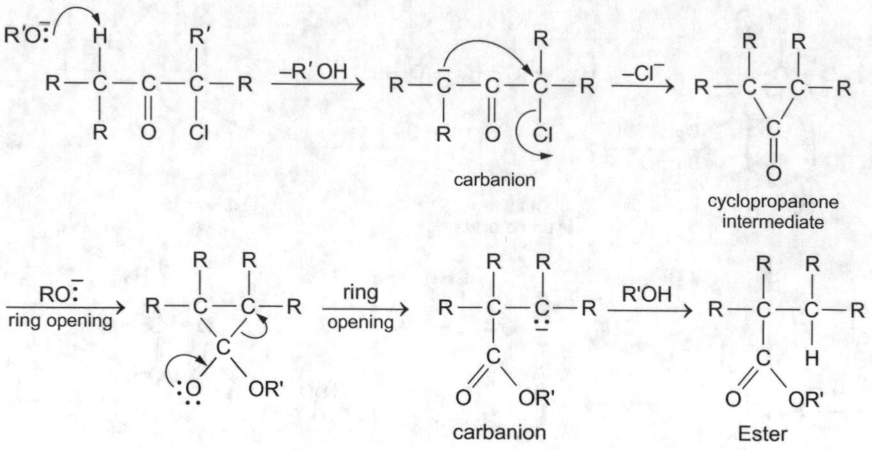

**Fig. 13.60**  Mechanism of Favorskii rearrangement

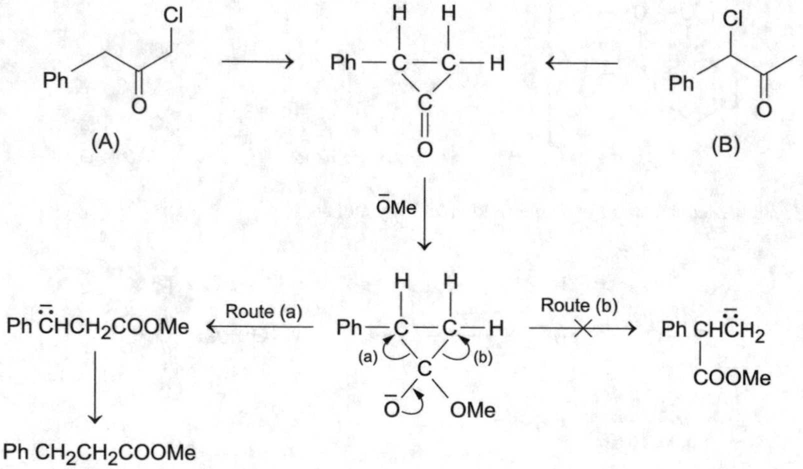

In symmetrical cyclopropanone intermediate, the two α-carbons being equivalent, the ring opening can occur by route a or b to give the same carbanion. However, in case of unsymmetrical cyclopropanone intermediate (as in the above case), the ring opening takes place in such a way so as to form more stable carbanian. For instance, the cyclopropanone ring in the above case opens via route a to give more stable carbanion (which is stabilised by resonance) and so the preferred product is Ph $CH_2CH_2COOCH_3$ (as given above).

In case of cyclic ketones, Favorskii rearrangement involves ring contraction. As an example, 2-bromocyclohexanone on treatment with sodium methoxide gives cyclopentane carboxylic acid (Fig. 13.61).

In a similar way, Favorskii rearrangement of 2-chlorocyclobutanone gives cyclopropane carboxylic acid (Fig. 13.62).

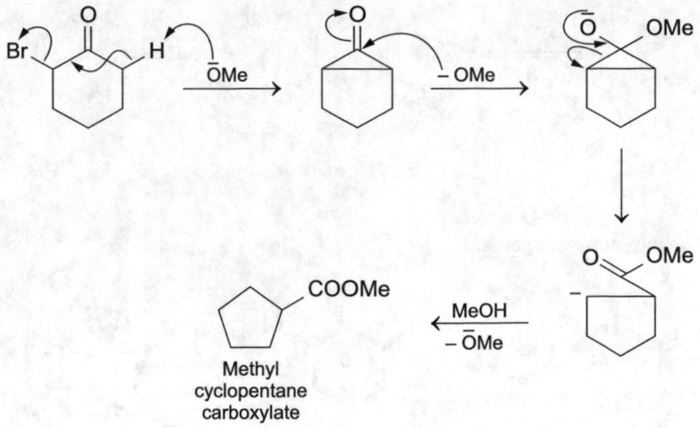

**Fig. 13.61**  Favorskii rearrangement of 2-bromocyclohexanone

**Fig. 13.62**  Favorskii rearrangement of 2-chlorocyclobutanone

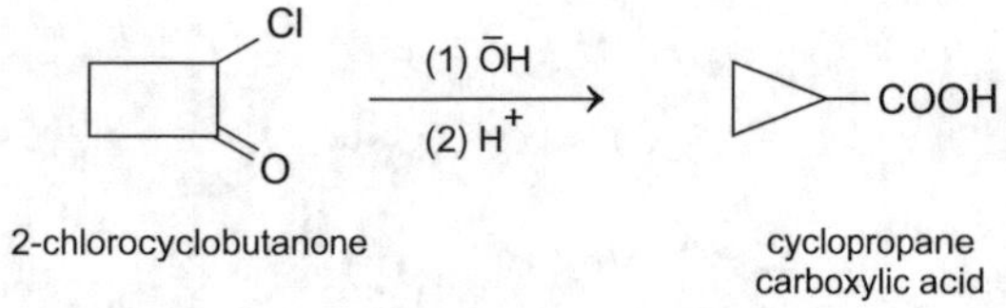

## References

1. A.E. Favorskii, J. Prakt. Chem., 1913, **88 (2)**, 658; O. Wallach, Ann. 1918, **414**, 296.
2. A.S. Kende, Org. Reactions, 1960, **11**, 261; Organic Synthesis, Coll. Vol., 1963, **4**, 594, N.J. Turro, Accts. Chem. Res., 1969, **2**, 25.

### 13.2.5.9  Jacobsen Rearrangement[1]

The migration of alkyl group from one position to another in polyalkyl benzenes during sulfonation is known as Jacobsen rearrangement. Thus, as an example, sulfonation of durene (1,2,4,5-tetramethylbenzene) gives prehnitenesulfonic acid (in which a methyl group has migrated), which on desulfonation yields prehnitene[2] (1,2,3,4-tetramethylbenzene) (Fig. 13.63).

The reaction involves ipso-sulfonation, followed by intramolecular migration of $CH_3$ group without pair of electrons giving a stable arenium ion. This is followed by the removal of a proton to regenerate the aromatic ring. The reaction is intramolecular (Fig. 13.64).

It is appropriate to state that in a number of Friedel–Crafts alkylation, some rearrangements have been observed leading to the change in relative position of the substituents in the benzene ring. Some such examples are given below (Fig. 13.65).

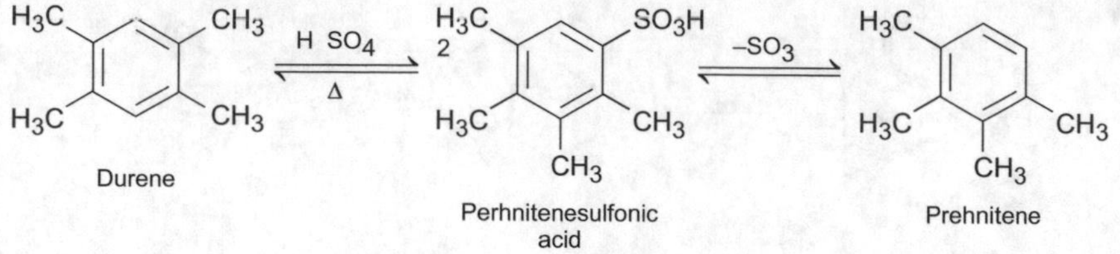

**Fig. 13.63**  Jacobsen rearrangement

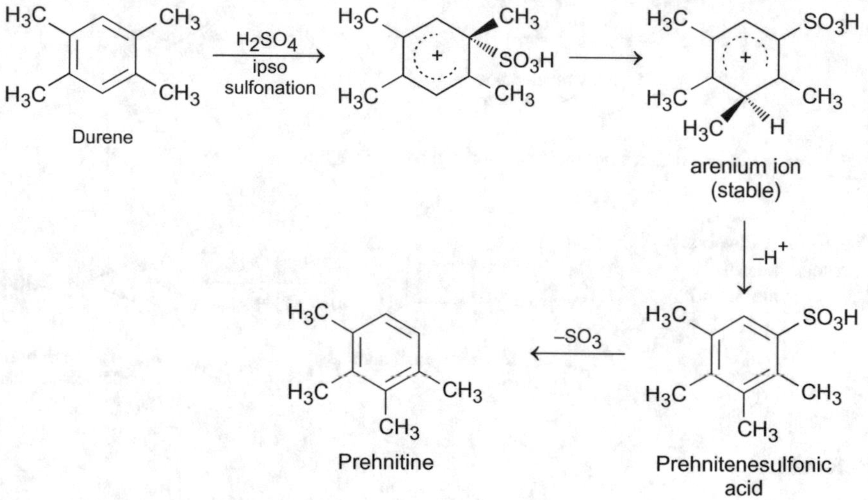

**Fig. 13.64**  Mechanism of Jacobsen rearrangement

In Jacobsen rearrangement, halogenated polyalkyl benzenes undergo isomerisation during sulfonation due to migration of halogen group. Some examples are given below (Fig. 13.66).

## References

1.  O. Jacobsen, Ber., 1886, **19**, 1209; L. I. Smit, Org. Reac. 1942, **I**, 370.
2.  H. Hart, J. F. Janssen, J. Org. Chem., 1970, **35**, 3637.


Neber Rearrangement[1]

It is closely related to Beckmann rearrangement and is the base-catalysed rearrangement of oxime tosylate to α-amino ketones via azirine. As an example, oximes of acetophenone and acetone are converted into the corresponding α-aminoketones (Fig. 13.67).

Aldoximes do not undergo Neber rearrangement.

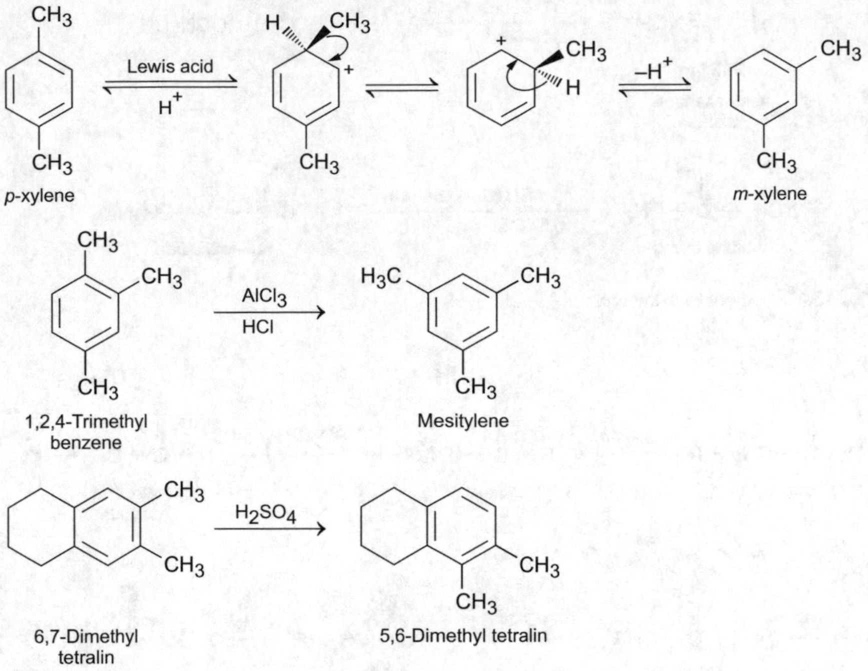

Fig. 13.65 Some rearrangements

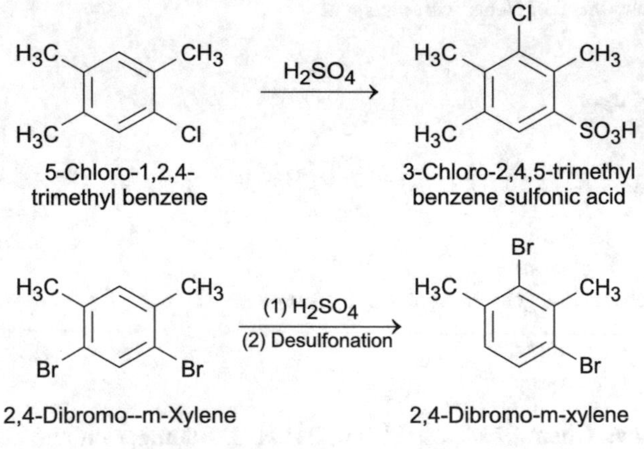

Fig. 13.66 Jacobsen rearrangements of halogenated polyalkylbenzenes

The reaction, as has already been stated, proceeds via the formation of azirine[2] (Fig. 13.68).

It has, however, been suggested that the azirine formation may involve the formation of nitrene as an intermediate (Fig. 13.69).

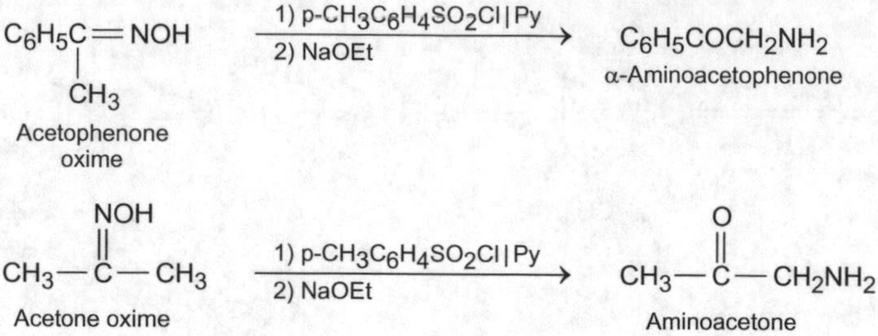

$$C_6H_5\underset{\underset{CH_3}{|}}{C}=NOH \xrightarrow[\text{2) NaOEt}]{\text{1) p-CH}_3C_6H_4SO_2Cl\,|\,Py} C_6H_5COCH_2NH_2$$

Acetophenone
oxime

α-Aminoacetophenone

$$CH_3-\underset{\overset{\|}{\underset{}{}}}{\overset{NOH}{C}}-CH_3 \xrightarrow[\text{2) NaOEt}]{\text{1) p-CH}_3C_6H_4SO_2Cl\,|\,Py} CH_3-\underset{\overset{\|}{O}}{C}-CH_2NH_2$$

Acetone oxime

Aminoacetone

**Fig. 13.67**   Neber rearrangement

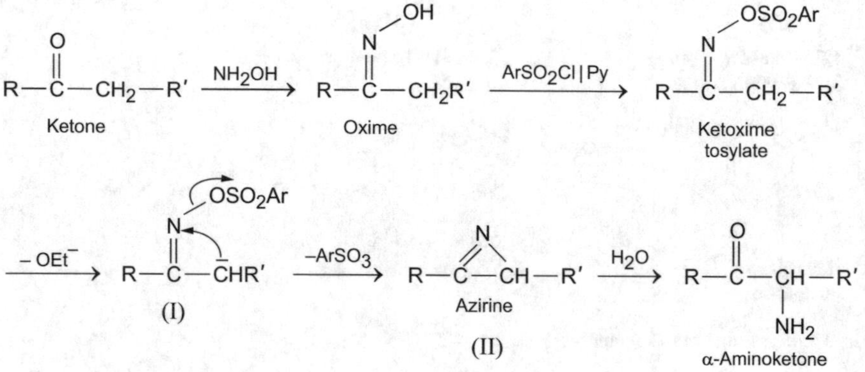

$$R-\underset{\overset{\|}{O}}{C}-CH_2-R' \xrightarrow{NH_2OH} R-\underset{\overset{\|}{N}\diagdown OH}{C}-CH_2R' \xrightarrow{ArSO_2Cl\,|\,Py} R-\underset{\overset{\|}{N}\diagdown OSO_2Ar}{C}-CH_2-R'$$

Ketone

Oxime

Ketoxime
tosylate

$$\xrightarrow{-\,OEt^-} R-\underset{(I)}{\overset{\overset{\displaystyle\frown OSO_2Ar}{N}\|}{C}}-CHR' \xrightarrow{-ArSO_3} \underset{(II)}{R-\overset{N}{\underset{\diagdown}{C}}-CH-R'} \xrightarrow{H_2O} R-\underset{\overset{\|}{O}}{C}-\underset{\underset{NH_2}{|}}{CH}-R'$$

Azirine

α-Aminoketone

**Fig. 13.68**   Mechanism of Neber rearrangement

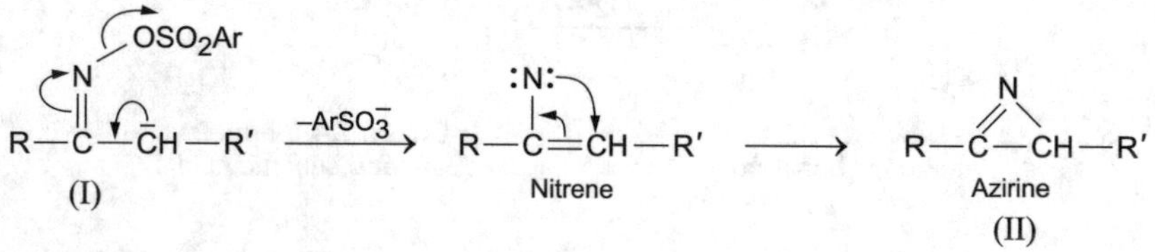

$$\underset{(I)}{R-\overset{\overset{\displaystyle\frown OSO_2Ar}{N}\|}{C}-\overset{\frown}{C}H-R'} \xrightarrow{-ArSO_3^-} \underset{\text{Nitrene}}{R-\overset{:N:}{\underset{|}{C}}=CH-R'} \longrightarrow \underset{\substack{\text{Azirine}\\(II)}}{R-\overset{N}{\underset{\diagdown}{C}}-CH-R'}$$

**Fig. 13.69**   Neber rearrangement via azirine formation

## References

1.  C. O'Brien, Chem. Rev., 19 64, **64**, 81; H.E. Baumgarten and F.A. Bower, J. Am. Chem. Soc., 1954, **76**, 4561.
2.  D. J. Cram and M.J. Hatch, J. Am. Chem. Soc., 1953, **75**, 38.

Sommelet–Hauser Rearrangement[1]

Benzhydryl trimethyl ammonium hydroxide on heating with concentrated sodium hydroxide rearranged to give *o*-benzyl benzyldimethyl amine (Fig. 13.70).

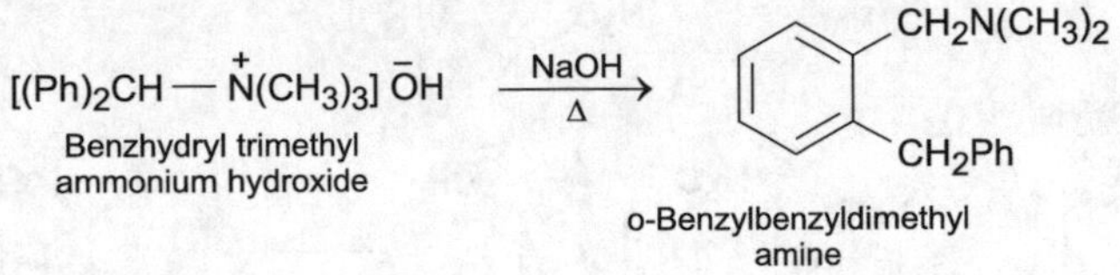

Fig. 13.70 Sommelet–Hauser rearrangement

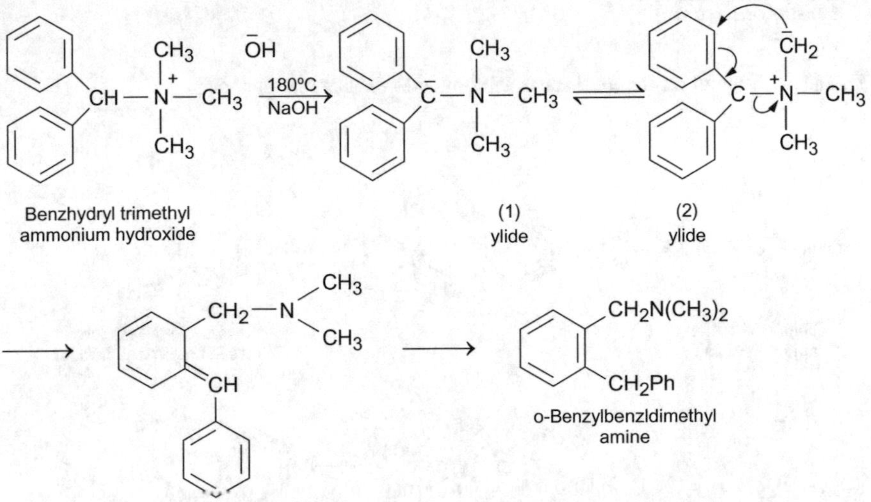

Fig. 13.71 Mechanism of Sommelet–Houser rearrangement

The above reaction occurs by initial deprotonation of the benzylic position to give the ylide (1), which is in equilibrium with a second ylide (2), which undergoes 2,3-sigmatropic rearrangement followed by aromalisation to give the final product (Fig. 13.71).

The proposed mechanism (as given above) finds support[2,3] in the observation that 2,4,6-trimethyl benzyl trimethyl ammonium iodide on treatment with sodamide in liquid ammonia gives a product with is non-aromatic (Fig. 13.72).

It was, however, found (Wittig, 1948)[2] that the same product, viz., o-benzylbenzyldimethyl amine was obtained during Stevens rearrangement when benzhydryl trimethyl ammonium bromide was treated with phenyl lithium in ether.

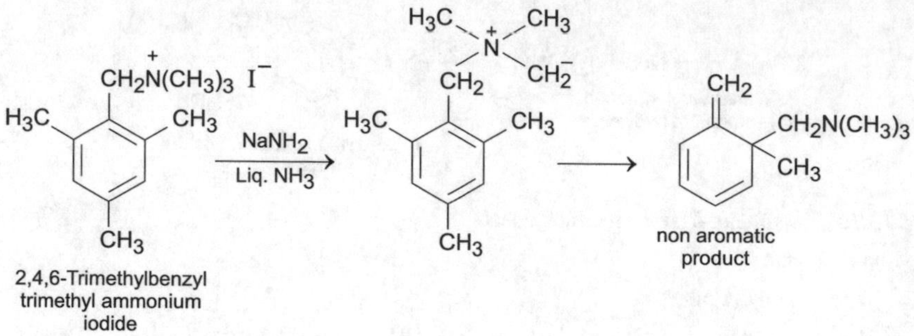

**Fig. 13.72**   Support for the mechanism of Sommelet–Hauser rearrangement

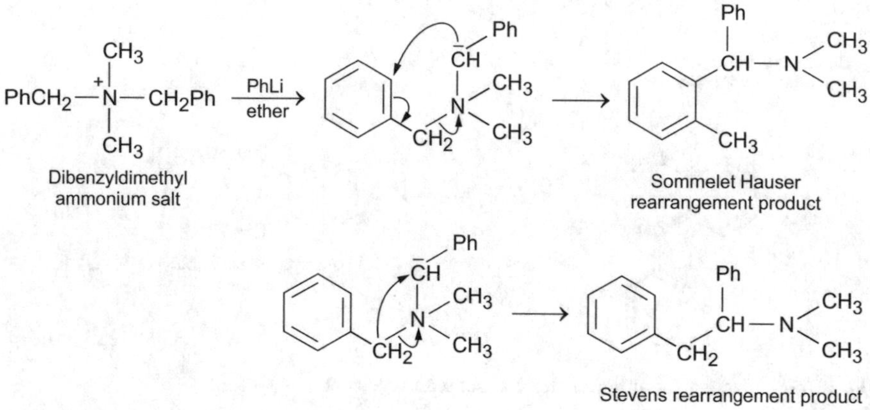

**Fig. 13.73**   The reaction of dibenzyldimethyl ammonium salt with phenyl lithium in ether

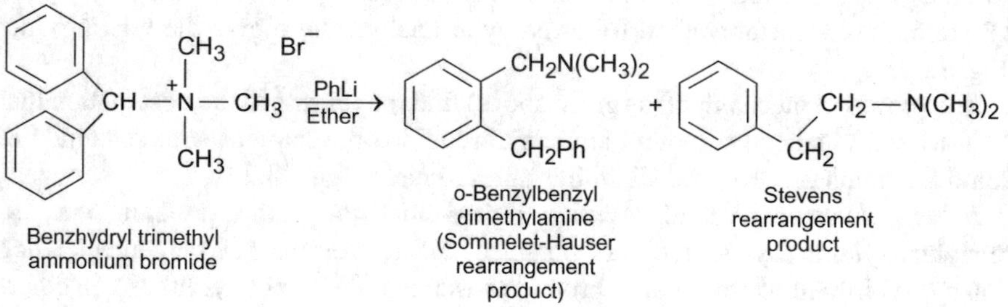

Thus, as an example, dibenzyldimethyl ammonium salt on treatment with phenyl lithium in ether gave both the Sommelet–Hauser rearrangement product and Stevens rearrangement product. It was found that at higher temperature, Steven's rearrangement product is preferably obtained, and at lower temperature, Sommelet rearrangement products obtained as the major product (Fig. 13.73).

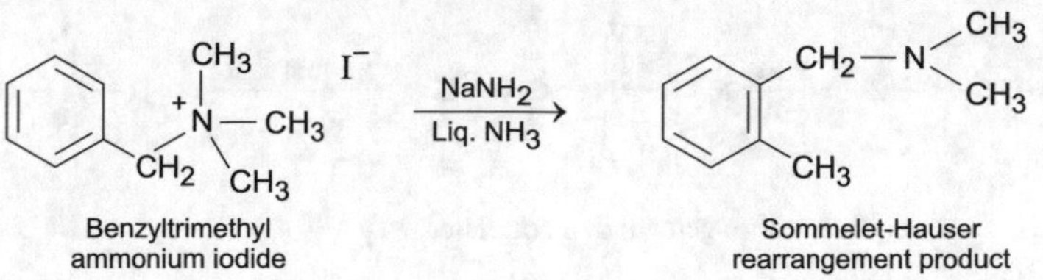

Benzyltrimethyl
ammonium iodide

Sommelet-Hauser
rearrangement product

**Fig. 13.74** Sommelet–Hauser type rearrangement using $NaNH_2$ in liquid ammonia

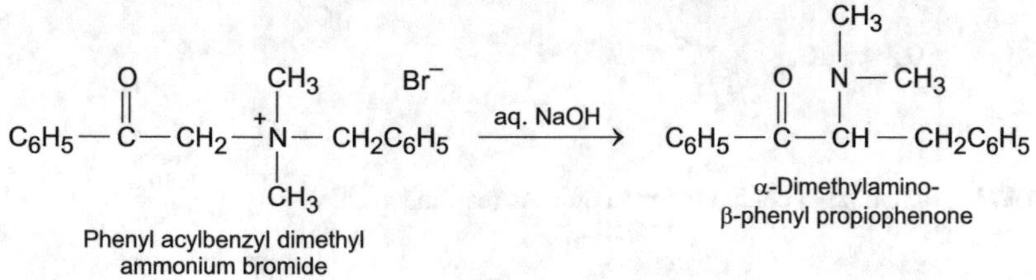

Phenyl acylbenzyl dimethyl
ammonium bromide

α-Dimethylamino-
β-phenyl propiophenone

**Fig. 13.75** Stevens rearrangement

Only Sommelet type rearrangements were found (Hauser, 1951) by using sodium or potassium amide[4]. Thus, treatment of benzyl trimethyl ammonium salt with sodamide in liquid ammonia gave only the Sommelet–Hauser rearrangement product, viz., 2-dimethylaminomethyl toluene (Fig. 13.74)

## References

1.  M. Sommelet, Compt. Rend., 1937, **205**, 56; Bull. Soc. Chem. France, 1918, **23**, 93.
2.  G. Wittig, R. Mangold and G. Fellets, Chem. Ann., 1948, **560**, 117.
3.  G. Wittig, H. Tenhaeff, W. Schoch and G. Koentnig, Ann. 1951, **572**, 1.
4.  S.W. Kantor and C.R. Houser, J. Am. Chem. Soc., 1951, **73**, 4122.

Stevens Rearrangement

Base-catalysed rearrangement[1] of a quaternary ammonium salt (in which none of the alkyl group has a β-hydrogen atom but one of the alkyl group has an electron-withdrawing group β- to the N atom) gives a tertiary amine. The rearrangement involves migration of a group (without pair of its electrons) from N to carbon having a negative charge. Thus, as an example, phenylacylbenzyl dimethyl ammonium bromide on treatment with alkali gives α-dimethylamino-β-phenyl propiophenone (Fig. 13.75).

It was earlier believed that this rearrangement occurred in a concerted way. The base abstracts a H from ammonium salt to give an ylide, which rearranges to a tertiary amine (Fig. 13.76).

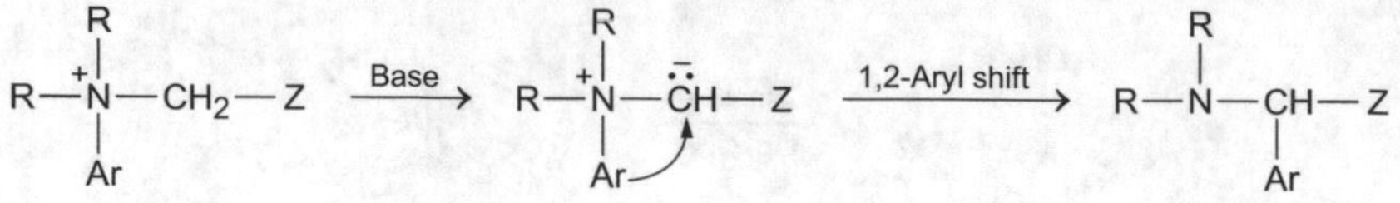

**Fig. 13.76**   Stevens ylide rearrangement in a concerted way

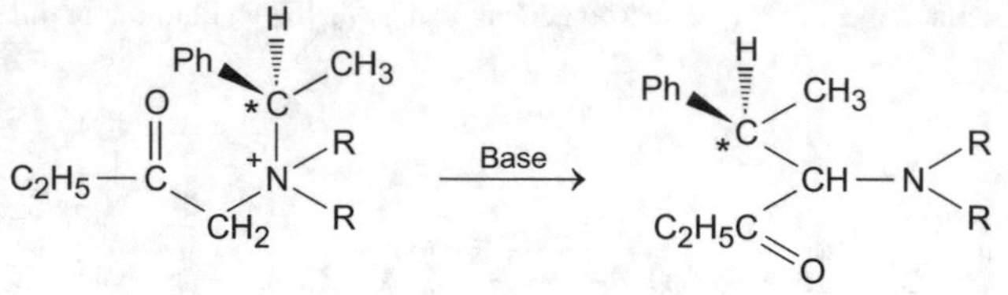

**Fig. 13.77**   Retention of configuration in Stevens rearrangement

The ylide is stabilised by the presence of an electron-withdrawing group like COOH.

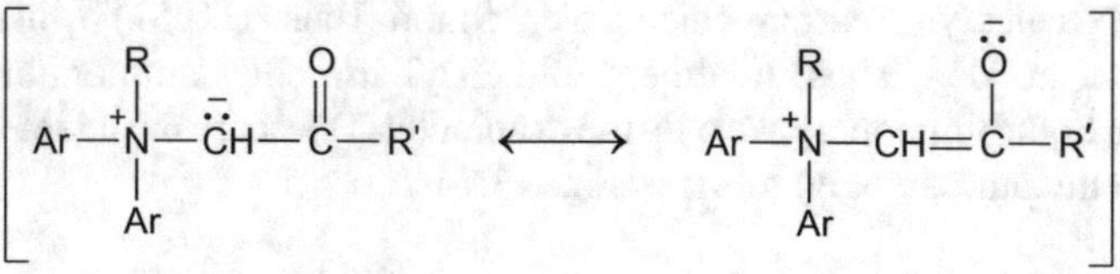

It has also been observed that the reaction occurred with retention of absolute stereochemistry at the migrating centre as shown by the following example (Fig. 13.77).

According to orbital symmetry rules, concerted rearrangement (as shown above) is not allowed. So a radical pair mechanism has been proposed. It involves deprotonation followed by homolytic fragmentation of the ylide yields a pair of radicals. The rapid recombination of the pair of radicals (which remain together in a tight solvent cage) yields the final product. A small amount of the coupling product, R–R formation supports the proposed mechanism (Fig. 13.78).

### Reference

1.   T. Thomson and T. S. Stevens, J. Chem. Soc., 1932, 55; T. S. Stevens, E. M. Creighton, A. B. Gorden, M. MacNicol, J. Chem. Soc., 1928, 3193.

Wittig Rearrangement[1]

This rearrangement is related to Stevens rearrangement and involves rearrangement of benzyl and allylethers with strong base like alkyllithium. As an example, benzyl phenylether on heating with sodium at 100 °C gives benzhydrol (Fig. 13.79).

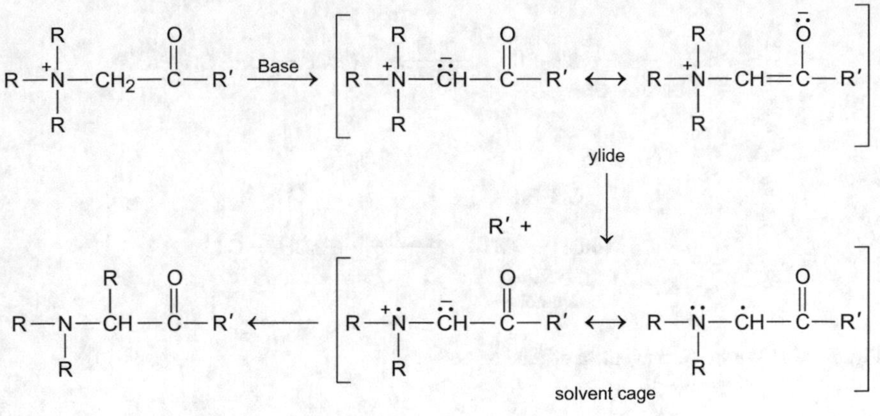

**Fig. 13.78** Radical mechanism for Stevens rearrangement

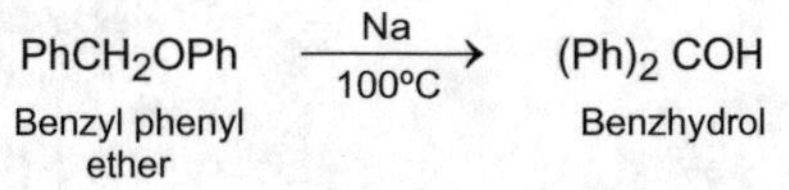

$$PhCH_2OPh \xrightarrow[100°C]{Na} (Ph)_2\ COH$$

Benzyl phenyl ether       Benzhydrol

**Fig. 13.79** Wittig rearrangement

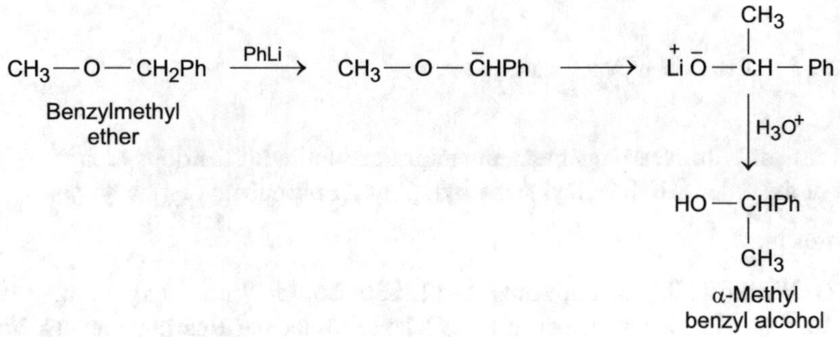

α-Methyl benzyl alcohol

**Fig. 13.80** Wittig rearrangement of benzylmethyl ether using PhLi

In case of benzylmethyl ether, a strong base like phenyl lithium is required to abstract a proton to give a conjugate base. This is followed by 1, 2-methyl shift from O (without pair of electrons) to adjacent carbon to give a stable alkoxide anion. The final workup gives the corresponding alcohol (Fig. 13.80).

In a similar way, dibenzyl ether gives 1,2-diphenylethanol (benzylphenyl methanol (Fig. 13.81).

The radical mechanism for Wittig rearrangement where migrating group is aryl or alkyl is given in Fig. 13.82.

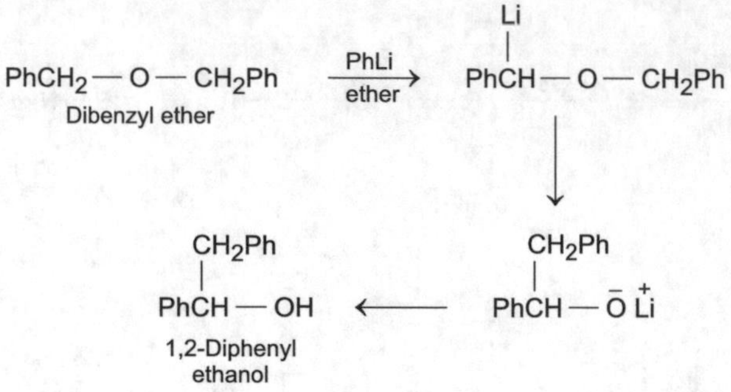

$$PhCH_2 - O - CH_2Ph \xrightarrow[\text{ether}]{\text{PhLi}} \overset{\overset{\displaystyle Li}{\displaystyle |}}{PhCH} - O - CH_2Ph$$

Dibenzyl ether

$$\underset{\text{1,2-Diphenyl ethanol}}{\overset{\overset{\displaystyle CH_2Ph}{\displaystyle |}}{PhCH} - OH} \longleftarrow \overset{\overset{\displaystyle CH_2Ph}{\displaystyle |}}{PhCH} - \bar{O} \overset{+}{Li}$$

**Fig. 13.81**   Wittig rearrangement of dibenzyl ether

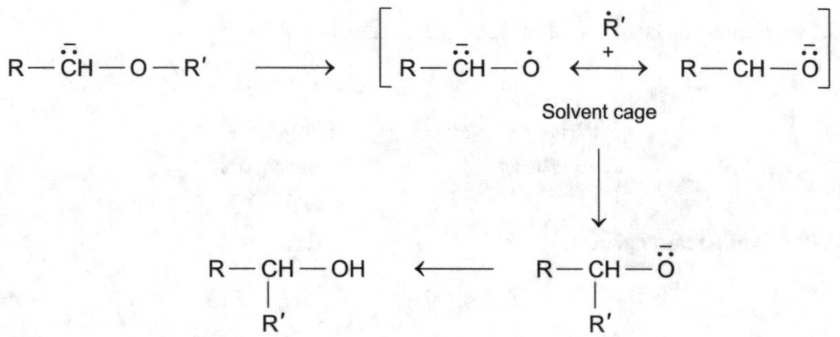

$$R - \overset{..}{\underset{..}{C}H} - O - R' \longrightarrow \left[ R - \overset{..}{\underset{..}{C}H} - \overset{.}{O} \overset{\overset{\displaystyle \overset{.}{R}'}{+}}{\longleftrightarrow} R - \overset{.}{C}H - \overset{..}{\underset{..}{O}} \right]$$

Solvent cage

$$\underset{\overset{|}{R'}}{R - CH - OH} \longleftarrow \underset{\overset{|}{R'}}{R - CH - \overset{..}{\underset{.}{O}}}$$

**Fig. 13.82**   Mechanism of Wittig rearrangement

In case of Wittig rearrangement, the migratory aptitude of various groups is found to be of the order. Allyl, benzyl > methyl, ethyl, p-nitrophenyl > phenyl.

**Reference**

1.   G. Wittig, L. Löhmann, Ann., 1942, **550**, 26; G. Wittig, Experientia, 1958, **14**, 389; H.E. Zimmerman in P. de Mayu, Molecular Rearrangements, Vol. 1 (Interscience, New York, 1963), p. 372.

*See* also Molecular Rearrangements (Sect. 7.2.3.3).

## Key Concepts

- **Anionic Rearrangement**: Rearrangement involving rearrangement of carbanions.
- **Arndt–Eistert Synthesis**: A procedure for converting a ketene (obtained from carboxylic acids) into esters, carboxylic acid or amides.

- **Baeyer–Villiger oxidation**: Oxidation of open chain ketones to esters and cyclic ketones to lactones by reaction with peracids.
- **Baker–Venkataraman Rearrangement**: Base-catalysed rearrangement of o-acylketones to β-diketones, an important intermediate for the synthesis of chromones and flavones.
- **Beckmann Rearrangement**: Conversion of Ketoximes into substituted amides by acidic reagents. Oximes of cyclic ketones undergo ring enlargements.
- **Benzidine Rearrangement**: Acid-catalysed rearrangement of hydrazobenzenes to 4, 4′-diaminobiphenyls (p-benzidine). In case the hydrazobenzenes contains a *para* substituent, the product is *p*-amino diphenylamine (Semidine rearrangement).
- **Benzil–benzilic acid Rearrangement**: Rearrangement of benzil to benzilic acid on treatment with base. A general reaction for the conversion of α-diketones to α-hydroxy acids.
- **Carbon–Carbon Rearrangement**: Rearrangement of molecules in which C–C bond is broken in one part of the molecule and reformed again at another part.
- **Carbon–Nitrogen Rearrangements**: Rearrangements in which a C–C bond is broken in one part of the molecule and a C–N bond is formed at another part.
- **Carbon Oxygen Rearrangement**: Rearrangement of molecules involving the breaking of a C–C bond in one part of the molecule and forming a C–O bond in another part.
- **Cationic Rearrangement**: Rearrangements involving a change in carbon skeleton via rearrangement of carbocation as intermediate.
- **Claisen Rearrangement**: Thermal rearrangement of allyl ethers of phenols to *o*-allyl phenols.
- **Cope Rearrangement**: Thermal rearrangement of 1,5-dienes by a 3,3-sigmatropic shift.
- **Curtius Rearrangement**: Thermal rearrangements of acyl azides to isocyanates, important intermediates for the synthesis of 1° amine.
- **Cycloaddition reaction**: Formation of a ring from two π systems.

- **Diazoamino Rearrangement**: Rearrangement of diazoamino benzene to *p*-ammo azobenzene by treatment with HCl at 40 °C.
- **Diels–Alder Reaction**: The [4 + 2] cycloaddition reaction between a conjugated diene (4π-electron system) and a compound containing a double or triple bond (dienophile) (2π-electron system) to form a cyclic adduct.
- **Electrocylic Reaction**: Reactions in which a single bond is formed between the termini of a conjugated polyene system.

- **Favorskii Rearrangement**: Base-catalysed rearrangement of α-haloketones (bromo or chloro) to carboxylic acid derivatives. In case of α-halo cyclic ketone ring, contraction occurs to give cylopentane carboxylic acid derivative.
- **Fischer–Hepp Rearrangement**: Acid-catalysed rearrangement of N-alkyl-N-nitroso arylamine to *p*-nitroso-N-alkylaniline.
- **Fries Rearrangement**: Rearrangement of phenylacetate to *o*- and phydroxyacetophenone by heating with anhyl $AlCl_3$.
- **Hofmann–Martius Rearrangement**: Thermal rearrangement of N-alkylaniline hydrochlorides to *o*- and *p*-alkylanilines.
- **Hofmann Rearrangement**: Rearrangement of primary amide into primary amine by reaction with sodium hypohalite (NaOH + halogen) via intermediate isocyanate.
- **Hydroperoxide Rearrangement**: Acid-catalysed rearrangement of hydroperoxide to give phenol. As an example, cumene hydroperoxide gives phenol.
- **Intermolecular Rearrangement**: Rearrangement reaction in which the atoms or groups that undergo migration are completely detached from the substrate and subsequently get attached at some other reactive site in the molecule.
- **Intramolecular Rearrangement**: Rearrangement reaction in which the migrating group is not completely detached from the system during migration.
- **Jacobsen Rearrangement**: Rearrangement of polymethylbenzenes with concentrated $H_2SO_4$ to give rearranged polymethyl benzene sulfonic acids, which on desulfonation gives rearranged polymethyl benzenes. The halogenated polyalkyl benzenes undergo isomerisation during sulfonation due to the migration of halogen group.
- **Neber Rearrangement**: Base-catalysed rearrangement of oxime to sylate to α-amino ketones via azirine.
- **Orton Rearrangement**: Rearrangement it N-chloroacelanilide to *o*- and *p*-chloro toluenes by treatment with dilute HCl.
- **Oxy-Cope Rearrangement**: Rearrangement of hydroxyl-substituted 1,5-dienes.
- **Pericyclic Rearrangements**: Reactions occurring via a cyclic transition state by the movement of electrons in a concerted way.
- **Pinacol–Pinacolone Rearrangement**: Acid-catalysed rearrangement of 1, 2-diols to ketones.
- **Rearrangement Reaction**: Reaction in which a synthon molecule undergoes reorganisation of its parts leading to change in molecular skeleton.
- **Sigmatropic reaction**: Reaction in which there is formation of a single bond adjacent to one or more π systems.

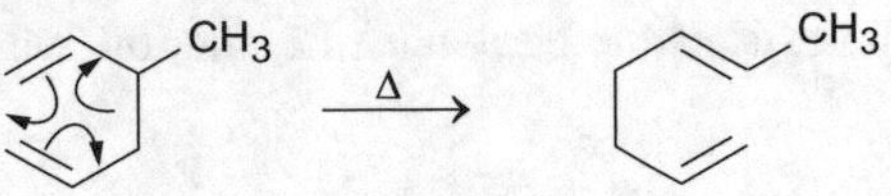

- **Sommelet Hauser Rearrangement**: Rearrangement of benzyl quaternary ammonium salts to ortho-substituted benzyldialkylamines on treatment will alkali metal amides.
- **Stevens Rearrangement**: Base-catalysed rearrangement of an appropriately substituted quaternary ammonium salt (in which none of the alkyl group has a β-hydrogen atom but one alkyl group as an electron-withdrawing group β to the N atom) gives a tertiary amine.
- **Wagner–Meerwein Rearrangement**: Carbon to carbon migration of alkyl, aryl or hydride ions. These rearrangements involve a change in carbon skeleton.
- **Wittig Rearrangement**: Base catalyst rearrangement of benzyl and allyl ether to give benzhydrol.
- **Wolff Rearrangement**: Rearrangement of diazoketones to ketene by action of heat, light or silver oxide.

## Problems

1.  What products are expected to be obtained in the following rearrangements? Give mechanisms.

(a)

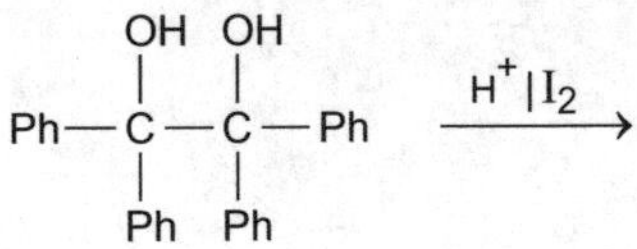

(b)

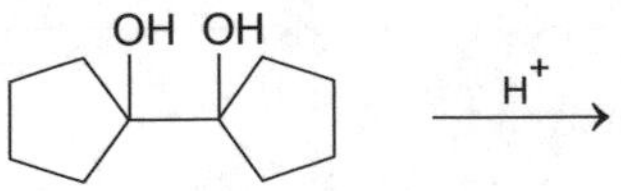

(c)

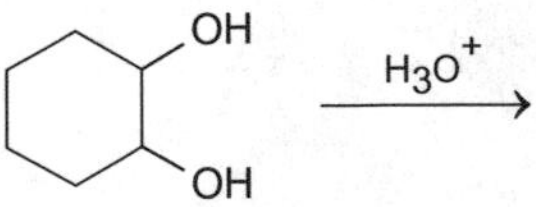

2. What products are expected to be obtained in the following anionic rearrangements?

(a)

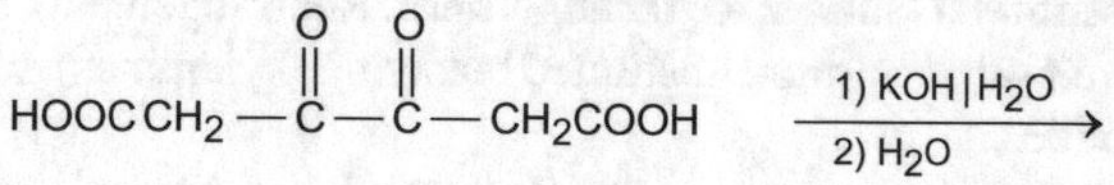

$$HOOCCH_2-\overset{\overset{O}{\|}}{C}-\overset{\overset{O}{\|}}{C}-CH_2COOH \xrightarrow[\text{2) H}_2\text{O}]{\text{1) KOH}|\text{H}_2\text{O}}$$

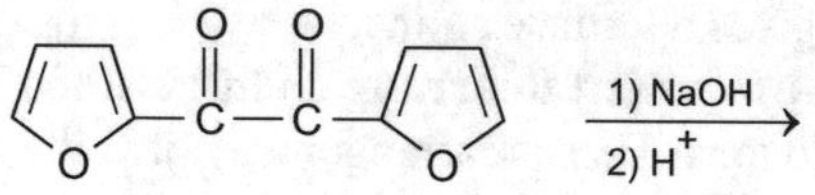

(b)

$\xrightarrow[\text{2) H}^+]{\text{1) NaOH}}$

(c)

$\xrightarrow[\text{2) H}^+]{\text{1) NaOH}}$

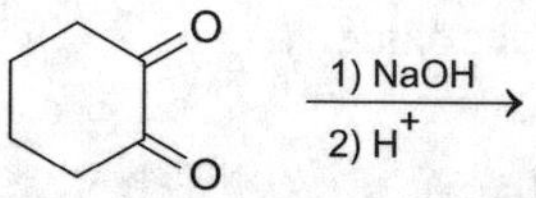

3. What products are expected to be obtained in the Hofmann rearrangement of the following:

(a)

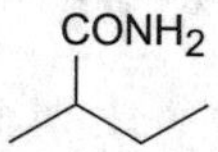

(b)

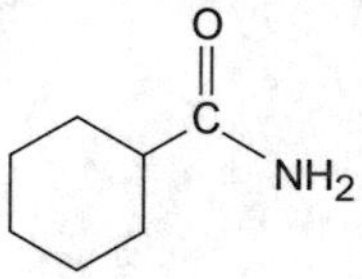

(c)

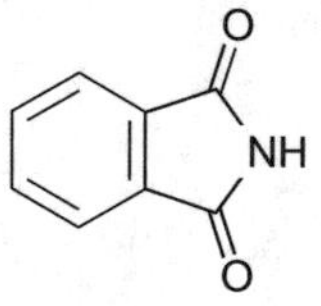

(d)

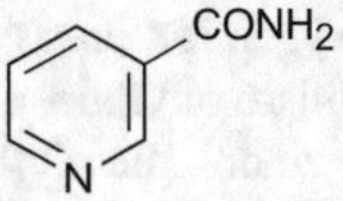

(e)

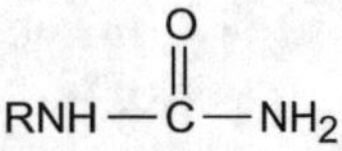

(f)

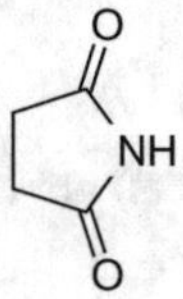

4.  How do you justify that Hofmann rearrangement is stereospecific?
5.  Explain Baeyer–Villger oxidation. What products are expected to be obtained in the Baeyer–Villeger oxidation of the following compounds:

(a)

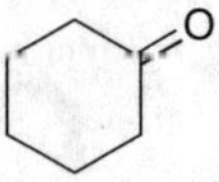

(b)

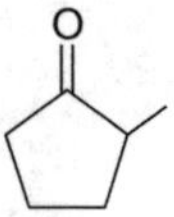

(c)

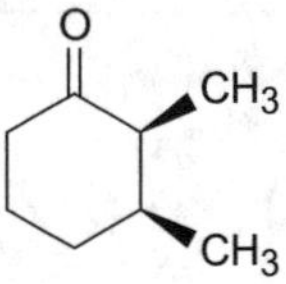

(d)

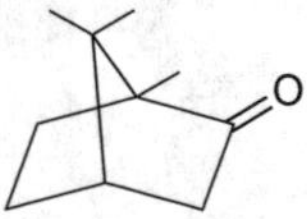

6. Bacyer–Villiger oxygen is stereospecific. Explain.
7. Discuss the mechanism of Claisen rearrangement.
8. In Claisen rearrangement, the allyl group is inverted. Comment.
9. What product is expected to be obtained in the Claisen rearrangement of the following:

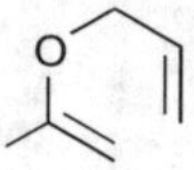

**Ans.**

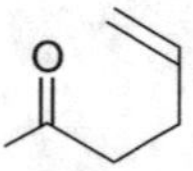

10. What product (products) are obtained in the following reactions:

(a)

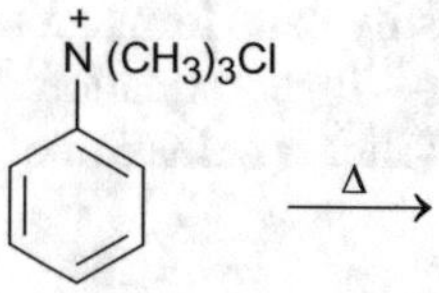

(b)

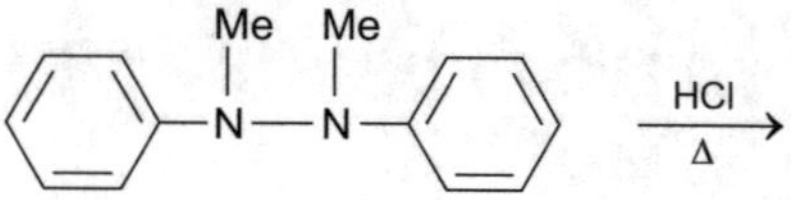

(c)

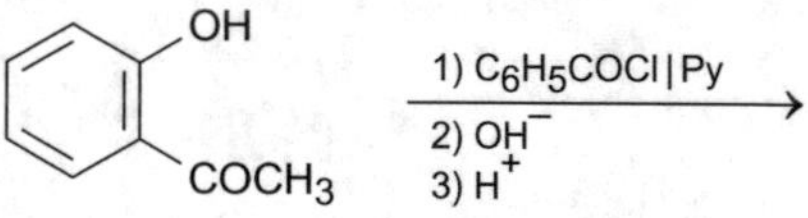

**Ans.** (*a*)

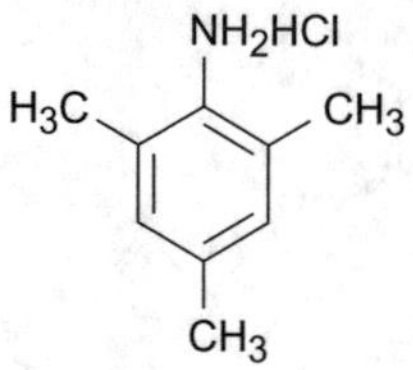

(*b*)  MeHN —⟨ ⟩—⟨ ⟩— NHMe

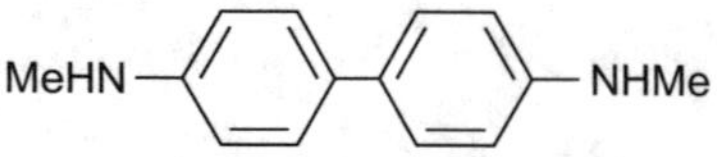

(c)

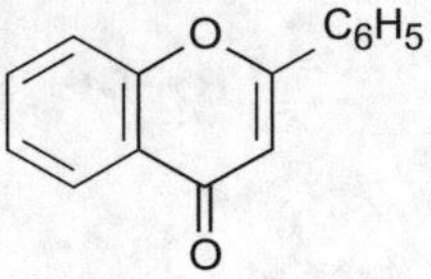

11. Give mechanism of the following:

    (a)    Beckmann rearrangement
    (b)    Fries rearrangement
    (c)    Benzil–benzilic acid rearrangement
    (d)    Favorskii rearrangement
    (e)    Wolff rearrangement

12. What products are obtained in the following reactions:

(a)

$$PhCH_2 - O - CH_2Ph \xrightarrow[\text{ether}]{\text{PhLi}}$$

(b)

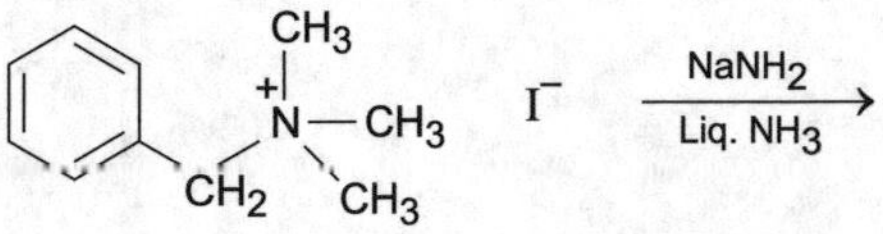

(c)

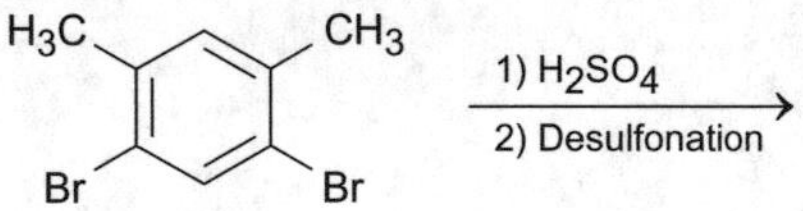

(d)

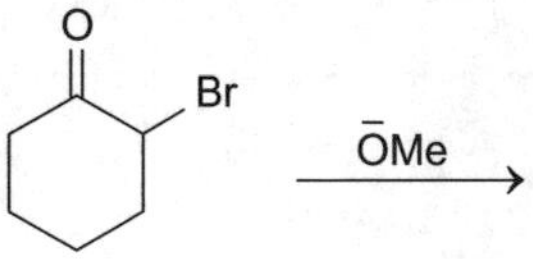

(e)

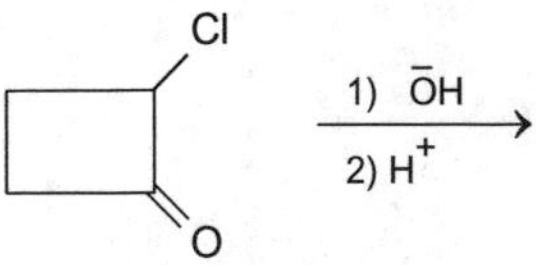

(f)

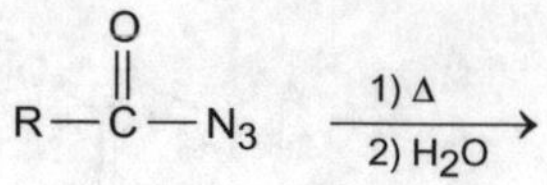
O
R—C—N3
1) Δ
2) H2O

(g)

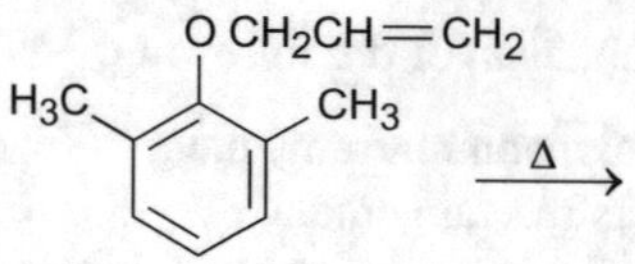
O CH2CH=CH2
H3C
CH3
Δ

(h)

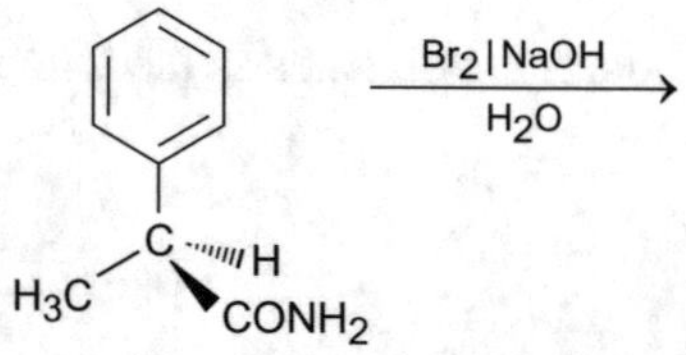
Br2 | NaOH
H2O
C
H
H3C
CONH2

(i)

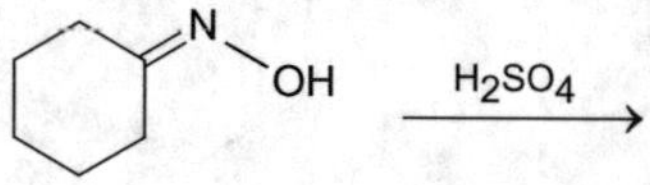
N
OH
H2SO4

# Chapter 14
# Stereochemistry of Pericyclic Reaction

## 14.1 Introduction

It is well known that most of the organic reactions like addition reactions, substitution reactions and elimination reactions proceed stepwise involving more than one step; the only exceptions are the $S_N2$ reactions and E2 reactions. However, a number of reactions proceed in a single step (concerted step) involving the formation of a cyclic transition state. Such reactions because of their cyclic transition states are called **pericyclic reactions.** Such reactions are generally initiated by heat or light and are highly stereospecific. Both processes (thermal and photochemical) give products with different stereochemistry. Such reactions do not involve any intermediates (ionic or free radical) and solvent and reagents have no effect on the formation of products.

## 14.2 Types of Pericyclic Reactions

The most common pericyclic reactions are cycloaddition reactions, sigmatropic rearrangement and electrocyclic reactions.

The **cycloaddition** reaction involves formation of a ring from the reaction of two π systems (Fig. 14.1).

In **sigmatropic rearrangement**, there is migration of a single bond adjacent to one or more π system (Fig. 14.2).

In **electrocyclic reaction**, a single bond is formed between the termini of a conjugated polyene system (Fig. 14.3) (see also Sect. 13.2.2.3).

© The Author(s), under exclusive license to Springer Nature Switzerland AG 2022    397
V. K. Ahluwalia, *Stereochemistry of Organic Compounds*,
https://doi.org/10.1007/978-3-030-84961-0_14

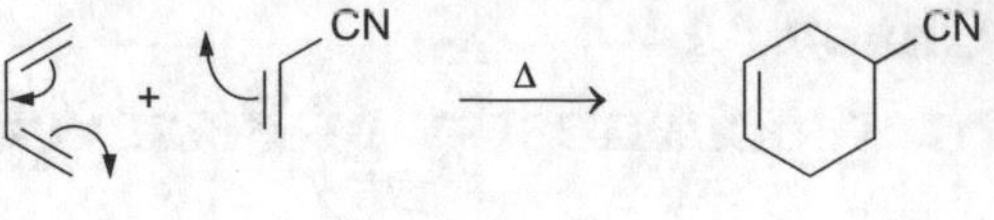

**Fig. 14.1**   Cycloaddition
reaction

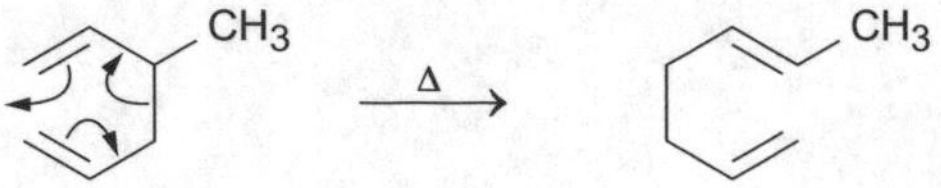

**Fig. 14.2**   Sigmatropic
rearrangement

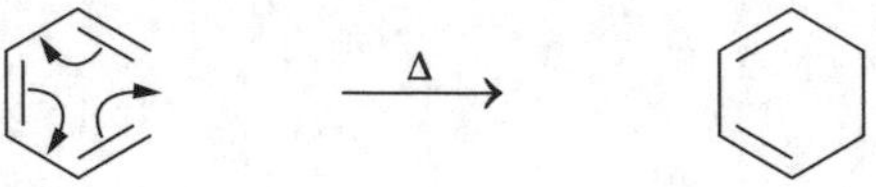

**Fig. 14.3**   Electrocyclic
reaction

## 14.3   Stereochemistry of Pericyclic Reactions

As already stated, pericyclic reactions are highly sterospecific and different products are obtained depending on whether the reaction is carried under thermal or photochemical conditions. Under both conditions, the reaction is stereospecific. It is also possible that a particular reaction may be possible only under one set of conditions. As an example the Diels Alder reaction (a cycloaddition reaction) is normally possible under thermal condition not under photochemical conditions. In a similar way, the cycloaddition of two moles of the alkene (I) is possible only under photochemical condition (Fig. 14.4).

In some cases, as already been stated different products are obtained under thermal and photochemical condition. Thus, as an example ring closure of *tran, trans*-2, 4-hexadiene under thermal and photochemical conditions gives only *cis*-3, 4-dimethyl cyclobutene and *trans*-3, 4-dimethyl cyclobutene, respectively (Fig. 14.5).

On the contrary, ring closure of *cis, cis*-2, 4-hexadiene gives cyclised products with opposite stereochemistry (compared to the cyclisation of *trans, trans*-2, 4-hexadiene, as shown in (Fig. 14.6).

The differential stereochemical effect is also seen in the cyclisation of *trans, cis, trans*-2,4,6-octatriene (Fig. 14.7).

According to Woodward and Hofmann (1965), in pericyclic reactions, there is conservation of orbital symmetry. This implies that in cyclic concerted reaction (reaction involving cyclic transition state), the orbitals in the reactant can transform in the product having the same symmetry properties with respect to the elements

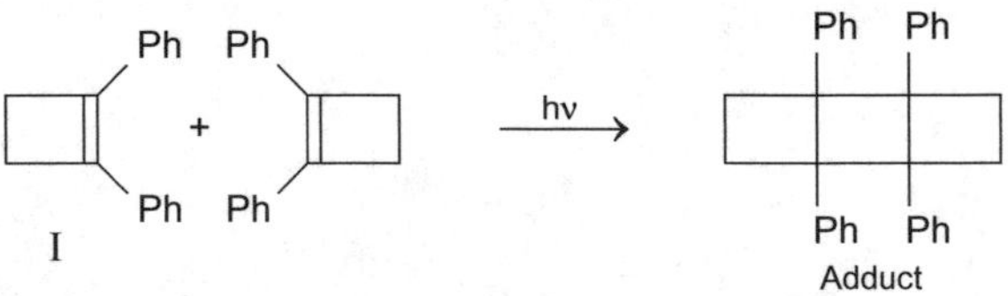

**Fig. 14.4**   Cycloaddition
reaction takes place only
under photochemical
conditions

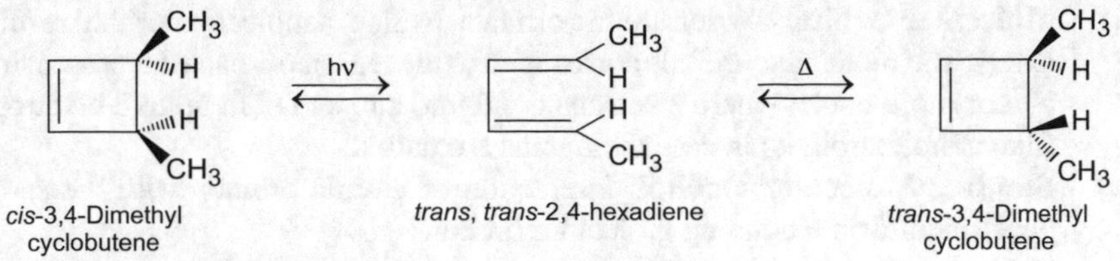

Fig. 14.5   Cyclisation of *trans, trans*-2, 4-hexadine

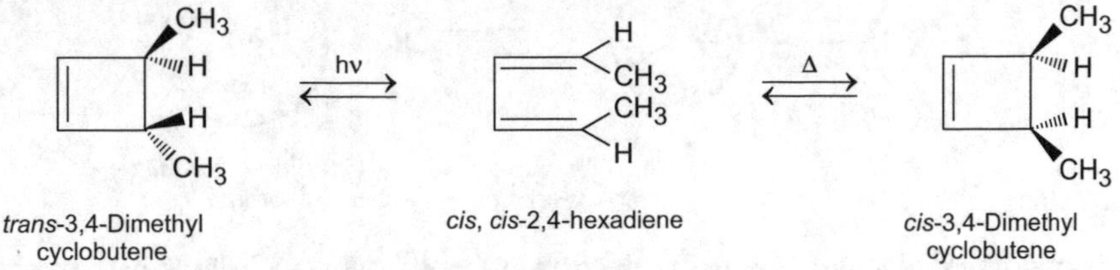

Fig. 14.6   Cyclisation of *cis, cis*-2, 4-hexadiene

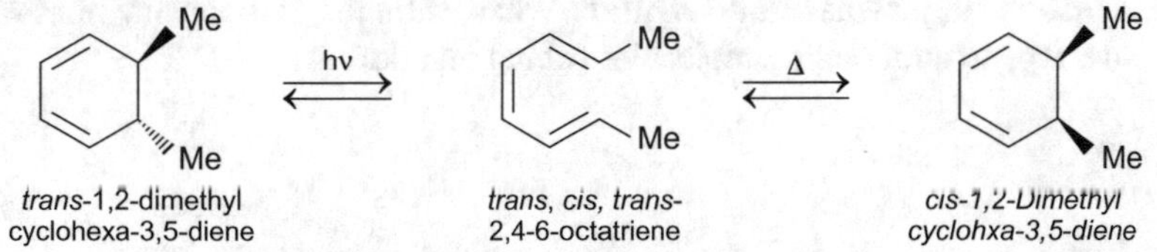

Fig. 14.7   Cyclisation of *trans, cis, trans*-2, 4, 6-octatriene

of symmetry being preserved in the reaction. As per quantum mechanical principle (*see* also Sect. 14.4), the energy level of the transition state of a symmetry allowed process is always lower than that of the alternative symmetry-forbidden path.

## 14.4   Some Useful Concepts that Come from Quantum Mechanics

(i)   **Atomic Orbital (AO)**: It corresponds to a region of space about the nucleus of a single atom where the probability of finding an electron is maximum. Atomic orbitals called $s$ orbitals are spherical in shape and $p$-orbitals are like two almost-tangent spheres. Orbitals can hold a maximum of two electrons when their spins are paired. The orbitals are described by a wave function, $\psi$. Each orbital has a characteristic energy. The phase signs of the orbitals are designated (+) and (−).

(ii)  **Molecular Orbitals**: When atomic orbitals overlap, a molecular orbital results (MO). The molecular orbitals correspond to the region of space encompassing two or more nuclei where electrons are found and can hold up to 2 electrons (like atomic orbitals) in case their spins are paired.

(iii)  **Bonding Molecular Orbital**: Interaction of atomic orbitals with the same phase sign form a bonding molecular orbital.

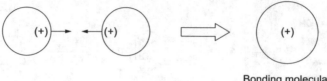

Bonding molecular
orbital

In bonding molecular orbital, the electron probability density is large in the region of space between the two nuclei where negative electrons hold the positive nuclei together.

(iv)  **Antibonding Molecular Orbital**: When orbitals of opposite phase sign overlap, an antibonding molecular orbital results.

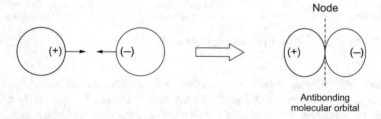

Antibonding
molecular orbital

Antibonding molecular orbital has higher energy compared to a bonding orbital. The electron probability density of the region between the two nuclei is small and it contains a node (a region where $\psi = 0$). The electrons in the antibonding orbital are not helpful to hold the nuclei together. In fact, the electrons fly apart due to intermolecular repulsion.

(v)  **Energy of Electrons**. The energy of electrons in the bonding molecular orbital is comparatively less than the energy of electrons in their separate atomic orbitals.

(vi)  **Number of Molecular Orbitals**. The number of molecular orbitals is equal to the number of atomic orbitals from which they are formed. The combination of two atomic orbitals will invariably yield two molecular orbitals (one bonding and one antibonding).

## 14.5 Molecular Orbital Theory

For an understanding of pericyclic reactions, it is useful to understand the basic features of molecular orbital theory. According to this theory, linear combination of atomic orbitals (LCAO) results in the formation of molecular orbitals.

It is known that in hydrocarbons like methane or ethane, the carbon atoms are $sp^3$ hybridised and form two types of σ-bonds, one between carbon atoms (by the overlap of $sp^3$-$sp^3$ atomic orbitals) and the other between carbon and hydrogen atoms (by the overlap of $sp^3$-$s$ atomic orbitals). Both the σ-bonds have a bonding orbital and a antibonding orbital.

As already stated, formation of a σ-bond between carbon and carbon involves the $sp^3$ hybrid orbitals of two carbons to combine (or overlap). In this case either the orbitals of the same phase can interact with each other resulting in the formation of a bonding molecular orbitals (σ-molecular orbital) or the orbitals of opposite phase interact with each other resulting in the formation of an antibonding molecular orbital (σ*-molecular orbital) these are explained in Sect. 14.4.

The phase (as already stated) is the orbital sign (+) and (–) or by designating by shading and no shading.

According to molecular orbital theory, the bonding orbitals are lower in energy compared to separate atomic orbitals and the antibonding orbitals. The antibonding orbital is of high energy and does not contribute to the bond formation (Fig. 14.8).

### 14.5.1 Molecular Orbitals of Ethene

In case of ethene, the carbon–carbon double bond consists of two kinds of bonds, a σ bond (resulting from overlapping of two $sp^2$ orbitals end to end and is symmetrical about an axis linking the two carbon atoms) and a π bond (resulting from sideways

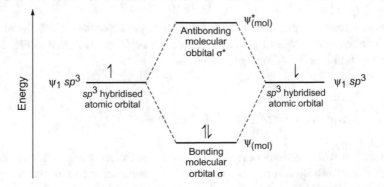

**Fig. 14.8** Energy diagram of molecular orbitals ($\psi_{(mol)}$ and $\psi^*_{(mol)}$ represent the wave functions for bonding and antibonding molecular orbitals)

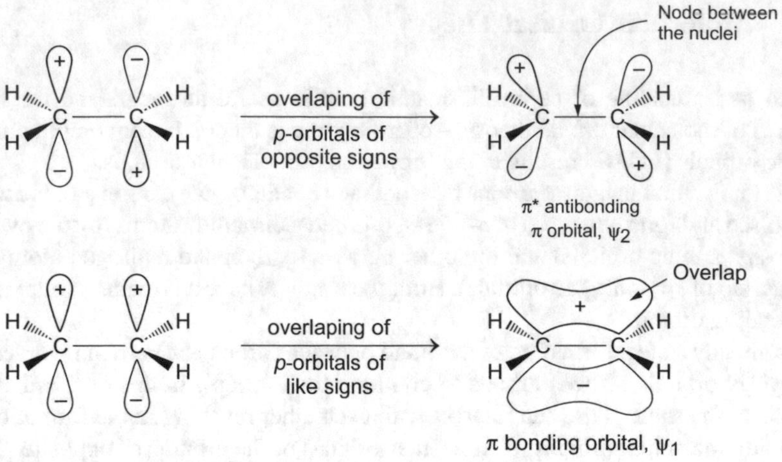

Fig. 14.9    π-Bonding and antibonding orbitals of ethene

Fig. 14.10    Relative energies of σ and π molecular orbitals in ground state of C = C in ethene

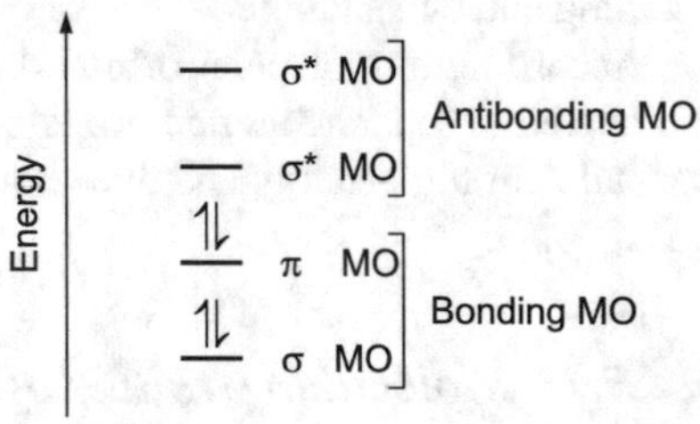

overlap of two p-orbitals. According to molecular orbital theory, both bonding and antibonding π molecular orbitals are formed when p-orbitals interact in this way to form a π bond. The bonding π orbital is formed when p-orbital lobes of like signs overlap and the antibonding π orbital results when p-orbital lobes of opposite signs overlap (Fig. 14.9).

The electrons in the π bond have greater energy compared to electrons of the σ bond. The relative energies to σ and π molecular orbitals are shown in Fig. 14.10. (The σ* orbital is antibonding sigma orbital).

## 14.5.2   Molecular Orbitals of Butadiene

In connection with the study of pericyclic reactions, the molecular orbitals of butadiene is important. Butadiene has four p-orbitals, which give rise to four molecular π orbitals having different energies and wave functions $\psi_1$, $\psi_2$, $\psi_3$ and $\psi_4$. Of these, $\psi_1$ and $\psi_2$ and π-bonding molecular orbitals (formed by overlap of orbitals of same phase) and $\psi_3$ and $\psi_4$ are π-antibonding molecular orbitals (formed by overlap of

**Fig. 14.11** Molecular
orbitals of butadiene

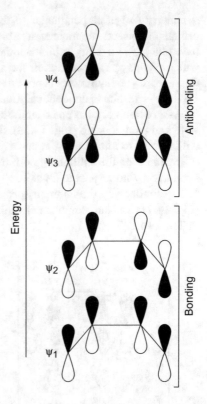

orbitals with different phase). The energy of $\psi_1$ is considerably lower than any other. Figure 14.11 shows the four $\pi$ molecular orbitals of butadiene in terms of increasing energy.

### 14.5.3 Molecular Orbitals of 1, 3, 5-Hexatriene

In case of 1,3,5-hexatriene (as in the case of butadiene), the six p-electrons are accommodated in the first three molecular $\pi$-orbitals ($\psi_1$, $\psi_2$, $\psi_3$) which are bonding orbitals and the remaining three higher energy molecular $\pi$ orbitals ($\psi_4$, $\psi_5$, $\psi_6$) remain unoccupied in the ground state (Fig. 14.2).

### 14.5.4 Symmetry Properties of Orbitals

The molecular orbitals can be classified according to their two independent symmetry properties, viz., plane of symmetry ($m$) and axis of symmetry ($C_2$). The plane of symmetry ($m$) is perpendicular to the plane of the atoms forming the molecular orbital and bisects the molecular orbital. However, molecular orbitals which do not have plane of symmetry exhibit another type of symmetry about a two-fold axis ($C_2$) which passes at right angles in the same plane and through the centre. The two-fold axis ($C_2$) of symmetry is present in molecules whose rotation about the axis by $180° \left( \frac{360°}{2} \right)$ results in an identical molecular orbital.

As an example, let us consider the symmetry properties of ethylene both in ground state and excited state (Fig. 14.13). The ground state ($\pi$) orbital is symmetric (S) with respect to mirror plane $m$ and asymmetric (A) with respect to the two-fold axis ($C_2$). However, the antibonding orbital ($\pi^*$) is antisymmetric with respect to m and symmetric with respect to $C_2$ axis.

In a similar way, the symmetry properties of four $\pi$ molecular orbitals of butadiene (Fig. 14.11) and six $\pi$ molecular orbitals of 1,3,5-hexatriene (Fig. 14.12) are given in Table 14.1.

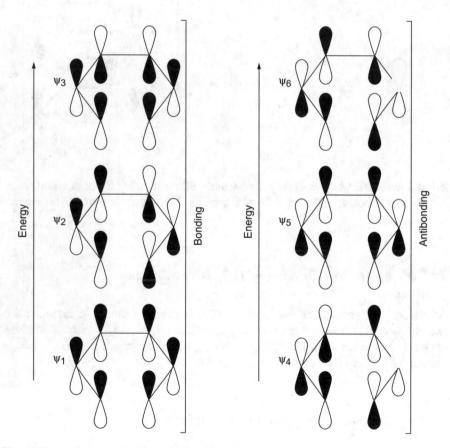

**Fig. 14.12**   Molecular orbitals of 1,3,5-hexatriene

**Table 14.1**   Symmetry properties of π molecular orbitals of butadiene and 1,3,5-hexatrienene

| Orbital | Butadiene | | Orbital | 1, 3, 5-Hexatriene | |
| --- | --- | --- | --- | --- | --- |
| | $m$ | $C_2$ | | $m$ | $C_2$ |
| $\psi_4$ | A | S | $\psi_6$ | A | S |
| $\psi_3$ | S | A | $\psi_5$ | S | A |
| $\psi_2$ | A | S | $\psi_4$ | A | S |
| $\psi_1$ | S | A | $\psi_3$ | S | A |
| | | | $\psi_2$ | A | S |
| | | | $\psi_1$ | S | A |

**Fig. 14.13**   Symmetry properties of ethylene in ground and excited state

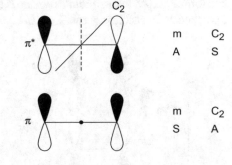

| | m | $C_2$ |
| --- | --- | --- |
| $\pi^*$ | A | S |
| $\pi$ | S | A |

**Fig. 14.14**   Symmetry properties of sigma orbitals of a C—C covalent bond

| | m | $C_2$ |
| --- | --- | --- |
| $\sigma^*$ | A | S |
| $\sigma$ | S | S |

In case of carbon–carbon covalent bond, the sigma orbitals have both mirror plane symmetry (since a rotation of 180° through its midpoint gives the same sigma orbital) and a $C_2$ symmetry (Fig. 14.14).

## 14.6   Electrocyclic Reactions

We know that in electrocyclic reaction, a single bond is formed between the termini of a conjugated polyene system and that the reaction can take place photochemically or thermally. We have seen that the cyclisation of *trans, cis, trans*-2,4,6-octatriene under thermal conditions yields *cis*-1,2-dimethyl cyclohexa-3,5-diene (as the sole product) and under photochemical conditions, the corresponding *trans* form (viz., *trans* 1,2-dimethyl cyclohexa-3,5-diene) as the sole product (*see* Fig. 14.7). The streochemistry of the cyclisation is so great that under thermal conditions less than one percent of the *trans* product is obtained though it is thermodynamically more stable.

The formation of different products (*cis* and *tran*) on cyclisation of *trans, cis, trans*-2,4,6-octatriene under thermal and photochemical conditions can be explained in the following way:

In case of 2,4,6-octatriene there are six molecular orbitals ($\psi_1$ to $\psi_6$) arising from six atomic orbitals (Fig. 14.15).

In case of 2,4,6-octatriene, $6\pi$ electrons (two per orbital) have to be accommodated. The HOMO (the electrons in the Highest Occupied Molecular orbital) will be $\psi_3$ molecular orbital (1). In order to form C–C sigma bond on cyclisation, the orbital lobes on the terminal carbons (carbon atoms carrying the methyl substituent) must rotate through 90° in case mutual overlap is to occur (in this case re-hybridisation must also occur). The necessary rotation can occur in the following two ways (Fig. 14.16).

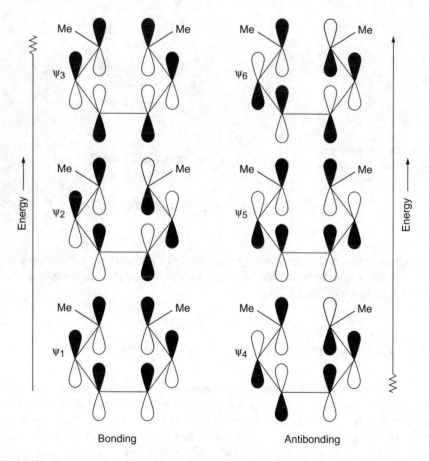

Bonding                                                     Antibonding

**Fig. 14.15**   Six molecular orbitals ($\psi$, to $\psi_6$) of 2,4,6-octatriene

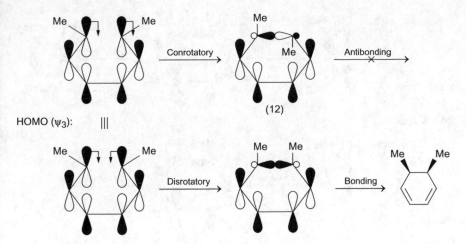

HOMO ($\psi_3$):    |||

**Fig. 14.16**  Conrotatory and disrotatory

– The two C–C sigma bonds rotate in the same direction, either clockwise or anticlockwise. Such a rotation is called **conrotatory**.
– The two C–C sigma bonds rotate in opposite directions. Such a rotation is called **disrotatory**.

As seen (Fig. 14.16), conrotatory movement results in orbital lobes on the terminal carbons being in opposite phase as in antibonding situation and disrotatory movement results in orbital lobes on the terminal carbons being in the same phase as in bonding situation. The later (disrotatory movement) leads to the formation of *cis*-1,2-dimethyl cyclohexa-3,5-diene.

On the other hand, irradiation during photochemical ring closure results in promotion of an electron into orbital of next higher energy level, *i.e.* $\psi_3 \xrightarrow{h\upsilon} \psi_4$. So in this case the HOMO which needs consideration is $\psi_4$ (Fig. 14.17).

In the above photochemical cyclisation (Fig. 14.17), the conrotatory movement results in the opposition of orbital lobes with the same phase (the bonding situation) given the *trans* isomer (*trans* 1,2-dimethyl cyclohexa-3,5-diene).

Interesting results are obtained in the cyclisation of hexa-2, 4-diene to 3,4-dimethyl cyclobutene. In the case exactly opposite stereochemistry is observed compared to cyclisation of 2,4,6-octatriene. Thus, cyclisation of *trans*, *trans* hexa-2,4-diene under thermal conditions yields *trans* 3,4-dimethylcyclobutene and under photochemical conditions yields the *cis* isomer (*cis* 3,4-dimethyl cyclobutene) (Fig. 14.18).

In case of thermal cyclisation (Fig. 14.19), the HOMO of the diene will be $\psi_2$ (Fig. 14.19). In this case four $\pi$ electrons are to be accommodated.

In the above thermal cyclisation (Fig. 14.19), the conrotatory movement gives, the diene in a bonding situation resulting in the formation of *trans* 3,4-dimethylcyclobutene (Fig. 16.19).

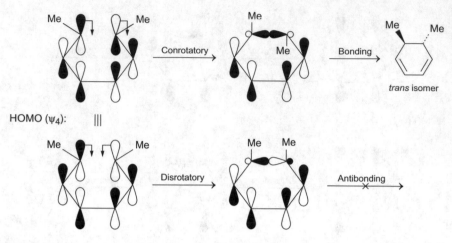

HOMO ($\psi_4$):    |||

**Fig. 14.17**    Photochemical cyclisation of 2,4,6-octatriene

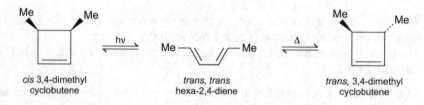

*cis* 3,4-dimethyl
cyclobutene

*trans, trans*
hexa-2,4-diene

*trans,* 3,4-dimethyl
cyclobutene

**Fig. 14.18**    Cyclisation of *trans, trans* hexa-2,4-diene

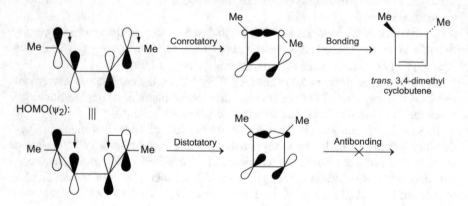

HOMO($\psi_2$):    |||

*trans,* 3,4-dimethyl
cyclobutene

**Fig. 14.19**    Conrotatory movement in thermal cyclisation of *trans, trans* hexa-2, 4-diene

However, in the photochemical cyclisation of *trans, trans* hexa-2,4-diene, an electron in the $\psi$ orbital is transformed into the next higher energy level, *i.e.*, $\psi_2 \rightarrow^{hv} \psi_3$ and it is the $\psi_3$ molecular orbital which takes part in the cyclisation (Fig. 14.20). In

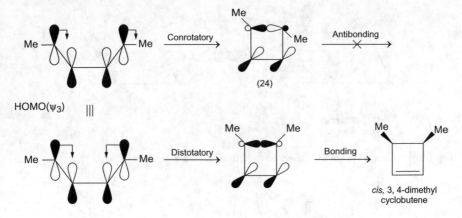

**Fig. 14.20**  Disrotatory movement in the photochemical cyclisation of *trans, trans* hexa-2,4-diene

**Table 14.2**  Type of bonding in some compounds

| No. of π electrons | Type of reaction | Motion for bonding |
|---|---|---|
| 4n | Thermal | Conrotatory |
| 4n | Photochemical | Disrotatory |
| 4n + 2 | Thermal | Disrotatory |
| 4n + 2 | Photochemical | Conrotatory |

this case, the disrotatory movement resulting in a bonding situation resulting in the formation of *cis*3,4-dimethyl cyclobutene.

On the basis of results obtained it is found that in thermal and photochemical cyclisation reactions of compounds containing different number of π electrons, the bonding of the type is given in Table 14.2.

Because of the rigid stereospecificity in electrocyclic reactions, these are of considerable interest in the formation of carbon–carbon bond formations.

## 14.6.1  *Frontial Molecular Orbital (FMO) Method*

The prediction of the outcome in an electrocyclic cyclisation is also possible by an alternate method known as Frontier Molecular Orbital (FMO) method. This method is very simple, the only guide being the symmetry of the highest occupied molecular orbital (HOMO) of the open chain unsaturated compound in the electrocyclic reaction. In case the orbital (HOMO) has a $C_2$ symmetry, the reaction follows a conrotatory mode. However, if the orbital (HOMO) has a mirror plane symmetry, the reaction follows a disrotatory mode.

As an example, in the electrocyclic cyclisation of butadiene to cyclobutene, $\psi_2$ is the highest occupied molecular orbital in butadiene and since it displays $C_2$ symmetry,

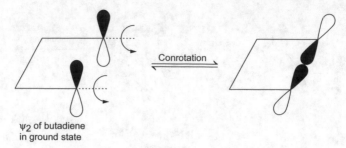

**Fig. 14.21**  Conrotation of $\psi_2$ molecular orbital of butadiene in ground state

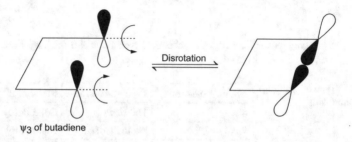

**Fig. 14.22**  Disrotation of $\psi$ molecular orbital of butadiene

ring closure is a conrotatory process. However, on irradiation, an electron from $\psi_2$ is transformed to $\psi_3$, which becomes the highest occupied molecular orbital and because $\psi_3$ has an $m$ symmetry, disrotation is necessary to effect ring closure. Similar approach can be used for other electrocyclic reactions. The FMO method is based on the fact that overlapping of wave functions of same sign is necessary for the formation of bond. As an example, butadiene molecule (in ground state), HOMO is $\psi_2$ and so cyclisation is possible only via conrotation (Fig. 14.21).

However, irradiation of butadiene promotes an electron to $\psi_3$ which in turn becomes the HOMO. The formation of bond takes place by disrotation of the $\psi_3$ HOMO (Fig. 14.22).

In a similar way, in the electrocyclic cyclisation of hexatriene to cyclohexadiene, the HOMO of hexatriene under thermal and photochemical conditions are $\psi_3$ and $\psi_4$, respectively. The reaction proceeds, as expected on heating by disrotation and by conrotation under photochemical conditions (Fig. 14.23).

## 14.7  Cycloaddition Reactions

Cycloaddition reactions involve reaction between two unsaturated molecules to form a cyclic product. The $\pi$ electrons of two $\pi$ systems lead to the formation of a ring.

**Fig. 14.23**  Thermal and photochemical cyclisation of hexatriene

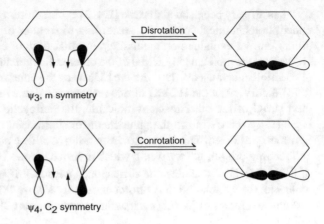

$\psi_3$, m symmetry

Disrotation

Conrotation

$\psi_4$, $C_2$ symmetry

These reactions are of three types, viz., $[2+2]$ cycloadditions, $[4+2]$ cycloadditions and 1, 3-dipolar cycloadditions.

## 14.7.1  *[2 + 2] Cycloadditions*

These reactions involve reaction of two molecule of an alkene to give cyclobutane derivatives. The cycloaddition reactions take place only under photochemical conditions but not thermally (Fig. 14.24).

In the above cycloaddition reaction, the orbitals of two molecules of the alkene may overlap in two modes. In case the bond formation takes place between two alkene molecules on the same side of the π-orbital, the addition is known as **suprafacial** and in case the bond formation is from opposite side of the π bond, the addition is known as **antarafacial** (Fig. 14.25).

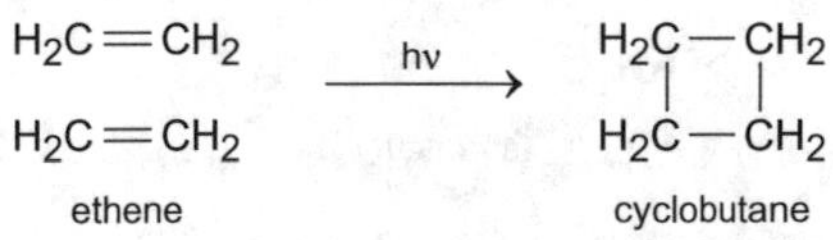

**Fig. 14.24**  [2 + 2] Cycloaddition

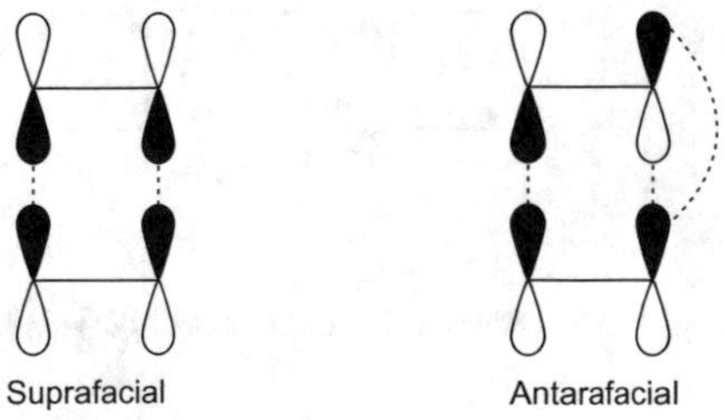

$$H_2C = CH_2$$
$$H_2C = CH_2$$
ethene

$\xrightarrow{h\nu}$

$$H_2C - CH_2$$
$$\phantom{H_2C} |\phantom{--}|$$
$$H_2C - CH_2$$
cyclobutane

**Fig. 14.25**  Suprafacial and antarafacial

Suprafacial

Antarafacial

It has already been stated that the [2 + 2] cycloaddition reactions take place only under photochemical conditions. According to frontier orbital theory, the cycloadditions can be explained on the assumption that flow of electrons occurs from the HOMO of one molecule to LUMO of the other molecule. In other words, the HOMO of one molecule must overlap with the LUMO of the other molecule. Also, according to FO theory when the HOMO of one reactant molecule has the same symmetry as the LUMO of the other reactant molecule, the pericyclic reaction is allowed; this is generally referred to as the photochemical suprafacial mode of cycloaddition is **Symmetry allowed**. This is because irradiation of ethene promotes an electron to the antibonding orbital, $\pi^*(\pi \xrightarrow{h\nu} \pi^*)$, which in turn becomes HOMO. As the symmetry of HOMO ($\pi^*$) of one ethylene corresponds with LUMO ($\pi^*$) of another ethylene molecule, the cycloaddition is **photochemically allowed** (Fig. 14.26).

Some examples of [2 + 2] cycloaddition reactions are shown below (Fig. 14.27).

HOMO   ($\xrightarrow{h\nu} \pi^*$)

$\xrightarrow{h\nu}$   cyclobutane

LUMO        ($\pi^*$)

**Fig. 14.26**  [2 + 2] cycloaddition of ethene under photochemical conditions

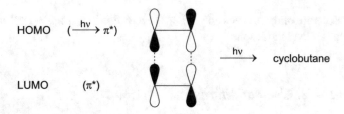

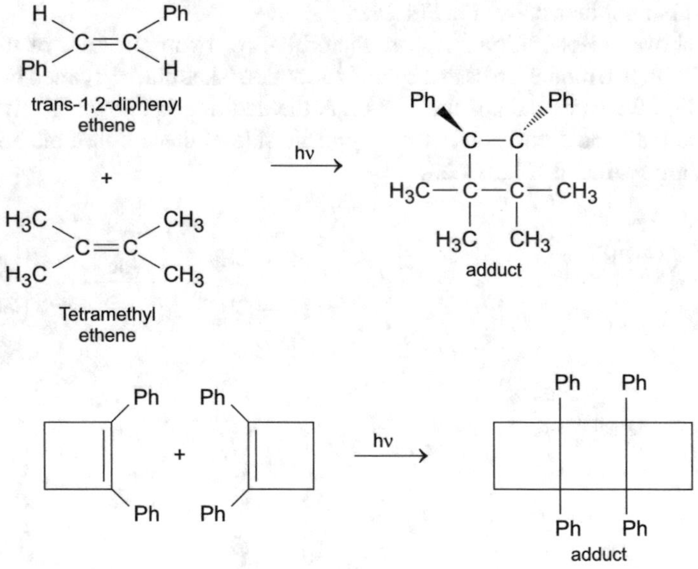

**Fig. 14.27**  Some other examples of [2 + 2] cycloaddition reactions

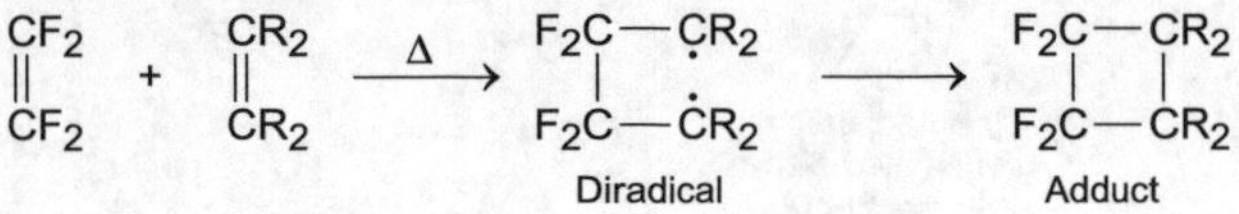

Fig. 14.28 [2 + 2] cycloadditions under thermal conditions

Fig. 14.29 π bonding and π* antibonding orbitals of ethene

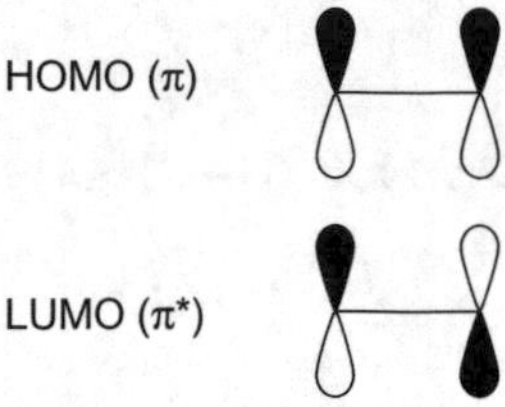

HOMO (π)

LUMO (π*)

The (2 + 2) cycloadditions reactions so far discussed are concerted reactions. However, certain dienes of the type $R_2C == CF_2$ or $CH_2 = CH–X$ (X = –COR, –CN, –COOR, etc.) proceed on heating under thermal conditions. However, these reactions proceed via radical mechanism (Fig. 14.28).

It has already been stated that ethene does not undergo [2 + 2] cycloaddition reaction under thermal conditions. This is explained as follows. Ethene is known to have two π molecular orbitals (π and π*) with wave functions $\psi_1$ and $\psi$. The π bonding orbital in the ground state is π bonding orbital and π* is the antibonding orbital and is LUMO as shown below (Fig. 14.29).

Under thermal conditions, ethene on heating the π electrons are not promoted (to π*) and remain in the ground state. It is due to this reason that there is no reaction of ethene under thermal conditions. Such a reaction is **symmetry forbidden**. As already stated, such a reaction can take place under different conditions, but the reaction will take place in a stepwise manner via radical intermediate (*see* Fig. 14.29).

### 14.7.2 [4 + 2] Cycloadditions

These reactions involve the addition of an alkene (called a dienophile, with 2π electrons) to a conjugated diene (with 4π electrons) to form the adduct (a cyclohexene derivative). The well-known Diels–Alder reaction is the thermal (4 + 2) cycloaddition reaction (Fig. 14.30).

In the Diels–Alder reaction, the diene (butadiene) must be cisoid (referring to the geometry around the single bond). The two conformations of butadiene are as shown in (Fig. 14.31).

The [4 + 2] cycloaddition of butadiene with ethene is thermally allowed only; such additions are not possible under photochemical conditions. This can be explained on the basis of the following. We know that a new sigma bond in [4 + 2] cycloaddition is formal by the HOMO–LUMO interaction of the p-orbitals in the diene and

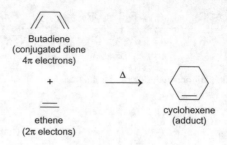

**Fig. 14.30**  [4 + 2] cycloaddition reaction (Diels–Alder reaction)

**Fig. 14.31**  Two conformations of butadiene

dienophile. In the thermally induced reaction, consider the π-electrons flowing from HOMO ($\psi_2$) of the diene (butadiene) to the LUMO ($\psi_2$) of the dienophile (ethene) (Fig. 14.32). As seen (Fig. 14.32) the reaction is symmetrically allowed.

However, under photochemical conditions, the diene (butadiene) on excitation by light, its $\psi_2$ HOMO becomes $\pi_3^*$-orbital (Fig. 14.33), which cannot overlap with

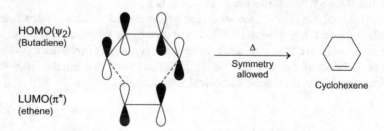

**Fig. 14.32**  Thermal Diels–Alder reaction (symmetry allowed)

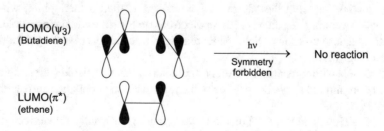

**Fig. 14.33**  Photochemical Diels–Alder reaction (symmetry forbidden)

the LUMO of the dienophile (ethene) as the two are not in phase. So the [4 + 2] photo-induced [4 + 2] cycloaddition is symmetry forbidden.

For more details about Diels–Alder reaction *see* Sect. 19.2.

### 14.7.3 1,3-Dipolar Cycloadditions

The [4 + 2] cycloadditions involve the reaction of a diene (as in butadiene) and the only requirement for a concerted pathway is the symmetry requirements for HOMO | LUMO. However, there are some non-dienic $4\pi$ electron systems involving three atoms and having one or more dipolar cononical structures and so the name 1,3-dipolar addition. Such structures, however, need not possess a residual dipole, as in the case of diazomethane (*a* ↔ *b*).

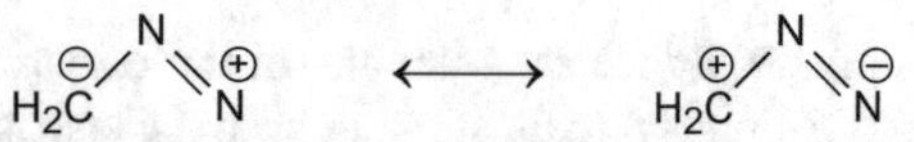

(canonical structures of diazomethane).

The initial addition of ozone to an alkene to form molozonide is regarded as 1,3-dipolar cycloaddition (Fig. 14.34).

In the above case (Fig. 14.34), the fragments, the carbonyl compound and a peroxy zwitter ion undergo 1,3-dipolar cycloaddition to yield the ozonide some other examples of 1,3-dipolar cycloadditions are (*i*) the reaction of phenyl azide with alkenes to give dihydrotriazoles and (*ii*) the reaction of diazomethane with ethyl acrylate to give dihydropyrazole (Fig. 14.35).

## 14.8 Sigmatropic Rearrangements

Sigmatropic rearrangements involve migration of a single bond (sigma bond) adjacent to one or more π systems. Due to the rearrangement of a sigma bond, these

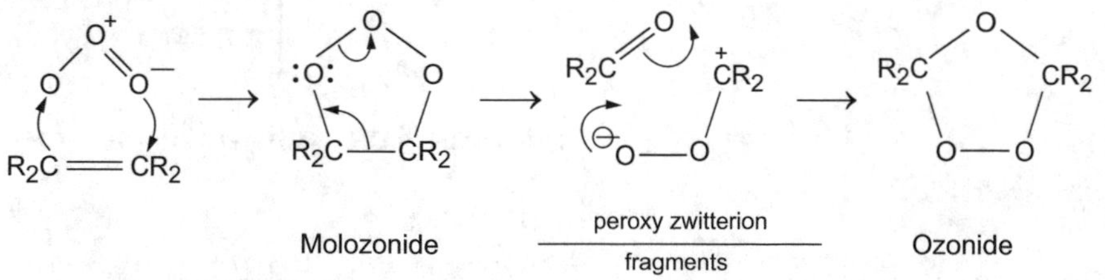

**Fig. 14.34** Formation of ozonide

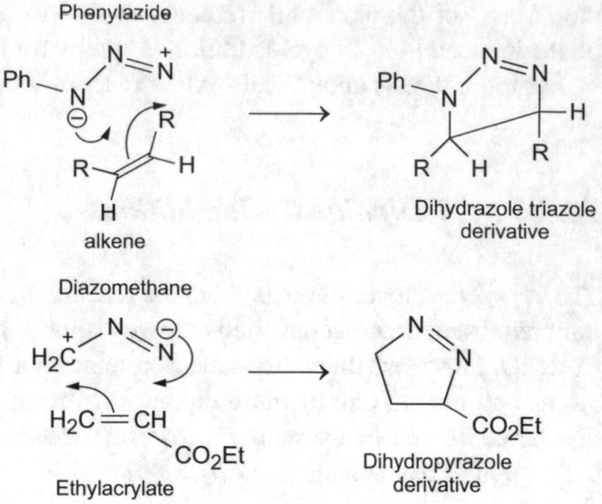

**Fig. 14.35**  Some examples
of 1,3-dipolar cycloaddtiions

Phenylazide

alkene

Dihydrazole triazole
derivative

Diazomethane

Ethylacrylate

Dihydropyrazole
derivative

reactions are called sigmatropic rearrangements. In these rearrangements, the total
number of σ-bonds and π-bonds remain same in both the starting and final product.

Sigmatropic rearrangements are generally of two types, viz., migration of a σ-
bond that carries a hydrogen atom and migration of a σ-bond that carries a carbon
substitutent.

## 14.8.1  Hydrogen Shifts

These are designated with two numbers, $i$ and $j$ set in brackets $(i, j)$ and refer to the
relative positions of the atoms which are involved in the migration. Some examples
of sigmatropic rearrangements involving shift of H atoms are [1, 3], [1, 5] and [1, 7]
rearrangements (Fig. 14.36).

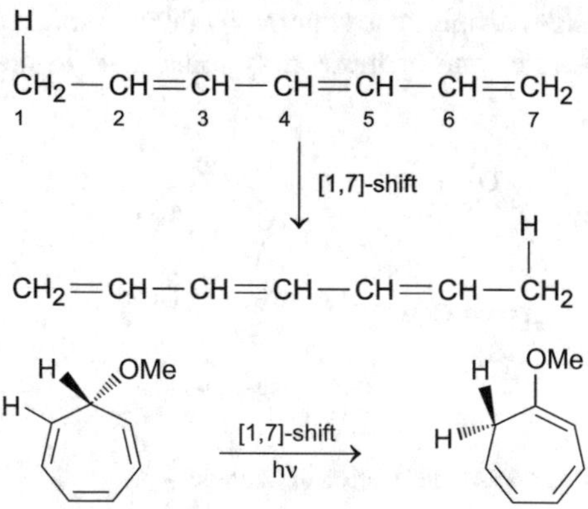

**Fig. 14.36**  Some examples
of sigmatropic
rearrangements involving
shift of H atoms

*[1, 3]-Rearrangement.*

$$\underset{1}{CH_2}\overset{\displaystyle H}{|}-\underset{2}{CH}=\underset{3}{CH_2} \xrightarrow{\text{[1,3]-shift}} CH_2=CH-\overset{\displaystyle H}{\underset{|}{CH_2}}$$

*[1, 5]-Rearrangement.*

$$\underset{1}{\overset{\displaystyle H}{\underset{|}{CH_2}}}-\underset{2}{CH}=\underset{3}{CH}-\underset{4}{CH}=\underset{5}{CH_2} \xrightarrow{\text{[1,5]-shift}} CH_2=CH-CH=CH-\overset{\displaystyle H}{\underset{|}{CH_2}}$$

*[1, 7]-Rearrangement.*

In the sigmatropic rearrangements (as referred above) there is migration of a σ bond across the π-electron system. This can occur by two different stereochemical courses. In case, the migrating σ bond moves across the same face of a conjugated system, it is known as **suprafacial process**. On the other hand, if the migrating σ bond is formed on the opposite side of π-electron face of the conjugated system, it is known as **antarafacial process**. Both the stereochemical courses (using [1, 5]-sigmatropic shift) are shown in (Fig. 14.37).

The suprafacial migrations are more common (due to steric reasons) than the antarafacial migrations. However, in systems containing long chain conjugated systems, the antarafacial migration is possible.

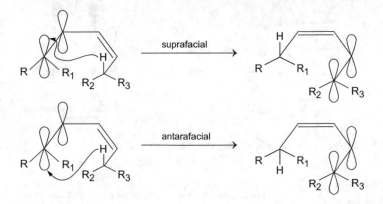

**Fig. 14.37** Suprafacial and antarafacial process in [1, 5] sigmatropic shifts

## 14.8.2  Analysis of Sigmatropic Rearrangements

It is assumed that the migrating bond undergoes homolytic cleavage to form a hydrogen radical and pentadienyl radical (Fig. 14.38).

The pentadienyl radical (Fig. 14.38) contains five π-electrons and so there are five π-molecular orbitals as shown in Fig. 14.39. Since in the ground state the HOMO is $\psi_3$, the hydrogen shift is governed by the symmetry of $\psi_3$ of pentadienyl radical.

The $\psi_3$ orbital has similar sign on the terminal lobes. So 1,5 hydrogen shifts is **thermally allowed** and occurs in suprafacial mode.

In case the above sigmatropic rarrangement is carried out photochemically, the HOMO is $\psi_4$ and it is unsymmetrical. So [1, 5] suprafacial migration is not possible and is symmetry forbidden.

On the basis of a number of sigmatropic rearrangements, selection rules have been formulated (Table 14.3).

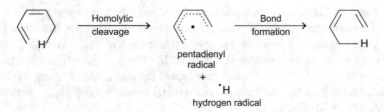

**Fig. 14.38**  Homolytic cleavage of the migrating bond

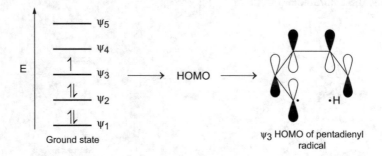

**Fig. 14.39**  Molecular orbitals of pentadienyl radical

**Table 14.3**  Selection rules for sigmatropic rearrangements

| Number of electrons $(i + j)$ | Ground state (Thermally allowed) | Excited state (Photochemically allowed) |
|---|---|---|
| $4n$ | Antarafacial | Suprafacial |
| $4n + 2$ | suprafacial | Antarafacial |

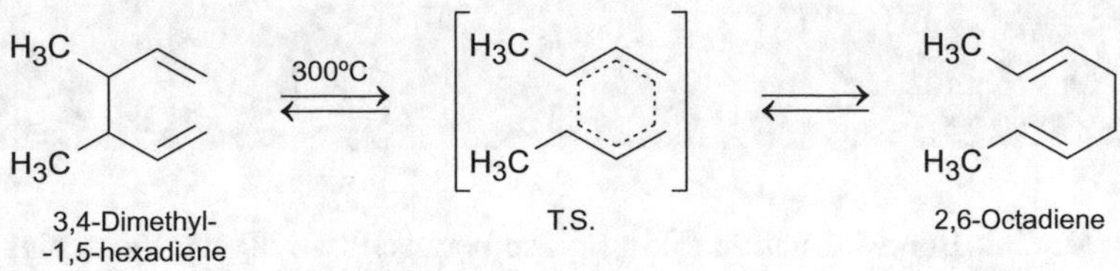

Fig. 14.40    Cope rearrangement

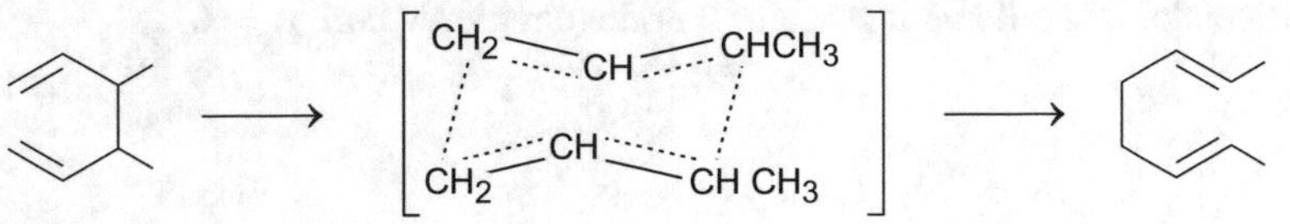

Fig. 14.41    Mechanism of cope rearrangement

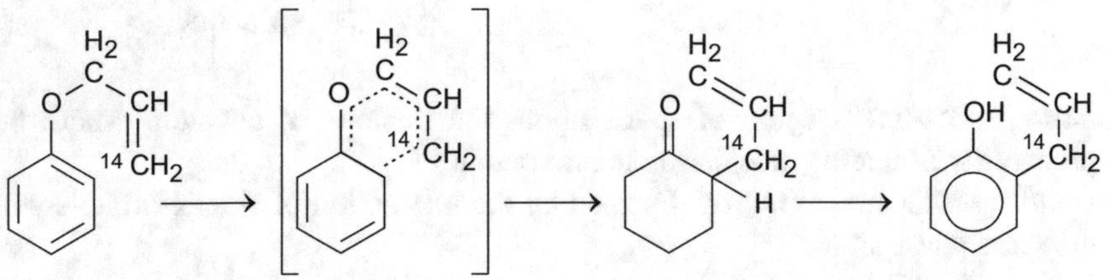

Fig. 14.42    Claisen rearrangement

## 14.8.3    Carbon Shifts

A typical example, involving a carbon moiety, is the shift from one carbon to another as in the cope rearrangement of 1,5-dienes, viz., 3,4-dimethyl-1,5-hexadiene to 2,6-octadiene (Fig. 14.40).

The cope rearrangement proceeds via a chair-like transition state, which is preferred over the boat-like transition state by about 5.7 kcal/mol. In case, the chair transition is inaccessible then the rearrangement proceeds via boat-like transition state (Fig. 14.41).

The chemical shift can also be from oxygen to carbon as in the Claisen rearrangement of allyl aryl ethers (Fig. 14.42).

The claisen rearrangement is intramolecular. [14]C labelling indicates that the position of the labelled carbon atom in the allyl group is 'inverted' during migration (Fig. 14.42).

For more details about Claisen rearrangement *see* Sect. 13.2.4.2.

## Key Concepts

- **Antrafacial**: Bond formation taking place between two alkene molecules from opposite side of the π-bond.
- **Antibonding Molecular Orbital**: Formed by the overlap of atomic orbitals with opposite phase sign. The electron probability density in the region between the two nuclei is small and it contains a node (a region where $\psi = 0$).

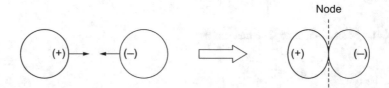

- **Atomic Orbital**: Region of space about the nucleus of an atom where the probability of finding an electron is maximum.
- **Bonding Molecular Orbital**. Formed by the interaction of atomic orbitals with the same phase sign.

- **Conrotatory**: Rotation of two C–C sigma bonds in the same direction, either clockwise or anticlockwise.
- **Cycloaddition Reactions**: Reactions involving formation of a ring from the reaction of two two π systems.

- **Disrtatory**: Rotation of two C–C sigma bonds in opposite direction.
- **Electrocyclic Reactions**: Reactions involving formation of a single bond between the termini of a conjugated system.

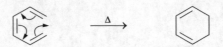

- **Molecular Orbital**: Formed by the overlap of atomic orbitals. These correspond to the region of space encompassing two or more nuclei where electrons are found and can hold up to 2 electrons (like atomic orbitals) in case their spins are paired.
- **Molecular Orbital Theory**: Deals with linear combination of atomic orbitals (LCAO) resulting in the formation of molecular orbitals.
- **Pericyclic Reactions**: Reactions involving the formation of a cyclic transition state and can be initiated by heat or light and are highly stereospecific
- **Suprafacial**: Bond formation taking place between two alkene molecules on the same side of the $\pi$-orbital.
- **Symmetry Allowed**: In a reaction involving a cyclic transition state (*i.e.*, percyclic reaction) if the orbitals in the reactants are the same (having same symmetry properties), the reaction is symmetry allowed, failing which it is **symmetry forbidden**.
- **Symmetry Forbidden**: *See* symmetry allowed.
- **Sigmatropic Rearrangements**: Reactions involving migration of a single bond (sigma bond) adjacent to one or more $\pi$ systems.

- **1, 3-Dipolar Additions**: Reactions of non-dienic $4\pi$-electron systems involving three atoms and having one or more dipolar cononical structures with alkenes as in formation of ozonide.
- **[2 + 2] Cycloaddition Reactions**: Reactions involving the reaction of two molecule of an alkene give cyclobutane derivatives. These reactions take place only under photochemical conditions and not thermally.

$$H_2C=CH_2 \atop H_2C=CH_2 \xrightarrow{h\nu} \begin{matrix} H_2C-CH_2 \\ | \quad | \\ H_2C-CH_2 \end{matrix}$$

- **[4 + 2] Cycloaddition Reactions**: Reactions involving addition of an alkene (dienophile with $2\pi$ electrons) to a conjugated diene (with $4\pi$ electrons in cisoid form) to form an adduct.

## Problems

1.  What are pericyclic reactions? How many types of pericyclic reactions are there?
2.  Discuss the stereochemistry of pericyclic reactions.
3.  Explain the terms, atomic orbital, molecular orbital, bonding molecular orbital, antibonding molecular orbital.
4.  Discuss briefly the molecular orbital theory.
5.  Discuss the symmetry properties of orbitals.
6.  Explain the following:

    (a)  Suprafacial and antarafacial
    (b)  Conrotatory and disrotatory.
    (c)  HOMO and LUMO
    (d)  Symmetry allowed and symmetry forbidden.

7.  Discuss Frontial Molecular Method for analysis an elctrocyclic reaction.
8.  What products are obtained in thermal and photochemical cyclisation of hexatriene?
9.  Discuss different types of cycloaddition reactions giving examples.
10. Discuss different types of sigmatropic rearrangements.

# Part IV
# Stereochemistry of Heterocyclic Compounds

# Chapter 15
# Stereochemistry of Some Compounds Containing Heteroatoms

Besides carbon, a number of other elements exhibit optical isomerism. The organic compounds that contain heteroatoms (like nitrogen, phosphorus, sulphur etc.) are important as far as their stereochemistry is concerned.

## 15.1  Stereochemistry of Nitrogen Compounds

Nitrogen compounds that are of interest include amines, quaternary ammonium salts, tertiary amine oxides, oximes, some other tetravalent nitrogen compounds containing a double bond and heterocyclic compounds containing nitrogen.

### 15.1.1  Stereochemistry of Amines

The nitrogen in an amine is surrounded by three atoms and one non-bonded electron pair making the nitrogen atom $sp^3$-hybridised and trigonal pyramidal; the bond angles are approximately 109.5°.

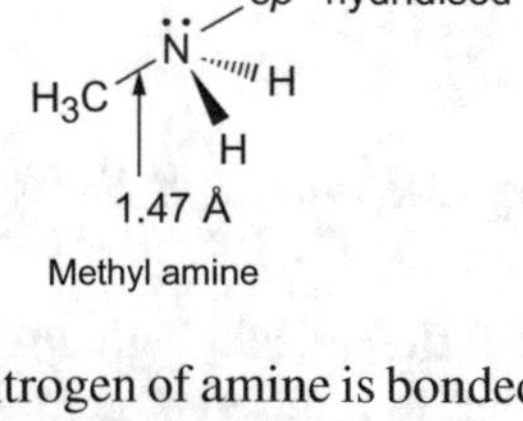

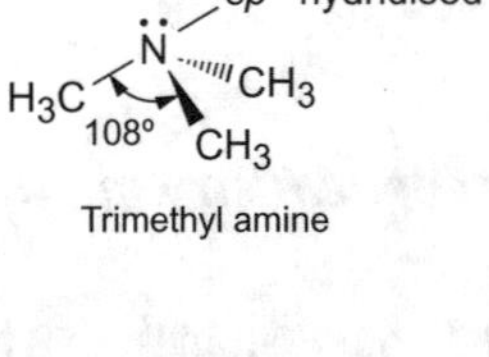

Methyl amine

Trimethyl amine

In case, nitrogen of amine is bonded to three different alkyl groups and an electron pair, it can be said that nitrogen is a stereogenic centre; in such a case, two non-superimposable trigonal pyramids can be drawn as shown ahead (Fig. 15.1).

© The Author(s), under exclusive license to Springer Nature Switzerland AG 2022
V. K. Ahluwalia, *Stereochemistry of Organic Compounds*,
https://doi.org/10.1007/978-3-030-84961-0_15

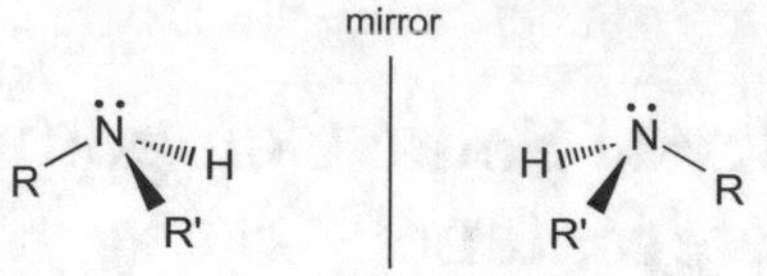

mirror

**Fig. 15.1** Non-superimposable mirror image of an amine with four different groups around N

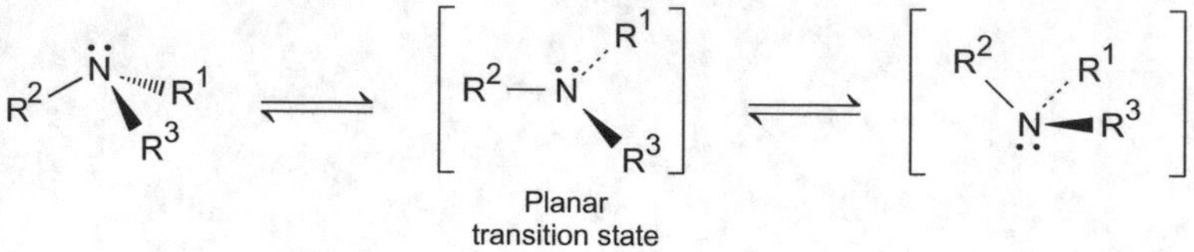

Planar
transition state

**Fig. 15.2** Conversion of one enantiomer into another via a planar transition state

**Fig. 15.3** Tröger's base

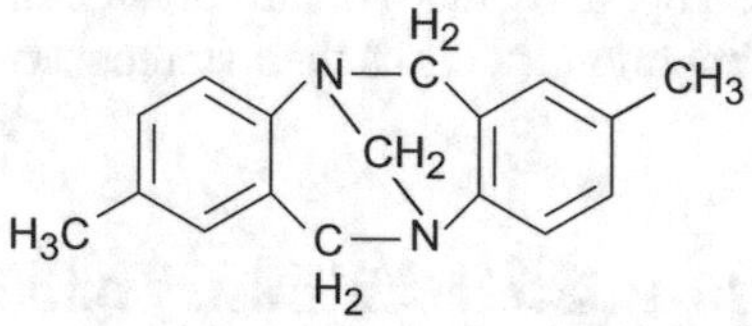

However, it does not mean that such an amine (Fig. 15.1) exists in two different enantiomers. This is because, at room temperature, one form is rapidly converted into the other form by an inversion process involving a planar transition state (Fig. 15.2).

It is found that in case of aliphatic tertiary amines, the inversion of this type (Fig. 15.2) occurs $10^3$ to $10^5$ times per second. In view of this, resolution of such amines into optical isomers is not possible by any of the available techniques. Since the two enantiomers interconvert, the chirality of the amine nitrogen can be ignored. However, in case the nitrogen atom is a part of the ring system, then the compound will be reasonably optically stable and can be isolated. This has been confirmed (Prelog and Wieland, 1944) by resolving Tröger's base (Fig. 15.3) by chromatographic adsorption on D-lactose.

The stereochemistry of heterocyclic compounds containing nitrogen forms the subject matter of Sect. 15.1.6.

## 15.1.2   Stereochemistry of Quaternary Ammonium Salts

In contrast to tertiary amines (having three different alkyl groups attached to nitrogen), the quaternary ammonium salts having four different groups attached to nitrogen are chiral. In this case, there is no non-bonded electron pair on the nitrogen atom, interconversion is not possible and the nitrogen atom is like carbon with four

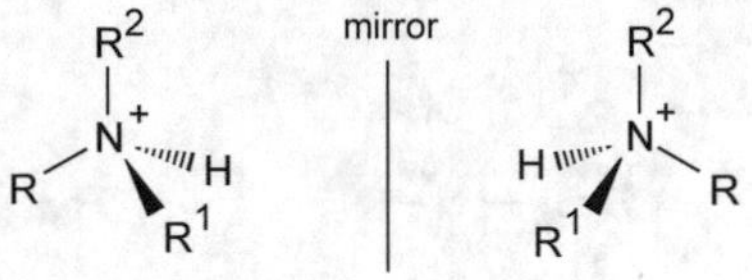

**Fig. 15.4** Two enantiomers of a quaternary ammonium salt

different groups around it. Thus, the nitrogen atom of a quaternary ammonium salt is a stereogenic centre when it is surrounded by four different groups (Fig. 15.4).

The first quaternary ammonium salt was resolved by Pope and Peachey (1899), who resolved allybenzyl methyl phenyl ammonium iodide by means of (+)-bromocamphorsulphonic acid. Subsequently, Jones (1905) resolved benzyl ethyl phenyl ammonium iodide (Fig. 15.5).

It was shown that the configuration of the quaternary ammonium salts is tetrahedral (Mills and Warren, 1925).

A quaternary ammonium salt, 4-carbethoxy-4'-phenylbispiperidinium-1, 1'-spiran bromide was prepared (Mills and Warren) and resolved. The spiran had no elements of symmetry and so it is resolvable. Since it could be resolved, it must be tetrahedral (Fig. 15.6).

It was found (Hanby and Rydon, 1945) that the diquaternary salts of dimethyl piperazine exhibited geometrical isomerism; this could be explained by the tetrahedral configuration of the nitrogen (Fig. 15.7).

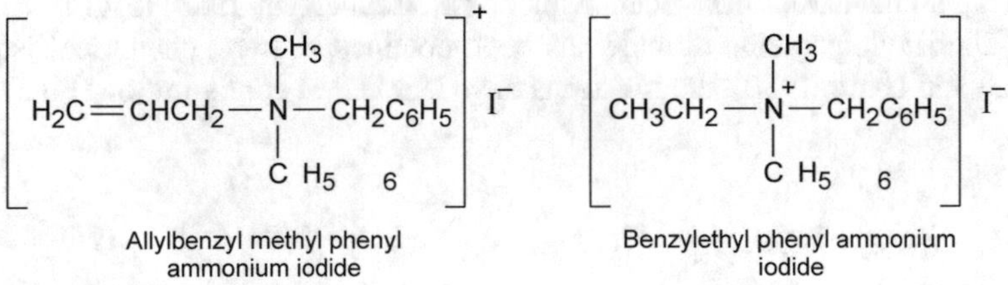

Allylbenzyl methyl phenyl ammonium iodide

Benzylethyl phenyl ammonium iodide

**Fig. 15.5** Quaternary salts resolved into enantiomers

**Fig. 15.6** 4-Carbethoxy-4'-phenylbisperidinium-1,1'-spiran bromide

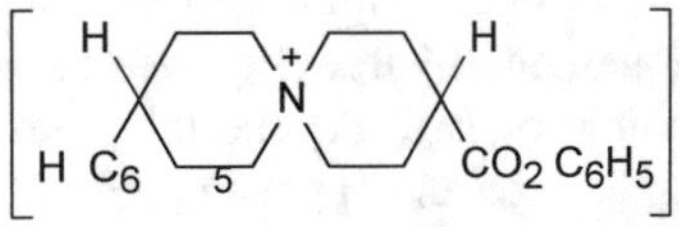

**Fig. 15.7** *cis* and *trans* forms of diquaternary salts of dimethyl piperazine

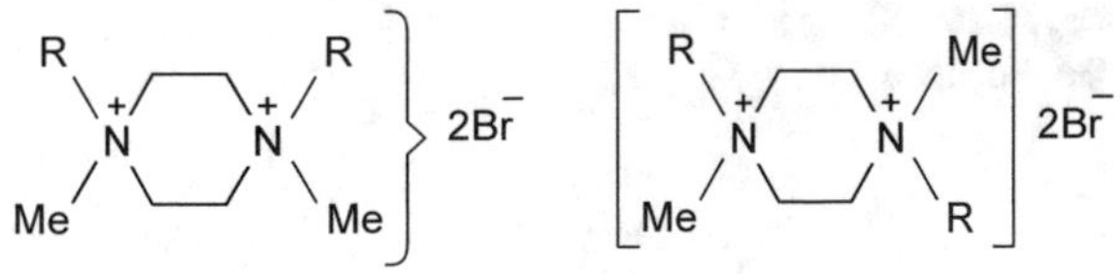

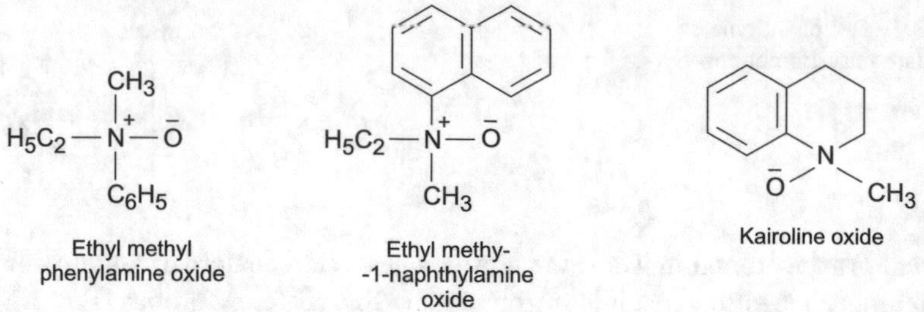

Ethyl methyl
phenylamine oxide

Ethyl methy-
-1-naphthylamine
oxide

Kairoline oxide

**Fig. 15.8**  Some tertiary amine oxides have been resolved

**Fig. 15.9**
1,4-Diphenylpeperazine
dioxide

## 15.1.3   Stereochemistry of Tertiary Amine Oxides

Tertiary amine oxides are prepared by the reaction of tertiary amines with $H_2O_2$. The nitrogen atom in tertiary amine oxides is joined into four different groups. On the basis of this, these have tetrahedral configurations. Such compounds can be resolved. Following amine oxides have been resolved (Meisenheimer, 1908) (Fig. 15.8).

1,4-Diphenylpiperazine dioxide has been obtained in two geometrical isomers (Bennett and Gynn, 1950), the nitrogens have tetrahedral configuration (Fig. 15.9).

## 15.1.4   Stereochemistry of Oximes

The oximes are prepared by the action of hydroxylamine on aldehydes or ketones and contain $C==N$ bonds. As an example, benzaldehyde gives two oximes (Fig. 15.10).

The oximes exhibit geometrical isomerism. The aldoxime in which both the hydrogen and hydroxyl groups are on the same side is called the *syn* form (corresponding to *cis* form) and the aldoxime in which both the hydrogen and hydroxyl groups are on the opposite side is called anti-form (corresponding to trans form).

**Fig. 15.10**   Oximes of
benzaldehyde

*cis* or *syn*
$\alpha$-Benzaldoxime

*trans* or *anti*
$\beta$-Benzaldoxime

Oximes are also named by E-Z system of nomenclature. The *syn* oxime is named benzaldehyde (E)-oxime or (E)-benzaldehyde oxime. The anti-oxime is named benzaldehyde (Z)-oxime or (Z)-benzaldehyde oxime. The group with higher priority (phenyl) is taken *cis* with respect to the hydroxyl group. In case of *syn-p*-tolyl phenyl ketoxime or anti-phenyl *p*-tolyl ketoxime, the *p*-tolyl has a priority over phenyl, the oxime is named (Z)-*p*-tolyl phenyl ketoxime (Fig. 15.11).

The configuration of the oxime is determined as follows. In case of aldoximes (*e.g.*, benzaldoxime), it is converted first into the corresponding acetyl derivative and then treated with aqueous $Na_2CO_3$. In case the original oxime is regenerated, the oxime is designated syn isomer or E isomer (Figs. 15.12 and 15.13).

However, if the acetate of the oxime on treatment with *aq.* $Na_2CO_3$, a cyanide is formed, then the oxime is Z or anti-oxime (Fig. 15.13).

The configuration of keto oximes (aromatic oximes) can be determined by Beckmann rearrangement (Sect. 13.2.3.1). Thus, the keto oximes on treatment with acidic

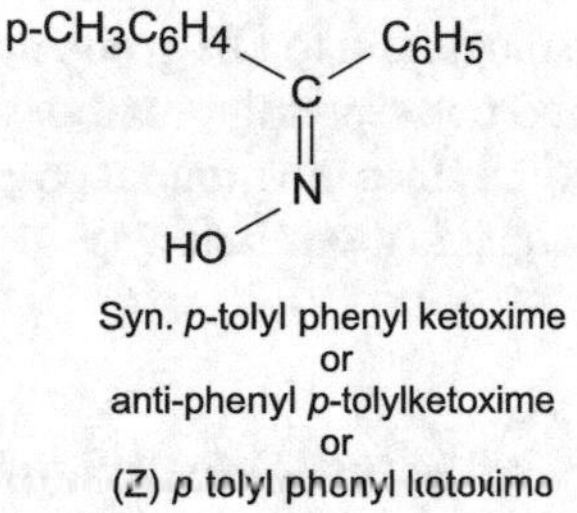

Syn. *p*-tolyl phenyl ketoxime
or
anti-phenyl *p*-tolylketoxime
or
(Z) *p* tolyl phenyl ketoxime

**Fig. 15.11**   Oxime of *p*-tolyl phenyl ketone

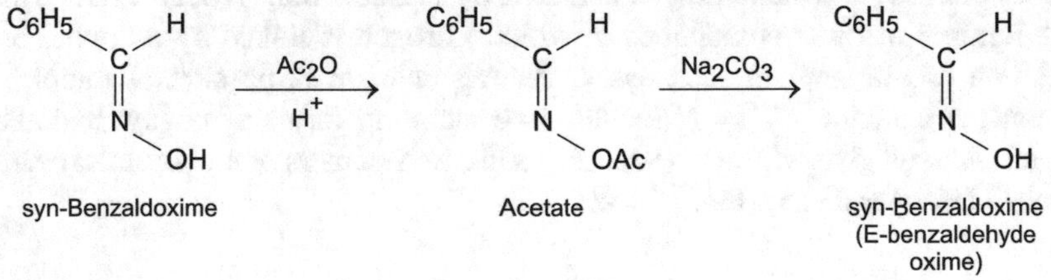

**Fig. 15.12**   Determination of configuration of syn-benzaldoxime

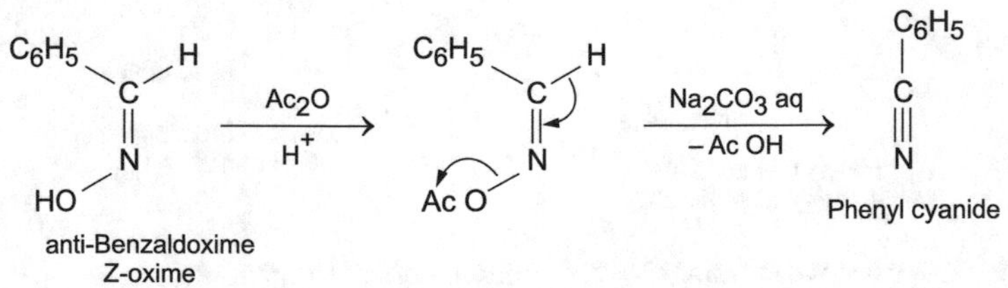

**Fig. 15.13**   Determination of configuration of anti-benzaldoxime

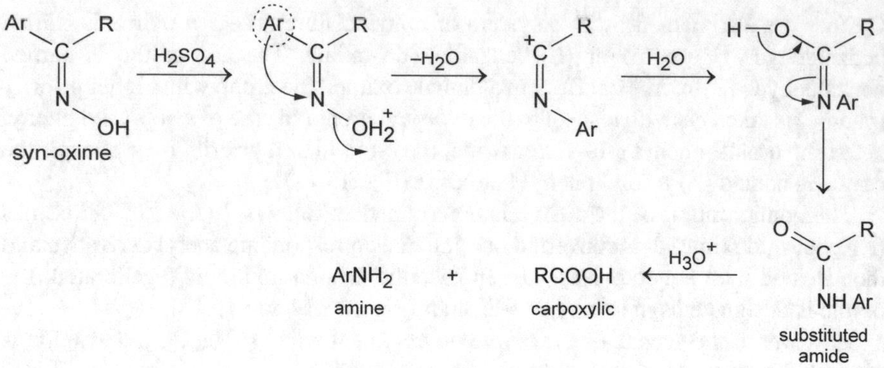

**Fig. 15.14**   Beckmann rearrangement

reagents ($H_2SO_4$, $P_2O_5$ etc.) undergo rearrangement to give a substituted amide. In this rearrangement, the group that is anti to OH group migrates; its structure is determined by its hydrolysis to the corresponding carboxylic acid. This indicates the configuration of the parent oxime since the group incorporated in the oxime must be anti to the OH group. An example is given below (Fig. 15.14):

## 15.1.5   Stereochemistry of Some Tetravalent Nitrogen Compounds Containing a Double Bond

Besides oximes (in which nitrogen is linked by a double bond) (Sect. 15.1.4), there are a number of other compounds in which nitrogen is linked by a double bond to carbon or nitrogen. In such cases, stereoisomerism is possible. Examples of compounds containing C = N include two isomeric forms of phenyl hydrazone of o-nitrophenyl glyoxylic acid (Krause, 1890), two isomers of monosemicarbazone of benzil (Hopper, 1925) (Fig. 15.15).

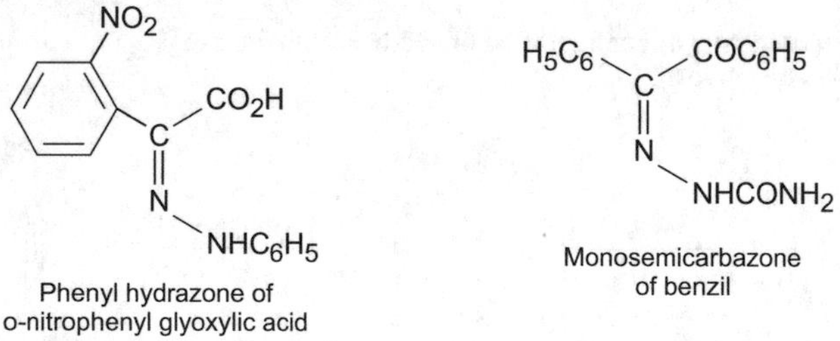

Phenyl hydrazone of
o-nitrophenyl glyoxylic acid

Monosemicarbazone
of benzil

**Fig. 15.15**   Compounds containing C = N exhibiting stereoisomerism

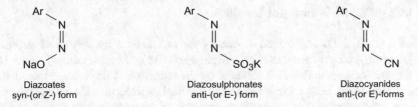

**Fig. 15.16**   Compounds containing N = N exhibiting geometrical isomerism

**Fig. 15.17**   Geometrical isomersism in azobenzene and azoxy benzene

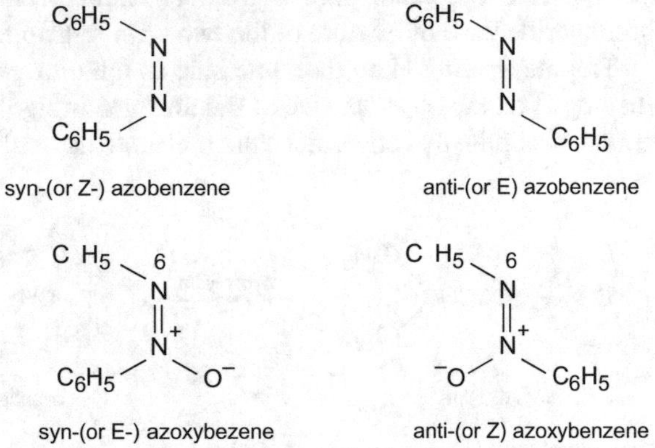

C$_6$H$_5$—N‖N—C$_6$H$_5$

syn-(or Z-) azobenzene

C$_6$H$_5$—N‖N—C$_6$H$_5$

anti-(or E) azobenzene

C H$_5$ 6 —N‖N$^+$—C$_6$H$_5$  O$^-$

syn-(or E-) azoxybezene

C H$_5$ 6 —N‖N$^+$— $^-$O  C$_6$H$_5$

anti-(or Z) azoxybenzene

The two forms (syn and anti) in the examples given above (Fig. 15.15) can be distinguished on the basis of their NMR spectra.

Geometrical isomersim is also possible in compounds containing N = N. Examples include diazoates, diazosulphonates and diazocyanides (Fig. 15.16).

Other compounds that contain N = N and exhibit geometrical isomerism are azoxybenzene (the ordinary azobenzene is the anti-isomer) and azoxybenzene (in which one N is tetracovalent and the other quadricovalent) (the ordinary azoxy benzene is anti-isomer) (Fig. 15.17).

## 15.1.6   Stereochemistry of Some Heterocyclic Compounds Containing Nitrogen

Most of the heterocyclic compounds containing nitrogen are alkaloids. Stereochemistry of such compounds is discussed below.

### 15.1.6.1    Stereochemistry of Tropines

Tropine ($C_8H_{15}NO$) is obtained as one of the products of hydrolysis of Atropine ($C_{17}H_{23}NO_3$), which occurs in deadly nightshade (*Atropa belladonna*) (Fig. 15.18).

Tropine (or tropanol) is also obtained by the reduction of tropinone by oxidation with chromic acid. In fact, tropinone can be reduced to give a mixture of two alcohols, tropine and $\psi$-tropine (pseudo-tropine), the ratio of the two alcohols depends on the nature of the reducing agent. Catalytic hydrogenation (Pt), electrolytic reduction, Zn dust and hydriodic acid gave tropine. However, reduction with sodium amalgam and sodium in ethanol gave $\psi$-tropine. Lithium aluminium hydride and sodium borohydride gave of mixture of the two with $\psi$-tropine predominating (Fig. 15.19).

Tropine (having H on the same side as the nitrogen bridge) and pseudo-tropine (having H on the opposite side of the nitrogen bridge) are epimers. Neither of these epimers is optically active, since the molecule has a plane of symmetry. C-1 and C-5

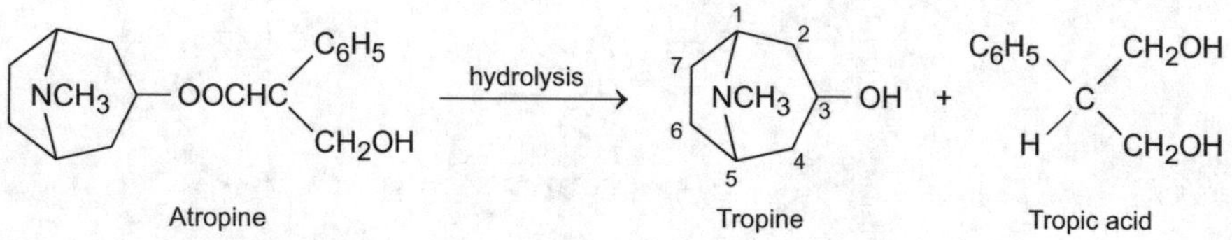

Fig. 15.18   Formation of tropine from atropine

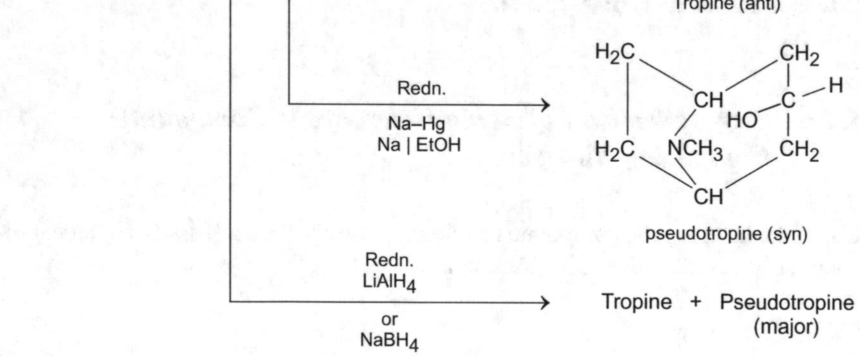

Fig. 15.19   Formation of tropine and pseudo-tropine

**Fig. 15.20** Boat conformation of piperidine ring in pseudo-tropine and tropine

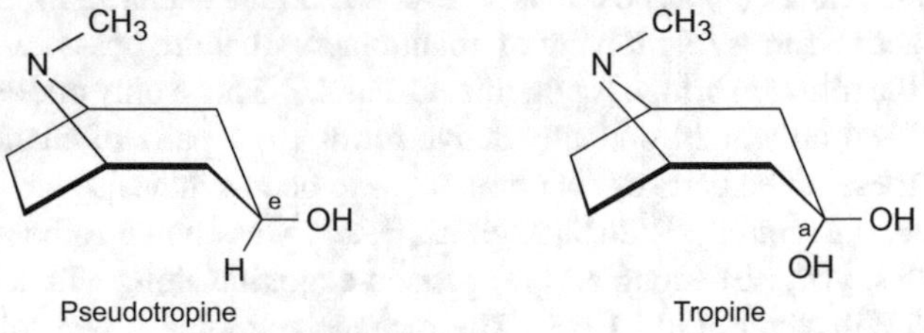

**Fig. 15.21** Chair conformation of piperidine ring in pseudo-tropine and tropine

are chiral centres. The optical inactivity is believed to be by internal compensation and thus each isomer is a *meso* form, C-3 being pseudo-asymmetric.

On the basis of evidences, Fodor (1953) showed that pseudo-tropine is the *syn* compound (OH group and N bridge are in *cis* position) and tropine is anti-compound (OH group and nitrogen bridge are in *trans* position). Besides, what has been stated, the stereochemistry of piperidine ring (whether boat or chair conformation) has also to be considered. Fodor (1953) proposed boat conformation in both isomers and axial orientation of methyl group in both isomers. However, the hydroxyl is axial in pseudo-tropine and equatorial in tropine as shown in Fig. 15.20.

Subsequently, Bose et al. (1953) assumed the chair conformation for the piperidine ring on the basis of analogy with the chair conformation of cyclohexane compounds. On this basis, the structure of pseudo-tropine and tropine is represented as shown below (Fig. 15.21).

It was Sparke (1953) who suggested that the chair form can easily change into boat form. In pseudo-tropine, the chair and boat forms are in mobile equilibrium, the boat form is the predominant form. As seen, the equatorial OH in the chair form of pseudo-tropine becomes axial in the boat form.

### 15.1.6.2 Stereochemistry of Cocaine

(−)-Cocaine ($C_{17}H_{21}NO_4$) occurs in COCa leaves. On hydrolysis, it gives ecgonine and benzoic acid (Fig. 15.22).

The structure of ecgonine has been confirmed by synthesis (Willstätter et al., 1901). From the stereochemical point of view, it is seen that in ecgonine, there are

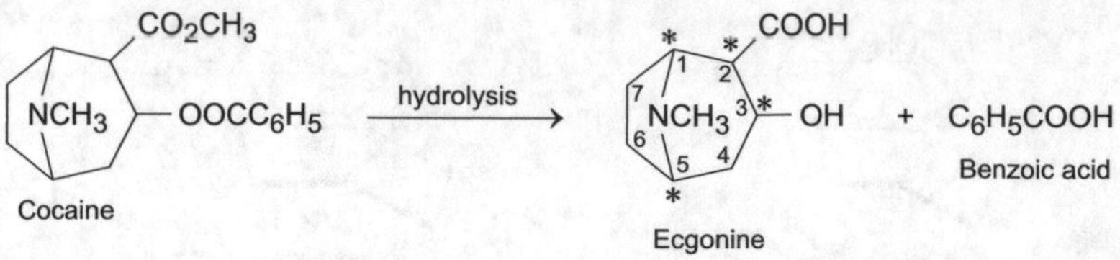

**Fig. 15.22**  Hydrolysis of cocaine to give ecgonine

four dissimilar centres (*) at positions 1, 2, 3 and 5 (*see* Fig. 15.22). So, there are 15 optically active forms (eight pairs of enantiomers) that are possible. As only the *cis*-fusion of the nitrogen bridge is possible, C-1 and C-5 have only one configuration (*cis*-form) and so only eight optically active forms (four pairs of enantiomers) are possible. Of these, three pairs of enantiomers have been synthesised.

The synthetic ecgonine (Willstätter et al., 1921) was shown to be a mixture of three racemates, viz., (±)-ecgonine, (±)-pseudo-ecgonine along with a third pair of enantiomers (Willstätter et al., 1923). The racemic ecgonine was resolved and the (–)-form esterified with methanol give ester and then benzoylated to give (–)-cocaine, identical with the natural product (Fig. 15.23).

Similarly (+)- and (–)- pseudo-cocaines were obtained from the corresponding pseudo-ecgonines.

The conformations of ecgonine and pseudo-ecgonine and the corresponding cocaines were established by Fodor et al. (1953, 1954) and Findlay (1953, 1954) (Fig. 15.24).

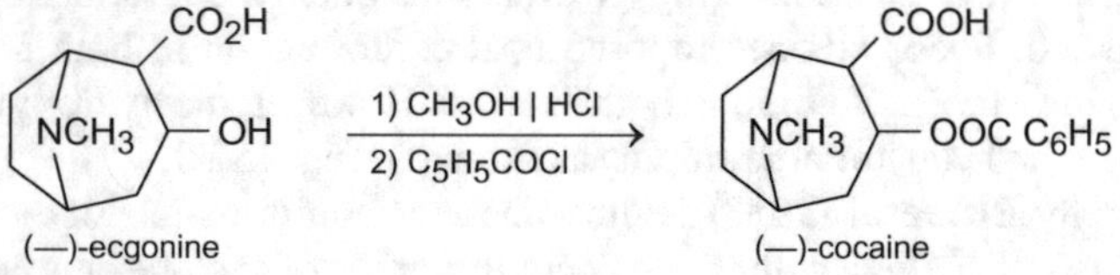

**Fig. 15.23**  Synthesis of (–)-cocaine

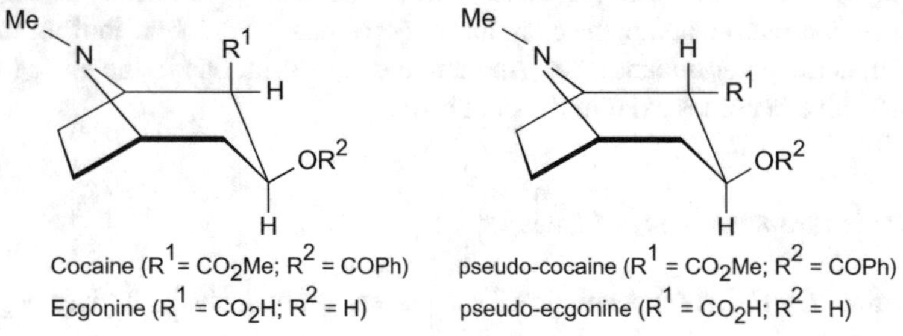

Cocaine ($R^1$ = $CO_2Me$; $R^2$ = COPh)     pseudo-cocaine ($R^1$ = $CO_2Me$; $R^2$ = COPh)
Ecgonine ($R^1$ = $CO_2H$; $R^2$ = H)          pseudo-ecgonine ($R^1$ = $CO_2H$; $R^2$ = H)

**Fig. 15.24**  Conformations of ecgonine and pseudo-ecgonine and the corresponding cocaines

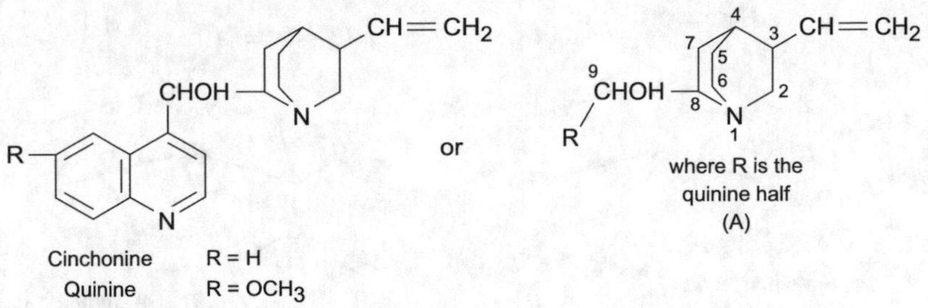

**Fig. 15.25** Structure of cinchonine and quinine

### 15.1.6.3 Stereochemistry of Cinchona Alkaloids

Cinchonine and quinine along with many other alkaloids are found to occur in the bark of various species of cinchona. Cinchonine is the parent substance in cinchona alkaloids though quinine is the most important member. The structures of both cinchonine ($C_{19}H_{22}N_2O$) and quinine ($C_{20}H_{22}N_2O_2$) are given below (Fig. 15.25).

As seen the formula (A) (Fig. 15.25) contains four chiral centres (at C-3, C-4, C-8 and C-9). As nitrogen is tertiary and all its (three) valences are part of the ring system, it (nitrogen atom) is chiral and cannot oscillate. So the structure (A) has five chiral centres. Since the bridge in structure (A) must be a *cis* fusion, the atoms 1 and 4 behave as 'one chiral unit'.

On the basis of extensive work by a number of investigators, it has been found that the configuration of C-3 and C-4 is same for all the compounds (*cis* configuration in the alkaloids). The hydrogens at C-3 and C-8 are *cis* with respect to each other. Since C-4 and C-8 are *cis* oriented, the hydrogen atoms at C-3, C-4 and C-8 are all *cis*-oriented in both cinchonine and quinidine, but in cinchonidine and quinine, the hydrogen atoms at C-3 and C-4 are *cis* but at C-3 and C-8 are *trans*.

The absolute configuration of the cinchona alkaloids is as given below (Fig. 15.26).

### 15.1.6.4 Stereochemistry of Morphine and Codeine

Both morphine and codeine belong to the phenanthrene group. (–)-Morphine is the chief alkaloid in opium. On the basis of extensive work, the structures of both these alkaloids are as given below (Fig. 15.27).

Both morphine and codeine contain five chiral centres (at C-5, C-6, C-9, C-13 and C-14). However, as the bridged ring system (across positions 9, 13) being *cis*, eight pairs of enantiomers are possible for each compound. On the basis of extensive work carried out to deduce the stereochemistry of codeine, it has been established that hydrogen atoms at C-5, C-6 and C-14 are *cis* and the bridge at C-9 and C-13 is also *cis*. This has been confirmed on the basis of X-ray analysis (Mackay et al. 1955). The conformational formula of both morphine and codeine is as given below (Fig. 15.28). The chair form for the cyclohexene ring has been used; the rings I, II

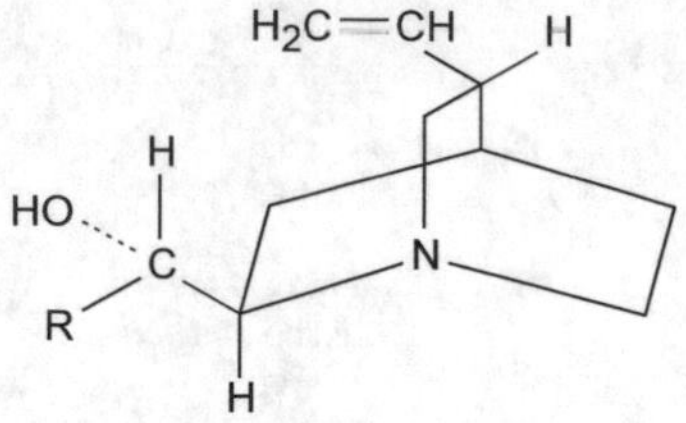

(+)-cinchonine
(+)-quinidine

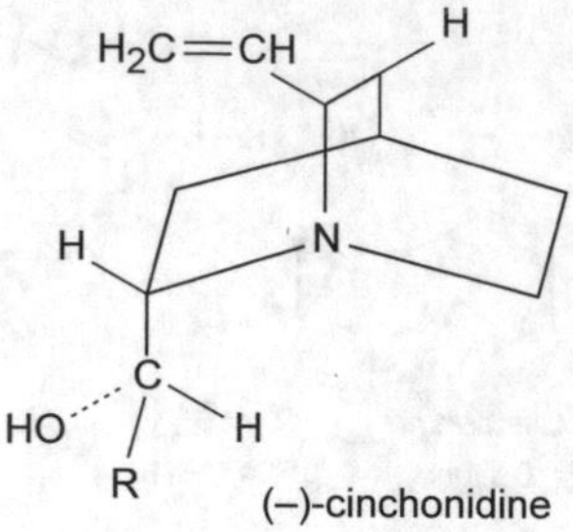

H₂C=CH H

(−)-cinchonidine
(+)-quinine

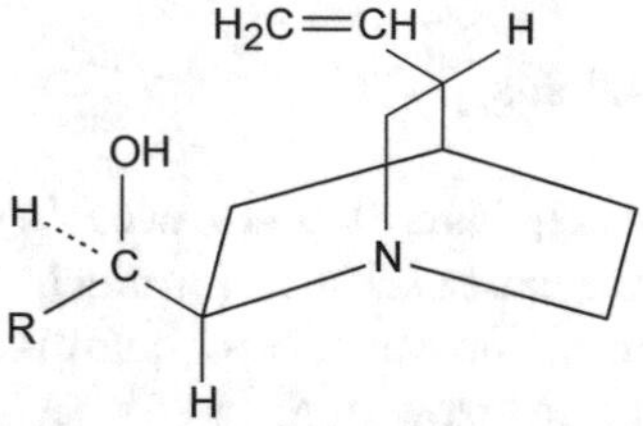

epiquinidine

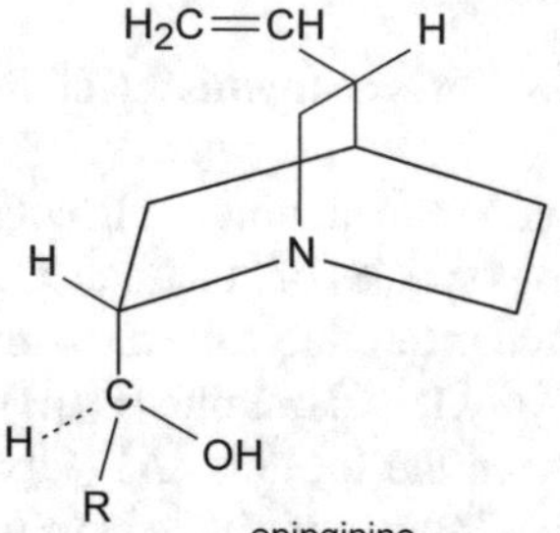

epipqinine

**Fig. 15.26**   Absolute configuration of cinchona alkaloids

**Fig. 15.27**   Structure of
Morphine and Codeine

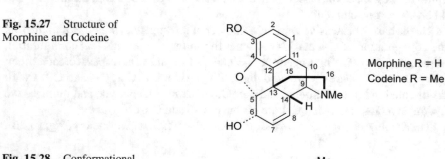

Morphine R = H
Codeine R = Me

**Fig. 15.28**   Conformational
formula of morphine and
codeine

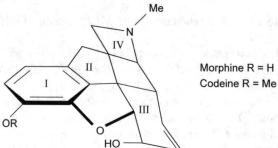

Morphine R = H
Codeine R = Me

and the oxide bridge are approximately in the plane of the paper, and the rings III and IV are approximately perpendicular to the plane of the paper.

Stereochemistry of some heterocyclic compounds (containing nitrogen), viz., pyridine, piperidine, deahydroquinoline has been discussed in Sect. 16.4 in).

## 15.2 Stereochemistry of Organophosphorus Compounds

Unlike nitrogen, which exhibit covalencies of 3 and 4, phosphorus can exhibit covalencies of 3, 4, 5 and 6 and so can give rise to more possible configurations than nitrogen. Organophosphorus compounds that are of interest from a stereochemical point of view include tertiary phosphines, quaternary phosphonium salts, tertiary phosphine oxides and some cyclic compounds.

### 15.2.1 Stereochemistry of Tertiary Phosphines

Unlike tertiary amines (bonded to three different alkyl groups, which cannot be resolved (*see* Sect. 15.1.1) as they undergo inversion too rapidly, the tertiary phosphines in which phosphorus is bonded to three different alkyl or aryl groups can be obtained in an optically pure state. An example is methylphenylpropyl phosphine, which can be obtained from optically active phosphonium salt, benzyl methyl phenyl propyl phosphonium bromide by electrolytic reduction (Fig. 15.29).

The phosphines (unlike amines) do not invert their configuration readily. In fact, an optically active phosphine like methyl propyl phenyl phosphine has to be heated in boiling toluene for about 3 h for complete racemisation.

Treatment of an optically active phosphine with benzyl bromide gives the phosphonium bromide with retention of configuration (*see* Fig. 15.29).

Since the absolute configuration of (+)-benzyl methyl phenyl propyl phosphonium bromide has been shown to be (S), so the phosphine obtained has also the (S) configuration.

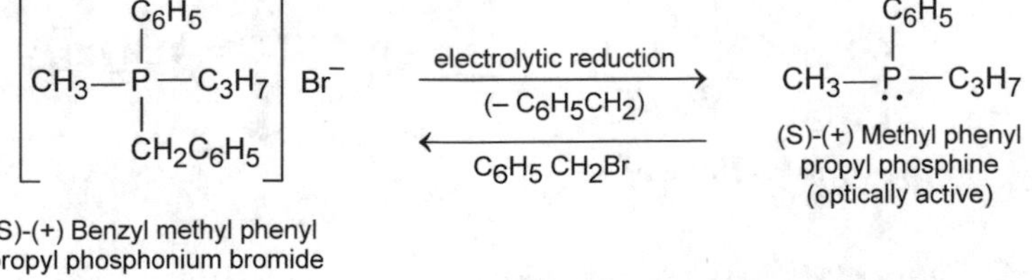

$$
\left[ \begin{array}{c} C_6H_5 \\ | \\ CH_3 - P - C_3H_7 \\ | \\ CH_2C_6H_5 \end{array} \right]^{+} Br^{-}
$$

(S)-(+) Benzyl methyl phenyl propyl phosphonium bromide

$$\xrightarrow[\text{(– }C_6H_5CH_2)]{\text{electrolytic reduction}}$$

$$\xleftarrow[\text{C}_6\text{H}_5\ \text{CH}_2\text{Br}]{}$$

$$
\begin{array}{c} C_6H_5 \\ | \\ CH_3 - \underset{\cdot\cdot}{P} - C_3H_7 \end{array}
$$

(S)-(+) Methyl phenyl propyl phosphine (optically active)

**Fig. 15.29** Preparation of optically active methyl propyl phenyl phosphine

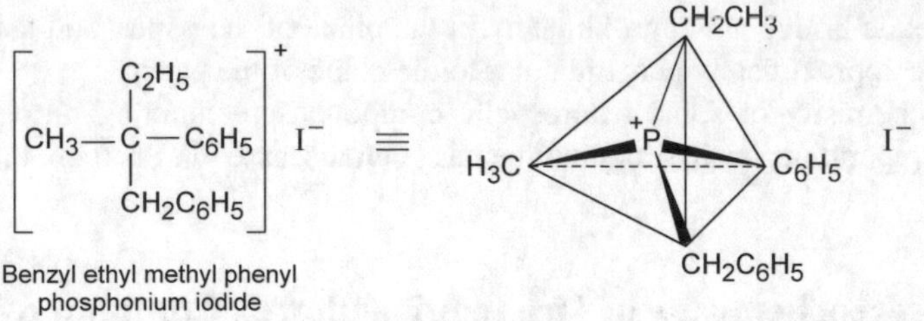

Benzyl ethyl methyl phenyl
phosphonium iodide

**Fig. 15.30**   Benzyl ethyl methyl phenyl phosphonium iodide

### 15.2.2   Stereochemistry of Quaternary Phosphonium Salts

The quaternary phosphonium salts resemble quaternary ammonium salts in their stereochemistry. The quaternary phosphonium salts of the type $R^1 R^2 R^3 R^4 P^+$ $X^-$ in which all the R groups are different can be resolved. One such quaternary phosphonium salt that can be resolved is benzyl methyl ethyl phenyl phosphonium iodide (Fig. 15.30).

### 15.2.3   Stereochemistry of Tertiary Phosphine Oxides

Tertiary phosphine oxides of the type $R^1 R^2 R^3 R^4 P = O$ in which all the four groups attached to phosphorus are different have non-planar configuration and so can be resolved into enantiomeric forms. An example of ethyl methyl phenyl phosphine oxide is given below (Fig. 15.31).

The phosphine oxides are conveniently obtained from phosphines by oxidation with $H_2O_2$ with retention of configuration. So it is possible to prepare an enantiomer of phosphine oxide with known absolute configuration. An example is given below (Fig. 15.32).

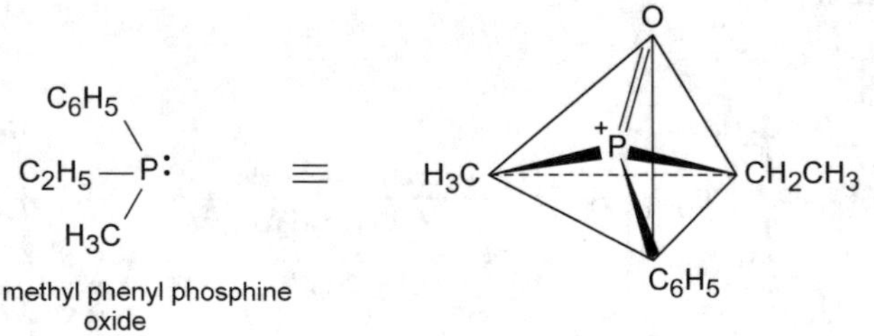

Ethyl methyl phenyl phosphine
oxide

**Fig. 15.31**   Ethyl methyl phenyl phosphine oxide

**Fig. 15.32** Preparation of
phosphine oxides

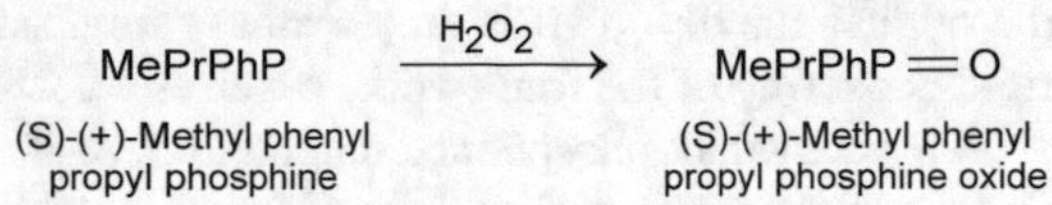

$$MePrPhP \xrightarrow{H_2O_2} MePrPhP = O$$

(S)-(+)-Methyl phenyl
propyl phosphine

(S)-(+)-Methyl phenyl
propyl phosphine oxide

## 15.3 Stereochemistry of Cyclic Phosphorus Compounds

Only a few cyclic phosphorus compounds have been synthesised. Some of such compounds are given below.

2-p-Hydroxyphenyl-2-phenyl-1, 2, 3, 4-tetrahydro-isophosphinolinium bromide (I) was prepared (Holliman and Mann, 1947) and resolved into enantiomers. Another spiro compound, P-spirobis-1,2,3, 4-tetrahydrophosphinolinium iodide (II) was synthesised (Mann et al., 1955) and resolved into (+) and (–) forms with high optical stability.

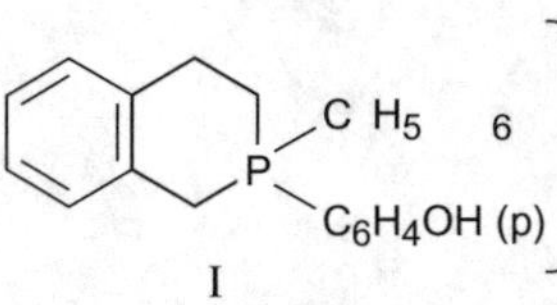

I
2-p-Hydroxyphenyl-2-phenyl-
-1,2,3,4-tetrahydro-
-isophosphinolinium bromide

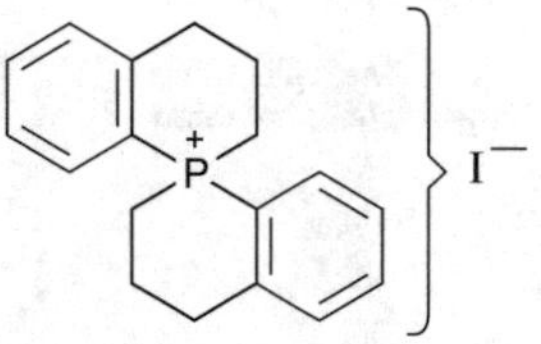

II
P-Spirobis-1,2,3,4-tetrahydro
phosphinolinium iodide

A number of azophosphaphenanthrenes (III) (e.g., $R^1 = H$; $R^2 = NMe_2$) were prepared (Campbell et al., 1966) but could not be resolved. However, the corresponding phosphine oxide (IV) (obtained by the oxidation of III with $H_2O_2$) could be resolved.

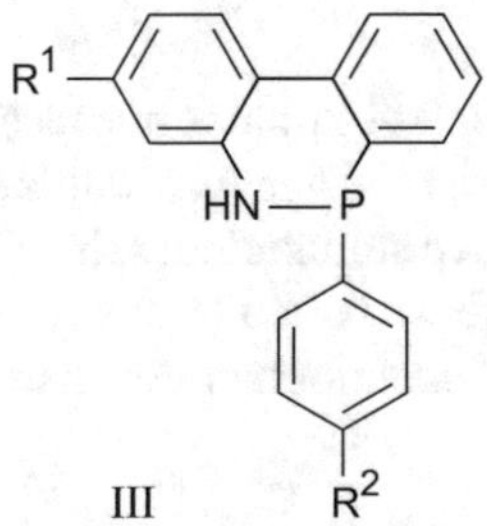

III
Azophosphaphenanthrene
$R^1$ = H; $R^2$ = NMe$_2$

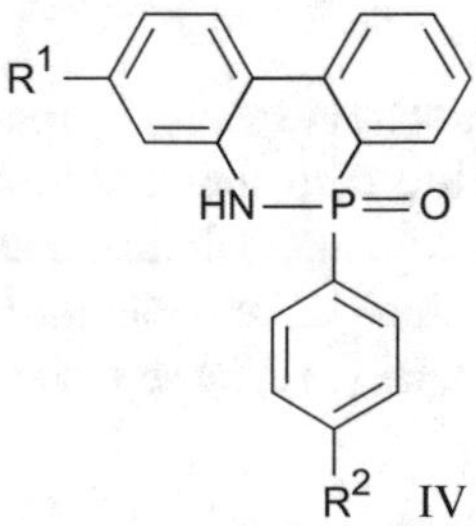

IV
Azophosphaphenanthrene oxide
$R^1$ = H; $R^2$ = NMe$_2$

The (+)-oxide (IV) on reduction with LiAlH$_4$ gave (–)-phosphine (III) and reduction of (–)-oxide (IV) gave the (+)-phosphine (III). It is, however, not certain, whether the optical activity in III is due to an asymmetric tetravalent phosphorus or due to rigid packering of the framework, which is a 2,2'-bridged biphenyl.

A tetracovalent phosphone compound 5,10-diethyl-5,10-dihydro-phosphanthren (V) did not exhibit optical isomerism. However, it exists as geometrical isomers (Mann et al., 1962). Salt of (VI) could be resolved (Hellwinkel, 1965). The anion contained hexavalent phosphorus and has an octahedral configuration ($sp^3d^2$).

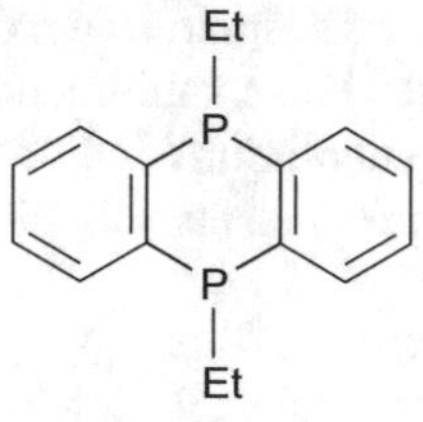

Et
|
P

P
|
Et
V
5,10-Diethyl-5,10-
dihydrophosphanthren

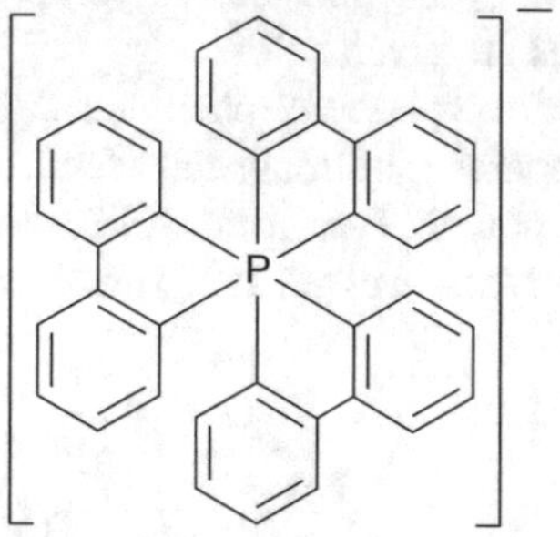

VI

## 15.4   Stereochemistry of Sulphur Compounds

The sulphur compounds that are of interest from a stereochemical point of view include sulphonium salts, sulphinic esters, sulphoxides, sulphilimines and sulphines.

### 15.4.1   Stereochemistry of Sulphonium Salts

The sulphonium salts are generally prepared from dialkyl sulphide by reaction with an alkyl halide. In this case, the sulphides act as nucliophilic agents towards substances that readily undergo nuclophilic displacement to give sulphonium salts (Fig. 15.33).

A special feature of sulphonium salts is that these salts if substituted by three different groups, they can be separated into optical enantiomers. As an example, the

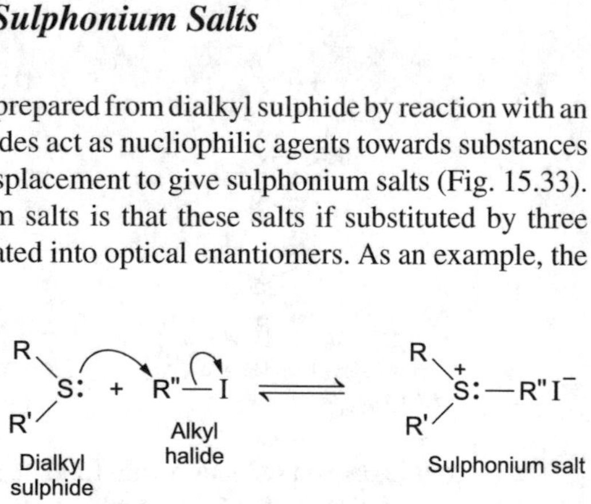

**Fig. 15.33**   Preparation of sulphonium salt

R
  \
   S:   +   R''—I   ⇌   R\+
  /                      S:—R''I⁻
R'                    R'/

Dialkyl        Alkyl
sulphide       halide        Sulphonium salt

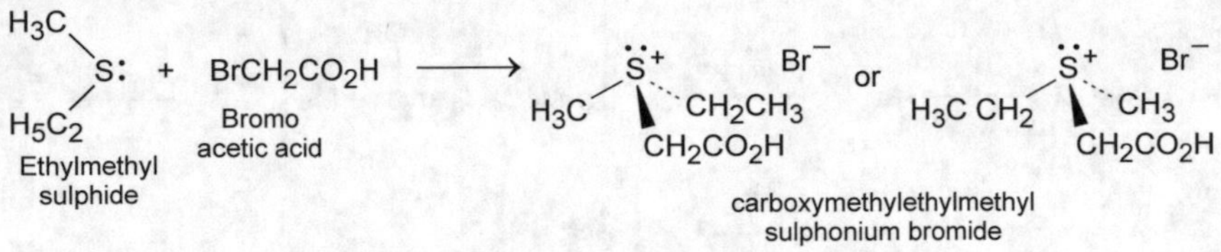

**Fig. 15.34** Synthesis of carboxymethylethylmethyl sulphonium bromide

reaction of methylethylsulphide with bromoacetic acid gives carboxymethylethyl-methyl sulphonium bromide. This molecule is not superimposable on its mirror image and so exists in two optically active forms (Fig. 15.34).

The sulphonium salt (prepared above, Fig. 15.34) could be separated into its optical enantiomers by treatment with (+)-camphorsulphonate and the salt obtained crystallised fractionally from a mixture of ethanol and ether (Pope and Peachey, 1900). The (+)-sulphonium camphorsulphonate being less soluble separated out ($M_D$ + 60°; the rotation of (+)-camphorsulphonate ion is about + 52°, so the difference (+16°) is the contribution of the sulphonium ion to the rotation].

Similarly, ethylmethyl phenacyl sulphonate picrate (I) was prepared (Smiles, 1900) in two optically active forms [[$\alpha$]$_D$ + 8.1° and −9.2° for the two forms]. Another sulphonium salt (II) in which sulphur atom is in a ring was also obtained (Mann and Halliman, 1946).

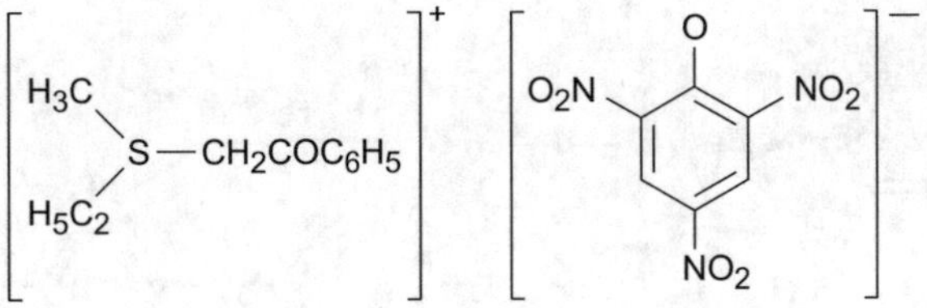

Ethylmethyl phenacyl sulphonate picrate.

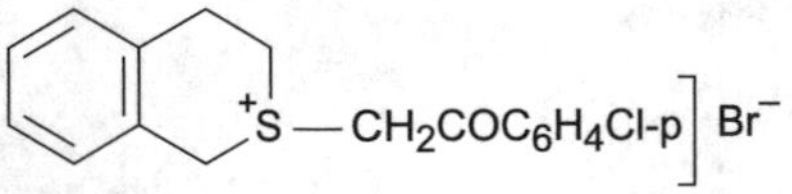

Optically active sulphonium salts are conveniently obtained from optically active sulphoxides (Anderson, 1971) as shown below (Fig. 15.35).

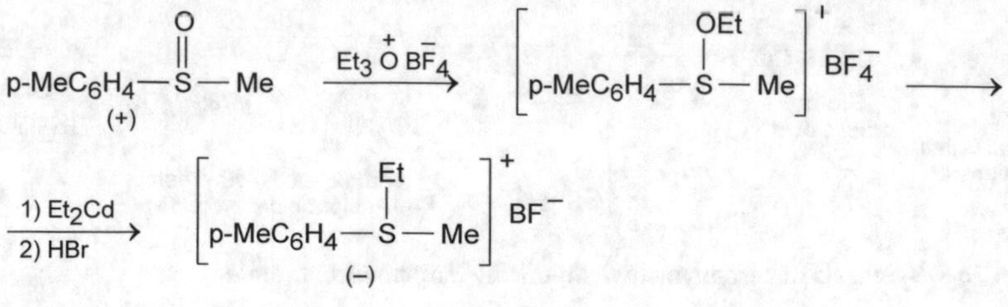

**Fig. 15.35**  Preparation of optically active sulphonium salts

## 15.4.2  Stereochemistry of Sulphinic Esters

Sulphinic esters are prepared from Grignard reagent by reacting with $SO_2$ followed by esterification of the formed sulphinic acids. Alternatively, these are prepared by the reduction of sulphonyl chlorides with zinc or sodium sulphite followed by esterification (Fig. 15.36).

The optical activity of sulphinic esters is explained if the structure is as given below in which the sulphur is $sp^3$ hybridised.

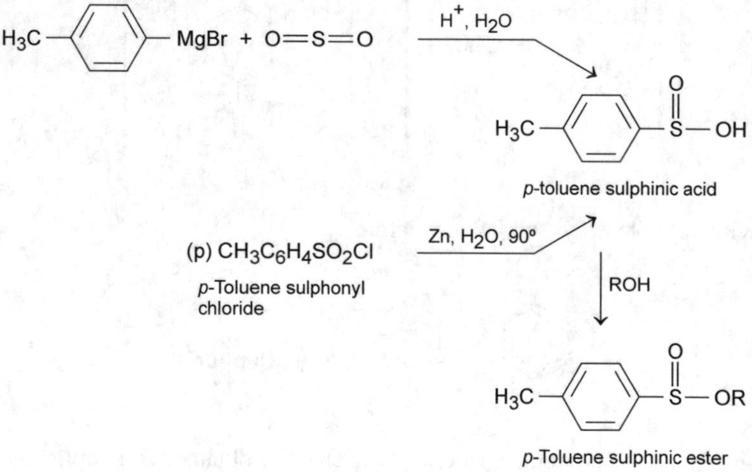

**Fig. 15.36**  Preparation of Sulphinic esters

The sulphinic esters could be resolved (Phillips, 1925) by alcoholysis. The method consists of heating two molecules of the sulphinic ester with one molecule of (−)-menthyl alcohol or (−)s-octyl alcohol. In case the sulphinic ester is a racemic modification, then the (+) and (−) forms will react at different rates with the optically active alcohol. It has been found (Phillips, 1925) that (+)-ester reacted faster than the (−)-ester.

## 15.4.3  Stereochemistry of Sulphoxides

Sulphoxides are prepared by the oxidation of sulphides, preferably 1.5 mol of $H_2O_2$ in acetic acid (Fig. 15.37).

In case excess of $H_2O_2$ (3.2 mols) is used, the product is the corresponding

$$\text{sulphone, } CH_3\overset{\overset{O}{\|}}{\underset{\underset{O}{\|}}{S}} - CH_2CH_2\overset{\overset{NH_2}{|}}{C}HCO_2H.$$

Sulphoxides of the types I and II have been resolved (Phillips et al., 1951).

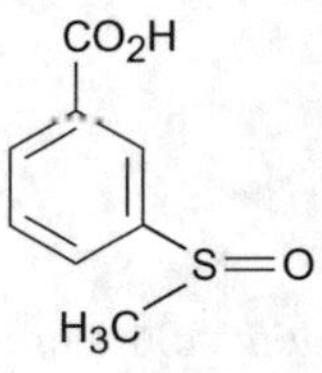

I 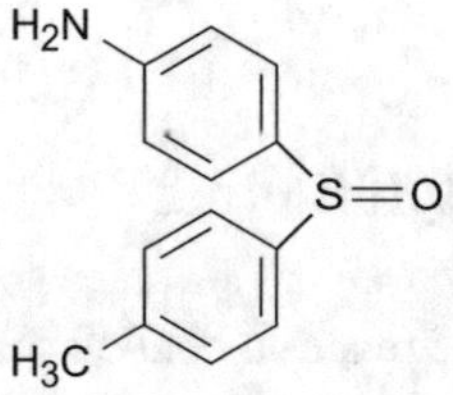  II 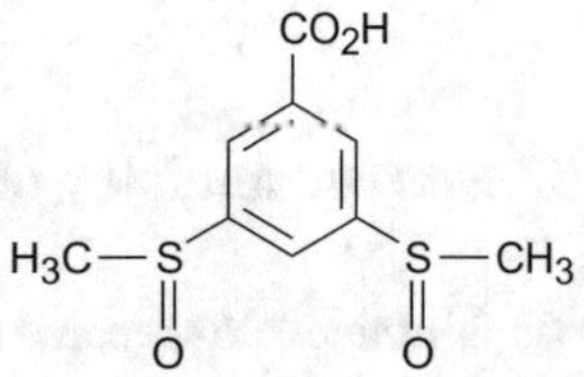  III

Disulphoxides of the type $CH_3SOCH_2CH_2SOCH_3$ could not be resolved (Bell and Bennett). However, the disulphoxide (III) could be resolved.

Diastereoisomeric sulphoxides have been prepared (Cram et al., 1963) by the oxidation of dialkyl sulphide with $t$-butyl hydroperoxide (an achiral reagent) (Fig. 15.38).

It has been shown (Henbest et al., 1966) that alkyl aryl sulphides are stereoselectively oxidised to sulphoxides (4–98% optical purity) in the presence of growing aerobic cultures of *Aspergillus niger*.

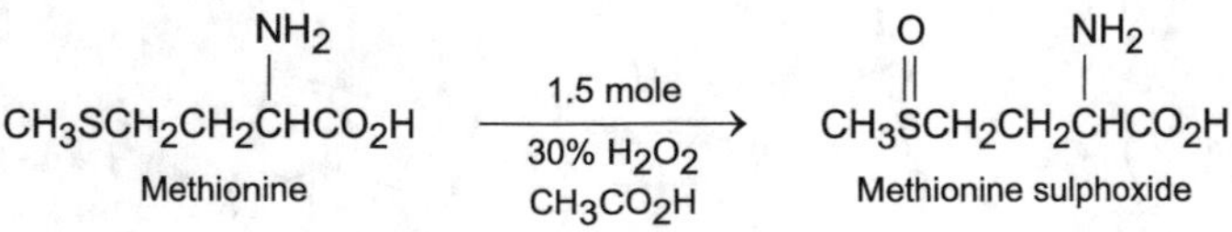

**Fig. 15.37**  Synthesis of sulphoxides

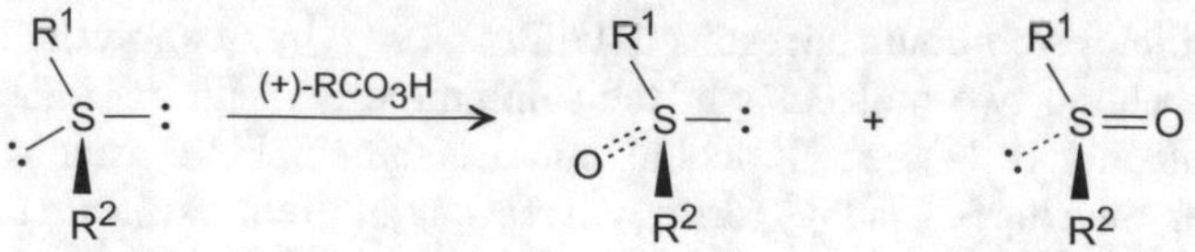

**Fig. 15.38**　Preparation of diastereoisomeric sulphoxides

In some compounds, the sulphoxide group can be inverted by heating. Thus, (+)-benzylp-tolyl sulphoxide is racemised on heating in declin at 160 °C (Henbest et al., 1964). It has also been shown that *cis*- and *trans*-cyclic sulphoxides (IV) on heating separately in declin at 190 °C gave the same *cis–trans* mixture containing 25% *cis* form. It should, however, be noted that t-butyl group is invariably equatorial and that the 1*e*, 4*e* (*trans*) is more stable than the 1*a*, 4*e* (*cis*).

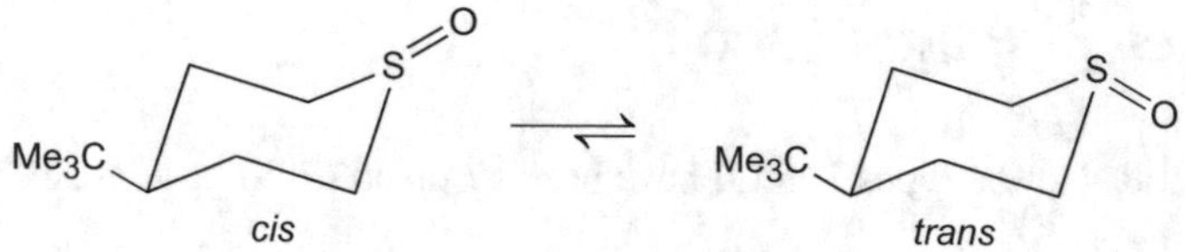

## 15.5　Stereochemistry of Sulphilimines

The sulphilimines are prepared by the reaction of alkyl sulphides with chloramine T (N-chloro-4-methyl benzenesulfonamide sodium salt) (Fig. 15.39).

The sulphilimines (iminosulphuranes) can be represented either as I or II (Fig. 15.39). In both the structures, sulphur atom is asymmetric. So sulphilimine can be resolved (Kenyon et al., 1972).

It is believed that sulphilimines are a resonance hybrid of the above two contributing structures I and II. On the basis of NMR spectroscopic studies (Lambert et al., 1971), it was shown that the parent compound, cyclic six-membered sulphilimine, thian-1-imine is preferentially in the equatorial position, whereas its N-benzene sulphonyl and N-tosyl derivatives are preferentially in axial position.

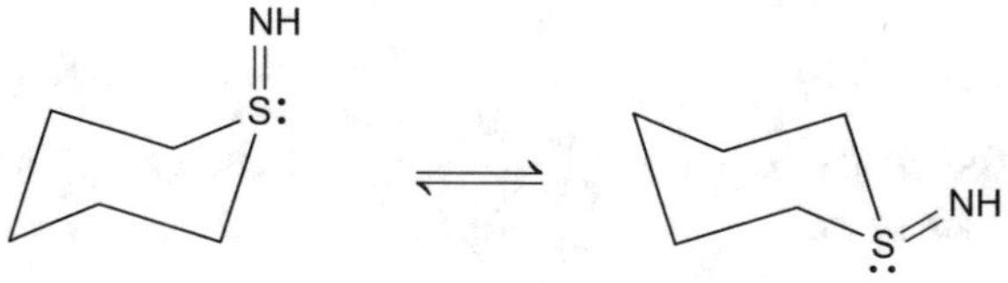

**Fig. 15.39** Preparation of sulphilimines

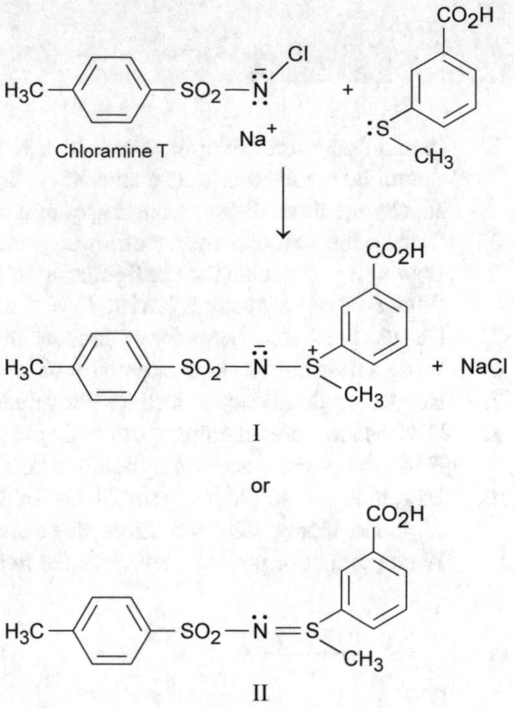

**15.5.1 Stereochemistry of Sulphines**

On the basis of NMR spectroscopes studies, it is found that in sulphines, the C = S = O system is a rigid non-linear group. In fact, due to this, syn and anti-isomers are possible in sulphines. It has been shown that the sulphine (I) prepared by the oxidation of thioketone with peroxy acid exists in two isomeric forms (E and Z). These could be isolated by column chromatography (Fig. 15.40).

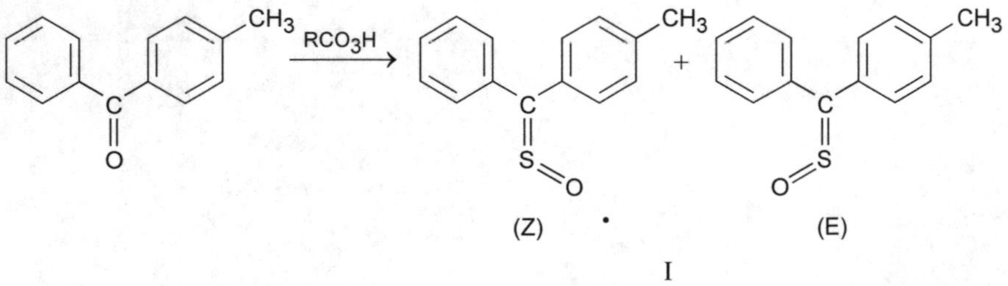

**Fig. 15.40** Preparation of sulphines

**Problems**

1. Explain why tertiary amines in which N is attached to different alkyl group cannot be resolved into its enantiomers but tertiary phosphines in which P is attached to three different alkyl groups can be resolved?

2. Discuss the stereochemistry of tertiary amine oxides and oximes.

3. How will you decide the configuration of oximes?

4. What type of compounds having $N = N$ can be resolved?

5. Discuss the stereochemistry of cocaine and quinine.

6. Write a note on the stereochemistry of morphine and codeine.

7. How are optically active tertiary phosphines prepared?

8. Discuss the stereochemistry of cyclic phosphorus compounds.

9. How are optically active sulphonium salts prepared?

10. Oxidation of sulphides with $H_2O_2$ in $CH_3COOH$ gives sulphoxide and sulphone. Under what conditions these are obtained in pure state?

11. What product or products are obtained in the following?

(a) 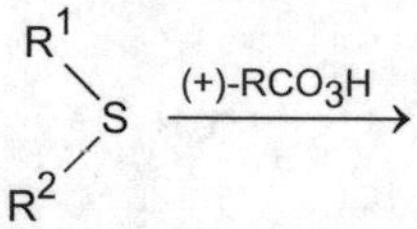
$$\begin{array}{c} R^1 \\ \diagdown \\ \phantom{xx}S \xrightarrow{\;(+)\text{-RCO}_3\text{H}\;} \\ \diagup \\ R^2 \end{array}$$

(b) Chloromine T + alkylsulphide $\rightarrow$

(c) 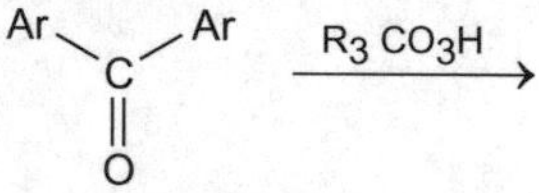
$$\begin{array}{c} Ar \diagdown \phantom{x} \diagup Ar \\ C \xrightarrow{\;R_3\,CO_3H\;} \\ \parallel \\ O \end{array}$$

# Chapter 16
# Stereochemistry of Some Heterocyclic Compounds

## 16.1 Three-Membered Heterocyclic Compounds

The three-membered heterocyclic compounds may contain one or two heteroatoms and can be either saturated or unsaturated. The saturated analogues are epoxides (oxiranes), aziridines and thiiranes and the unsaturated analogues are oxirene, azirines and thiirenes (Fig. 16.1).

Besides these, the three-membered heterocyclic compounds may also contain two heteroatoms.

The three-membered heterocyclic compounds are strained due to bond angle distortion. In general, these have shorter C–C bonds than in cyclopropane (thiirane is an exception). The bond angles, bond lengths and diploe moments of the three heterocycles are given ahead (Fig. 16.2).

### *16.1.1 Three-Membered Saturated Heterocyclic Compounds*

Among the three-membered saturated heterocyclic compounds, the three saturated analogues, epoxides (oxiranes), aziridines and thiirenes are of considerable importance.

#### 16.1.1.1 Epoxides (Oxiranes)

Epoxides are the simplest types of oxygen-containing heterocycles and are most conveniently prepared by the epoxidation of alkenes with a peracid like *m*-chloroperbenzoic acid (Fig. 16.3).

Appropriately substituted epoxides exist in both geometrical and optical isomers. An example is stilbene oxide (Fig. 16.4).



V. K. Ahluwalia, *Stereochemistry of Organic Compounds*,
https://doi.org/10.1007/978-3-030-84961-0_16

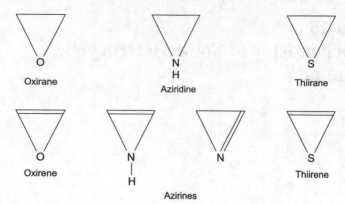

**Fig. 16.1**  Three-membered heterocyclic compounds

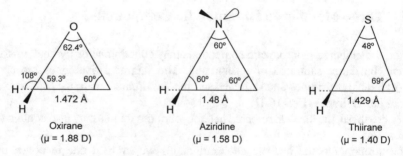

**Fig. 16.2**  Parameters of oxirane, aziridine and thiirane

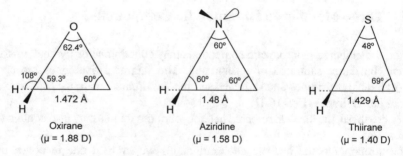

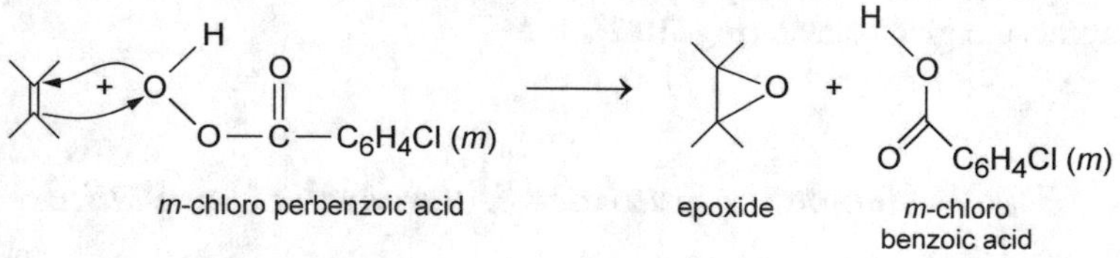

**Fig. 16.3**  Synthesis of epoxide

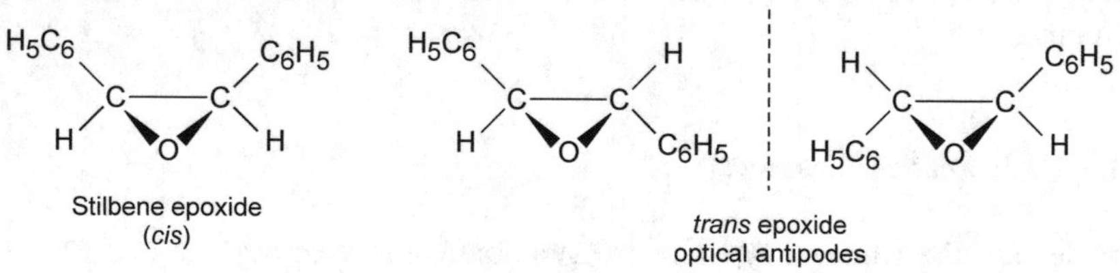

**Fig. 16.4**  Geometrical and optical forms of substituted epoxides

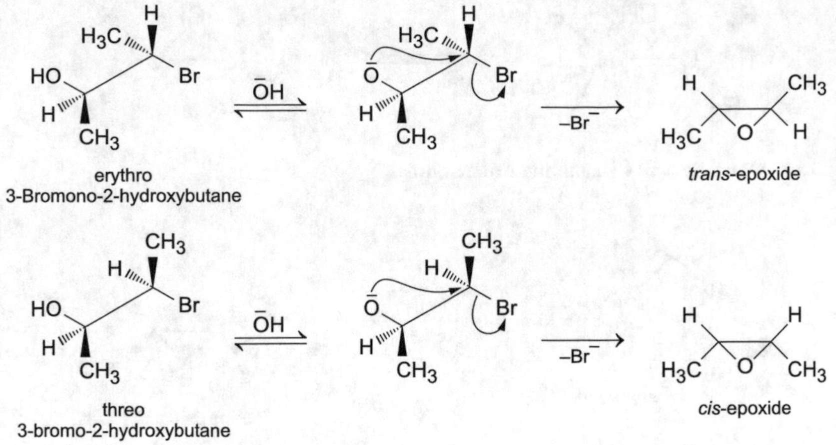

erythro
3-Bromono-2-hydroxybutane

*trans*-epoxide

threo
3-bromo-2-hydroxybutane

*cis*-epoxide

**Fig. 16.5** Synthesis of *trans* and *cis* epoxides

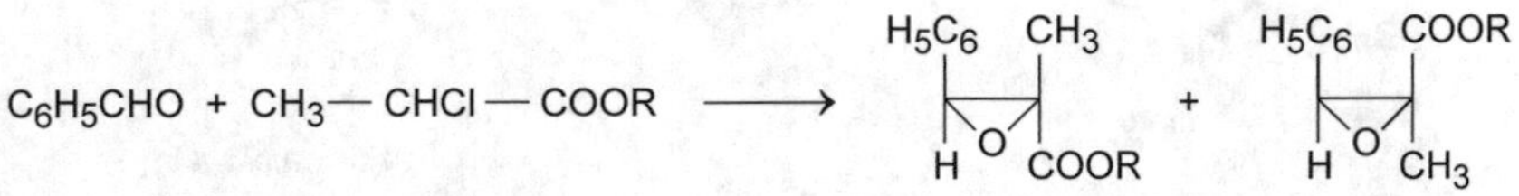

$C_6H_5CHO + CH_3\!-\!CHCl\!-\!COOR \longrightarrow$

**Fig. 16.6** Darzen condensation

Even in monosubstituted epoxides like propylene epoxide, optical isomerism is observed.

A convenient route for the preparation of epoxides involves the removal of hydrogen halide from halohydrins. Thus, as an example, cyclisation of erythro and threo-3-bromo-2-hydroxybutanes yield *trans* and *cis*-epoxides, respectively (R.E. Carter and T. Darkenberg, Chem. Commun, 1972, 582) (Fig. 16.5).

The stereoisomeric epoxides are also obtained by the **Darzen condensation** (M.C Roux-Schmitt, J. Seydon-Penne and S. Wolfe, Tetrahedran, 1972, **28**, 4965–4979). An example is given below (Fig. 16.6).

### 16.1.1.2 Aziridines

The aziridines are interesting as the substituent on the trivalent nitrogen occupies a different plane than that of the ring. Thus, it is possible to resolve appropriately substituted aziridines into optically active enantiomers (Fig. 16.7).

In this case, nitrogen undergoes inversion of configuration rapidly even at room temperature. So, the resolution can be achieved at very low temperatures.

As in the case of epoxides, aziridines (appropriately substituted) can exist in *trans* and *cis* forms. A convenient procedure for the synthesis *cis* and *trans* aziridines is Gabrial ring closure method (A. Weissperger and H. Bach, Chem. Ber., 1931, **64**,

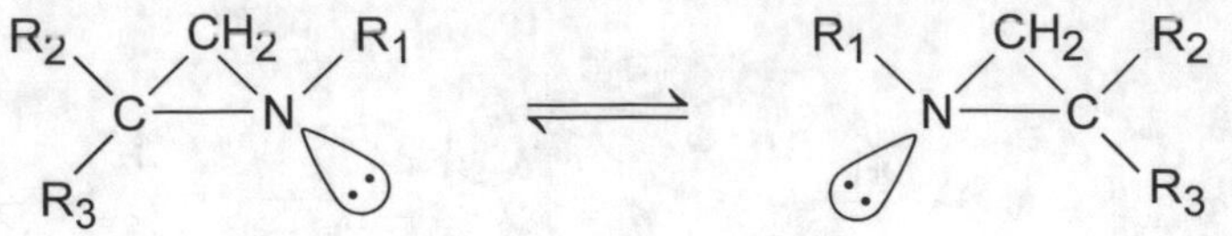

**Fig. 16.7**  Optically active enantiomers of aziridines

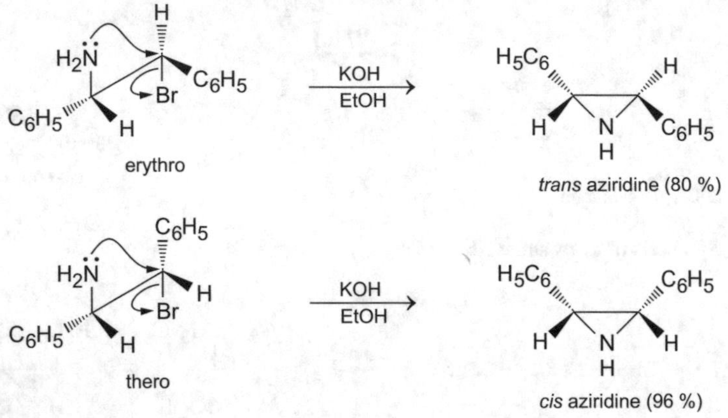

**Fig. 16.8**  Synthesis of *trans* and *cis* aziridines

1095; 1932, **65**, 631). It involves intramolecular displacement of halogen atom by an amino group. The cyclisation is stereospecific and occurs at with inversion at the carbon bearing the leaving group (Fig. 16.8).

### 16.1.1.3   Thiiranes

Also known as episulphides or thiacyclopropanes, these are three-membered sulphur containing saturated heterocycles. Among the three-membered heterocycles, thiirane has the lowest strain energy (83.26 kJ/mol). Its dipole moment (1.66 D) is lower than that of oxirane (1.88 D); the difference is attributed to smaller polarity of the C–S bond than that of the C–O bond. The carbon–carbon bond length (1.429 Å) is intermediate between the normal C–C (1.54 Å) and C $=$ C (1.34 Å) bond lengths suggesting partial double bond character in thiiranes. The C–S bond length (1.829 Å) is of the same order as in dialkyl sulfides (1.810 Å). The H–C–H and C–S–S bond angles are 60° and 48.5°, respectively.

As in the case of oxiranes, the thiiranes are obtained by the cyclisation of 2-halomercaptans with dilute alkali (Fig. 16.9).

Also, *trans*-2-chlorocyclohexane thiol in the presence of sodium bicarbonate gives the corresponding thiirane (Fig. 16.10).

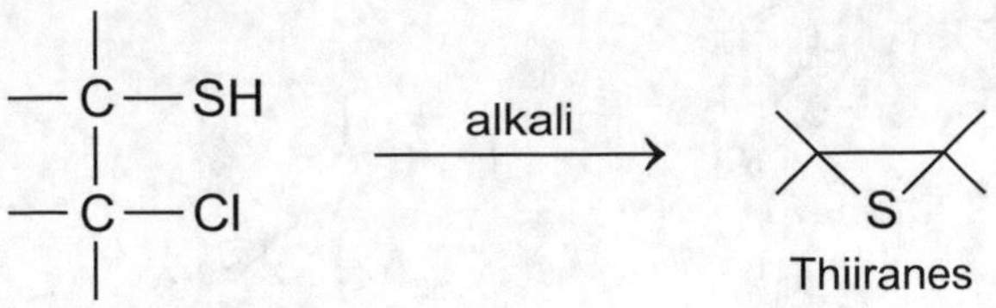

**Fig. 16.9** Synthesis of thiiranes

Trans-2-chloro
cyclohexane thiol

Cyclohexene
sulphide

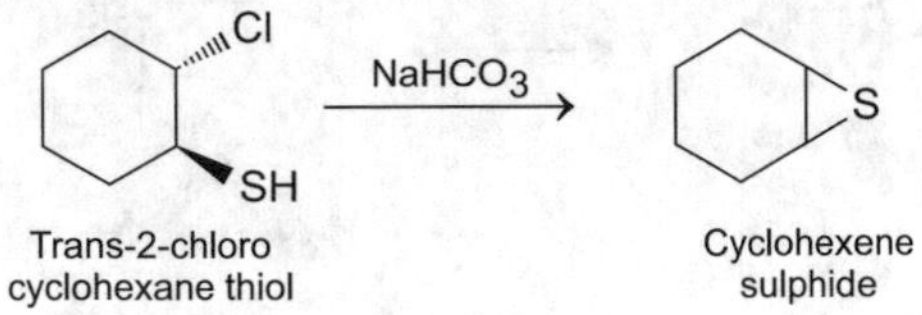

**Fig. 16.10** Synthesis of cyclohexene sulphide

A widely used method for the synthesis of thiiranes is the reaction of epoxides with thiourea or thiocyanate salts (Fig. 16.11).

By using appropriately substituted *cis* or *trans* epoxides, the corresponding *cis* or *trans* thiiranes can be obtained.

## 16.1.2 Three-Membered Unsaturated Heterocyclic Compounds

As already mentioned, among the three-membered unsaturated compounds, the three analogues are oxirenes, azirines and thiirenes (Fig. 16.12).

### 16.1.2.1 Oxirenes

These are resonance stabilised anti-aromatic heterocycles with four π electrons. Oxirene can not be isolated; it exists as an unstable intermediate due to destablising electronic configuration. On the basis of molecular orbital calculations, it has been suggested that oxirene is less stable than its acyclic analogue. Oxirenes are obtained as intermediate in the oxidation of alkynes with peracids (E. Lewars and G. Morrison, Tetrahedron Lett, 1977, 501) and also in the photochemical **Wolf rearrangement** of α-diazoketones (K.P. Zeller, Tetrahedron Lett., 1977, 707) (Fig. 16.13).

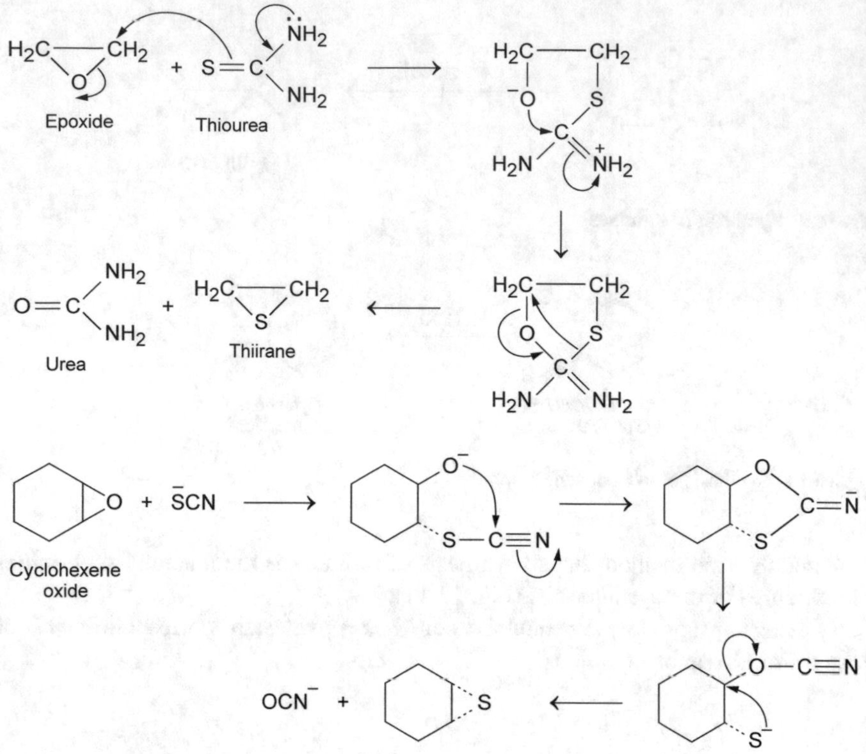

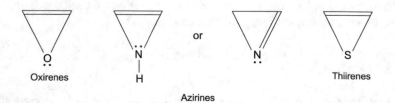

**Fig. 16.11**  Synthesis of thiiranes from epoxides

**Fig. 16.12**  Three-membered unsaturated heterocycles

### 16.1.2.2  Azirines

These are less stable compared with their saturated counterparts. This is due to the introduction of unsaturation (in small ring) for $sp^2$-hybridised atoms, which introduces ring strain. These are difficult to prepare. As already stated, these can exist in two isomeric forms, viz., 1H-azirines (enamines or 2-azirines) and 2H-azirines (imines or 1-azirines) (Fig. 16.14).

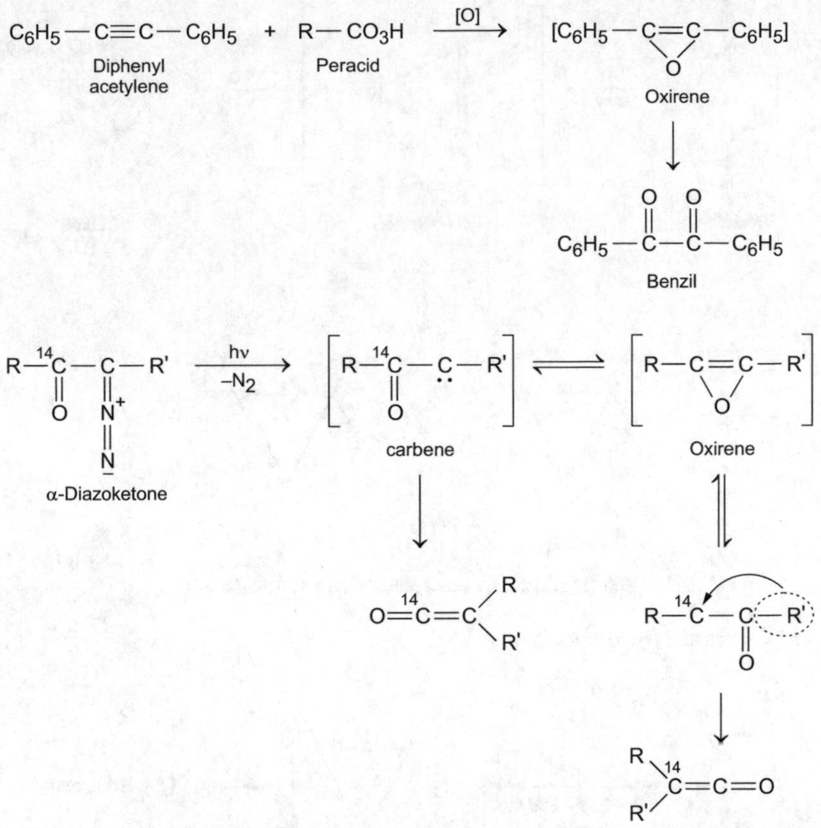

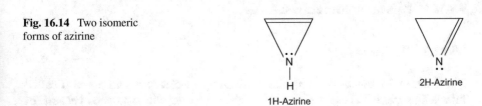

$$C_6H_5\!-\!C\!\equiv\!C\!-\!C_6H_5 \quad + \quad R\!-\!CO_3H \xrightarrow{\;[O]\;} [C_6H_5\!-\!C\!=\!C\!-\!C_6H_5]$$

Diphenyl acetylene    Peracid    Oxirene

$$C_6H_5\!-\!\overset{O}{\overset{\|}{C}}\!-\!\overset{O}{\overset{\|}{C}}\!-\!C_6H_5$$

Benzil

Fig. 16.13 Formation of oxirenes as intermediates

Fig. 16.14 Two isomeric forms of azirine

1H-Azirine

2H-Azirine

1H-Azirine is an anti-aromatic and resonance destablised. Its instability is due to not only the large-angle strain in the unsaturated three-membered ring but also due to the potential overlapping of the lone pair of electrons at nitrogen with the olefinic π-electrons. Due to this, 1H-azirine is very difficult to synthesise. However, 1H-azirine is obtained as an intermediate during the pyrolysis of triazolines (I and II).

The intermediate 1-H-azirine gives a mixture of 2-H-azirines (III and IV) (Fig. 16.15).

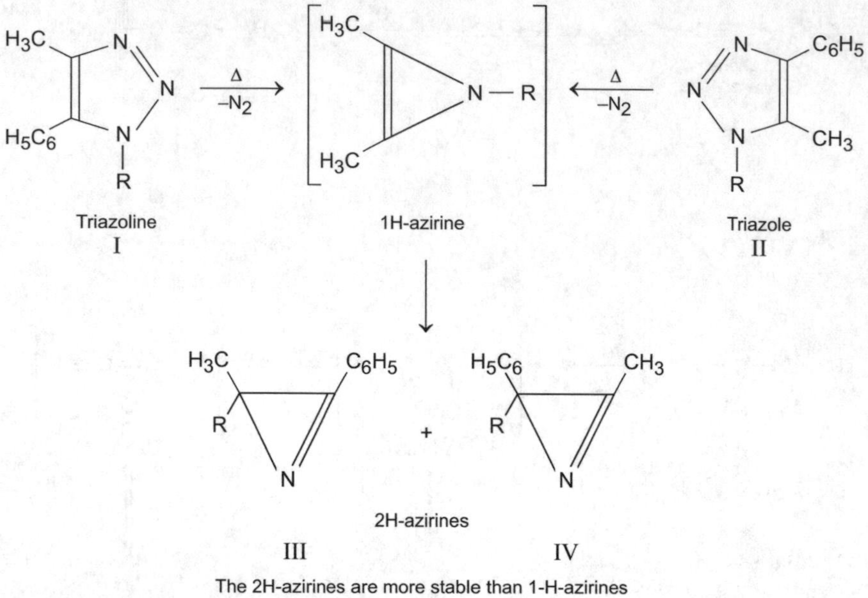

Fig. 16.15  Formation of azirines

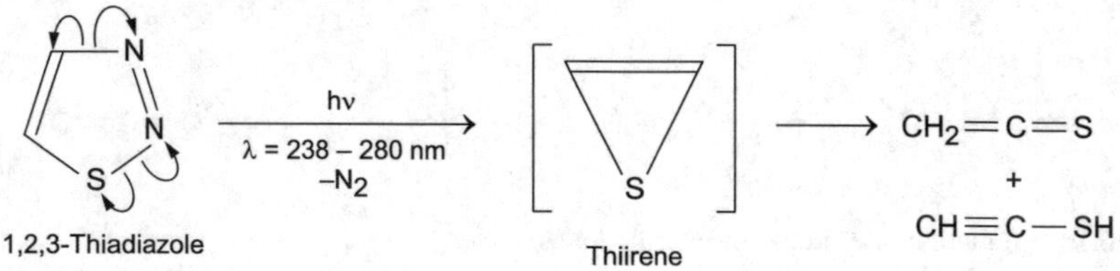

Fig. 16.16  Formation of thiirene as intermediate

### 16.1.2.3  Thiirenes

These are anti-aromatic heterocycles containing $4\pi$-electrons and are also called thiacyclopropenes. These are less stable than their acyclic analogues because of having resonance destablised $4\pi$-electron system. These are, however, formed as transient intermediates as in the photochemical decomposition of 1, 2, 3-thiadiazole (Fig. 16.16).

The instability of thiirenes is reduced with the formation of thiirenium salts, thiirene-1-oxides and thiirene-1,1-dioxide.

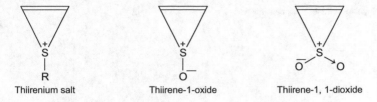

| Thiirenium salt | Thiirene-1-oxide | Thiirene-1, 1-dioxide |

## 16.1.3 Three-Membered Heterocyclic Compounds with Two Heteroatoms

The three-membered heterocyclic compound, with two heteroatoms can be saturated (like diaziridines and oxaziridines) or unsaturated (like diazirines) (Fig. 16.17).

### 16.1.3.1 Diaziridines

These are weak bases and form salts. Due to the increased rotational barrier resulting from the lone pair-lone pair interations on nitrogen atoms, the nitrogen inversion in diaziridine is retarded. So, diaziridine can be resolved into enantiomers.

Diaziridines can be obtained from diazirines by reaction with Grignard reagent (E. Schnaitz, R. Ohme and D.R. Schmitt, Chem. Ber., 1962, **95**, 2714) (Fig. 16.18).

The required diazirines can be obtained as given in Sect. 16.1.3.2.

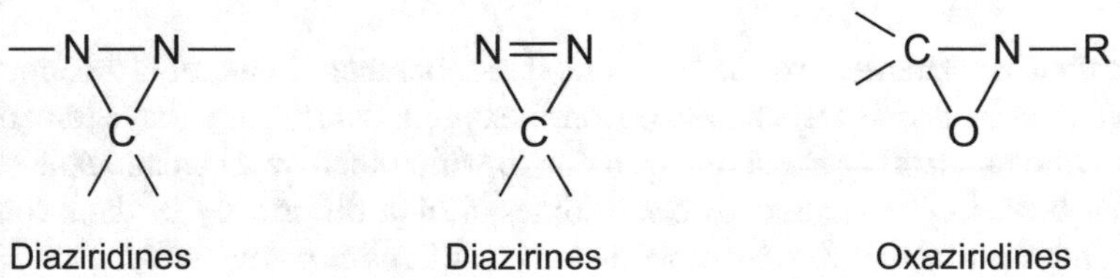

| Diaziridines | Diazirines | Oxaziridines |

**Fig. 16.17** Three-membered heterocyclic compounds with two heteroatoms

Diazirines  →(RMgX)→  →(HOH)→  Diaziridines

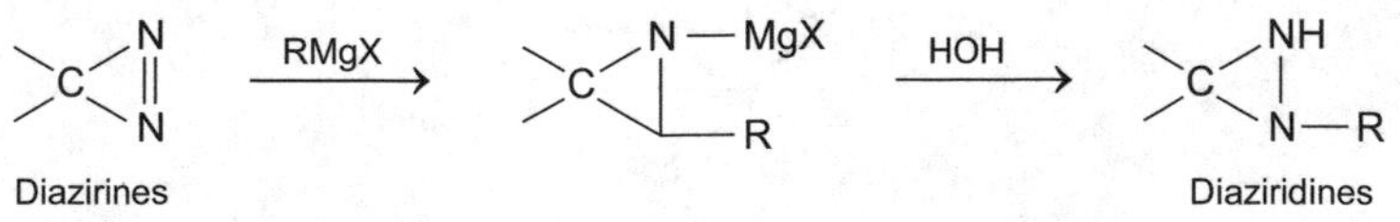

**Fig. 16.18** Synthesis of diaziridines

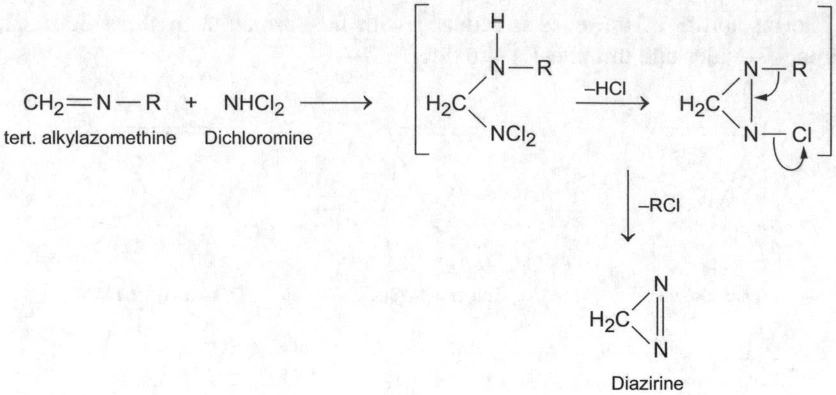

$$CH_2{=}N{-}R \quad + \quad NHCl_2 \longrightarrow$$

tert. alkylazomethine    Dichloromine

**Fig. 16.19** Synthesis of diazirine

### 16.1.3.2  Diazirines

These are three-membered heterocycles having two nitrogens and are the cyclic isomers of diazoalkanes.

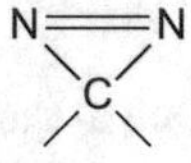

Diazirine

The reaction of tert-alkylazomethine with dichloramine gives diazirine (Fig. 16.19).

### 16.1.3.3  Oxaziridines

Also called oxaziranes, oxaziridines are three-membered saturated heterocycles containing one of each carbon, nitrogen and oxygen atoms. These have remarkable conformational stability about nitrogen due to which there is a considerable higher nitrogen barrier. The isomers of oxaziridines, which differ only in their configuration about nitrogen, are resolvable and are configurationally stable (D.R. Boyd, Tetrahedran Lett., 1968, 4561).

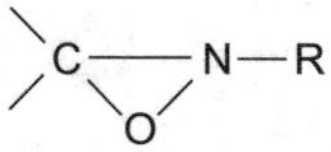

Oxaziridines

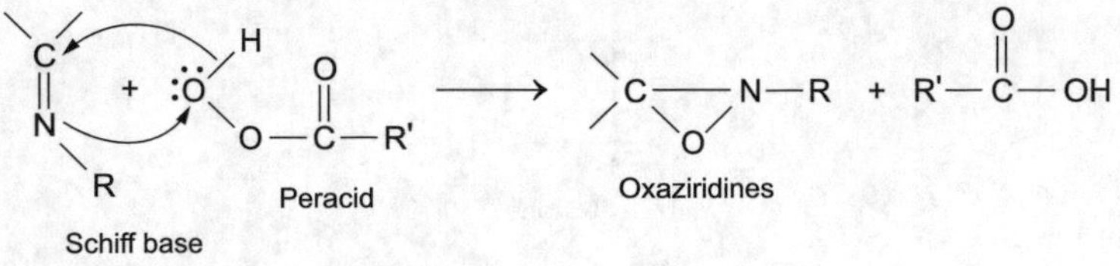

**Fig. 16.20** Synthesis of oxaziridines

Oxaziridines can be conveniently obtained by the oxidation of Schiff bases (prepared by the reaction of 1° amines with aldehydes and ketones) with peracids (Fig. 16.20).

## 16.2 Four-Membered Heterocyclic Compounds

These are heterocyclic analogues of cyclobutane and are derived by replacing a methylene group ($-CH_2$) by a heteroatom (NH, O or S). Depending on the heteroatom (O, NH or S), these are known as oxetanes, azetidines and thietanes (Fig. 16.21).

The four-membered heterocyclic compounds are less strained (compared with three-membered heterocyclic compounds) and so are more stable. Oxetane and thietane molecules are planar but not square because of large size of oxygen and sulphur atoms than the carbon atoms. The planarity of oxetane and thietane as compared with cyclobutane (which is puckered) is due to the reduction in the number of non-bonded interactions between methylene groups.

### 16.2.1 Oxetanes

Also called oxacyclobutanes (oxygen is assigned position 1), oxetanes, as already stated are planar molecules compared to the puckered cyclobutane ring, due to the replacement of $CH_2$ groups by a divalent oxygen atom which causes a reduction in the number of non-bonding interactions between the neighbouring hydrogen atoms.

**Fig. 16.21** Four-membered heterocyclic compounds

Oxetane      Azetidine      Thietane

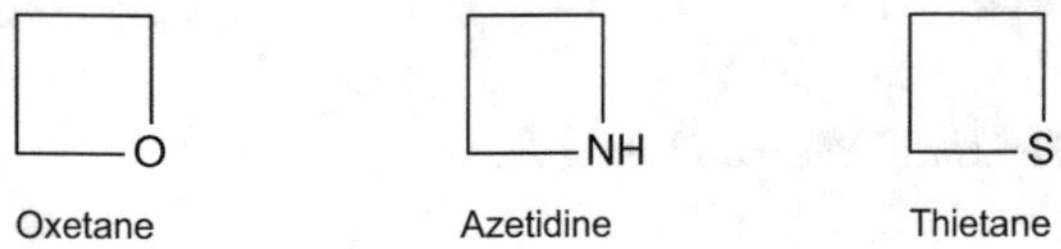

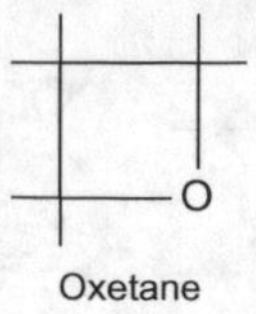

Oxetane

In oxetane, the C–C bond length (1.54 Å) is greater than C–O bond length (1.46 Å) indicating that the molecule is not a perfect square. The calculated value of the dipole moment of oxetane (2.01D) is higher than that of ether (1.22D) or dimethyl ether (1.31D). This indicates that the electron density at oxygen atom in oxetane is more than in acyclic aliphatic ethers.

Oxetanes are prepared by the cyclisation of 1,3-halohydrins in presence of a base (Fig. 16.22).

Oxetanes are synthesised in high yield by the well-known **Paterno-Büchi reaction** involving photochemical [2 + 2] cycloaddition of carbonyl compounds to olefins (Fig. 16.23).

The stereochemistry of the above photocyclisation is believed to involve the formation of a diradical intermediate (Fig. 16.24).

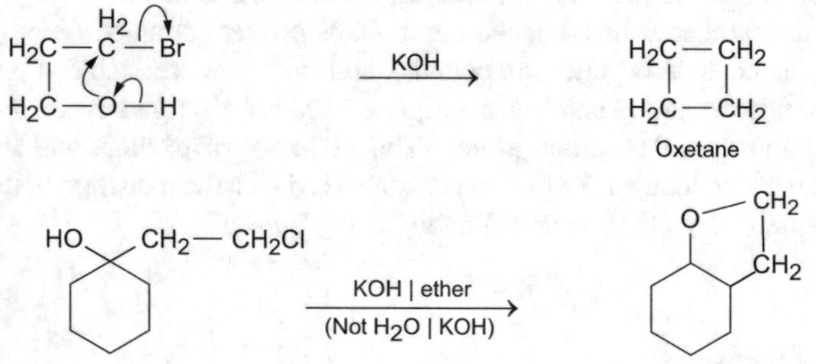

**Fig. 16.22**  Synthesis of Oxetanes

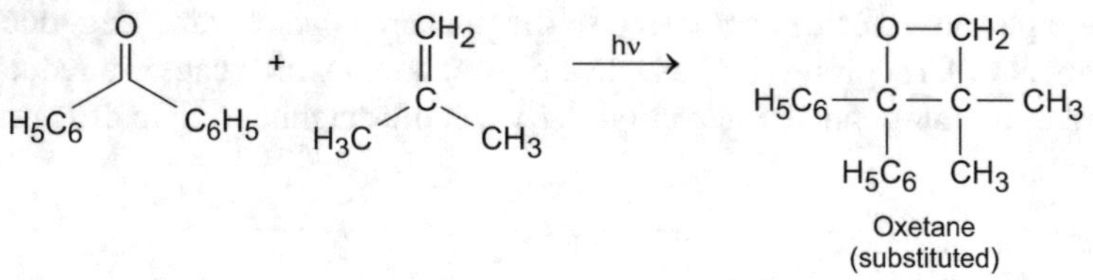

**Fig. 16.23**  Paterno-Büchi reaction

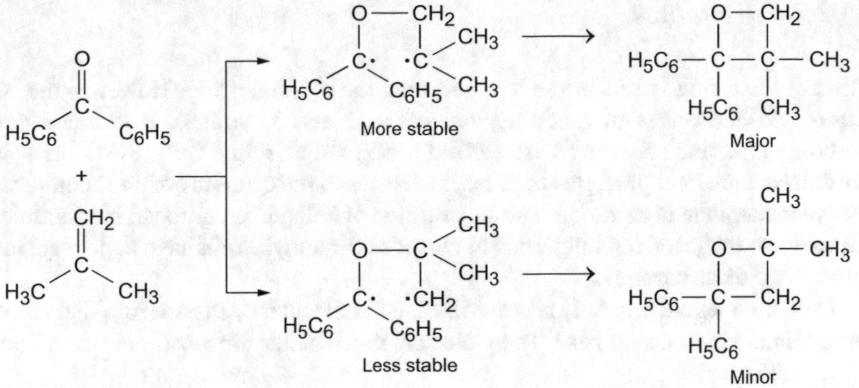

**Fig. 16.24** Mechanism of Paterno-Buchi reaction

## 16.2.1.1 Oxetanones (β-Lactones)

Oxetanones, also called oxetan-2-ones (2-oxetanones) or β-lactones, are carbonyl derivatives of oxetane. These are strained internal esters. The simplest β-lactam is β-propiolactone or β-oxetanone is a planar molecule having the following parameters:

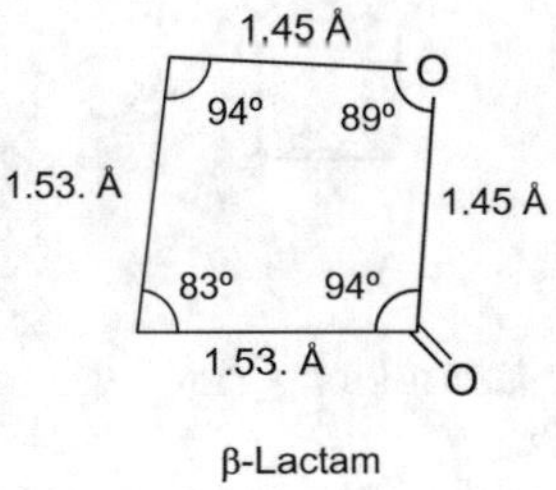

β-Lactam

The IR spectra of β-lactones exhibit carbonyl stretching frequency of 1840 cm$^{-1}$, higher than in cyclic esters (1736 cm$^{-1}$) and γ-lactone (1770 cm$^{-1}$). This is due to the presence of ring strain in the molecule and deformation of bond angles.

β-Lactones are best obtained by the intramolecular cyclisation of β-halo acids with a base under controlled conditions (Fig. 16.25).

**Fig. 16.25** Synthesis of β-lactone

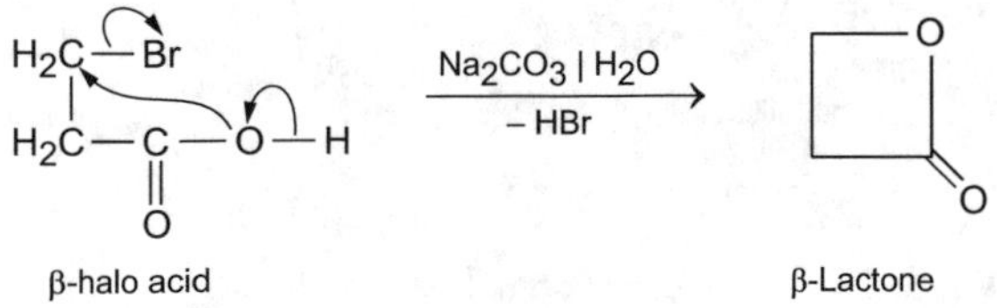

β-halo acid          β-Lactone

## 16.2.2   Azetidines

The azetidine ring is much less strained than the aziridine ring. However, the N-substituted derivatives of azetidines behave as 2 and 3° amines. It is found that azetidine is a much stronger base (*pKa* 11.29) than aziridine (*pKa* 8.04). Also in azetidines, the effect of angle strain on the barrier to nitrogen inversion is comparatively smaller than in aziridine. The substitution of halogen on nitrogen in azetidines slows down the inversion rate. Thus N-chlororo-2-methyl azetidine can be resolved into its two diastereomers.

The azitidines are generally prepared by interamolecular cyclisation of γ-haloalkyl amines in the presence of base. The yields are much better for substituted azetidines (Fig. 16.26).

### 16.2.2.1   Azetidinones (β-Lactams)

Also known as 2-azetidinones (carbonyl derivatives of azetidines containing carbonyl group at position 2), the azetidinones (β-lactams) are of great importance due to their use as anti-bacterial agents.

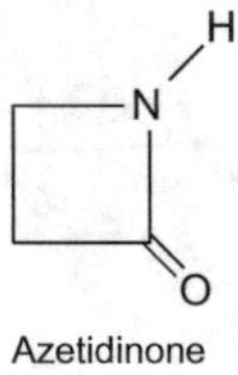

Azetidinone

It is well known that the azetidinone ring system is part of two antibiotics, penicillins and cephalosporins.

These are commonly prepared by the reaction of β-amino esters with Grignard reagents. It is advantageous to use mesityl magnesium bromide since its reaction with the formed β-lactam at the carbonyl site is prevented due to steric hindrance (Fig. 16.27).

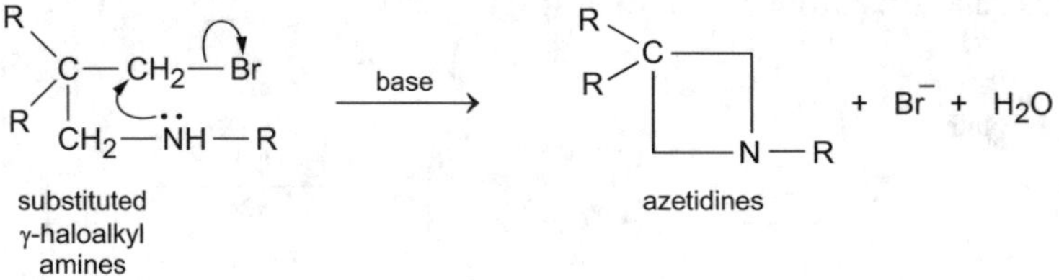

Fig. 16.26   Preparation of azetidines

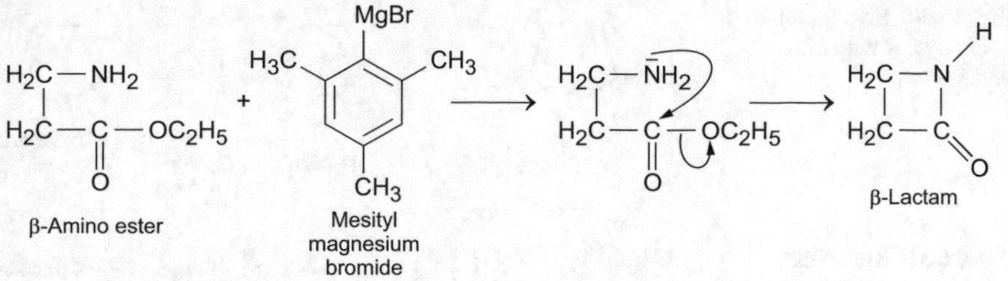

Fig. 16.27 Preparation of β-lactam

## 16.2.3 Thietanes

It is a four-membered saturated heterocycle containing sulphur as heteroatom. The thietane molecule is planar due to a reduction in the number of non-bonding inter-actions and lower torsional energy barrier about the C–S bond. The parameters of thietane are given below (Fig. 16.28).

Thietanes are best prepared by the reaction of 1, 3-dihaloalkanes with anhydrous sodium sulphide in ethanol in the presence of a phase transfer catalyst (hexadecyltriethyl ammonium chloride) (Fig. 16.29).

## 16.3 Five-Membered Heterocyclic Compounds with One Heteroatom

The simplest five-membered heterocyclic compounds containing one heteroatom are furan, pyrrole and thiophene (Fig. 16.30).

A close examination of the structures (given above) gives an impression that these compounds should have characteristic properties of diene and an ether (in case

Fig. 16.28 Parameters of thietane

$$C—S = 1.847 \text{ Å}$$

$$C—S = 1.549 \text{ Å}$$

$\varphi_1 = 76.8°$

$\varphi_2 = 90.6°$

$\varphi_3 = 95.6°$

$\varphi_4 = 90.6°$

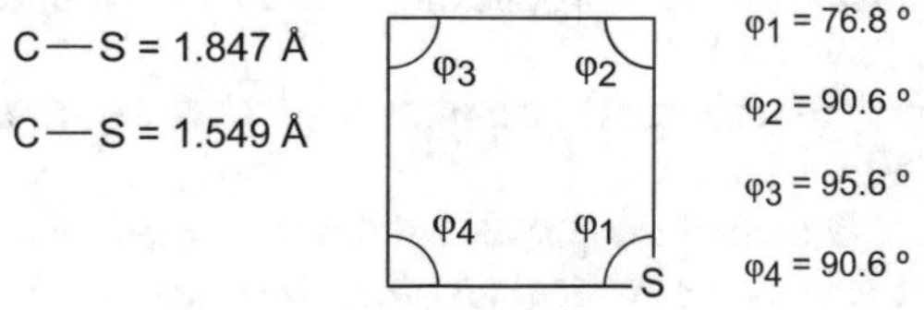

Fig. 16.29 Preparation of thietane

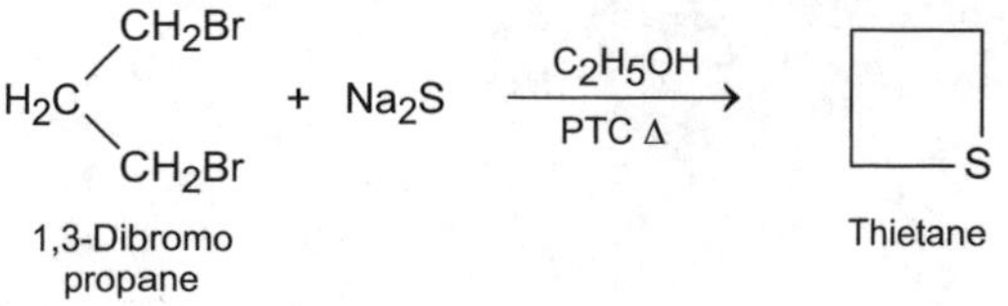

1,3-Dibromo propane

Thietane

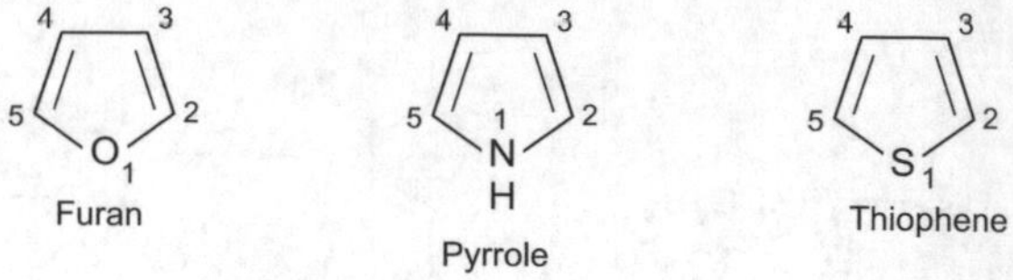

**Fig. 16.30** Five-membered
heterocycles with one
heteroatom

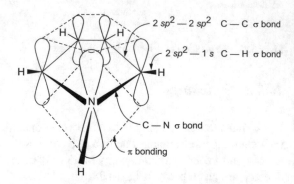

Furan          Pyrrole          Thiophene

**Fig. 16.31** Molecular
orbital picture of pyrrole

$2 sp^2 - 2 sp^2$  C — C  σ bond

$2 sp^2 - 1 s$  C — H  σ bond

C — N  σ bond

π bonding

of furan), an amine (in case of pyrrole) or an sulphide (in case of thiophene). It has, however, been found that none of these compounds possesses such properties but possesses aromatic character since they undergo electrophilic substitution reactions like nitration, sulphonation, Friedel–Crafts reaction etc. In fact, the aromatic character of all these compounds is in analogy with benzene.

The aromatic character of these compounds, as we know, arises due to the delocalisation of four carbon π-electrons and two paired electrons donated by heteroatom (O, N or S) thus forming a sextet of electrons (characteristic of aromatic system). These heterocycles, in terms of molecular orbitals, are planar pentagons consisting of $sp^2$-hybridised carbon atoms. There remain four carbon $p$-orbitals with a π electron, each to overlap with doubly filled $p$-orbital of the heteroatom (see Fig. 16.31).

The fully reduced products of pyrrole, furan and thiophene are tetrahydropyrrole tetrahydrofuran (THF, used as an industrial solvent) and tetrahydrothiophene. The bond lengths $s$ and bond angles in all the five-membered heterocycles are given below (Table 16.1).

The structural parameters of pyrrole, furan and thiophene are given below (Fig. 16.32).

In case of thiophene, sulphur being less electronegative than oxygen or nitrogen, it is mono aromatic than other five-membered heterocyclic compounds.

**Table 16.1**  Bond lengths and bond angles in five-membered heterocycles

| Heterocycles | Bond lengths (Å) | | | Bond angles | | |
|---|---|---|---|---|---|---|
| | C–X | $C_2$–$C_3$ | $C_3$–$C_4$ | $C_2$–X–$C_3$ | X–$C_2$–$C_3$ | $C_2$–$C_3$–$C_4$ |
| Pyrrole | 1.370 | 1.382 | 1.417 | 109.80 | 107.70 | 107.40 |
| Pyrroledine | 1.470 | 1.540 | 1.540 | | | |
| Furan | 1.362 | 1.361 | 1.430 | 106.50 | 110.65 | 111.47 |
| Tetrahydro furan | 1.430 | 1.540 | 1.540 | 106.40 | 103.70 | 100.30 |
| Thiophene | 1.714 | 1.370 | 1.423 | 92.17 | 111.47 | 112.45 |
| Tetrahydro thiophene | 1.820 | 1.540 | 1.540 | 93.42 | 106.10 | 105.00 |

$\theta_1 = 109.8°$

$\theta_2 = 107.7°$

$\theta_3 = 107.4°$

$\theta_4 = 121.5°$

$\theta_5 = 125.5°$

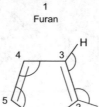

Pyrrole

| Bond angles (degrees) | 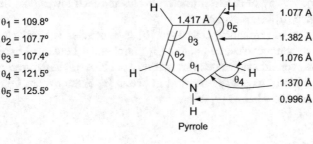 | Bond lengths (Å) |
|---|---|---|
| $C_2$ — O — $C_5$ = 106.50 | | O — $C_2$ = 1.362 |
| O — $C_2$ — $C_3$ = 110.65 | | $C_2$ — $C_3$ = 1.361 |
| $C_2$ — $C_3$ — $C_4$ = 106.77 | | $C_3$ — $C_4$ = 1.430 |
| O — $C_2$ — H = 115.98 | | $C_2$ — H = 1.075 |
| $C_2$ — $C_3$ — H = 127.83 | | $C_3$ — H = 1.077 |

Furan

| Bond angles (degrees) | | Bond lengths (Å) |
|---|---|---|
| $C_2$ — S — $C_5$ = 92.17 | | $C_2$ — S = 1.714 |
| S — $C_2$ — $C_3$ = 111.47 | | $C_2$ — $C_3$ = 1.370 |
| $C_2$ — $C_3$ — $C_4$ = 112.45 | | $C_3$ — $C_4$ = 1.423 |
| $C_2$ — $C_3$ — H = 123.28 | | $C_2$ — H = 1.078 |
| | | $C_3$ — H = 1.081 |

Thiophene

**Fig. 16.32**  Structural parameters of pyrrole, furan and thiophene

## 16.3.1  Benzo-Fused Five-Membered Heterocyclic Compounds with One Heteroatom

The fusion of benzene ring with five-membered heterocyclic compounds (viz., pyrrole, furan and thiophenes) gives benzo [*a*], benzo [*b*] and benzo [*c*]-fused heterocycles depending on the annelation of benzene ring onto the 'face *a*' (1, 2-bond), 'face *b*' (2,3-bond) and 'face *c*' (3,4-bond), respectively (Fig. 16.33).

It is found that fusion of benzene ring to 3,4-bond (face *c*) (to give benzo [*c*] heterocycles) results in less stable bicyclic heterocyclic compounds than those resulting from 2,3-bond fusion (face *b*) (to give benzo [*b*] heterocycles). In fact, benzo [*b*] heterocycles are stable but benzo [*c*] heterocycles are stable only at low temperatures. The resonance energy of benzo [*c*] heterocycles is much lower than the resonance energy of benzo [*b*] heterocycles.

The sequence of aromaticity in benzo [*b*] heterocycles is similar to that in the five-membered heterocycles, i.e., benzo [*b*] thiophene > benzo [*b*] pyrrole > benzo [*b*] furan. However, this sequence in aromaticity of benzo [*b*] heterocycles is different from the aromaticity sequence observed in benzo [*c*] heterocycles, i.e., isoindole > benzo [*c*] thiophene > benzo [*c*] furan.

### 16.3.1.1  Indole

It is a planar molecule with $sp^2$-hybridised carbons and nitrogen atom. The $sp^2$-hybridised orbitals of carbon and nitrogen overlap axially with each other and $s$ orbitals of hydrogen bond resulting in the formation of C–C, C–N, C–H and N–H

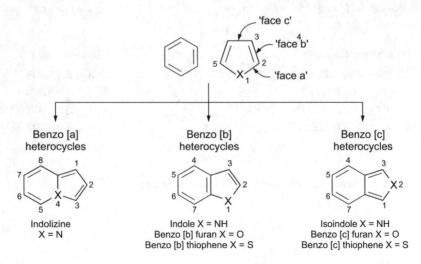

**Fig. 16.33**  Benzo-fused five-membered heterocyclic compounds with one heteroatom

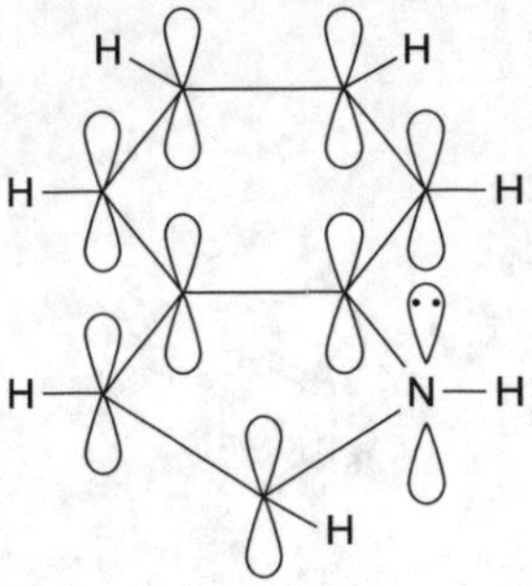

**Fig. 16.34** Molecular
orbital structure of indole

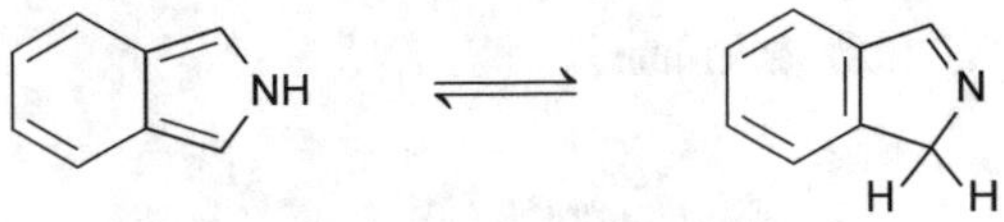

**Fig. 16.35** Two tautomeric
forms of isioindole

2H-Isoindole
(Isoindole)

1H-Isoindole
(Isoindolenine)

sigma bonds. The unhybridised orbital on C and N overlaps laterally forming a π
molecular orbital with 10 π-electrons (Fig. 16.34).

### 16.3.1.2 Isoindoles

Isoindole, also called benzo [c] pyrroles, are 10π electron systems retaining
appreciable aromatic character. Isoindole exists in two tautomeric forms, viz., 2
H-isoindole (isonidole) and 1 H-isoindole (isoindolenine); the energy difference
between the two tautomers is 33.0 kJ/mol (Fig. 16.35).

### 16.3.1.3 Indolizines

Also known as pyrrocolines, indolizine is an aromatic heterocycle with 10 π-
electrons delocalised. The nitrogen atom in indolizine is present at a ring junction (as
bridgehead nitrogen). Indolizine exhibits characteristics of both pyrrole and pyridine
(Fig. 16.36).

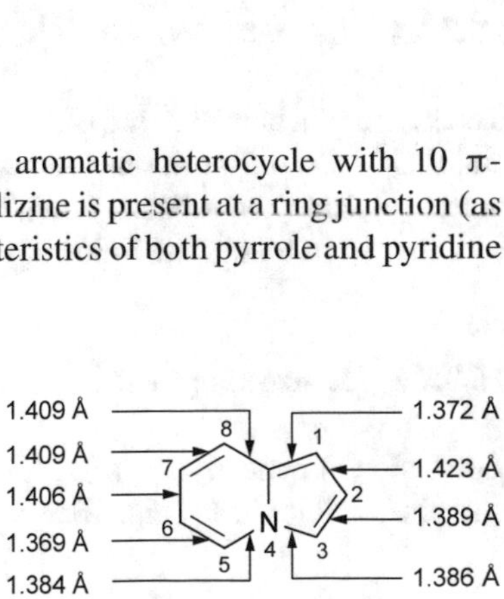

**Fig. 16.36** Bond lengths in
indolizines

1.409 Å
1.409 Å
1.406 Å
1.369 Å
1.384 Å

1.372 Å
1.423 Å
1.389 Å
1.386 Å

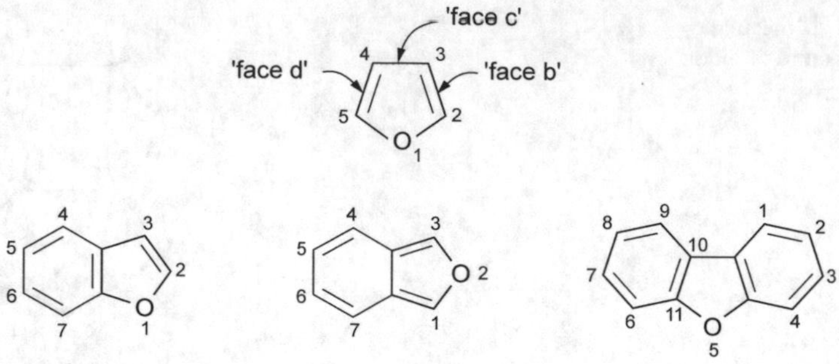


Benzo [b] furan (Benzofuran) | Benzo [c] furan (isobenzofuran) | Dibenzo furan

**Fig. 16.37** Benzofurans

| **Bond lengths (Å)** | **Bond angles (°)** |
|---|---|
| $C_1 - C_2 = 1.389$ | $C - O - C = 104.1$ |
| $C_2 - C_3 = 1.393$ | $C - O_{11} - C_{10} = 112.3$ |
| $C_3 - C_4 = 1.388$ | $C_{11} - C_6 - C_7 = 116.7$ |
| $C - O = 1.404$ | $C_6 - C_7 - C_8 = 120.9$ |
| $C_{10} - C_{12} = 1.393$ | $C_7 - C_8 - C_9 = 121.9$ |
| | $C_8 - C_9 - C_{10} = 117.9$ |

**Fig. 16.38** Structural parameters of dibenzofurans

The structural parameters of indolizine are shown in the structure given above (Fig. 16.36).

### 16.3.1.4 Benzofurans

The fusion of a benzene ring in 2,3-positions (or face '*b*'); 3,4-positions (or face '*c*') and at 2,3- and 4,5-positions (on both sides) of the furan ring gives benzo [*b*] furan, benzo [*c*] furan and dibenzo furan, respectively (Fig. 16.37).

Dibenzo furan has the following structural parameters (Fig. 16.38).

### 16.3.1.5 Benzothiophenes

The fusion of benzene ring (*s*) to thiophene gives [*b*] thiophene, benzo [*c*] thiophene and dibenzothiophene (Fig. 16.39).

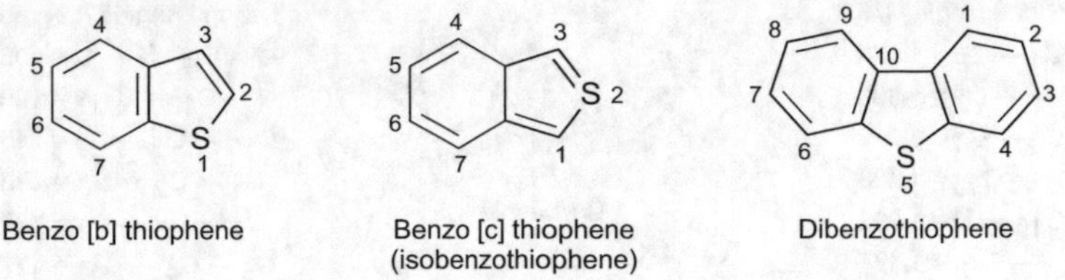

Benzo [b] thiophene  Benzo [c] thiophene (isobenzothiophene)  Dibenzothiophene

**Fig. 16.39** Benzothiophenes

## 16.3.2 Dibenzoheterocyclic Compounds with One Heteroatom

These are obtained by the fusion of benzene rings on 2,3- and 4,5-bonds of five-membered heterocycles (pyrrole, furan and thiophene). The dibenzoheterocycles are carbazole, dibenzofuran and dibenzothiophene (Fig. 16.40).

In these heterocycles, there is an appreicable increase in carbon–heteroatom bond due to the annelation. All these heterocycles are slightly bow-shaped and have small dihedral angles between the planes of the five- and six-membered rings (Table 16.2).

Carbazole  Dibenzofuran  Dibenzothiophene

**Fig. 16.40** Dibenzoheterocyclic compounds with one heteroatom

**Table 16.2** Comparison of bond lengths of dibenzoheterocycles with parent heterocycles

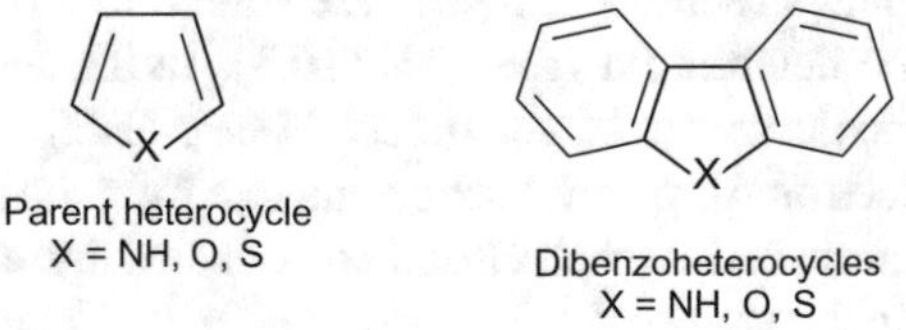

Parent heterocycle
X = NH, O, S

Dibenzoheterocycles
X = NH, O, S

| Parent heterocycle | C–X (Å) | Dibenzoheterocycles | C–X(Å) | $C_6$–$C_{6'}$(Å) | Dihedral angles (°) |
|---|---|---|---|---|---|
| Pyrrole (X = NH) | 1.383 | Dibenzo pyrrole | 1.414 | 1.477 | 1.0 |
| Furan (X = O) | 1.362 | Dibenzo furan | 1.404 | 1.481 | 1.12 |
| Thiophen (X = S) | 1.714 | Dibenzo thiophene | 1.740 | 1.441 | 0.5–1.2 |

Bond lengths (Å)

$C_1 - C_2 = 1.372$
$C_2 - C_3 = 1.393$
$C_3 - C_4 = 1.392$
$C_1 - C_{10} = 1.403$
$C_{10} - N = 1.393$
$C_{11} - C_{12} = 1.477$
$C_{10} - C_{11} = 1.408$
$N - H = 1.020$

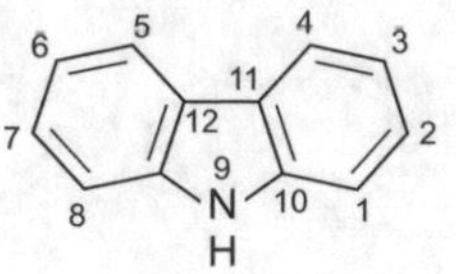

Carbazole

Bond angles (degrees)

$C - N - C = 108.3$
$N - C_{10} - C_{11} = 109.7$
$C_{10} - C_1 - C_2 = 115.6$
$C_1 - C_2 - C_3 = 123.9$
$C_2 - C_3 - C_4 = 120.1$
$C_3 - C_4 - C_{11} = 117.9$
$C_4 - C_{11} - C_{10} = 120.6$
$C_{11} - C_{10} - C_1 = 121.9$

**Fig. 16.41**  Structural parameters of carbazole

The above dibenzoheterocycles behave as diphenylamine, diphenyl ether and diphenyl sulphide in their reactions.

### 16.3.2.1   Carbazole

Carbazole is benzo [*b*] indole and is analogous to anthracene. The structural parameters of carbazole are as given below (Fig. 16.41).

## 16.3.3   Five-Membered Heterocyclic Compounds with Two Heteroatoms

These are considered to be derived from pyrrole, furan or thiophene by the replacement of –CH = group by $sp^2$-hybridised azomethine nitrogen (pyridine type nitrogen) at position-2 or 3. These heterocycles can be of two types, viz., 1, 2-azoles (the azomethine nitrogen (–N = ) is substituted at position-2 or 1,3-azoles (the azine nitrogen (–N = N) is substituted at position 3. The 1,2- and 1,3-azoles are of three types, each as given below (Fig. 16.42).

The molecular orbital structures of azoles are similar to those of five-membered heterocycles with one heteroatom (see Sect. 16.3). In azoles, each atom is $sp^2$-hybridised. The aromatic sextet of six delocalised π-electrons is believed to be contributed by one electron from each carbon and one from the azomethine nitrogen (pyridine type nitrogen), and two electrons are donated by the heteroatom as in pyrrole, furan and thiophene.

The lone pair on azomethine nitrogen does not participate in the formation of aromatic sextet. The molecular orbital structures of 1, 2 and 1, 3 azoles are as given below (Fig. 16.43).

The structural parameters of imidazole, oxazole, thiazole, pyrazole and isothiazole are given below. These are helpful to exactly imagine the shape of the heterocycles (Fig. 16.44).

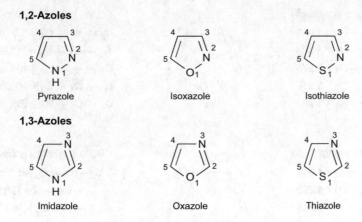

**Fig. 16.42** Five-membered heterocyclic compounds with two heteroatoms

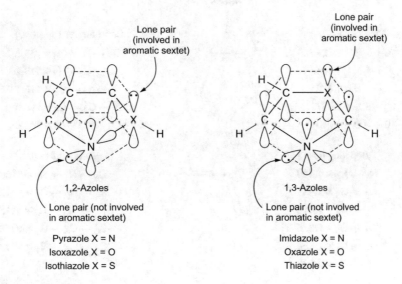

**Fig. 16.43** Molecular orbital structures of 1,2- and 1,3-azoles

## 16.4 Six-Membered Heterocyclic Compounds

Typical six-membered heterocyclic compounds are α-pyran, γ-pyran and pyridine (Fig. 16.45).

The present discussion is restricted to pyridine, its reduced product, piperidine, decahydroquinolines, 1,3-dioxans and 1,4-dioxans.

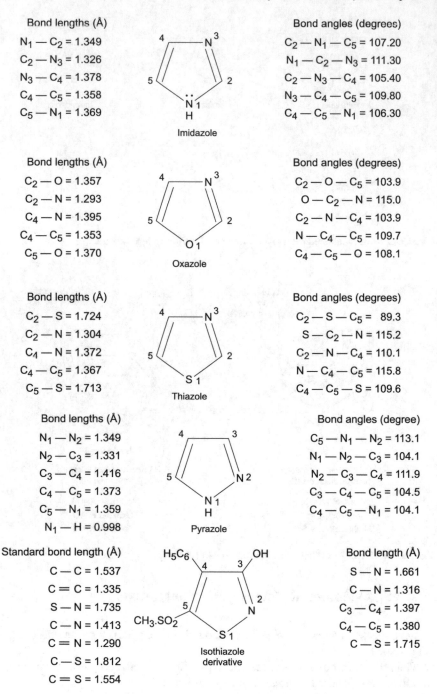

Bond lengths (Å)

$N_1 — C_2 = 1.349$
$C_2 — N_3 = 1.326$
$N_3 — C_4 = 1.378$
$C_4 — C_5 = 1.358$
$C_5 — N_1 = 1.369$

Imidazole

Bond angles (degrees)

$C_2 — N_1 — C_5 = 107.20$
$N_1 — C_2 — N_3 = 111.30$
$C_2 — N_3 — C_4 = 105.40$
$N_3 — C_4 — C_5 = 109.80$
$C_4 — C_5 — N_1 = 106.30$

Bond lengths (Å)

$C_2 — O = 1.357$
$C_2 — N = 1.293$
$C_4 — N = 1.395$
$C_4 — C_5 = 1.353$
$C_5 — O = 1.370$

Oxazole

Bond angles (degrees)

$C_2 — O — C_5 = 103.9$
$O — C_2 — N = 115.0$
$C_2 — N — C_4 = 103.9$
$N — C_4 — C_5 = 109.7$
$C_4 — C_5 — O = 108.1$

Bond lengths (Å)

$C_2 — S = 1.724$
$C_2 — N = 1.304$
$C_4 — N = 1.372$
$C_4 — C_5 = 1.367$
$C_5 — S = 1.713$

Thiazole

Bond angles (degrees)

$C_2 — S — C_5 = 89.3$
$S — C_2 — N = 115.2$
$C_2 — N — C_4 = 110.1$
$N — C_4 — C_5 = 115.8$
$C_4 — C_5 — S = 109.6$

Bond lengths (Å)

$N_1 — N_2 = 1.349$
$N_2 — C_3 = 1.331$
$C_3 — C_4 = 1.416$
$C_4 — C_5 = 1.373$
$C_5 — N_1 = 1.359$
$N_1 — H = 0.998$

Pyrazole

Bond angles (degree)

$C_5 — N_1 — N_2 = 113.1$
$N_1 — N_2 — C_3 = 104.1$
$N_2 — C_3 — C_4 = 111.9$
$C_3 — C_4 — C_5 = 104.5$
$C_4 — C_5 — N_1 = 104.1$

Standard bond length (Å)

$C — C = 1.537$
$C = C = 1.335$
$S — N = 1.735$
$C — N = 1.413$
$C = N = 1.290$
$C — S = 1.812$
$C = S = 1.554$

Isothiazole
derivative

Bond length (Å)

$S — N = 1.661$
$C — N = 1.316$
$C_3 — C_4 = 1.397$
$C_4 — C_5 = 1.380$
$C — S = 1.715$

**Fig. 16.44** Structural parameters of imidazole, oxazole, thiazole, pyrazole and isothiazole

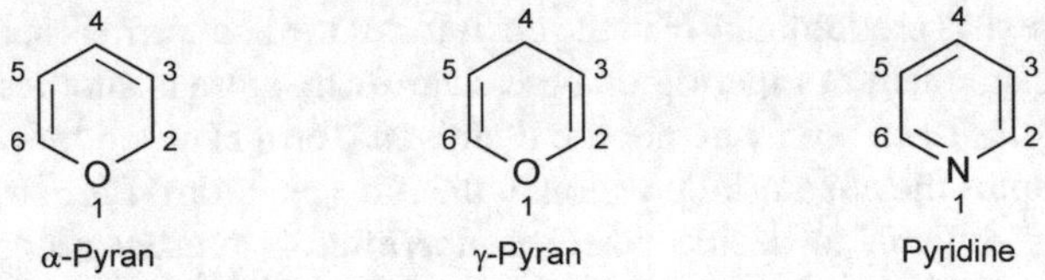

4
5   3
6   2
1
α-Pyran

4
5   3
6   2
1
γ-Pyran

4
5   3
6   2
N
1
Pyridine

**Fig. 16.45** Six-membered heterocyclic compounds

**Fig. 16.46** Molecular
orbital structure of pyridine

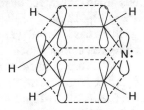

## 16.4.1 Pyridine

The pyridine molecule is flat with bond angles of 120°. The C–C and C–N bond lengths are intermediate between those of C–C single bond and $C = N$ double bond lengths.

| | |
|---|---|
| C — C bond length in pyridine 1.39Å | C — C bond length 1.54 Å |
| | C = C bond length 1.34 Å |
| C — N bond length in pyridine 1.37 Å | C — N bond length 1.47 Å |
| | C = N bond length 1.28 Å |

All the carbon atoms and the nitrogen atom in pyridine are $sp^2$-hybridised leaving one $p$-orbital on each atom. The unhybridised $p$-orbital at each atom is perpendicular to the plane of the ring atoms, overlapping each other to form a π-electron cloud above and below the plane of the ring. The molecular orbital structure of pyridine is given below (Fig. 16.46).

## 16.4.2 Piperidine

It is a saturated six-membered heterocycle containing a nitrogen atom. Its stereochemistry is more or less similar to that of cyclohexene (see Sect. 2.5.5. in Chap. 5).

Piperidine exists predominantly in the chair form; the boat form is less favourable. Besides the assignment of piperidine in the chair form, there is another aspect to be considered. It has to be seen whether the hydrogen (form I) or the free electron pair (form II) occupies the equatorial position at the nitrogen atom (Fig. 16.47).

It was very difficult to decide whether piperidine is present predominantly in form I or II as different data are available. On the basis of dipole moments and IR spectra, it was concluded that the form I with an equatorial hydrogen atom is favourable. The NMR data indicated that the axial orientation of the hydrogen of the NH group is regarded as reliable (W.J. Le Noble and Y. Ogo, Tetrahedran, 1970, **26**, 16, 4119–4124).

In the case of N-alkyl substituted piperidine, it was found that the alkyl group has equatorial conformation. This was assigned on the basis of dipole moments.

In the case of N-chloropiperidine, it has been concluded on the basis of spectral data (O.S. Anisimova, Yu. A. Pentin and L.G. Yudin, Zhur. Prikladnoi Spektroskopii, 1968, **8**, 6, 1027–1030), it was found that at 0 °C, 94% of N-chloropiperidine has chlorine in the equatorial position and 6% in the axial conformation. However, on freezing, the whole of the molecule has the equatorial conformation of chlorine.

The conformations of 4-chloro-N-alkyl piperidines have been found on the basis of NMR spectroscopy (M.-L. Stien et al., Tetrahedran, 1971, **27**, 2, 411–423) as given below (Fig. 16.48).

As already stated, piperidine exists predominantly in the chair conformation. However, in some substituted piperidine (as in III), boat conformation has been found; its IR spectrum shows a strong intramolecular hydrogen bond. On the other hand, the compound IV, which has no gem-dimethyl group, exists in chair conformation; its IR spectra show the presence of a free OH group (Fig. 16.49).

**Fig. 16.47**  Piperidine

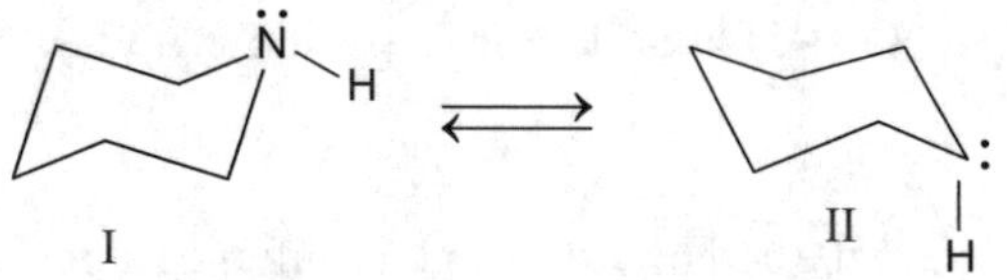

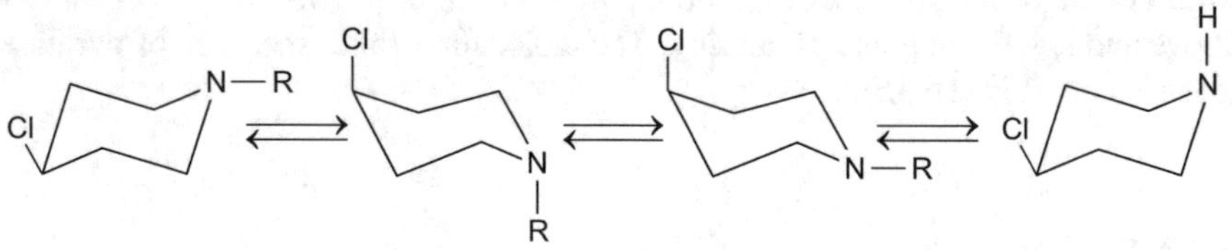

| R = CH$_3$ | 54 % | 7 % | 21 % | 18 % |
| R = C(CH$_3$)$_2$ | 70 % | 0 % | 30 % | 0 % |

**Fig. 16.48**  Conformations of 4-chloro-N-alkyl piperidine

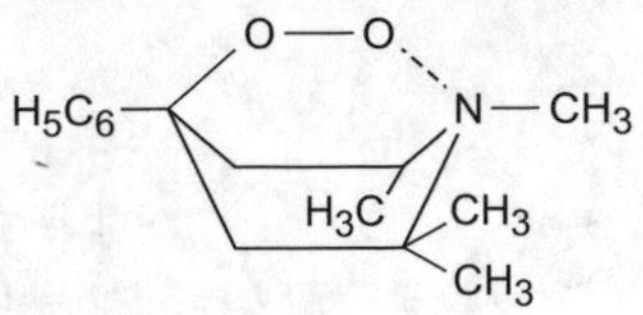

**Fig. 16.49** Boat and chair conformations in some substituted piperidines

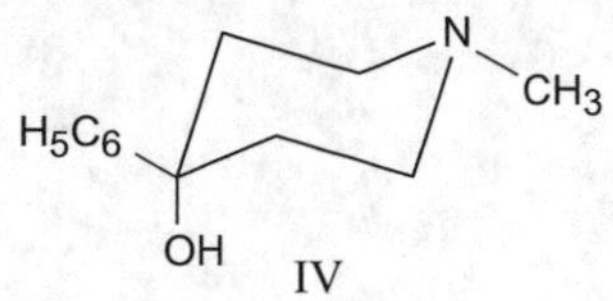

## 16.4.3 Decahydroquinolines

The fusion of piperidine ring with benzene ring gives decahydroquinoline. As in the case of decalins (see Sect. 8.2.1, Fig. 8.10, Chap. 8), the decahydroquinoline also exists in *cis* and *trans* forms (Fig. 16.50).

The decahydroquinolines are examples of bicyclic nitrogen-containing heterocycles. The *cis*- and *trans*-decahydroquinolines are both stable isomeric forms, which can exist separately. In contrast to this, quiniolisidene, analogous forms of a related cyclic structure are readily interconvertible (F.G. Riddell, Quart. Rev., 1967, **21**, 3, 364–378). In this case, the *cis–trans* forms are conformers that cannot be isolated individually (Fig. 16.51).

Stereochemistry of some heterocyclic compounds (containing nitrogen), viz., tropines, cocaine, cinchona alkaloids, morphine and cocaine has been discussed in Chap. 15.

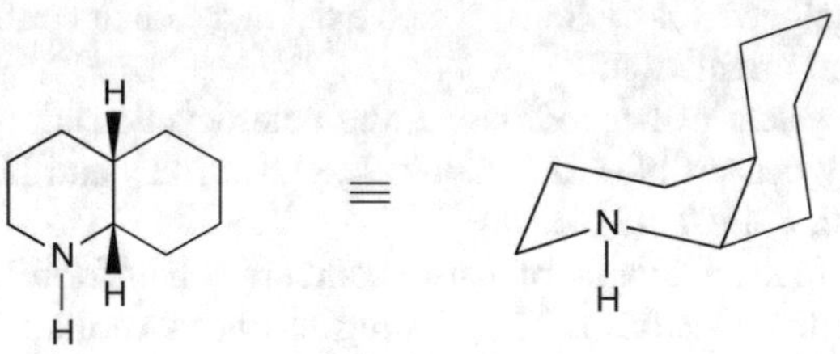

*cis*-decahydroquinoline

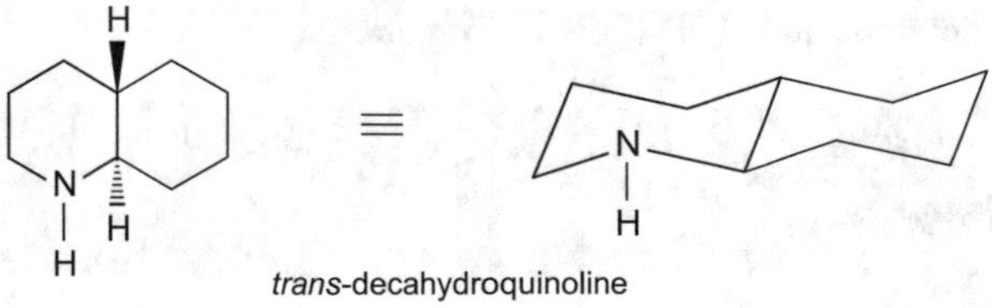

*trans*-decahydroquinoline

**Fig. 16.50** *cis* and *trans* decahydroquinolines

**Fig. 16.51**  Quinolimes

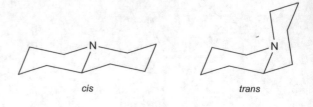

*cis*                                                         *trans*

**Fig. 16.52**  1,3-Dioxans

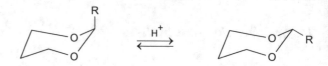

## *16.4.4   1,3-Dioxans*

The 1, 3-dioxans have been found to exist in thermodynamic equilibrium between axial and equatorial forms in acid medium on the basis of their NMR spectra (E.L. Eliel, Angew. Chem., 1972, **84**, 17, 779–791) (Fig. 16.52).

## *16.4.5   1,4-Dioxans*

As in the case of other six-membered rings, 1,4-dioxane has the chair form. In *trans*-2,3- and *trans*-2,5-dihalogen-1,4-dioxans, which exist in the chair conformation, the halogens have the axial orientation.

In the fused-ring system of benzodioxane, the heterocyclic ring has half chair form, which is readily convertible (M.J. Cook, A.R. Katritzky and M.J. Sewell, J. Chem. Soc., (B), 1970, **6**, 1207–1210).

In piperazine, a nitrogen analogue of 1,4-dioxane, the chair form is more stable (by 16 kJ/mole) than the boat form. This finding has been verified in case of N, N'-dichloropiperazine on the basis of electron diffraction method. In this case, the chlorine atoms are found to occupy equatorial positions. Similar results have been obtained for piperazine and N,N'-dimethyl piperazine.

**Key Concepts**

- **Darzen Reaction:** Condensation of aldehydes or ketones with esters of α-haloacids gives thermally unstable glycidic acids.
- **Paterno-Buchi Reaction:** Formation of oxetanes by photochemical cycloaddition of carbonyl compounds to olefins.

- **Wolf rearrangement:** Rearrangement of diazoketones by heat or light to give ketenes via the formation of intermediate oxirenes.

## Problems

1.  Three-membered heterocyclic compounds are strained. Comment.
2.  How are *trans* and *cis* epoxides and *trans* and *cis* aziridines obtained?
3.  How are epoxides converted into thiiranes?
4.  Azirines exist as 1 H and 2 H azirine. How they are obtained?
5.  Explain why oxetanes are planar molecules.
6.  Discuss the molecular orbital picture of pyrrole, indole, pyrazole and thiazole.
7.  Discuss the stereochemistry of piperidine, decahydroquinolines and quinolidene.

# Chapter 17
# Stereochemistry of Biomolecules

Biomolecules are complex organic molecules, which form the basis of life. Most of the biomolecules are of natural origin and are optically active. These include carbohydrates, proteins, nucleic acids, etc. The problems of biochemistry can only be solved on the present-day level only by consideration of their stereochemistry.

## 17.1  Carbohydrates

These are naturally occurring compounds and include compounds like sugars, starches and cellulose and are composed of carbon, hydrogen and oxygen. They are the main sources of energy for human and animal body and also support plant tissues. The carbohydrates, as we know, are divided into various groups. These include monosaccharides (examples, glucose, fructose), disaccharides (examples sucrose, maltose), trisaccharides (examples, raffinose), tetrasaccharides (example stachyose) and polysaccharides (examples, starch, cellulose).

The optically active carbohydrates include tetroses, pentoses and hexoses. Their optical isomerism has been discussed in Sect. 4.4.6 (Chap. 4).

The configuration of the carbohydrates (or its stereochemistry) defines the arrangement of H and OH groups on the asymmetric carbon atoms. It is generally determined on the basis of known configuration of the carbohydrate from which these are obtained. As an example, the configuration of glucose and mannose is determined on the basis of known configuration of D-arabinose. For depicting the configuration of carbohydrates, the sign (+) and (−) are used. These refer to the direction of rotation of the polarised light. In case the compound rotates the plane of the polarised light to the right, it is said to be dextrorotatory and is represented by the sign (+). On the other hand, if the plane of polarised light is rotated towards the left, it is said to be laevorotatory and is designated by the sign (−). The letters D- and L- refer to the absolute configuration around the asymmetric carbon atom. If the hydroxy group on

© The Author(s), under exclusive license to Springer Nature Switzerland AG 2022
V. K. Ahluwalia, *Stereochemistry of Organic Compounds*,
https://doi.org/10.1007/978-3-030-84961-0_17

the asymmetric carbon atom farthest from the aldehyde or ketone group projects to the right, the compound belongs to D-series. On the other hand, if the hydroxy group on the asymmetric carbon atom farthest from aldehyde or ketonic group projects to the left, the compound belongs to the L-series. The D and L concept has been used in case of tetroses, pentoses and hexoses (*see* Sect. 4.4.6 in Chap. 4).

Following are given the stereochemical formulae (configurations) of some typical, carbohydrates.

## 17.1.1   Glucose

It can be represented in open-chain form (Fischer projection) or in the cyclic form (Haworth projection), which is normally used. The cyclic form of glucose can be six-membered (pyranose structure) or five-membered (furanose structure). In fact most of the glucose is present in pyranose form as a mixture of α-glucose (m.p 146 °C) and β-glucose (m.p. 150 °C) having different specific rotations (+111° and +19.2°, respectively). It is known that if the two forms are separately allowed to stand in water solution, the specific rotation of the solution changes gradually until a final value +52.5° is obtained. The equilibrium is attained faster in the presence of a catalyst ($H^+$ or $^-OH$).

$$\alpha-(D)\text{-glucose} \rightleftharpoons \text{Equilibriumn} \rightleftharpoons \beta\text{-D-glycose}$$

sp-rotation+111°                              mixture                              +19.2°

+52.5°

A change in the specific rotation of an optically active compound is called **mutarotation**.

Besides α, and β glucose, a small amount of glucose is also present in furanose form. The stereochemical assignments of all the forms of glucose are given below.

D-Glucopyranose
Fisher projection

α-D-Glucose

β-D-Glucose

Haworth Projection

In the Haworth projection (as shown in above), the lower thickened edge of the ring is assumed to be nearest to the observer. The groups to the right, in the Fisher's Projection, go below the plane of the ring, while those to the left side go above the ring.

On the basis of X-ray studies, it has been shown that the cyclic forms of glucose (as shown above) preferentially exist in the non-planar chair conformation like those of cyclohexanes.

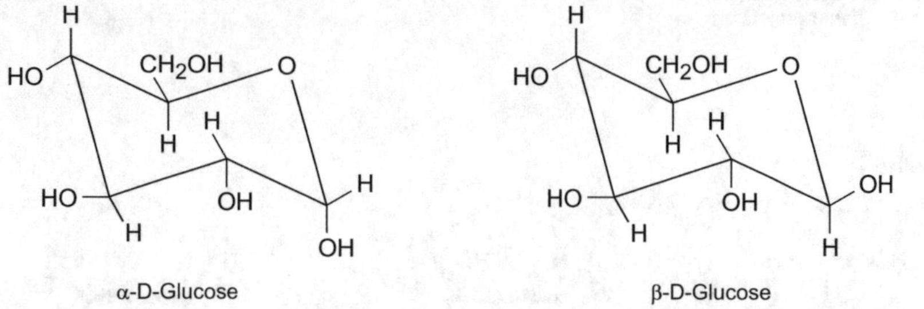

α-D-Glucose          β-D-Glucose

Chair conformations of glucose

α- and β-D Glucose (Furanose form)

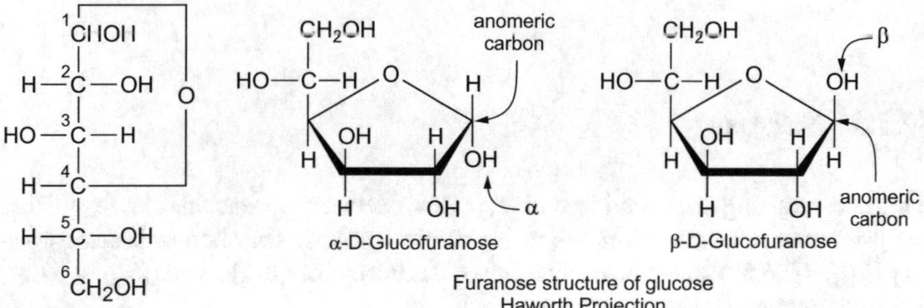

D-Glucose furanose
Fisher Projection

α-D-Glucofuranose          β-D-Glucofuranose

Furanose structure of glucose
Haworth Projection

## 17.1.2 Fructose

Like glucose, fructose is also present as α- and β-D-fructopyranose and α- and β-D-fructose furanose as shown below.

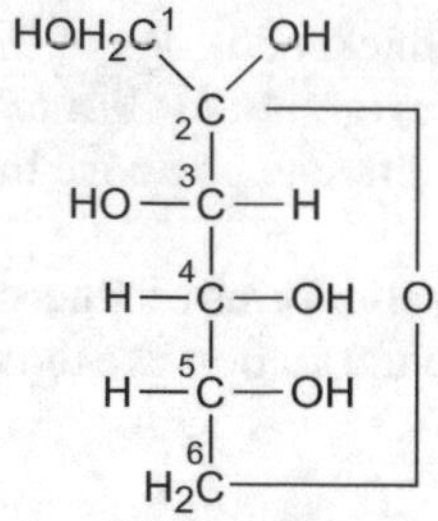

D-Fructopyranose
Fisher projection

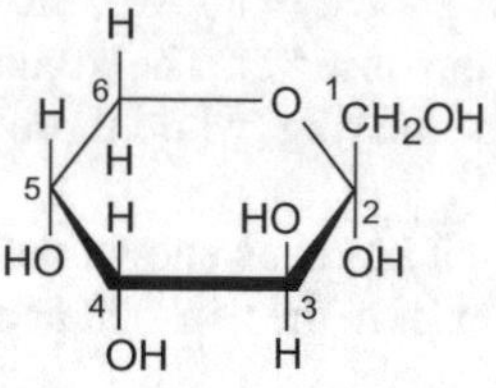

α-D-Fructopyranose

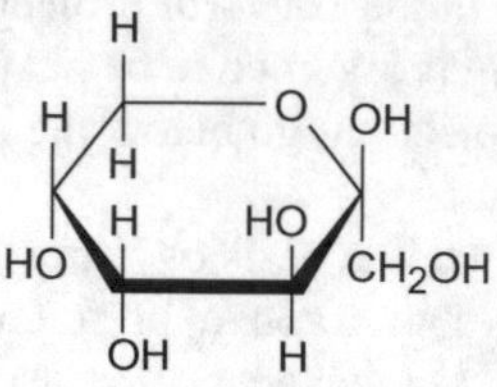

β-D-Fructopyranose

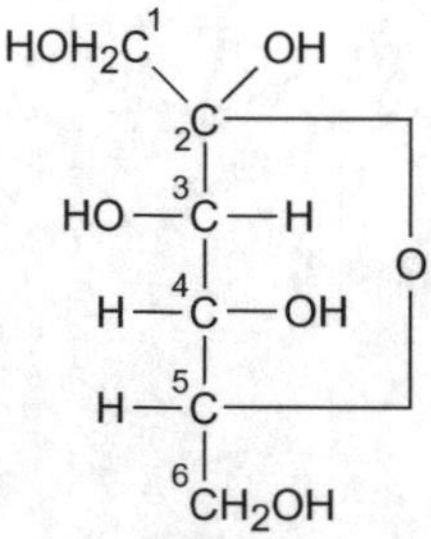

D-Fructofuranose

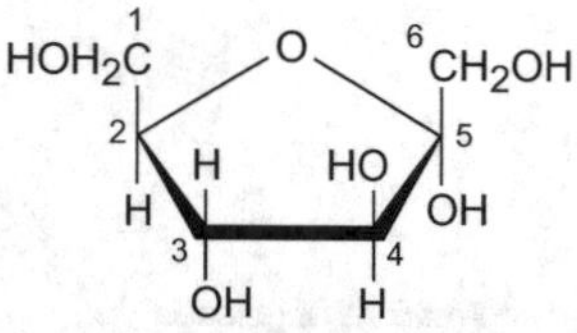

α-D-Fructofuranose

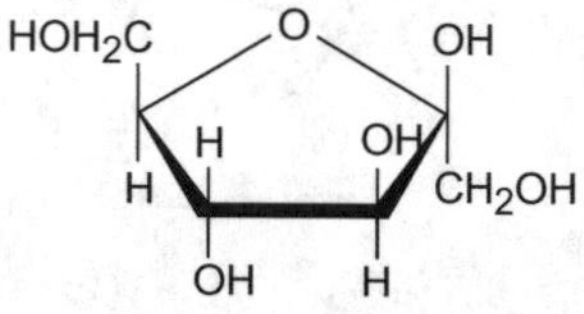

β-D-Fructofuranose

## 17.1.3  *Sucrose*

It is a non-reducing disaccharide, and so, in this, the two monosaccharides (α-glucose and β-fructose) are linked through their reducing groups. The sucrose is dextrorotatory ($[\alpha]_D + 66.5°$) and its hydrolysis gives dextrorotatory α-glucose ($[\alpha]_D + 52.5°$) and laevorotatory β-fructose ($[\alpha]_D - 92.4°$).

$$C_{12}H_{22}O_{11} \quad +H_2O \xrightarrow{H^+} \quad C_6H_{12}O_6 \quad + \quad C_6H_{12}O_6$$

$$\text{sucrose} \qquad\qquad\qquad \alpha - \text{Glucose} \qquad \beta\text{-Fructose}$$

$$[\alpha]_D = +66.5° \qquad\qquad [\alpha]_D = +52.5° \quad [\alpha]_D = -92.4°$$

It has been shown that in sucrose, D(+) glucose contained a pyranose ring while D(−) fructose is present in furanose form. Since sucrose is non-reducing sugar, so both the reducing groups, i.e., C–1 of glucose and C–2 of fructose are involved in the linkage. Also in sucrose, α-glucose is linked to fructose (since sucrose can be hydrolysed with maltase and fructose is linked to glucose in the β-form (since sucrose

could be hydrolysed by an enzyme specific for the hydrolysis of β-fructofuranoside).
On the basis of results cite α, sucrose is α-D-glucopyranosyl-β-D-fructofuranoside.

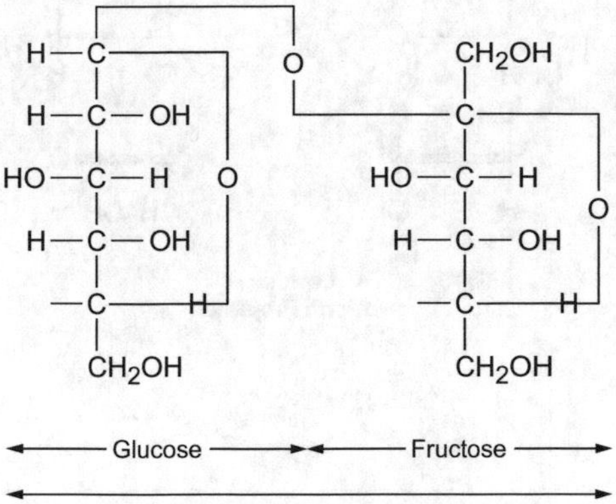

Sucrose
Fisher Projection

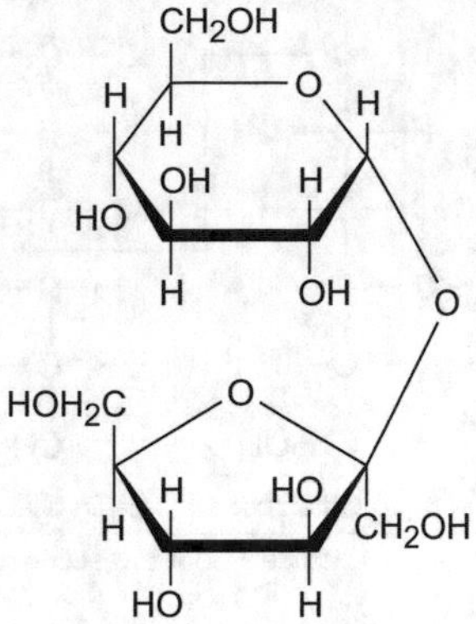

Sucrose
Haworth Projection

On the basis of X-ray analysis, sucrose has the following structure:

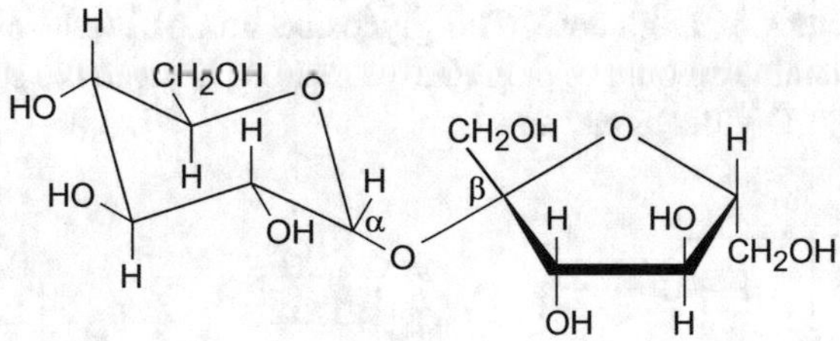

Sucrose

## 17.1.4 *Lactose*

It is a reducing type of disaccharide. On hydrolysis with emulsin, it gives glucose and
galactose, since emulsin is a β-glycosidic splitting enzyme, lactose is β-glucoside

$$\underset{(+)\ Lactose}{C_{12}H_{22}O_{11}} + H_2O \xrightarrow{emulsin} \underset{(+)\ Glucose}{C_6H_{12}O_6} + \underset{(+)\ Galactose}{C_6H_{12}O_0}$$

On the basis of degradative studies, lactose is 4-O-(β-D-galactopyranosyl)-D-glucopyranose.

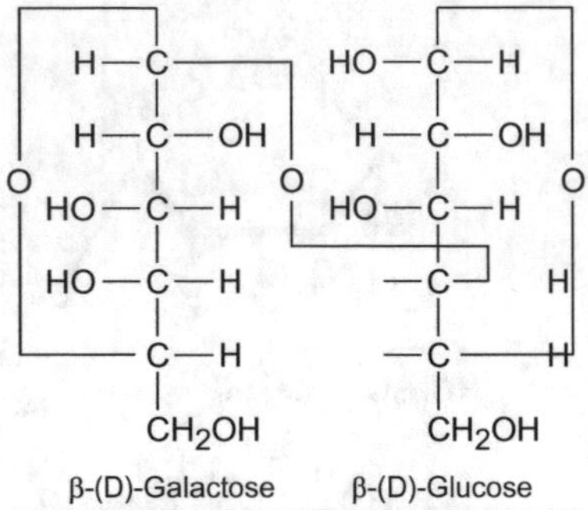

Lactose Fisher Projection

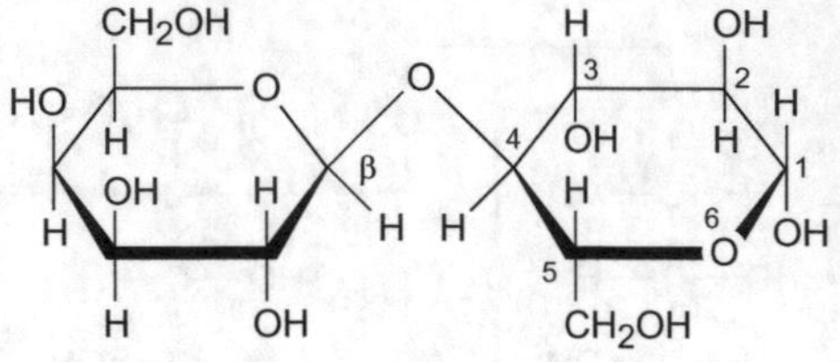

Lactose
Haworth Projection

### 17.1.5  *Maltose*

It is a reducing type of disaccharide and on hydrolysis with the enzyme emulsin gives two molecules of D-glucose. The glycoside link of D-glucose is α. On the basis of a considerable amount of degradative maltose is given the structure as 4-O-D-glucopyranosyl-D-glucopyranose.

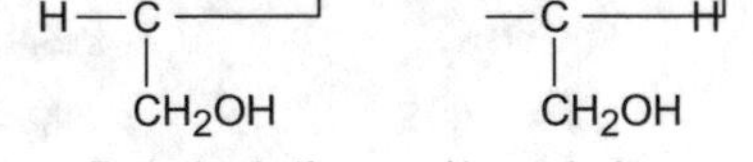

Reducing half
Maltose

Non-reducing
half

Fisher Projection

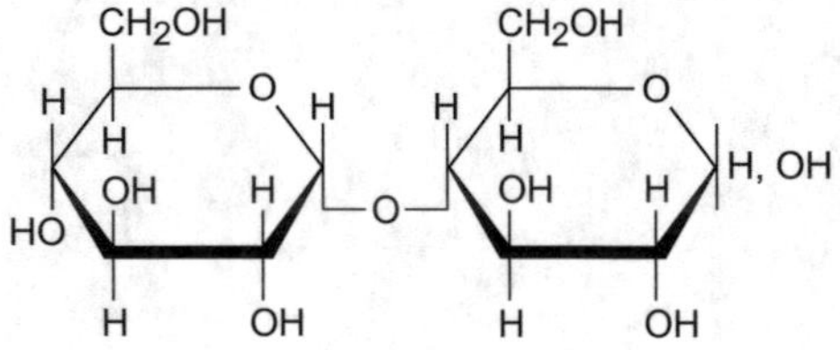

Non-reducing
half

Reducing
half

Maltose
Haworth Projection

## 17.1.6 Trehalose

A non-reducing disaccharide trehalose is α-D-glucopyranosyl-α-D-glucopyranoside. On hydrolysis, two molecules of glucose are obtained. The glycosidic link was found to be as α-α due to its high positive rotation.

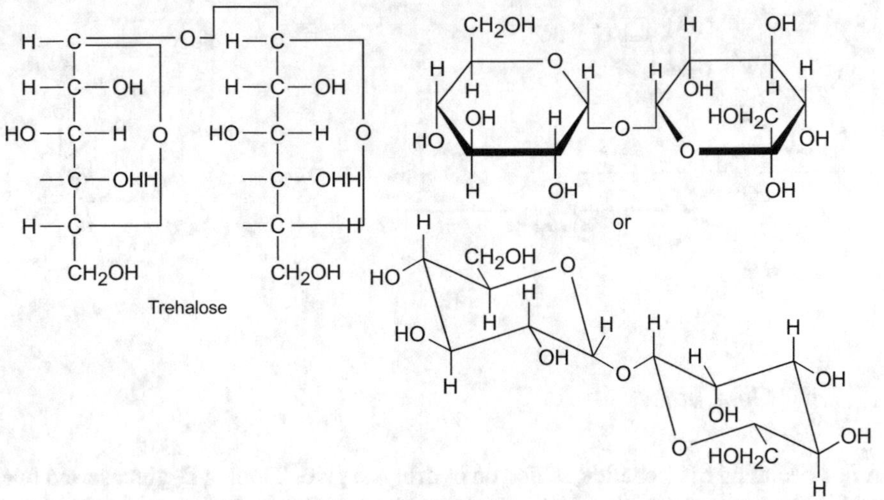

Trehalose

## 17.1.7 Raffinose

A non-reducing trisaccharide is giving one molecule each of D-fructose, D-glucose and D-galactose on vigorous hydrolysis. However, on mild hydrolyse with acids gives D-fructose and the disaccharide mellibiose and enzymatic hydrolysis with invertase yields D-fructose and mellibiose and enzymatic hydrolysis with α-glycosidase yields galactose and sucrose. On the basis of the hydrolysis products, it is found that the linkage of the three monosaccharides is of the order: galactose–glucose–fructose. Thus, raffinose has the structure O-α-D-galactopyranosyl-(1 → 6)-O-α-D glucopyranosyl-(1 → 2)-β-D-fructofuranoside.

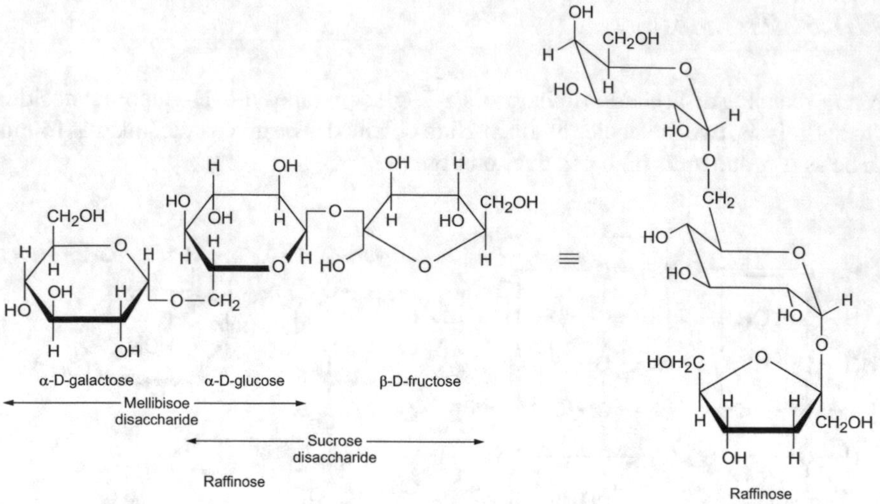

α-D-galactose        α-D-glucose        β-D-fructose

Mellibiose
disaccharide

Sucrose
disaccharide

Raffinose

Raffinose

## 17.1.8   *Gentibiose*

A non-reducing trisaccharide, which on hydrolysis gives 2 mol of D-glucose and one mole of D-fructose. On hydrolysis with invertase, it yields D-fructose and gentibiose, and with emulin, it yields D-glucose and sucrose. The arrangement of the three monosaccharides is glucose–glucose–fructose.

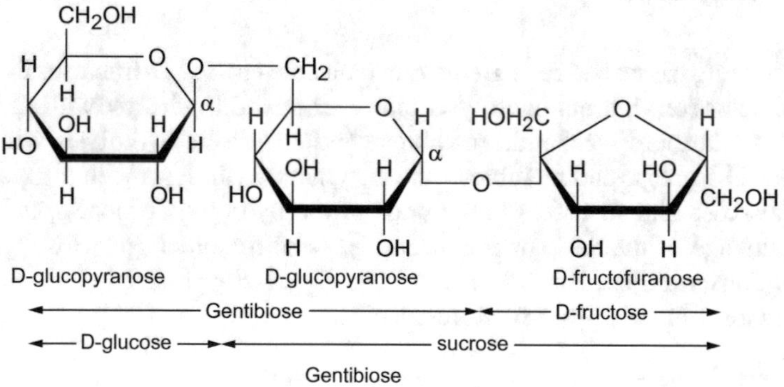

D-glucopyranose          D-glucopyranose          D-fructofuranose

Gentibiose                    D-fructose

D-glucose                    sucrose

Gentibiose

## 17.1.9 *Cellulose*

A polysaccharide of high molecular weight (2,50,000 to 1,000,000) on hydrolysis gives only glucose. X-ray analysis indicates that cellulose is a linear molecule. It is found that cellulose is made up of chains of D-glucose, each unit being joined by a glycoside linkage to C-4 of the next glucose unit. On treatment with acetic anhydride and sulphuric acid, cellulose yields octa-O-acetyl cellobiose. Thus, cellulose is considered as a polymer of cellobiose. Also enzymic hydrolysis cellulose gives cellobiose, which has β-glycosidic linkage. This implies that all glycosidic linkages in cellulose are β-linkages.

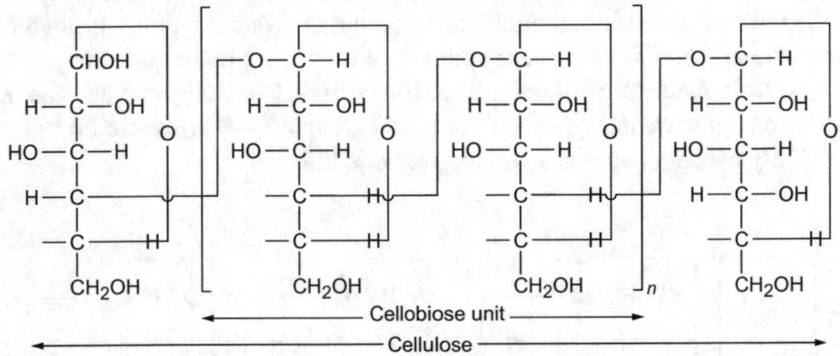

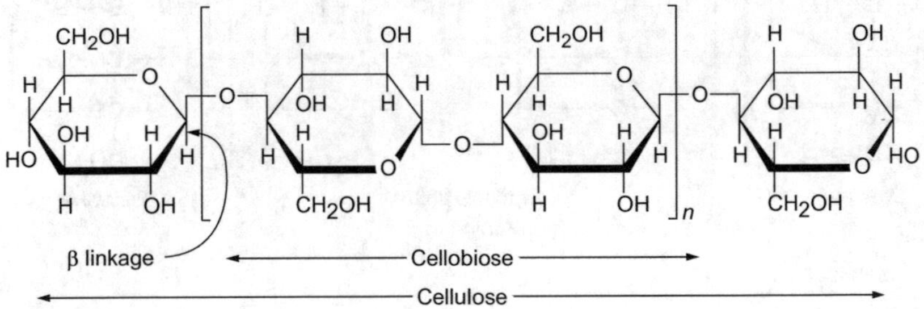

The cellulose molecule contains about 1500 glucose units. The long chains lie side by side in bundles and are held together by hydrogen bonds between the neighbouring hydroxyl groups. Finally, the bundles are twisted to form rope-like structures, which, in turn, form the fibres. It is worth noting that humans cannot digest cellulose, since the digestive enzymes of humans cannot attack its β, 1:4 linkages. However, cows etc. can use the cellulose of grass as a food since their digestive juices contain β-glucosidases, which can attack the β, 1: 4 linkages.

## 17.1.10   Starch

Starch is not a pure compound. It is a mixture of two polysaccharides, viz., amylose (20%, water-soluble fraction and gives a blue colour with iodine) and amylopectin (80%, water insoluble and gives no colour with iodine).

### 17.1.10.1   Amylose

On hydrolysis, amylose gives maltose (as the only disaccharide) and glucose (as the only monosaccharide). It is found that glucose unit is joined by α-glycosidic linkage to C–4 of the next molecule of glucose. In fact each D-glucose unit in amylose is attached to two other D-glucose units, one through C–1 and the other through C–4. Also amylose is found to be a linear polymer of D-glucose units joined together by α-linkages (as in maltose) with one end having a free aldehyde group and the other end having a D-glucose unit, having a free –OH group at C–4. Amylose differs from cellulose only in stereochemistry of the glycoside linkage.

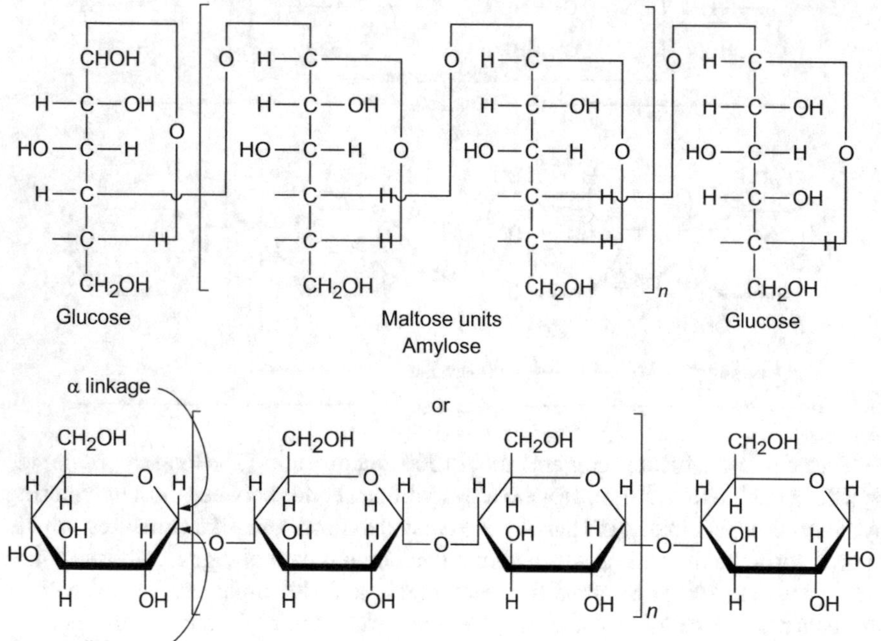

Glucose                Maltose units                              Glucose
                       Amylose

α linkage                              or

α linkage                            Amylose

### 17.1.10.2 Amylopectin

Amylopectin on hydrolysis gives maltose as the only disaccharide. In fact amylopetin is made up of chains of D-glucose and that each unit is joined by a α-glycoside linkage to C–4 of the next unit. It has been found that amylopectin has a highly branched structure consisting of several hundred short chains 20–25 D-glucose units; one end of each chain is joined through C–1 to C–6 of next chain. Amylopectin is a branched α-1, 4′ polymer of glucose with a α-1,6′ linkage, which provides an attachment point for other chains.

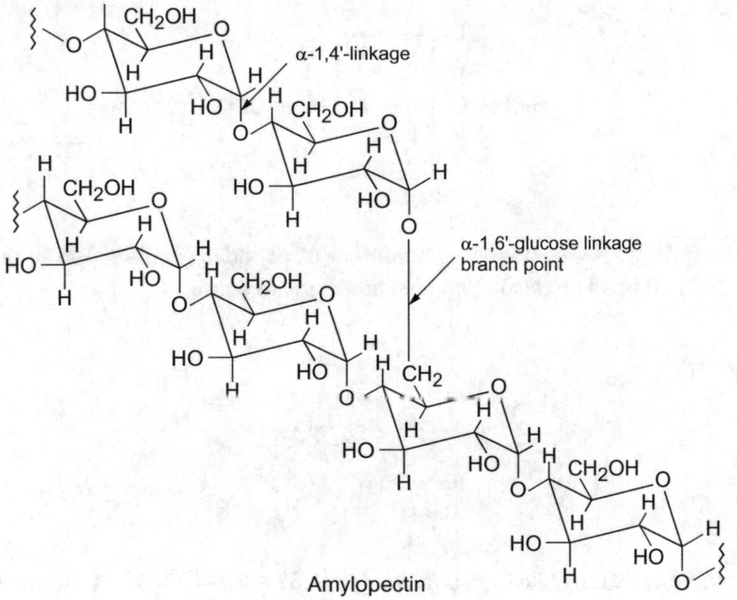

Amylopectin

## 17.2 Proteins

The basic unit of proteins is the amino acids, which combine to form peptide or polypeptide. The term peptide and proteins are used interchangeably. The term peptide is normally used if the number of amino acids in a chain is about 50 and for chains more than 50 are called proteins. The optical activity in proteins is due to the optical activity of amino acids that make up the proteins. Also the amino acids contained in proteins belong to the L series. For a discussion on D- and L system of nomenclature, *see* Sect. 4.4.4 (in chapter on stereochemistry of organic compounds containing asymmetric C). The absolute configuration of amino acids (R and S system of nomenclature) forms the subject matter of Sect. 4.4.5 (Chap. 4).

As already stated, the amino acids in proteins are joined in a linear fashion, to form a peptide bond (–CONH–group). In the formation of a peptide bond, the COOH group of one amino acid combines with $NH_2$ group of the other amino acid. On the basis of this, the protein molecule is a linear polymer of amino acid.

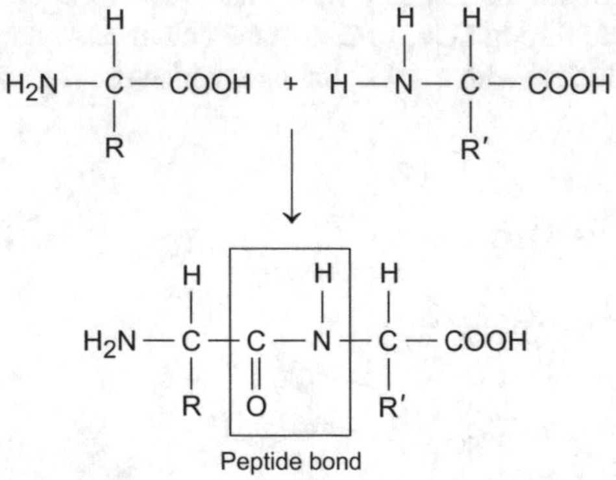

Peptide bond

On the basis of X-ray studies of a number of peptides (Pauling 1953), the bond lengths (Å) and bond angles in peptides are as given below:

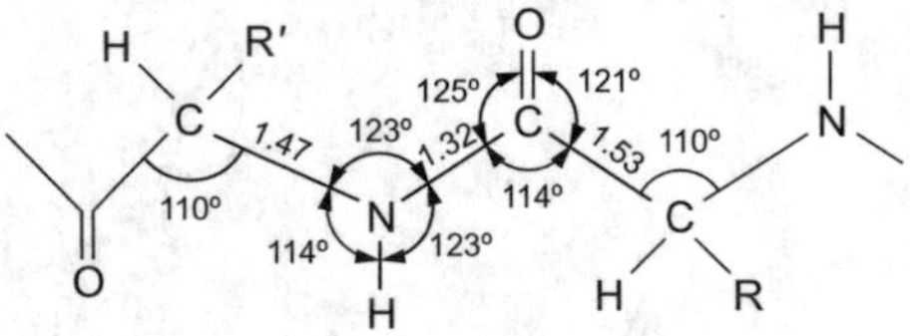

According to Pauling, the C–N bond length (1.32 Å) of –CO–NH is shorter than the usual C–N bond length (~1.47 Å) and so this bond has some double-bond character. This was explained by saying that C–N bond permits the possibility of geometrical isomerism; the *trans* form is more likely due to the much larger steric repulsion operating in the *cis* isomer. Also, the rotation can occur about the $R^1CH$–CO and $R^2CH$–NH bond as shown below:

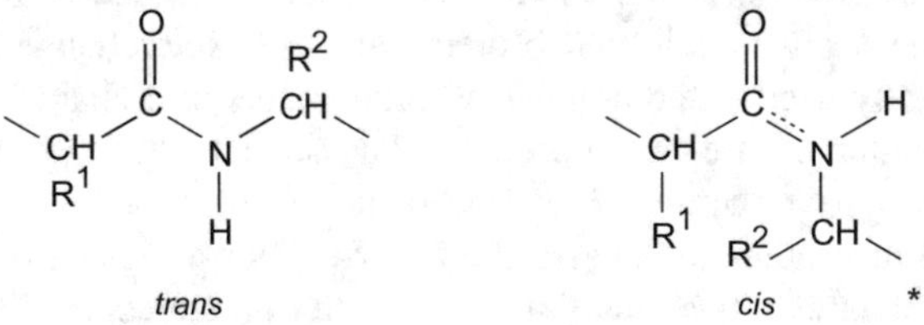

*trans*                                                          *cis*                    *

Before discussing the stereochemistry of protein, it is of interest to study the stereochemistry of some biologically important peptides.

## 17.2.1 Biologically Important Peptides

These include oxytocin, insulin and gramicidin S.

### 17.2.1.1 Oxytocin

It is a peptide hormone containing nine amino acids. It contains a disulphide linkage between cysteine residues at positions 1 and 6. Its structure including three-dimensional structure is given ahead (Fig. 17.1).

**Fig. 17.1**
Three-dimensional structure
of oxytocin

Structure of oxytocin

### 17.2.1.2    Insulin

The peptide hormone, insulin, well known for controlling sugar level in humans has been found to have the following structure:

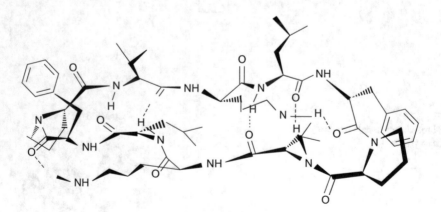

Structure of human insulin

### 17.2.1.3    Gramicidin S

A peptide antibiotic has the following three-dimensional structure (Fig. 17.2).

**Fig. 17.2**  Three-dimensional structure of a peptide antibiotic

## 17.2.2   Stereochemistry of Proteins

As already stated, the proteins are believed to be polymers containing a large number of amino acids joined to each other and have definite a three-dimensional shape. In fact the stereochemistry of proteins is considered at four different levels, viz., primary, secondary, tertiary and quaternary structures. The *primary structure* deals with the spatial arrangement of the polypeptide in a protein molecule. In other words, it deals with the amino acid sequence in the protein.

The *secondary structure* deals with the confirmation that the polypeptide chain assumes.

The *tertiary structure* deals with the way, the protein molecule folds in order to provide a specific shape. Finally, the tertiary structure is the arrangement and ways in which the subunits of proteins are held together.

### 17.2.2.1   Secondary Structure of Proteins

It was Pauling (1951), who, on the basis of theoretical grounds, proposed a $\alpha$-helix model for the conformation of proteins. This was subsequently verified by experimental evidences. According to Pauling, the peptide group is planar and the hydrogen bonding stabilises the conformation of the protein. The strength of the hydrogen bond is a maximum when the atoms concerned (C==O–N–H) are collinear. Pauling proposed a helix in which each turn contained either 3.7 or 5.1 amino-acid residues. Stereochemical considerations showed that $\alpha$-helix with 3.7 residues per turn was most stable. This is represented below (Fig. 17.3).

As seen above, each hydrogen bond is formed between the CO group of one residue and the NH group of the fourth residue in the chain. The hydrogen bonding prevents free rotation indicating that the helix is rigid.

The $\alpha$-helix may be left or right handed. As already stated, the common amino acids, except glycine are optically active and have the L-configuration. According to Moffit (1956), the right-handed helix (for L-amino acids) is more stable than the left-handed helix; this was deduced on the basis of theoretical considerations. In view of this, it was expected that the right-handed helix is expected to occur naturally.

Pauling (1951) also proposed that proteins exist in the $\beta$-conformation (or pleated sheet). In this conformation, the polypeptide chain is extended and the chains are held together by intermolecular hydrogen bonds. According to Pauling, two types of pleated structures are possible. These are parallel (I) in which all chains run in the same direction and anti-parallel(II), in which chains run alternately in the opposite directions.

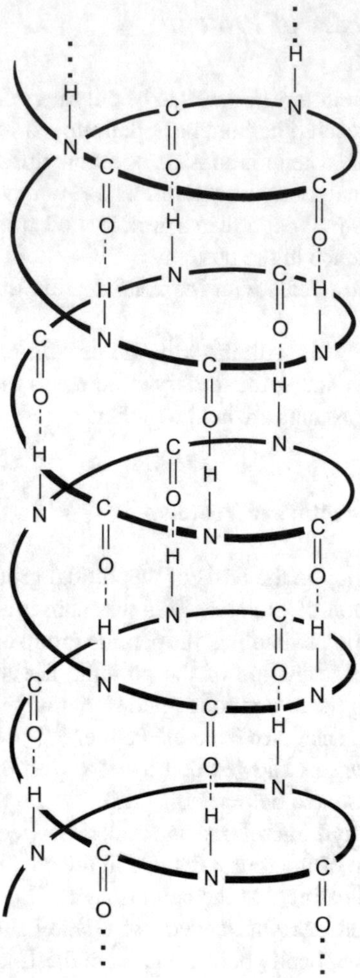

**Fig. 17.3**  Representation of α-helix

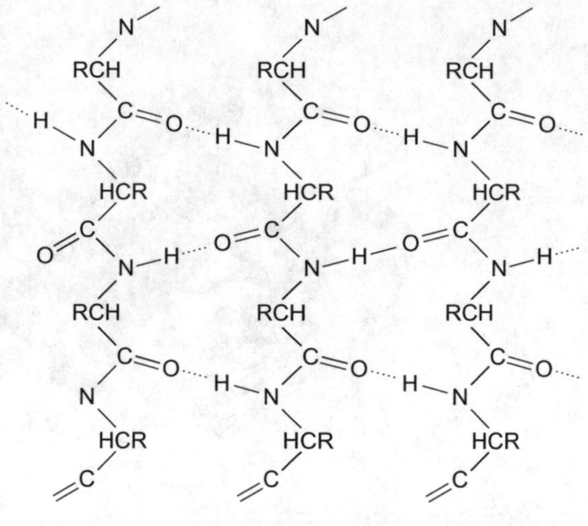

I (Parallel)

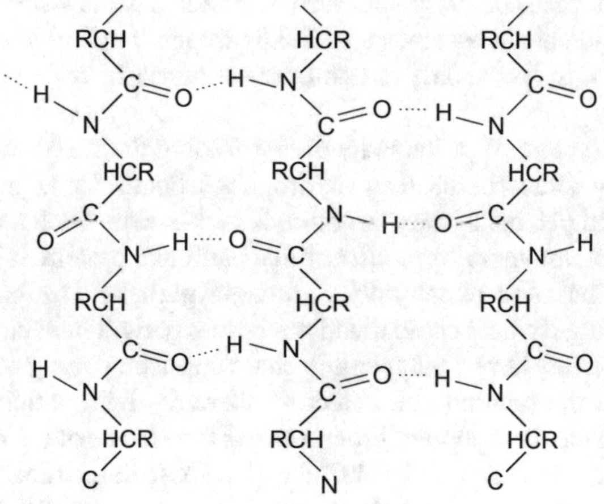

II (Anti-parallel)

On the basis of X-ray analysis, it was believed that the α-helix in proteins is present in solid state. The X-ray data also revealed that the α-helix has two types of repeat units, viz., one in which the distance between two successive turns is about 5.0–5.5 Å and the other being the distance in the direction of the helical axis between two like atoms in the chain (1.5 Å), i.e., there is the 'rise' from the first N to the second N in NHCOCHRNH. It is estimated that the diameter of the helix is about 10 Å. It is found that all polypeptide chains cannot form the α-helix, as the stability of the helix is dependent on the nature and sequence of the side chains in the polypeptide chain.

The X-ray analysis is also helpful to establish the existence of the pleated sheet structure in solid proteins. For example, the chains are parallel in Keratin and anti-parallel in fibroin.

**Fig. 17.4** Tertiary structure
of the proteins lysozyme

## 17.2.2.2    Tertiary Structure of Proteins

As already stated, the secondary structure of proteins deals with the confirmation
that the polypeptide chain assumes and the hydrogen bonding is of importance in
the stabilization of the secondary structure of proteins (the α-helix and the pleated
sheets).

The tertiary structure of proteins involving folding the entire molecule involves
hydrogen bonding, ionic, chemical and hydrophobic bonds. Under normal conditions
of temperature and pH, the tertiary structure is most stable. Such a tertiary structure
is referred to as the *native conformation* of that particular protein. It is found that two
major molecular shapes occur naturally, fibrous and globular. The former (i.e., fibrous
proteins) have a large helical content and are basically rigid molecules and the latter
(i.e., globular proteins) have a polypeptide chain consisting partly of helical sections
and folded about the random coil selection thereby giving a spherical shape. In
globular proteins, the most popular groups are on the surface of the molecule and the
most hydrophobic side chains are inside the molecule. A diagrammatic representation
of the tertiary structure of a protein lysozyme is given below (Fig. 17.4).

## 17.2.2.3    Quaternary Structure of Proteins

According to IUPAC-IuB, the quaternary structure of a protein molecule is the
arrangement of its subunits in space and the ensemble of its intersubunits contacts
and interactions, with regard to the internal geometry of the subunits.

Both the fibrous and globular proteins consist of only one polypeptide chain. In
case, a number of chains are present, the globular protein is said to be oligomeric.
The individual chains are called protomers or subunits, which may or may not be
identical. The subunits are held together by hydrogen bonds.

**Fig. 17.5** Structure of
haemoglobin showing four
subunits and the haeme
moiety

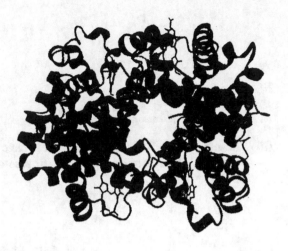

In haemoglobin, each of the four chains binds a prosthetic group called haeme, which consists of four chains that bind a prosthetic group called haeme, which consists of an iron atom held in a cage by protoporphyrin. The structure of haemoglobin is given below (Fig. 17.5).

## 17.3   Nucleic Acids

Like proteins and polysaccharides, nucleic acids are biological polymers or biomolecules. Nucleic acids are associated with the transfer of genetic information (hereditary characteristics). These are of two types, *viz.*, DNA (Deoxyribonucleic acid) and RNA (Ribonucleic acid).

The nucleic acids are made up of (*i*) sugars (ribose in case of RNA and 2-deoxyribose in case of DNA), (*ii*) four bases (three of these bases are adenine, guanine and cytosine; the fourth base is uracil in case of RNA and thiamine in case of DNA), and (*iii*) phosphoric acid. The combination of sugars, bases and phosphoric acid gives nucleoside (sugar + base), nucleotide (sugar + base + phosphoric acid) and nucleic acid [(sugar + base + phosphoric acid)$_n$]. The nucleic acids, in fact, are called polynucleotides.

The structures of various constituents (viz., sugars, bases and phosphate group) are given below:

(i)   **Sugars:** As already stated the sugars can be either D-Ribose (in case of RNA) or deoxyribose (in case of DNA).

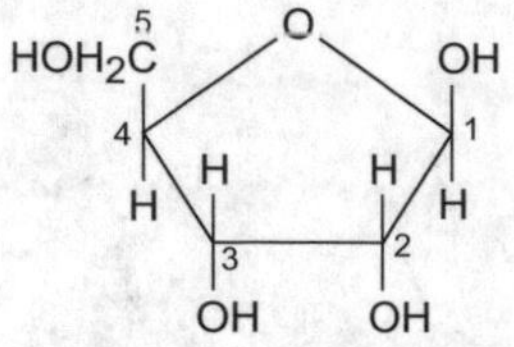

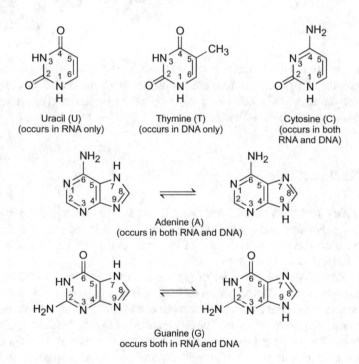

D-(—)Ribose
(in case of RNA)

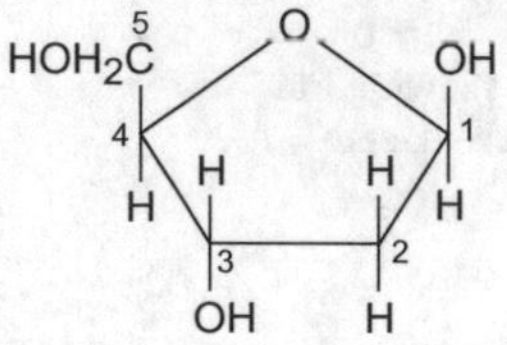

2-Deoxy-D-(—)Ribose
(in case of DNA)

(ii)   **Bases:** Three of the bases present in nucleic acids are adenine, guanine and cytosine. The fourth base is uracil (in case of RNA and thymine in case of DNA.

Uracil (U)
(occurs in RNA only)

Thymine (T)
(occurs in DNA only)

Cytosine (C)
(occurs in both
RNA and DNA)

Adenine (A)
(occurs in both RNA and DNA)

Guanine (G)
occurs both in RNA and DNA

(iii)   **Phosphate Group Linkage:**

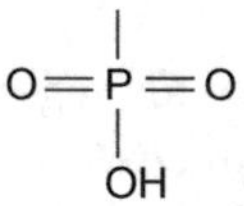

Phosphate linkage
(present in nucleic acids)

**Nucleosides of RNA and DNA**

In the nucleoside of RNA, the base uracil is bonded to a sugar, ribose and in the nucleoside of DNA, the base thymine is bonded to the sugar deoxy ribose.

|                        |                        |
|------------------------|------------------------|
| Base<br>uracil          | Base<br>Thymine         |
| Sugar<br>Ribose         | Sugar<br>2-Deoxyribose  |

Nucleoside of RNA            Nucleoside of DNA<br>
(Ribonucleoside)            (Deoxyribonucleoside)

In the nucleosides, the nitrogen base (uracil or thymine) is attached to $C_1$ of the sugar (ribose or 2-deoxyribose) as shown ahead.

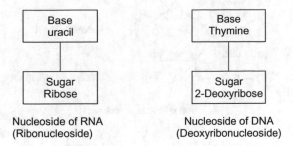

Uracil — Ribose — Uradine (uracil +ribose)

Thymine — 2-deoxyribose — Thymidine (Thymine + 2-deoxyribose)

Other nucleosides are guanosine (guanine + ribose), cytidine (cytosine + ribose) and adenosine (adenine + ribose).

## 17.3.1   *Nucleotides*

The nucleotides are phosphorylated nucleotides. The phosphate linkage is attached to position 3 of the sugar. An example is adenylic acid (a nucleotide), which contains adenine, ribose and phosphate group.

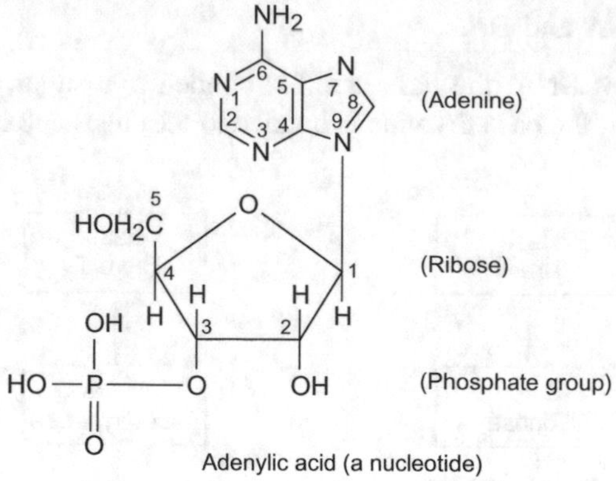

Adenylic acid (a nucleotide)

## 17.4   Nucleic Acids

The structure of nucleic acids is discussed at three levels, viz., primary, secondary and tertiary structures.

### 17.4.1   Primary Structure

As stated earlier, the nucleic acids are polynucleotides and are formed by the polymerisation of a large number of molecules of nucleotides. During polymerisation, the 5-$CH_2OH$ group of sugar moiety of one nucleotide combines with the phosphoric acid group present at C-3 position of the other nucleotide forming a long chain of polynucleotide chain. Such an arrangement gives the primary structure of nucleic acid as given below. As already stated, common bases in RNA are adenine, guanine, uracil and cytosine and the common bases in DNA are same as in RNA except that uracil is replaced by thymine. Also the sugar present in RNA is ribose while in DNA it is 2-deoxyribose (Fig. 17.6).

### 17.4.2   Secondary Structure of Nucleic Acids (DNA)

The secondary structure is exhibited by DNA. It is known that the common bases in DNA are adenine (A), guanine (G), thymine (T) and cytosine (C). On the basis of X-ray analysis (James Wilson and Francis Crick, 1953), it was found that DNA consists of two polynucleotide chains or strands, which run in opposite direction (i.e., the free phosphate residue at 3′ (or 5′ positions of the two strands lie in opposite side of the α-helix giving a double helix structure to DNA (see Fig. 17.7). The sugar-

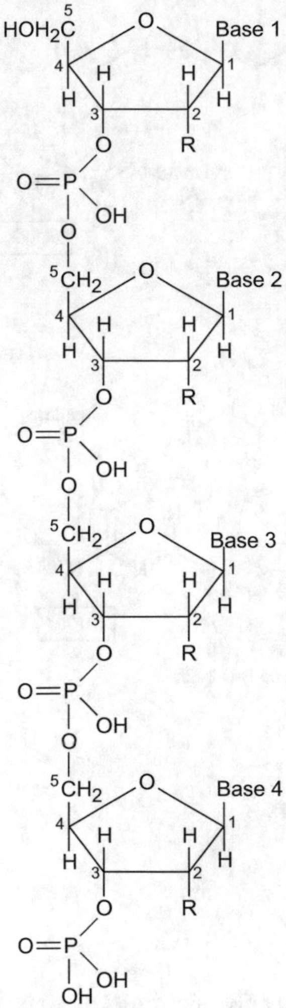

RNA Base 1 – Guanine
Base 2 – Uracil
Base 3 – Adenine
Base 4 – Cytosine

Sugar = R = OH (Ribose)

DNA Bases present are:
Guanine
Thymine
Adenine
Cytosine

Sugar = R = H (2-Deoxyribose)

**Fig. 17.6** Primary structure of nucleic acids (RNA and DNA)

phosphate units constitute the backbone of each strand. The base units of each strand are pointed into the interior of the helix. The two strands of the double helix are held together at fixed distances through H-bonds, which are specific between a purine and a pyrimidine base pair. Thus, guanine (G) pairs with cytosine (C) through three H-bonds and adenine (A) pairs with thymine (T) through two H-bonds. The two strands of the double helix are not identical but complementary as the base sequence of one strand fixes automatically that of the other due to base-pairing principle as stated above. The paired bases are stacked together one above or below the other and the distance between two base pairs is 3.4 Å. It has been found that the distance between any two turns of the helix is 34 Å indicating that base pairs in each turn. The

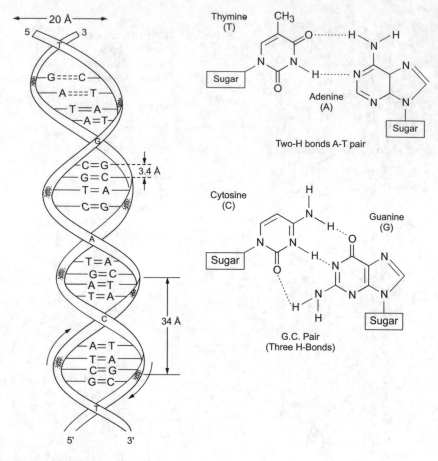

**Fig. 17.7**  Secondary Structure of DNA

diameter of the helix is about 20 Å. In RNA, there is only one strand as compared with DNA, which has two strands.

## 17.4.3   Tertiary Structure of DNA

The DNA can also exist in tertiary structure form. In this, the double helix of DNA is further coiled into a compact three-dimensional structure, known as supercoiling, which is an intrinsic aspect of DNA tertiary structure.

## Key Concepts

- **Amylopectin:** A polysaccharide on hydrolysis gives maltose as the only disaccharide. It is made up of D-glucose and each unit is joined by a α-glycoside linkage to C–4 of the next unit.
- **Amylose:** A polysaccharide on hydrolysis gives maltose as the only disaccharide and glucose (as the only monosaccharide). The glucose unit is joined by α-glycosidic linkage to C–4 of the next molecule of glucose.
- **Biomolecules:** Complex organic molecules form the basis of life.
- **Cellulose:** A polysaccharide of high molecular weight (2,50,000–1,000,000) on hydrolysis gives only glucose and each unit is joined by a glycoside linkage to C–4 of the next glucose.
- **DNA:** *See* nucleic acids
- **Gentibiose:** A non-reducing trisaccharide on hydrolysis gives two moles of glucose and one mole of D-fructose.
- **Gramicidin:** S A peptide antibiotic.
- **Insulin:** A well-known peptide hormone controls sugar levels in humans.
- **Lactose:** A reducing type of disaccharide ($C_{12}H_{22}O_{11}$) on hydrolysis with emulsin gives (+) glucose and (+) galactose.
- **Maltose:** A reducing type of disaccharide ($C_{12}H_{22}O_{11}$) on hydrolysis with emulsin gives two molecules of α-glucose.
- **Mutarotation:** A change in the specific rotation of an optically active compound.
- **Nucleic acids:** Biological polymers associated with transfer of genetic information (hereditary characteristics). These are of two types, viz., DNA and RNA. These are made up of sugars (ribose in case of RNA and 2-deoxyribose in case of DNA), four bases (three of these are adenine, guanine and cytocine; the fourth base is uracil in case of RNA and thiamine in case of DNA) and phosphonic acid.
- **Nucleoside:** The combination of sugars and bases gives nucleoside (sugar + base).
- **Nucleotide:** Combination of sugar + base + phosphoric acid.
- **Oxytocin:** A peptide hormone containing nine amino acids and containing a disulphide linkage between cysteine residues at positions 1 and 6.
- **Protein:** Nitrogenous substances occurring in the protoplasm of all animal and plant cells.
- **Raffinose:** A non-reducing trisaccharide on vigorous hydrolysis gives one molease of each D-fructose, D-glucose and D-galactose. It is designated as O-α-D-galacto pyranosyl-(1 → 6)-O-α-D-glucopyranosyl-(1 → 2)-β-D-fructofuranoside.
- **RNA:** *See* nucleic acids
- **Starch:** A mixture of two polysaccharides, viz., amylose (20%, soluble in water and gives blue colour with iodine) and amylopectin (80%, water insoluble and gives no colour with iodine).

- **Sucrose:** A non-reducing disaccharide ($C_{12}H_{22}O_{11}$) on hydrolysis gives 1:1 mixture of α-glucose and β-fructose.
- **Trehlose:** A non-reducing disaccharide and is α-D-glucopyranosyl-α-D-glucopyranoside.

## Problems

1. What are carbohydrates? Give Haworth Projection of glucose (pyranose and furanose forms).
2. Explain the term mutarotation.
3. Discuss the configuration of cellulose and starch.
4. What is a polypeptide? How does it differ from proteins?
5. Discuss primary, secondary and tertiary structures of proteins and nucleic acids.
6. What are nucleic acids?
7. What is the difference between RNA and DNA?

# Part V
# Stereoselective Synthesis and Organic Reactions

# Chapter 18
# Stereoselective Synthesis

## 18.1 Introduction

If in a substance, there are two different functional groups and the reagent reacts preferentially with one of the functional group, such a reaction is called **chemoselective**. As an example 4-nitrobutanal on reduction with sodium borohydride gives 4-nitro-1-butanol (Fig. 18.1).

On the other hand, if in a reaction an unequal mixture of enantiomers is produced the reaction is called **enantioselective**. However, if the reaction gives only one enantiomer, the reaction is called **enantiospecific** or **stereoselective**. As an example, the reduction of 2, 4-hexadione with Baker's yeast gives 2-(S) hydroxy-4-hexanone with 99% ee (Fig. 18.2).

## 18.2 Importance of Stereoselective Synthesis

Stereoselective synthesis has a special relevance in pharmaceutical industry. It is known that one enantiomer of a drug is useful for the effective treatment of a disease compared to the other enantiomer which may be ineffective. As an example, the well known drug ibuprofen contains one stereogenic centre and exists as a pair of enantiomer. However, only (S)-Ibuprofen is effective as an anti-inflammatory agent. The (R)-Ibuprofen shows no anti-inflammatory activity, but is slowly converted into the S-enantiomer in vivo.

(S)-Ibuprofen

 
V. K. Ahluwalia, *Stereochemistry of Organic Compounds*,
https://doi.org/10.1007/978-3-030-84961-0_18

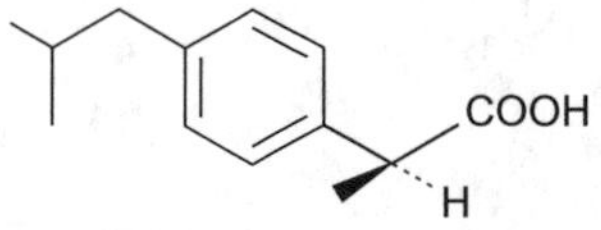

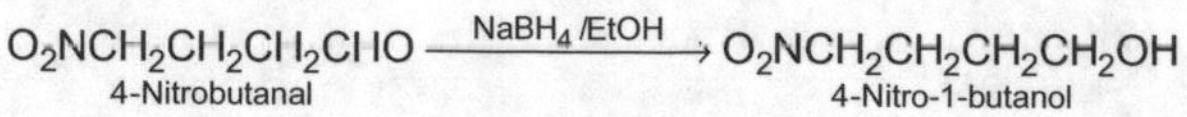

$$O_2NCH_2CH_2CH_2CHO \xrightarrow{\text{NaBH}_4\text{/EtOH}} O_2NCH_2CH_2CH_2CH_2OH$$

4-Nitrobutanal                                    4-Nitro-1-butanol

**Fig. 18.1** Reduction of 4-nitrobutanal to 4-nitro-1-butanol

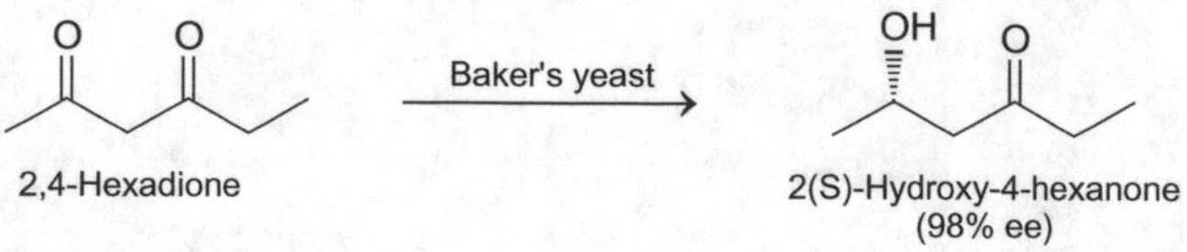

2,4-Hexadione                          2(S)-Hydroxy-4-hexanone
                                                      (98% ee)

**Fig. 18.2** Reduction of 2,4-hexadione to 2(S)-hydroxy-4-hexanone

Another drug, fluoxetine is used as an antidepressant. In this case only the R-enantiomer is active component.

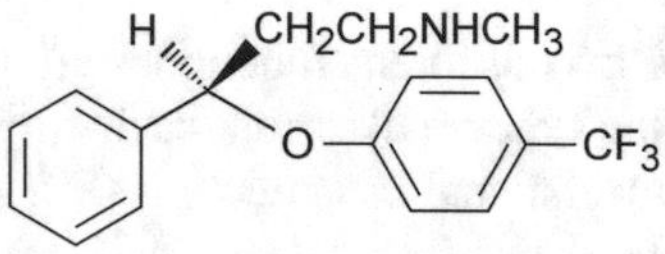

(R)-Fluoxetine

The anti-inflammatry agent, **naproxen** owes its activity due to (S)-enantiomer. The (R) enantioner is extremely harmful and is a liver toxin.

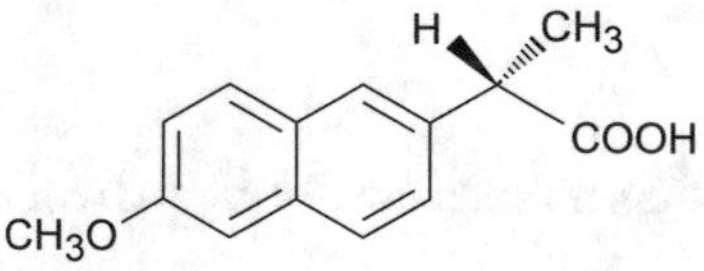

(S)-naproxen

The antihypertensive drug methyldopa (Aldomet) is effective in the (S) isomer. Penicillamine, the (S) isomer is a potent therapeutic agent for chronic arthrits; the (R) isomer is highly toxic and has no therapeutic action.

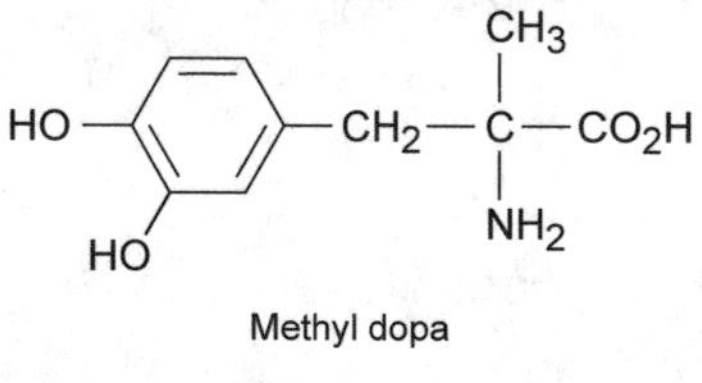

Methyl dopa 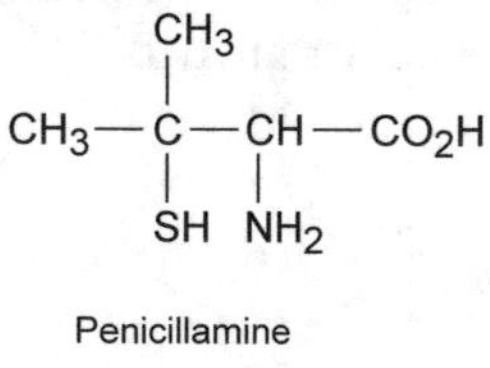                                      Penicillamine

An interesting example of a biologically active anti-nausea drug for expectant mothers in Europe and Great Britan (1959–1962), thalidomide. It was used as a sedative but was withdrawn subsequently since it caused catastropic birth defects to children born to women who took thalidomide during pregnancy. Investigations showed that this drug was sold as a mixture of its two enantiomers and each of these has a different biological activity. The (R)-enantiomer of thalidomide had the desired therapeutic effect, the other isomer (S) was responsible for birth defects and is highly mutagenic.

(R)-Thalidomide

A large number of other examples exist, in which only one enantiomer is active (For details *see* Chem. Eng. News, 1998, **76**, 83–104).

In view of what has been stated, it is useful to carry out stereoselective synthesis. Though a racemic mixture can be resolved into its enantiomeric forms, this procedure is cumbersome and time consuming (For details about resolution of racemic mixture *see* Sect. 4.5.2).

## 18.3   Enantioselective Synthesis

A number of procedures are available for enantioselective synthesis. Some of these methods are discussed below.

### 18.3.1   Using a Chiral Starting Synthon

A large number of naturally occurring compounds are chiral. Some examples include L-alanine (α -amino acid), (+)-tartaric acid, (+)-Malic acid, R-(+)-α-pinene, D-glyceraldehyde, (–)-ephedrine, (+)-conine, (–)-nicotine (Fig. 18.3).

A straight forward way to make a chiral substance is to start with a known chiral substance which reacts in a specific way. As an example (R)-2-hydroxy propionic acid (D-lactic acid) on treatment with phosphorus tribromide ($S_N2$ reaction) gives (S)-2-bromo propionic acid (Fig. 18.4).

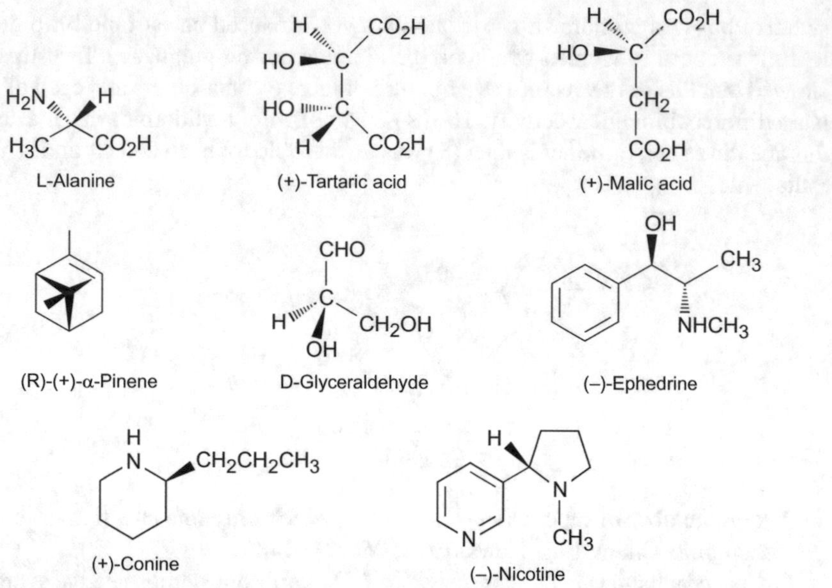

Fig. 18.3  Some naturally occurring chiral compounds

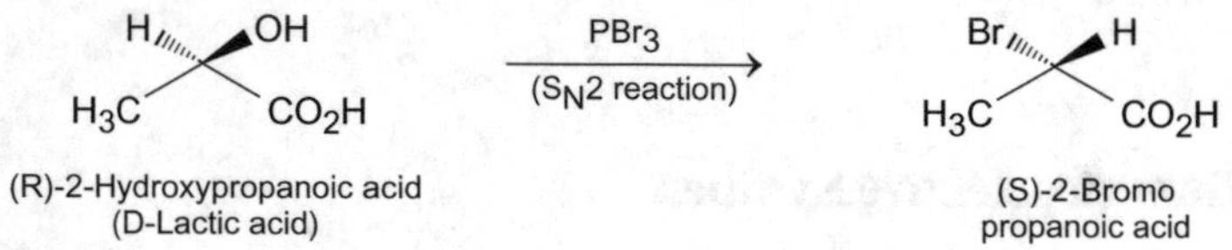

Fig. 18.4  Conversion of D-Lactic acid into the corresponding bromo compound

Another example is the conversion of L-tyrosine into L-DOPA, which is used for the treatment of Parkinson's disease (Fig. 18.5).

In both the above synthesis, the existing stereocentre is not affected.

## 18.3.2  *Enantioselectie Epioxidations*

It is known that *cis*-alkenes on epoxidation with per acids give *cis*-epoxides and *trans* alkenes give *trans* epoxides in Fig. 18.6.

In place of peracids, **chiral oxidants** have been successfully used. A chiral oxidant is a combination of an oxidising agent with a chiral ligand and a transition metal catalyst. The chiral reagent is capable of controlling the stereochemistry of epoxidations. In a chiral oxidant, some of the metallic compounds used are vanadyl

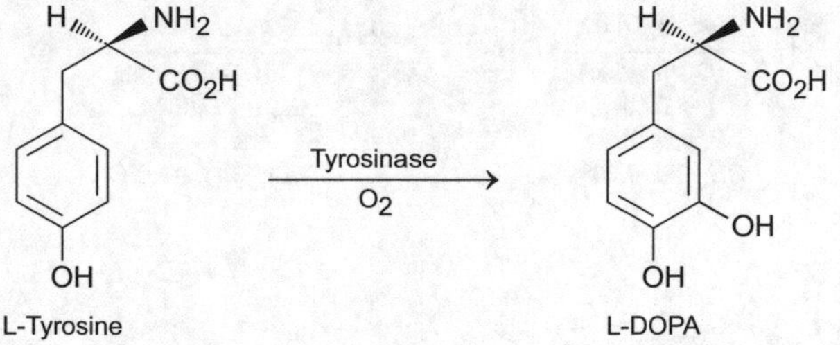

**Fig. 18.5**  Conversion of L-tyrosine into L-DOPA

**Fig. 18.6**  Formation of *cis* and *trans* epoxide

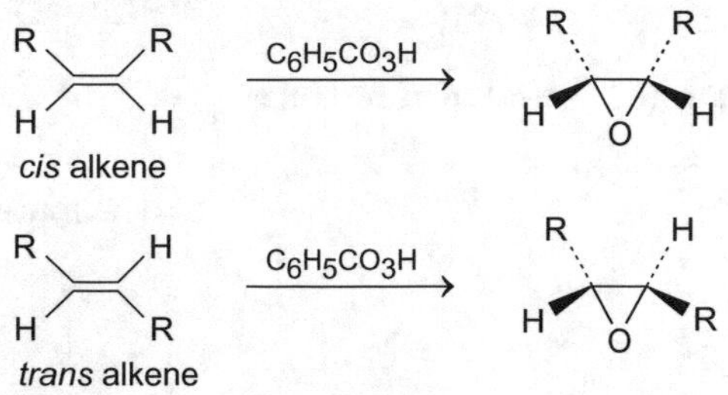

acetylacetone, Vo(acac), molybdenyl acetyl acetonate, $MoO_2(acac)_2$ and especially titanium tetraisopropoxide $[Ti[(OCH(CH_3)_2]_4$. Best results are obtained by **Sharpless reagent**, which is a mixture of 1 mol of titanium isopropoxide, 1–2 mol of tert butyl hydroperoxide and 2 mol of (R, R)-or (S, S)-diethyl tartarate as the chiral component.

Thus, the epoxidation of allylic alcohols with *t*-butyl hydroperoxide in presence of either (+) or (−) diethyl tartarate and titanium tetraisopropoxide yields the corresponding asymmetric epoxidation product in high optical yield[1,2]. In this chiral epoxidation, there is uniformly high asymmetric induction and the absolute configuration of the formed epoxide can be predicted. It has been established that for a given tartarate, the system delivers the epoxide oxygen from the same face of the olefin regardless of the substitution pattern. If allyl alcohol is represented as shown below (so that the $CH_2OH$ group is at the lower right), oxygen is delivered[1,2] from the top face in the presence of D-(−) diethyl tartarate and from the bottom face in the presence of L-(+)-diethyl tartarate.

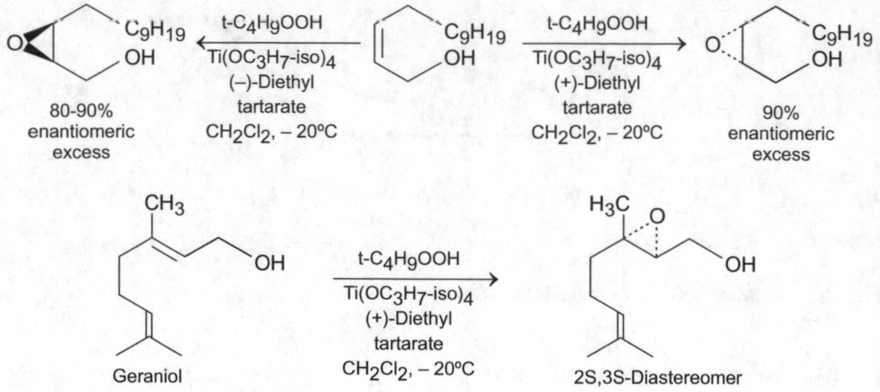

Fig. 18.7   Asymmetric epoxidation

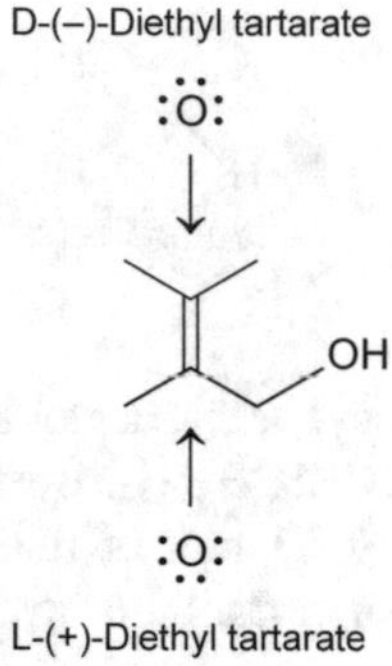

A number of examples of highly enantioselective epoxidations of allylic alcohols have been reported. Some examples are given below (Fig. 18.7).

The asymmetric epoxides are important synthons for organic synthesis (*see* Sect. 3.3).

### References

**1.** T. Kasturi and K.B. Sharpless, J. Am. Chem. Soc., 1980, **102**, 5975.

**2.** B.E. Rossiter, T. Kasturi and K.B. Sharpless, J. Am. Chem. Soc., 1981, **103**, 464.

## 18.3.3   *Epoxides as Synthons for Stereoselective Sysnthesis*

The epoxide ring is a strained three-membered ring. The ring can be easily opened up by nucleophilic attack. The ring opening occurs readily with strong nucleophiles (Fig. 18.8).

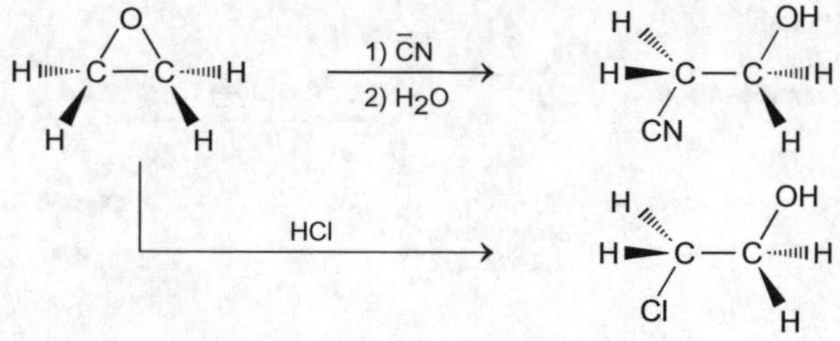

**Fig. 18.8**  Ring opening of epoxides with nucleophiles

In the above ring opening reaction (Fig. 18.8), the nucleophile attacks an electron-deficient carbon of the epoxide, cleaving a C—O bond to from an alkoxide. Subsequent treatment with water gives a neutral product with two functional groups on different carbons (Fig. 18.9).

Common nucleophiles that normally open the epoxide ring include ⁻OH, ⁻OR, ⁻CN, ⁻SR, NH₃. The reaction occurs via S_N2 mechanism. The nucleophile attacks the ring from back side (Fig. 18.10).

In case the epoxide is unsymmetrical, the nucleophile attacks at the less substituted carbon atom (Fig. 18.11).

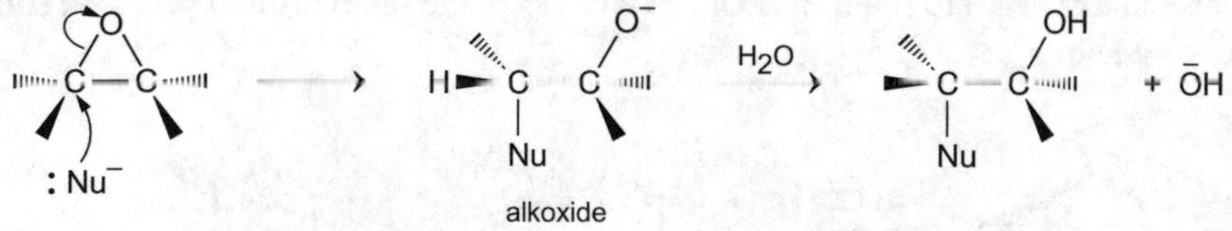

**Fig. 18.9**  Mechanism of epoxide ring opening

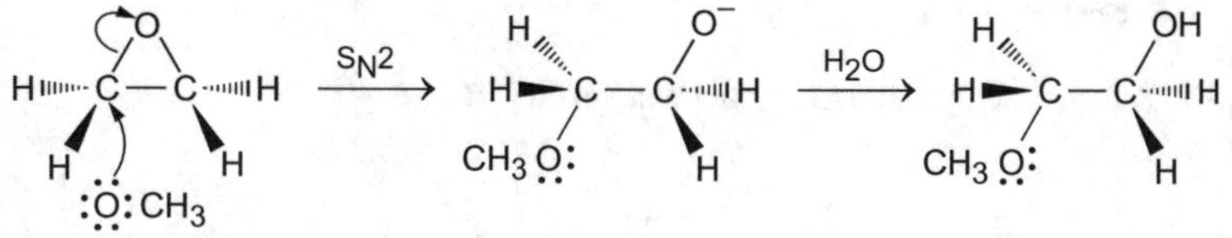

**Fig. 18.10**  Back side approach of the nucleophile

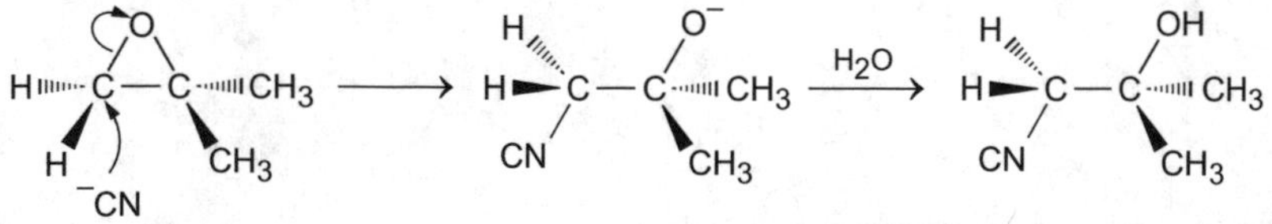

**Fig. 18.11**  Attack of nucleophile at the less substituted carbon

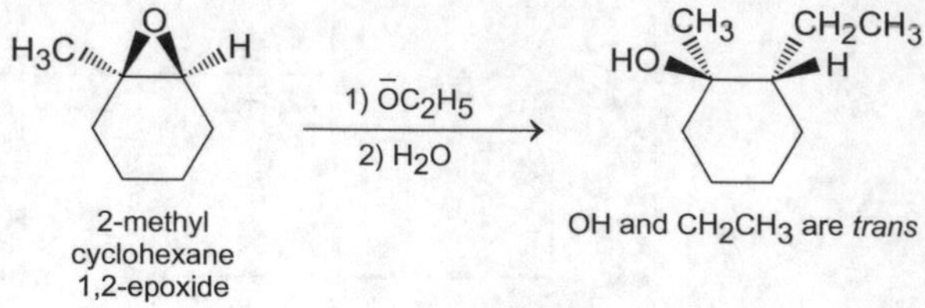

**Fig. 18.12** Reaction of
2-methylcyclohexane 1,
2-epoxide with ethoxide ion

2-methyl
cyclohexane
1,2-epoxide

OH and CH₂CH₃ are *trans*

As an example, 2-methylcyclohexane, 1,2-epoxide reacts with a nucliophile in the following way (Fig. 18.12).

In case of cyclohexane 1, 2-epoxide reaction with $^-OCH_3$ gives two *trans*-1, 2-disubstituted cyclohexanes, which are enantiomeric, each having two stereogenic centres (Fig. 18.13). It should, however, be noted that cyclohexane 1, 2-epoxide is achiral, since it possesses a plane of symmetry.

As a general rule, the reaction of an achiral substrate invariably yields a product, which is achiral (meso) or racemic.

Contrary to the ring opening of the epoxide moiety with strong nucleophiles (which attacks the less substitued carbon atom (Fig. 18.11), reaction with acids (HZ, containing nucleophilic Z) ring opening takes place as shown below (Fig. 18.14).

In this case (Fig. 10), the epoxide oxygen gets protonated followed by the attack of the nucleophile (Z) from backside thereby ring opening. Acids like HCl, HBr, HI all react in this fashion. Even $H_2O$ and ROH react in the same fashion provided some acid is added.

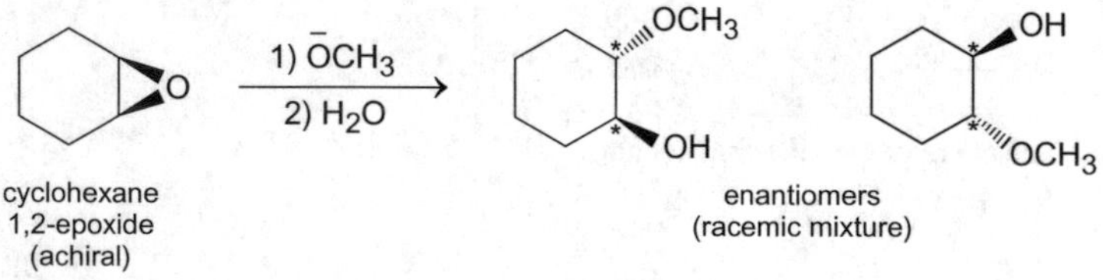

cyclohexane
1,2-epoxide
(achiral)

enantiomers
(racemic mixture)

**Fig. 18.13** Reaction of Cyclohexane 1, 2-epoxide with $-OCH_3|H_2O$

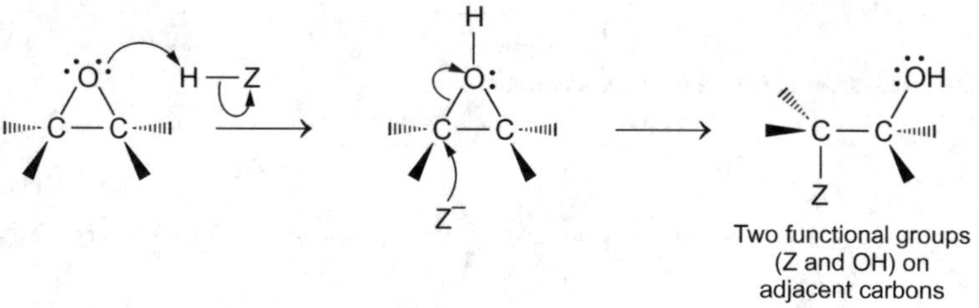

Two functional groups
(Z and OH) on
adjacent carbons

**Fig. 18.14** Epoxide ring opening by acid, H–Z

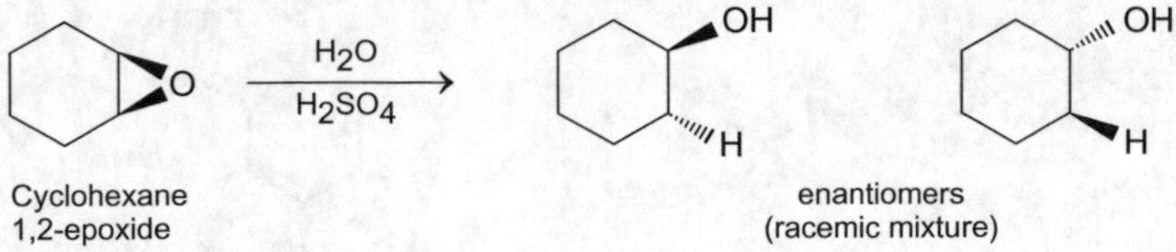

Fig. 18.15 Reaction of cyclohexane epoxide with $H_2O \mid H^+$

When the epoxide ring is fused to a ring (as in the case of cyclohexane-1,2-expoxide), the products obtained are 1, 2-disubstituted cycloalkanes (Fig. 18.15).

With unsymmetrical epoxide, the nucleophilic attack (using acids H–X) occurs at the more substituted carbon atom [this is contrary to nucleophic attack with strong nucleophiles (Fig. 18.11)]. As an example, 2,2, dimethyloxirane reacts with HCl as shown below (Fig. 18.16).

The ring opening of epoxide with a strong nucleophile ($Nu^-$) or an acid (HZ) is **regioselective** as only one product is obtained in major amount or exclusive product. The difference in the two is that the site of attack in these two reactions is opposite. The mechanism of the reaction of 2,2-dimethyl oxirane with HCl is shown below (Fig. 18.17).

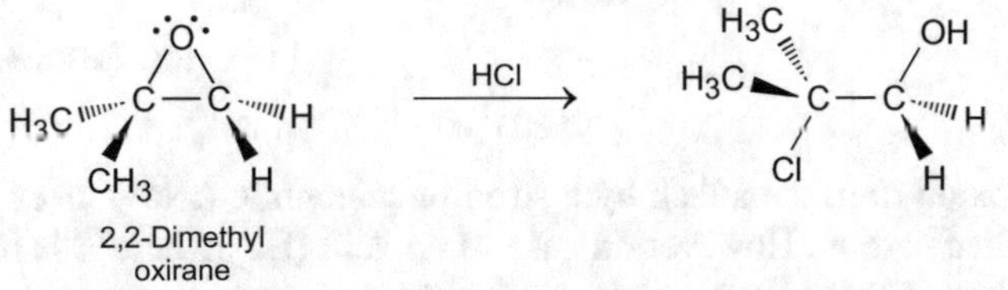

Fig. 18.16 Reaction of 2,2-dimethyl oxirane with HCl

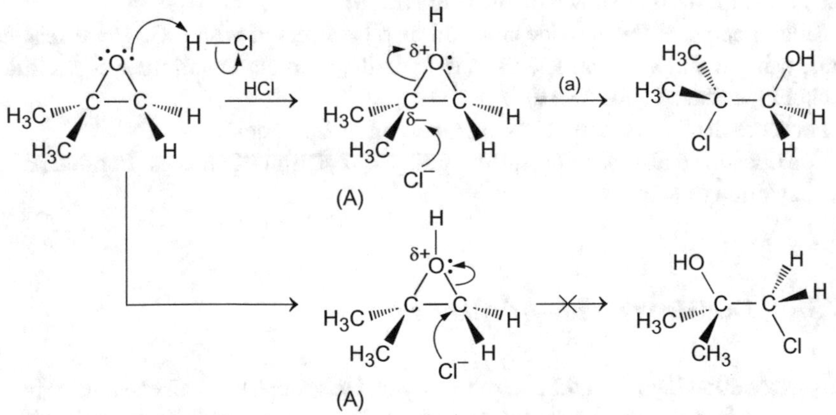

Fig. 18.17 The nucleophile attack takes place at the more substituted carbon (path a)

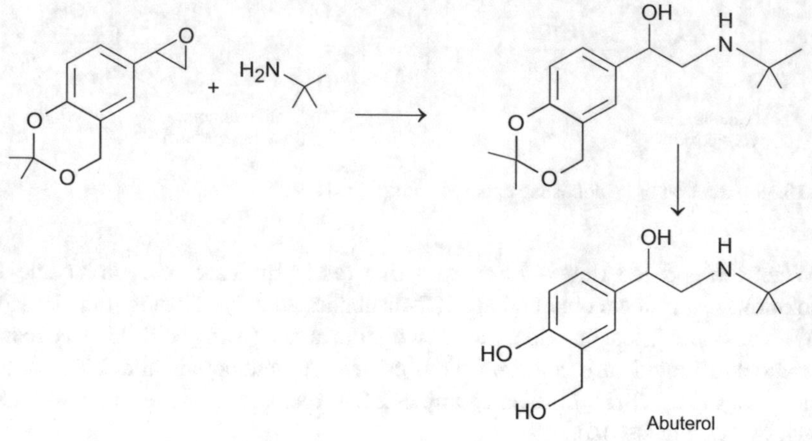

**Fig. 18.18** Synthesis of abuterol

**Fig. 18.19** Reaction of asymmetric epoxide with Grignard reagent

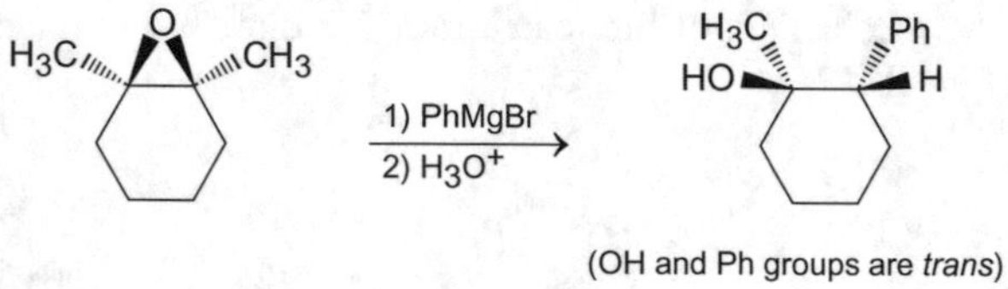

(OH and Ph groups are *trans*)

In an epoxide ring, the attack by a strong nucleophile (:Nu⁻) takes place at the less substituted carbon. However, in case of an acid (HZ, where Z is nucleophile), the nucleophile (Z) attacks at the more substituted carbon.

The epoxide ring opening can also be affected by a nitrogen nucleophile resulting in the formation of a new C–N bond. As an example, a drug abuterol (bronchodialator drug) is synthesised as shown below (Fig. 18.18).

Ring opening of the epoxide ring can also be affected with a Grignard reagent. Thus, using a single enantiomer of the epoxide, a single enantiomer of a tertiary alcohol is produced (Fig. 18.19).

For more details *see* section "Ring Opening of Epoxides".

Epoxides have also been useful for antihydroxylation of alkenes. For details *see* hydroxylation of alkenes.

## 18.3.4   Dihydroxylation of Alkenes

Dihydroxylation involves addition of two hydroxyl groups to an alkene double bond. Depending on the nature of the reagent the addition be syn addition or anti addition to give *cis* or *trans*; diols respectively (Figs. 18.20 and 18.21).

**Fig. 18.20**  Anti and syn addition to alkene double bond

## 18.3.4.1  Anti Dihydroxylation

In this process, the alkene is epoxidised with peracids followed by ring opening of the epoxide ring with ⁻OH or $H_2O$. As an example cyclopentene is converted into a racemic mixture of two *trans* 1, 2-cyclopentane diols. The reaction involves anti addition of two hydroxyl groups (Fig. 18.21) (*see* also Sects. 18.3.2 and 18.3.3).

The stereochemistry of the products formed is represented in (Fig. 18.22).

As seen (Fig. 18.22), the expoxidation of cyclopentene adds an O atom either from below or above the plane of the double bond to give a achiral epoxide (only one representation is shown in the Fig. 18.22). Finally, the epoxide ring opens up with backside attack of the nucleophile (⁻OH) at either of the C–O bond [path (*a*)

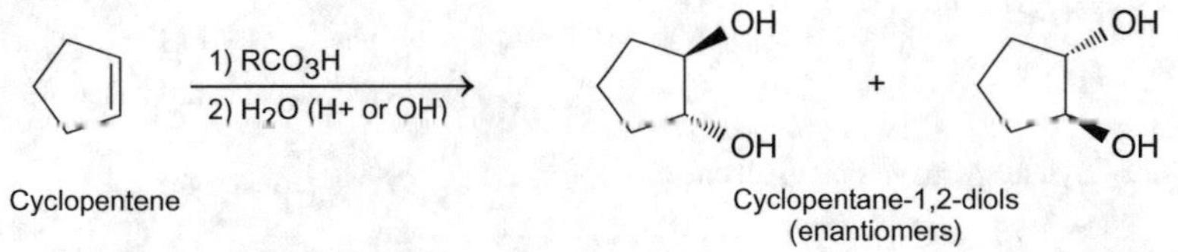

**Fig. 18.21**  Anti dihydroxylation of cyclopentene

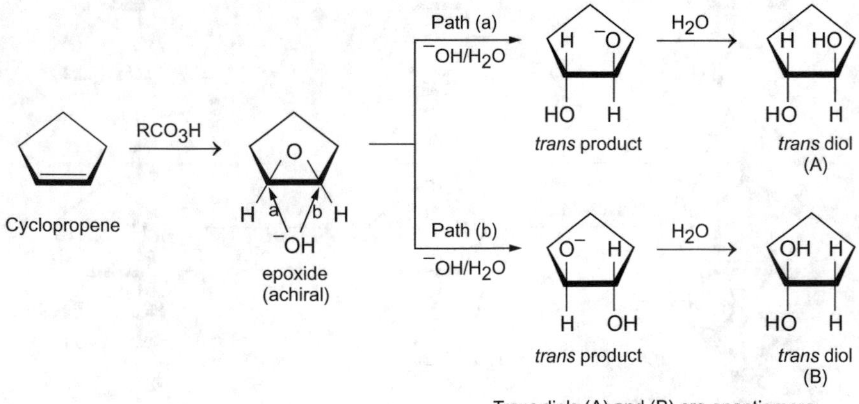

**Fig. 18.22**  Mechanism of formation of anti hydroxylation

and (*b*)]. Since one OH group of the diol is from epoxide and the other OH group is from the nucleophile (‾OH), the result is anti addition of the two OH groups to an alkene.

### 18.3.4.2  Syn Dihydroxylation

Treatment of an alkene with $KMnO_4$ or $OsO_4$ results in syn dihydroxylation (Fig. 18.23).

In the above case (Fig. 18.23), each of the reagents ($KMnO_4$ or $OsO_4$) adds two oxygen atoms from the same side of the double bond (syn addition) giving *cis*-1, 2-diol. In case $OsO_4$ is used, sodium bisulphite ($NaHSO_3$) is also added in the final hydrolysis step. The mechanics of syn dihydrooxylation is shown in Fig. 18.24.

Though $KMnO_4$ is a cheaper reagent and is radily available, its use is limited due to its insolubility in organic solvents. In order to prevent further oxidation of the diol, the reaction mixture must be kept basic by addition of ‾OH. The problem of insolubility of $KMnO_4$ in organic solvents (like t-butanol, benzene, toluene) has

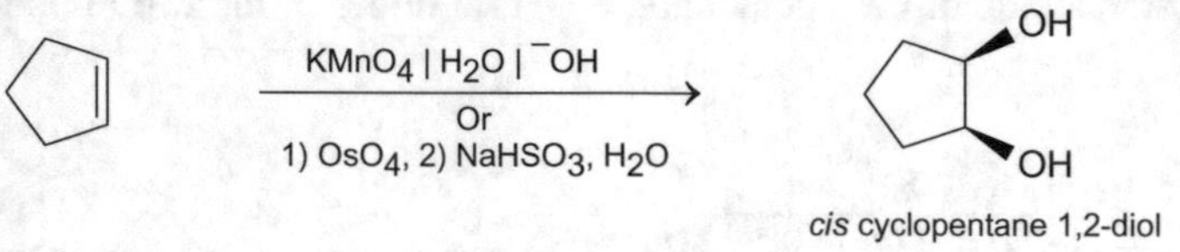

Fig. 18.23  Syn dihydroxylation of alkene

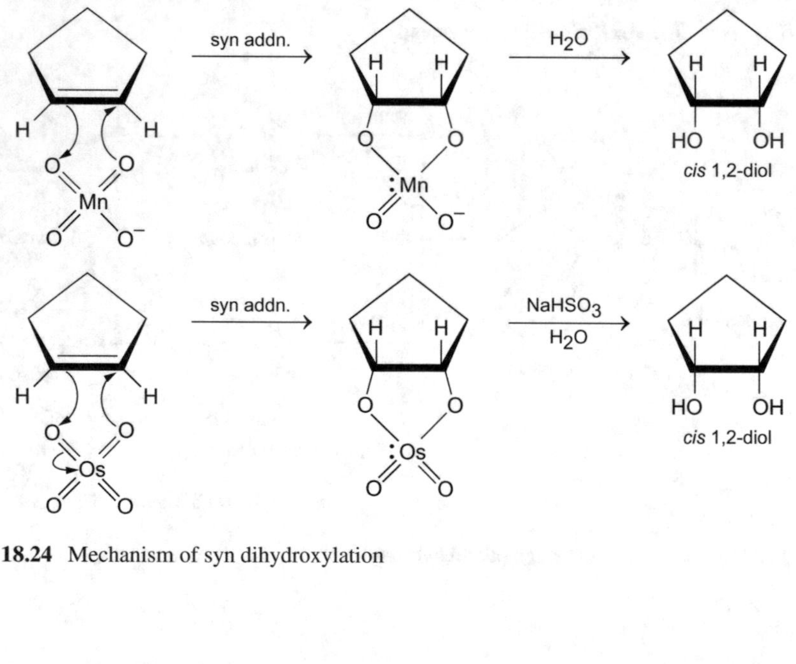

Fig. 18.24  Mechanism of syn dihydroxylation

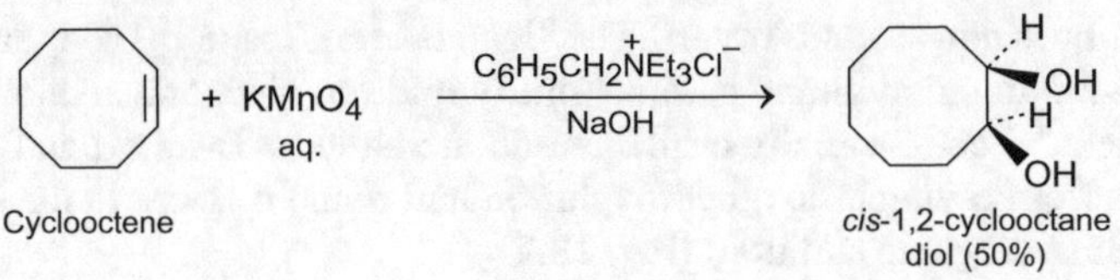

**Fig. 18.25**  Syn dihydroxylation of cyclooctene with KMnO$_4$ in presence of PTC

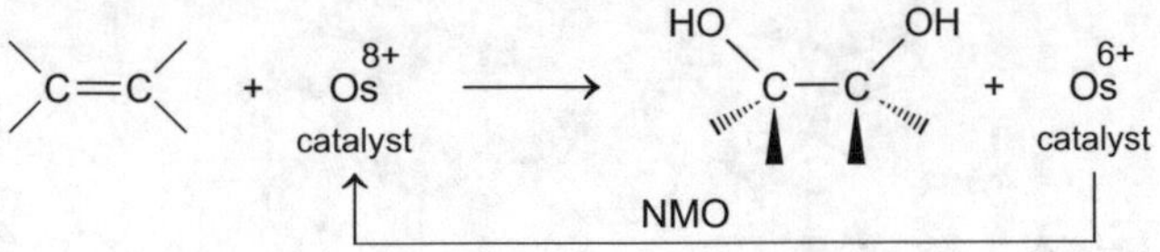

**Fig. 18.26**  Syn hydroxylation of alkene with catalytic amount of OsO$_4$ in presence of NMO

been overcome by using only a small amount of KMnO$_4$ in presence of a phase transfer catalyst (C$_6$H$_5$CH$_2$N$^+$Et$_3$Cl$^-$) [W.B. Weber and J.P. Shephard, Tetrahedron Lett., 1972, 4907].

Under alkaline conditions, the yield of the 1,2-diol is 50% compared to about 7% by the classical technique (Fig. 18.25).

Compared to KMnO$_4$, osmium tetraoxide is much more expensive and is also toxic. This problem has been overcome by using a catalytic amount of OsO$_4$ in presence of N-methylmorpholine N-oxide (NMO).

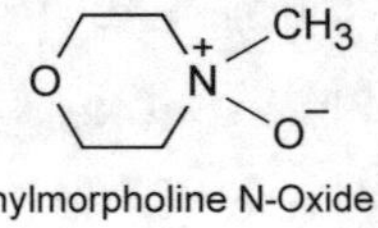

N-Methylmorpholine N-Oxide (NMO)

During dihydroxylation of the alkene, the Os$^{8+}$ oxidant is convertd into Os$^{6+}$, which is reoxidised by NMO to Os$^{8+}$, which is used again and again (Fig. 18.26).

### 18.3.4.3  Asymmetric Dihydroxylation of Alkenes Using OsO$_4$ and Chiral Amine Ligands

It has already been stated (Sect. 18.3.4.2) that dihydroxylation of alkenes with OsO$_4$ gives vicinal diols. It was, however, found that if the oxidiation of alkene with OsO$_4$ is carried out in presence of chiral amine ligand. It was K.B. Sharpless (2001), who found that addition of a chiral amine ligand $\pi$ a reaction mixture, containing potassium osmate and potassium ferricyanide (a cooxidant that generates OsO$_4$ in situ), asymmetric dihydroxylation takes place. For this discovery and development of the catalytic reagents K.B. Sharpless was awared 2001 Nobel Prize in chemistry which was shared by Knowles and Yoyori. The ligands were made by coupling a linker molecule with an chiral amine guinine and quinidine (cinchona alkaloids). These reagents are now commercially available and are called AD-mix α and AD-mix β [AD stands from asymmetric dihydroxylation and α and β refere to the face of the double bond undergoing addition] (Fig. 18.27).

In the dihydroxylation using sharpless chiral reagents, a good enantioselectivity is obtained. Some examples are given below (Fig. 18.28).

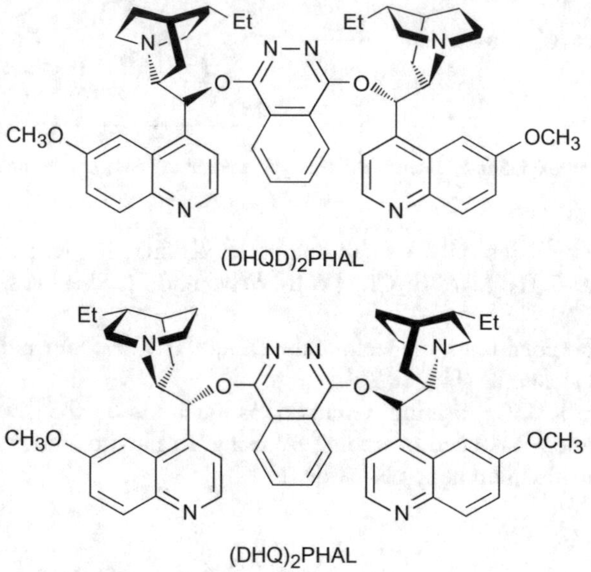

(DHQD)₂PHAL

(DHQ)₂PHAL

AD-mix α: K₂OsO₂(OH)₄, K₃Fe(CN)₆, K₂CO₃, (DHQ)₂PHAL

AD-mix β: K₂OsO₂(OH)₄, K₃Fe(CN)₆, K₂CO₃, (DHQD)₂PHAL

**Fig. 18.27**  Sharpless chiral reagents

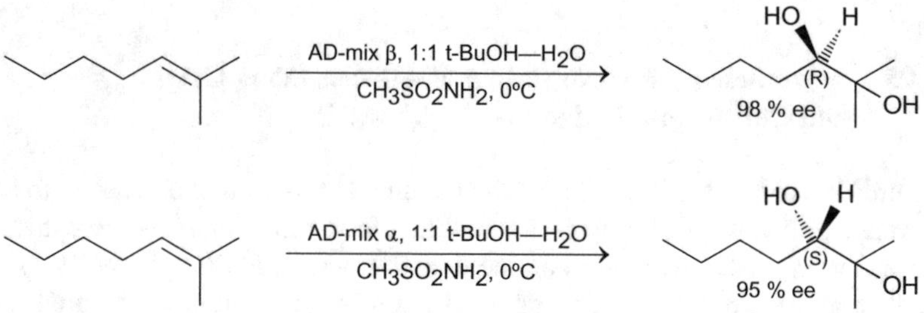

**Fig. 18.28**  Enantioselective hydroxylation of some alkenes

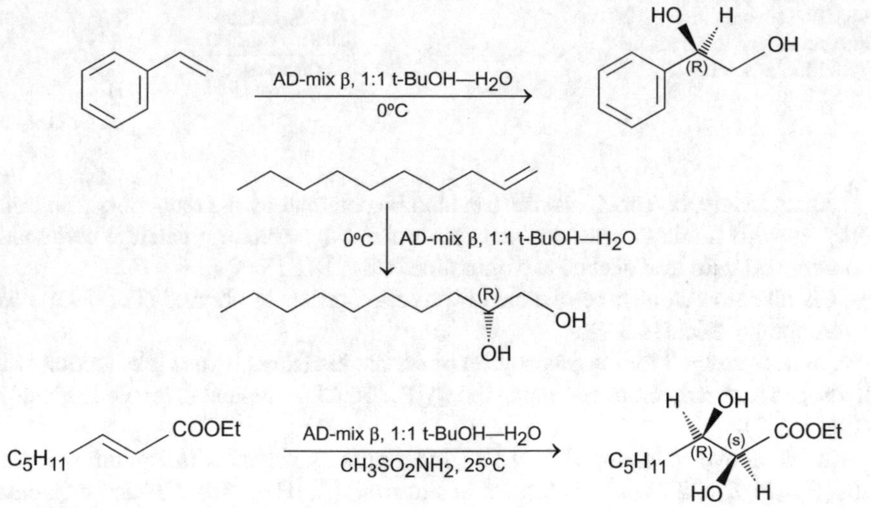

**Fig. 18.29** Enantioselective dihydroxylation of some terminal alkenes

Terminal alkenes do not react enantioselectively. However, using more recently developed reagents have been used for dihydroxylation of terminal alkenes. Some examples are given in Fig. 18.29.

It has also been possible to carry out hydroxylation of aromatic rings to give *cis* diols (for details *see* Sect. 19.3.12).

## 18.3.5 Stereoselective Reduction of Alkynes

Alkynes can be reduced to Z(or *cis*) or E(or *trans*) alkenes depending on the catalyst and other conditions. Thus, *cis* alkenes can be obtained by reduction of an internal alkyne in presence of nickel boride) also called P-2 catalyst (which is prepared by the reduction of nickel acetate with sodium borohydride) (Fig. 18.30).

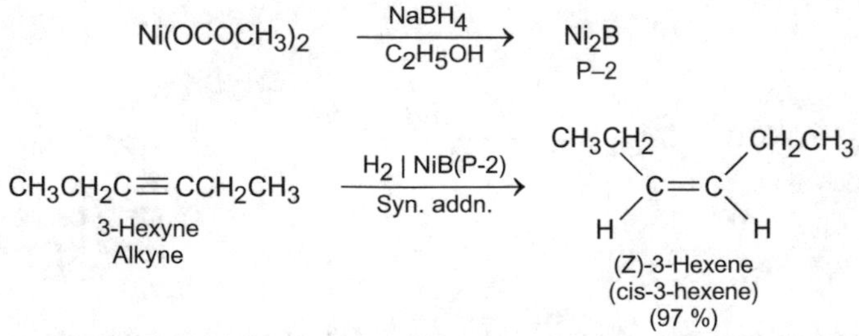

**Fig. 18.30** Reduction internal alkyne to *cis* alkene using P-2 catalyst

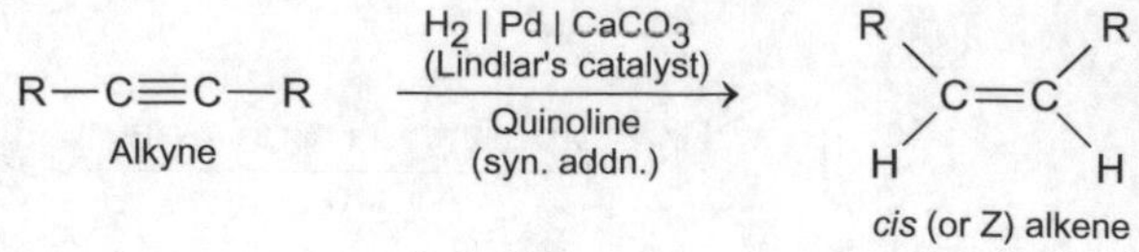

**Fig. 18.31** Reduction of internal alkyne to *cis* alkene with Lindlar's catalyst

$$R-C\equiv C-R \quad \xrightarrow[\substack{\text{Quinoline} \\ \text{(syn. addn.)}}]{\substack{H_2 \mid Pd \mid CaCO_3 \\ \text{(Lindlar's catalyst)}}} \quad \begin{array}{c} R \\ \diagdown \\ \diagup \\ H \end{array} C=C \begin{array}{c} R \\ \diagup \\ \diagdown \\ H \end{array}$$

Alkyne

*cis* (or Z) alkene

Alternatively, *cis* (or Z) alkenes can also be obtained by the reduction of internal alkynes with Lindlar's catalyst (palladium metal deposited on calcium carbonate, conditioned with lead acetate and quinoline) (Fig. 18.31).

*Cis* alkenes can also be obtained by hydroboration of alkynes (For details *see* hydroborane, Sect. 19.3.7).

Another method for the preparation of *cis* olefins from alkynes is reduction with hydrogen and ruthenium complex $[C_6H_5)_3P]_3$ RhCl. A typical example is given in (Fig. 18.32).

In the above reduction (Fig. 18.32), the actual catalyst is ruthenium complex $[(C_6H_5)_3P]_3RuClH$, which is formed in situ from $[(C_6H_5)_3P]_3RuCl_2$ and molecular hydrogen in benzene in presence of a base, triethylamine, *trans*-alkenes are prepared by the reduction of internal alkynes with sodium or lithium in ammonia or an amine (Fig. 18.33).

The above reduction takes place by the addition of an electron to the triple bond to give a radical anion; subsequent protonation with the solvent ammonia yields a radical. This is followed by the addition of a second electron to the radical (formed above) to give a carbanion, which gets protonated to give *trans* alkene (Fig. 18.34).

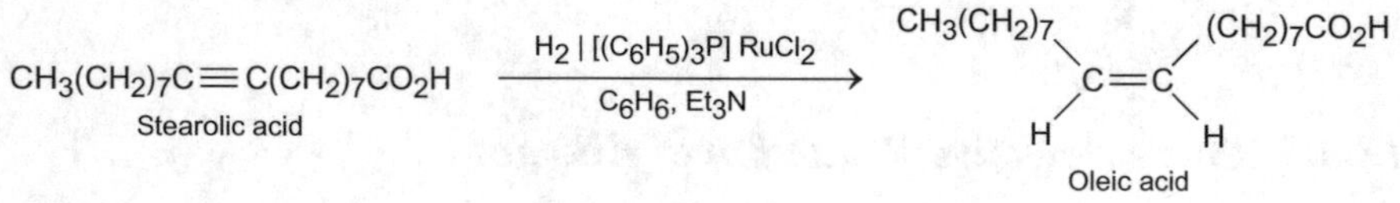

$$CH_3(CH_2)_7C\equiv C(CH_2)_7CO_2H \quad \xrightarrow[C_6H_6, Et_3N]{H_2 \mid [(C_6H_5)_3P]\, RuCl_2} \quad \begin{array}{c} CH_3(CH_2)_7 \\ \diagdown \\ \diagup \\ H \end{array} C=C \begin{array}{c} (CH_2)_7CO_2H \\ \diagup \\ \diagdown \\ H \end{array}$$

Stearolic acid

Oleic acid

**Fig. 18.32** Partial reduction of stearolic acid to oelic acid

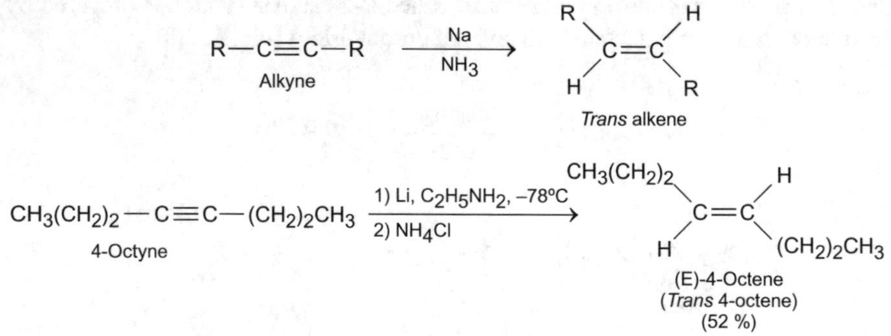

$$R-C\equiv C-R \quad \xrightarrow[NH_3]{Na} \quad \begin{array}{c} R \\ \diagdown \\ \diagup \\ H \end{array} C=C \begin{array}{c} H \\ \diagup \\ \diagdown \\ R \end{array}$$

Alkyne

*Trans* alkene

$$CH_3(CH_2)_2-C\equiv C-(CH_2)_2CH_3 \quad \xrightarrow[\text{2) NH}_4Cl]{\text{1) Li, } C_2H_5NH_2, -78°C} \quad \begin{array}{c} CH_3(CH_2)_2 \\ \diagdown \\ \diagup \\ H \end{array} C=C \begin{array}{c} H \\ \diagup \\ \diagdown \\ (CH_2)_2CH_3 \end{array}$$

4-Octyne

(E)-4-Octene
(*Trans* 4-octene)
(52 %)

**Fig. 18.33** Reduction of internal alkyne with Na or Li in liquid ammonia or an amine

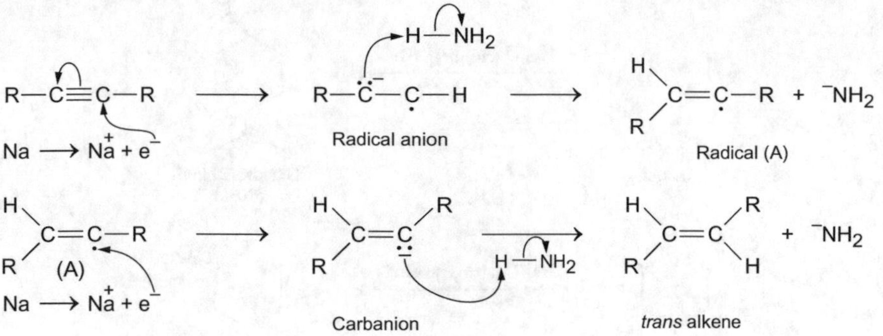

**Fig. 18.34** Mechanism of reduction of alkyne to *trans* alkene

## 18.3.6 *Enantioselective Hydrogenations of Alkenes*

Enantioselective hydrogenation of appropriately substituted alkenes can be carried out by using homogeneous catalysts. These catalysts are soluble catalysts and can be used for enatioselective hydrogenations in homogeneous solution. A number of soluble catalysts have been discovered, the most effective are the rhodium and ruthenium complexes [tris (triphenylphosphine) chloro rhodium [$(C_6H_5)_3$ P]$_3$ RhCl and hydridochloroltris (triphenyl phosphine) ruthenium [$(C_6H_5)_3$P]$_3$ RuClH. The former catalyst, [$(C_6H_5)_3$P]$_3$ RhCl is commonly known as **Wilkinson catalyst** and is made by reacting rhodium chloride with excess of triphenyl phosphine in boiling alcohol.

$$RhCl_3 \cdot 3H_2O + (C_6H_5)_3P \xrightarrow[\text{heat}]{C_2H_5OH} [(C_6H_5)_3P]_3\ RhCl$$

Rhodium chloride    Triphenyl phosphene                Wilkinson catalyst

Using Wilkinson catalysts, mono- and disubstituted double bonds can be reduced more rapidly, permitting partial hydrogenation of compounds containing different kind of double bonds[2]. As an example, linalool is reduced to dihydro compound in 90% yield, the reduction of vinyl group taking place selectively. In a similar way, carvone is converted into carvotanacetone (Fig. 18.35).

In the above cases, hydrogenation takes place by *cis* addition to the double bond. This has been shown by the catalysed reaction of deuterium with 1, 4-androstadien-3,17-dione to give the dideutro compound by *cis* addition to the α-face of the disubstituted double bond (Fig. 18.36).

A novel catalyst[3] designated as $\dot{P} - \dot{P}$ catalyst (structure shown in Fig. 18.37) was developed by William S. Knowles. Using this catalyst L-DOPA (used for the treatment of Parkinson's diesease) was synthesised (Fig. 18.37).

In the $\dot{P} - \dot{P}$ complex, the phosphorus is chiral and contains rhodium (*see* Fig. 18.37). For the discovery of the chiral catalyst for the enantioselective hydrogenation of alkene, Knowles was awarded 2001 Nobel Prize, which was shared by Ryoji Noyori, who developed still better catalysts. A disadvantage of using $\dot{P} - \dot{P}$

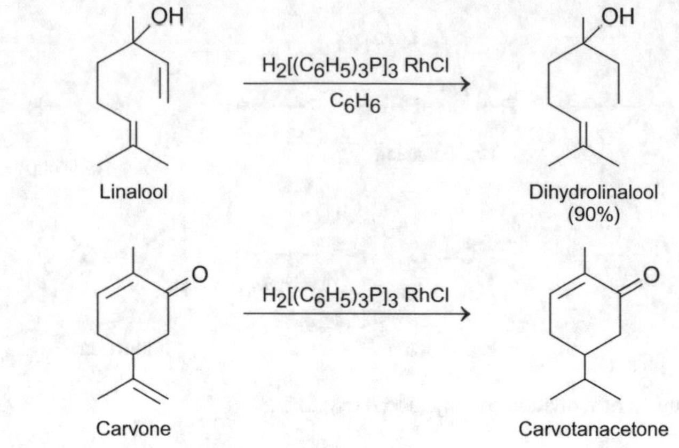

**Fig. 18.35** Selectivity in the reduction of alkenes

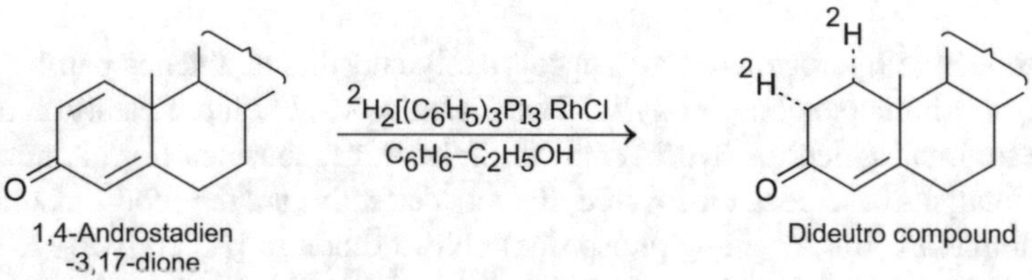

**Fig. 18.36** Reduction of 1,4-androstadiene with deuterium and catalyst

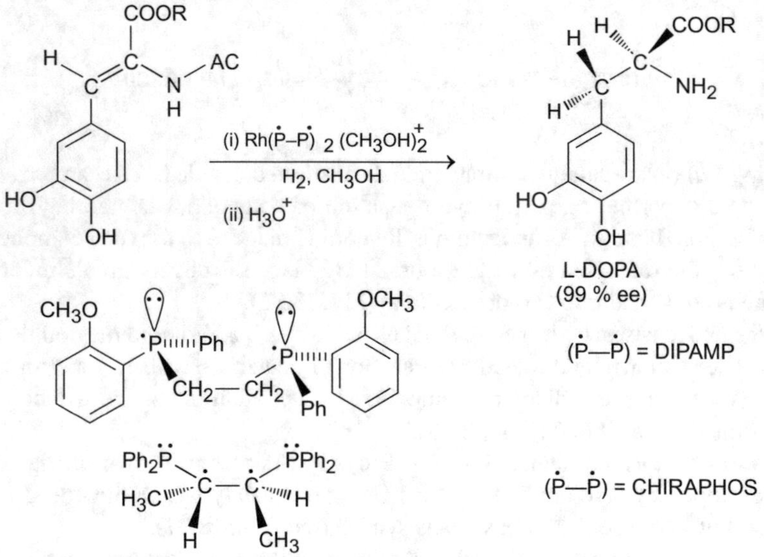

**Fig. 18.37** Synthesis of L-DOPA

catalyst is that only alkenes with aryl ring and amido group attached to the double bond are reduced enantioselectively.

The noval catalyst[4] developed by Ryoji Noyori contained ruthenium and have phosphine ligands. In these catalyst dessignated BINAP(S and R) two naphthalene rings containg diphenyl phosphine group (PPh$_2$) in ortho positions are used. These are chiral since the rings do not lie coplanar. Any of the isomer of BINAP is bound to ruthenium (II) ion and an ancillary ligand (like carboxylate, halide and amine), designated as L$_x$ in the structure of (S) or (R) BINAP (Fig. 18.38).

The complex (S)-BINAP]RuLx is a useful catalyst for enantioselective hydrogenation of alkenes. Some examples are given below (Fig. 18.39).

### References

**1.** R.E. Harmon, S.K. Gupta and D.J. Brown, Chem. Rev., 1973, **73**, 21.

**2.** J.F. Biellmann, Bull. Soc. Chem. Fr., 1968, 3055.

**3.** T.N. Sowell, Organic chemistry, Viva Books, second Edn. page 539.

**4.** Ref 3. Page 540.

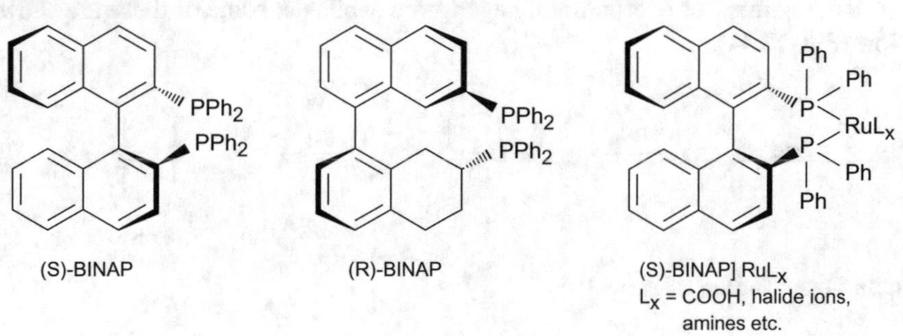

(S)-BINAP                          (R)-BINAP                          (S)-BINAP] RuL$_x$
                                                                      L$_x$ = COOH, halide ions,
                                                                      amines etc.

**Fig. 18.38**  Isomers of BINAP

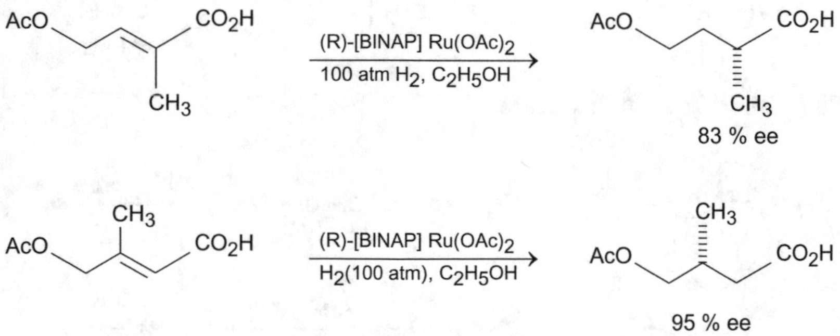

**Fig. 18.39**  Some enantioselective hydrogenations of alkenes

## 18.3.7  *Enantioselective Hydroboration*

Alkenes and alkynes add on to borane (which exist as dimer diborane, $B_2H_6$) to yield organoboranes. This process is commonly known as hydroboration. This reaction was first reported by Herbert C. Brown (1955), who was awarded Nobel Prize for chemistry in 1979 due to tremendous synthetic applications of hydroboration.

Hydroboration involves addition of $BH_3$ to the $\pi$ bond of the alkene to give an alkyl borane as an intermediate, which on oxidation ($H_2O_2$ I NaOH) is converted into the corresponding OH ($BH_2$ is replaced by OH) (Fig. 18.40).

Hydroboration is a concerted addition of H and $BH_2$ from the same side of the planar double bond. In this process, the $\pi$ bond and H–$BH_2$ bonds are broken and two new sigma bonds are formed. The transition state is said to be four centred because four atom are involved (Fig. 18.41).

The alkyl borane (Fig. 18.42) is obtained by the addition of one equivalent of alkene to $BH_3$, has still two B–H bonds. So it can react with two more equivalents of alkene to form trialkyl borane (Fig. 18.42).

Hydroboration is regioselective. Thus, with unsymmetrical alkenes, the boron atom gets bonded to the less substituted carbon atom as shown below (Fig. 18.43).

A large number of hydroborating agents are available, some of these are shown below (Fig. 18.44).

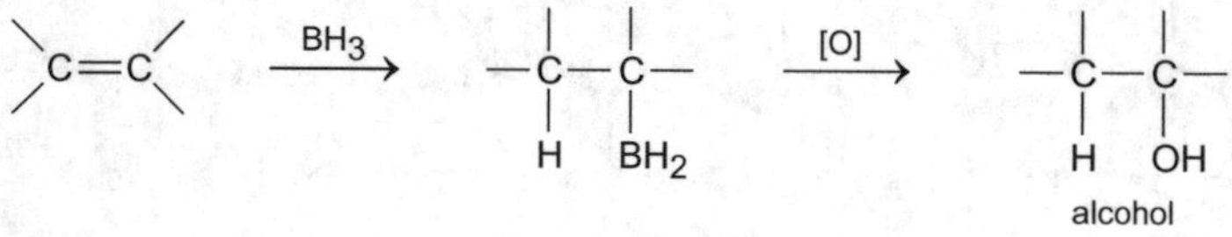

**Fig. 18.40**  Addition of alkene to $BH_3$

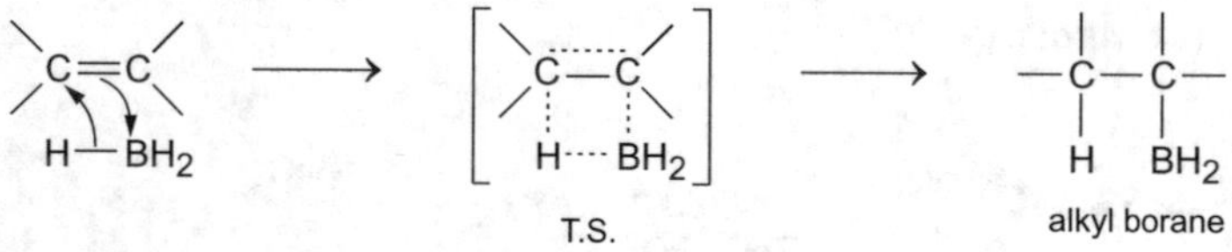

**Fig. 18.41**  Mechanism of hydroboration

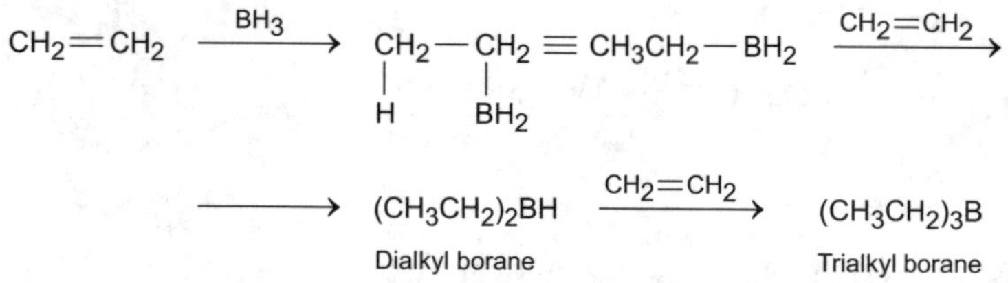

**Fig. 18.42**  Formation of trialkyl borane

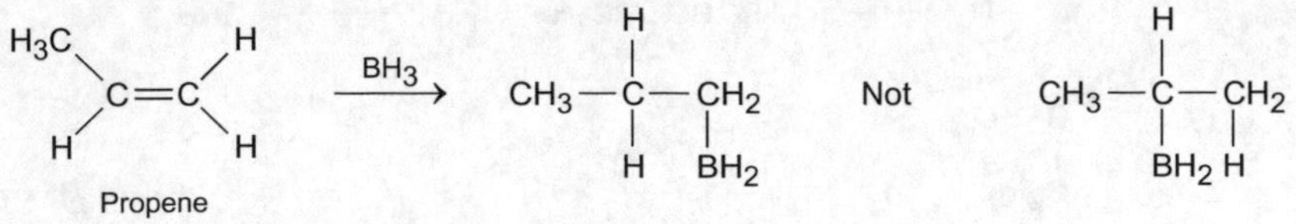

**Fig. 18.43** Hydroboration of an unsymmetrical alkene

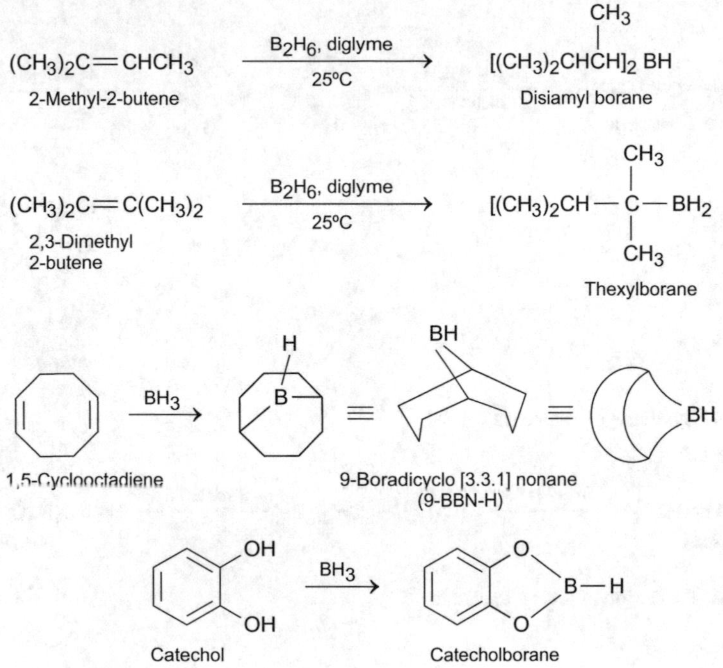

**Fig. 18.44** Some useful hydroborating agents

Though hydrobration has a number of synthetic application, but in the present section only two important applications are discussed. These are protonolysis and oxidation of organoboranes.

### 18.3.7.1 Protonolysis of Organoboranes

The organoboranes undergo hydrolysis (protonolysis) on treatment with organic carboxylic acids. This provides a convenient method for the reduction of alkenes (carbon–carbon double bond) and alkynes (carbon–carbon triple bond). The carboxylic acids cleave the C–B bond. This process involving replacement of a heteroatom with a proton is called **protonolysis**. The mechanism of protonolysis is shown below (Fig. 18.45).

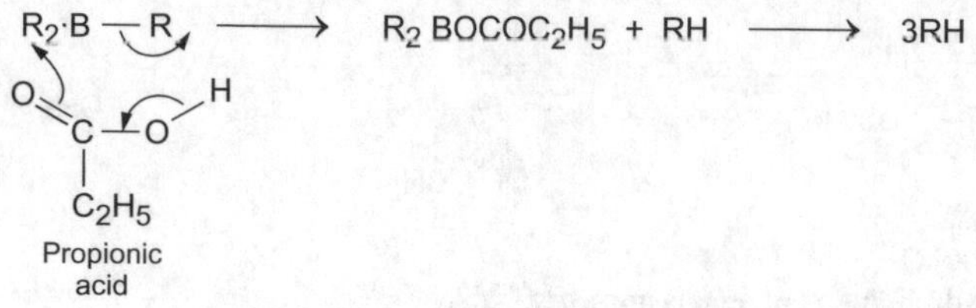

$$R_2 \cdot B \overset{\curvearrowright}{\frown} R \longrightarrow R_2\,BOCOC_2H_5 + RH \longrightarrow 3RH$$

Propionic
acid

**Fig. 18.45**  Mechanism of protonolysis of organoboranes with propionic acid

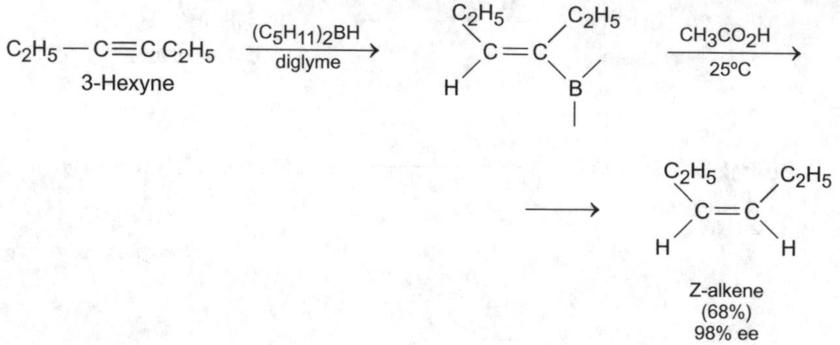

$$C_2H_5 - C \equiv C\ C_2H_5 \xrightarrow[\text{diglyme}]{(C_5H_{11})_2BH}$$

3-Hexyne

$$\underset{H}{\overset{C_2H_5}{>}}C = C \underset{B}{\overset{C_2H_5}{<}} \xrightarrow[25°C]{CH_3CO_2H}$$

$$\underset{H}{\overset{C_2H_5}{>}}C = C \underset{H}{\overset{C_2H_5}{<}}$$

Z-alkene
(68%)
98% ee

**Fig. 18.46**  Synthesis of Z-alkenes

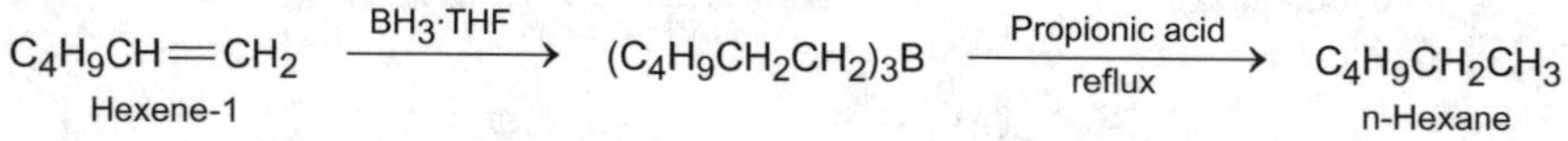

$$C_4H_9CH = CH_2 \xrightarrow{BH_3 \cdot THF} (C_4H_9CH_2CH_2)_3B \xrightarrow[\text{reflux}]{\text{Propionic acid}} C_4H_9CH_2CH_3$$

Hexene-1                                                                n-Hexane

**Fig. 18.47**  Reduction of alkene to alkane

Thus, an internal alkyne, 3-hexyne on treatment with disiamyl borane in diglyme gives the corresponding organoborane, which on treatment with acetic acid at 25 °C gives Z-alkene in 68% yield (Fig. 18.46).

On the other hand, an alkene is reduced to alkane (Fig. 18.47).

Thus, the hydroboration reaction can be considered as reduction reaction. It is the replacement of one electropositive element (B) with another (H).

### 18.3.7.2  Oxidation of Organoboranes: Asymmetric Synthesis of Optically Active Secondary Alcohols

Organoboranes on oxidation with alkaline hydrogen peroxide give alcohols. In this procedure many functional groups are unaffected under the reaction conditions and a number of different alkenes can be converted into alcohols. The reaction results in anti-Markownikoff addition of water to the double bond or triple bond. During hydroboration the boron atom adds to the less substituted carbon of the multiple bonds. As an example terminal alkynes give aldehydes (Fig. 18.48).

The mechanism of the oxidation of organoborane with alkaline hydrogen peroxide is shown in (Fig. 18.49).

An interesting example is the hydroboration-oxidation of 1-methylcyclohexene. In this case there is syn addition of $BH_3$ from above and below the planar double bond resulting in the formation of two enantiomeric alkylboranes, which on oxidation with alkaline hydrogen peroxide gives a racemic mixture of alcohols. The two alcohols obtained are enantiomers (Fig. 18.50).

Asymmetric synthesis of optically active secondary alcohols can be achieved by hydroboration of alkenes with an optically active alkyl borane followed by oxidation. The optically active alkyl boranes used are diisopinocamphenl borane (I) ($Ipc_2BH$) (H.C. Brown, M.C.Desai and P.K. Jadhav., J. Org. Chem., 1982, **47**, 5065) and mono-isopinocamphenylborane (II) (Ipc $BH_2$) (H.C. Brown, A.K. Mandal, N.M. Yoon, B. Singaram, J.R. Schwier and P.K. Jadhav, J. Org. Chem., 1982, **47**, 5069, 5074).

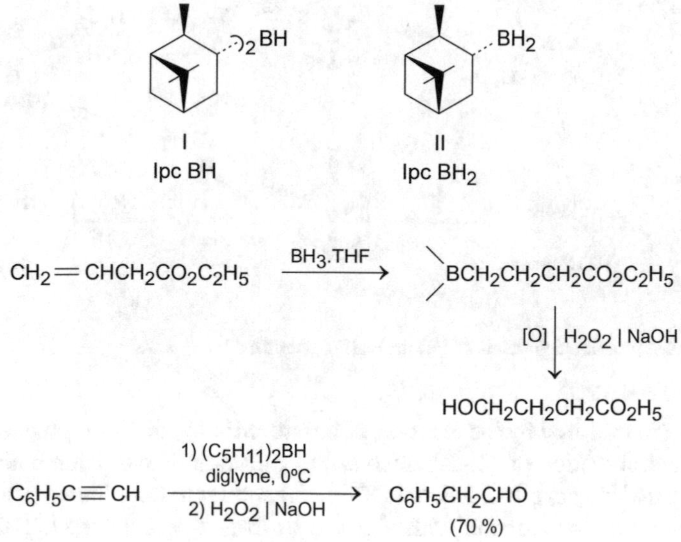

$$CH_2 = CHCH_2CO_2C_2H_5 \xrightarrow{BH_3.THF} \diagup BCH_2CH_2CH_2CO_2C_2H_5$$

$$[O] \downarrow H_2O_2 \mid NaOH$$

$$HOCH_2CH_2CH_2CO_2C_2H_5$$

$$C_6H_5C \equiv CH \xrightarrow[\text{2) } H_2O_2 \mid NaOH]{\begin{array}{c}\text{1) } (C_5H_{11})_2BH \\ \text{diglyme, 0°C}\end{array}} C_6H_5CH_2CHO \\ (70\%)$$

**Fig. 18.48** Hydroboration followed by oxidation of alkenes and alkynes

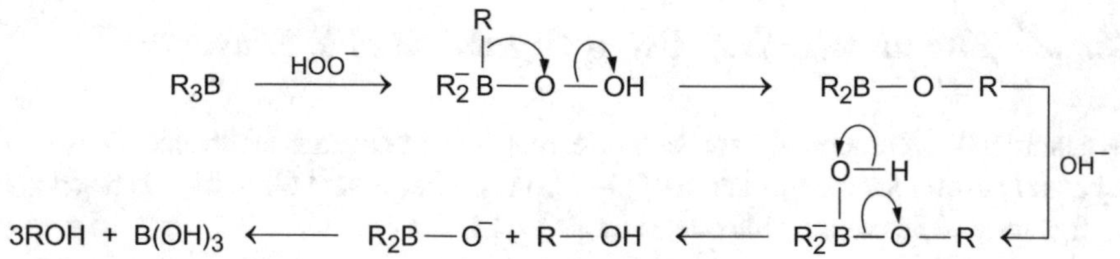

**Fig. 18.49** Mechanism of the oxidation of organoborane

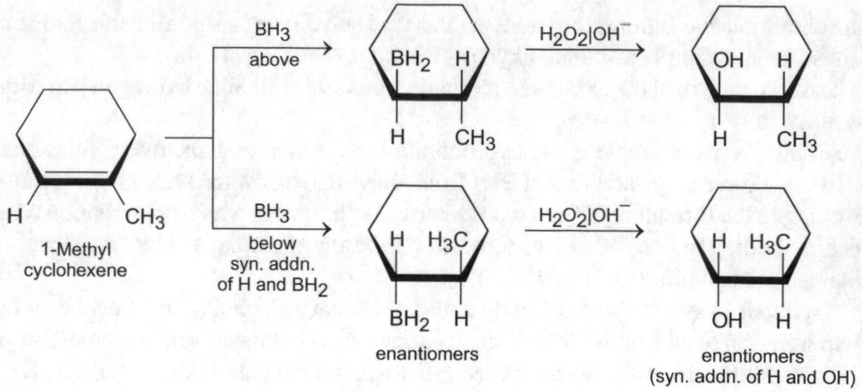

**Fig. 18.50**  Hydroboration-oxidation of 1-methylcyclohexene

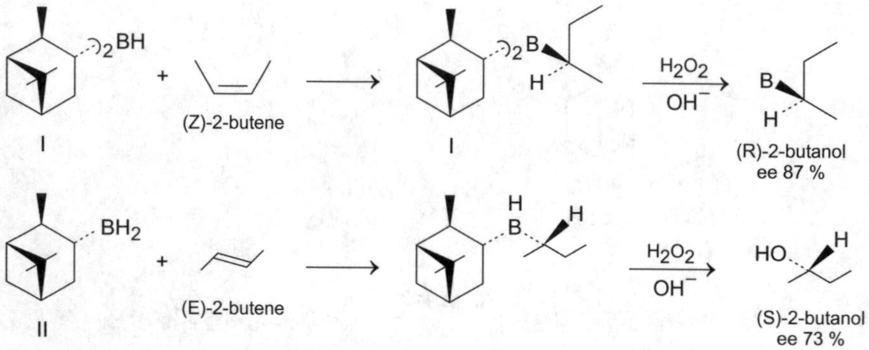

**Fig. 18.51**  Asymmetric Synthesis of (R) and (S)-2-butanol

I and II are prepared by the reaction of borane with (+) or (–)-α -pinene.

Thus, hydroboration of (Z)-2-butene with I followed by oxidation with alkaline hydrogen peroxide gave R-2-butanol (87% ee.) and hydroboration of (E)-2-butene with II followed by oxidation with alkaline hydrogen peroxide gave S-2-butanol (73% ee.) (Fig. 18.51).

## 18.3.8  *Enantioselectivity Using Organometallic Reagents*

Organometallic compounds are versatile reagents in organic synthesis. These are especially useful for the formation of C–C bonds. The persent discussion is restricted to organometallic compounds containing Mg, Li, Cu.

### 18.3.8.1 Organomagnesium Halides (Grignard Reagents)

These are commonly known as **Grignard reagents**. These were discovered by Victor Grignard, who demonstrated their termendous synthetic potential. For this work, Victor Grignard was awarded Nobel Prize in 1912. These are represented as RMgX, where R is alkyl, alkenyl, alkynyl or aryl group and X is Cl, Br, I. These are prepared by the action of magnesium on alkyl halide in anhydrous ether.

$$R–X + Mg \xrightarrow[\text{reflux}]{\text{ether}} RMgX$$

Organic halides which are unreactive can be made to react with magnesium by converting it into active form. This can be achieved by sonication (J.L. Luche and J.C. Damianu, J. Am. Chem. Soc., 1980, **102**, 7926). However, there is a serious limitation on the structures of the organic halides that can be used. In fact any organic halides which has an acidic proton (like alcohol phenol, carboxylic acid, amine etc.) or a reactive functional group (like keto, nitrile, nitro or epoxide) will not give Grignard reagent, since the formed Grignard reagent will react with the group present. In such cases, the functional group has to be protected.

Some of the important reactions of Grignard reagent are given below.

Reaction of Grignard Reagent with Compounds Containing Acidic Protons

Any substance that has a acidic proton (*pKa* < 40) reacts with as acid towards a Grignard reagent.

$$CH_3Mg\,I \xrightarrow{H_2O} CH_3 - H + Mg(OH)I$$

This reaction has been used for the synthesis of (Z)– and (E)– phenylpropene by the reaction of (Z)- and (E)- β -bromostyrene with methyl magnesium iodide in ether in presence of $NiCl(PPh_3)_2$. In the absence of $NiCl(PPh_3)_2$, the yields are poor due to intervening side reactions like α- and β-eliminations (M. Kumada, Pure and Appl. Chem. 1980, **52**, 669).

Reaction of Grignard Reagent with Ketones

This is a convenient procedure for the generation of a chiral centre.

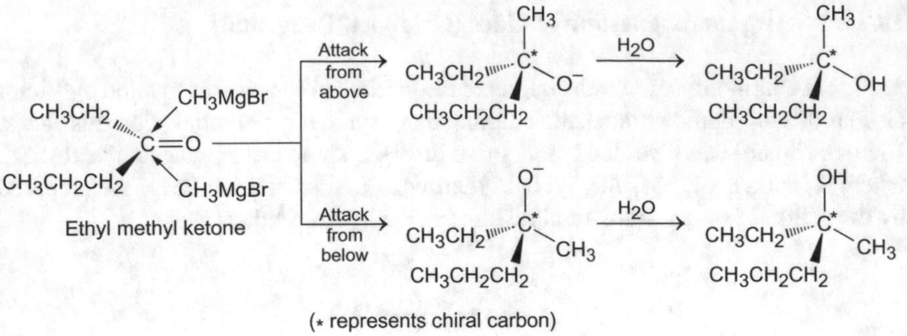

(∗ represents chiral carbon)

**Fig. 18.52** Synthesis of Chiral alkanes

**Fig. 18.53** Reaction of cyclic ketones with Grignard reagents

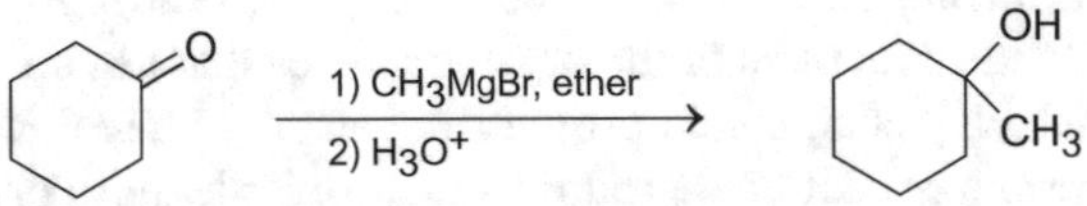

In the above reaction, addition of organometallic reagent converts an $sp^2$ hybridised carbonyl carbon to a tetrahedral $sp^3$ hybridized carbon. The addition of Grignard reagent occurs from both side of the trigonal planar carbonyl group. Thus, an achiral starting material gives an equal mixture of enantiomers. As an example, the reaction of ethyl methyl ketone with Grignard, reagent is represented as shown below (Fig. 18.52).

Cyclic ketones (e.g., cyclohexanone) reacts in the following way (Fig. 18.53).

## Ring Opening of Epoxides

The reaction of ethylene oxide with Grignard reagent gives primary alcohol (Fig. 18.54).

However, unsymmetrical epoxides react with Grignard reagent to give 2° and 3° alcohols (depending on the substitution pattern). In this case, Grignard reagent reacts of the less hindered carbon atom of the epoxide ring (Fig. 18.55).

## Synthesis of Higher Homologues of Alkenes and Alkynes

Alkenes are obtained by the reaction of Grignard reagents with unsaturated alkyl halides (Fig. 18.56).

Higher alkynes are obtained by the reaction of terminal alkynes with Grignard reagent followed by the treatment of the formed alkynyl magnesium halide with alkyl halides (Fig. 18.57).

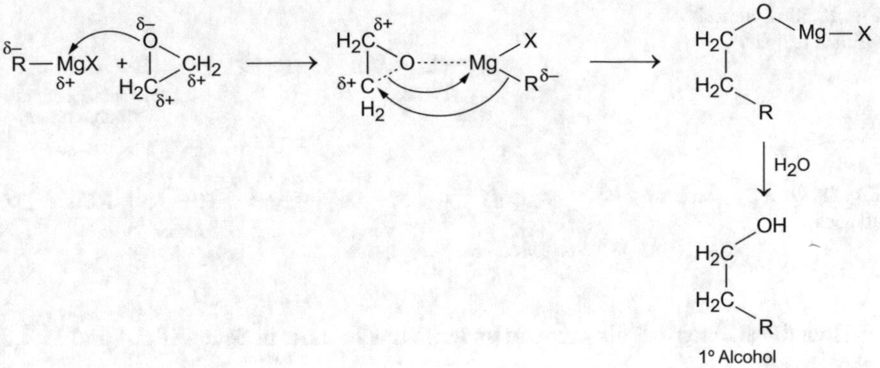
R—MgX
H2O
1° Alcohol

**Fig. 18.54** Synthesis of 1° alcohols

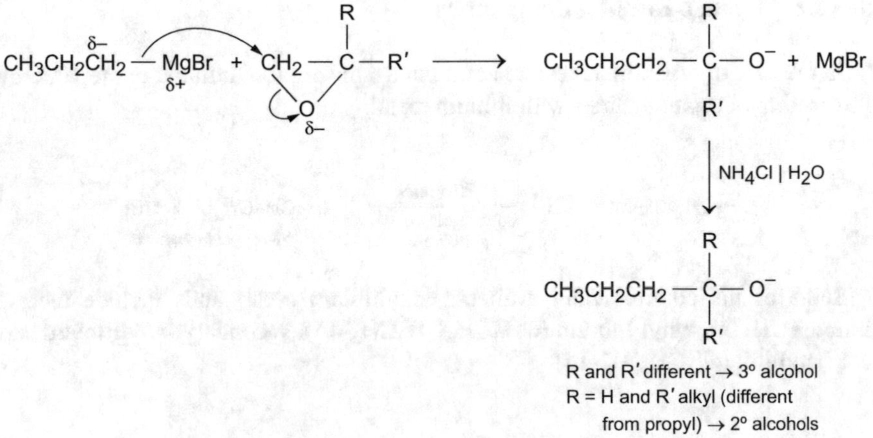
NH4Cl | H2O
R and R′ different → 3° alcohol
R = H and R′ alkyl (different
from propyl) → 2° alcohols

**Fig. 18.55** Ring opening of epoxides. Synthesis of 2° and 3° alcohols

$$CH_3MgI + CH_2{=}CHCH_2Br \longrightarrow CH_2{=}CHCH_2CH_3 + Mg(Br)I$$

**Fig. 18.56** Synthesis of higher alkenes

$$\underset{\text{alkyne}}{R{-}C{\equiv}C{-}H} + CH_3MgI \longrightarrow \underset{\text{alkynyl magnesium iodide}}{RC{\equiv}C{-}MgI} \xrightarrow[S_N2]{R'I} R{-}C{\equiv}C{-}R' + MgI_2$$

**Fig. 18.57** Synthesis of higher alkynes

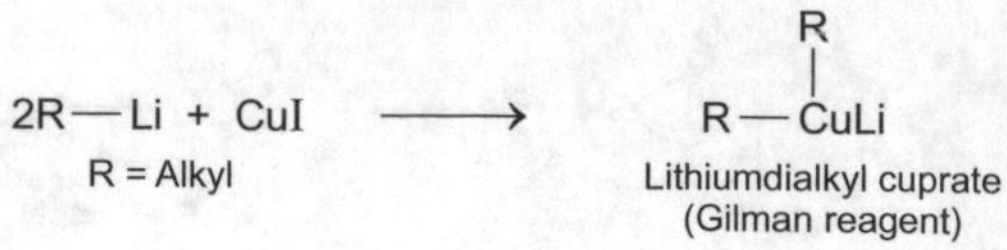

**Fig. 18.58**  Synthesis of
Gilman reagent

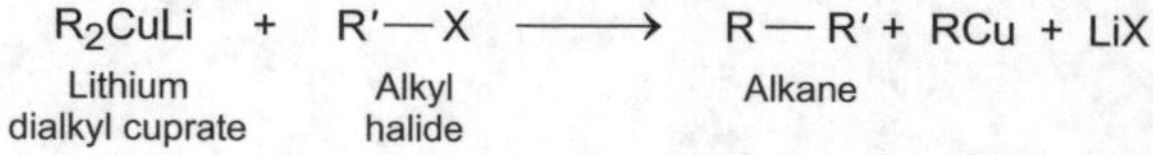

$$2R—Li + CuI \longrightarrow R—\overset{\displaystyle R}{\underset{\displaystyle |}{Cu}}Li$$

R = Alkyl

Lithiumdialkyl cuprate
(Gilman reagent)

**Fig. 18.59  a** Synthesis of
alkanes

$$R_2CuLi + R'—X \longrightarrow R—R' + RCu + LiX$$

Lithium
dialkyl cuprate

Alkyl
halide

Alkane

Both the alkenes and alkynes can undergo reactions as in Sects. 18.3.4 and 18.3.5 respectively.

### 18.3.8.2   Organo Lithium Compounds

As in the case of organomagnesium compounds, the organo lithium are prepared by the reaction of organohalides with lithium metal.

$$CH_3CH_2CH_2Br + 2Li \xrightarrow[\substack{\text{anhydrous}\\\text{conditions}}]{\text{diethylether}} CH_3CH_2CH_2Li + LiBr$$

Propyl lithium

Some of the commerically available alkyllithium compounds include methyl lithium ($CH_3Li$), butyl lithium ($CH_3CH_2CH_2CH_2–Li$), secondary butyllithium and tert. butylithium [$(CH_3)_3C–Li$].

$$(CH_3CH_2—\overset{\displaystyle Li}{\underset{\displaystyle |}{CH}}—CH_3) \text{ and tert. butylithium } [(CH_3)_3C\text{-}Li].$$

Organo lithium compounds react in the same way as Grignard reagents. Some of important reations of organo lithium reagents are given below.

Synthesis of Lithium Dialkyl Cuprate

These are commonly known as **Gilman reagents** and are prepared by the reaction of organo lithium compounds with cuprous iodide (CuI) (H. Gilman and J. M. Straley, Rect. Trave. Chem. Pays. Bas., 1936, **55**, 821; H. Gilman, R.G. Jones and L. A. Woods, J. Org. Chem., 1952, **17**, 1930) (Fig. 18.58).

The Gilman reagents react with alkyl halide to give alkanes (Fig. 18.59a).

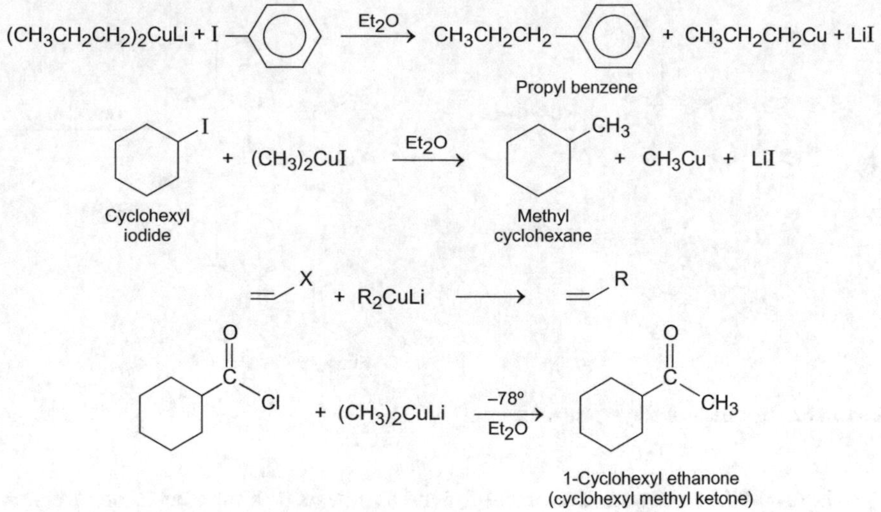

**Fig. 18.60**  Synthesis of some typical compounds

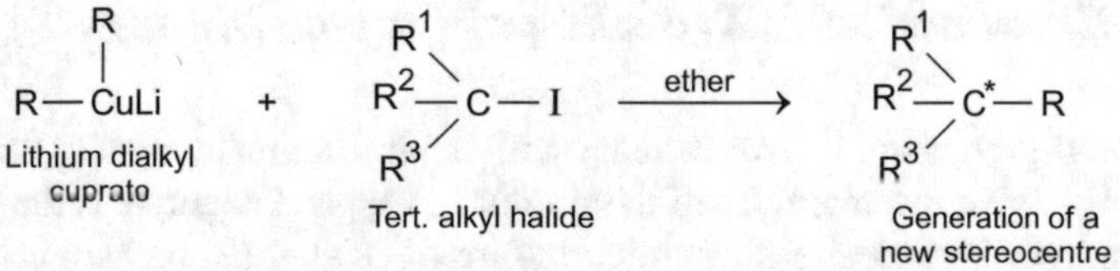

**Fig. 18.61**  Reaction of tert. alkyl halide with Gilman reagent

The above method is known as the Covery-Postner, Whitesides-House synthesis. This method is useful for the synthesis of alkanes which cannot be synthesised by other methods. Some examples are given below (Fig. 18.60).

A special application of Gilman reagents is the generation of a new stereocentre by the reaction of a tertiary halide with Gilman reagent (Fig. 18.61).

Synthesis of Alkynyllithium Reagents

These are prepared by the reaction of terminal alkynes with alkyl lithium.

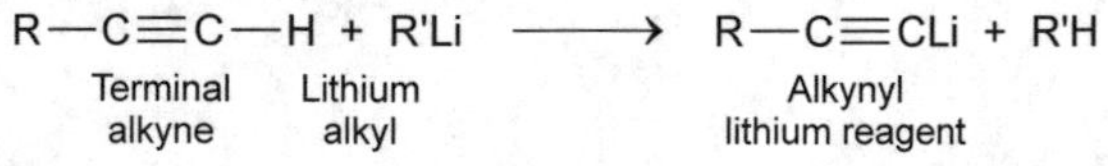

$$R—C≡C—H + R'Li \longrightarrow R—C≡CLi + R'H$$

Terminal   Lithium               Alkynyl
alkyne      alkyl               lithium reagent

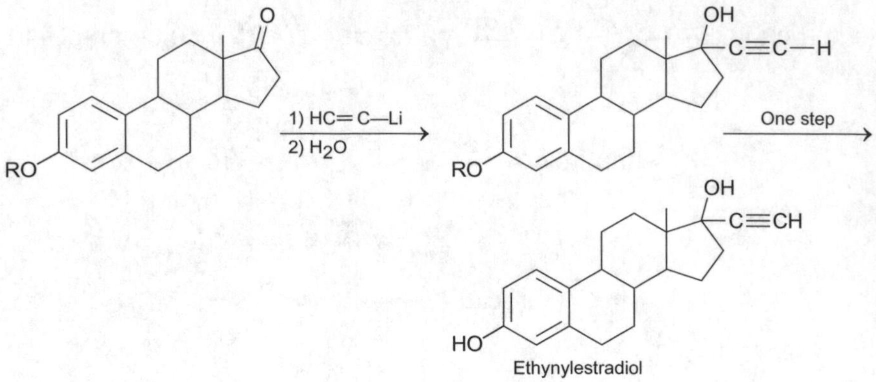

**Fig. 18.62** Synthesis of ethynylestradiol

Alkynyllithium reagents are useful for the synthesis of a number of products. A typical example is the synthesis of ethynylestradiol, an oral contractive (Fig. 18.62).

## 18.3.9 *Enantioselective Reduction of Carbonyl Groups*

The carbonyl groups are known to react with achiral reducing agents like $NaBH_4$ and $LiAlH_4$ to give racemic form of the product. A typical example is the reduction of ethylmethylketone to give a racemic mixture of (R)-(–)-(2)-butanol and (S)-(+)-(2)-butanol. In this case the rate of reaction by path (*a*) is equal to the rate reaction by path (*b*) (Fig. 18.63).

However, reduction with chiral reagents like enzymes give one enantiomeric form of a chiral product. Such a synthesis is called **enantioselective**. Enzymatic reduction are straight forward and highly stereoselective. Prelog was the first to study the reduction of carbonyl compounds using a number of enzymes. The reduction of

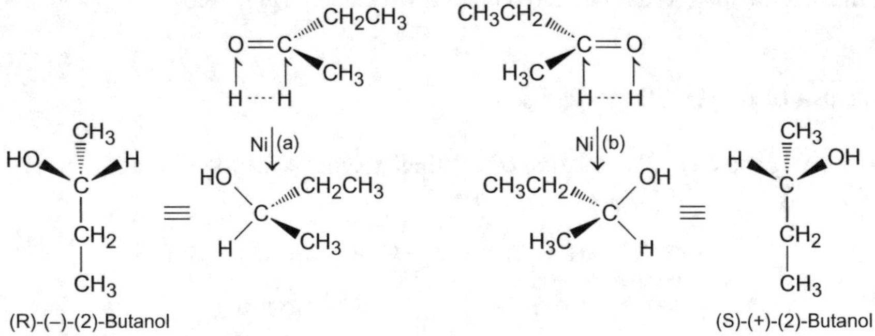

**Fig. 18.63** Reduction of ethylmethylketone with $H_2|Ni$

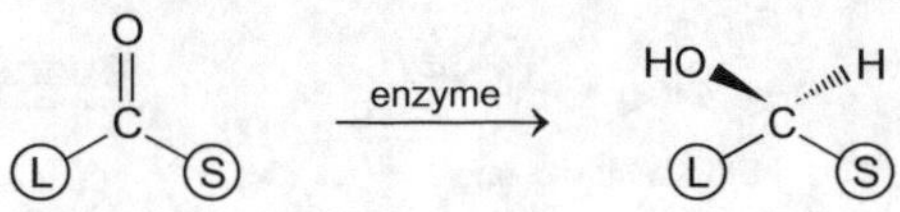

**Fig. 18.64** Reduction of carbonyl group with enzymes

ketones with *Curvularia fulcatta* gave a stereochemical induction based on Prelog's rule (V. Prelog, Pure and Applied Chem., 1964, **9**, 179).

According to this rule, if the steric difference between large (L) and small (S) groups attached to the carbonyl group is large enough, the enzyme attacks from the less hindered face (over S) to give the corresponding alcohol as shown in Fig. 18.64.

Carbonyl group can be reduced by *Thermoanaerobium brockii*. Thus, 2-butanone on reduction gives the R-alcohol (2-butanol) in 12% yield and 48% ee. However, larger ketones like 2-hexanone are reduced to the S-alcohol (85% yield, 96% ee.) (E. Kinan, E.K. Mafeli, K. K. Seth and R. Lamed, J. Am. Chem. Soc. Chem. Common, 1980, 1026). Thus, the selectively of reduction is found to depend on the size and the nature of the carbonyl group (Fig. 18.65).

A common enzyme, **Baker's yeast** (*Saccharomyces cerevisiae*) is useful for the selective reduction of β-ketoesters. Thus, the reduction of ethyl acetoacetate with Baker's yeast gave the S-alcohol. However, ethyl β-Ketovalerate on reduction.

gave the R-alcohol. Thus in this case, also, the selectivity of reduction changed from S (with small chain esters) to R selectivity (will long chain esters (B. Zhou, A.S. Gopalan, F. Van Middlesworth, W. –R. Shieh and C. Sih, J. Am. Chem. Soc., 1983, **105**, 5925) 1, 3-Diketones are reduced to β-ketoalcohol and 2, 4-hexanedione is reduced to S alcohol (Fig. 18.66).

Simple ketones can also be reduced by Baker's yeast. Thus, the ketonic moiety in the side chain of (A) was reduced selectively to R alcohol (B) with Baker's yeast (J.K. Lieser, Syn. Commun., 1982, **13**, 765) (Fig. 18.67).

Another enzyme *Beanverea sulfurescens* has been used for the reduction of conjugated carbonyls. Thus *trans*-crotonaldehyde on reduction gives 2-buten-1-ol in 80% yield. However, 2-methyl-1-pentanal gave a mixture of partially and completely

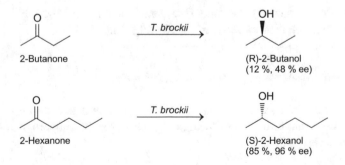

2-Butanone → *T. brockii* → (R)-2-Butanol (12 %, 48 % ee)

2-Hexanone → *T. brockii* → (S)-2-Hexanol (85 %, 96 % ee)

**Fig. 18.65** Enzymatic reduction butanone and 2-hexanone

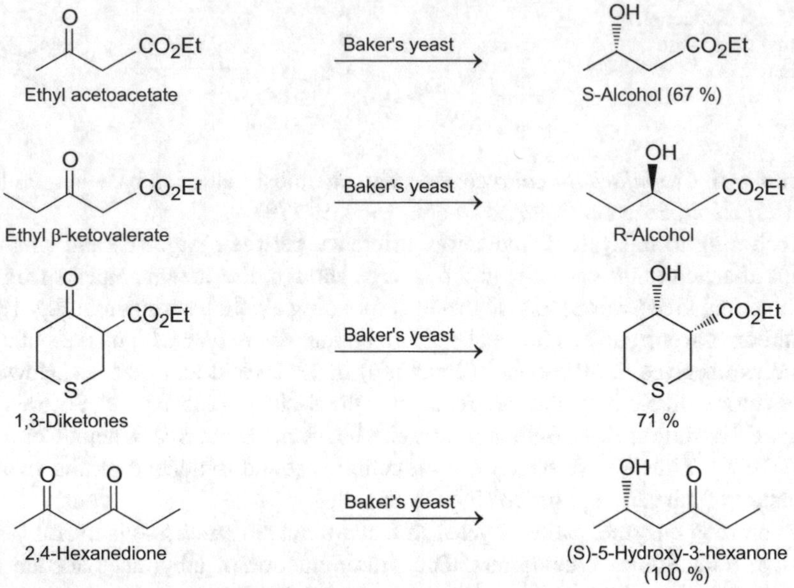

**Fig. 18.66**  Reduction with Baker's yeast

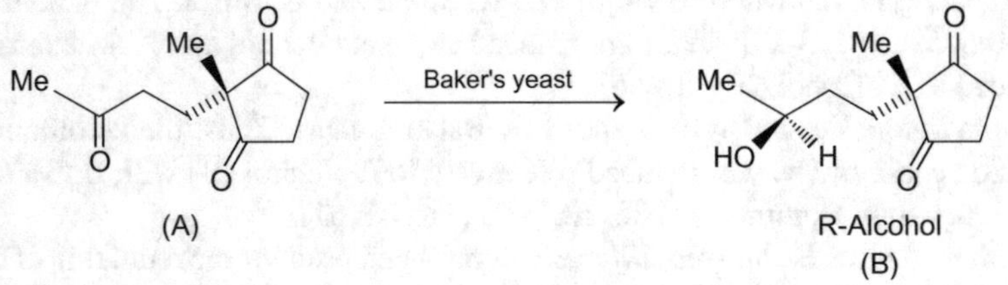

**Fig. 18.67**  Reduction using Baker's yeast

reduced product, 2-methyl-1-pentanol (M. Desert, A. Kergomard, M. F. Renard and H. Veschambre, Tetrahedron 1981, **37**, 3825) (Fig. 18.68).

## 18.3.9.1   Stereoselective Synthesis of Amino Acids

Since most of the naturally occurring amino acids are L-amino acids, the best course is to synthesise L-amino acids. Following are given some of the important procedures, for the synthesis of L-amino acids.

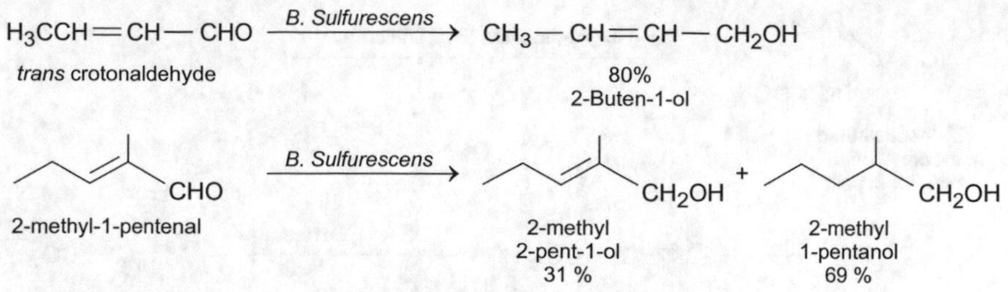

$$H_3CCH=CH-CHO \xrightarrow{\text{B. Sulfurescens}} CH_3-CH=CH-CH_2OH$$

*trans* crotonaldehyde

80%
2-Buten-1-ol

2-methyl-1-pentenal $\xrightarrow{\text{B. Sulfurescens}}$ CH_2OH + CH_2OH

2-methyl
2-pent-1-ol
31 %

2-methyl
1-pentanol
69 %

**Fig. 18.68** Reduction using B. Sulfurescens

### Reduction of 2-acetylaminopropeonic Acid

A number of chiral hydrogenation catalysts derived from transition metals have been used for the reduction of 2-acetylaminopropeonic acid. One such chiral catalyst developed by Bosnich is derived from a rhodium complex with (R)-1, 2-bis(diphenyl phosphino) propanol. It is commonly called (R)-prophos.

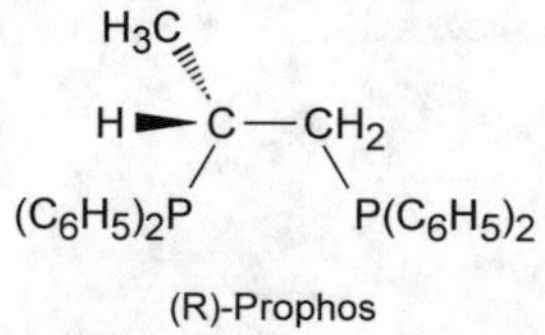

$$H_3C$$
$$H \blacktriangleright C-CH_2$$
$$(C_6H_5)_2P \qquad P(C_6H_5)_2$$

(R)-Prophos

Treatment of (R)-prophos with rhodium complex of norbornadiene (NBD) results in the replacement one of the molecules of norbornadiene surrounding the rhodium giving the chiral rhodium complex.

$$[Rh(NBD)_2]ClO_4 + (R) - propos \rightarrow \begin{array}{c} [Rh(R) - prophos)(NBD)]ClO_4 + NBD \\ Chiral\ rhodium\ complex \end{array}$$

Reduction of (Z)-3-substituted 2-acetyl aminopropeonic acid with the chiral catalyst in a suitable solvent gave N-acetyl derivative of L-aminoacids in 90% enantiomeric excess. Hydrolysis of the N-acetyl group yields L-amino acids. Since the hydrogenation catalyst is chiral, it transfers stereoselectively its hydrogen atoms. This type of reaction is called asymmetric synthesis or enantioselective synthesis (Fig. 18.69).

### Corey's Method

This method is useful for the preparation of optically pure amino acids from achiral molecules using chiral reagents having an asymmetric centre. Thus, α-keto acids

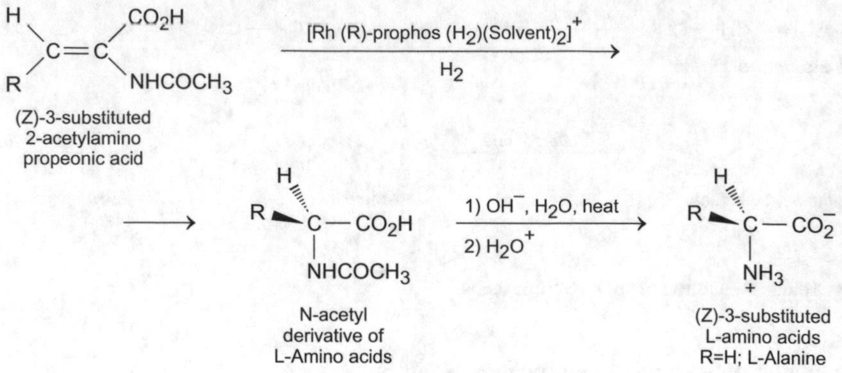

**Fig. 18.69**  Synthesis of L-aminoacids

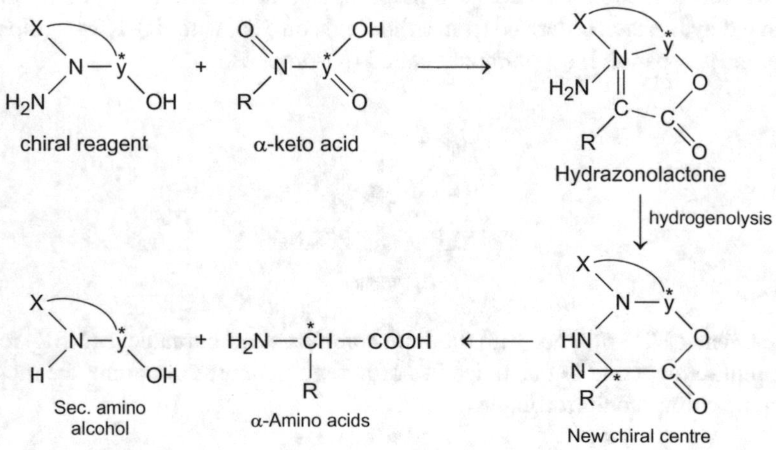

**Fig. 18.70**  Synthesis of α-Amino acids

which act as precursors of α-amino acids) on reaction with a chiral reagent form a hydrazonolactone ring. Various steps involved are given in Fig. 18.70.

In the above method, the stereospecific reduction of $N = C$ double bond produces the chiral carbon of the amino acid. The original chiral reagent is regenerated from the chiral secondary amino alcohol obtained.

Using the above method D-alanine (optical purity 80%) is obtained from methylacetoformate ($CH_3COCOOCH_3$) using a chiral reagent with a bicyclic indoline structure (Fig. 18.71).

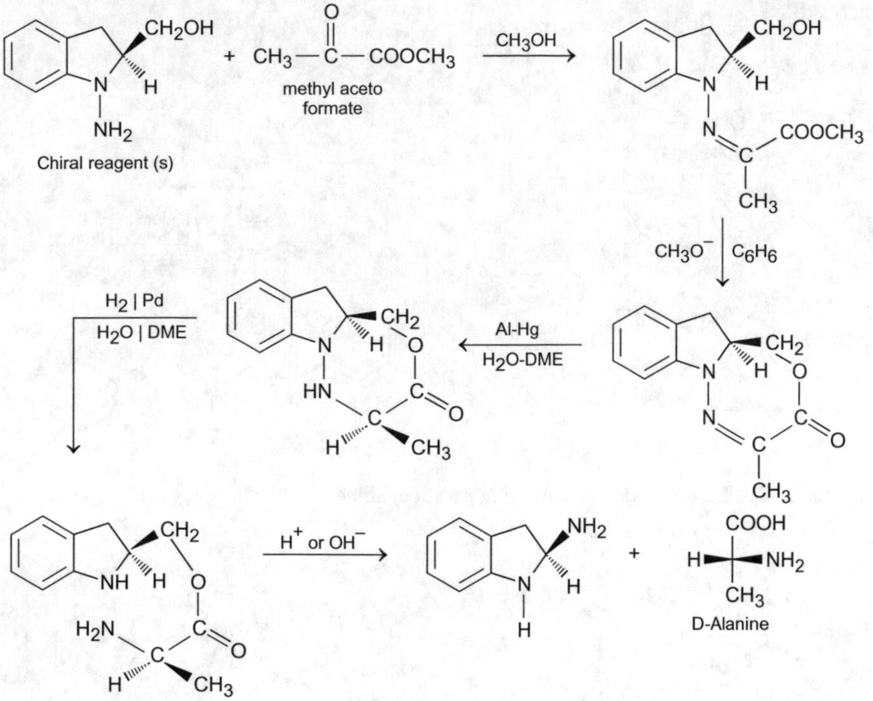

**Fig. 18.71**  Synthesis of D-Alanine

## Using Pseudoephedrine as Chiral Auxiliary

In this method, pseudoephedrine (a readily available and inexpensive chiral auxiliary) is reacted with glycine methyl ester to give pseudoephedrineglycinamide. In this reaction, the secondary amino group of pseudoephedrine forms an amide bond with the carboxyl group of glycine methyl ester. The reaction of the formed pseudoephedrine glycinamide with a number of electrophiles gives excellent diastereoselectivity with good yields (Fig. 18.72).

## Schollkopf's Method

In this method, an amino acid, valine is used as chiral auxiliary. Thus, condensation of valine with glyane yields a *bis*-lactim ether, which on enolisation and electrophilic reaction with an aldehyde or a ketone gives an intermediate with high level of distereoface selectivity. Various steps involved are shown in (Fig. 18.73).

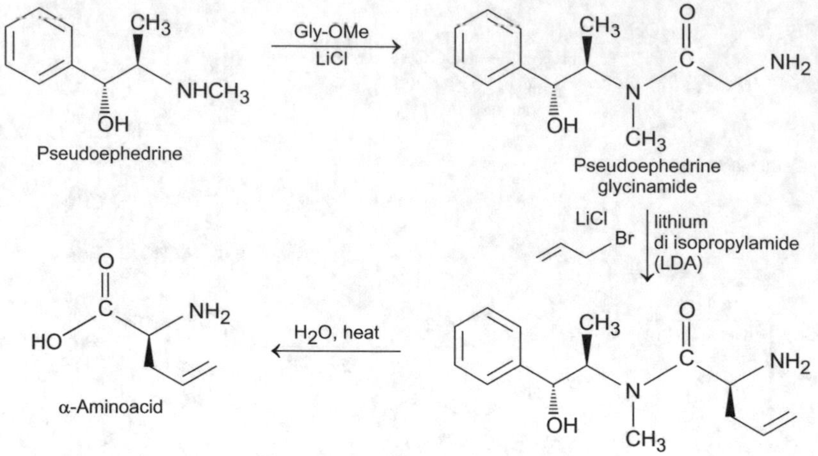

**Fig. 18.72** Synthesis of a-aminoacids using pseudoephirine

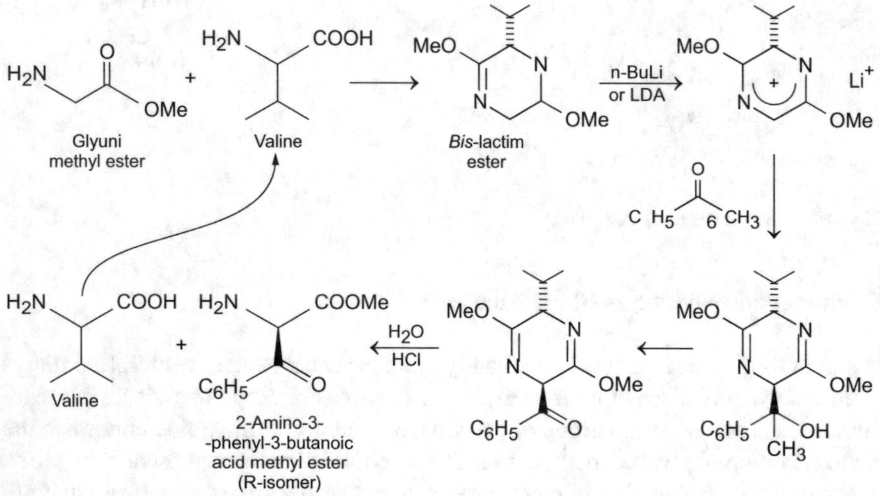

**Fig. 18.73** Synthesis of (R) 2-amino-3-phenyl-3-butenoic acid methyl ester

## 18.3.9.2   Slectivity in Hydride Reductions

Reactions which are known to proceed by transfer of hydride ions are commonly used in organic synthesis. A number of hydrides are available for reduction. The most common are aluminium isopropoxide, lithium aluminium hydride and sodium borohydride.

Aluminium isopropoxide has commonly been used under the name of **Meerwein-Pondorff-Verley reduction** for the reduction of carbonyl compounds to alcohols.

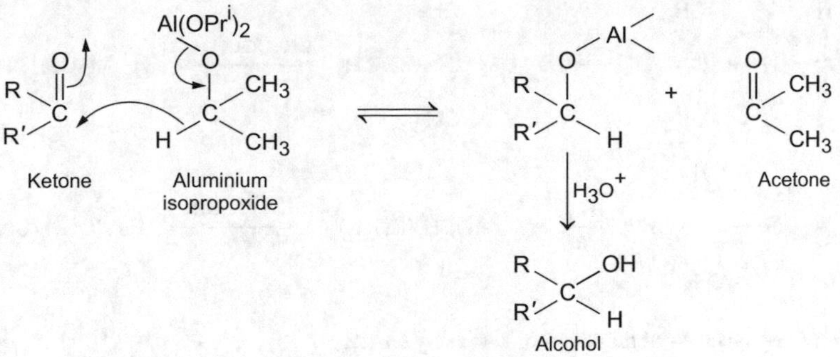

**Fig. 18.74** Mechanism of Meerwein-Pondorff-Verley reaction

This reaction consist in heating the components in solution in isopropanol. An equilibrium is set up in the reaction and the product of reduction (alcohol) is obtained by using either excess of the reagent or by distilling the acetone as it is formed. The reaction is known to proceed by transfer of a hydride ion from isopropoxide to the carbonyl compound via a six membered transition state. A special feature of this reaction is that many unsaturated groups are not affected thus permitting selective reduction of the carbonyl group. Thus, cinnamaldehyde is converted into cinnamyl alcohol and o-nitrobenzaldehyde into o-nitrobenzyl alcohol. The mechanism of reduction of carbonyl group to alcohol is given in (Fig. 18.74).

Lithium aluminium hydride and sodium borohydride are the most commonly used amongst metal hydrides. The anions of these two hydrides are believed to be derived from lithium or sodium borohydride and are nucleophilic reagents and normally attack polarised multiple bonds like $C = O$ or $C \equiv N$ by the transfer of hydride ion to the more positive atom. Isolated carbon–carbon double or triple bonds are not reduced.

$$LiH + AlH_3 \rightarrow Li^+ AlH_4^-$$

$$NaH + BH_3 \rightarrow Na^+ BH_4^-$$

Both the reagents ($LiAlH_4$ and $NaBH_4$) transfer all the four hydrogen atoms for reduction as shown below for the reduction of a ketone with $LiAlH_4$ (Fig. 18.75).

Reduction with $LiAlH_4$ can be carried out only in non-hydroxylic solvents (like ether, THF) since it reacts with water and other compounds containing active hydrogen atoms. However $NaBH_4$ can be used in water or alcohol, since it reacts only slowly in these solvent. The selectivity of reduction of both $LiAlH_4$ and $NaBH_4$ is shown below (Fig. 18.76).

Lithium aluminium hydride reduces triple bond of propargylic alcohols to *trans* double bonds (B. Grant and C. Djerassi, J. Org. Chem., 1974, **39**, 968). Only the triple bonds flanked by hydroxyl groups are reduced (Fig. 18.77).

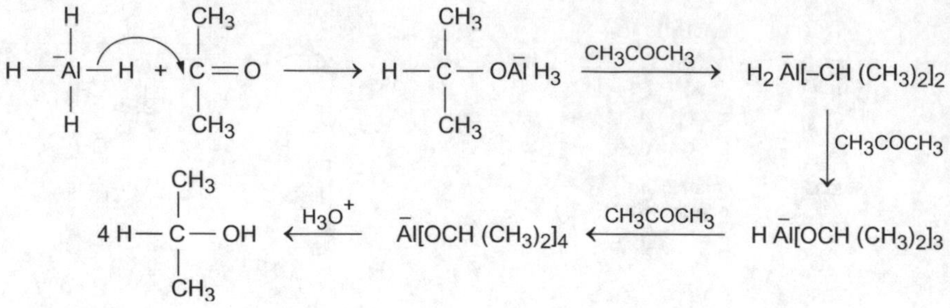

**Fig. 18.75** Mechanism of reduction of ketone by LiAlH$_4$

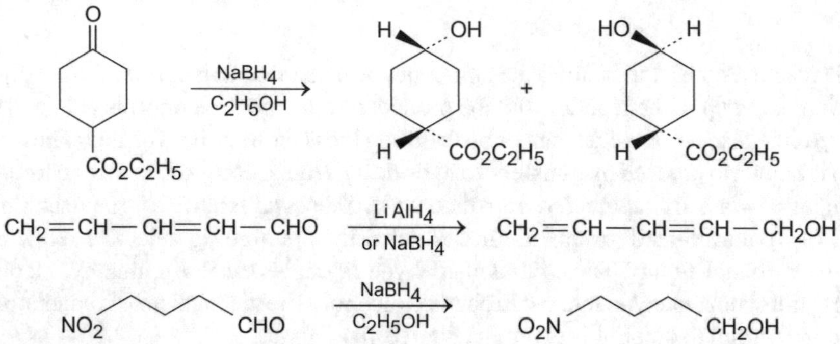

**Fig. 18.76** Selectivity of the reduction by LiAlH$_4$ and NaBH$_4$

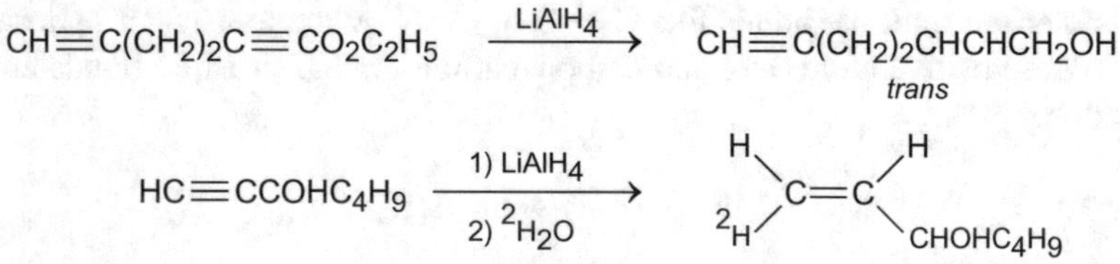

**Fig. 18.77** Reduction of triple bond to *trans* double bond

In case of unsymmetrical ketones (like ethyl methyl ketone) the reduction give only racemic alcohol. This is because carbonyl is a **prochiral centre**, addition of hydride can occur from either face of the carbonyl to generate a stereogenic centre. As both the faces are equally susceptible to attack, the alcohol produced is racemic (Fig. 18.78).

In case of a ketone which has an asymmetric centre, the two forms of the corresponding alcohol are not produced in equal amounts. As an example reduction of the ketone (A) with lithium aluminium hydride the threo (anti) (B) form of the alcohol is obtained as predominant product (Fig. 18.79).

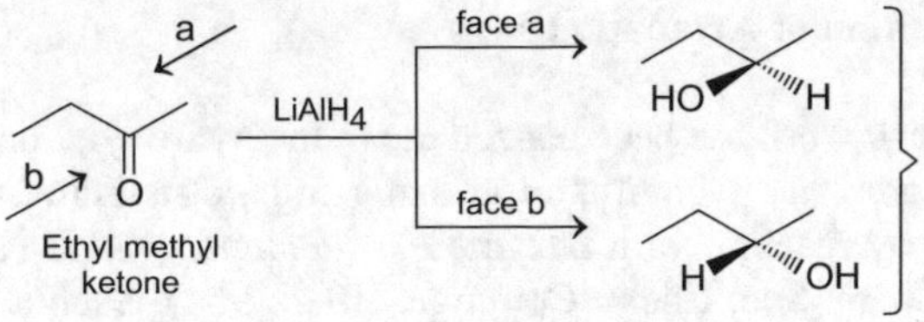

**Fig. 18.78** Reduction of ethyl methyl ketone gives racemic alcohol

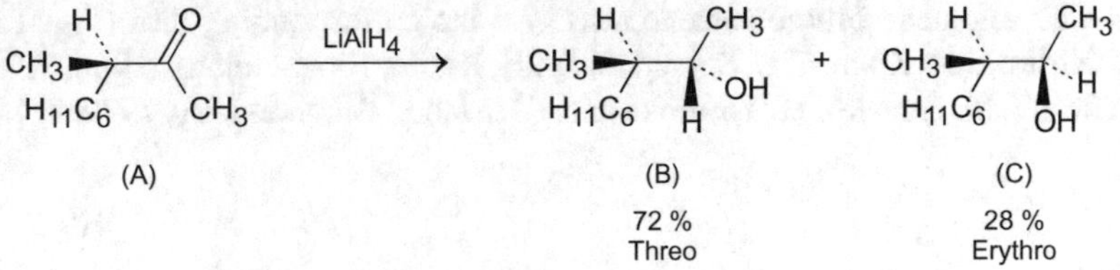

(A)                          (B)                          (C)

72 %                         28 %
Threo                        Erythro

**Fig. 18.79** Reduction of a ketone having an asymmetric centre

**Fig. 18.80** Cram's rule

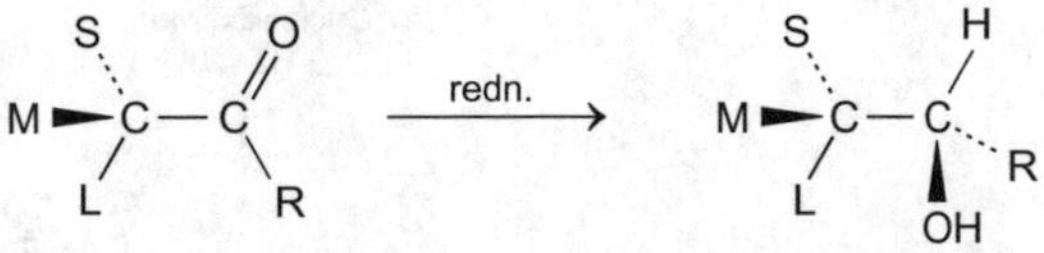

In the above reduction, the formation of main product (B in this case) can be predicted on the basis of Cram's rule (D.J. Cram and F.A. Abd Elhafez, J. Am. Chem. Soc., 1952, **74**, 4828). According to Carm's rule that distereromer predominates in the product which is formed by the approach of the reagent to the less hindered side of the carbonyl group. This may be represented by using Newmann's projection formula (Fig. 18.80) where S, M and L represent small, medium and large substituents.

The selectivity of reduction has also been found by using $NaBH_4$ and DIBAL (di isobutylaluminium hydride). In all these cases the predominate product is obtained in the ratio 90–97: 2–5. The ratio refers to predominant and minor product. (K. E. Strickland and J. B. Pommerville, J. Org. Chem., 1988, **53**, 4877).

Exclusive formation of one enatiomer can, however be obtained by using chiral reducing agents like organoboranes (*see* Sect. 19.3.7) and organo magnesium halides (*see* Sect. 19.3.8).

Enzymes are also chiral reagent and have been used for the reduction of carbonyl compounds to give only one enantiomer. For details *see* Sect. 19.3.9.

### 18.3.9.3   Hydroxylation of Aromatic Rings

Considerable amount of work has been carried out in the hydroxylation of aromatic rings. It is possible to convert a pi bond of an aromatic ring such as benzene into cyclo-hexadiene 1, 2-diols by reaction with enzyme *Pseudomonas putida* (L.M. Shirley and S.C. Taylor, J. Chem. Soc. Chem. Commun., 1983, 954). Such a biochemical transformation is not possible by conventional chemical means. The diol obtained has been used for the synthesis of conduritol F and pinitol (Fig. 18.81).

As in the case of benzene, other aromatic hydrocarbons like toluene chlorobenzene and 4-Chlorotoluene bromobenzene also give the corresponding diol (Fig. 18.82) (D.T. Gibson, J.R. Koch, C.L. Schuld and R.E. Kallio, Biochemistry, 1968, **7**, 3795; D.T. Gibson, M. Hensley, H. Yoshioka and T.J. Mabry, Biochemistry, 1970, **9**, 1626).

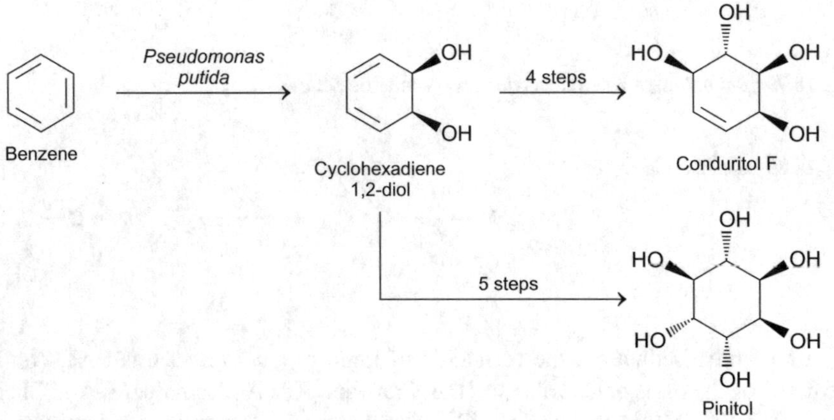

**Fig. 18.81**  Hydroxylation of benzene

Toluene, X = CH$_3$, R = H
Chlorobenzene, X = Cl, R = H
4-Chlorotoluene, X = CH$_3$, R = Cl

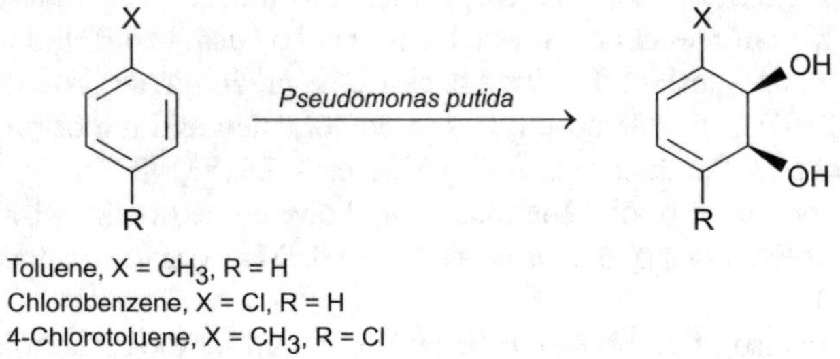

**Fig. 18.82**  Some examples of dihydroxylation of benzene derivatives

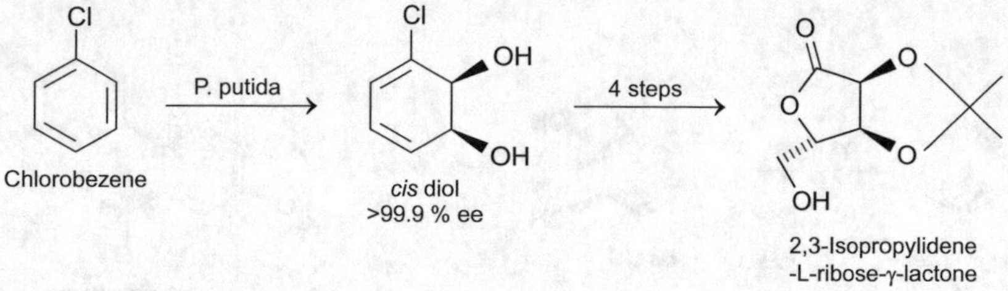

**Fig. 18.83** Synthesis of 2, 3-isopropylidene-L-ribose-γ-lactone

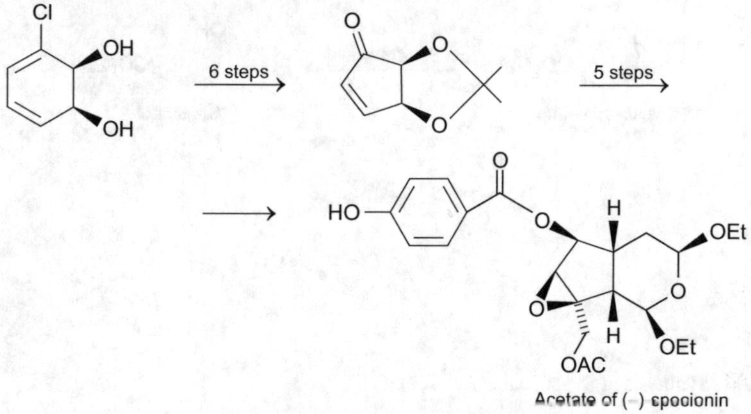

**Fig. 18.84** Synthesis of acetate of (–)-specionin

## 18.3.9.4 Some Typical Enantioselective Synthesis

A large number of enantioselective synthesis have been reported. Some of the typical synthesis are given below.

Synthesis of 3,3-Isopropylidene-L-Ribose-γ-Lactone and Acetate of (–)-Specionin

The *cis* diol from chlorobenzene was converted (H. Tudlicky and J.D. Price, Synlett, 1990, 159; T. Hudlicky, H. Luna, J. Price and F. Rulin, Tetradron Lett., 1989, **30**, 4053) into 2,3-isopropylidene-L-ribose-γ-lactone in four steps (overall yield 22%) (Fig. 18.83).

The same diol has also been used for the synthesis of (–)-specionin via a syntheic intermediate (M. Natchus, J. Org. Chem., 1992, **57**, 4740) (Fig. 18.84).

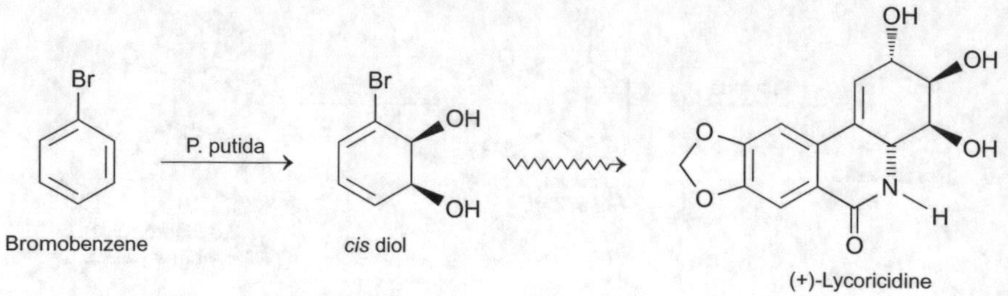

**Fig. 18.85**  Synthesis of (+)-lycoricidine

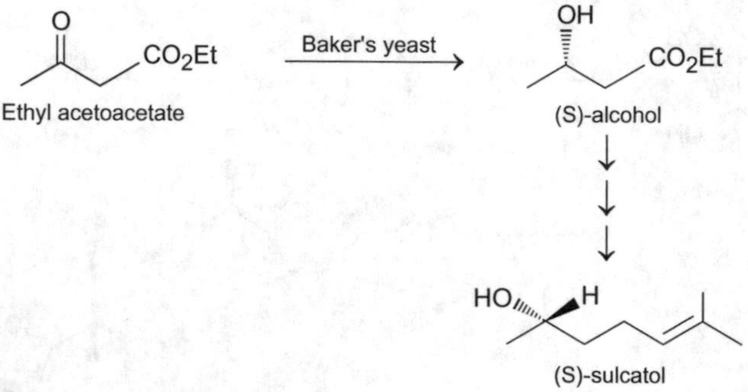

**Fig. 18.86**  Synthesis of (S)-sulcatol

## Synthesis of (+)-Lycoricidine

The *cis* diol obtained from bromobenzene (T. Hudlicky and H.F. Oliva, J. Am. Chem. Soe., 1992, **114**, 9694 and the references cited therein) is converted into (+)-lycoridicine (Fig. 18.85).

## Synthesis of (S)-(+)-Sulcatol

The (S) hydroxy ester obtained by the reduction of ethylacetoacetate has been used for the synthesis of (S)-(+)-sulcatol (K. Mori, Tetrahedron, 1981, **37**, 1341) (Fig. 18.86).

## Synthesis of S-Citronellol

Reduction of 7-methyl-3-oxo-oct-6-enoate by Baker's yeast gave (M. Hirma, M. Shimitzu and M. Iwashita, J. Chem. Soc. Chem. Commun., 1983, 599) the corresponding (R) acid, which is a synthon for the synthesis of S-citronellol (Fig. 18.87).

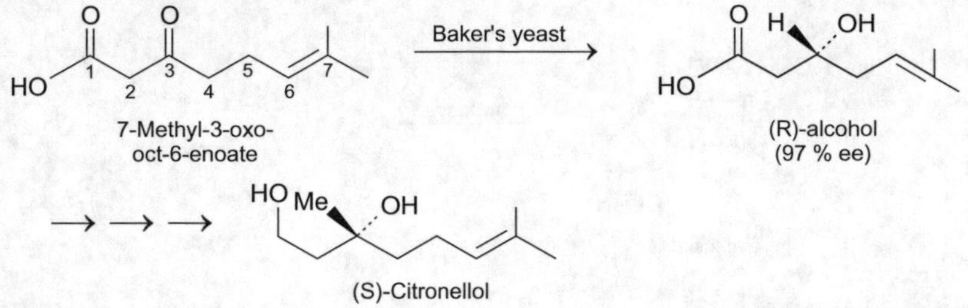

**Fig. 18.87** Syntehsis of (S)-citronellol

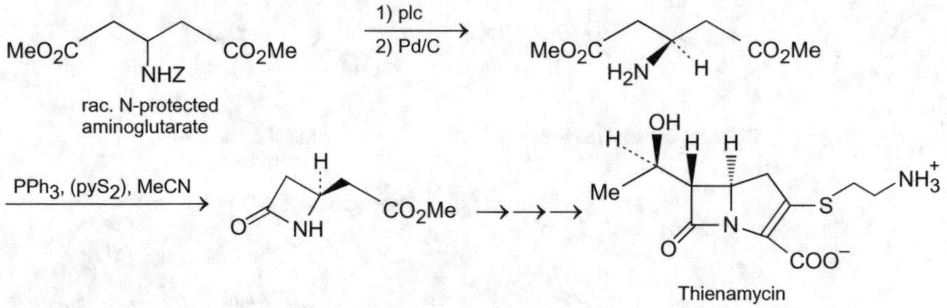

**Fig. 18.88** Synthesis of thienamycin

Synthesis of Thienamycin

Rac. N-Protected aminoglutarate on hydrolysis by procane liver esterase (plc) followed by deprotection gave the (S)-amino acid which was converted via the macrocyclic β-lactam into thienamycin (S. Kobayashi, T. Imori, T.I. Zawa and M. Ohnu, J. Am. Chem. Soc., 1981, **103**, 2406; R. Rossi, A Carpita and M. Chini, Tetrahedron, 1985, **41**, 627) (Fig. 18.88).

Synthesis of 6-Aminopenicillic Acid (6APA)

It was obtained (D.L. Regan, P. Dunnil and M.D. Lilley, Biotech, Bioeng, 1974, **16**, 333) by the cleavage of 6-acetyl group from Penicillin G by using an isolated immobilized amidase (Fig. 18.89).

Synthesis of Aspartame

It was industrially prepared (K. Oyama, Chirality in Industry, eds. A.N. Collins, G.N. Sheldrake and J. Crosby, Wiley, Chichester, 1992, Page 237) as given in (Fig. 18.90).

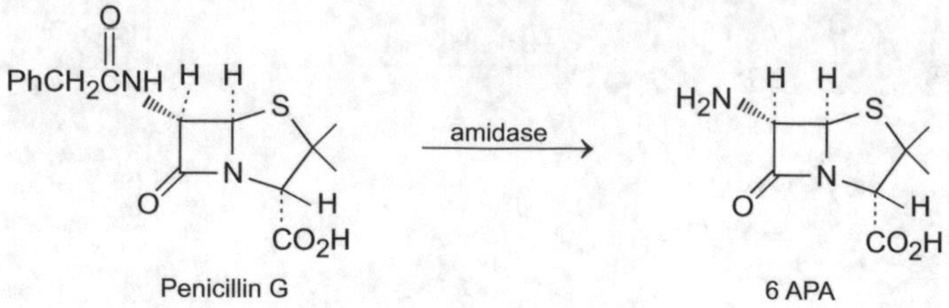

**Fig. 18.89** Synthesis of 6 APA

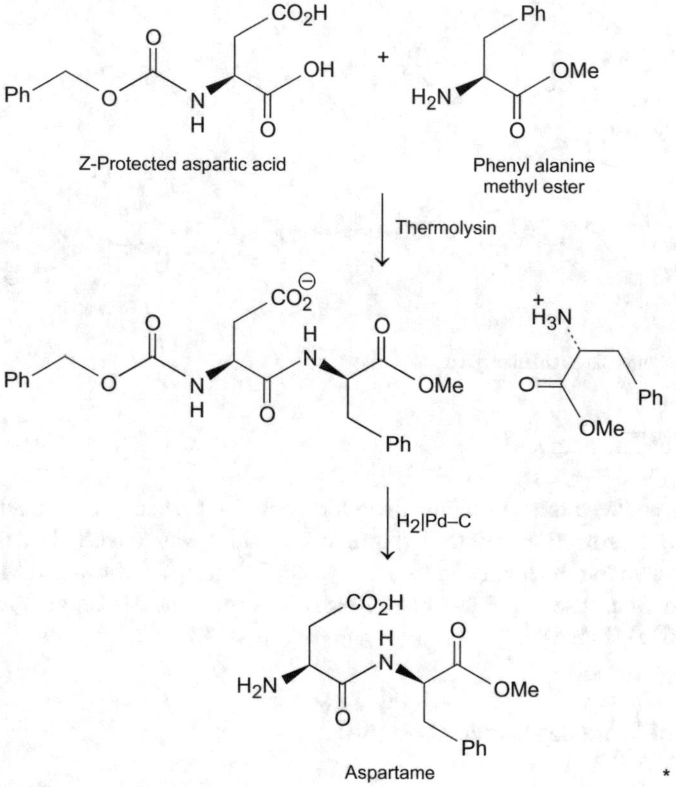

**Fig. 18.90** Synthesis of aspartame

## Synthesis of Either D or L-amino Acids

This synthesis has been achieved by the dynamic resolution of amino acid using the enzyme hydantiomase followed by treatment with a second hydrolyticenzyme, carbamoylase, to release the free amino acid of desired configuration. This process has been used on industrial scale (A.S. Bommarius, K. Drauz, U. Groegar and C.

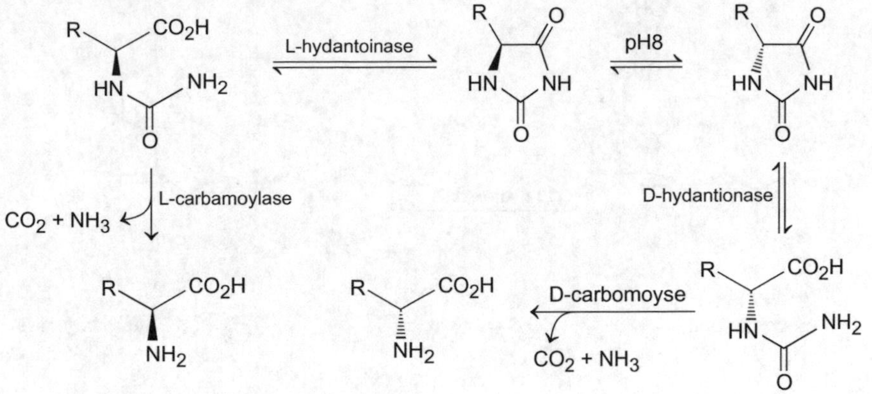

**Fig. 18.91**  Synthesis of D-and L amino acids

Wandrey in Chirality in Industry, eds. A.N. Collins, G. N. Sheldrake and J. Crosby, Wiley Chichester, 1972, Page 371) (Fig. 18.91).

Synthesis of Prednisolone

Prednisolone, a cortisone analog, used as a drug against rheumatoid arthritis is obtained on a commercial scale from Reichstein's compound, a steroid precursor (K. Meyers and E.D. Mihlich, Angew. Chem. Internal. Edn, 1976, **15**, 270). This compound is passed through a series of two different columns, each containing a specific enzyme attached to a polymer support (Fig. 18.92).

Besides the examples of enantioselective synthesis using enzymes given above, a large number of other compounds have also been synthesised using enzymes. For more details see V.K. Ahluwalia, Enzymes for green organic synthesis, Narosa Publishing House, 2010 and the references cited there in.

*Synthesis of Cholesterol*

The main problem in the synthesis of steroids is due to stereochemical problems. Cholesterol molecule contains eight chiral centres and so there are 256 optical isomers. Due to this each step in the synthesis, which produces a new chiral centre the formation and isolation of the desired product has to be isolated by using suitable procedure including resolution of the racemic modification. Another problem is to develop specific reagents so that the required enantiomer can be obtained in maximum amount. Following is described in the outline of Woodward's synthesis. Following steps are involved:

1.  The required intermediate, 4-methoxy-2,5-toluquinone is obtained as follows

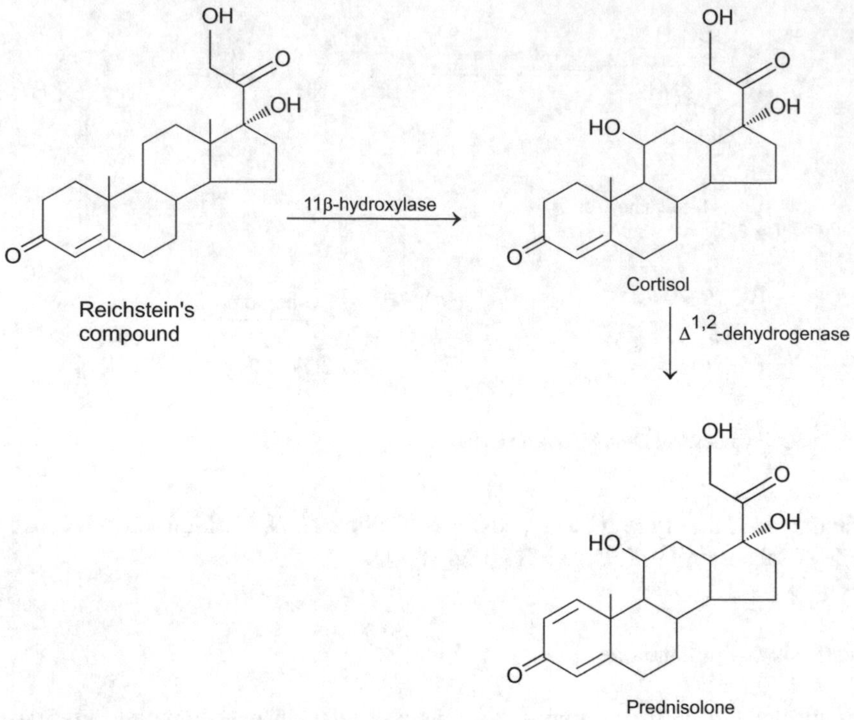

Reichstein's
compound

11β-hydroxylase

Cortisol

$\Delta^{1,2}$-dehydrogenase

Prednisolone

**Fig. 18.92** Synthesis of prednisolone

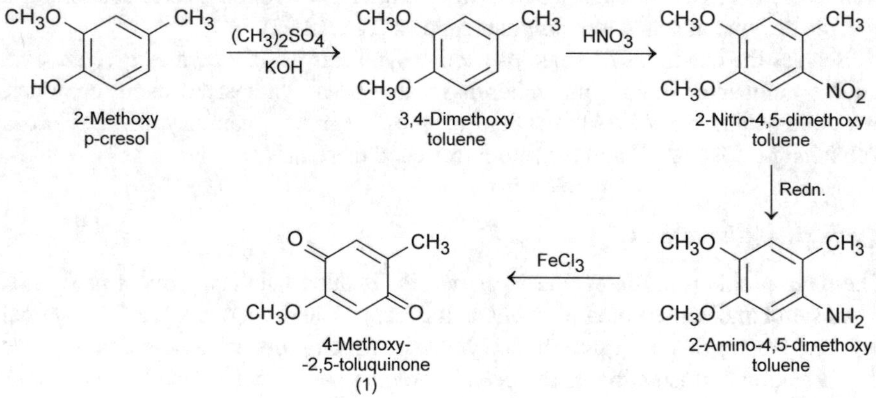

2-Methoxy
p-cresol

(CH₃)₂SO₄
KOH

3,4-Dimethoxy
toluene

HNO₃

2-Nitro-4,5-dimethoxy
toluene

Redn.

2-Amino-4,5-dimethoxy
toluene

FeCl₃

4-Methoxy-
-2,5-toluquinone
(1)

2.    Condensation of (1) with butadiene gave the adduct (2) and had the *cis* configu-
      ration. It was isomerised quantitatively to the *trans*-isomer (3) by dissolving in
      aqueous alkali, seeding with a crystal of *trans*-isomer followed by acidification.

3. Reduction of (3) with lithium aluminium hydride gave the quinol (4),which being a vinyl ether of a glycol, on treatment with aqueous acid underwent hydrolysis (demethylation) to give a β-hydroxyketone (5).

4. Removal of hydroxyl group from (5) is achieved by heating the acetate with zinc in acetic anhydride. The formed (6) on reaction with ethyl formate in presence of sodium methoxide gave the hydroxymethylene ketone (7) (Claisen condensation).

5. Treatment of (7) with ethyl vinyl ketone in presence of potassium t-butoxide (Michael condensation) gave (8), which on cyclisation by potassium hydroxide in dioxane gave (9) as the sole product. In this case, the cyclisation is stereospecific and occurs by an intramolecular Aldol condensation followed by dehydration.

6. Oxidation of (9) with osmium tetroxide gave the *cis* glycol (10) (glycol formation occurs easily at the isolated double bond; the other two double bonds are conjugated and so have less double bond character). The two glycols were separated and the desired isomer (the one insoluble in benzene) on treatment with acetone in presence of anhydrous copper sulphate gave the isopropylidene derivative (11).

7. Catalytic reduction of (11) with $H_2$-Pd/$SrCO_3$ gave the reduced product (12), which on condensation with ethyl formate in presence of sodium methoxide gave (13).

8. The reaction of (13) with methylaniline gave (14). In this way regiospecific control was achieved by blocking the position 4 and leaving 3-keto group for subsequent reaction.

9. Condensation of (14) with vinyl cyanide (cyanoethylation) followed by hydrolysis of the formed product with alkali a mixture of two keto acids were obtained. These were separated and the stereoisomer (15) (methyl group in front and propionic acid group behind the plane of the ring) was converted into the enol lactone (16), which on treatment with grignard reagent (methyl magnesium bromide) gave (17).

10. Ring closure of (17) with alkali gave (18), which on oxidation with periodic acid in aqueous dioxan, gave the dialdehyde (19) (obtained by hydrolysis of the diol).

11. The dialdehyde (19) on heating in benzene solution in presence of a small amount of piperidine acetate gave (20) along with a small quantity of an isomer. This cyclisation involved an intramolecular aldol condensation in presence of the base,piperidine acetate. The isomer (20) is formed due to steric reasons.

12. The cyclised product (20) on oxidation gave the corresponding acid, which on esterification with diazomethane gave the methyl ester (21). The methyl ester (21) was a racemate and was resolved by reduction of the keto group with sodium brohydride to the hydroxy esters (($\pm$)-3α- and ($\pm$)-3β). The (+) form of the 3β-alcohol was preferentially precipitated by digitonin, and this isomer (21) on oppenauer oxidation gave the desired stereoisomer (+)-(21).

13.   Catalytic reduction ($H_2$/Pt) of (+)-(21) to (22) and its subsequent oxidation give a mixture of stereoisomers (23) which were separated, reduced (sodium borohydride) and hydrolysed. The β-isomer (24) was converted into methyl ketone by first acetylation, then treating with thionyl chloride and finally with dimethyl cadmium. The acetylated hydroxyketone (25) on treatment with isohexylmagnesium bromide gave (26). This (26) was a mixture of isomers, since a new chiral centre was introduced at position 20.

14.   Dehydration of (26) gave one product (27), which on catalytic hydrogenation ($H_2$/Pt) gave a mixture of 5α-cholesteryl acetates (the chiral C-20 has been reintroduced). These acetates were separated and the desired isomer, on hydrolysis, gave 5α-cholestan-3β-ol (28), which was identical with natural cholestanol.

15.   Finally, conversion of cholestanol (28) into cholesterol (33) was carried out by a series of reactions. Thus, oxidation of (28) followed by bromination of the formed (29) with bromine in acetic acid in the presence of hydrogen bromide gave 2α-bormo derivate (30). Treatment of (30) with pyridine gave (31), which on heating with acetyl chloride in presence of acetic anhydride gave the enol acetate (32). Final reduction of (32) with lithium aluminium hydride followed by acidification gave cholesterol (33).

All the above steps have been shown in Fig. 18.93.

## Key Concepts

- **Alkynyl Lithium Reagents**: R–C≡CLi. These are prepared by the reaction of terminal alkynes with alkyl lithium.
- **Chemoselective**: A reagent which reacts with only one functional group (in a compound containing two or more functional group) is said to be chemoselective.
- **Chiral Synthon**: An optically active starting material.
- **Enantioselective Reaction**: A reaction in which there is predominance of one enantiomer.
- **Enantiospecific Reaction**: Also known as regioselective reaction and is a reaction giving only one enantiomer.
- **Enantioselective Epoxidation**: Expoxidation reaction of alkenes which give one (*trans* or *cis*) epoxide.
- **Enzymes**: These are proteins and are capable of catalytic activity and are referred to as biological catalysis or bio catalysts. These have been extensively used in organic synthesis. These are enantioselective reagents.
- **Gilman Reagents**: These are lithium dialkyl cuprates and are prepared by the reaction of organo lithium compounds with cuprous iodide

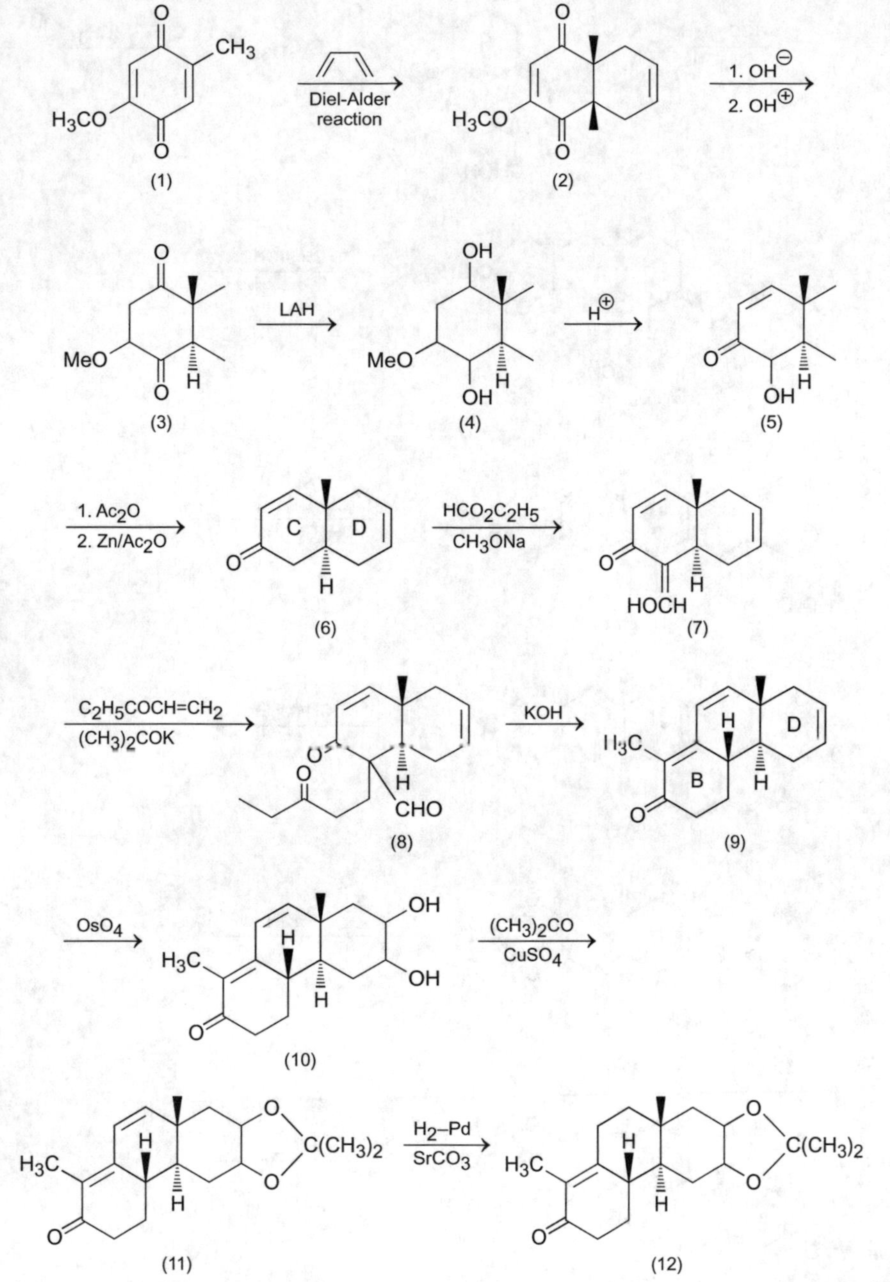

**Fig. 18.93** Synthesis of Cholesterol

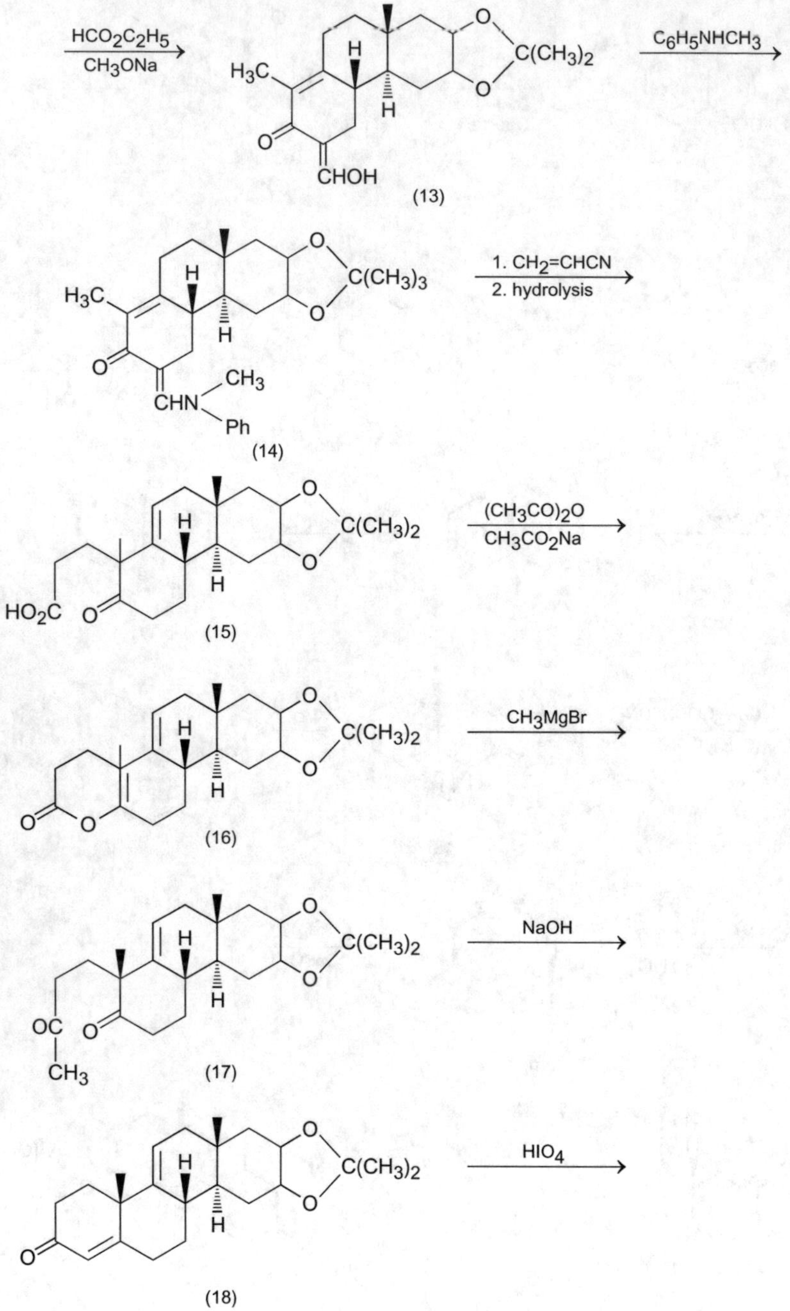

**Fig. 18.93** (continued)

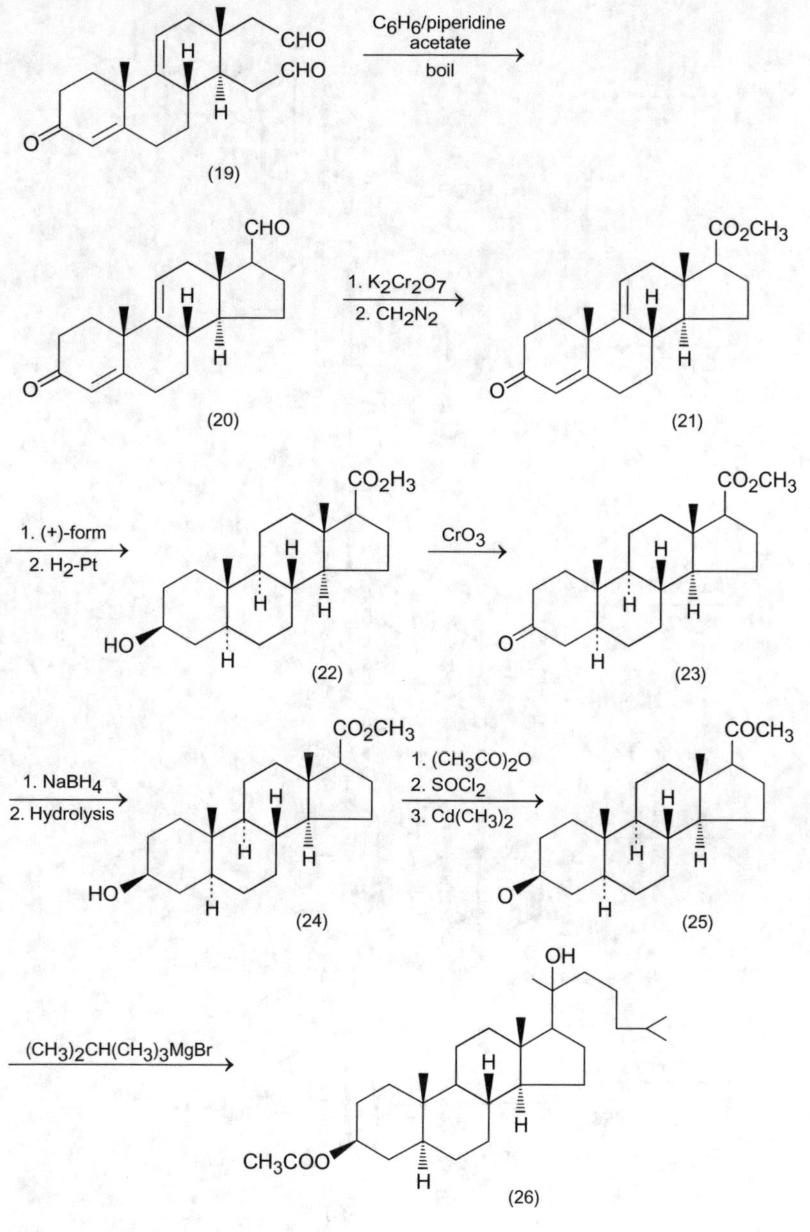

**Fig. 18.93** (continued)

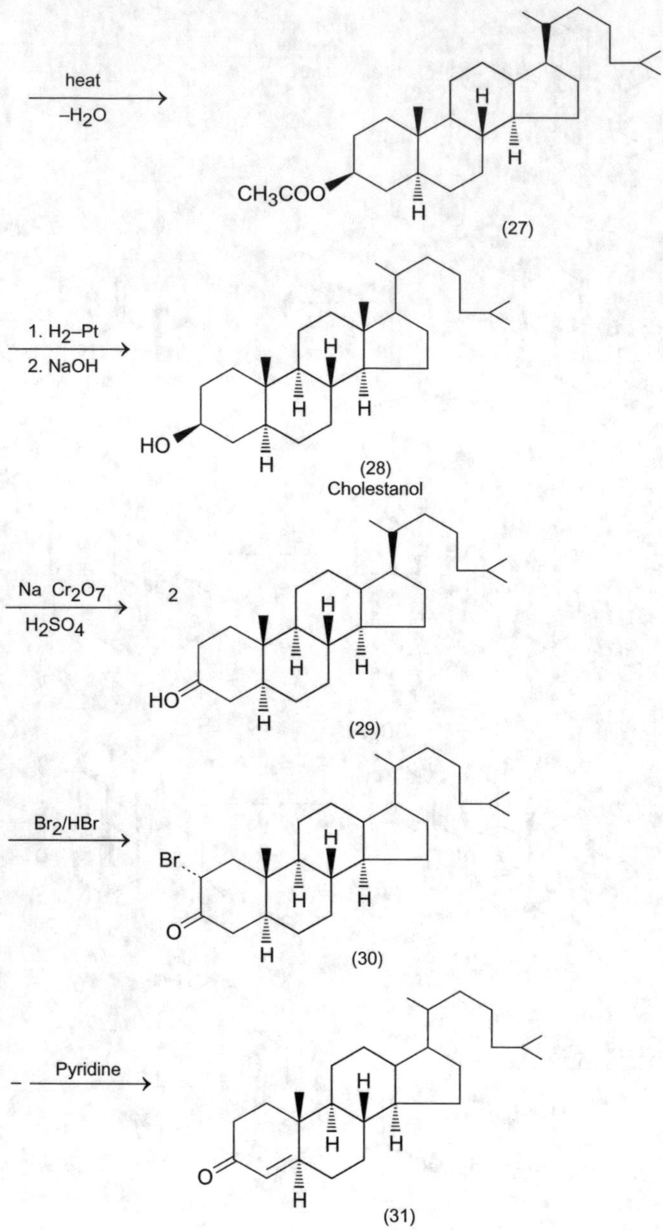

**Fig. 18.93** (continued)

**Fig. 18.93** (continued)

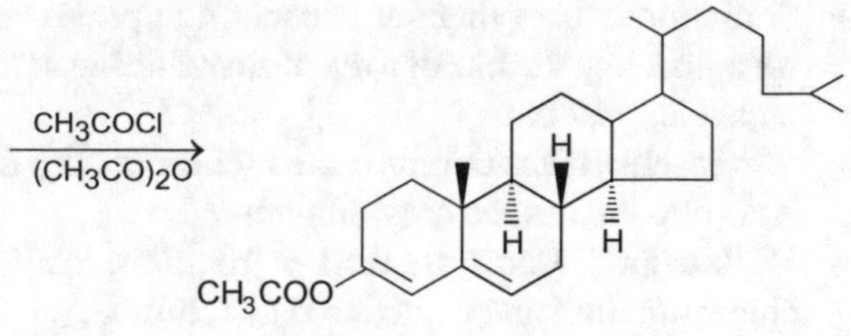
CH3COCl
(CH3CO)2O
H
H
H
CH3COO

(32)

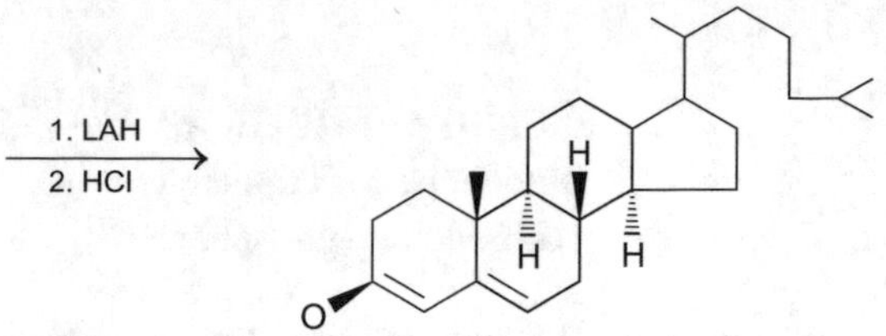
1. LAH
2. HCl
H
H
H
O

(33) Cholesterol

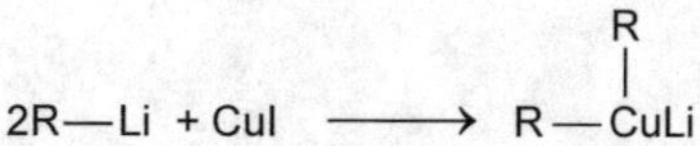
2R—Li + CuI
R
R—CuLi

- **Grignard Reagents**: These are organo magnesium halides and are prepared by the reaction of an organic halide with magnsium in presence of ether.
- **Hydroboration**: Formation of organoboranes by the reaction of alkenes and alkynes with hydroboration reagent like diborane.
- **Lindlar's Catalyst**: Palladium metal deposited on calcium carbonate and conditioned with lead acetate and quinoline. It reduces alkynes to Z-alkens.
- **Protonolysis**: A term used in reaction of organoboranes and refers to cleavage of the C–B bond; the boron atom is replaced with proton. Usually, carboxylic acids are used for this purpose.
- **Prochiral Center**: A number of molecules which do not have a chiral center but are such that on reaction with a reagent, a chiral centre in generated. The group that is involved in the generation of a chiral center is called a prochiral center.
- **(R)-Prophos**: These are chiral hydrogenation catalysts derived from transition metals. A typical example is a rhodium complex, (R)-1, 2-bis (diphenyl phosphino) propanol, called (R)-prophos

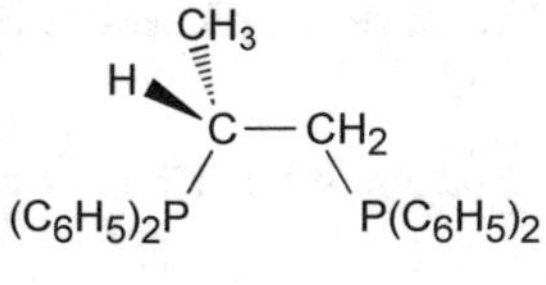
CH3
H
C—CH2
(C6H5)2P
P(C6H5)2

- **Sharpless Reagent:** It is titanium tetraiso propoxide, $[Ti[OCH(CH_3)_2]_4$.

- **Sonication:** Carrying out a reaction in presence of ultrasonic waves. It is useful for enhancing the rate of a reaction including activation of metals like magnesium, nickel, copper etc.
- **Stereoselective reduction:** Reduction of alkynes in which only one alkene (E or Z) is obtained as the only product.
- **Wilkinson's Catalyst:** It is a rhodium complex, tris (triphenyl phosphine) chlororhodium and is prepared as follows

$$RhCl_3 3H_2O + (C_6H_5)_3P \xrightarrow[\text{boil}]{C_2H_5OH} [(C_6H_5)P]_3RhCl$$

| Rhodium | Triphenyl | Wilkinson |
| chloride | phosphene | catalyst |

It is used for the homogeneous hydrogenation of non conjugated alkenes and alkynes.

## Problems

1. What product is obtained by the reaction of 2,2-dimethyl oxirane with

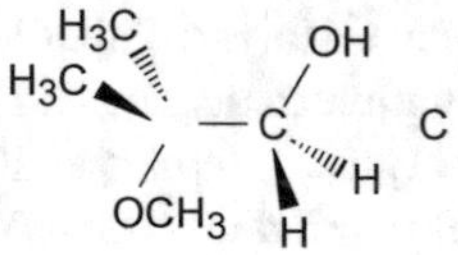

(i) $^-OCH_3$ | $H_2O$ and (ii) $CH_3OH$ | $H_2SO_4$

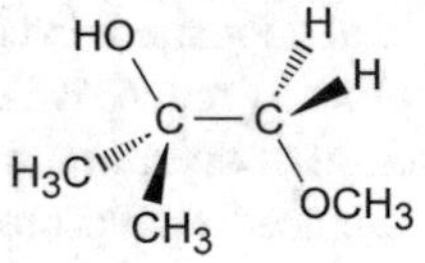

2. What products are obtained by the addition to hydrogen to an alkene as shown below:

    (i)   Using 2 equivalent hydrogen.
    (ii)  Using 1 equivalent hydrogen.
    (iii) Using 1 equivalent hydrogen in presence of Lindlar's catalyst.
    (iv)  Using 1 equivalent hydrogen in presence of P-2 catalyst.
    (v)   Using 1 equivalent hydrogen in presence of sodium and liquid **ammonia** (give the mechanism).

3. What product is expected to be obtained in the folowing reduction under different conditions:

    (i)   $HC{\equiv}CCH_2CH_2CH_3$
    (ii)  $CH_3C{\equiv}CCH_2CH_3$

4.  What product is expected to be obtained in the reduction of the following compound with Na–NH$_3$?

$$CH_3CH = CH - CH_2 - C \equiv C - CH_3$$

5.  In the following conversion, what reagent can be used?

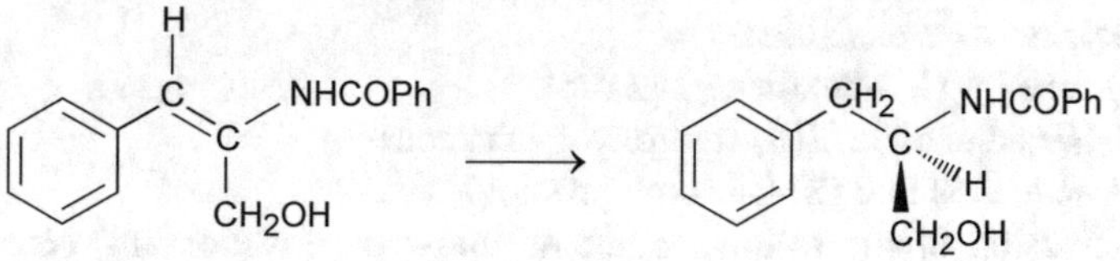

**Hint:** The alkene is substituted with both aryl and amido group and since it is allylic alcohol, the catalyst used in Rh(·P–·P)$_2$ (CH$_3$OH$_2$)$_2^+$ | H$_2$ | CH$_3$OH. However the catalyst (BINA·P)Ru(OAC)$_2$ | H$_2$ | CH$_3$OH can also be used.

6.  For the preparation naproxen (a pain killer) (Structure given below), what is the starting material and the catalyst used?

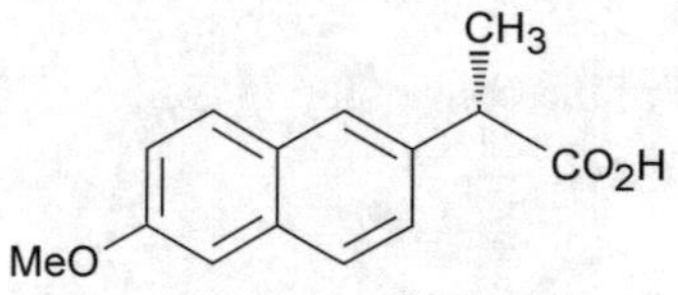

(Naproxen)

**Hint:** The starting material and the catalyst are

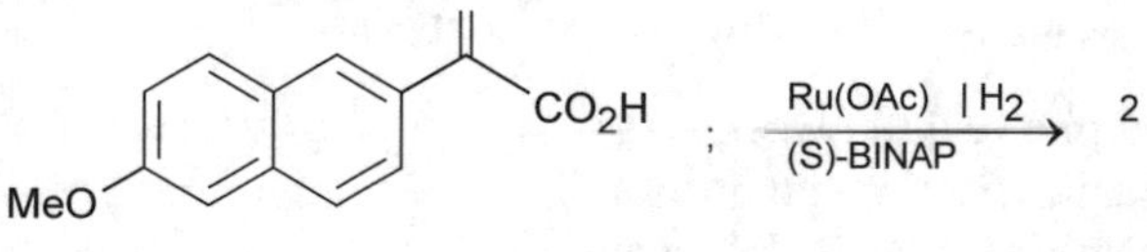

7.  Using hydroboration-oxidation, how will you carry out the following conversion:

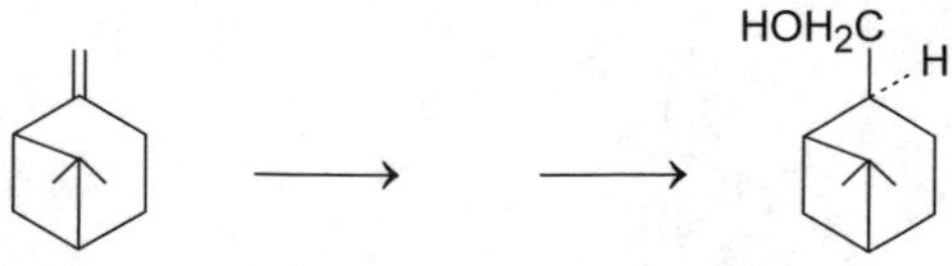

8.    What products are obtained is the hydroboration-oxidation of:

(*i*)  $CH_3CH=CH_2$  (*ii*) 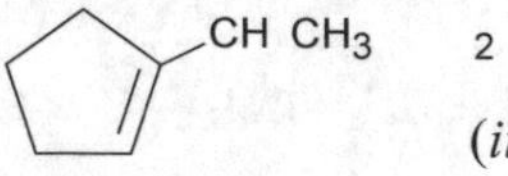    (*iii*)

9.    How following conversion are effected

(i)     Alkynes into *cis* alkenes:
(ii)    Alkynes into *trans* alkenes.
(iii)   Benzene into 2,4-hexdiene 1, 2-diol.
(iv)    2,4-Hexadiene into 2(S)-hydroxy-1-hexanone
(v)     D-Lactic acid into (S)-2-bromo propionic acid
(vi)    Conversion of the following alkene into the corresponding *cis* and *trans* epoxides

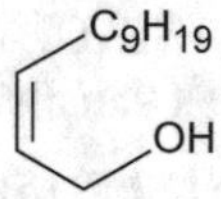

(g)

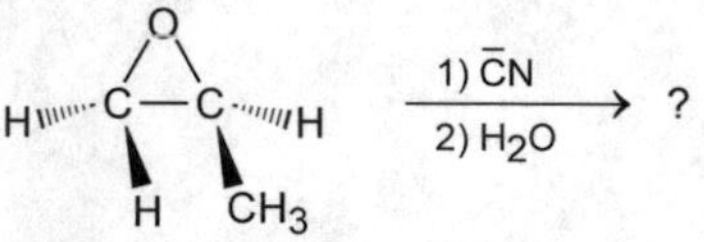

(h)    Give the mechanism of the conversion in (g).

(i)    $Ni(OCOCH_3)_2 + NaBH_4 \xrightarrow{\text{alcohol}}$ ?

(j)

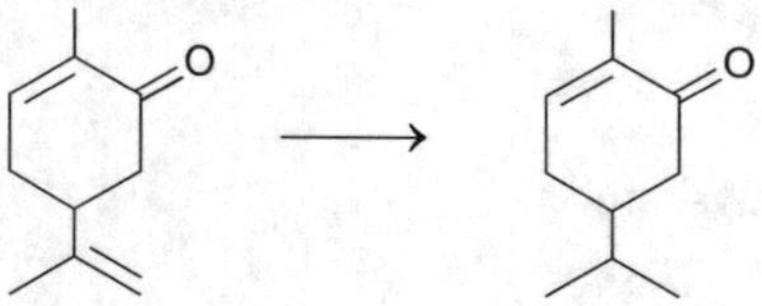

10.   Write notes on:

(a)    Sharpless chiral reagents
(b)    Enantioselective epoxidation
(c)    Enantioselective hydrogenation
(d)    Enantioselective hydroboration
(e)    Gilman's Reagents
(f)    Enantioselective reduction of carbonyl groups

    (g)    Prelog's rule

    (h)    Selectivity in hydride reductions

    (i)    Prochiral Centre

    (j)    Hydroxylation of aromatic rings.

# Chapter 19
# Enantioselective-Stereoselective Organic Reactions

Most of the organic reactions give normal products. However, there are certain reactions which give enantioselectivity in the formed product or there is induction of asymmetry. Some of the important such reactions are discussed below.

## 19.1 Aldol Reaction

In simple form, aldol reaction involves self-condensation of aldehyde molecules in presence of base to give $\beta$-hydroxy aldehydes called aldol. Thus, acetaldehyde gives $\beta$-hydroxy butyraldehyde (aldol) (Fig. 19.1).

However, aldol reaction between two different aldehydes gives a mixture of 4 products and such a reaction has no synthetic utility.

The aldol reaction between two different molecules of an aldehyde or ketone in a protic solvent (like water or alcohol) also has no synthetic utility. Such a reaction is called crossed-aldol reaction. In such a reaction a mixture of products arising from self-condensation of aldehyde and ketone is obtained. Another reason is that a reaction between an aldehyde and a ketone gives a mixture of $\alpha$-alkyl-$\beta$-hydroxycarbonyl compound as a mixture of syn (erythro) and anti (threo isomers), which have to be separated (Fig. 19.2).

However, in mixed aldol reaction, if an aromatic aldehyde (having no $\alpha$-proton) is used, a single product is obtained (Fig. 19.3).

In the above aldol reaction (Fig. 19.3), the aldol formed undergoes rapid dehydration by the E1cb mechanism to give $\alpha$, $\beta$-unsaturated product. Since the ketone forms only a single enolate ion, then only one crossed aldol product is obtained. Another example is given in Fig. 19.4a.

© The Author(s), under exclusive license to Springer Nature Switzerland AG 2022
V. K. Ahluwalia, *Stereochemistry of Organic Compounds*,
https://doi.org/10.1007/978-3-030-84961-0_19

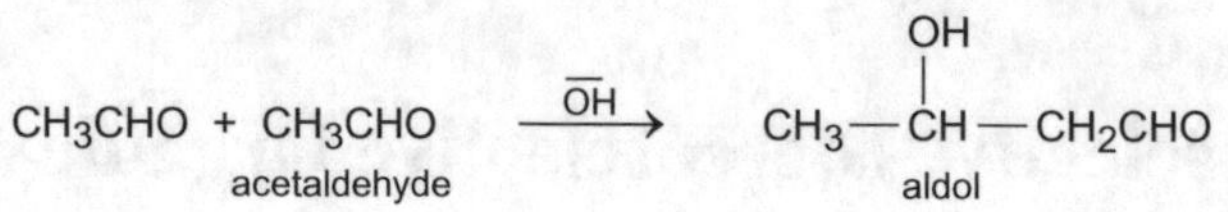

$$CH_3CHO + CH_3CHO \xrightarrow{\overline{O}H} CH_3-\underset{\underset{aldol}{|}}{\overset{\overset{OH}{|}}{CH}}-CH_2CHO$$

acetaldehyde

**Fig. 19.1**  Aldol reaction of acetaldehyde

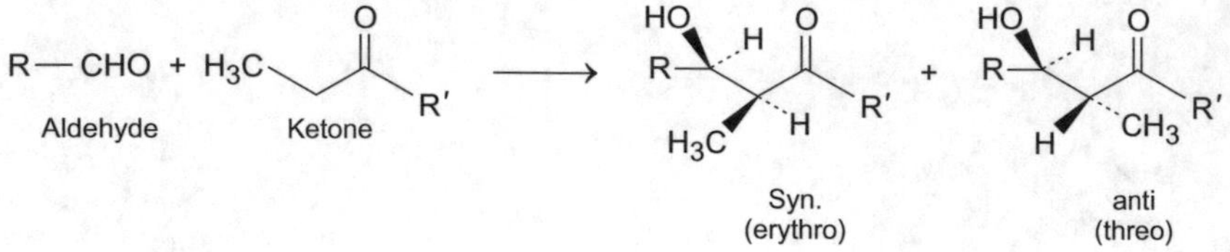

**Fig. 19.2**  Mixed aldol reaction

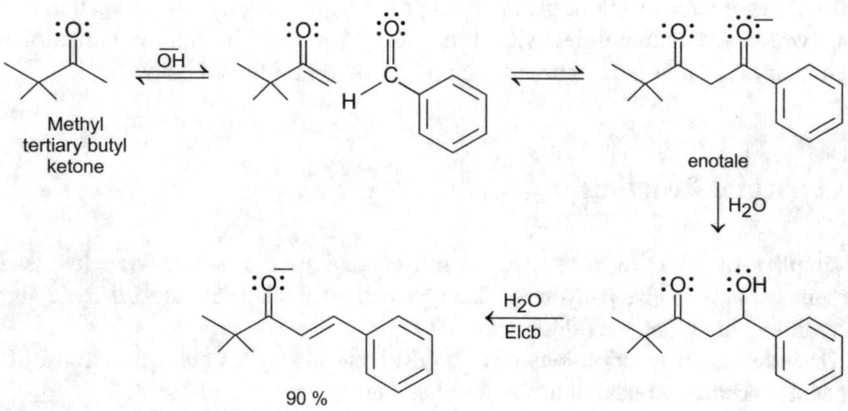

**Fig. 19.3**  Mixed aldol reaction between an aromatic aldehyde and a ketone

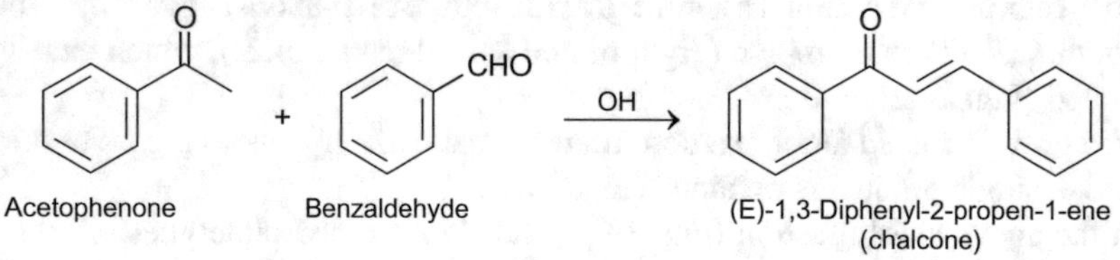

**Fig. 19.4  a** Synthesis of chalcone by mixed aldol reaction

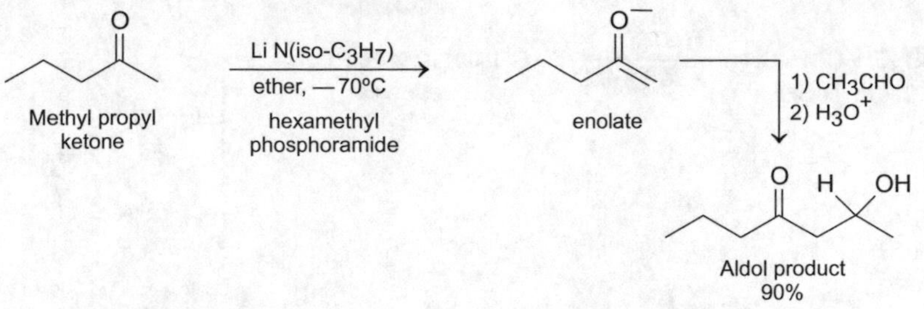

**Fig. 19.5** Directed aldol reaction

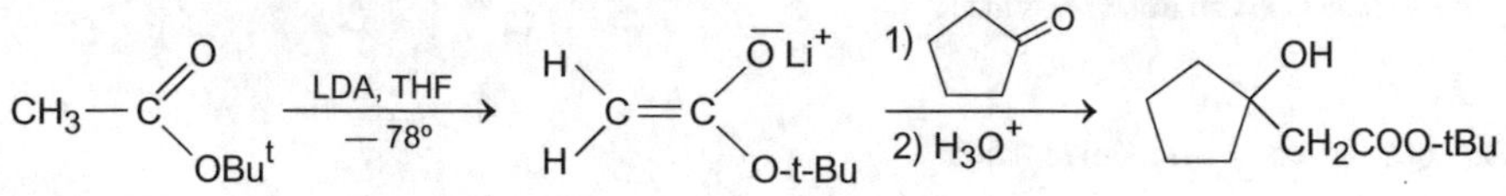

**Fig. 19.6** Directed aldol reaction between an ester and cyclic ketone

## 19.1.1 Directed Aldol Reaction

An interesting varian of aldol reaction is **directed aldol reaction** between two different carbonyl compounds to give the corresponding aldol product. This reaction is useful for the formation of a single product. It consists of the treatment of the ketone with lithium diisopropylamide to give the enolate, which reacts with acetaldehyde to give a mixed aldol product in 90% yield (Fig. 19.5).

Another example of directed aldol reaction is the reaction of an ester with a cyclic ketone (Fig. 19.6).

## 19.1.2 Stereoselective Aldol Reaction

A **stereoselective aldol reaction** is known as **Mukaiyama reaction** or **Mukaiyama aldol reaction** (T. Mukariyama, Angew. Chem. Int. Ed., Engl., 1977, **16**, 817; Organic React. 1982, **28**, 203). This reaction is carried out in presence of titanium tetrachloride. Thus, as an example, diethylketone on treatment with LDA in presence of $(CH_3)SiCl$ gives the silyl enol ether which on treatment with an aldehyde in presence of $TiCl_4$ gave the aldolate product, which on hydrolysis yields the aldol (Fig. 19.7).

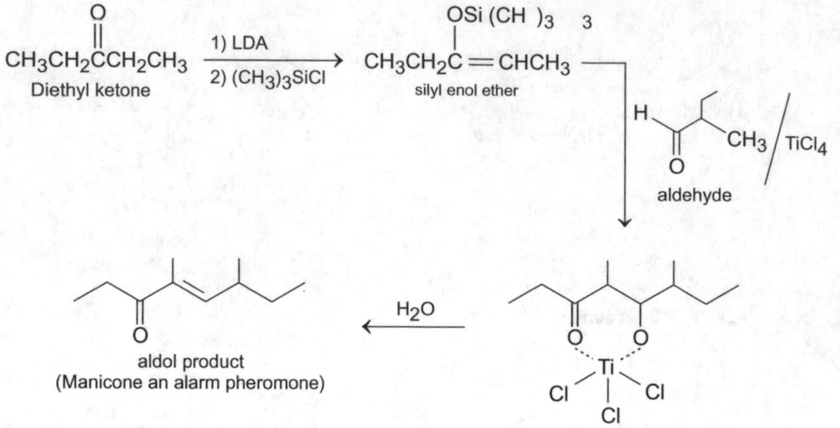

**Fig. 19.7** Mukaiyama aldol reaction

## 19.1.3 Enantioselective Aldol Reaction

It is known that the aldol reaction involving an achiral aldehyde and an achiral ketone gives a mixture of syn and anti-aldol products (Fig. 19.8).

However, the aldol reaction involving a chiral aldehyde and an achiral enolate gives aldol products (C. Heathcock, C.T. White, J.J. Morrison and D. Van Derveer, J. Org. Chem. 1981, **46**, 1296; P. Fellmann and J. E. Dubois, Tetrahedron, 1978, 34, 1349; J. E. Dubois and P. Fellmann, Tetrahedron Lett., 1975, 1225). The chiral centre in aldehyde having S configuration is retained as the aldol product. The aldol products are SRS diasterermer and SRR diastaromer (Fig. 19.9).

The aldol condensation of an achiral aldehyde with a chiral enolate give the aldol product. If the chiral centre has R configuration, the aldol products will be RSR diasteromer and SRR diasteromer (Fig. 19.10).

In case both the aldehyde and the enolate are chiral, configuration of the stereocentre of the aldehyde and the enolate is retained. The aldol products will be S RSR and SSRR diasteromers (Fig. 19.11).

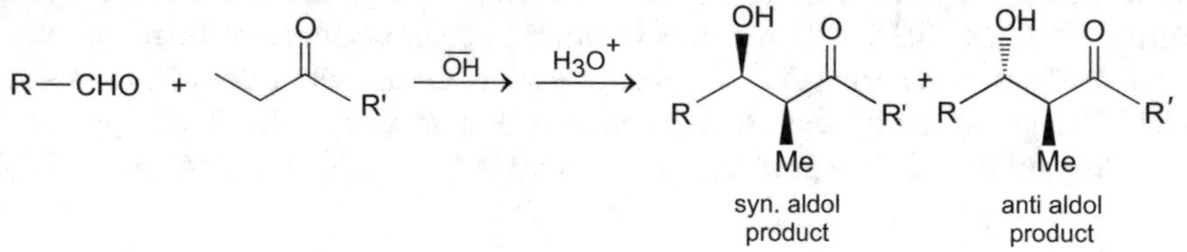

**Fig. 19.8** Aldol reaction of an achiral aldehyde and an achiral ketone

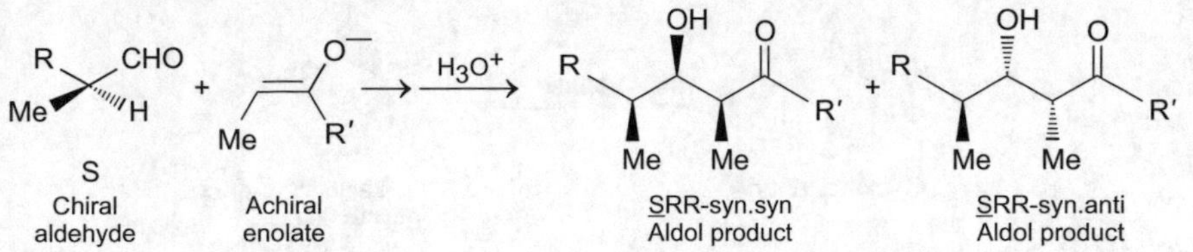

**Fig. 19.9** Aldol reaction between a chiral aldehyde and an achiral enolate

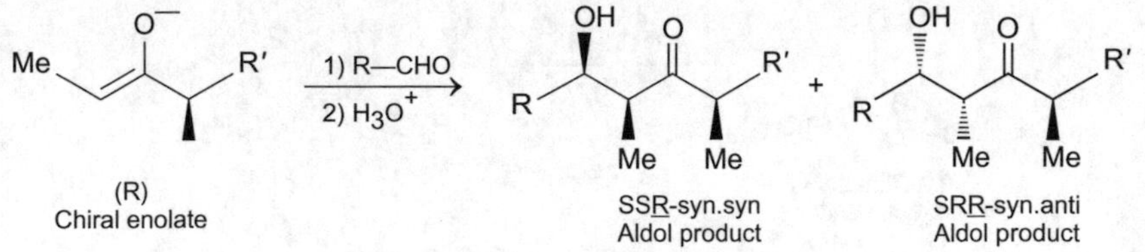

**Fig. 19.10** Aldol reaction between an achiral aldehyde with chiral enolate

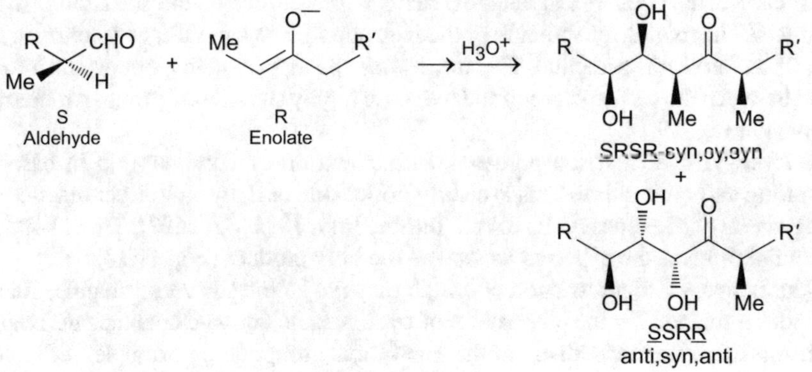

**Fig. 19.11** Aldol reaction between a chiral aldehyde and chiral enolate

## 19.2 Baeyer–Villiger Reaction

Normally this reaction involves the conversion of a ketone to an ester (Fig. 19.12) by treatment with peracid.

Baeyer Villiger reaction is especially useful for the conversion of cyclic ketones to lactones (cyclic esters) (Fig. 19.13).

**Fig. 19.12** Simple form of Baeyer–Villiger reaction

$$C_6H_5COCH_3 \xrightarrow[25°C]{C_6H_5CO_3H, \ CHCl_3} C_6H_5OCOCH_3$$

Acetophenone                    Phenyl acetate

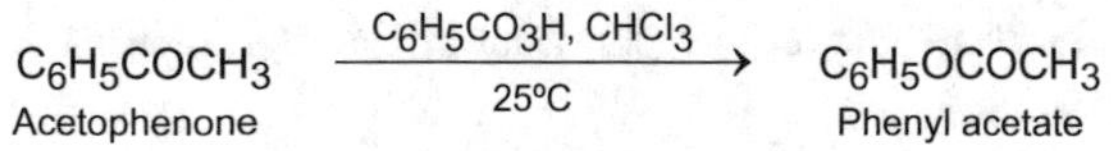

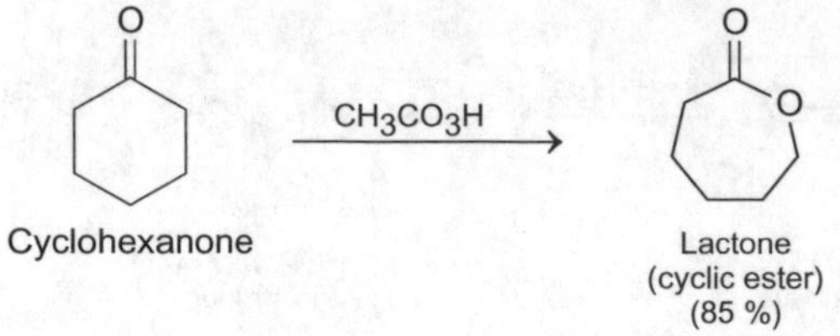

**Fig. 19.13**  Conversion of cyclic ketones into lactones

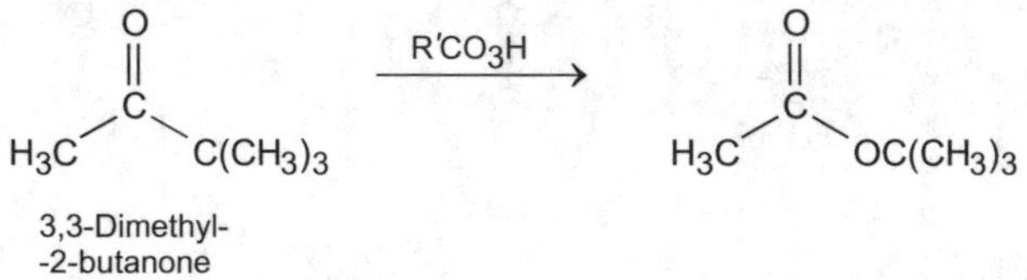

**Fig. 19.14**  Baeyer–Villiger oxidation of 3,3-dimethyl-2-butanone

In case of unsymmetrical ketones, there is possibility of the formation of two products. The migratory aptitutde of the R group in Baeyer–Villiger reaction is H > 3° > 2° and aryl > 1° > methyl. This means that the new oxygen atom inserts into the bond between the carbonyl group and the more highly substituted group. An example is given in Fig. 19.14.

Baeyer–Villiger reaction proceeds with retention of configuration in case the migrating carbon is chiral. This is clearly noticeable in Baeyer–Villiger reactions of bridged bicyclic ketones (G.R. Know, Tetrahedron, 1981, **37**, 2697). Thus, 1-methyl norcamphor gives the expected lactone as the only product (Fig. 19.15).

The Baeyer–Villiger reaction of bridge bicyclic ketones is very valuable, since it provides a method for the preparation of cyclohexane and cyclopentane derivatives with control of stereochemistry of the substituent group. As an example, the lactone (A), which is an important synthon in the synthesis of prostaglandins, is obtained by the following sequence (Fig. 19.16).

An interesting example of Baeyer–Villiger reaction of a compound containing a chiral centre is given below (Fig. 19.17).

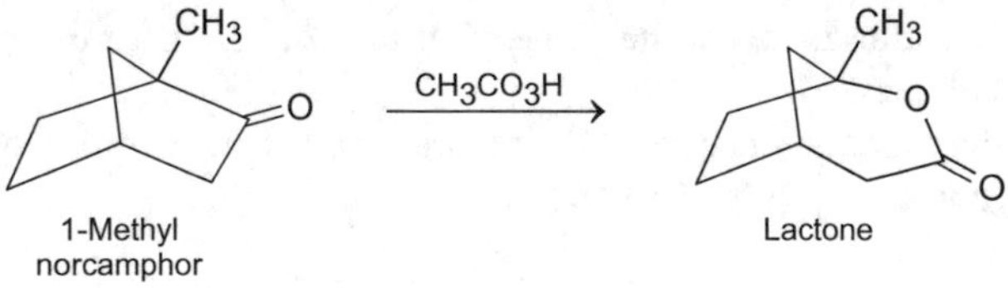

**Fig. 19.15**  Baeyer Villiger reaction of 1-methyl norcamphor

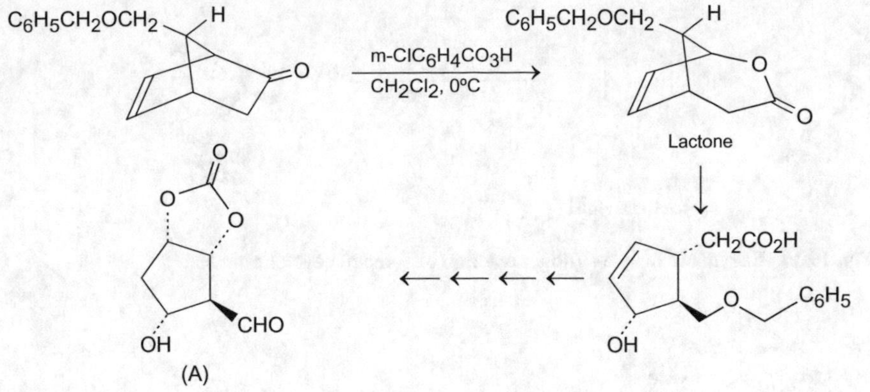

**Fig. 19.16** Baeyer–Villiger reaction of bridged bicyclic ketone

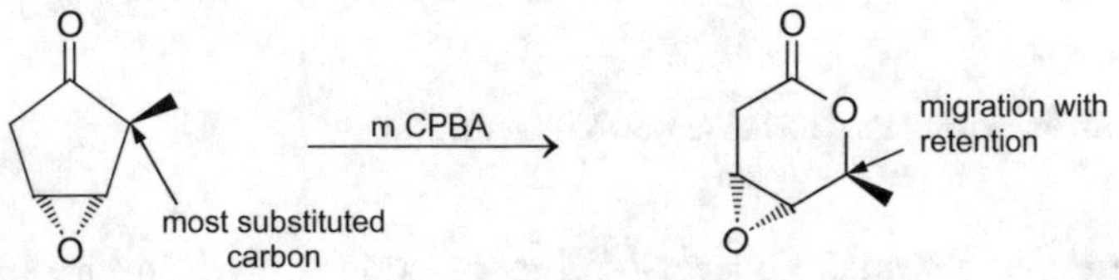

**Fig. 19.17** Baeyer–Villiger reaction involving retention of the configuration of the chiral centre

## 19.2.1 Enzymatic Baeyer–Villiger Reaction

The enzymatic conversion of ketones to esters is commonly achieved by using enzymes (C.J. Sih and J.P. Rosazza). In applications of biochemical systems in organic chemistry (J.B. Jones, C.J. Sih and D. Perlan, Eds. Wiley, New york, 1976, Part II, PP 100–102), a typical transformation is conversion of cyclohexanone using a purified cyclohexanone oxygenase to give lactone (C.C. Ryerson, D.P. Ballou and C. Walsh, Biochemistry, 1982, **21**, 2644; N. A. Donoghue, D.B. Norris and P. W. Trudgill, Eur. J. Biochem., Biochem., 1976, **63**, 175) (Fig. 19.18).

However, 4-methyl cyclohexanone on enzymatic Baeyer–Villiger reaction gave the lactone in 80% yield with 97% ee (M.J. Taschner and Q. –Z. Chem. Biorganic and Medicinal Chem. Lett., 1991, **1**, 535) (Fig. 19.19).

**Fig. 19.18** Enzymatic Baeyer–Villiger reaction of cyclohexanone

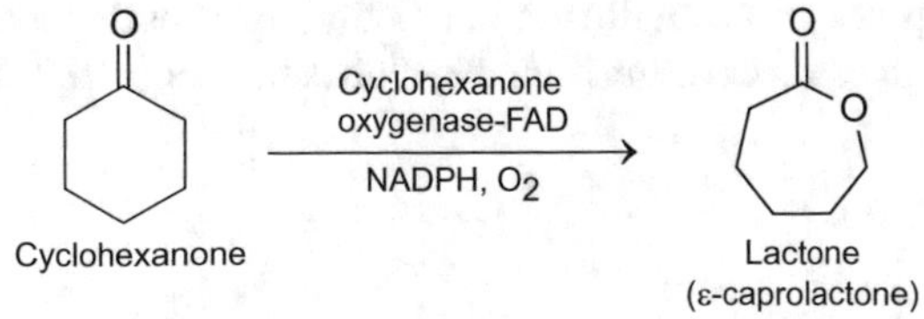

Cyclohexanone

Cyclohexanone oxygenase-FAD

NADPH, O$_2$

Lactone ($\varepsilon$-caprolactone)

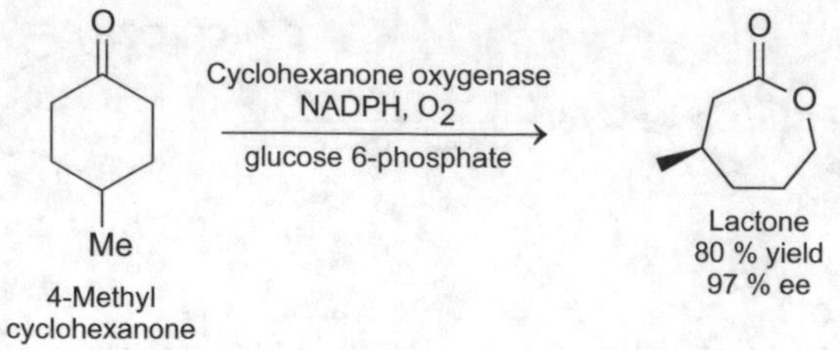

**Fig. 19.19**   Enzymatic Baeyer–Villiger reaction of 4-methyl cyclohexanone

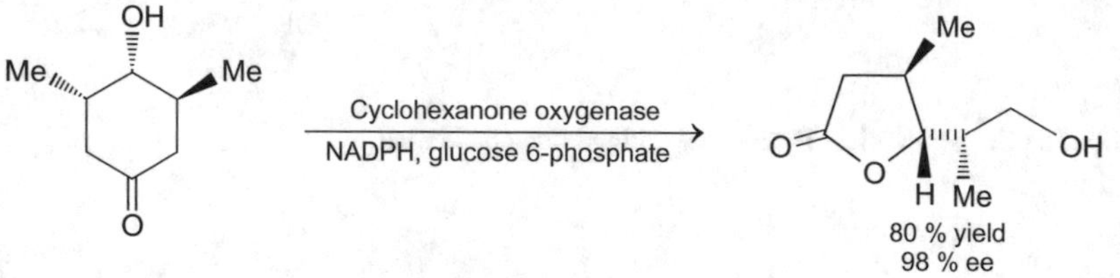

**Fig. 19.20**   An interesting Enzymatic Baeyer–Villiger reaction

An interesting enzymatic Baeyer–Villiger reaction is given in Fig. 19.20 (M.J. Taschner and D.J. Black, J. Am. Chem. Soc., 1998, **110**, 6892).

## 19.3   Diels Alder Reaction

Diels Alder reaction is a **Cycloaddition reaction** (a type of **pericyclic reaction**), discovered by Otto Diels and Kurt Alder is the 1920s, who were award 1950 Nobel Prize in Chemistry. It is one of the most versatile method for the preparation of six-membered rings. The Diels Alder reaction consist in the reaction of a conjugated diene and a dienophile to give a cyclohexene derivative. Some examples are given below (Fig. 19.21).

The presence of electron-withdrawing group as a substituent in the dienophile gives much better yields (Fig. 19.22).

The Diels–Alder reaction is also called (4 + 2) cycloaddition reaction (the numbers 4 and 2 represent the number of $\pi$ electrons in each reaction) and is a stereospecific reaction. The stereochemistry of the dienophile is retained in the formed product. As an illustration diethyl *trans* butendioate on reaction with 1, 3-butadiene gives *trans* disubstituted cyclohexene (Fig. 19.23).

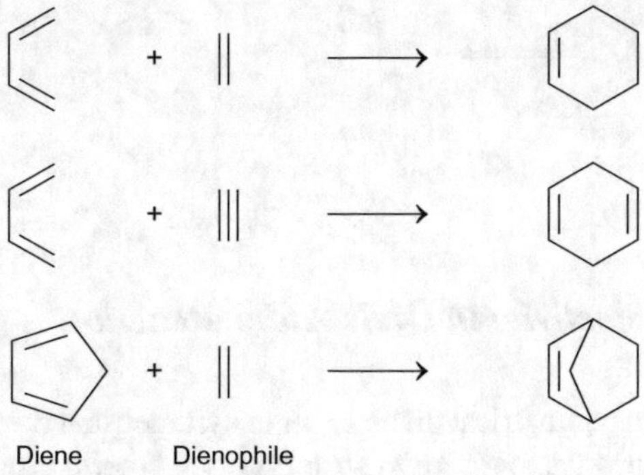

**Fig. 19.21** Some examples of Diels–Alder reaction

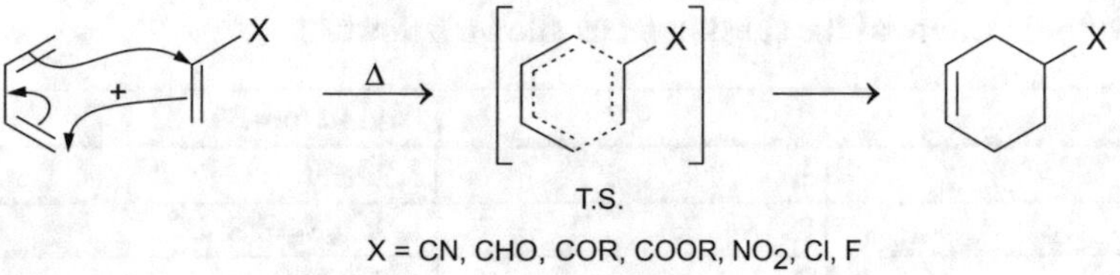

T.S.

X = CN, CHO, COR, COOR, NO$_2$, Cl, F

**Fig. 19.22** Diels Alder reaction of conjugated diene and dienophile containing electron-withdrawing substituent

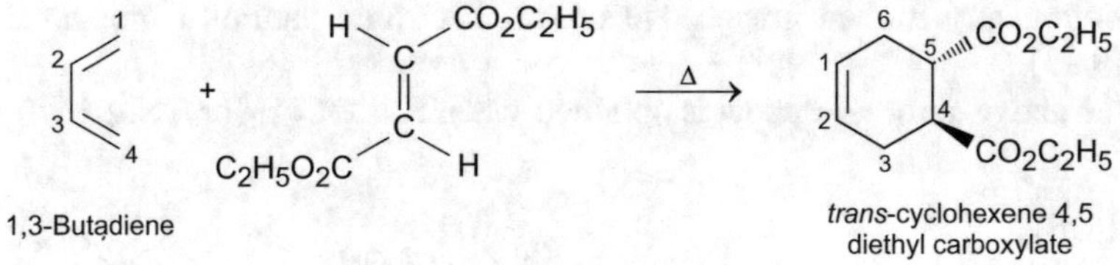

1,3-Butadiene

trans-cyclohexene 4,5 diethyl carboxylate

**Fig. 19.23** Stereospecific Diels–Alder reaction

Some of excellent dienophile include maleic anhydride $\left( \begin{array}{c} HC-C\overset{O}{\underset{O}{\diagdown}} \\ \| \quad \quad \diagup O \\ HC-C\underset{O}{\diagup} \end{array} \right)$, ,

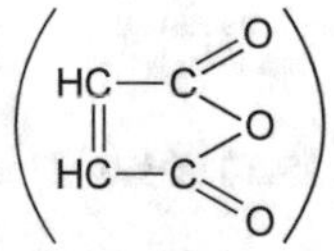

acrylonitrile (CH$_2$=CH—CN), methyl acrylate (H$_2$C=CH—$\overset{O}{\overset{\|}{C}}$—OCH$_3$) and dimethyl acetylene dicarboxylate (CH$_3$OOC—C≡C—COOCH$_3$).

The diene to be used in Diels Alder reaction must be in the S-*cis* conformation or should be able to adopt the S-*cis* conformation. This is because in S-*cis* conformation there is free rotation about the single bond. In case the diene is locked into transoid conformation as in (A), the reaction does not take place.

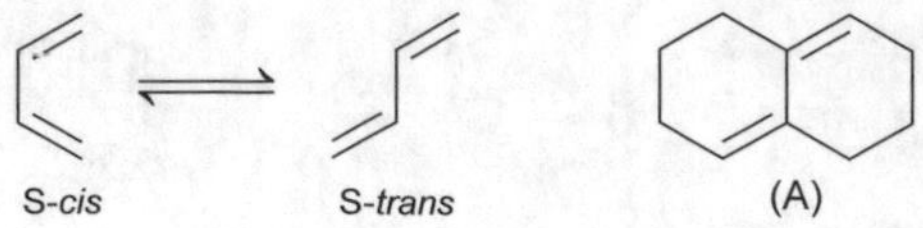

S-cis          S-trans          (A)

## 19.3.1   Regio Selectivity in Diels Alder Reaction

The addition of an unsymmetrical diene to an unsymmetrical dienophile in principle can take place in two ways to give two structurally isomeric products (Fig. 19.24).

In practice, it is found that the formation of only one of the isomers is favoured (J. Saver, Angew. Chem. internal. edn., 1967, **6**, 16). Thus, in the addition of acrylic acid derivative to 1-substituted butadiens, the 1, 2-adduct is favoured irrespective of the electronic nature of the substituent as shown below:

| R | R' | Product formed |
|---|---|---|
| NEt$_2$ | H | 1, 2 only |
| CH$_3$ | H | 1, 2: 1, 3 = 18: 1 |
| CO$_2$H | H | 1, 2 only |
| CMe$_3$ | CMe$_2$ | 1, 2: 1, 3 = 0.9: 1 |

In the addition of methyl acylate to 2-substituted butadiene, the 1,4-adduct is predominantly formed irrespective of the electronic nature of the substitutent (Fig. 19.25).

In the above, only 1,4-adduct is obtained when R is OC$_2$H$_5$ or CN.

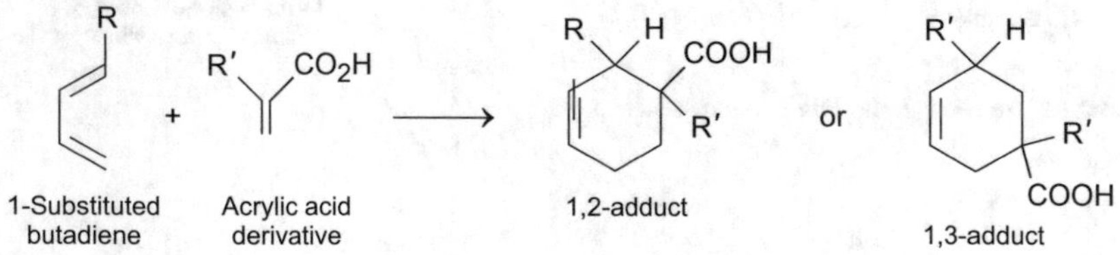

1-Substituted      Acrylic acid      1,2-adduct                      1,3-adduct
butadiene          derivative

**Fig. 19.24**  Diels–Alder reaction of 1-substituted butadiene- and acrylic acid derivatives

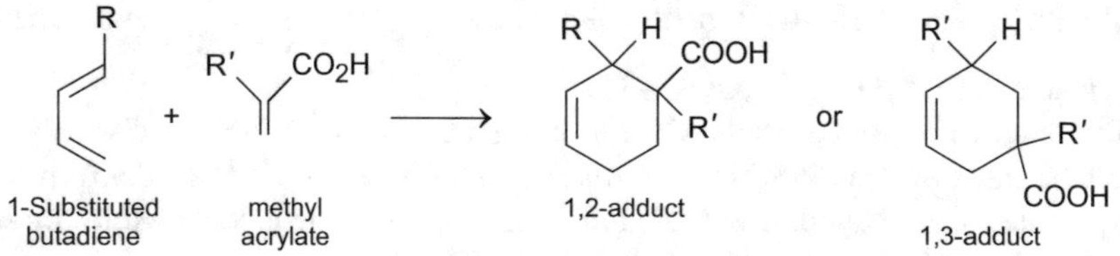

1-Substituted      methyl         1,2-adduct                      1,3-adduct
butadiene          acrylate

**Fig. 19.25**  Diels–Alder reaction of 2 substituted butadiene and methyl acrylate

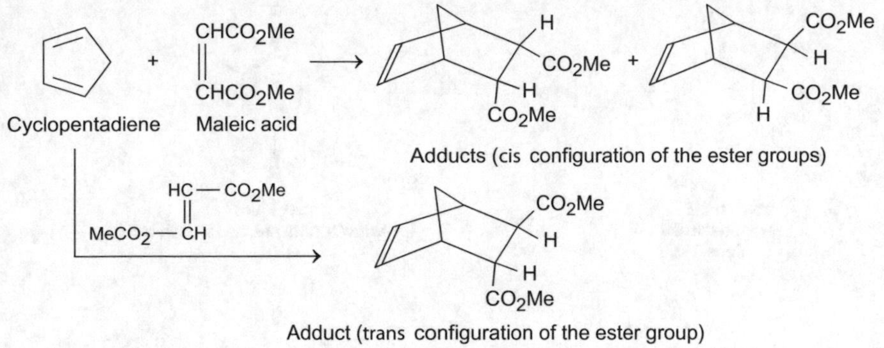

Fig. 19.26  Stereospecific Diels–Alder reaction

## 19.3.2  Stereoselectivity in Diels–Alder Reaction

The usefulness of Diels–Alder reaction is due to its remarkable stereoselectivity. The stereochemistry of the adduct obtained in a number of Diels–Alder reaction can be determined on the basis two emprical rules (the *cis* principle and the *endo* addition rule) formulated by Alder and Stein (1937).

### 19.3.2.1  *Cis* Principle

According to the '*cis* principle', the relative stereochemistry of substituent in both the diene and dienophile is retained in the adduct. As an example, the reaction of cyclopentadiene with dimethyl maleate, the *cis* adducts are formed. However, in the reaction with dimethyl fumarate, the *trans* configuration of the ester groups is retained in the adduct (Fig. 19.26).

In a similar way, the relative configuration of the diene is retained in the adduct. Thus, a *trans, trans* disubstituted dienes give adduct in which substituents in 1- and 4 positions are *cis* to each other and a *cis, trans*-disubstituted diene give adducts with *trans* substitutents (Fig. 19.27).

The *cis* principle is widely applicable and is followed.

### 19.3.2.2  Endo Addition Rule

In the case of Diels–Alder reaction of cyclopentadiene and maleic anhydride, two different products, the *endo* and the *exo*, are possible to be formed in principle. According to the endo addition rule, in the Diels–Alder reaction of maleic anhydride and cyclopentadiene, the *endo* product is formed from orientation of the two reactants with maximum accumulation of double bonds. Thus, the *endo* adduct is produced

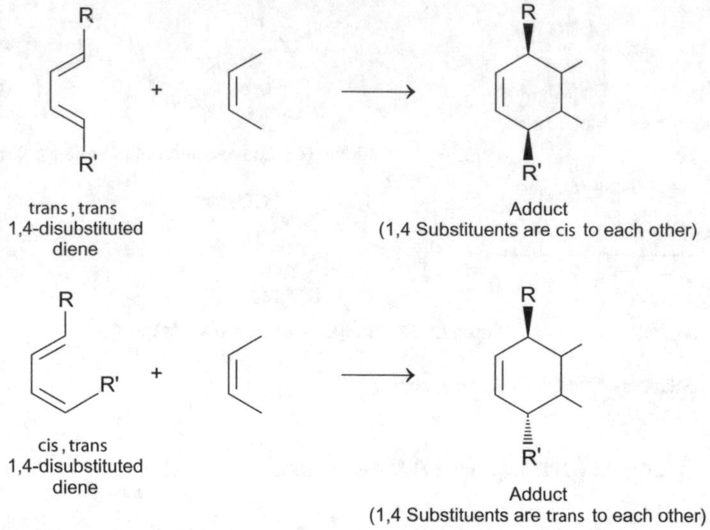

**Fig. 19.27** Diels–Alder reaction of *trans, trans* and *cis, trans*-1, 4-disubstituted dienes with alkenes

exclusively. However, the thermodynamically more stable *exo* adduct is formed in less than 1.5% (Fig. 19.27).

The *endo* rule is strickly applicable only to the Diels–Alder reaction of cyclic dienophiles and cyclic dienes. However, it is also a useful guide in many other cases.

### 19.3.3  Catalystic Diels–Alder Reaction

A number of Diels–Alder reactions are found to be remarkably accelerated in presence of aluminium chloride and other Lewis acids like boron trifluoride and tin (IV) chloride (P. Yates and P. Eaton, J. Am. Chem. Soc., 1960, **82**. 4436; J. Sauer, Angew. Chem. Internal. Edn., 1967, **6**, 16). As an example, the Diels–Alder reaction of anthracene and maleic anhydride in methylene chloride solution in presence of aluminium chloride gave quantitative yield of the adduct in 90 s at room temperature; the reaction in the absence of catalyst required 4800 h. for 95% conversion. Similarly, butadiene and methyl vinyl ketone reacted in 1 h at room temperature in presence of SnCl$_4$ giving 75% yield of the adduct (acetylcyclohexene). No reaction takes place in the absence of a catalyst.

A number of other examples have been recorded in the use of catalysts in Diels–Alder reaction (W. Kreiser, W. Haumesser and A. F. Thomas, Helv. Chim. Acta, 1974, **57**, 164). Two such examples are given in Fig. 19.29.

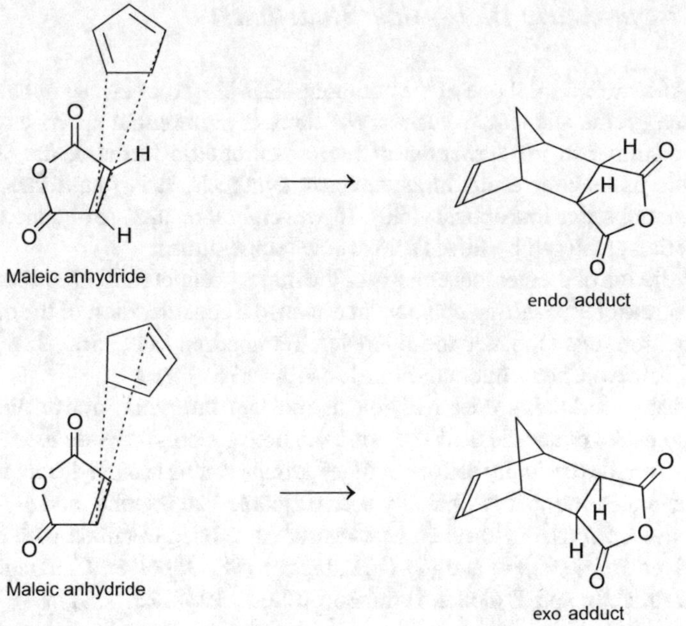

**Fig. 19.28** Formation of endo and exo adducts in Diels Alder reaction of maleic anhydride and cyclopentadiene

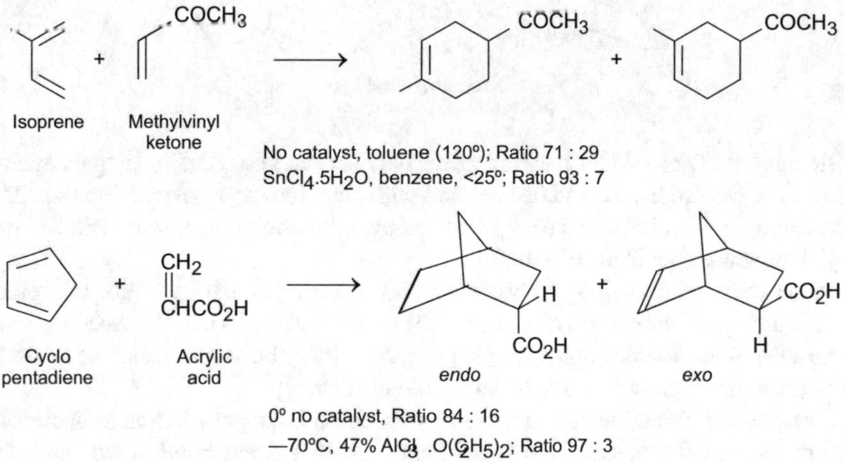

**Fig. 19.29** Some examples of Diels–Alder reaction in presence of catalysts

## 19.3.4  Asymmetric Diels–Alder Reaction

In Diels–Alder reaction, if one of the components is chiral, a new asymmetric centre is generated (in the adduct). In such a case there is preferential approach from one direction resulting in the formation of two enantiomeric forms of the new chiral centre. This procedure results in asymmetric synthesis. It is found that the usual thermal reactions give low optical yields. However, best results are obtained in Diels–Alder reaction catalysed by Lewis acid at low temperatures.

The prediction of the stereochemistry of the major product formed can be predicted from that of the chiral starting component and also a consideration of the orientation of the transition state (J. Sauer and J. Kredel, Tetrahedron Lett, 1966, 731, 6359; W. Oppolzer, Angew. Chem. Internal. Edn. 1984, **23**, 876).

In Asymmetric Diels–Alder reaction in most of the work, active dienophiles, particularly esters of acrylic acid and optically active alcohols, were used. After the reaction is over, the optically active auxiliary groups can be removed from the adduct and used again. A number of optically active alchols like menthol and (–)-8-phenyl menthol have been used. However, best results have been obtained with neopentyl esters derived from (R)-(+)-and (S)-(–) camphor (W. Oppolzer, C. Chapius, G.M. Dao, D.R. Reichlin and T. Godel, Tetrahedron Lett., 1982, **23**, 4781).

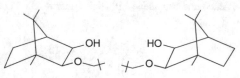

Neopentyl esters derived from
(R)-(+) and (S)-(–) camphor

In catalytic Diels–Alder reaction of derived acrylate ester and cyclopentadiene, the (2R) or (2S) adducts could be obtained with complete asymmetry induction. The chiral auxiliary alcohol could be regenerated by reduction of the adduct with LiAlH$_4$ to give pure endo alcohol (Fig. 19.30).

In the above reaction (Fig. 19.30) it is believed the asymmetric reaction take place by the addition of the diene to the ester in the conformation (A). In this case access to the re-face of the double bond is hindered by the tertiary butyl group and the addition takes place preferentially from the back side exclusively.

Asymmetric induction also take place with chiral ketols in addition to cyclopentadiene even in absence of a Lewis acid catalyst. If R is tert butyl in the ketol, the reaction occurs at –20 °C to give endo product with diastereotopic selectivity of more than 100: 1 (W. Choy, L. A. Reed and S. Masamune, J. Org. Chem., 1983, **48**, 1137) (Fig. 19.31).

Dienes containing optically active auxiliary group have not been widely used. An example is the reaction of (S)-O-methylmendelyl ester of 1-hydroxy-1, 3-butadiene with a quinone (Juglone) in presence to boron triacetate to give an adduct in 98%

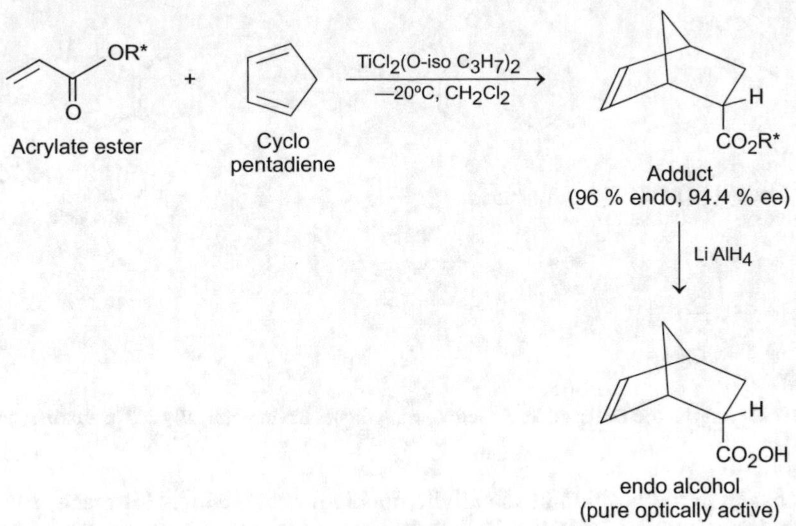

**Fig. 19.30** Asymetric Diels–Alder Reaction

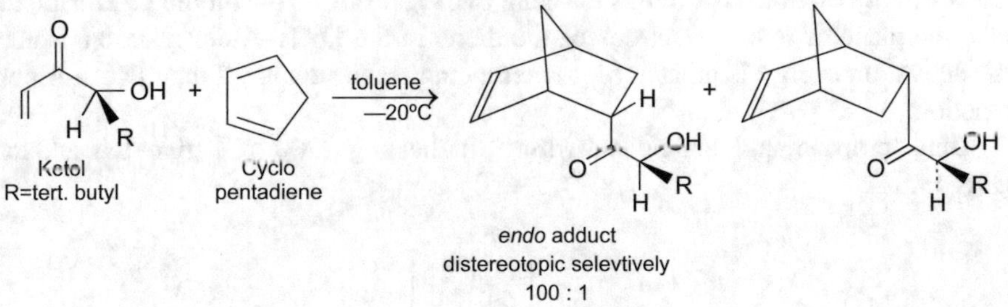

**Fig. 19.31** Assymetric Diels–Alder Reaction of an asymmetric ketol and cyclopentadiene

yield; in this case there is complete induction (B.M. Trost, D. O'Krongly and J.L. Bellatire, J. Am. Chem. Soc., 1980, **102**, 7595) (Fig. 19.32).

See also cycloaddition reaction (Sect. 14.7) in stereochemistry of pericyclic reactions.

## 19.4 Ene Reaction

It is related to Diels–Alder reaction and involves the addition of an alkene bearing an allylic hydrogen atom (instead of a diene used in Diels–Alder reaction) with a dienophile (now called an **enophile**) resulting in the formation of a new $\sigma$ bond to the terminal carbon of the allyl group; there is 1,5-migration of the allylic hydrogen

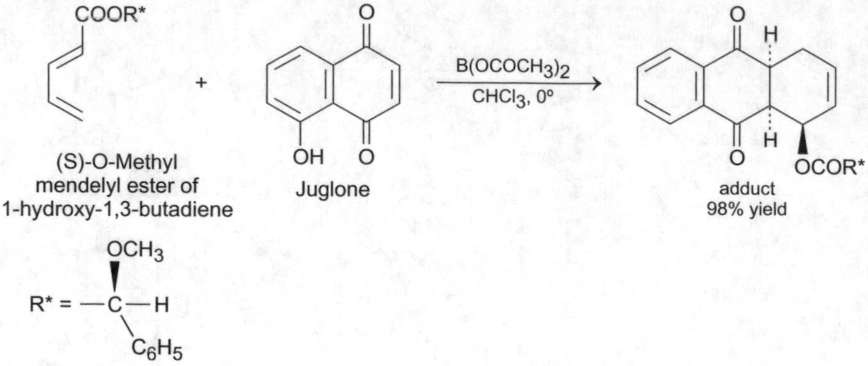

(S)-O-Methyl
mendelyl ester of
1-hydroxy-1,3-butadiene

Juglone

B(OCOCH3)2
CHCl3, 0°

adduct
98% yield

$$R^* = -\underset{\underset{C_6H_5}{|}}{\overset{\overset{OCH_3}{|}}{C}}-H$$

**Fig. 19.32** Assymetric Diels–Alder Reaction using dienes having optically active auxiliary group

and change in the position of the allylic double bond (H.M.R. Hoffmann, Angew. Chem. Internal. Edn., 1969, **8**, 556; E.C. Keung and H. Alper, J. Chem. Educ., 1972, **49**, 97) (Fig. 19.33).

The ene reaction is also a $6\pi$-electron electrocyclic reaction (as in the case of Diels–Alder reaction), but in this case the two electrons of the allylic C–H $\sigma$ bond take the place of two $\pi$-electrons of the diene in the Diels–Alder reaction. Since the activation energy is greater, higher temperatures are required than Diels–Alder reaction.

Thus, propene and maleic anhydride on heating (200 °C) give the adduct (Fig. 19.34).

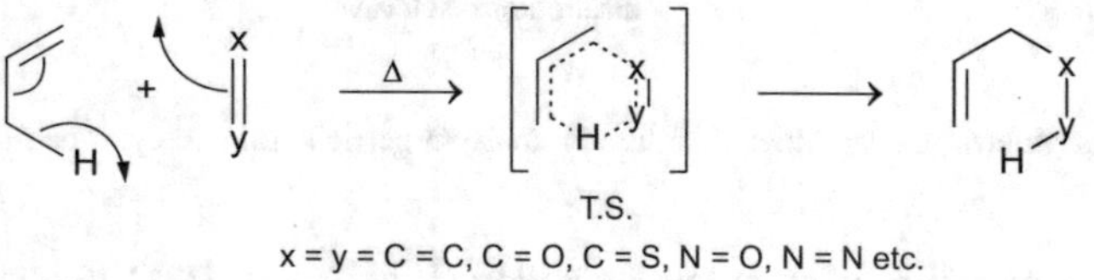

T.S.

x = y = C = C, C = O, C = S, N = O, N = N etc.

**Fig. 19.33** The representation of ene reaction

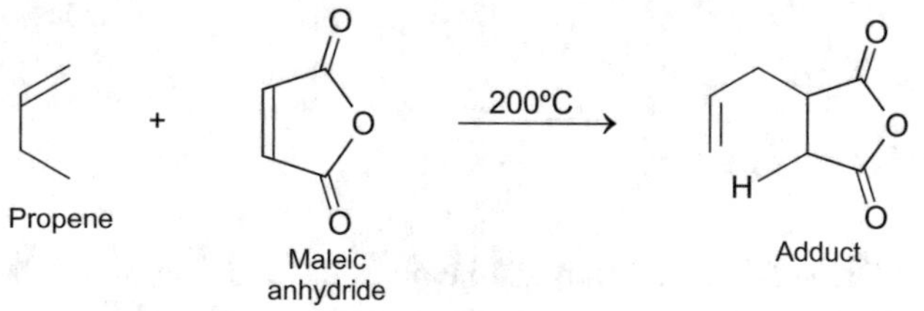

Propene

Maleic
anhydride

200°C

Adduct

**Fig. 19.34** Ene reaction of propene and maleic anhydride

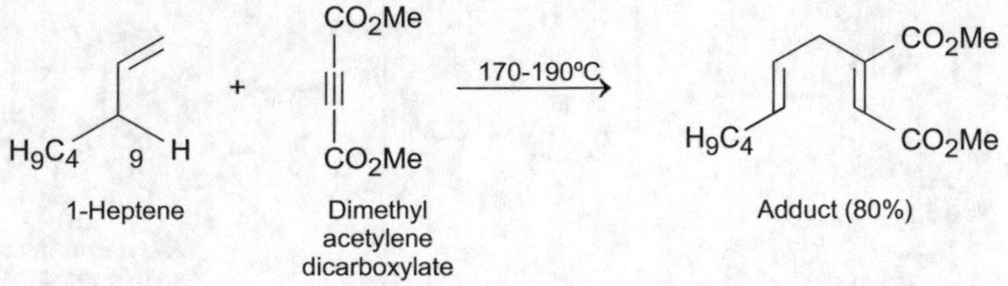

**Fig. 19.35** Ene reaction of 1-heptene with dimethyl acetylene dicarboxylate

## 19.4.1 Catalytic Ene Reaction

Like Diels–Alder reaction, the catalytic ene reaction can also take place easily at much lower temperature to give good yields of the adducts. The catalyst used are $AlCl_3$, $SnCl_4$ and $TiCl_4$. However, best results are obtained with alkylaluminium halides, which act as a proton scavengers and as Lewis acids and also help to prevent the formation unnecessary byproducts (B.B. Snider, Acc. Chem. Res., 1980, **13**, 426). As an example the reaction of methyl acrylate with 2-methyl-2-propene in presence ethylaluminium dichloride gives good yields of the adduct at 25°, without catalyst a temperature of 230 °C is required.

## 19.4.2 Stereoselectivity in Ene Reaction

The stereoselectivity in Ene reaction is similar to that in Diels–Alder reaction, and there is preference for the formation of *endo* products.

As an example Ene reaction of 1-heptene with dimethyl acetylenedicarboxylate gives the adduct (Fig. 19.35). In this case, the hydrogen atom and the alkyl residue add to the same side of the triple bond of the endophile.

## 19.4.3 Intramolecular Ene Reaction

Intramolecular ene reaction has great utility for the synthesis of cyclic compounds, particularly for synthesis of five membered rings. These reactions in a number of cases are highly stereoselective. As an example, the *cis* diene (A) on heating gave the *cis* disubstituted cyclopentane (B) with complete stereoselectivity by way of *exo* transition state (the *endo* transition state is highly strained in this case). However the *trans* diene (C) gave mainly the *cis*-cyclopentane (D); in this case the transition state is *endo* (Fig. 19.36).

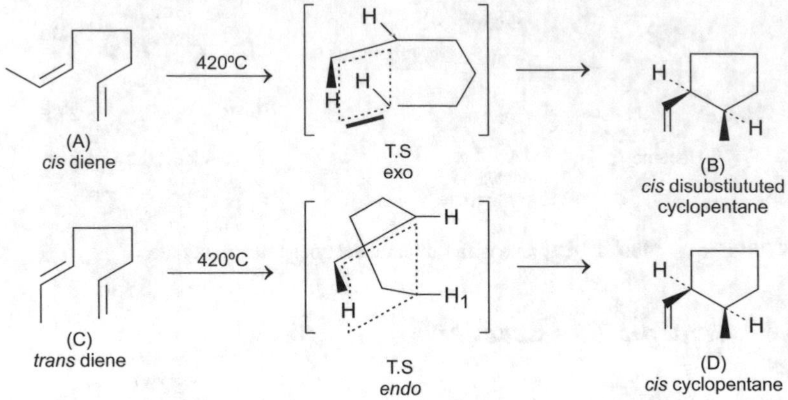

**Fig. 19.36**  Intramolecular Ene reaction

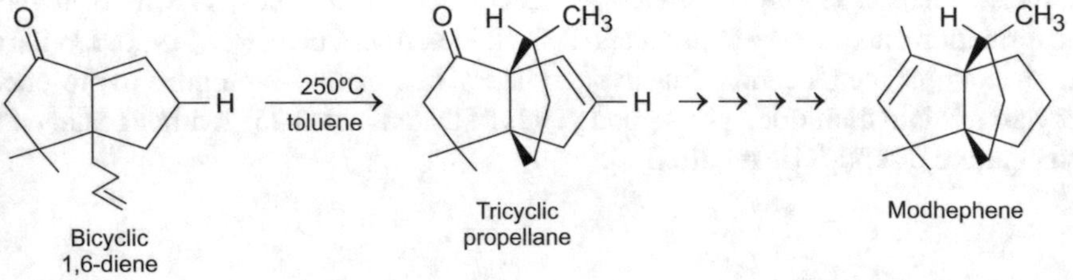

**Fig. 19.37**  Synthesis of modhephene, a sesquiterpene

The intramolecular ene reaction has been used for the synthesis of a number of natural products (W. Oppolzer, Pure and Appl. Chem., 1981, **53**, 1181). As an example, a natural product, a sequiterpene modhephene, has been synthesised by intramolecular ene reaction of bicyclic 1,6-diene to the tricyclic propellane, which has been used as a synthon to obtain modhephene (Fig. 19.37).

Besides 5-membered rings, the intramolecular ene reaction has also been useful for the synthesis of 6-membered rings by the cyclisation of 1,7-dienes. However, lower yields are obtained and higher temperature are required.

The intramolecular ene reaction has also been used to make fused, spiro and bridged ring compounds. It has been found that dienes, in which the ene component is in a ring give sterospecifically bicyclic systems in high yield. An example is synthesis of sesquiterpene alcohol (±)-acorenol; in this case the spiro[5,4] decane system is obtained as an intermediate by an ene reaction of appropriate 1,6-diene (W. Oppolzer, Helv. Chem. Acta, 1973, **56**, 1812) (Fig. 19.38).

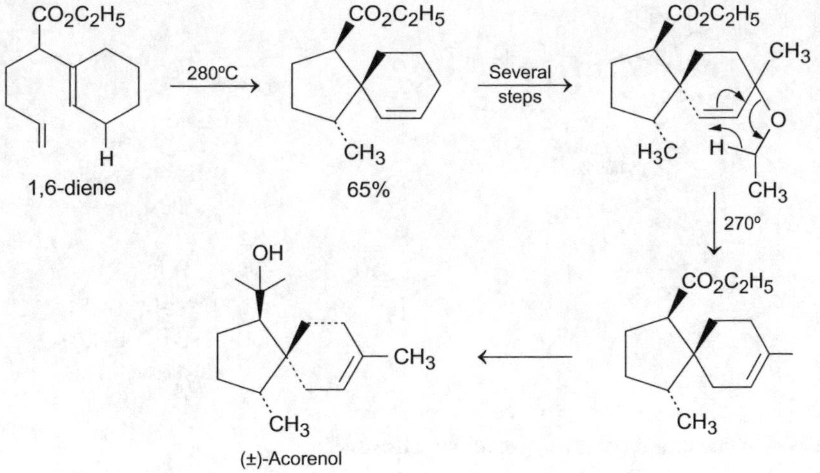

**Fig. 19.38** Synthesis of acorenol

## 19.4.4 *Chiral Ene Reaction*

Optically active products are obtained in ene reactions in which hydrogen atom transferred is attached to a chiral centre. As an example, the reaction of maleic anhydride with an optically active alkenes (A) gave adducts which are optically active (R.K. Hill and M. Rabinovitz, J. Am. Chem. Soc., 1964, **86**, 965) (Fig. 19.39).

Using optically active esters as the enophile component also gave optically active adducts. Thus, acid catalysed ene reaction of 8-phenylmethyl glyoxalate with 1-hexene give (S) alcohol in 98% diastereomeric excess (J.K. Whitesell, A. Bhattacharya, D.A. Aguilor and K. Henkle, J. Chem. Soc. Chem. Commun, 1982, 989) (Fig. 19.40).

As intramolecular ene reaction of 8-phenyl menthyl ester give an optically active adduct, which was used (as a synthon for the synthesis (+)-$\alpha$-allokainic acid (W. Oppolzer, Pure and Appl. Chem., 1981, **53**, 1181) (Fig. 19.41).

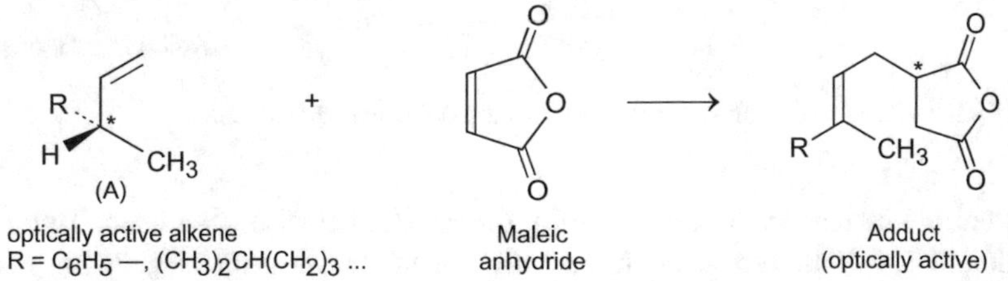

optically active alkene
R = C$_6$H$_5$ —, (CH$_3$)$_2$CH(CH$_2$)$_3$ ...

Maleic anhydride

Adduct (optically active)

**Fig. 19.38** Chiralene reaction of optically active alkene

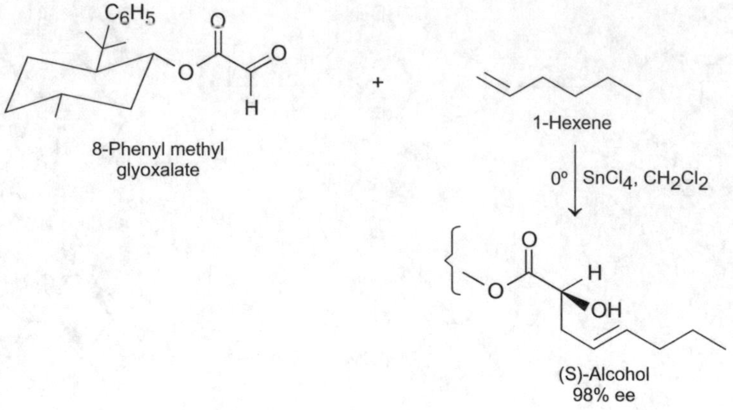

**Fig. 19.40** Chiral ene reaction using optically active esters

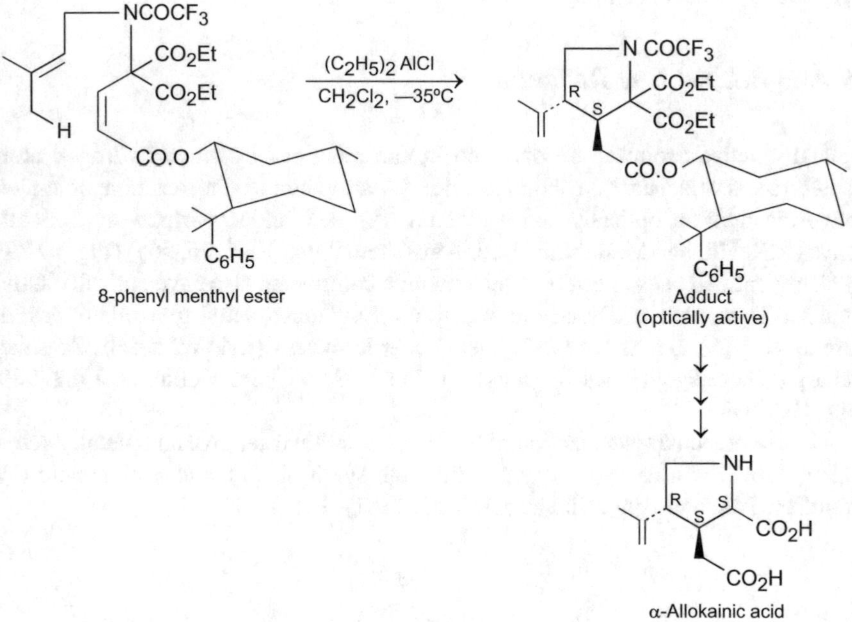

**Fig. 19.41** Synthesis of $\alpha$-allokainic acid by an intramolecular ene reaction

In chiral ene reactions, chiral aluminium catalyst has also been used. Thus, ene reaction of 2, 6-dichlorobenzaldehyde with 2-phenylthio-1-propene gave the adduct in 89% yield (ee 65%) (Fig. 19.42) (O. Achmatowicz, and O. Achmatowicz, Jr., Roczniki Chem., 1962, **36**, 1791; Chem. Abstr, 1963, **59**, 8610b).

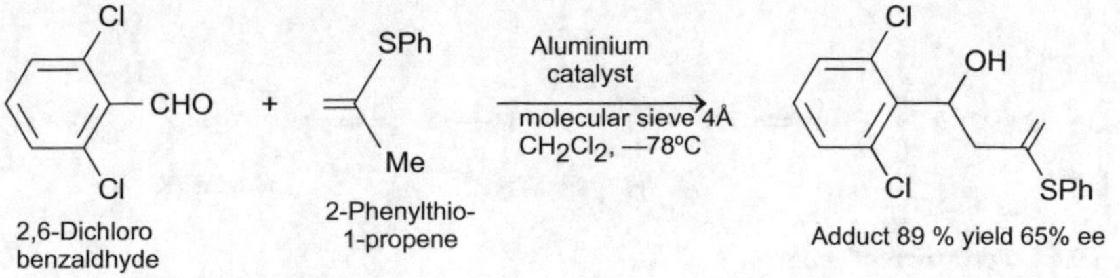

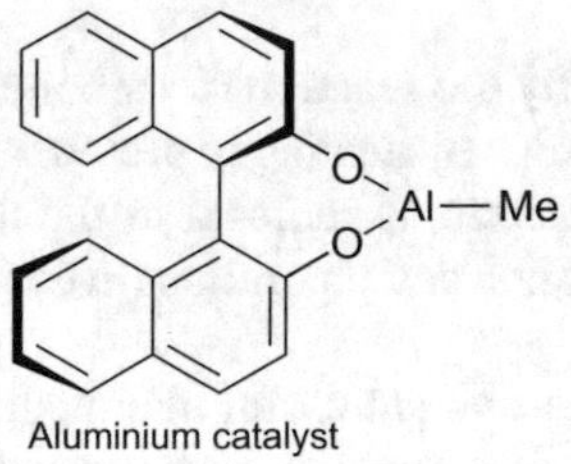

**Fig. 19.42** Chiral ene reaction using chiral aluminium complex

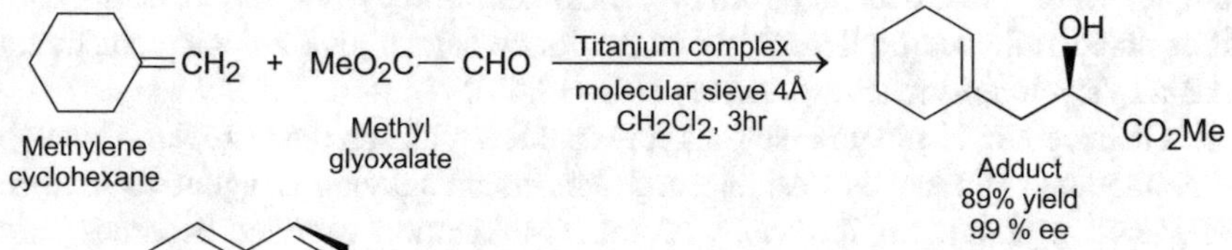

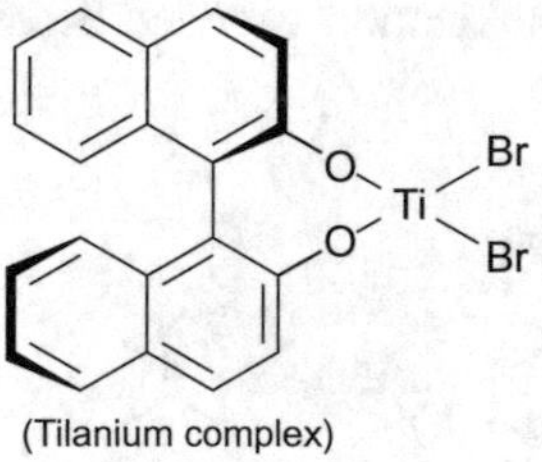

**Fig. 19.43** Chiral ene reaction using chiral titanium complex

Another example of using a chiral complex in chiral ene reaction is the reaction of methylene cyclohexane and methyl glyoxalate giving the adduct in 89% yield (98% ee). The chiral complex used is a titanium complex, which was prepared in situ by the reaction of (R)-bisisnaphthyl and bis-diisopropoxytitanium dibromide (K. Mikaini, M. Terada and T. Nakai's J. Am. Chem. Soc., 1989, **111**, 1940) (Fig. 19.43).

## 19.5 Enamine Reaction

The enamine reaction was introduced by Stork et al. (G. Stork, and S.R. Dowel, J. Am. Chem. Soc., 1963, **85**, 2178; P.W. Hickmott, Tetrahedron, 1982, **38**, 1975, 3363; J. W. Whitesell and M. A. Whitesell, Synthesis, 1983, 517). Enamines are $\alpha$,

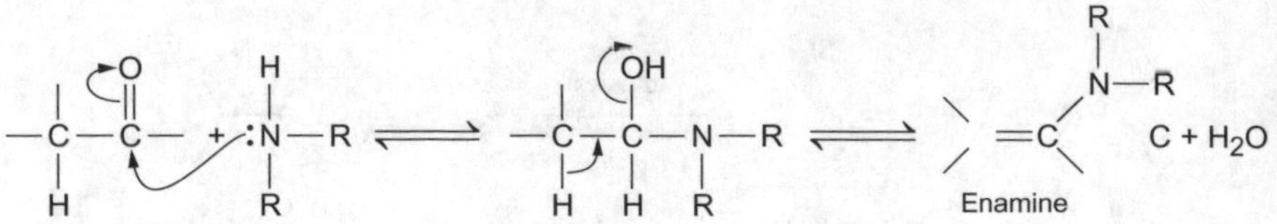

**Fig. 19.44** Formation of enamine

$\beta$-unsaturated amines and can be obtained by the reaction of carbonyl compounds (aldehydes and ketones) with a secondary amine by heating in presence of a catalytic amount of toluene-p-sulphonic acid with azeotropic removal of the formed water. Commonly, amines like pyrrolidine, morpholine and piperidine (in decreasing order of reactivity) are used (Fig. 19.44).

The enamines behave as nitrogen enolates and are useful intermediates for alkylation or acylation of aldehydes and ketones at $\alpha$-position, and the method is known as **Strok Enamine Synthesis**. The enamine reaction provides a valuable alternative method for the selective alkylation and acylation of aldehydes and ketones. As an example, the enamine obtained from cyclohexanone and pyrolidine can be alkylated and acylated followed by hydrolysis of the imine salt to give 2-acylcylohexanone and 2-alkylcyclohexanone respectively (Fig. 19.45).

It is believed that in some cases, alkylation takes place on the nitrogen. However, the N-alkylated product is unstable and acts as an alkylation agent to give the C-alkylated product. So the yield of the C-alkylation product becomes more (Fig. 19.46).

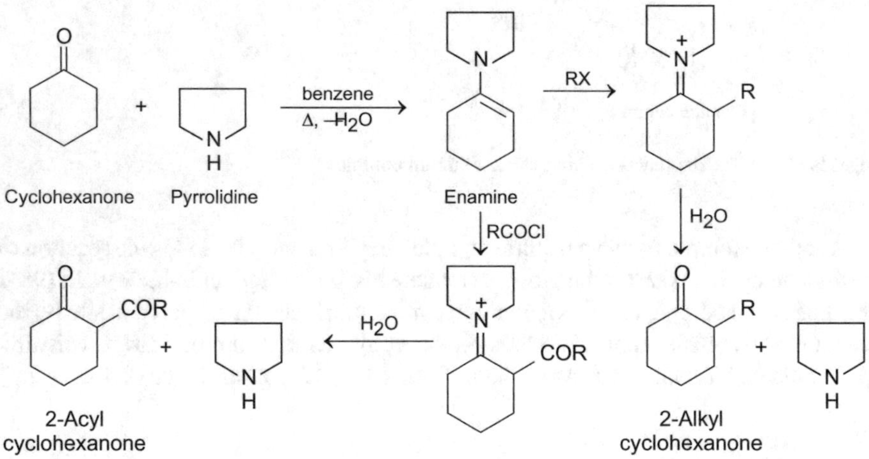

**Fig. 19.45** The alkylation and acylation of cyclohexanone

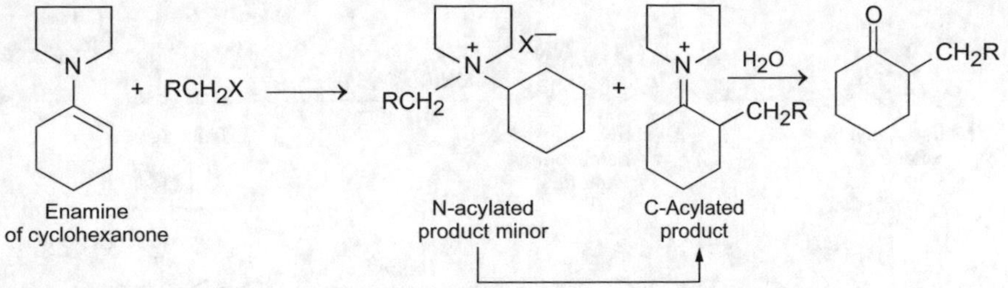

**Fig. 19.46** Formation of N-alkylated product in enamine reaction, which acts as an alkylation agent

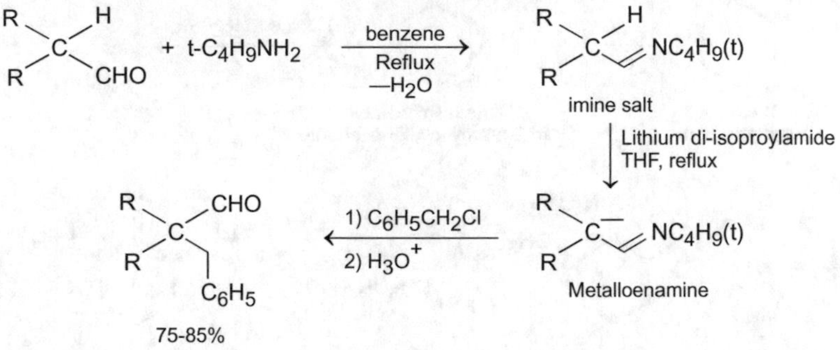

**Fig. 19.47** Formation of metalloenamines

It is found that imines formed from enolizable aldehydes and ketones on reaction with aliphatic primary amines can be deprotonated on treatment with lithium diiso-proylamide or Grignard reagent metalloenamine give the metal salts called **metalloe-namines**. The metalloenamines on reaction with primary or secondary alkyl halides give monoalkylated carbonyl compounds (Fig. 19.47).

The alkylation of aldehydes and ketones via the formation of metalloen amine is a valuable alternative to the usual enamine synthesis, since N-alkylation does not take place.

A convenient and useful alternative to the metalloenamine procedure to alkyl derivatives proceed from dimethylhydrazone of an aldehyde or ketone (instead of an imine in metalloenamine procedure) (E. J. Corey and D. Enders, Tetrahedron Lett., 1976, 11). The dimethyl hydrazones on treatment with lithium diisopropylamide or n-butyl lithium give the lithium derivatives, which can be alkylated with alkyl halides. Finally, the dimethylhydrazine group is removed by oxidation with sodium periodate to give the alkylated product. Such reactions take place with high positional and stereoselectivity under mild conditions. The alkylation generally occurs at the less substituted $\alpha$-position of the unsymmetrical ketone. With cyclohexane derivative, axil methylation is favoured. Thus, methyl pentyl ketone gives ethyl pentyl ketone in 95% yield. Similarly 2-methyl cyclohexanone gives *trans*-2,6-dimethyl cyclohexanone (Fig. 19.48).

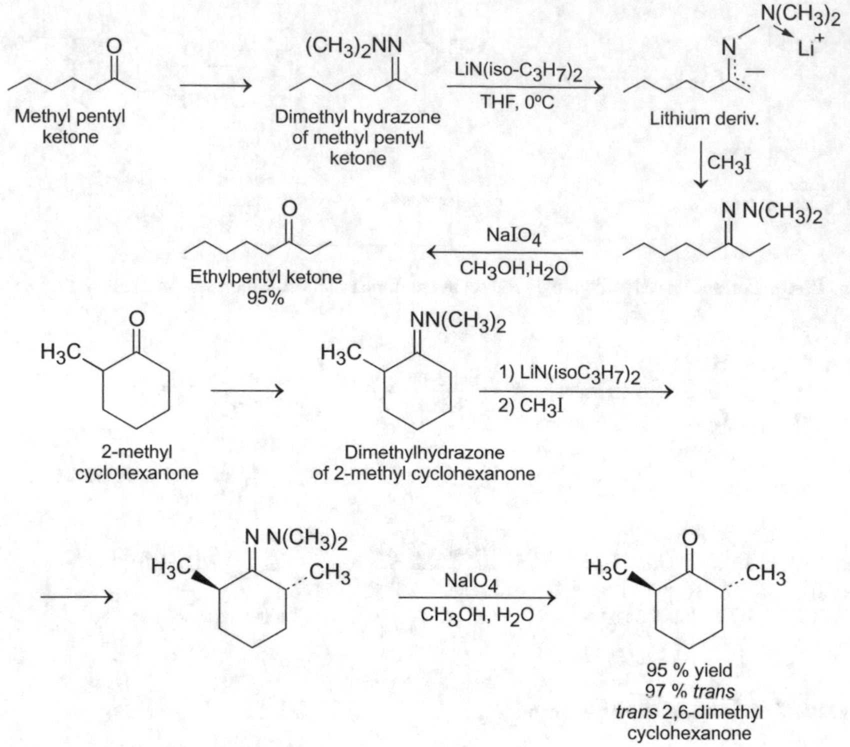

**Fig. 19.48** An alternative to metalloenamine procedure using dimethylhydrazones of aldehydes or ketones

An advantage of lithiated dimethylhydrazone (Fig. 19.48) over metalloe-namines (Fig. 19.47) is that these can be easily converted into organocuprates, which take part in carbon–carbon bond formation. Thus, as an example, acetaldehyde dimethyl hydrazone on reaction with CuI (iso-$C_3H_7$)$_2$ gives the organocuprate, which on treatment with cyclohexenone gives the keto aldehyde (Fig. 19.49).

## 19.5.1   Asymmetric Enamine Synthesis

The metalloenamine reaction has been used for the enantioselective alkylation of ketones using optically active amine. A typical example is given in Fig. 19.50.

In the above synthesis, the methoxy group in the optically active amine plays an important role in asymmetric induction. In the absence of methoxy groups, the optical yields are low.

Asymmetric acylation of aldehydes and ketones can also be affected by the reaction of ketones with an optically active hydrazine as chiral auxiliary group. The

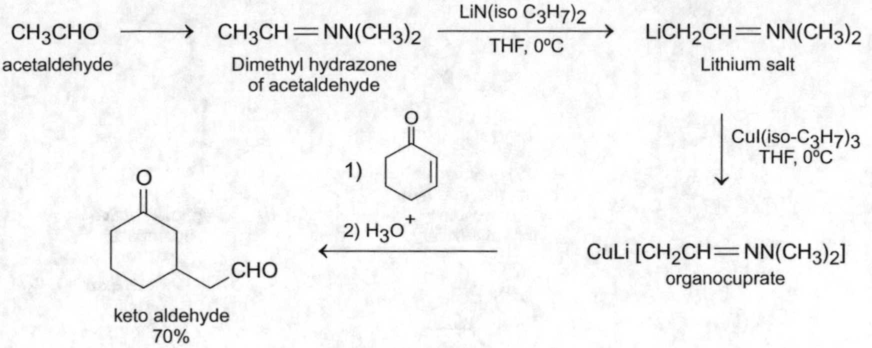

CH$_3$CHO $\longrightarrow$ CH$_3$CH$=$NN(CH$_3$)$_2$ $\xrightarrow[\text{THF, 0°C}]{\text{LiN(iso C}_3\text{H}_7)_2}$ LiCH$_2$CH$=$NN(CH$_3$)$_2$

acetaldehyde      Dimethyl hydrazone                Lithium salt
                 of acetaldehyde

CuI(iso-C$_3$H$_7$)$_3$
THF, 0°C

1)

2) H$_3$O$^+$

CHO

CuLi [CH$_2$CH$=$NN(CH$_3$)$_2$]
organocuprate

keto aldehyde
70%

**Fig. 19.49** Synthesis of ketoaldehyde via organocuprate

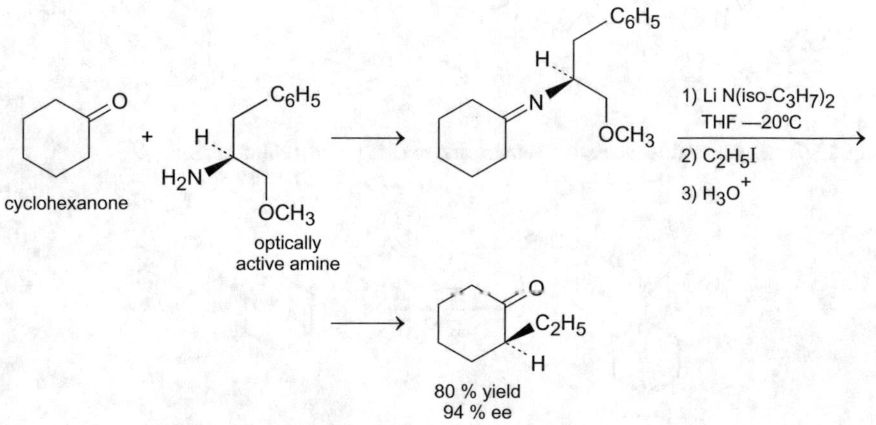

cyclohexanone    +    H$_2$N, C$_6$H$_5$, OCH$_3$

optically
active amine

1) Li N(iso-C$_3$H$_7$)$_2$
    THF —20°C
2) C$_2$H$_5$I
3) H$_3$O$^+$

C$_2$H$_5$
H

80 % yield
94 % ee

**Fig. 19.50** Asymmetric enaming synthesis using metalloenamine

formed hydrazone is treated with lithium isopropyl amide and alkyl halide (C$_3$H$_7$I) and finally with acid gives the asymmetrically acylation product in 60% yield (95.5% ee) (Fig. 19.51).

The above procedure has been used for the asymmetric synthesis of a number of natural products (D. Enders, H. Eichenauer, U. Basu, H. Schubert and K.A.M. Kremer, Tetrahedron, 1984, **40**, 1345).

In asymmetric enamine synthesis, chiral enamines have also been used. The chiral amines must be carefully chosen. Two such examples are given below (J.K. Whitesell, and S.W. Felman, J. Org Chem., 1977, **42**, 1663) (Fig. 19.52).

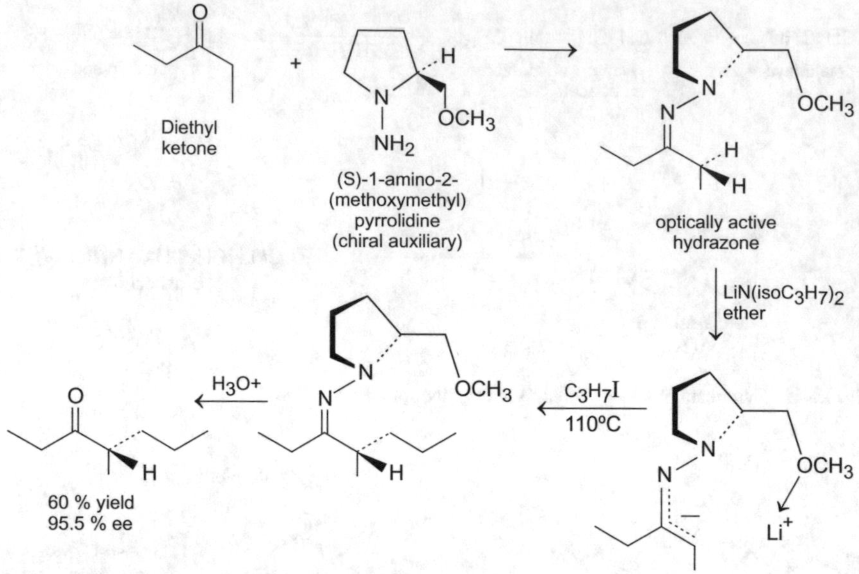

**Fig. 19.51**  Asymmetric acylation of ketones and reacting with a chiral auxiliary

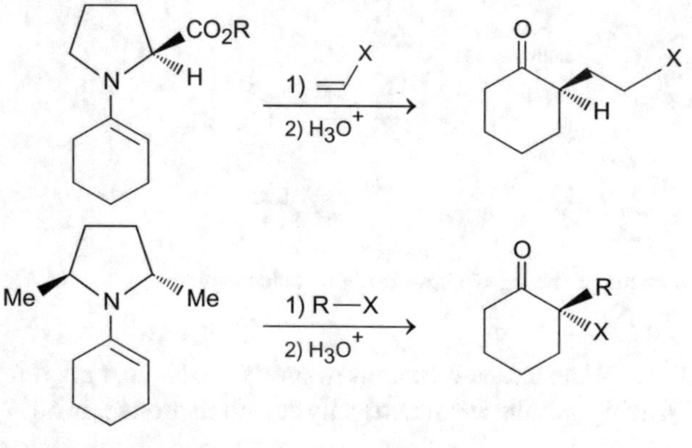

**Fig. 19.52**  Asymmetric synthesis using chiral amines

## 19.6   Friedel Crafts Reaction

Friedel Crafts alkylation and acylation are two types of Friedel Crafts reactions. (Fig. 19.53).

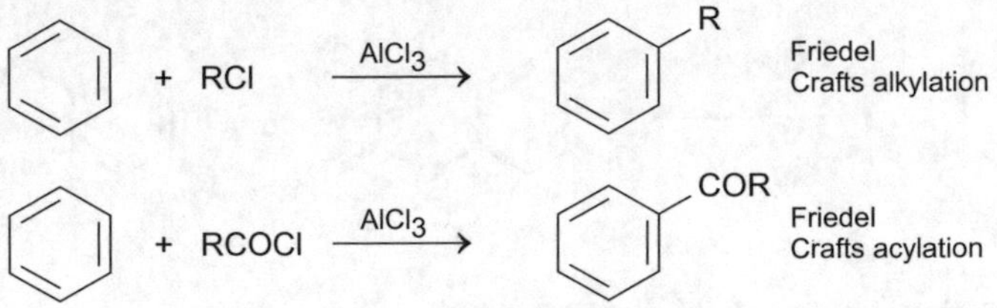

**Fig. 19.53**  Friedel Crafts alkylation and acylation

## 19.6.1  Asymmetric Induction in Friedel Crafts Reaction

The Friedel Craft reaction is known to proceed via a cation and so the reaction with chiral halides is expected to give racemic products. It has however been found (S. Suga, M. Segi, K. Kitano, S. Masuda and T. Nakajima, Bull. Chem. Soc. Jpn., 1981, **54**, 3611) that some asymmetric induction can be achieved in presence of mild Lewis acid at low temperatures for a short duration of time. Thus, the reaction of benzene with (S)-2-chlorobutane at 0 °C for 4 min. in presence of $FeCl_3$ give the product in 70% yield with 24% ee (Fig. 19.54).

Another approach for the asymmetric induction in Friedel Crafts reaction is to use appropative chloro compound like 1-phenyl-2-chloropropane (S. Masuda, T. Nakajima and S. Suga, Bull. Chem. Soc. Jpn. 1983, **56**, 1089) (Fig. 19.55).

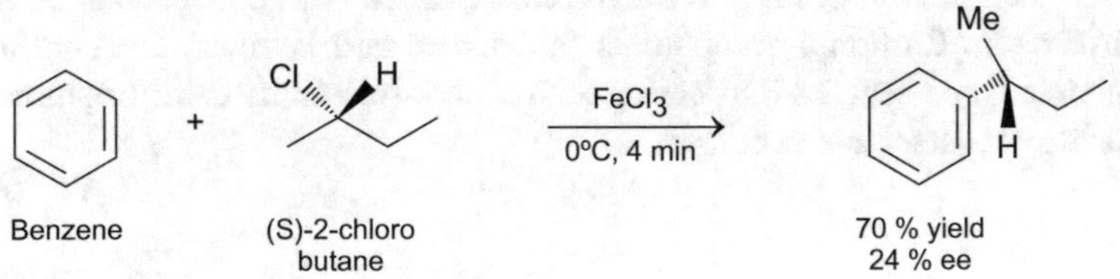

**Fig. 19.54**  Asymmetric induction in Friedel Crafts reaction

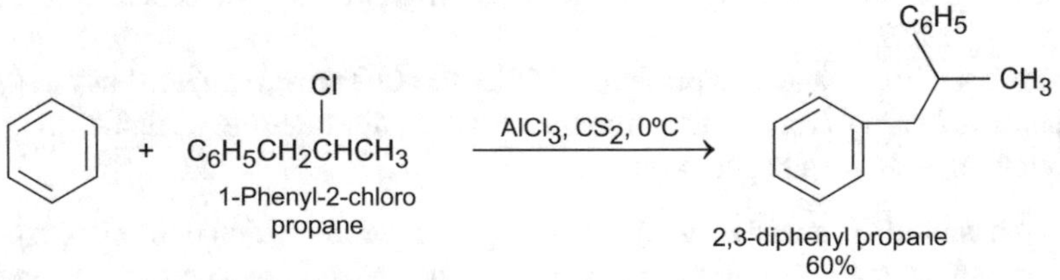

**Fig. 19.55**  A symmetric induction using 1-phenyl-2-chloropropane

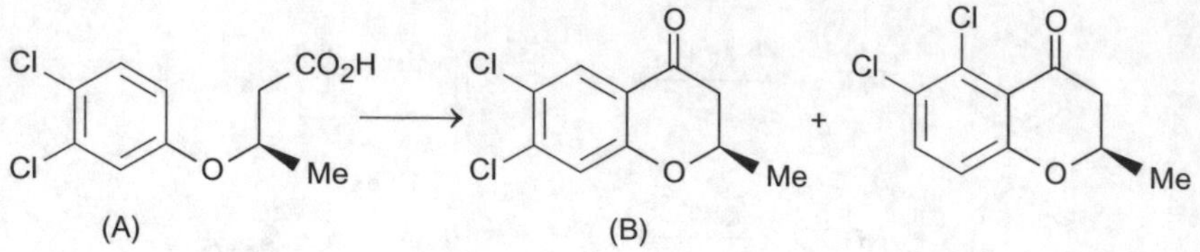

**Fig. 19.56** Regioselectivity in Friedel crafts reaction

## 19.6.2   Regioselectivity in Friedel Crafts Synthesis

The intramolecular Friedel Crafts acylation are important in organic synthesis. The product formed depends on the reagent used. Thus, Friedel–Crafts acylation of (A) gave a mixture of products (B) and (C) (cited in Organic Synthesis, Michael B. Smith, Mcgraw-Hill International Edn. Page 1321). Use of conc. $H_2SO_4$ gave a mixture of B and C favouring B (Ratio 35: 65). However, using a mixture of phosphores tribromide and aluminium chloride gives high regioselectivity for (B) (ratio of B and C is 95: 5) (Fig. 19.56).

## 19.7   Grignard Reaction

It involves the reaction of organo magnesium halides with compounds containing acidic hydrogen. Carbonyl compounds (aldehydes and ketones, terminal alkynes etc.). For details see Sect. 18.3.8 dealing with enantioselectivity using organometallic reagents in stereoselective synthesis.

## 19.7.1   Asymmetric Induction in Grignard Reaction

The reaction of unsymmetrical ketones with Grignard reagent generates a chiral centre (Fig. 19.57).

In case a chiral reagent is present in either the Grignard reagent (but not at the carbon atom bearing Mg) or the carbonyl substrate, diastereomers result.

Following four possibilities are:

(i)    An achiral Grignard reagent + carbonyl compound which does not possess a substituent attached to a chiral centre → one new chiral centre is generated.

**Fig. 19.57** Asymmetric induction in Grignard reaction

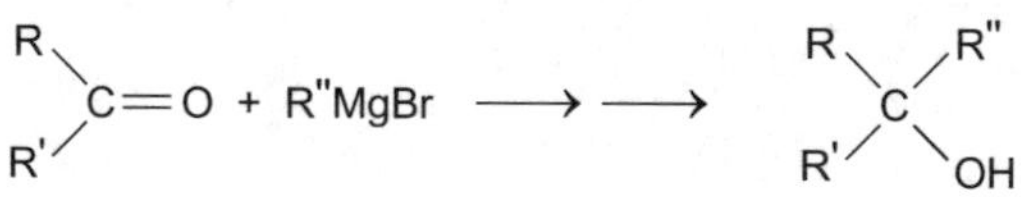

$$CH_3CH = CHBr \xrightarrow[\text{ether}]{Mg} CH_3CH = CHMgBr$$

Vinyl bromide                    Vinyl manesium halide
                                 (alkenyl Grignard reagent)

**Fig. 19.58** Formation of vinyl magnesium halide

$$CH_3CH = CHMgBr + CO_2 \longrightarrow CH_3CH = CH-COOH$$

**Fig. 19.59** Reaction of vinyl magnesium bromide with $CO_2$

(ii) A chiral Grignard reagent + carbonyl compound that does not possess a substituent attached to a chiral centre (as in (i) above), → a diastereomer is generated.

(iii) A achiral Grignard reagent + a carbonyl compound that has a substituent attached to a chiral centre → diastereomers (with possible diastereoselectivity) is generated.

(iv) A chiral Grignard reagent + a carbonyl compound having a substituent attached to a chiral centre → good diastereoselectivity results.

## 19.7.2 Stereoselectivity in the Formation of Vinyl Magnesium Halides from Appropriate Alkenes

Vinylhalides react with magnesium to give vinyl magnesium halides an alkenyl Grignard reagent (Fig. 19.58).

The vinyl magnesium halide on treatment with $CO_2$ gives the corresponding acids (Fig. 19.59).

It has been found that the reaction of pure E or Z alkenyl halide with magnesium (Fig. 19.58) gives a mixture at E and Z vinyl Grignard reagents (G.J. Martin, B. Mechin and M. L. Martin, C. R. Acad. Sci. Ser. C, 1968, 207, 986). Also the reaction of vinyl Grignard reagents with $CO_2$ gives an E-Z mixture of final product (Fig. 19.59).

The table below shows the ratio of Z: E ratio of the vinyl magnesium halide (from stereochemically pure 1-bromo-1-alkene) and the corresponding Z: E ratio of the carboxylic acid.

1-Bromo-1-propene

$$CH_3CH = CH/Br \xrightarrow[\text{ether}]{Mg} CH_3CH = CHMgBr \xrightarrow{CO_2} CH_3CH = CHCOOH$$

| Z : Eratio | Z : E | Z : E |
|---|---|---|
| Z (98.5 : 1) | 80-90 : 20-10 | 85-95 : undected |
| E (5 : 95) | 30-40 : 70-60 | 15-5 : 85-95 |

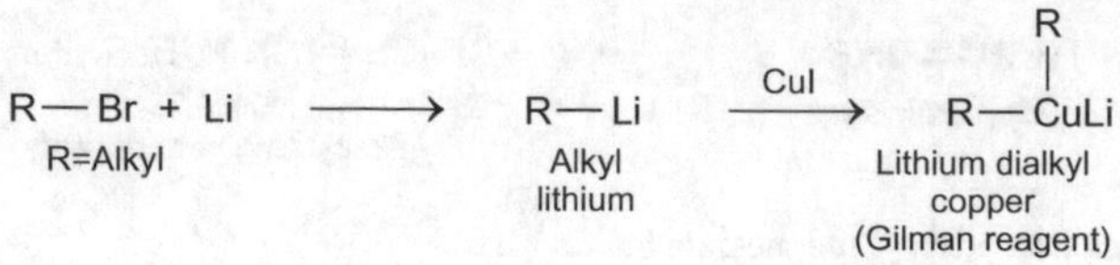

$$R\!-\!Br + Li \longrightarrow R\!-\!Li \xrightarrow{CuI} \underset{R}{\overset{R}{R\!-\!CuLi}}$$

R=Alkyl          Alkyl       Lithium dialkyl
                     lithium       copper
                           (Gilman reagent)

**Fig. 19.60** Preparation of Gilman reagents

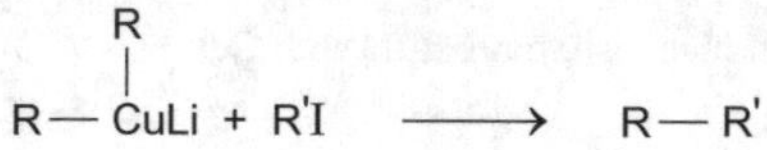

$$\underset{R}{\overset{R}{R\!-\!CuLi}} + R'I \longrightarrow R\!-\!R'$$

**Fig. 19.61** Preparation of alkanes

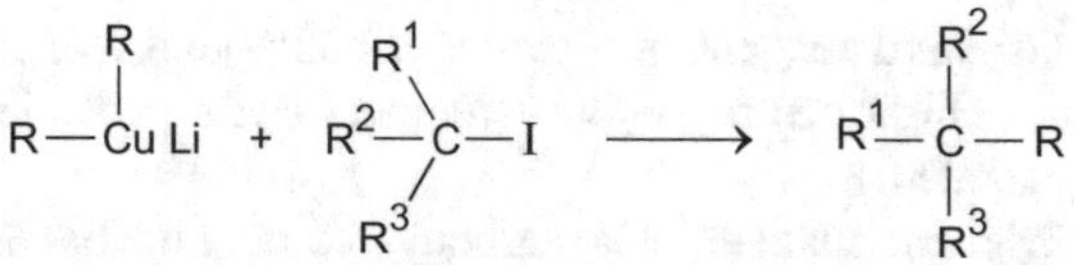

$$\underset{R}{\overset{R}{R\!-\!CuLi}} + R^2\!-\!\underset{R^3}{\overset{R^1}{C}}\!-\!I \longrightarrow R^1\!-\!\underset{R^3}{\overset{R^2}{C}}\!-\!R$$

**Fig. 19.62**  Generation of a new stereocentre using Gilman reagent

## 19.8   Corey-Posner, Whites-House Synthesis

The synthesis involves the reaction of alkyl halides with organo copper reagents, known as Gilman reagents (H. Gilman and J. M. Straley, Rect. Trave. Chem. Pays. Bae., 1936, 55, 821; H. Gilman, R.G. Jones and L. A. Woods, J. Org. Chem., 1952, **17**, 1630). The Gilman reagents are prepared as follows (Fig. 19.60).

The Gilman reagents react with other alkyl halides to generate an alkane (Fig. 19.61).

Use of tertiary alkyl halide generates a new stereocentre (Fig. 19.62).

## 19.9   Sharpless Epoxidation Reaction

It is a enantioselective reaction that converts an alkene to epoxides using sharp-less reagent, which is a mixture of tert. butyl hydroperoxide[$(CH_3)_3C\!-\!OOH$], a titanium catalyst-usually titanium isopropoxide [$Ti(OCH(CH_3)_2)_4$] and diethyl tartarate (DET). There are two different chiral diethyl tartarates labelled as (+) DET or (−) DET, viz., (+)-(R, R)-diethyl tartarate and (−)(S, S)-diethyl tartarate (Fig. 19.63).

The nature of the DET isomer (i.e., (+) or (−)) determines which enatiomer is obtained as the major product in the epoxidation of an allylic alcohol with sharpless reagent. It is found that in the major product by using (+)-DET, O is added from

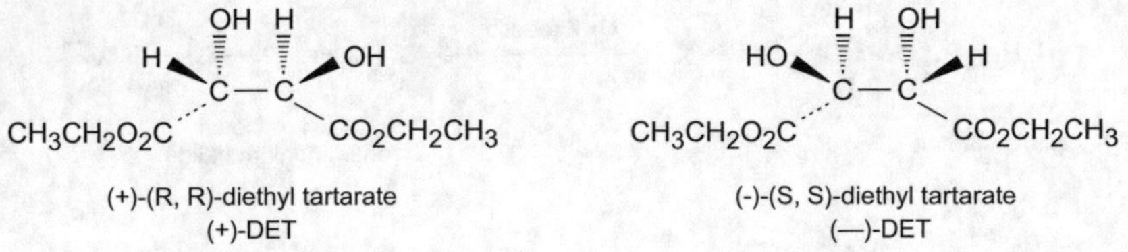

OH H

H⸜⸜⸜C—C⸝⸝⸝OH

CH₃CH₂O₂C          CO₂CH₂CH₃

(+)-(R, R)-diethyl tartarate
(+)-DET

H  OH

HO⸜⸜⸜C—C⸝⸝⸝H

CH₃CH₂O₂C          CO₂CH₂CH₃

(-)-(S, S)-diethyl tartarate
(—)-DET

**Fig. 19.63**  (+)-DET and (–)-DET

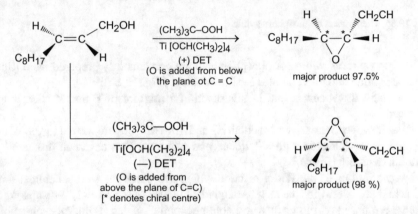

**Fig. 19.64**  Epoxidation of an allylic alcohol with (+)-DET and (–)-DET

above the plane of $C == C$, and in the major product by using (–)-DET, O is added from the bottom (Fig. 19.64).

Also see Sect. 18.3.2 in stereoselective synthesis.

## 19.10  Wittig Reaction

It involves the reaction of a phosphorus ylide (or phosphorane) with aldehydes and ketones to give alkenes and triphenyl phosphine oxide (Fig. 19.65).

R\
  C=O  +  (C₆H₅)₃P—C⟨R″/R‴  ⟶  R\C=C/R″  +  O=P(C₆H₅)₃
R′/                                    R′/    \R‴

Aldehyde or       Phosphorus ylide              Alkene          Triphenyl
ketone            (or phosphorane)          (E and Z isomers)   phosphine
                                                                oxide

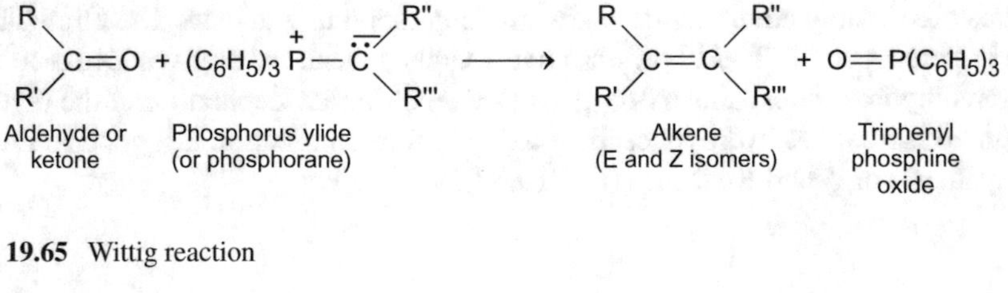

**Fig. 19.65**  Wittig reaction

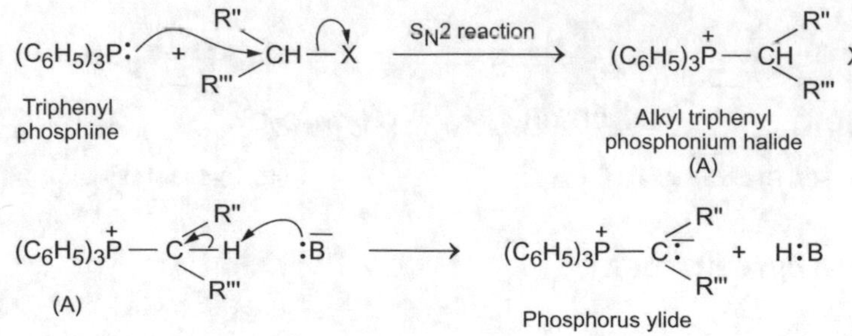

**Fig. 19.66** Preparation of phosphorus ylide

The phosphorus ylides (or phosphoranes) are conveniently prepared from triphenyl phosphine and alkyl halides as shown below (Fig. 19.66).

The main drawback is that in this reaction a mixture of E and Z alkenes are obtained.

It has however been shown that both *cis* and *trans* alkenes are formal by the Wittig reaction of an aldehyde with alkylidene phosphorane. This has been discussed in detail in Sect. 3.6 (Fig. 3.17).

Two variations of the Wittig reaction which overcome the stereochemical limitation and are useful for the stereoselective synthesis of Z and E-1, 2-disubstituted alkenes are based on the readily available phosphine oxides (I) and phosphonobis-N,N-dialkylamides (II).

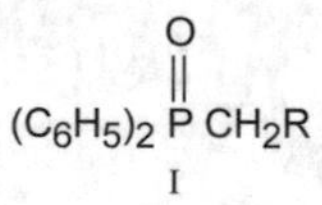

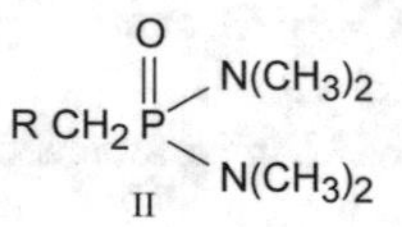

$$
\underset{\text{I}}{(C_6H_5)_2 \overset{\displaystyle O}{\overset{\|}{P}} CH_2R} \qquad\qquad \underset{\text{II}}{R\,CH_2\,\overset{\displaystyle O}{\overset{\|}{P}} \underset{\displaystyle N(CH_3)_2}{\overset{\displaystyle N(CH_3)_2}{}}}
$$

The reactions involving the anions derived from phosphine oxides (the **Horner-Wittig reaction**) are synthetically useful finding increasing applications (B. Lythgoe, T.A. Moran, M.E.N. Nambudiry, and S. Rustan, J. Chem. Soc. Perkin, 1976, I, 2386; J.M. Clough and G. Pattenden, Tetrahedron, 1981, **37**, 3911).

The oxides are conveniently obtained by the quaternisation of triphenyl phosphene and hydrolysis of the phosphonium salts. The lithio derivative of the phosphine oxide (I) on reaction with aldehydes or ketones give $\beta$-hydroxy phosphine oxide, which on treatment with sodium hydride give the corresponding alkene. The elimination step is stereospecific. Thus, E erythro hydroxyphosphene oxide gives Z-alkene and the threo hydroxyphosphine oxide gives E-alkene. In fact, separation of the erythro and threo isomers of hydroyphosphine oxide before elimination is a good route for the synthesis of Z-and E-alkenes (Fig. 19.67).

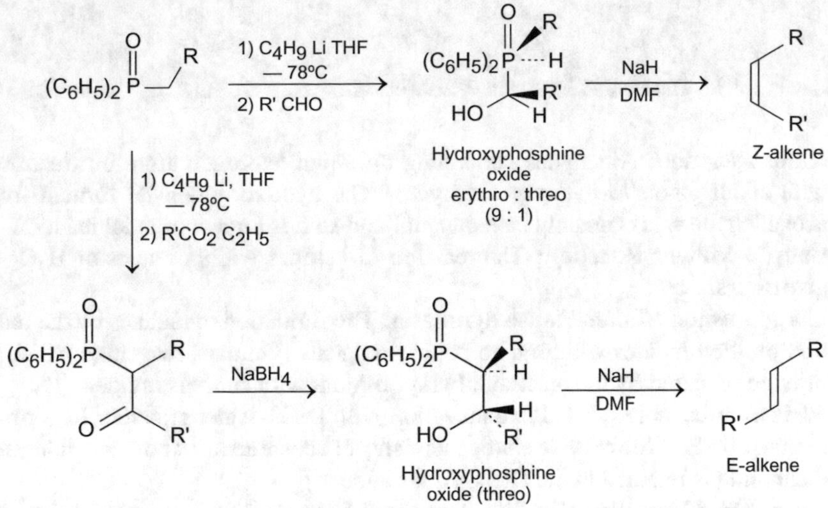

**Fig. 19.67**  Synthesis of Z and E alkenes by Horner-Wittig reaction

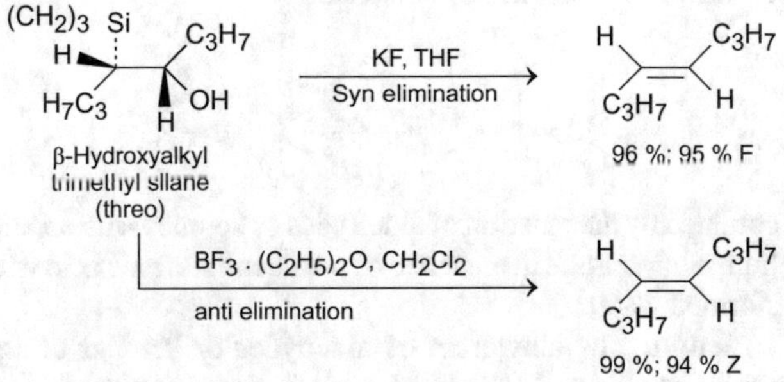

**Fig. 19.68**  Peterson reaction, synthesis of E and Z alkenes

Both Z- and E-alkenes can be individually obtained by **Peterson reaction** (D. Peterson, J. Org. Chem., 1968, **33**, 781; D.J. Ager, Synthesis, 1984, 384). This is a silicon version of the Wittig reaction and involves elimination of trimethylsilanol, $(CH_3)_3SiOH$ (volatile and easier to remove than triphenyl phosphine oxide in Wittig reaction) from a $\beta$-hydroxyalkyltrimethyl silane. In this reaction, the steric course of the reaction is easily controlled. Both Z- and E- alkenes can be separately obtained from a single stereomer of the hydroxysilane depending on whether elimination is affected under acidic or basic conditions (Fig. 19.68).

## Key Concepts

- **Aldol Reaction:** Aldehydes containing an alpha hydrogen atom on treatment with alkali giving $\beta$-hydroxy aldehydes. The hydroxy aldehyde formed from acetaldehyde was originally called alchol and so it is known as aldol reaction.
- **Baeyer–Villiger Reaction:** The reaction of ketones with per acids or $H_2O_2$ to give esters.
- **Corey-Posner, Whites-House Synthesis:** The formation of alkanes by the reaction of alkyl halides with organo copper reagents (Gilmans reagents), which in turn are prepared by the reaction of alkyl lithium with cuprous iodide.
- **CisPrinciple:** A term used in connection with Diels–Alder reaction. According to this principle, the relative stereochemistry of substituents in both the diene and dienophile is retained in the Diels–Alder adduct.
- **Diels–Alder Reaction:** The [4 + 2] cycloaddition reaction between a conjugated diene ($4\pi$-electron system) and a compound containing a double or triple bond (dienophile) ($2\pi$-electron system) to form a cyclic adduct.
- **Enamines:** These are $\alpha$, $\beta$-unsaturated amines

$$\left[ \overset{|}{\underset{}{C}}=\overset{|}{\underset{}{C}}-\ddot{N}\overset{}{\underset{}{<}} \longleftrightarrow \overset{-}{\underset{}{C}}=\overset{|}{\underset{}{C}}-\overset{+}{N}\overset{}{\underset{}{<}} \right)$$

and are obtained by the reaction of aldehydes or ketones with an amine (pyrrolidine, morpholine, piperidine in presence of p-toluene sulfonic acid with azeotropic removal of formed water.
- **Enamine Reaction:** The alkylation of aldehydes or ketones at alpha position by reaction with enamines followed by hydrolysis to give alkylated or acylated carbonyl compounds.
- **Enantioslective Reaction:** Also known as stereoselective reaction and is a reaction giving only one enatiomer.
- **Endo Addition Rule:** A term used in connection with Diels–Alder reaction. According to this rule in Diels–Alder reaction *endo* product is formed from orientation of the two reactants with maximum accmulation of double bonds.
- **Ene Reaction:** Related to Diels–Alder reaction and involves the addition of an alkene bearing an allylic H atom (instead of a diene used in Diels–Alder reaction) with a dienophile (now called an enophile) resulting is the formation of a new sigma bond to the terminal carbon of the allyl group and this is 1,5-migration of the allylic H and change in the position of the allylic double bond.
- **Friedel Crafts Reaction:** The reaction of aromatic compounds (like benzene) with alkyl halides or acylhalides in presence of $AlCl_3$ to give alkyl benzenes (Friedel Crafts alkylation) or acylbenzenes (Friedel Crafts acylation).

- **Gilman Reagents:** Organo copper reagents and are prepared by treating alkyl lithium with cuprous iodide.

$$R—Li + CuI \longrightarrow R—\overset{\overset{\displaystyle R}{|}}{Cu} Li$$

- **Grignard Reagent:** Organo magnesium halides and are prepared by the reaction of magnesium on alkyl halides in ether.
- **Grignard Reaction:** The reaction of Grignard reagent with various substrates like aldehydes and ketones to give 1° or 2° alcohols.
- **Horner-Wittig Reaction:** A modification of Wittig reaction involving reaction of phosphine oxide with *n*-BuLi followed by reaction of the formed anion with an aldehyde and treatment of the formed hydroxyphosphene oxide with a base to give Z-alkene.
- **Peterson Reaction:** A silicon version and Wittig reaction involving elimination of trimethyl silanol from a $\beta$-hydroxyalkyl trimethyl silane. The reaction under appropriate conditions gives both E and Z alkanes.
- **Sharpless Reagent:** A mixture tertiary butyl hydroperoxide, $(CH_3)_3C\text{-}OOH$, Titanium isoproxide $[Ti(OCH(CH_3)_2)_4]$ and diethyl tartarate (DET).
- **Sharpless Epoxidation Reaction:** An enantioselective reaction that converts an alkenes to epoxides using sharpless reagent.
- **Stereoselective Reaction:** A reaction producing one stereoisomer.
- **Wittig Reaction:** The reaction of a phosphorus ylide (or phosphorane) with carbonyl compounds to give alkenes.

## Problems

1. What product is expected to be obtained in the Baeyer–Villiger oxidation of t-butylethyl ketone?

**Ans.** 5-tert.butyl migration is observed.

2.  Can cyclopentadiene undergo Diels–Alder reaction? If yes, what product is obtained.

    **Ans.** Yes, Dicyclopentadiene

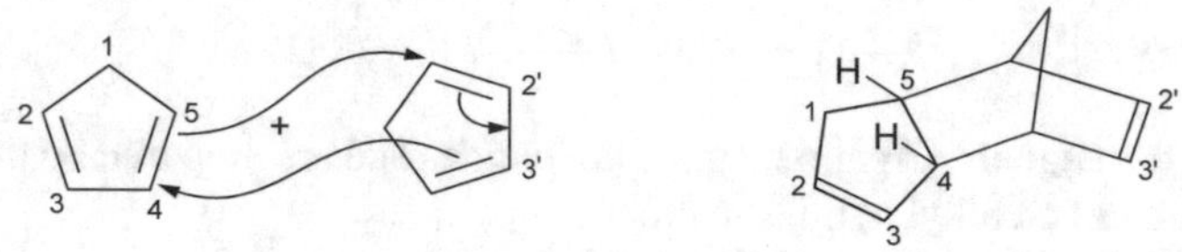

3.  Predict the major product in the epoxidations.

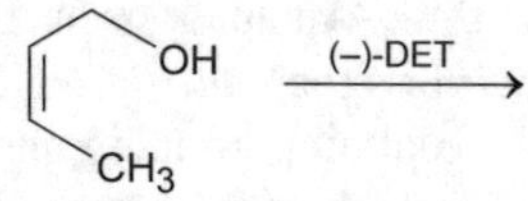

4.  What product is expected to be obtained in the aldol reaction of benzaldehyde and methyl tertiary butyl ketone?

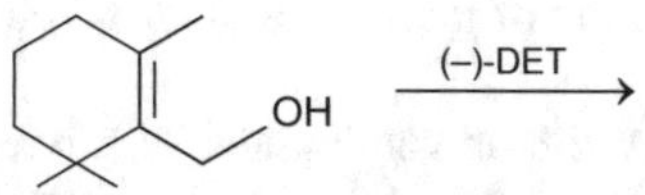

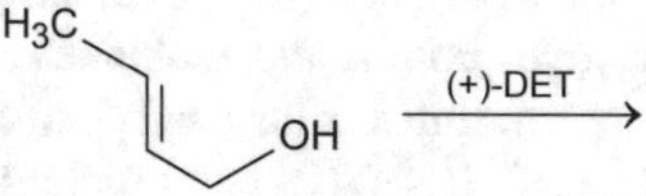

    What is directed aldol reaction. Explain giving one example.

5.  Write notes on:

    (a)  Makaiyama aldol reaction
    (b)  Aldol reaction between a chiral aldehyde and chiral enolate
    (c)  Enzymatic Baeyer–Villiger reaction
    (d)  Stereopsecific Diels–Alder reaction
    (e)  *Endo* addition rule
    (f)  Asymmetric enamine synthesis

6.  What is the difference between Ene reaction and Diels–Alder reaction?
7.  Give an example of intramolecular ene reaction.
8.  What are enamines? How they are useful in organic synthesis?
9.  What are metalloenamines? What are their uses?
10. Starting with 2-methyl cyclohexanone, how will you obtain *trans*-2,6-dimethyl cyclohexanone?
11. What are organo cuprates? Give their applications.
12. How asymmetric induction can be achieved in Friedel Crafts reaction?

13.   Explain Corey-Posner, White-House synthesis.
14.   Write a note on sharpless epoxidation reaction.
15.   Using Horner-Wittig reaction and Peterson reaction how will you obtain E and Z alkanes?

# References

Advanced inorganic chemistry J (1991) March, 4 th edn, Willy, New york.
Ahluwalia VK (2013) Organic chemistry, fundamental concepts. Narosa Publishing House
Ahluwalia VK, Kumar LS, Kumar S (2006) Chemistry of Natural Products: Amine Acids, Peptides and Enzymes, Ane Books India
Ahluwalia VK, Natural products, Vishal Publishing Co. Jalandhar
Francis A Carey and Richard J. Sundbery, Advanced Organic Chemistry, Vol. I and II. Sponiger.
Eliel EL, Stereochemistry of carbon compounds. Tata McGraw Hill Education Pvt. Ltd., New Delhi
Eliel EL, Wilin SH (1994) Stereochemistry of organic compounds. Willy, Inc, New York
Findr IL, Organic chemistry, Volume II. Longman Groun
Marye Anne Fox and Tames K. Whitesell, Organic Chemistry. Jones and Barllett Publishers
Gupta RR, Kumar M, Gupta V, Heterocyclic compounds, Vol. I and II. Springer, Berlin
Janice Gorzynski Smith, Organic chemistry, The McGraw Hill companies.
Sykes P, A guide book to mechanisms in organic chemistry, 6th edn, Orient Longman
Potapow VM (1979) Stereochemistry. Mir Publishers, Moscow
Thomas N. Sovrell, Organic Chemistry, Viva Books, New Delhi.

# Index

**A**

© The Author(s), under exclusive license to Springer Nature Switzerland AG 2022
V. K. Ahluwalia, *Stereochemistry of Organic Compounds*,
https://doi.org/10.1007/978-3-030-84961-0

601

Printed in the United States
by Baker & Taylor Publisher Services